SCIENTIFIC PRINCIPLES OF ADIPOSE STEM CELLS

SCIENTIFIC PRINCIPLES OF ADIPOSE STEM CELLS

Edited by

LAUREN KOKAI
Departments of Plastic Surgery and Bioengineering, University of Pittsburgh, Pittsburgh, PA, United States
McGowan Institute for Regenerative Medicine, University of Pittsburgh, Pittsburgh, PA, United States

KACEY MARRA
Departments of Plastic Surgery and Bioengineering, University of Pittsburgh, Pittsburgh, PA, United States
McGowan Institute for Regenerative Medicine, University of Pittsburgh, Pittsburgh, PA, United States

J. PETER RUBIN
Departments of Plastic Surgery and Bioengineering, University of Pittsburgh, Pittsburgh, PA, United States
McGowan Institute for Regenerative Medicine, University of Pittsburgh, Pittsburgh, PA, United States

Academic Press is an imprint of Elsevier
125 London Wall, London EC2Y 5AS, United Kingdom
525 B Street, Suite 1650, San Diego, CA 92101, United States
50 Hampshire Street, 5th Floor, Cambridge, MA 02139, United States
The Boulevard, Langford Lane, Kidlington, Oxford OX5 1GB, United Kingdom

Library of Congress Cataloging-in-Publication Data
A catalog record for this book is available from the Library of Congress

British Library Cataloguing-in-Publication Data
A catalogue record for this book is available from the British Library

ISBN 978-0-12-819376-1

For information on all Academic Press publications
visit our website at https://www.elsevier.com/books-and-journals

Publisher: Stacy Masucci
Acquisitions Editor: Elizabeth Brown
Editorial Project Manager: Pat Gonzalez
Production Project Manager: Kiruthika Govindaraju
Cover Designer: Mark Rogers

Typeset by SPi Global, India

Dedication

This book is dedicated to my wonderful colleagues and mentors, without whom this text would not have been possible. I further thank my family for their constant support and love: Aaron, Adelaide, and Avery; you make every day brighter.

– Lauren Kokai, PhD

This book is dedicated to my wonderful family for their constant support and to my amazing colleagues in the laboratory. This book would not have been possible without them.

– Kacey Marra

This work is dedicated to my family, whose bedrock support has been a source of strength along my career path; to the amazing people in our laboratory at the University of Pittsburgh; and to all the patients who inspire us to explore new approaches in regenerative medicine.

– J. Peter Rubin, MD

Contents

5. Mechanisms of adipose-derived stem cell aging and the impact on therapeutic potential

Xiaoyin Shan and Ivona Percec

6. Human pluripotent nontumorigenic multilineage differentiating stress enduring (Muse) cells isolated from adipose tissue: A new paradigm in regenerative medicine and cell therapy

Karen L. Leung and Gregorio D. Chazenbalk

7. Adipose stem cell homing and routes of delivery

Ganesh Swaminathan, Yang Qiao, Bhavesh D. Kevadiya, Lucille A. Bresette, Daniel D. Liu, and Avnesh S. Thakor

8. Bioreactors and microphysiological systems for adipose-based pharmacologic screening

Mallory D. Griffin and Rosalyn D. Abbott

3

Adipose cell therapy and regenerative medicine

9. Adipose stem cells and donor demographics: Impact of anatomic location, donor sex, race, BMI, and health

Adam Cottrill, Yasamin Samadi, and Kacey Marra

10. Immunomodulatory properties of adipose stem cells in vivo: Preclinical and clinical applications

Matthias Waldner, Fuat Baris Bengur, and Lauren Kokai

11. Clinical experience with adipose tissue enriched with adipose stem cells

Shawn Loder, Danielle Minteer, and J. Peter Rubin

12. Adipose-derived stem cells for wound healing and fibrosis

Yasamin Samadi, Francesco M. Egro, Ricardo Rodriguez, and Asim Ejaz

13. Oncologic safety of adipose-derived stem cell application

Hakan Orbay and David E. Sahar

14. FDA regulation of adipose cell use in clinical trials and clinical translation

Mary Ann Chirba, Veronica Morgan Jones, Patsy Simon, and Adam J. Katz

4 Engineering with adipose stem cell

15. Biomaterial control of adipose-derived stem/stromal cell differentiation

John Walker and Lauren Flynn

16. Genetic modification of adipose-derived stem cells for bone regeneration

Harsh N. Shah, Abra H. Shen, Sandeep Adem, Ankit Salhotra, Michael T. Longaker, and Derrick C. Wan

17. Adipose-derived stromal/stem cells for bone tissue engineering applications

Nathalie Faucheux, Fabien Kawecki, Jessica Jann, François A. Auger, Roberto D. Fanganiello, and Julie Fradette

18. The hematopoietic potential of stem cells from the adipose tissue

Béatrice Cousin and Louis Casteilla

19. Adipose stem cells for peripheral nerve engineering

Benjamin K. Schilling, George E. Panagis, Jocelyn S. Baker, and Kacey Marra

Contributors

Rosalyn D. Abbott Biomedical Engineering, Carnegie Mellon University, Pittsburgh, PA, United States

Sandeep Adem Department of Surgery, Division of Plastic and Reconstructive Surgery, Stanford University School of Medicine, Stanford, CA, United States

Sara I. Al-Ghadban Center for Stem Cell Research and Regenerative Medicine, Tulane University School of Medicine, New Orleans, LA, United States

François A. Auger The Tissue Engineering Laboratory (LOEX), a Laval University Research Center, and Regenerative Medicine Division, CHU of Quebec—Laval University Research Center; Department of Surgery, Faculty of Medicine, Laval University, Quebec, QC, Canada

Jocelyn S. Baker Department of Bioengineering, School of Engineering, University of Pittsburgh, Pittsburgh, PA, United States

Fuat Baris Bengur Department of Plastic Surgery, University of Pittsburgh, Pittsburgh, PA, United States

Lucille A. Bresette Interventional Regenerative Medicine and Imaging Laboratory, Department of Radiology, Stanford University School of Medicine, Palo Alto, CA, United States

Aaron C. Brown Center for Molecular Medicine, Maine Medical Center Research Institute, Scarborough, ME; School of Biomedical Sciences and Engineering, The University of Maine, Orono, ME; Tufts University School of Medicine, Boston, MA, United States

Bruce A. Bunnell Center for Stem Cell Research and Regenerative Medicine, Tulane University School of Medicine, New Orleans, LA, United States

Louis Casteilla RESTORE, Toulouse University, INSERM-1301, CNRS-5070, EFS, ENVT, University Toulouse III - Paul Sabatier (UPS), Toulouse, France

Gregorio D. Chazenbalk Department of Obstetrics and Gynecology, University of California Los Angeles (UCLA), Los Angeles, CA, United States

Mary Ann Chirba Boston College Law School, Newton Centre, MA; Tufts Medical School, Boston, MA; New York University Law School, New York, NY, United States

Adam Cottrill Department of Plastic Surgery, University of Pittsburgh, Pittsburgh, PA, United States

Béatrice Cousin RESTORE, Toulouse University, INSERM-1301, CNRS-5070, EFS, ENVT, University Toulouse III - Paul Sabatier (UPS), Toulouse, France

Christian Dani Université Côte d'Azur, iBV, UMR CNRS/INSERM, Faculté de Médecine, Nice, France

Vincent Dani Université Côte d'Azur, iBV, UMR CNRS/INSERM, Faculté de Médecine, Nice, France

Francesco M. Egro Department of Plastic Surgery, University of Pittsburgh Medical Center, Pittsburgh, PA, United States

Asim Ejaz Department of Plastic Surgery, University of Pittsburgh Medical Center, Pittsburgh, PA, United States

Roberto D. Fanganiello Group of Oral Ecology Research (GREB), Faculty of Dentistry, Laval University, Quebec, QC, Canada

Nathalie Faucheux Laboratory of Cell-Biomaterial Biohybrid Systems, Department of Chemical and Biotechnological Engineering, University of Sherbrooke; CHUS Research Center, Sherbrooke, QC, Canada

Lauren Flynn Department of Anatomy and Cell Biology, Schulich School of Medicine and Dentistry; Department of Chemical and Biochemical Engineering, Thompson Engineering Building, The University of Western Ontario, London, ON, Canada

Julie Fradette The Tissue Engineering Laboratory (LOEX), a Laval University Research Center, and Regenerative Medicine Division, CHU of Quebec—Laval University Research Center; Department of Surgery, Faculty of Medicine, Laval University, Quebec, QC, Canada

Mallory D. Griffin Biomedical Engineering, Carnegie Mellon University, Pittsburgh, PA, United States

Mark A.A. Harrison Center for Stem Cell Research and Regenerative Medicine, Tulane University School of Medicine, New Orleans, LA, United States

Jessica Jann Laboratory of Cell-Biomaterial Biohybrid Systems, Department of Chemical and Biotechnological Engineering, University of Sherbrooke; CHUS Research Center, Sherbrooke, QC, Canada

Veronica Morgan Jones Department of Surgery, Wake Forest, School of Medicine, Winston-Salem, NC, United States

Adam J. Katz Department of Plastic Surgery, Wake Forest University, Winston-Salem, NC, United States

Fabien Kawecki The Tissue Engineering Laboratory (LOEX), a Laval University Research Center, and Regenerative Medicine Division, CHU of Quebec—Laval University Research Center; Department of Surgery, Faculty of Medicine, Laval University, Quebec, QC, Canada

Bhavesh D. Kevadiya Interventional Regenerative Medicine and Imaging Laboratory, Department of Radiology, Stanford University School of Medicine, Palo Alto, CA, United States

Lauren Kokai Departments of Plastic Surgery and Bioengineering; McGowan Institute for Regenerative Medicine, University of Pittsburgh, Pittsburgh, PA, United States

Karen L. Leung Department of Obstetrics and Gynecology, University of California Los Angeles (UCLA), Los Angeles, CA, United States

Daniel D. Liu Interventional Regenerative Medicine and Imaging Laboratory, Department of Radiology, Stanford University School of Medicine, Palo Alto, CA, United States

Shawn Loder Department of Plastic Surgery, University of Pittsburgh, Pittsburgh, PA, United States

Michael T. Longaker Department of Surgery, Division of Plastic and Reconstructive Surgery; Institute for Stem Cell Biology and Regenerative Medicine, Stanford University School of Medicine, Stanford, CA, United States

Kacey Marra Departments of Plastic Surgery and Bioengineering; McGowan Institute for Regenerative Medicine, University of Pittsburgh, Pittsburgh, PA, United States

Danielle Minteer Plastic & Reconstructive Surgery, University of Pittsburgh Medical Center, Pittsburgh, PA, United States

Omair A. Mohiuddin Center for Stem Cell Research and Regenerative Medicine, Tulane University School of Medicine, New Orleans, LA, United States

Benjamen T. O'Donnell Center for Stem Cell Research and Regenerative Medicine, Tulane University School of Medicine, New Orleans, LA, United States

Hakan Orbay Department of Surgery, Crozer Chester Medical Center, Upland, PA, United States

George E. Panagis Department of Plastic Surgery, School of Medicine, University of Pittsburgh, Pittsburgh, PA, United States

Ivona Percec Department of Surgery, Perelman School of Medicine, University of Pennsylvania, Philadelphia, PA, United States

Yang Qiao Interventional Regenerative Medicine and Imaging Laboratory, Department of Radiology, Stanford University School of Medicine, Palo Alto, CA; Department of Interventional Radiology, The University of Texas MD Anderson Cancer Center, Houston, TX, United States

Ricardo Rodriguez CosmeticSurg, LLC, Luthersville, MD, United States

J. Peter Rubin Departments of Plastic Surgery and Bioengineering; McGowan Institute for Regenerative Medicine, University of Pittsburgh, Pittsburgh, PA, United States

David E. Sahar University of California Davis Medical Center, Sacramento, CA, United States

Ankit Salhotra Department of Surgery, Division of Plastic and Reconstructive Surgery; Institute for Stem Cell Biology and Regenerative Medicine, Stanford University School of Medicine, Stanford, CA, United States

Yasamin Samadi Department of Plastic Surgery, University of Pittsburgh Medical Center; Department of Plastic Surgery, University of Pittsburgh, Pittsburgh, PA, United States

Benjamin K. Schilling Department of Bioengineering, School of Engineering, University of Pittsburgh, Pittsburgh, PA, United States

Harsh N. Shah Department of Surgery, Division of Plastic and Reconstructive Surgery; Institute for Stem Cell Biology and Regenerative Medicine, Stanford University School of Medicine, Stanford, CA, United States

Xiaoyin Shan Department of Surgery, Perelman School of Medicine, University of Pennsylvania, Philadelphia, PA, United States

Abra H. Shen Department of Surgery, Division of Plastic and Reconstructive Surgery, Stanford University School of Medicine, Stanford, CA, United States

Patsy Simon Department of Plastic Surgery, Regulatory and Clinical Affairs, Center for Innovation in Restorative Medicine, University of Pittsburgh, Pittsburgh, PA, United States

Brianne N. Sullivan Center for Stem Cell Research and Regenerative Medicine, Tulane University School of Medicine, New Orleans, LA, United States

Ganesh Swaminathan Interventional Regenerative Medicine and Imaging Laboratory, Department of Radiology, Stanford University School of Medicine, Palo Alto, CA, United States

Avnesh S. Thakor Interventional Regenerative Medicine and Imaging Laboratory, Department of Radiology, Stanford University School of Medicine, Palo Alto, CA, United States

Matthias Waldner Department of Plastic Surgery and Hand Surgery, University Hospital Zurich, Zurich, Switzerland

John Walker Department of Anatomy and Cell Biology, Schulich School of Medicine and Dentistry, The University of Western Ontario, London, ON, Canada

Derrick C. Wan Department of Surgery, Division of Plastic and Reconstructive Surgery, Stanford University School of Medicine, Stanford, CA, United States

Rachel M. Wise Center for Stem Cell Research and Regenerative Medicine, Tulane University School of Medicine, New Orleans, LA, United States

Xi Yao Université Côte d'Azur, iBV, UMR CNRS/INSERM, Faculté de Médecine, Nice, France

Preface

Adipose stem cells (ASCs) are a critical, multipotent adult stem cell population, utilized for decades in exciting fields such as tissue engineering, regenerative medicine, and medical research. Since 1995, the number of journal publications regarding ASCs has increased every consecutive year, with over 2000 new publications in 2020 alone. As research with ASCs has encompassed a broad spectrum of ideas, sciences, and technologies, there is significant diversity in represented perspectives in literature, including those of clinicians, scientists, industry developers, and federal regulators. This book, *The Scientific Principles of Adipose Stem Cells*, intends to provide readers with indepth and expert knowledge in ASC theory and practice and combines knowledge from many pioneers in ASC research, of diverse background, into a cohesive and balanced educational tool. The chapter topics range from basic principles regarding adipose developmental origins, critical signaling mechanisms, and immunomodulatory properties, to clinical insights into ASC therapeutic applications. It is our hope that new and experienced readers will discover new facets of ASCs that promote scientific advancement of this important and rapidly expanding research field.

As the scope of ASC research is multidisciplinary and complex, the contents of this book have been developed with specific focus toward scientists and clinicians who are broadly familiar with basic cell biology principles and are seeking advanced information specific to ASCs. As such, the material is expansive and ranges from developmental biologic origins to current governmental regulatory considerations for clinical utilization. As ASC research continues to diversify into fields such as biochemistry, genetics, and molecular biology as well as new and exciting arenas such as environmental science, energy, and outer space, we hope that this book will provide valuable and expert ASC knowledge applicable to readers in a variety of fields. Further, we have included information for those moving ASCs from the benchtop to the bedside or those working in industry. Finally, we have reviewed ASC applications in cutting edge applications such as microphysiologic "organ-on-a-chip" systems, which will undoubtedly have important applications in drug discovery and for tissue homeostasis and pathology. Due to ease of access, potency, and safety, ASCs are well positioned to make substantial contributions to the future of cell therapy, and this book aims to benefit new students and established investigators driving this process forward.

We could not have created this collective without expert knowledge from leaders in the field, including many members of the International Federation of Adipose Therapeutics and Science, or IFATS, to whom we are forever indebted. To this end, we humbly thank our many authors who found time in their busy schedules to contribute in-depth content on complex and sophisticated topics imperative for holistic education on ASCs. Further, we are grateful for our publisher, Elsevier, for their outstanding administrative and editorial assistance and ensuring that we were *nonsolus* during the long process of bringing this book to fruition. Finally, we are forever grateful to our families for their support and patience as we devoted long nights and weekends to this work, taking time from our all-encompassing careers and burning the midnight oil toward this labor of love. Given the 2019–21 SARS-CoV-2 pandemic, it has been an extremely interesting time to attempt such a feat and required many expert interviews to be held virtually from the inside of a coat closet, hiding from our children. However, this perspective also enabled us increased reverence on the power of scientific advancement to transform societal health and medicine. As we posit what the future may hold for ASCs, we know that the knowledge contained within these pages is only the beginning of the story, that the deeper mechanistic understanding of ASCs' regenerative capacity is an evolving field, and that through perseverance and fortitude, we will collectively find a path toward ASC clinical utilization that is safe, efficacious, and transformative.

PART 1

Introduction and front matter

CHAPTER

1

Plastic surgery, fat, and fat plasticity: How adipose tissue changed the landscape of stem cell therapeutics

Lauren Kokai[a,b] *and J. Peter Rubin*[a,b]

[a]Departments of Plastic Surgery and Bioengineering, University of Pittsburgh, Pittsburgh, PA, United States [b]McGowan Institute for Regenerative Medicine, University of Pittsburgh, Pittsburgh, PA, United States

Pittsburgh, Pennsylvania, is a famous point of confluence. The city is positioned at the head of the Ohio River, where the tortuous and powerful Monongahela and Allegheny rivers come together, creating a hill-laden landscape that is highly dissected by numerous bridges and tunnels. Pittsburgh is also a famous city of rebirth. What was once an ash-covered capital of the coal industry is now a beacon of medical and technological innovation. It therefore seems appropriate that it was there, at the University of Pittsburgh Medical Center (UPMC) in the mid-1990s, where an immunologist, a plastic surgeon, and an aspiring tissue engineer came together and discovered a new purpose for discarded adipose tissue. At that time, the field of tissue engineering was just beginning to flourish, and Pittsburgh was one of the earliest hotbeds, thanks in large part to Peter Johnson, MD, and the founding of the Pittsburgh Tissue Engineering Initiative. At the same time, Thomas Starzl, MD, PhD, a pioneer and driving force in solid organ transplantation science was recruiting many top immunologists, including Ramon Llull, MD, PhD, to relocate to Pittsburgh for research experience. And finally, William (Bill) Futrell, MD, chair of the Plastic Surgery Division, was intensely building a research program and simultaneously had entirely too much discarded fat from the operating room on his hands.

The field of tissue engineering loosely began in the late 1980s when clinicians and engineers worked together with the goal of manipulating living tissues and potentially combining them with prosthetic materials [1]. However, a critical moment in the birth of a new, multidisciplinary field came in the form of a publication in *Science* in 1993 by Langer and Vacanti, where the authors defined tissue engineering as "applying the principles of biology and engineering to the development of functional substitutes for damaged tissue" [2]. The

Scientific Principles of Adipose Stem Cells
https://doi.org/10.1016/B978-0-12-819376-1.00009-3

development of the PTEI in 1996 was one of the first significant efforts outside of Boston to explore the potential of this new field. At that time, one of the long-term goals of tissue engineers was to make whole human organs for transplantation into patients, and to do this, scientists quickly learned that you needed cells—a lot of cells. To make a new muscle, for example, a scientist required a scaffold and a cell population that could either be expanded on the scaffold before inducing maturation or inoculated as functional cells. While efforts in the laboratory with in vitro and animal experiments showed initial success, clinical translation of engineered tissues was frustratingly out of reach. Reading the literature and listening to presentations, Ramon Llull (and many others) could only conclude that it was a lack of cell density that resulted in the failure to transform engineered tissues into a beneficial therapeutic effect. Thus began a search for an adequate tissue source that continued for decades, as evident by a 2006 article by Charles Vacanti, one of the fathers of tissue engineering, who wrote, "To ultimately be effective in humans, it will be necessary to generate relatively large volumes, starting with very few cells. Mature cells, expanded in vitro, lose efficacy…To be effective, cells should be easily procured, effectively expanded *in vitro*, survive the initial implantation, be accepted as self; that is, not recognized as foreign, and function normally and not become malignant. In addition, it would also be quite convenient if no moral concerns or questions were generated as a result of the cell type used" [1].

Adam Katz, MD, arrived in Pittsburgh in 1993 by way of the University of Michigan and Duke University as a young resident in the Division of Plastic Surgery at UPMC. He grew up playing varsity sports and has a strong "team player" mentality; thus, he fit in well with Dr. Futrell and the University of Pittsburgh Plastic Surgery family. Adam was part of the new 3-2-3 model for plastic surgery residents, where he first completed 3 years of general surgery, then 2 years of dedicated research, and finished with 3 years of plastics. For his research project, Adam decided to create a new tissue-engineered breast to be used in reconstructive surgery, which, given the volume of the tissue being replaced, would require an enormous source of cell inoculum. At this same time, Bill Futrell was using departmental budget to dispose of human adipose tissue, and he was not happy about it. Plastic surgeons in the department were frequently performing a number of procedures that resulted in adipose medical waste, including liposuction, body contouring (abdominoplasty, panniculectomy), etc., and the resulting medical waste, i.e., human adipose tissue, required packaging into large red bags marked "medical waste hazard" and processing according to the local and state statutes regarding the method of disposal. Thus, Adam was given the task by his Chief of figuring out how to repurpose adipose tissue destined for the biohazard bin and extract valuable cells to generate new tissues.

As Pittsburgh is a city of bridges, so too does the discovery of adipose stem cells (ASCs) require a bridge in the form of Ramon Llull, MD, PhD. Ramon received his medical degree in Pamplona, Spain, prior to completing a PhD in immunology at the University of California at Irvine. A brilliant immunologist with interest in transplantation, Ramon was recruited to Pittsburgh by Thomas Starzl for a 1-year post-doc position in the Pittsburgh Transplantation Institute (PTI). Dr. Starzl met Ramon during a Paris meeting of the Transplantation Society and was impressed. As Ramon's time was coming to an end in the PTI laboratory, Dr. Starzl connected Ramon to Bill Futrell and endorsed him for a resident position in Plastics. To this day, Dr. Futrell speaks glowingly of Ramon's impact on adipose research and recounts past conversations with Dr. Starzl whereupon Dr. Starzl admitted that "Ramon's the smartest guy I've ever seen."

Though Adam was coming into the laboratory as Ramon was transitioning into the clinic, the two quickly became partners in research. Adam was the driving force at the bench. He designed experiments by building upon past work in adipose literature and by absorbing insights from publications by Arnold Caplan, PhD, that utilized bone marrow mesenchymal stem cells. Ramon assisted as often as possible, mentoring Adam in the lab on nights and weekends and aiding with interpretation of experimental outcomes. Together Adam and Ramon discovered that cultured adipose tissue yielded fibroblast-like cells, that those cells looked very similar to bone-marrow stem cells, and that they could in fact be stimulated to differentiate into adipocytes and bone. Donning the sheen of positive results that only a young scientist can, one morning Adam brought with him a T25 flask with adipose cells stained black with Von Kossa, evidence of calcium mineralization and osteogenesis, to a morbidity and mortality (M&M) conference and informed Ramon and Bill Futrell that he had succeeded in his quest; he had learned to culture and manipulate progenitor cells from fat.

Plastic surgery and fat plasticity

In the late 1970s and into the early 1980s, the race to identify the adipocyte precursor cell was in full swing. Thanks to improvements in the Coulter counter, an apparatus used to count and size particles or cells, scientists were able to observe how adipocyte size and number related to lipid deposition in the body, and the frequency distribution of adipocyte sizes during adipose expansion was coming into focus [3]. At the time, there was limited knowledge of the mechanisms through which white adipose tissue expanded by preadipocyte hypertrophy, and there was an urgent need to understand hormonal and genetic drivers of adipogenesis to combat the growing obesity epidemic. Cell populations suggested as precursors to adipocytes included free endothelial cells, fibroblast-like perivascular reticulum cells, and macrophages [4]. By the late 1980s, protocols for isolating stromal-vascular cells and fat cells from human adipose tissue for culture were well established [5]. Further, it was known that ceiling-cultured mature adipocytes could dedifferentiate into mitotically active cells [6]. However, though adipocyte progenitor cells could be isolated from adipose tissue and cultured on tissue culture plastic, the capacity of these cells to undergo maturation into mature adipocytes was highly variable. In some reports, roughly only 6.5% of the originally inoculated cell population exhibited an adipocyte morphology after stimulation [7]. This was in part because, at that time, little was known regarding adipoblasts (i.e., ASCs), and markers to distinguish between fat cell precursor populations were lacking. Because the stromal vascular fraction of enzymatically digested adipose contains only a small percentage of progenitor cells ($\sim$1%), to adequately explore and characterize ASCs, researchers needed a much larger quantity of starting material.

At the same time that scientists were exploring adipose progenitor cells in the laboratory, dermatologists and plastic surgeons were rapidly expanding the use of liposuction for body contouring. Improvements in the surgical technique by Jeffrey Klein, MD, which combined local anesthesia with tumescent solution [8] and improved instrumentation [9], brought about an increased level of comfort and confidence with the procedure, enabling large volumes (<5 L) of adipose tissue to be removed with minimally invasive procedures. Initially, many physicians perceived the liposuction product, i.e., lipoaspirate, containing residual tumescent fluid and tissue fragments as nonviable. The procedure to remove excess adipose involved pressing tissue into a hole in a cannula and shearing the tissue particle with back-and-forth motion before it is sucked into a collection cannister with negative pressure. In most instances, the lipoaspirate was discarded as medical waste with little thought. Sydney Coleman, MD, however, thought differently. Dr. Coleman learned of autologous fat grafting, or transferring a patient's own adipose to a new location by way of liposuction and injection, while in Plastic Surgery residency in the 1980s [10, 11] and independently worked to refine and advance the technique for over a decade before publishing the now widely used technique of adipose condensation with centrifugation prior to injection, commonly referred to as the "Coleman Technique" [12]. Using this technique, Dr. Coleman reported that grafted fat obtained through liposuction achieved long-term results, perhaps, he hypothesized, even permanent. Though fat grafting was first decried by some in the surgical community as tantamount to malpractice [13], Dr. Coleman's vision of "lipoinfiltration" withstood the test of time, and the applications in aesthetic and reconstructive surgeries were indeed profound.

As the impact of fat grafting on aesthetic and reconstructive surgery took hold in the late 1990s, many surgeons travelled to New York City to attend Dr. Coleman's lectures and/or courses to learn his technique. Adam Katz, then a resident at UPMC Plastic Surgery working in the laboratory with fat, together with Dennis Hurwitz, MD, an attending plastic surgeon, were two such surgeons who heard Dr. Coleman speak on "structural fat grafting." During his talk, Dr. Coleman reflected on the regenerating effects he had observed in many of his patients. The fact that lipoaspirate tissue maintained volume after injection into new tissue beds and appeared to impact the local tissue strongly implied that lipoaspirate contained viable cells, and perhaps, as Katz, Hurwitz, and Coleman postulated in a 2001 article, even multipotential stem cells [14]. Upon returning to Pittsburgh, Adam and Ramon successfully showed what no one had previously shown, that lipoaspirate contained progenitor cells capable of differentiating into multiple mesenchymal cell lineages. Due to the novelty of this discovery and concern of peer criticism, Adam and Ramon's work was verified prior to publication through a collaboration with the Regenerative Bioengineering and Research ("REBAR") laboratory at University of California at Los Angeles (UCLA) [15]. Less than 20 years later, this original article has been cited close to 9000 times and stands as a pillar in the subsequent field of ASC research that emerged.

The Adipose Stem Cell Center in Pittsburgh and the International Federation of Adipose Therapeutics and Science

As Adam Katz and Ramon Llull left Pittsburgh to build laboratories in Charlottesville, VA, and Mallorca, Spain, respectively, a young plastic surgeon-scientist from Harvard and Massachusetts General Hospital arrived in Pittsburgh. In Boston, J. Peter Rubin, MD, experienced laboratory training in both tissue engineering and transplant immunology in the early 1990s and developed a strong interest in adipose biology and tissue engineering. In 2002, Peter joined the Plastic Surgery faculty at the University of Pittsburgh with the intent of starting an ASC laboratory and sourcing adipose tissue from his clinical work in body contouring after weight loss. He had the good fortune to collaborate with Kacey Marra, PhD, a biomaterials scientist and polymer chemist, and together they founded the Adipose Stem Cell Center at the University of Pittsburgh in 2005. Peter also developed strong collaborations with Adam Katz,

Ramon Llull, and Bill Futrell. Recognizing that this was a new and rapidly evolving field, the four of them founded The International Federation for Adipose Therapeutics and Science (IFATS), a nonprofit scientific foundation, serving as the world's only interdisciplinary fat tissue society, with a focus on ASCs and the regenerative applications of ASCs and adipose tissue. IFATS was designed to provide an annual forum for the presentation of cutting-edge science relating to ASCs and to help facilitate collaboration and networking within this growing scientific community. The first meeting was convened in beautiful Mallorca, Spain, in 2003, hosted by Ramon Llull. The second meeting was held in Pittsburgh in 2004, hosted by Peter Rubin, and the third meeting convened in Charlottesville, VA, in 2005, hosted by Adam Katz. Over the years, many significant leaders and contributors in the field of ASC biology have hosted the meeting across the globe (Table 1). The 17th Annual IFATS Meeting in Marseille, France, drew participants from 37 countries.

TABLE 1 Historical International Federation for Adipose Therapeutics and Science conferences from inception to date (2020).

Year	Location	Host
2003	Mallorca, Spain	Ramon Llull, MD, PhD
2004	Pittsburgh, PA, USA	J. Peter Rubin, MD
2005	Charlottesville, VA, USA	Adam Katz, MD
2006	Baton Rouge, LA, USA	Jeffrey Gimble, MD, PhD
2007	Indianapolis, IN, USA	Keith March, MD, PhD
2008	Toulouse, France	Anne Bouloumie, PhD,
		Louis Casteilla, PhD
2009	Daegu, South Korea	Sang Hong Baek, MD, PhD
		Jae-Ho Jeong, MD, PhD
2010	Dallas, TX, USA	Spencer Brown, PhD
2011	Miami Beach, FL, USA	Stuart Williams, PhD
2012	Quebec City, QC, Canada	Julie Fredette, PhD
2013	New York City, NY, USA	Kacey Marra, PhD
		Sydney Coleman, MD
2014	Amsterdam, The Netherlands	Marco Helder, PhD
2015	New Orleans, LA, USA	Bruce Bunnell, PhD
2016	San Diego, CA, USA	Ricardo Rodriguez, MD
2017	Miami Beach, FL, USA	Brian Johnstone, PhD
2018	Las Vegas, NV, USA	Kotaro Yoshimura, MD
2019	Marseille, France	Guy Magalon, MD
2021	Fort Lauderdale, FL, USA	Ivona Percec, MD, PhD

Since 2001, the number of journal publications regarding ASCs has linearly increased in both basic foundational science and in clinical research. While developing primarily out of the field of plastic surgery, ASCs are now used for tissue engineering applications, drug discovery, and developmental models and as therapeutics. Due to the incredible international interest, ASCs have been utilized in a significant number of published clinical studies, for a variety of applications. Further, multiple corporations have been formed around ASC processing devices or use as a cell therapy. With this deepening scientific and clinical experience with ASCs, there is now an incredible breadth of new information available in multiple scientific journal publications. In this book, we present a compendium of the scientific principles of ASCs into one encompassing and thorough resource. It is our intent to provide readers with in-depth and expert knowledge on ASCs in regard to developmental biologic origins, foundational research on signaling mechanisms and immunomodulatory properties, and clinical insights into applications in regenerative medicine. We are confident that continued exploration of these easily extractable mesenchymal stem cells will lead to valuable new therapies, and we are continuously humbled by the complexity with which ASCs impact health.

References

[1] C. Vacanti, The history of tissue engineering, J. Cell Mol. Med. 10 (3) (2006) 569–576.
[2] R. Langer, J.P. Vacanti, Tissue engineering, Science 260 (1993) 920–926.
[3] P.J.H. Björntorp, Size, number and function of adipose tissue cells in human obesity, Horm. Metab. Res. (1974) 77.
[4] G.J. Hausman, D.R. Campion, R.J. Martin, Search for the adipocyte precursor cell and factors that promote its differentiation, J. Lipid Res. 21 (6) (1980) 657–670.
[5] D. Dixon-Shanies, et al., Observations on the growth and metabolic functions of cultured cells derived from human adipose tissue, Exp. Biol. Med. 149 (2) (1975) 541–545.
[6] D.A. Roncari, S. Kindler, C.H. Hollenberg, Excessive proliferation in culture of reverted adipocytes from massively obese persons, Metabolism 35 (1) (1986) 1–4.
[7] P. Pettersson, et al., Adipocyte precursor cells in obese and nonobese humans, Metabolism 34 (9) (1985) 808–812.
[8] J.A. Klein, The tumescent technique for lipo-suction surgery, Am. J. Cosmet. Surg. 4 (4) (1987) 263–267.
[9] L.S. Toledo, Liposculpture of the face and body, in: Annals of the international symposium recent advances in plastic surgery, 1989.
[10] M. Bircoll, B.H. Novack, Autologous fat transplantation employing liposuction techniques, Ann. Plast. Surg. 18 (4) (1987) 327–329.
[11] R. Ellenbogen, Free autogenous pearl fat grafts in the face—a preliminary report of a rediscovered technique, Ann. Plast. Surg. 16 (3) (1986) 179–194.
[12] S. Coleman, The technique of periorbital lipoinfiltration, Oper. Tech. Plast. Reconstruct. Surg. 1 (3) (1994) 120–126.
[13] S.R. Coleman, A.P. Saboeiro, Fat grafting to the breast revisited: safety and efficacy, Plast. Reconstr. Surg. 119 (3) (2007) 775–785.
[14] D. Hurwitz, S.R. Coleman, A. Katz, Structural fat grafting of the face: lessons from a teacher and his student, Key Issues Plast. Cosmet. Surg. 17 (2001) 14–38.
[15] P.A. Zuk, et al., Multilineage cells from human adipose tissue: implications for cell-based therapies, Tissue Eng. 7 (2) (2001) 211–228.

PART 2

Basic biology of adipose stem cells

CHAPTER

2

Developmental origins of adipocytes: What we learn from human pluripotent stem cells

Xi Yao, Vincent Dani, and Christian Dani

Université Côte d'Azur, iBV, UMR CNRS/INSERM, Faculté de Médecine, Nice, France

Introduction: Three types of adipocytes in mammals

In mammals, three types of adipocytes coexist, i.e., white, brown, and beige/brite. White adipose tissue (WAT) is dispersed throughout the body and is mainly involved in energy storage. The two largest depots of WAT in human and rodents are subcutaneous and visceral WAT. These two types of WAT differ in important aspects during obesity. Whereas increased subcutaneous fat depots present little or no cardiovascular risk, increased white visceral fat depots correlate with adverse metabolic outcomes of obesity. In contrast to WAT, brown adipose tissue (BAT) has an abundance of mitochondria and utilizes stored energy to produce heat. BAT represents a minor fraction of adipose tissue in adult humans and disappears from most areas with age, persisting only in the anterior neck, in thoracic paravertebral location, and around deeper organs [1]. Beige/brite adipocytes, the third type of adipocytes identified more recently, are brown-like adipocytes (BAs) that reside in clusters scattered within subcutaneous WAT depots.

BAs are recruited in WAT upon stimulation such as cold. It has been demonstrated that white and brown adipocytes are generated from distinct adipose progenitors (APs) that originate from different developmental pathways. Different cellular sources for BAs have been reported, and BAs may derive from transdifferentiation of white adipocytes [2]. Alternatively, distinct BA progenitors have been identified from stromal vascular cell populations of mouse and human subcutaneous WAT [3, 4]. Therefore, the three types of adipocytes are all involved in energy balance regulation while also having opposite functions. The question as to whether adipocytes with different properties originate from disparate developmental pathways has recently emerged.

Scientific Principles of Adipose Stem Cells
https://doi.org/10.1016/B978-0-12-819376-1.00015-9

Lineage tracing, the gold standard approach to identify the origin of tissue progenitors, has revealed that, for mice, the origin of adipocytes is more complex than first thought. Investigations on the developmental origins of adipocytes in humans have been initiated more recently, thanks to the use of human PSCs, i.e., ESCs and iPSCs. Analyses of PSC differentiation toward the adipogenic lineage revealed that human brown/beige adipocytes may derive from different developmental pathways. Understanding the mechanisms governing the commitment of human PSCs toward adipocytes should help to characterize new therapeutic targets to reduce or promote an adipocyte type in patients. For regenerative medicine, knowing the embryonic origin of adipocytes of different fat depots should allow surgeons to better match the donor and the host fat site to improve postoperative outcomes and safety of fat grafting.

Developmental origins of white adipocytes

Adipocytes are generally described to derive from mesenchymal stem cells (MSCs), which themselves are thought to arise from mesoderm. However, it is worth noting that during development of higher vertebrates, the mesoderm is not the only germ layer source of mesenchymal cells. In the head, for instance, the facial bones have been shown to derive from the neural crest (NC). The NC is a vertebrate cell population that arises from the neuroectoderm. After neural tube closure, NC cells (NCCs) undergo an epithelial–mesenchyme transition and migrate to diverse regions in the developing embryo, where they differentiate into various cell types. In the head and neck, the NC also yields mesenchymal precursors differentiating into connective tissue cells (reviewed in Ref. [5]). ESCs from mouse were the first to be used to investigate the earliest steps of adipocyte development in vitro and revealed the surprising conclusions regarding the ontogeny of white adipocytes in the NC.

White adipocytes from mouse embryonic stem cells derive from neural crest

Mouse ESCs are proliferating PSCs that have been isolated from the epiblast of blastocyst-stage mouse embryos. They can be propagated indefinitely in an undifferentiated state in vitro. Furthermore, when transplanted into a mouse blastocyst, ESCs integrate into the embryo and contribute to all cell lineages, including germ cells. When aggregated to form embryoid bodies (EBs) in vitro, they undergo differentiation in ectodermal, mesodermal, and endodermal derivatives. In addition, ESCs are easy to modify genetically and can be produced in large numbers, thus offering a unique cell culture model to study the earliest steps of mammalian development [6].

Directed differentiation of mouse ESCs toward the adipocyte lineage was first accomplished in 1997 by Dani et al., who showed that functional adipocytes could be obtained when ESCs were exposed to appropriate extracellular cues [7]. In this system, the generation of adipocytes is dependent on an early and transient exposure of EBs to retinoic acid (RA) and a subsequent treatment with conventional adipogenic factors (e.g., insulin, triiodothyronine, and rosiglitazone). Thus, this model has provided a powerful system to address the different steps of adipocyte development [8, 9].

In a first attempt to unravel the events underlying the formation of mesenchymal derivatives in RA-treated mouse ESCs, Kawaguchi et al. examined the expression of various mesodermal and mesenchymal markers in early EBs. Surprisingly, they noticed that treatment with RA resulted in a sharp reduction in expression of several mesodermal markers such as *Brachyury, Flk1*, and *PDGFRα*, as well as in the suppression of cardiomyocyte formation, suggesting that RA reduces overall mesoderm formation [10]. Since RA at 10^{-6}M or 10^{-7}M was also shown to promote neural differentiation of mouse ESCs, these authors then analyzed the expression of various NC markers. They showed that *Sox9, Sox10, Foxd3*, and *Runx2*, which all play an important role in NC formation, were upregulated upon RA treatment.

Altogether, these studies suggest that adipocytes generated from RA-exposed mouse ESCs, arise from the NC, rather than from the mesoderm. It is important to note that RA commits ESCs toward the adipogenic lineage when ESCs are exposed to RA for 3 days at an early stage of development, i.e., during mesoderm and NC formation. The RA pathway is also important in the latest stage of adipocyte differentiation through stimulation of uncoupling protein 1 (UCP1) in brown adipocytes [11].

White adipocyte developmental origins in quail and mouse embryos

Billon et al. isolated primary cultures of quail NCCs from both the cephalic and thoracic level and maintained them in culture medium permissive to adipocyte differentiation. This analysis revealed that typical mature adipocytes could be readily produced from cephalic NCCs and, to a lesser extent, from truncal NCCs [12, 13]. Next, a lineage tracing approach was performed in mice to investigate the adipocyte lineage origin in vivo and provided direct evidence for the contribution of the NC. Because Sox10 is strongly and specifically expressed in the NC from early embryonic development and is not expressed in mesoderm, Sox10-cre/yfp transgenic mice were used to map NC derivatives in vivo. This study revealed adipocytes derived from NC in cephalic adipose depots, between the salivary gland and the ear area. In contrast, no NC-derived adipocytes could be detected in truncal adipose depots, including subcutaneous, perirenal, periepididymal, or interscapular tissues. Therefore, this observation provided new information about the ontogeny of the adipocyte lineage and demonstrates that, during normal development, a subset of adipocytes in the face originates from NC and not from mesoderm. The role of RA in the early steps of adipocyte development remains to be demonstrated in vivo in mice. Interestingly, RA has recently been shown to be required for differentiation of cephalic NCCs into adipocytes in developing zebrafish embryos [14], which is a reminiscence of the role of RA in mouse ESC adipogenesis.

Lineage tracing studies reveal a complex situation for the developmental origins of adipocytes

Subcutaneous and visceral adipocytes have different developmental origins

Regarding white adipocytes, the laboratories of Kahn [15] and of Kirkland [16] first reported differences in the expression of 96 genes related to embryonic development between visceral fat and subcutaneous fat, suggesting that these two fat depots have different embryonic origins. Lineage tracing has subsequently confirmed that visceral adipocytes derive from mesothelial Wilms tumor 1 (Wt1)-expressing cells, while subcutaneous and brown adipocytes originate from different pathways [17]. More recently, expression of Wt1 in visceral and not in subcutaneous adipose tissue has been confirmed in human samples [18], although Wt1 did not label all visceral adipocytes but was heterogeneous, suggesting that other lineages could be involved in the generation of visceral adipocytes. A hematopoietic origin of adipocytes has also been reported [19, 20]; however, further investigations are required to determine the contribution of the myeloid lineage in subcutaneous and visceral adipocytes.

White, beige, and brown adipocytes originate from different developmental pathways: Evidence of mosaic origins in individual fat depot

The pioneer work of Timmons and colleagues showed that APs isolated from mouse BAT express Myf5, a critical transcription factor for skeletal myoblast differentiation. In fact, brown Aps (BAPs) have a molecular signature more similar to skeletal myoblast than to white APs (WAPs), and this observation suggests that brown adipocytes, but not beige or white adipocytes, derive from the skeletal muscle developmental pathway [21]. A lineage tracing approach using the Myf5 Cre/Lox has further supported this hypothesis. In this animal model, Cre was expressed under the control of Myf5 promoter as driver, and mice were crossed with indicator mice that express green fluorescent protein (GFP) as reporter under the control of a ubiquitous promoter but blocked by a stop cassette. This cassette is eliminated when Cre is expressed, making Myf5 positive cells and their progenies indelibly marked (Fig. 1). From this Myf5 Cre/Lox model, it was assumed that brown adipocytes share a common precursor with muscle, while white adipocytes, at least inguinal and epididymal, arise from a different lineage.

However, other studies indicate that the situation is more complex. In subsequent lineage tracing tools with different promoters as drivers, examination of different fat depots has revealed that, in addition to BAT, some white adipocyte depots, such as interscapular, anterior subcutaneous, and retroperitoneal WAT, also originate from Myf5 precursors. In addition, the labeling of adipocytes in BAT is heterogeneous, indicating that not all brown adipocytes originate from Myf5 in an individual fat depot. Similar observations were made with Pax3-Cre. Pax3, similar to Myf5, is a myogenic factor. Pax3 and Myf5 adipose tissue lineages largely overlap, but also exhibit some significant differences such as the perigonadal WAT, which is marked by Pax3 but not by Myf5 (for review see Ref. [22]). Therefore, evidence suggests that there is more than one precursor cell origin of both brown and white, a subset of which can be defined by Myf5 and/or Pax3.

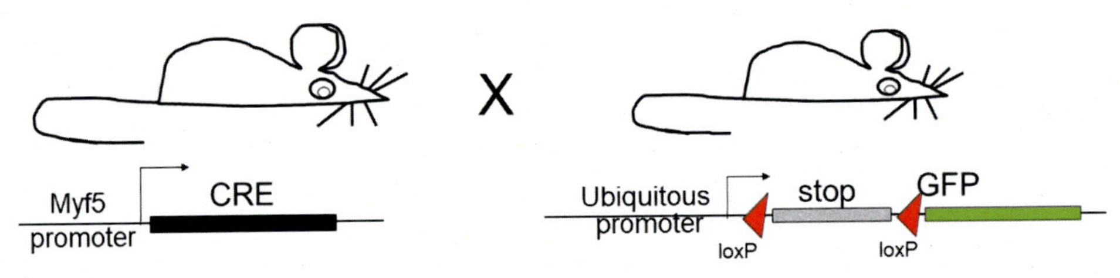

FIG. 1 A strategy for lineage tracing.

Human models to investigate the developmental origins of adipocytes

As lineage tracing cannot be performed in humans, in vitro models that recapitulate the embryonic development of human adipocytes are extremely useful for identifying human adipogenic pathways and development. We will present data generated from human PSCs, ESCs, and iPSCs, and show that these models enable one to analyze the developmental origins of human white and beige adipocytes.

Capacity of human pluripotent stem cells to generate adipocytes

PSCs, i.e., ESCs and iPSCs, display a quasi-unlimited self-renewal capacity and are an abundant source of multiple cell types of therapeutic interest. Some papers in the early 2000s reported the potential of human ESCs to generate adipocytes [23–25]. These observations suggested that PSCs could be a valuable tool to identify developmental pathways governing the generation of the different types of adipocytes. Then, Taura et al. demonstrated that human iPSCs (hiPSCs) have an adipogenic potential comparable to that of human ESCs [26]. However, these authors did not address the adipogenesis efficiency and the phenotype of adipocytes generated. Surprisingly, a cocktail of hematopoietic factors allowed Nishio and colleagues to report, for the first time, the capacity of hiPSCs to generate substantial BAs [27]. These findings support the idea that, as previously shown in mice [28], the bone morphogenetic protein (BMP) signaling pathway plays a critical role in human brown adipocyte generation. Ahfeldt et al. purified hiPSC-derived fibroblasts that were able to undergo differentiation into white adipocytes or BAs following forced expression of adipogenic master genes [29]. This strategy allowed the generation of human white adipocytes and BAs at a large scale, which may be a powerful tool for drug discovery, but the question arose as to whether these cells with ectopic expression of adipogenic master genes faithfully reflected physiological developmental pathways of adipocytes. More recently, a procedure to isolate expandable BAPs from hiPSCs and to generate high levels of functional BAs with no gene transfer was described [30–32], thus providing an opportunity to make effective use of hiPSC features to identify developmental pathways governing the generation of BAs.

Human pluripotent stem cell commitment toward the white and brown-like adipogenic lineage is regulated by the retinoic acid pathway

Mohsen-Kanson and colleagues, in our laboratory, investigated factors involved in the commitment of PSCs toward adipogenic lineages [30]. Adipogenic genes, including *UCP1*, *Dio2*, *PGC1a*, and *PRDM16*, were detected in differentiated cultures, indicating that cells having a BA gene program were spontaneously generated during differentiation. However, the adipogenesis efficiency was weak, as adipocytes represented only 2% of cells in the differentiated cultures. As previously reported in mouse (see above), the RA pathway appeared to promote hiPSC commitment toward the white adipogenic lineage by increasing the percentage to 15%. In contrast, expression of the brown adipocyte specific marker UCP1 was inhibited in RA-treated cultures. Together, these data support the hypothesis that white and brown adipocytes derive from different pathways. RA pathway activation at an early development stage dramatically promotes the differentiation of human PSCs toward the white adipocyte lineage, while inhibiting the brown adipocyte generation. This observation is reminiscent of the critical role of RA in the early steps of mouse ESC white adipogenesis [7, 10]. The identification of RA targets could provide a means to uncover genes involved in the development of adipocytes. The combination of computational and experimental approaches in mouse ESCs has revealed an extensive network of transcription factors that might coordinate the expression of genes essential for the acquisition of adipocyte characteristics [33]. This could represent a unique comprehensive resource that could be further explored to investigate human adipocyte development.

Adipocytes derived from human induced pluripotent stem cells display a brown-like adipocyte phenotype and originate from mesoderm and from neural crest

A molecular analysis of APs derived from hiPSCs in the absence of RA treatment, to generate UCP1-expressing adipocytes, indicates that they display a molecular brown-like phenotype. A number of earlier studies have proposed several specific markers for brown, beige/brite, and white adipocyte lineages in mouse. Not all of these markers are relevant in humans, but the following ones have been reported in the literature as informative: *ZIC1* (brown), *PAX3, DIO2, CIDEA* (brown/beige/brite), *CD137, TMEM26, TBX1* (beige/brite), *HOXC8, BMP4, HOXA5, HOXC9,* and *TCF21* (white). Hafner and colleagues investigated expression of these markers both in hiPSC-BAPs and in hiPSC-WAPs. *TMEM26, TBX1,* and *TCF21* were detected but with no differential expression between BAPs and WAPS. In contrast, *PAX3, DIO2, CD137, CIDEA,* and *ZIC1* were expressed more in BAPs than in WAPs. *HOXC8, BMP4, HOXA5,* and *HOXC9* were weakly, if at all, expressed in hiPSC-BAPs. Altogether, these data indicate that hiPSC-BAPs were expressed in both classical brown and beige/brite adipocyte markers. Interestingly, Mohsen-Kanson and colleagues reported that brown-like APs derived from hiPSCs expressed PAX3, which has a functional role in the generation of APs [30]. As discussed earlier, Pax3 has been reported in mice as a marker of the brown lineage. PAX3 is also associated with early human NC development and its derivatives in the differentiation of multipotent NC precursors [25, 26]. As Myf5 expression was not detected by the authors during the differentiation of hiPSCs, it has been proposed that human BAs may also be derived from NC rather than from the myogenic lineage origin [31]. Expression of other NC markers, such as *Sox10* and *p75*, precede the generation of PAX3-APs during the differentiation of hiPSCs, supporting the NC origin of BAs (Dani et al., unpublished data). However, this hypothesis remains to be conclusively demonstrated. More recently, Aaron Brown's laboratory described the generation of human beige/brite adipocytes from hiPSCs of defined origin using a developmental approach [32]. The first step consisted of the generation of mesoderm, then the selection of FOXF1+ mesoderm cells that were subsequently induced to undergo differentiation into MSCs that had the capacity to differentiate into beige adipocytes when maintained in the appropriate adipogenic cocktail. At the molecular level, progenitors expressed beige/brite markers such as *CD137* and *TMEM26* but not the brown-specific marker *Zic1*. Altogether, these studies strongly suggest that human BAs are derived both from mesoderm and from NC (Fig. 2). Whether the NC or mesoderm developmental origin has consequences on BA functional properties remains to be addressed.

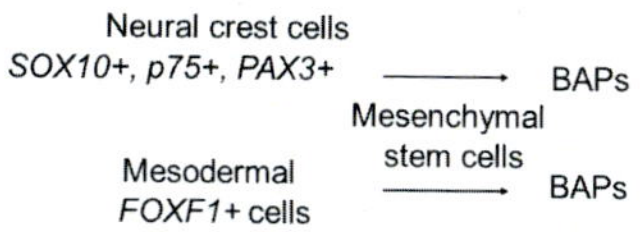

Neural crest cells
SOX10+, p75+, PAX3+ → BAPs
Mesenchymal stem cells
Mesodermal *FOXF1+* cells → BAPs

FIG. 2 Brown-like adipose progenitors can originate both from neural crest (NC) and mesoderm. Expression of specific markers for neural crest and for mesoderm is indicated.

hiPSC-3D adipospheres: The next steps toward gaining greater insight into the development of human brown-like adipocytes

In classic cell culture models, cells are conventionally grown as 2D monolayers, which does not reflect the in vivo situation. Adipose tissue exhibits a complex lobular architecture that plays a functional role in adipogenesis [18]. Indeed, adipose tissue is a highly vascularized tissue and made up of lobules, corresponding to clusters of adipocytes, separated from each other by a structured extracellular matrix [34]. Interestingly, it has been proposed that the adipocyte browning phenomenon specifically occurs in these lobules [35]. In an effort to improve the physiological relevance of in vitro studies, 3D culture technologies and bioengineering methods for seeding different cell types in an organoid-like structure are highly promising [36–38]. Three-dimensional cultures represent a bridge between traditional cell culture and live tissue [36], and 3D suspension cultures that are routinely used by stem cell biologists to differentiate PSCs toward the three germ lineages strongly suggest that the development of hiPSC-BAPs could be investigated in 3D culture conditions. As a first step toward the generation of brown-like adipose tissue-like organoids, our laboratory has formed 3D spheroids of hiPSC-BAPs enriched with endothelial cells that are able to differentiate into adipospheres expressing UCP1 (Fig. 3). Interestingly, hiPSC-adipospheres display an extracellular matrix structure resembling human subcutaneous adipose tissue (Fig. 4). Altogether, these observations highlight the physiological relevance of the in vitro 3D adiposphere model to analyze the earliest steps of adipocyte generation.

Importance of matching the embryonic origin between transplanted cells and the host environment for tissue regeneration: A lesson for the selection of the best donor fate site?

The embryonic origin plays an unsuspected role in the regeneration processes, as recent observations indicate that the match between the donor site and the host environment is crucial for the behavior of transplanted stem cells. Factors governing the fate of adipose stem cells after transplantation in a heterotopic site remain to be fully identified, but the embryonic origin appears to be among them [39]. It has been shown that transplantation of tibia-derived stem cells originating from mesoderm into the mandible environment of NC origin leads to aberrant bone regeneration, illustrating that matching the positional identity and the embryonic origins of transplanted cells with that of the host microenvironment appears as a critical parameter to achieve regeneration. Recent publications underlined that individual fat depots commonly used in reconstructive surgery exhibit distinct embryonic origins. As lineage tracing approaches are not feasible in humans, molecular studies using developmental markers have been investigated to determine the embryonic origin of human facial and limb fat

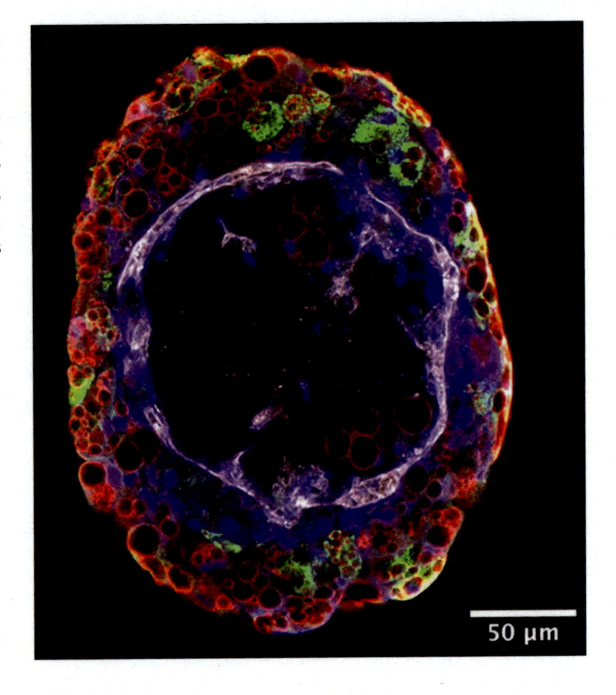

FIG. 3 Human induced pluripotent stem cell (hiPSC)-adipospheres containing a network of endothelial cells and UCP1-expressing adipocytes. hiPSC-spheroids were enriched in endothelial cells expressing a green fluorescent protein (GFP) reporter gene and then induced to differentiate for 24 days. After adiposphere clearing, adipocytes were stained with perilipin 1 (red) and UCP1 (green), and the endothelial cell network was revealed using anti-CD31 antibody.

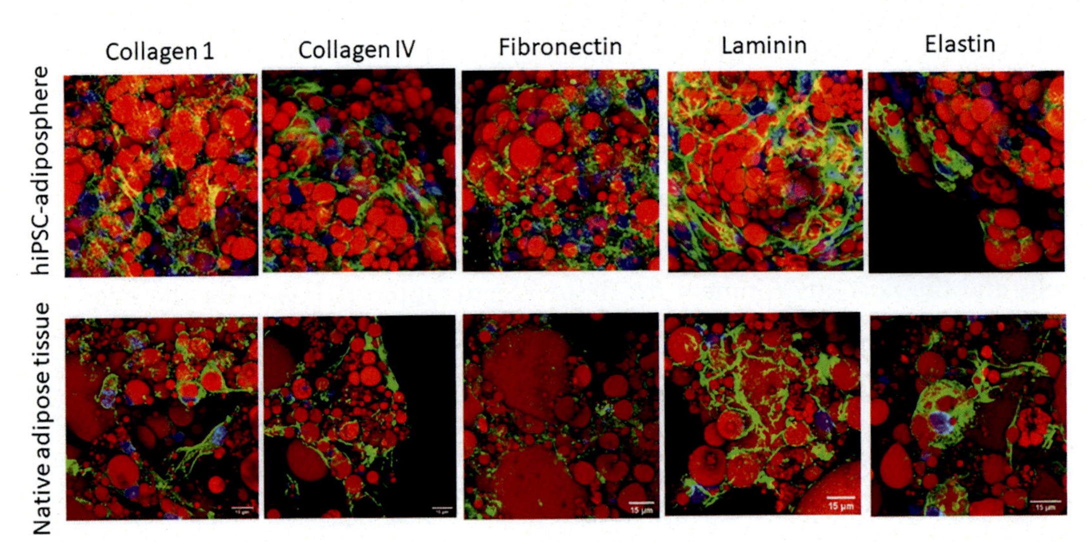

FIG. 4 Expression of extracellular proteins in human induced pluripotent stem cell (hiPSC)-adipospheres and in adipose tissue. hiPSC-adipospheres were maintained in differentiation for 24 days. Human subcutaneous adipose tissues were obtained via liposuction. Lipid droplets were stained with Oil Red O and laminin or fibronectin and revealed by immunofluorescence. DAPI marked cell nuclei.

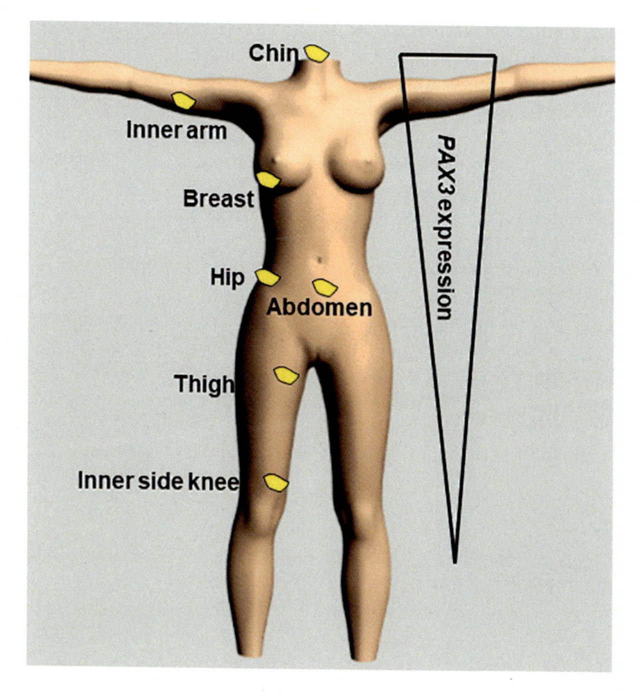

FIG. 5 The fat depots used in reconstructive surgery exhibiting different features. Fat tissue indicated in yellow on the scheme has been analyzed for the expression of PAX3, a marker on neural crest (NC) embryonic origin. As indicated, a gradient of PAX3 expression was revealed.

depots. These studies showed that facial adipose stem cells are of NC origin, whereas limb adipose stem cells are likely of mesodermal origin [40]. In addition to the mismatch of the embryonic origin between these two fat depots, adipose stem cells generate adipocytes presenting a different functional phenotype. In fact, fat depots in the knee and the face display a white and a brown-like phenotype, respectively [40]. These results are in agreement with the NC origin of BAs revealed from hiPSC studies. The molecular profile of several fat depots has been reported more recently [41]. The results, schematized in Fig. 5, show a gradient of PAX3 expression from the upper to the lower body. Altogether, these studies suggest that different fat depots used in clinical practice for plastic and reconstructive surgery for volume replacement have different embryonic origins. This further gives a reflection on the request to choose the most appropriate donor site for fat grafting, according to the host environment.

References

[1] H. Sacks, M.E. Symonds, Anatomical locations of human brown adipose tissue: functional relevance and implications in obesity and type 2 diabetes, Diabetes 62 (6) (2013) 1783–1790.

[2] G. Barbatelli, et al., The emergence of cold-induced brown adipocytes in mouse white fat depots is determined predominantly by white to brown adipocyte transdifferentiation, Am. J. Physiol. Endocrinol. Metab. 298 (6) (2010) E1244–E1253.

[3] R. Xue, et al., Clonal analyses and gene profiling identify genetic biomarkers of the thermogenic potential of human brown and white preadipocytes, Nat. Med. 21 (7) (2015) 760–768.

[4] J. Wu, et al., Beige adipocytes are a distinct type of thermogenic fat cell in mouse and human, Cell 150 (2) (2012) 366–376.

[5] E. Dupin, S. Creuzet, N.M. Le Douarin, The contribution of the neural crest to the vertebrate body, Adv. Exp. Med. Biol. 589 (2006) 96–119.
[6] A.G. Smith, Embryo-derived stem cells: of mice and men, Annu. Rev. Cell Dev. Biol. 17 (2001) 435–462.
[7] C. Dani, et al., Differentiation of embryonic stem cells into adipocytes in vitro, J. Cell Sci. 110 (11) (1997) 1279–1285.
[8] Q. Tong, et al., Function of GATA transcription factors in preadipocyte-adipocyte transition, Science 290 (5489) (2000) 134–138.
[9] Y. Takashima, et al., Neuroepithelial cells supply an initial transient wave of MSC differentiation, Cell 129 (7) (2007) 1377–1388.
[10] J. Kawaguchi, P.J. Mee, A.G. Smith, Osteogenic and chondrogenic differentiation of embryonic stem cells in response to specific growth factors, Bone 36 (5) (2005) 758–769.
[11] R. Alvarez, et al., A novel regulatory pathway of brown fat thermogenesis. Retinoic acid is a transcriptional activator of the mitochondrial uncoupling protein gene, J. Biol. Chem. 270 (10) (1995) 5666–5673.
[12] N. Billon, et al., The generation of adipocytes by the neural crest, Development 134 (12) (2007) 2283–2292.
[13] N. Billon, M.C. Monteiro, C. Dani, Developmental origin of adipocytes: new insights into a pending question, Biol. Cell. 100 (10) (2008) 563–575.
[14] N. Li, et al., Regulation of neural crest cell fate by the retinoic acid and Pparg signalling pathways, Development 137 (3) (2010) 389–394.
[15] S. Gesta, et al., Evidence for a role of developmental genes in the origin of obesity and body fat distribution, Proc. Natl. Acad. Sci. U. S. A. 103 (17) (2006) 6676–6681.
[16] T. Tchkonia, et al., Identification of depot-specific human fat cell progenitors through distinct expression profiles and developmental gene patterns, Am. J. Physiol. Endocrinol. Metab. 292 (1) (2007) E298–E307.
[17] Y.Y. Chau, N. Hastie, Wt1, the mesothelium and the origins and heterogeneity of visceral fat progenitors, Adipocyte 4 (3) (2015) 217–221.
[18] D. Esteve, et al., Lobular architecture of human adipose tissue defines the niche and fate of progenitor cells, Nat. Commun. 10 (1) (2019) 2549.
[19] Y. Sera, et al., Hematopoietic stem cell origin of adipocytes, Exp. Hematol. 37 (9) (2009) 1108–1120. 1120 e1–4.
[20] S.M. Majka, et al., De novo generation of white adipocytes from the myeloid lineage via mesenchymal intermediates is age, adipose depot, and gender specific, Proc. Natl. Acad. Sci. U. S. A. 107 (33) (2010) 14781–14786.
[21] J.A. Timmons, et al., Myogenic gene expression signature establishes that brown and white adipocytes originate from distinct cell lineages, Proc. Natl. Acad. Sci. U. S. A. 104 (11) (2007) 4401–4406.
[22] J. Sanchez-Gurmaches, C.M. Hung, D.A. Guertin, Emerging complexities in adipocyte origins and identity, Trends Cell Biol. 26 (5) (2016) 313–326.
[23] C. Xiong, et al., Derivation of adipocytes from human embryonic stem cells, Stem Cells Dev. 14 (6) (2005) 671–675.
[24] V. van Harmelen, et al., Differential lipolytic regulation in human embryonic stem cell-derived adipocytes, Obesity (Silver Spring) 15 (4) (2007) 846–852.
[25] N.R. Hannan, E.J. Wolvetang, Adipocyte differentiation in human embryonic stem cells transduced with Oct4 shRNA lentivirus, Stem Cells Dev. 18 (4) (2009) 653–660.
[26] D. Taura, et al., Adipogenic differentiation of human induced pluripotent stem cells: comparison with that of human embryonic stem cells, FEBS Lett. 583 (6) (2009) 1029–1033.
[27] M. Nishio, et al., Production of functional classical Brown adipocytes from human pluripotent stem cells using specific hemopoietin cocktail without gene transfer, Cell Metab. 16 (3) (2012) 394–406.
[28] Y.H. Tseng, et al., New role of bone morphogenetic protein 7 in brown adipogenesis and energy expenditure, Nature 454 (7207) (2008) 1000–1004.
[29] T. Ahfeldt, et al., Programming human pluripotent stem cells into white and brown adipocytes, Nat. Cell Biol. 14 (2) (2012) 209–219.
[30] T. Mohsen-Kanson, et al., Differentiation of human induced pluripotent stem cells into brown and white adipocytes: role of Pax3, Stem Cells 32 (6) (2014) 1459–1467.
[31] A.L. Hafner, T. Mohsen-Kanson, C. Dani, Differentiation of brown adipocyte progenitors derived from human induced pluripotent stem cells, Methods Mol. Biol. 1773 (2018) 31–39.
[32] S. Su, et al., A renewable source of human beige adipocytes for development of therapies to treat metabolic syndrome, Cell Rep. 25 (11) (2018) 3215–3228.e9.
[33] N. Billon, et al., Comprehensive transcriptome analysis of mouse embryonic stem cell adipogenesis unravels new processes of adipocyte development, Genome Biol. 11 (8) (2010) R80.

[34] F. Wassermann, The development of adipose tissue, in: Compr. Physiol. 2011, Supplement 15: Handbook of Physiology, vol. 87, American Physiological Society, 1965, p. 100.
[35] C. Barreau, et al., Regionalization of browning revealed by whole subcutaneous adipose tissue imaging, Obesity (Silver Spring) 24 (5) (2016) 1081–1089.
[36] P. Horvath, et al., Screening out irrelevant cell-based models of disease, Nat. Rev. Drug Discov. 15 (11) (2016) 751–769.
[37] S.A. Langhans, Three-dimensional in vitro cell culture models in drug discovery and drug repositioning, Front. Pharmacol. 9 (2018) 6.
[38] C. Liu, et al., Modeling human diseases with induced pluripotent stem cells: from 2D to 3D and beyond, Development 145 (5) (2018) 1–6.
[39] P. Leucht, et al., Embryonic origin and Hox status determine progenitor cell fate during adult bone regeneration, Development 135 (17) (2008) 2845–2854.
[40] M. Kouidhi, et al., Characterization of human knee and chin adipose-derived stromal cells, Stem Cells Int. 2015 (2015) 592090.
[41] R. Foissac, et al., Homeotic and embryonic gene expression in breast adipose tissue and in adipose tissues used as donor sites in plastic surgery, Plast. Reconstr. Surg. 139 (3) (2017) 685e–692e.

CHAPTER

3

Establishing the adipose stem cell identity: Characterization assays and functional properties

Mark A.A. Harrison, Sara I. Al-Ghadban, Benjamen T. O'Donnell, Omair A. Mohiuddin, Rachel M. Wise, Brianne N. Sullivan, and Bruce A. Bunnell

Center for Stem Cell Research and Regenerative Medicine, Tulane University School of Medicine, New Orleans, LA, United States

Introduction

Following their initial description by Zuk et al. in 2001, adipose stem cells (ASCs) rapidly gained interest as a potential therapeutic tool and became a topic of intense research [1]. In comparison with other mesenchymal stem cells (MSCs), the abundance, ease of both harvest and isolation, and proliferative abilities make ASCs ideal for research purposes. Adipose tissue, removed either by lipectomy or liposuction, can be subjected to enzymatic digestion and centrifugation to produce a cell pellet known as the stromal vascular fraction (SVF) [2]. This pellet contains numerous cell populations, including immune cells, endothelial cells, pericytes, and ASCs. Isolation of ASCs can be accomplished by plating the SVF onto standard cell culture plastic and maintaining only the plastic adherent cells. Following this, the appropriate characterization of the isolated cells is essential. Both the International Federation for Adipose Therapeutics and Science (IFATS) and the International Society for Cellular Therapy (ISCT) have released positional statements regarding the minimum characterization requirements necessary for identifying multipotent mesenchymal stromal/stem cells [3, 4]. In addition to these characterization standards, assessment of the immunomodulatory abilities of isolated ASCs may provide valuable information prior to use in downstream applications. The aim of this chapter is to describe the characterization and functional assays needed to

https://doi.org/10.1016/B978-0-12-819376-1.00002-0

appropriately determine the identity of ASCs. Where possible, comparisons with other common sources of MSCs will be made. Finally, the impact of donor-to-donor variation, focused on factors such as age, body mass, and health status, in relation to ASC function will be discussed.

Physical characterization

Morphological characterization

ASCs are a heterogeneous population characterized by their ability to adhere to tissue culture plastic, readily proliferate in culture, form colonies, and differentiate into multiple lineages in vitro and in vivo [5, 6]. ASCs display a spindle-shaped morphology with a distinctly fibroblast-like appearance in passages lower than 10 (p10), very similar to the appearance of bone marrow-derived mesenchymal stem cells (BMSCs) (Fig. 1). This morphology and the ability to adhere to plastic are both retained following cryopreservation [7, 8]. Following long-term culture, greater than p15, ASC morphology changes into large

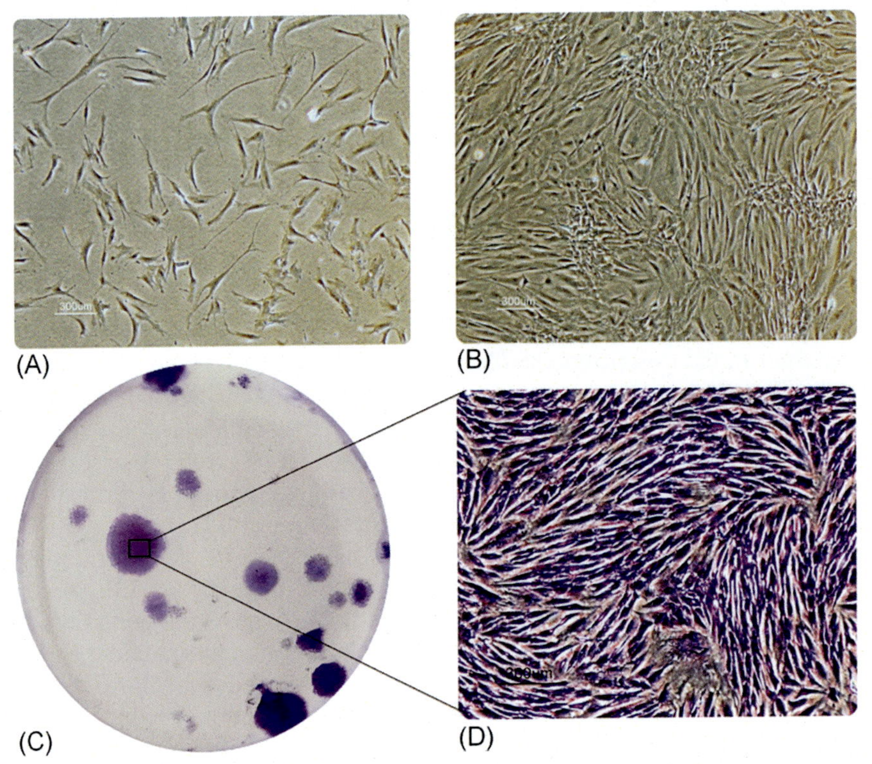

FIG. 1 Adipose stem cell (ASC) morphologic analysis and colony formation assay. Bright field imaging of ASCs at low (A) and high (B) confluence using a 4 × objective. Colony formation fibroblast assay plate following staining with crystal violet (C) and bright field image of the same staining using a 4 × objective (D).

trident-shaped senescent cells. Senescence can be verified by employing a staining assay in which a chromogenic substrate 5-bromo-4-chloro-3-indolyl-beta-D-galactopyranoside (X-gal) is incubated with cells to determine the presence of β-galactosidase [9–11]. Dmitrieva et al. demonstrated that senescence in BMSCs could be detected as early as p3, whereas ASCs did not show any evidence of cellular senescence prior to p6 [12]. These results were further corroborated by numerous studies indicating that ASCs exhibit decreased senescent features in comparison with BMSCs [13–15]. However, it has been demonstrated that the percentage of senescent cells in ASC cultures does increase significantly following extended passaging [9]. Thus, although several studies have encouraged the use of long-term culture-expanded ASCs for clinical trials, a careful morphological assessment prior to their use is crucial.

Assessment of proliferation

Human ASC proliferation rate is variable and primarily determined by passage number, donor characteristics such as age and body mass index (BMI), harvest site, surgical extraction procedure, and culture conditions [16–18]. Reported ASC doubling times between 20 and 40h are standard for low-passage cells, although some studies indicate times of up to 100h [9, 19, 20]. ASCs begin proliferating once they adhere to tissue culture plastic and can be assessed by Trypan Blue-assisted cell counting. In a study conducted by Di Battista et al., the ability of ASCs to proliferate increases in early passages (p2–p3), decreases at p4, but then increases again at p6 [21]. Similar studies have demonstrated decreases in proliferation at p10 followed by significant decreases after p15 without any changes in the expression of tumor-related genes or the development of chromosomal abnormalities [22, 23]. Cryopreservation of ASCs has not been demonstrated to affect the rate of their proliferation nor their differentiation potential when compared with noncryopreserved cells [24]. However, ASCs isolated from donors who are either older (>50 years of age), overweight (BMI >25), or both, exhibit a decreased proliferation potential in comparison with young (<35 years of age), lean (BMI <25) donor-derived ASCs [25, 26].

A significant factor in the ability of ASCs to adhere and proliferate is the choice of culture medium. Studies have shown that the addition of growth factors (fibroblast growth factor [FGF], epidermal growth factor [EGF], platelet-derived growth factor [PDGF]), cytokines (interleukin-6 [IL-6]), and ascorbic acid to serum-supplemented medium enhances their growth rate [27]. The source of serum is also a significant concern for the culture of ASCs. Potential risks involved in the use of fetal bovine serum (FBS) include the transmission of zoonotic microorganisms and induction of an immune response, which may result in the rejection of transplanted cells [28]. Many translational studies are now considering the use of human platelet lysate (hPL) as a safe, effective substitute for FBS owing to its rich protein and growth factor complement and minimal risk of xenogenic immune reaction [27, 29, 30]. ASCs cultured in hPL-supplemented medium have been shown to proliferate faster than those in FBS-supplemented medium, thus making hPL ideal for large-scale expansion of ASCs [30].

ASC proliferation rate can also be significantly affected by the harvest site of adipose tissue and the surgical procedure used to collect it. ASCs isolated from the hip/thigh and abdomen using resection and tumescent liposuction displayed increased yield and proliferation rate in comparison with ASCs obtained by ultrasound-assisted liposuction [16]. Breast tissue-derived ASCs have exhibited very low cell yield, decreased proliferation, and impaired differentiation abilities in comparison with subcutaneous fat, which was likely due to the significant presence of glandular and connective tissue characteristic of the harvest site [16]. Finally, it has been reported that adipose tissue yields roughly 5000 cells per mL in comparison with 100–1000 cells per mL of bone marrow [31]. In addition, donor-matched proliferation assays demonstrated that ASCs continue proliferating well through 21 days in culture, whereas BMSCs cease proliferating as early as day 14 [32].

Colony-forming unit–fibroblast assay

Self-renewal is one of the primary characteristics of stem cells and can be readily assessed using the colony-forming unit–fibroblast assay (CFU-F). In this assay, ASCs are plated on plastic cell culture dishes at a low density (i.e., 15–30 cells/cm^2) and their clonal expansion is assessed following 2 weeks of culture. Crystal-violet staining can be used to identify and visualize colonies under a bright field microscope (Fig. 1C). The number of colonies composed of more than 50 cells (2 mm in diameter) is counted and reported. Similar to the proliferation rate, the clonogenic potential of ASCs can be affected by donor characteristics as well as tissue harvest site. Frazier et al. have shown that ASCs isolated from lean donors (BMI <25) form significantly higher numbers of colonies than ASCs isolated from overweight (BMI 25–30) and obese donors (BMI >30) [17]. Likewise, ASCs isolated from subcutaneous abdominal adipose tissue have 50% higher clonogenic potential than those isolated from hip/thigh tissue [33, 34]. Compared with BMSCs, ASCs yield approximately four times more colonies per milliliter of harvested tissue sample, making them preferable in applications when high numbers of cells are required [18].

Phenotypic characterization

Cellular phenotype is the culmination of genetic makeup, epigenetic factors, and extracellular microenvironment. Phenotypic characterization allows for the grouping of cell populations with similar functions and is an essential part of defining the stem cell identity. It should be performed at different passages during in vitro expansion to ensure the maintenance of the stem cell identity. Flow cytometry is ideal for this characterization as most of the canonical markers for stem cells are cell surface cluster of differentiation (CD) proteins. Similar to BMSCs, ASCs express classical MSC markers and are negative for endothelial and hematopoietic markers as well as the human leukocyte antigen (HLA)-DR. According to IFATS and ISCT, ASCs are distinct from BMSCs in their expression of fatty acid translocase marker CD36, and the absence of vascular cell adhesion molecule (VCAM-1/CD106) [4, 35, 36]. At present, there is no single CD marker that in itself identifies an ASC. Generally, both positive and negative selection markers are employed to ensure that the isolated cells are stem cells and that they are not contaminated with other cell types (Table 1).

Positive selection markers

Human ASCs express the traditional MSC surface markers, including the cell adhesion molecules CD29, CD34, CD44, CD146, and CD166; the receptor molecules CD90 and

TABLE 1 Positive and negative selection markers used for the characterization of adipose stem cells.

CD markers	Descriptive names	Cell expression
Positive markers		
CD29	Integrin beta-1 (ITGB1)	MSCs, including ASCs
CD34	Hematopoietic progenitor cell antigen CD34	MSCs, including ASCs
CD44	Homing cell adhesion molecule (HCAM)	MSCs, including ASCs
CD73	5′-nucleotidase	MSCs, including ASCs
CD90	Thy-1 GPI-anchored glycoprotein	MSCs, including ASCs
CD105	Endoglin	MSCs, including ASCs
CD146	Melanoma cell adhesion molecule (MCAM)	MSCs, including ASCs
CD166	Activated leukocyte cell adhesion molecule (ALCAM)	MSCs, including ASCs
Negative markers		
CD3	T-cell receptor T3 complex	Hematopoietic cells, endothelial cells, T cells
CD11b	Integrin alpha M (ITGAM)	Monocytes
CD14	Monocyte differentiation antigen CD14	Macrophages
CD19	B-lymphocyte surface antigen B4	B cells
CD31	Platelet endothelial cell adhesion molecule (PECAM1)	Hematopoietic cells, endothelial cells, platelets
CD45	Protein tyrosine phosphatase receptor type C (PTPRC)	Leukocytes
CD163	Scavenger receptor cysteine-rich type 2 protein	Monocyte, macrophage
CD200	OX-2 membrane glycoprotein	B cells, activated T cells, endothelial cells, neuronal cells
CD271	Low-affinity nerve growth factor receptor	BMSCs
HLA-DR	Human leukocyte antigen DR-isotype	Activated T cells

Abbreviations: ASCs, *adipose stem cells;* BMSCs, *bone marrow-derived mesenchymal stem cells;* MSCs, *mesenchymal stem cells.*

CD105; and the GPI-anchored enzyme CD73. Although ASCs retain the classic mesenchymal characteristics at least up to p10, the expression of certain markers such as CD105 and CD166 can be decreased at early passages, particularly when the cells are committed toward particular lineages [35, 37]. Phenotypic characterization of ASCs isolated from young and old donors have demonstrated no significant differences [25]. However, ASCs obtained from obese donors exhibit a decrease in typical stemness markers as compared with those isolated from nonobese donors [38]. Moreover, cryopreserved ASCs did not elicit changes in cell surface marker expression when compared with noncryopreserved cells [7].

Negative selection markers

Ensuring that isolated ASCs are not contaminated with another cell type is critical. The SVF, from which ASCs are isolated, is rich in hematopoietic cells, endothelial cells, and immune cells [39]. Thus, the use of a wide array of surface markers is required to ensure purity. ASCs lack the expression of hematopoietic lineage markers, including CD11b, CD13, CD14, CD19, and CD45. They are also negative for the endothelial markers CD31 and HLA-DR [4]. In addition to these classical negative selection markers identified by ISCT, studies have shown that ASCs do not express CD163, CD200, CD271, or CD274, allowing them to be distinguished from chondrocytes and osteoblasts [40]. The same study suggested that CD163, a marker of monocytes and macrophages, could be used to differentiate ASCs from other MSCs. Finally, negative selection using CD54 permits ASCs to be distinguished from fibroblasts, and the $CD54^-$ ASC population demonstrated increased osteogenic and adipogenic differentiation abilities [41].

Donor-to-donor variability, culture medium, and growth factors have also been shown to impact the growth of distinct ASC subpopulations. These different ASC subpopulations, as characterized by their phenotypic expression patterns, may also possess varied functional characteristics. For instance, positive selection for $CD105^+$ ASCs resulted in a population of ASCs that displayed significantly greater osteogenic and chondrogenic differentiation potential than $CD105^-$ ASCs [42]. Nevertheless, $CD105^-$ ASCs demonstrated stronger immunomodulatory abilities than the $CD105^+$ population [43]. Additionally, populations of $CD34^+$ ASCs were shown to proliferate faster than $CD34^-$ ASCs, although the latter exhibited an increased ability to differentiate along several mesenchymal lineages [44]. Results such as these represent the potential for significant future research to identify and characterize optimal ASC populations for various in vitro and in vivo applications.

Multipotent differentiation

The ability to differentiate into a variety of lineage-specific cell types is a defining characteristic of stem cells. Like all MSCs, ASCs maintain the ability to differentiate into adipocytes, osteoblasts, and chondrocytes when exposed to specific differentiation factors and conditions. The ability of isolated ASCs to differentiate into these lineages is essential for proper characterization and demonstration of their stemness. ASCs have also been differentiated into cells of nonmesodermal lineage such as cardiomyocytes [45], endothelial cells [46], neurons [47], and hepatocytes [48]. However, when it comes to the characterization of ASCs, differentiation into the three primary mesenchymal cell types is sufficient according to ISCT and IFATS [3, 4].

Adipogenesis

In vitro, ASCs act as a reservoir of cells that can efficiently differentiate into mature adipocytes. Due to their location, many believe that ASCs have an increased adipogenic potential as compared with MSCs from other depots. Multiple studies have demonstrated that ASCs have an increased adipogenic potential and a corresponding decrease in osteogenic and chondrogenic potential [49, 50]. However, other studies have found no difference in the differentiation abilities between ASCs and BMSCs [51, 52]. These conflicting findings are potentially due to donor variability in age, BMI, and health status, as well as the adipose depot from which the cells were isolated. These factors all have an impact on the ability of ASCs to differentiate into adipocytes [25, 53–57].

Mechanism(s) of differentiation

Due to the obesity epidemic, significant research has been performed to elucidate the differentiation pathway(s) from ASCs to mature adipocytes in vivo. Still, it is a complicated pathway involving a myriad of genetic and biologic factors that are being updated constantly. Evidence has suggested that Wnt/β-catenin, Hedgehog, and transforming growth factor-β (TGF-β)/mothers against decapentaplegic homolog 3 (Smad3) signaling pathways all serve to inhibit adipogenesis, while extracellular matrix (ECM) stiffness plays a significant role in adipogenesis [58–60]. However, in vitro differentiation strategies often skip the early differentiation stages and focus instead on increasing late-stage differentiation regulators such as CAAT/enhancer-binding proteins (C/EBPs) and peroxisome proliferator-activated receptor gamma (PPARγ) (Fig. 2).

Adipogenic differentiation is a multistep process progressing from ASCs to preadipocytes and finally ending with mature adipocytes. Correspondingly, the genetic differentiation pathway is divided between early-, mid-, and late-stage genes. Understanding of the transition from ASC to preadipocyte is still in its infancy, but recent work suggests that transcriptional factors PPARγ and zinc-finger protein 423 (ZNF423) may play roles in this process [61]. While preadipocyte commitment is not well understood, maturation from preadipocyte to adipocyte has been well characterized. The C/EBP family of transcription factors are potent regulators of adipogenesis. In particular, C/EBPβ and C/EBPδ are significantly upregulated in the early stages of adipogenesis, during preadipocyte commitment, and maturation of adipocytes [62, 63]. The upregulation of C/EBPβ is largely mediated by glucocorticoid signaling and cAMP induction [63]. Binding of these transcription factors to their targeted promotors induces transcription of C/EBPα and PPARγ, widely considered the master regulators of adipocyte commitment and maturation [63, 64]. The activation of these regulators leads to a positive feedback loop that results in maturation and lipid accumulation. Work by Wu et al. has demonstrated that inhibition of PPARγ results in the loss of preadipocyte commitment, while the inhibition of C/EBPα still allows for lipid accumulation but also results in a loss of insulin sensitivity [65–67]. The effect of this is critical, as insulin receptor activation results in phosphatidylinositol 3-kinase (PI3K)/Akt signaling, which is the driving force behind lipid accumulation in maturing adipocytes [63]. Canonical Akt signaling has been shown to play a crucial role in adipocyte formation through PPARγ signaling [68, 69]. Akt knockout mouse models have demonstrated an inhibition of adipogenesis through the loss of PPARγ expression. This is mediated through the activation of mammalian target of rapamycin complex 1 (mTORC1) signaling, which directly regulates adipocyte protein synthesis and morphology [70]. Akt signaling also inhibits the expression of forkhead box O1 (FOXO1), which is an inhibitor of PPARγ [71, 72].

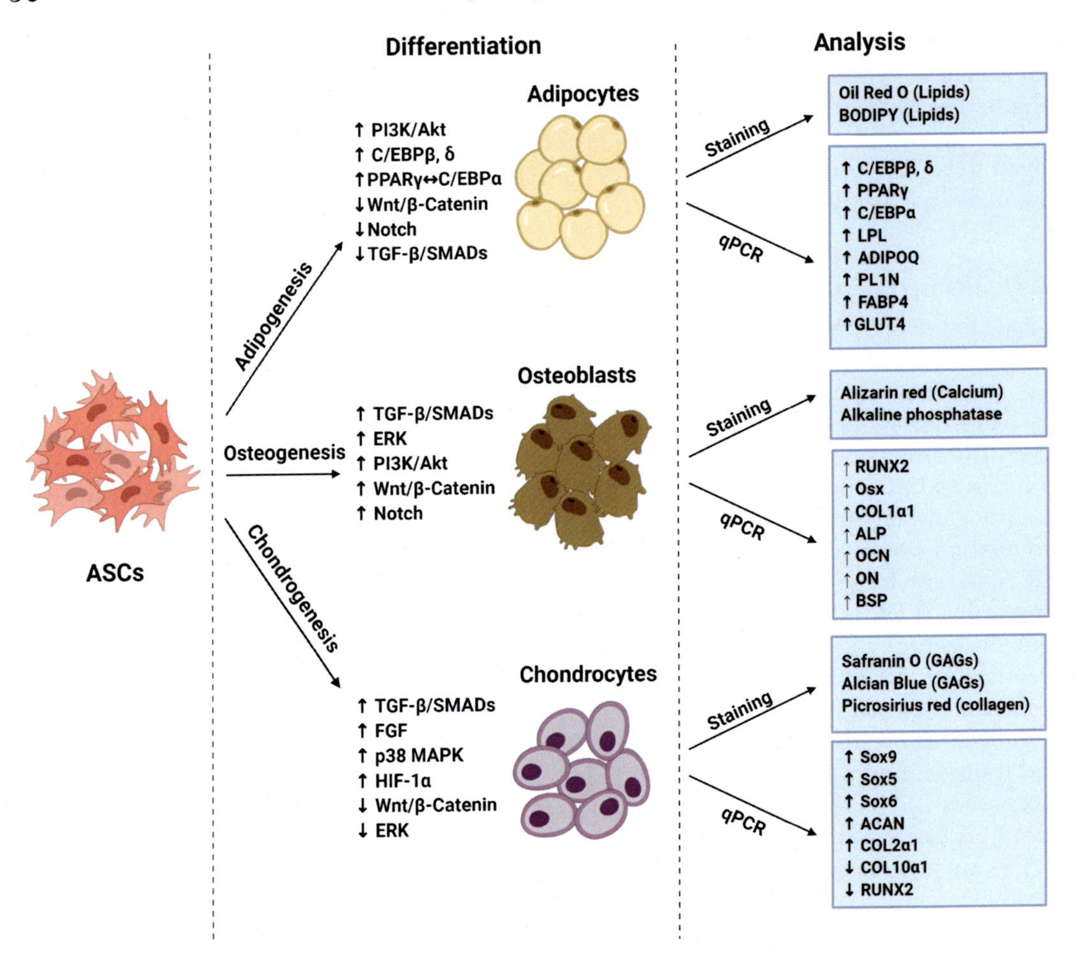

FIG. 2 Overview of differentiation of adipose stem cells to adipocytes, osteoblasts, and chondrocytes and methods of differentiation confirmation. Overlapping signaling pathways work in concert to induce differentiation and establish tissue line commitment. Various cellular stains and gene expression analysis using quantitative polymerase chain reaction are used to confirm differentiation. *ACAN*, aggrecan; *ADIPOQ*, adiponectin; *ALP*, alkaline phosphatase; *BSP*, bone sialoprotein; *COL1α1*, collagen 1-alpha-1; *COL2α1*, collagen 2-alpha-1; *COL10α1*, collagen 10-alpha-1; *ERK*, extracellular signal-regulated kinase *C/EPB*, CAAT/enhancer-binding protein; *FABP4*, fatty acid binding protein 4; *FGF*, fibroblast growth factor; *GAGs*, glycosaminoglycans; *GLUT4*, glucose transporter type 4; *HIF-1α*, hypoxia-inducible factor 1-alpha; *LPL*, lipoprotein lipase; *OCN*, osteocalcin; *ON*, osteonectin; *Osx*, osterix; *PI3K*, phosphatidylinositol 3-kinase; *PL1N*, perilipin; *PPARγ*, peroxisome proliferator-activated receptor gamma; *p38 MAPK*, p38 mitogen-activated protein kinase; *RUNX2*, runt-related transcription factor-2; *Smad*, TGFβ mothers against decapentaplegic; *SOX*, SRY-related HMG-box; *TGF-β*, transforming growth factor beta. *Created with BioRender.com*

Late-stage adipogenic genes are those required for the healthy maintenance of the mature adipocyte. These include adipokines such as adiponectin (APN) and leptin (LEP). Additionally, the expression of lipoprotein lipase (LPL), perilipin (PL1N), and fatty acid binding protein 4 (FABP4) are all critical for the production and maintenance of the classic lipid-filled vacuoles.

Differentiation methodology

Strategies for adipogenic differentiation often vary in the composition and concentration of their additives, but the overall strategy remains the same [73–77]. Cells are cultured in adipogenic medium consisting of standard stromal medium supplemented with several additives. Dexamethasone, a corticosteroid, is used to stimulate transcription of C/EBPβ. 3-isobutyl-1-methylxanthine (IBMX), an inhibitor of cyclic adenosine monophosphate (cAMP) phosphodiesterase, is used to increase the cytosolic concentration of cAMP, resulting in the transcription of C/EBPδ. The activity of C/EBPβ and C/EBPδ results in the expression of PPARγ and C/EBPα, the master regulators of adipogenesis, which initiate differentiation. Periodic exposure to high levels of PPARγ agonists such as indomethacin or rosiglitazone serves to maintain differentiation and push the cells towards a mature state by reinforcing the PPARγ-C/EBPα positive feedback loop. Finally, insulin is used to activate PI3K/Akt signaling, which encourages the storage of lipids and increases the size of neutral lipid vacuoles in the cytosol [73–77].

Confirmation of adipogenesis

In two-dimensional (2D) cultures, adipogenesis can be confirmed by microscopy, as neutral lipid droplets are readily visible in the cytosol of differentiated ASCs. However, appropriate staining of lipid droplets or lipid-droplet-associated proteins, as well as analysis of gene expression patterns of the differentiated cells, is preferred to ensure appropriate differentiation.

The staining of neutral lipid droplets can be achieved by several methods; however, it should be noted that the use of any fixative containing significant amounts of alcohols can disrupt the lipid droplets. Sudan Red and Oil Red O are diazo dyes that preferentially stain lipids and can be used to visualize adipogenesis on a bright field microscope. Quantification of Sudan Red and Oil Red O staining is possible since isopropanol can be used to extract the stain from the cells. The absorbance of the solution can be read at 507 nm and 584 nm, respectively [74, 75]. To make comparisons between ASC populations, it is necessary to normalize the absorbance values to the total protein content of a sample.

Fluorescent microscopy is another method by which ASC adipogenesis can be examined. Of the lipid-droplet-specific fluorescent probes, the most well-known is BODIPY. However, a wide range of fluorescent probes exists for visualizing lipid droplets in all different excitation and emission spectra, as expertly reviewed by Fam et al. [78]. The availability of these probes in varying spectra allows for the use of nuclear and cytoskeletal counterstains to visualize overall cell morphology. As ASCs differentiate, they gain a rounded morphology with spiky projections. The neutral lipid droplets will appear in a rough circle surrounding the nucleus before slowly gaining size. As differentiation progresses, these droplets will begin to push all of the other organelles to the edges of the cell as they combine and continue to grow [79–81].

Assessment of adipocyte differentiation of ASCs can also be performed by examination of essential differentiation genes via quantitative polymerase chain reaction (qPCR). Upregulation of C/EBPα and PPARγ in comparison with undifferentiated ASCs is considered sufficient to illustrate initiation of the positive feedback loop that leads to the commitment of ASCs to adipocyte lineages. While C/EBPβ and C/EBPδ upregulation precede induction of C/EBPα and PPARγ, evidence of this alone does not demonstrate adipocyte commitment.

Finally, upregulation of late-stage genes such as APN, LPL, PL1N, or FABP4 demonstrates the development of mature adipocytes (Fig. 2).

Adipose stem cell adipocyte secretome

Historically, adipose tissue was believed to be merely an energy depot in the form of triglyceride-packed vacuoles. Recent advances, however, have demonstrated that adipose tissue is a highly active endocrine organ whose interactions extend to many tissues, including the liver, brain, and vasculature. Collectively, the endocrine and paracrine effects of adipocytes are mediated by adipokines. The most common adipokines are involved in appetite suppression (LEP), glucose storage mediated by insulin sensitivity (APN), and insulin resistance (resistin) [82]. ASCs differentiated into adipocytes will secrete similar proteins, however, discrepancies may exist possibly attributable to the differentiation protocol employed [83]. Of increasing interest are the immunomodulatory adipokines that are thought to play a major role in the inflammatory state of obese and diabetic patients [84]. Undifferentiated ASCs produce antiinflammatory cytokines such as TGFβ and IL-10 [83, 85]. However, upon differentiation towards adipocytes, there is an increase in the expression of proinflammatory genes and concurrent decrease of most antiinflammatory genes [86]. Adipocyte differentiation and adipokine secretion also varies by adipose depot and donor obesity status [57, 87, 88]. Additionally, the expression of adipocyte proinflammatory cytokines IL-1β, IL-6, and IL-8 is elevated in obese patients [89, 90]. Age and sex of donor also has a role in the ability of ASCs to differentiate. It has been demonstrated that ASCs isolated from a more aged donor have a reduced ability to differentiate into adipocytes [25, 56].

Osteogenesis

BMSCs are typically the cells of choice for bone regeneration, with some studies suggesting that they exhibit osteogenic capacity superior to that of ASCs [91–94]. Recently, however, there has been an increased focus on ASC-based bone regeneration due to ASCs' ease of harvest, availability, reduced need for in vitro expansion, and proven osteogenic abilities [95]. Further characterization of the osteogenic potential of ASCs has revealed that higher donor BMI results in reduced differentiation [17, 96]. On the other hand, donor age and cryopreservation status have not been found to affect the osteogenic differentiation abilities of ASCs [19, 20, 97].

Mechanism(s) of differentiation

The capacity of ASCs to undergo osteogenic differentiation is well established; however, the underlying molecular mechanism(s) is not thoroughly understood. Several different signaling pathways have been found to play a role in the osteogenic differentiation of ASCs. These signaling pathways include extracellular signal-related kinase (ERK), Notch, TGF-β, Wnt/β-catenin, and phosphoinositide 3-kinase/Akt (PI3K/Akt) signaling (Fig. 2) [98–100].

Osteogenic differentiation of ASCs results in a significant upregulation of ERK signaling. A study by Liu et al. demonstrated that ASCs cultured in osteogenic medium display increased ERK levels by day 7, peak at day 10, and decrease by day 14 [101]. Additionally, pharmacologic inhibition of ERK activation by PD98059 reduced the osteogenic differentiation capabilities of ASCs [101, 102]. When these same ERK-inhibited ASCs were exposed to dexamethasone, they proceeded to differentiate into adipocytes [101, 102]. Similarly, active ERK signaling results in phosphorylation of PPARγ, which decreases its activity and produces and antiadipogenic effect on ASCs [99]. Thus, ERK represents a key regulatory pathway for the commitment of ASCs to either the osteogenic or adipogenic lineage.

The Notch signaling pathway also plays a significant role in the osteogenic differentiation of ASCs. Inhibition of Notch signaling results in reduced osteogenic differentiation, while transfection of the downstream Notch-1 intracellular domain (ICD) restored their osteogenic potential [103].

Bone morphogenic proteins (BMPs) and TGF-β regulate osteogenic differentiation via the TGF-β signaling pathway and result in upregulation and activation of downstream transcription factors including Smads, runt-related transcription factor-2 (RUNX2), and osterix (Osx), all of which are key promoters of osteogenesis [98, 104–106].

Activation of the PI3K/Akt signaling pathway and Wnt/β-catenin pathway have also been shown to upregulate osteogenesis in ASCs [107–109]. Wnt5a specifically promotes osteogenesis by activating the Wnt/β-catenin pathway and suppressing the activation of PPARγ [106].

Of particular interest is the role of nuclear factor kappa-light chain-enhancer of activated B cells (NF-κB). NF-κB is a transcription factor family that plays a significant role in the inflammatory responses of cells as well as exerting control over both proliferation and cell survival. It is rapidly activated in response to harmful stimuli such as bacterial lipopolysaccharides, reactive oxygen species, and proinflammatory cytokines such as TNFα and IL-1β. In the context of BMSCs, it has been demonstrated that activation of NF-κB resulted in potent inhibition of osteogenic differentiation [110–113]. This observation appears to remain true for ASCs as well. Wang et al. demonstrated that inhibition of prostaglandin G/H synthase 1 (PTGS1) caused suppression of NF-κB signaling, resulting in enhanced osteogenic differentiation [114].

The coordinated regulatory effects of the aforementioned signaling pathways lead to the activation of pro-osteogenic transcription factors RUNX2 and Osx that transform ASCs into osteoblasts. RUNX2 is one of the earlier markers of osteogenic differentiation and promotes the formation of osteoblast-specific ECM by regulating collagen I and osteopontin (OPN) expression [115, 116]. Collagen type I represents the principal structural protein of bone ECM [101, 117]. Osx, a later marker of osteogenic differentiation, is a zinc finger transcription factor that regulates several other key osteogenic genes such as osteocalcin (OCN), osteonectin (ON), and bone sialoprotein (BSP) [96, 116]. OCN, ON, and BSP are involved in the regulation of osteoclast activity, binding of collagen, and the calcification of the ECM [101, 117].

Differentiation method

Differentiation of ASCs begins with the formation of osteoblast precursor cells, which later mature to form osteoblasts capable of producing and depositing minerals in the ECM [115]. Thus far, ASCs have been encouraged to undergo osteogenic differentiation in vitro using differentiation medium, electromagnetic stimulation [118], and mechanical stimulation [119]. While the mechanisms behind electromagnetic and mechanical stimulation have not been fully elucidated, osteogenic differentiation medium is well established and is the method of choice for the characterization of ASCs.

Osteogenic differentiation medium is primarily comprised of standard stromal medium supplemented with dexamethasone, ascorbic acid 2-phosphate, and β-glycerophosphate [106, 115, 116, 120–122]. Dexamethasone, a glucocorticoid, promotes osteogenic differentiation by increasing the transcription of four and a half LIM domains protein 2 (FHL2), which then binds to β-catenin, translocates to the nucleus, and initiates transcription of RUNX2 [123]. Ascorbic acid 2-phosphate serves as a cofactor for the hydroxylation of proline and lysine molecules in collagens and results in the upregulation of procollagen, alkaline phosphatase (ALP), and OCN [76, 124]. Finally, β-glycerophosphate aids in the synthesis of hydroxyapatite by providing inorganic phosphate molecules [125]. The activation of these signaling pathways and the synthesis of hydroxyapatite lead to the osteogenic differentiation of ASCs and the deposition of calcium in the ECM.

Confirmation of osteogenesis

ASC osteogenic differentiation in vitro is typically confirmed by assessment of alkaline phosphatase (ALP) activity and examination of calcium deposition. This evidence is then further reinforced by gene expression profiling of the cells and analysis of secreted factors.

ALP is an enzyme that is used to detect early stages of osteogenesis. ALP expression is upregulated in osteoblasts in comparison with ASCs [115]. Staining of ALP can be accomplished with nitro blue tetrazolium/5-bromo-4-chloro-3-indolyl phosphate (NBT/BCIP), which can then be viewed under a bright-field microscope [101, 105, 108, 114, 118, 120]. For a quantifiable measurement of ALP, its activity can be examined following cell lysis. The lysate is mixed with *p*-nitrophenyl phosphate substrate solution and incubated. The reaction is stopped using a sodium hydroxide solution, and the absorbance of the solution can be measured at 405 nm for quantification of ALP [101, 118, 126, 127].

Extracellular calcium deposition is a defining characteristic of terminally differentiated osteoblasts and can be stained using alizarin red [96, 101, 118, 120, 128–130]. Calcium deposits are then visualized using a bright-field microscope. For the purposes of quantification, alizarin red can be extracted via incubation with cetylpyridinium chloride or acetic acid, and its absorbance can be measured at 584 nm and 405 nm, respectively [96, 105, 108, 120, 128].

Gene expression changes can also be used to verify the osteogenic differentiation of ASCs using RT-qPCR (Fig. 2). Commonly upregulated genes include *RUNX2*, Osx (*MID1*), collagen 1α1 (*COL1α1*), ALP (*ALPL*), OCN (*BGLAP*), ON (*SPARC*), and BSP (*IBSP*). The differentiation process can be broken down into early, middle, and late stages based on gene expression patterns. Typical early gene markers include *RUNX2* and *ALPL*, whereas *IBSP* and *BGLAP* are intermediate and late-stage osteogenic differentiation markers, respectively [115, 116, 118].

Adipose stem cell osteoblast secretome

A limited number of studies have focused on the quantification of cytokines such as interleukins and growth factors secreted by osteogenically differentiated ASCs. Mussano et al. used Bio-Plex analysis to compare the levels of IL-6, IL8, IL-10, IL-12, monocyte chemoattractant protein-1 (MCP-1), vascular endothelial growth factor 1 (VEGF-1), and hepatocyte growth factor (HGF) between undifferentiated and osteogenically differentiated ASCs. Analysis of conditioned medium revealed decreased levels of all the aforementioned cytokines except for IL-8 [127, 131]. Transcriptome profiling of osteogenically differentiated ASCs using RNA sequencing resulted in the upregulation of several transcripts, including vascular cell adhesion molecule-1 (VCAM-1), vascular endothelial growth factor-B (VEGF-B), insulin-like growth factor-1 (IGF-1), TNF-α, BMP-4, platelet factor-4, IL-16, and angiopoietin-1 [100]. Of particular interest, this same group found that two major classes of ECM remodeling enzymes, the matrix metalloproteinases (MMPs) 2, 15, and 28, and A disintegrin and metalloproteinase with thrombospondin motifs (ADAMTs) 8, 13, 15, and 18 were upregulated in osteogenically differentiated ASCs. Further analysis of osteogenically differentiated ASC medium based on the current transcriptomic findings would provide greater insight into the composition of the osteogenic ASC secretome and could provide valuable insight into their use for bone regeneration.

Chondrogenesis

Due to the prevalence of joint injuries and diseases such as osteoarthritis where cartilage is degraded, there is great interest in the processes that control chondrogenesis. Comparisons have been made surrounding BMSCs and ASCs' ability to differentiate. A study by Mohammad-Ahmed et al. used donor-matched BMSC and ASCs and claimed that BMSCs possess superior chondrogenic differentiation potential [32]. These results were predicated on increased aggrecan (*ACAN*) gene expression in BMSCs after 28 days of differentiation. However, Alcian Blue staining for the presence of glycosaminoglycans (GAGs) demonstrated few differences between the pellets at the same timepoint. Another study by Pagani et al. showed that ASCs produced superior chondrogenic differentiation abilities. However, when the same MSCs were differentiated in a proinflammatory environment, meant to mimic osteoarthritic conditions, the findings were reversed [132]. Research employing bisulfite sequence examination of the differences in methylation of CpG islands in important chondrogenic promotor regions found that there were no significant differences between BMSCs or ASCs [133]. Thus far, no studies have been conducted demonstrating differences in chondrogenic abilities of ASCs from high-BMI donors; however, it has been shown that increased age in donor results in decreased chondrogenic ability [25]. Finally, cryopreservation of ASCs does not seem to alter the chondrogenic differentiation abilities of ASCs when evaluated after use of standard freezing medium of 10% dimethyl-sulfoxide, 90% serum freezing medium [134].

Mechanism(s) of differentiation

The similarities between the osteogenic and chondrogenic differentiation often complicate the understanding of both, though there are a few notable exceptions (Fig. 2). Canonical Wnt/β-catenin signaling, which is essential for osteogenesis, has been shown to repress chondrogenesis. Functional loss of β-catenin also rendered cells unable to differentiate into osteoblasts; instead, the cells become chondrocytes [135]. The upregulation of ERK signaling is also important for osteogenesis. However, the pharmacologic inhibition of ERK signaling using PD98059 was shown to promote chondrogenesis and also suppress hypertrophic chondrocytes, as evidenced by decreased COL10A1 and RUNX2 expression [136].

In vitro chondrogenesis is mediated by both environmental factors and cellular signaling molecules. They act to upregulate the expression of the transcription factor SRY-related

HMG-box-9 (Sox9). Sox9 is the primary transcription factor regulating chondrogenic differentiation and, in combination with Sox5 and Sox6, is a direct regulator of COL2A1 the primary collagen species in cartilage [137].

Sox9 expression has been shown to be enhanced in transcription factor ZFN145 overexpressing cells, leading to chondrogenesis [138]. Signaling mediated by the TGF-β family of growth factors, including the BMPs, results in phosphorylation of Smad1/5/8, leading to expression and phosphorylation-mediated stabilization of Sox9 [139]. Additionally, the upregulation of yes-associated protein-1 (YAP) decreased Smad1/5/8 phosphorylation and inhibited chondrogenic differentiation [140].

TGF-β signaling also encourages chondrogenesis through inhibition of *RUNX2* and *COL10A1* transcription [141]. Many different combinations of TGF-βs and BMPs have been examined to assess their contribution to chondrogenesis. It has been separately shown that TGF-β1 and BMP2 both promote chondrogenesis and that, when combined, they produce a synergistic effect [142]. Similarly, TGF-β2 and BMP7 were shown to enhance chondrogenesis [143]. Finally, incorporation of FGF-18 and TGF-β3 also produced a synergistic prochondrogenic effect [144].

Micromass culture, the preferred mechanism for in vitro chondrogenesis, produces a dense ASC cell pellet that promotes a rounded cell morphology and mimics the initial cell condensation stage of in vivo chondrogenesis [145]. The culture of cells in this manner also induces hypoxic conditions reminiscent of the largely avascular end product of chondrogenesis, cartilage [146]. Hypoxia has been shown to preferentially promote chondrogenesis over osteogenesis by inducing p38 MAPK signaling, which results in transcription of *COL2A1* and *ACAN* [147]. Hypoxic conditions also stimulate early commitment to chondrogenesis by inducing HIF-1α signaling, leading to a twofold increase in Sox9 expression. The ablation of HIF-1α was found to reverse this finding [148].

While several studies have reported transdifferentiation of hypertrophic chondrocytes into osteoblasts, inhibition of chondrocyte hypertrophy is essential to avoid unwanted mineralization of the ECM [149, 150]. This collagen-to-bone process is known as endochondral bone formation. In hypertrophic chondrocytes, this transition can be examined and is mediated by the expression and activity of COL10, ALP, and MMP-13 [150]. The continued expression of Sox9 has been shown to actively repress transcription of *COL10A1* by inhibiting the activity of *RUNX2* [151]. Hypoxia also results in the downregulation of *COL10A1* and *MMP13*, thus inhibiting chondrocyte hypertrophy [152]. Parathyroid hormone-related peptide (PTHrP) supplementation has been shown to suppress chondrocyte hypertrophy [153].

Differentiation method

Unlike adipogenic and osteogenic differentiation of ASCs, chondrogenic differentiation is often not performed in a cell monolayer. Monolayer differentiation of ASCs can be achieved, and evidence suggests that coating cell culture plastic with fibroblast-derived ECM may aid the process [154]. However, an environment more conducive to chondrogenesis can be achieved by pelleting the cells in a conical tube or v-bottomed plate [145, 155, 156]. It has also been shown that ASCs cultured in a variety of scaffolds can also improve their chondrogenic differentiation [156–160].

Medium supplements and concentrations may vary between protocols, but the overall differentiation strategy remains the same. ASC stromal medium is supplemented with dexamethasone, L-ascorbic acid 2-phosphate, and members of the TGF-β family. Additionally, some labs include an insulin–transferrin–selenium mixture to help promote differentiation [74–76, 142–144, 155, 161]. Similar to adipogenic and osteogenic differentiation, dexamethasone is used to initiate differentiation and, in combination with other signaling molecules, has been shown to upregulate transcription of several collagens. Again, L-ascorbic acid 2-phosphate acts as a cofactor for the hydroxylation of prolines in collagens and induces transcription of COL2A1 [76, 143]. Importantly, the inclusion of different TGF-βs, BMPs, and FGFs has been shown to increase the differentiation of ASCs into chondrocytes that secrete the appropriate ECM [142–144, 161].

Confirmation of chondrogenesis

Confirmation of chondrogenesis is largely determined by the analysis of the ECM that was deposited around the cells in micromass culture. This can be accomplished by embedding, sectioning, and staining the pellet or by proteolytic enzyme-mediated digestion followed by component analysis (Fig. 2).

The Bern scoring system, introduced by Grogan et al., can be used for the qualification of in vitro engineered cartilaginous tissue. The Bern system uses a 0–9 score to describe cartilage quality [162]. It is assessed using a 0–3 scale along three axes: (1) Safranin O-fast green staining to examine GAG uniformity and intensity, (2) distance between cells as a measure of matrix deposition, and (3) assessment of typical rounded chondrocyte morphology.

Paraffin-embedded chondrogenic pellets can be sectioned and subjected to numerous histological stains to confirm chondrogenesis. The most common of these are Alcian Blue, Toluidine Blue, and Safranin O, all of which stain polysaccharides such as those found in GAGs [74, 156, 162]. The intensity and evenness of staining can be used to assess the degree of differentiation of ECM-producing chondrocytes. Picrosirius red is a linear anionic dye that can be used to stain collagen networks [163]. Additionally, collagen subtype-specific antibodies can be used to determine which types of collagens are present in the chondrogenic pellets.

Following digestion with a proteolytic enzyme, quantification of GAG content can be made using dimethylmethylene blue (DMMB). DMMB is a cationic dye that preferentially stains sulfated GAGs in solution and can be quantified by measuring its absorbance at 525 nm [144]. The amount of DNA that can be isolated from each pellet should be used to normalize the DMMB measurements to control for cell number. Similarly, a hydroxyproline assay can be used to determine collagen content indirectly. Briefly, hydroxyprolines are hydrolyzed, oxidized into pyrroles, and reacted with *p*-dimethylaminobenzaldehyde (DMAB, Ehrlich's reagent) to form a chromophore whose absorbance can be read at 550 nm [144, 164]. The absorbance can be used to calculate initial collagen content. Alternatively, the hydroxyproline to collagen ratio of 1:7.69 has been used to determine collagen content [144]. Again, these results should be normalized to the DNA content of the pellet.

Finally, the digestion of the pellet allows for analysis of gene expression of the cells following RNA isolation. Gene expression analysis can be used to examine the chondrocyte stages of maturation. Immature and mature chondrocytes can be determined through the expression of *SOX5*, *SOX6*, *SOX9*, *COL2A1*, and *ACAN* [165, 166]. Finally, hypertrophic chondrocytes can be assessed by the expression of *COL10A1*, *ALPL*, and *MMP13* [167, 168].

Adipose stem cell chondrocyte secretome

Chondrocytes' secretome is predominantly related to the production and maintenance of the ECM. This consists predominantly of GAGs and collagens supplemented by stabilizing proteins such as link protein, cartilage oligomeric protein, decorin, and fibromodulin [168]. Importantly, once chondrocytes become hypertrophic, they begin to secrete COL10, ALP, and various MMPs, which work in a concerted effort to initiate bone deposition. Thus, inhibition of chondrocyte hypertrophy is essential for the maintenance of clinically relevant chondrogenesis and cartilage formation. Future research optimizing chondrogenic differentiation and maintenance of mature chondrocytes will provide valuable information for both treatment and prevention of disease.

Characterization of in vitro immunomodulatory abilities

The immunomodulatory capabilities of MSCs were only recently reported by Bartholemew et al. in a landmark 2002 publication illustrating their suppressive effects on lymphocyte proliferation [169]. Since then, their potent immunomodulatory potential has made them the subject of intense research. Much of this research has focused on secreted molecules, such as cytokines and growth factors. However, recently researchers have started examining MSC-secreted extracellular vesicles carrying proteins and RNA species that also direct immune cell function. It has also been demonstrated that treatment of ASCs with proinflammatory factors such as lipopolysaccharide (LPS) [170, 171], and interferon-gamma (IFN-γ) [172] or exposure to hypoxic conditions [173], can enhance the secretion of immune-active compounds.

Regardless of the mechanism of action, the ability of MSCs, including ASCs, to modulate components of the immune environment in vitro and in vivo make them exciting potential therapeutics. For the purposes of this chapter, only the in vitro activity of ASCs relative to specific immune populations is discussed (see Chapter 10 for in vivo studies). In vitro, the immunomodulatory impact of ASCs can be assessed via direct cell coculture, indirect coculture, or treatment of immune cells with ASC-conditioned medium. Depending on the intended application of ASCs, characterization of their immunomodulatory capacity may or may not be necessary. However, analysis of their activity in vitro, as well as examination of their secreted immunoactive compounds may serve as a helpful tool to evaluate their immunomodulatory potential prior to use.

Adipose stem cells and the innate immune system

Innate immunity is one of two interconnected systems, along with adaptive immunity, used by most vertebrates to mount a response against foreign insults such as viruses, bacteria, and parasites. The innate immune system serves as the first line of defense against all foreign substances, providing both physical barriers and specialized immune cells that inhibit the invaders' spread throughout the body. Physical barriers such as the skin, gastrointestinal tract, and their respective acellular defense mechanisms help prevent the invasion of pathogens, while the specialized immune cells are responsible for rapidly and indiscriminately killing any pathogens that breach these barriers. The primary subsets of innate immune cells are phagocytes, including neutrophils and macrophages, dendritic cells (DCs), and natural killer (NK) cells.

Macrophages

Macrophages are highly plastic phagocytic cells derived from either the embryonic yolk sac or circulating monocytes [174]. They are present in specialized forms in nearly every tissue of the body and exert their critical immune function by engulfing and phagocytosing pathogens and apoptotic cells, acting as professional antigen-presenting cells (APCs), and secreting effector molecules such as chemokines and cytokines to direct the recruitment and activation of other immune cells. Primary monocytes and macrophages can be isolated from peripheral blood mononuclear cells (PBMCs), bone marrow, or peritoneal lavage for in vitro studies, and are typically selected or validated by their expression of CD14. Furthermore, immortalized cell lines such as U-937, THP-1 (human), and RAW264.7 (mouse) can be used to examine macrophages in vitro.

Activated macrophages are classified into two primary categories: classically activated, or "M1" macrophages, which elicit proinflammatory activity, and alternatively activated, or "M2" macrophages, which are antiinflammatory and proangiogenic. Polarization of macrophages into M1 phagocytes requires the use of any combination of activation compounds such as LPS, IFN-γ, TNF-α, and phorbol-12-myristate-13-acetate (PMA). M2 phagocyte polarization can be achieved through exposure to IL-4 and/or IL-13. Macrophage dynamics are characterized in vitro through a comprehensive examination of cell surface phenotype, secretome, morphology, and phagocytic capacity [175, 176].

Macrophage polarization state can be determined by measuring gene transcript and protein expression of cell surface markers such as the costimulatory molecules CD80 and CD86 (M1) and the pattern recognition receptors CD206 and CD163 (M2). Interestingly, exposure of both primary $CD14^+$ monocyte and M1-activated macrophages to ASCs consistently downregulates surface expression of M1 markers and upregulates M2 markers, and this is most likely due to soluble factors from ASCs, as conditioned medium, Boyden chamber systems, and isolated extracellular vesicles all produce equally robust effects when compared with direct coculture [177–181].

Canonical M1 secreted factors include TNF-α, IFN-γ, IL-6, nitric oxide (NO), IL-12, and IL-1β, which cumulatively promote a T-helper cell 1 (Th1) and Th17-driven immune response. In contrast, M2 macrophages commonly secrete IL-10, VEGF, and TGF-β, which promote a shift towards a Th2 and regulatory T-cell (Treg)-driven immune response [182, 183]. ASCs' paracrine activity has profound effects on the monocyte and macrophage secretome, as conditioned medium or Boyden chamber cocultures induce powerful downregulation of M1-secreted cytokines and chemokines and concomitant upregulation of M2 factors [178–180, 184–188]. The ASC-derived soluble factors suggested to be responsible for this influence include microRNA-packed extracellular vesicles [181], TNF-α-stimulated gene-6 (TSG-6) [189], or prostaglandin E2 (PGE2) [179, 180, 190]. Regardless, several groups have demonstrated robust suppression of NF-κB activation in macrophages following exposure to ASCs, which may explain these phenotypic shifts [181, 184, 185].

In addition to the modulation of their secretory profile, ASCs can also alter the morphology, migration ability, and phagocytic rate of macrophages. Following exposure to ASCs or their secretome, macrophages decrease their phagocytic and migratory abilities and adopt a ramified morphology [178, 181, 185]. The ASC secretome is powerful enough to induce these functional and phenotypic shifts even in macrophages from septic or colitic mice [187]. Overall, evidence suggests that ASCs are superior to BMSCs in their ability to shift the balance of macrophages in favor of the antiinflammatory M2 phenotype [184, 190, 191]. This effect is enhanced when ASCs are stimulated with either TNF-α or IFN-γ, or cultured in low-serum medium or hypoxic conditions to enhance their immunomodulation [177, 181, 189, 190].

Certain disease conditions may alter the ability of ASCs to influence macrophage activation and polarization. Interestingly, when ASCs isolated from equine metabolic syndrome animals were cocultured with a mouse macrophage cell line, RAW264.7, in a xenogeneic model, the macrophages exhibited increased gene expression and secretion of proinflammatory cytokines [192]. Similarly, an investigation of the effects of obesity on ASCs immunomodulatory functions found that, following coculture with ASCs from obese donors, RAW264.7 cells upregulated the expression of proinflammatory genes and increased inducible nitric oxide synthase (iNOS) signaling [193]. These proinflammatory changes did not occur in RAW264.7 cells that had been cocultured with ASCs from lean donors. Alternatively, ASCs isolated from patients with coronary heart disease were still able to suppress activation and lipid droplet accumulation in oxidized low-density lipoprotein (ox-LDL)-treated foamy allogeneic macrophages. In this study, ASCs also outperformed BMSCs in the ability to induce secretion of antiinflammatory mediators and protect against ox-LDL-mediated apoptosis [194]. When comparing lipoma-derived human ASCs with control, both cell populations similarly suppressed the secretion of proinflammatory mediators by mouse macrophage cell line, RAW264.7. However, lipoma-derived ASCs did result in a downregulation of IL-10 after coculture, suggesting they may be less immunosuppressive than controls [188].

Finally, the interaction between ASCs and macrophages is bidirectional, and crosstalk between these two cell types has been reported by several groups. As a result of coculture with monocytes and macrophages, ASCs upregulate expression of M2-priming cytokines IL-4, IL-13, and PGE2, and exhibit enhanced migration abilities [178, 179, 190].

These data point to the ability of ASCs to outperform BMSCs and regulate the activation, polarization, and immune function of macrophages. This results from their potent paracrine activity, and while this may be impaired in some disease states, it can also be restored and even enhanced with exogenous preconditioning strategies. Thus, ASCs are able to dynamically respond to macrophage signals to adapt their immunosuppressive vigor to meet the needs of a changing environment (Fig. 3).

Dendritic cells

DCs are phagocytes that primarily differentiate from circulating bone marrow-derived hematopoietic progenitors. They are the most potent antigen presenting cell (APC) in the innate immune system and also perform a variety of functions, including surveillance for pathogens or danger signals, phagocytosis, and processing of antigens for major histocompatibility complex (MHC)-II presentation, and trafficking to lymphoid organs to convey activation signals to adaptive immune cells. They form a crucial bridge between the innate and adaptive immune systems by communicating directly with B- and T-lymphocytes to initiate antigen-specific immune responses. After encountering an antigen, the immature DC (iDC) engulfs and processes the antigen and initiates maturation. They upregulate expression of chemokine

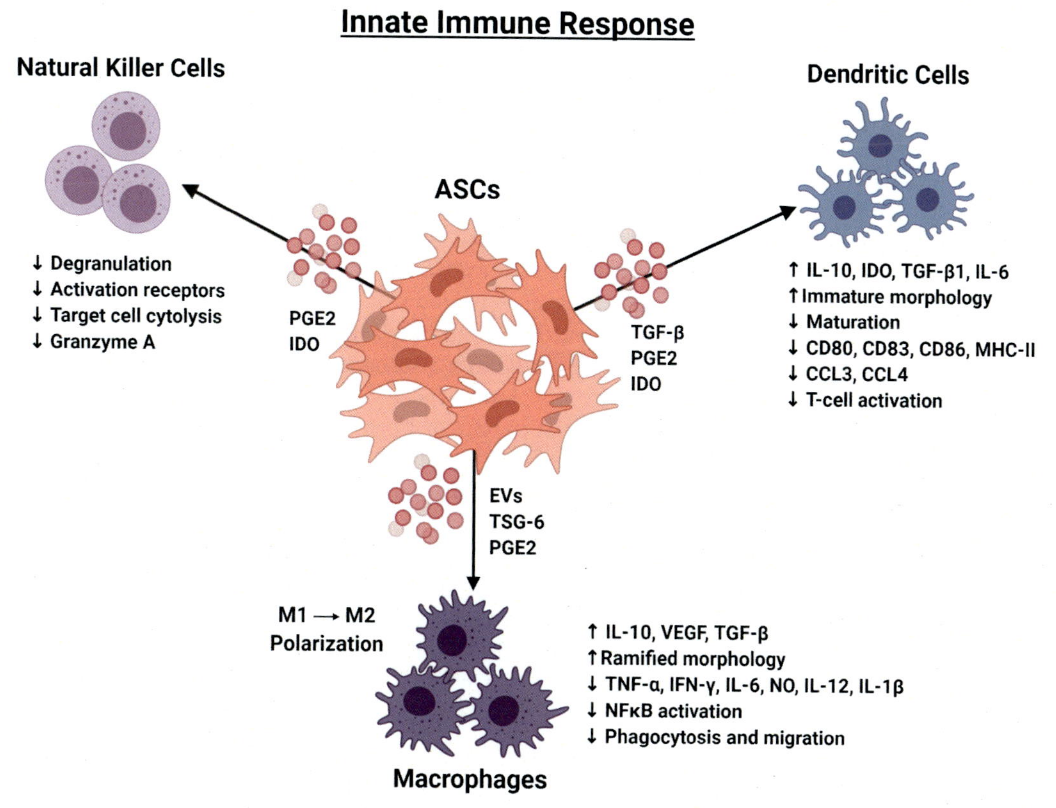

FIG. 3 In vitro effects of adipose stem cells (ASCs) on cell of the innate immune system. ASCs display robust contact-dependent and paracrine-mediated immunomodulatory effects on dendritic cells, macrophages, and NK cells in vitro. *ASCs*, adipose-derived stem cells; *CCL3*, C-C motif chemokine ligand 3; *CCL4*, C-C motif chemokine ligand 4; *EVs*, extracellular vesicles; *IDO*, indoleamine 2,3-dioxidase; *IFN-γ*, interferon-gamma; *IL-6*, interleukin 6; *IL-10*, interleukin 10; *IL-12*, interleukin 12; *IL-1β*, interleukin 1 beta; *NFκB*, nuclear factor kappa-light-chain-enhancer of activated B cells; *NO*, nitric oxide; *PEG2*, prostaglandin E2; *TGF-β*, transformation growth factor-beta; *TNF-α*, tumor necrosis factor alpha; *TSG-6*, TNF-α-stimulated gene-6; *VEGF*, vascular endothelial growth factor. *Created with Biorender.com*

receptors that direct their migration to draining lymph nodes, costimulatory and MHC molecules that allow for antigen presentation and priming of naïve T cells, and cytokines that direct recruitment and activation of adaptive immune cells [195, 196]. DCs are grouped into two main subsets based on their developmental pathways, surface marker expression, and molecular signature. The myeloid DCs (mDC) are derived from the common myeloid progenitor (CMP) that also gives rise to macrophages, while the plasmacytoid DCs (pDC) are derived from the common lymphoid progenitor (CLP) that also gives rise to B cells, T cells, and NK cells. Functionally, mDCs respond to immune stimulation by inducing either Th1 or Th2 cell activation, while pDCs produce the bulk of type I interferons and induce either Th2 or Tregs [197].

Generation of DCs for in vitro assays is typically accomplished by CD14$^+$ positive selection from PBMCs or bone marrow. This is followed by the differentiation of immature CD14$^+$ monocytes into either immature mDCs using a combination of IL-4, granulocyte-macrophage colony-stimulating factor (GM-CSF), and/or TNF-α, or immature pDCs using IL-3 [198]. This is then followed by maturation of iDCs using LPS or other proinflammatory stimuli [199].

Due to the heterogeneity of DC subsets, several in vitro assays have been developed to examine specialized cell functions. Some of the most commonly employed include characterization of maturation state, secretory activity, and the ability to induce activation and proliferation of CD4$^+$ T lymphocytes. Flow cytometric phenotyping is used to determine the maturation state of DCs based on surface expression of CD14 (immature only) and the mature DC markers CD83, CD80, CD86, and HLA-DR or MHC-II. Additionally, morphology can be used to discern maturation state, as iDCs have a smooth and round appearance while mDCs display the dendritic morphology that lends these cells their name [200]. Activated DCs exhibit a unique secretory signature that produces both paracrine and autocrine effects on naïve T cells and other iDCs, respectively [201]. Finally, T-cell proliferation and activation phenotype are used as indirect measures of mDC function in mixed lymphocyte reaction (MLR) assays.

The effect of MSCs on DC maturation and function has been well characterized. Comparisons between the effects of BMSCs and ASCs have repeatedly shown the latter to be more potent inhibitors of DC maturation in both direct and indirect coculture [198, 202–205]. These results are evidenced by elevated numbers of CD14$^+$ cells exhibiting smooth immature morphology and dampened expression of the mature DC markers CD83, CD80, CD86, and MHC-II. The secretory activity of DCs plays a role in how they recruit and activate adaptive immune cells and has thus been a major focus of investigation. In response to direct coculture with ASCs, levels of the secreted chemokine ligands CCL3 (macrophage inflammatory protein 1 alpha, or MIP1α) and CCL4 (macrophage inflammatory protein 1 beta, or MIP1β), both of which act as chemoattractants for CD8$^+$ T-effector cells, is nearly abolished in DCs [202]. ASCs also significantly upregulate secretion of the antiinflammatory cytokine IL-10 when cocultured with CD14$^+$ monocytes. This cytokine inhibits not only the maturation of iDCs but also the subsequent production of the Th1-priming cytokines, IL-12, TNF-α, and IFN-γ [198, 202, 204, 205].

An essential immune function of DCs is the ability to prime naïve CD4+ and CD8+ T cells against a particular antigen and promote their proliferation. Some groups purport that PGE2 plays an indispensable role in the ability of DCs to suppress activated T-cell proliferation. In coculture with MSCs, particularly ASCs, the ability of iDCs to induce T-cell proliferation in mixed lymphocyte assays is significantly suppressed [198]. This suppression is likely to be a result of the robust increase in the secretion of PGE2 by ASCs, as blockade of PGE2 but no other upregulated cytokine negates this outcome. In comparison with DCs alone, those that are directly cocultured with ASCs produce higher levels of the antiinflammatory and pleiotropic cytokines TGF-β1, IL-10, IL-6, and indoleamine 2,3-dioxidase (IDO) [198, 204, 205]. When these DCs are subsequently cultured with CD4$^+$ T cells, those that were treated with ASCs suppress T-cell proliferation and induce significantly more Tregs than those cultured alone. This effect was attributed to TGF-β-induced elevations in IDO, as ASCs pretreated with TGF-β1 siRNAs or the IDO inhibitor, indomethacin, did not produce this effect [198, 204, 205].

Overall, ASC paracrine activity demonstrates robust and reliable suppression of DC maturation and T-cell-activating function, making them promising agents for combating inflammation and autoimmune conditions (Fig. 3).

Natural killer cells

NK cells are cytolytic effector cells of lymphoid origin that derive their name from their ability to "naturally" kill defective or foreign cells without initial priming or antigen presentation. NK cytolytic activity is determined by the balance of activating or inhibiting ligands on the surface of target cells. Healthy cells expressing high levels of the "self" MHC-I molecules evade NK cytolysis, while those experiencing viral infection or malignant transformation downregulate MHC-I and shift the balance toward NK cell activation [206]. Once activated, NK cells perform their cell-killing function by either inducing Fas/FasL-dependent apoptosis or releasing lysosomal granules filled with perforin and granzymes to digest the target cell [207].

NK cells can be isolated from PBMCs using either positive selection for the canonical NK marker CD56 or negative selection to eliminate monocytes, other lymphocytes, granulocytes, and erythrocytes [208]. In vitro analysis of NK cell phenotype and function is performed using surface marker expression of the pathogen recognition receptor (PRR) CD56, degranulation assays measuring the expression of lysosomal-associated membrane proteins CD107a/b (LAMP1/2), and intracellular staining for cytokines such as IFN-γ in the presence of protein transport inhibitors [209].

Comparative studies examining the effects of different MSC sources on NK cell characteristics have yielded somewhat contradictory findings. Some reports detail an enhanced immunosuppressive effect of ASCs over other MSC sources on NK cell activation, proliferation, and cytolytic capacity [210–212]. Others, however, argue that ASCs perform similarly to BMSCs when it comes to suppressing activation, proliferation, or cytokine secretion and that BMSCs are more potent inhibitors of NK cytolysis than ASCs (Fig. 3) [203, 213]. These discrepancies may derive from different NK purification methods, culture conditions, and activation stimuli.

The enhanced potency of ASC-mediated suppression of NK cytolytic function in comparison with BMSCs has been demonstrated in both direct and indirect cocultures. Results have shown evidence of significantly less degranulation of IL-2 activated, allogeneic NK cells [210]. Incubation of NK cells with both ASCs and BMSCs significantly downregulated NK cell surface expression of the activation receptor DNAX accessory molecule-1 (DNAM-1), reduced degranulation capacity when subsequently cultured with K562 target cells, and downregulated levels of granzyme A, suggesting less NK cell activation [210]. However, both MSC types elicited a similar IFN-γ response as measured by intracellular staining. The ability of ASCs to suppress activation and proliferation of NK cells has been attributed to the downregulation of NK activation receptors such as NKp30, NKp44, and NKG2D through the activity of PGE2 and IDO, both of which are upregulated by coculture with ASCs [211, 212]. Other groups have demonstrated no significant differences between BMSCs and ASCs in immunosuppression of NK cells or susceptibility to their cytotoxic function [203, 213].

MSCs are proven immunomodulators capable of soothing inflammation and promoting an environment of repair and homeostasis. NK cells are primarily responsible for the rapid clearance of transplanted MSCs, and resilience to NK-mediated cytolysis is, therefore, a valuable measure of how long MSCs will survive in the posttransplant environment to exert their therapeutic benefit. Here, again, results appear contradictory, with some arguing that ASCs are equally vulnerable to NK cell-induced apoptosis as BMSCs. Others purport reduced susceptibility of ASCs to the cytolytic activity of NK cells due to lower surface expression of HLA class I molecules and negligible expression of activating ligands for NK cells [210]. This enhanced resilience is further confirmed when ASCs are used as target cells for activated NK cells, as they induce significantly less degranulation compared with BMSCs [210].

Evidence suggests that ASCs are at least comparable to BMSCs in their ability to suppress NK cell activation and evade cytolytic destruction and may even be superior to BMSCs in these respects. These abilities indicate that ASCs may be able to persist longer after transplantation and exert enhanced immunomodulatory effects by suppressing NK cell activation.

Adipose stem cells and the adaptive immune system

While the innate immune system is critical for the rapid identification and elimination of pathogens, the adaptive immune system provides a selective and specific mechanism for recognizing self- and non-self-antigens [214]. There are two main types of cells in the adaptive immune system: the T lymphocytes, which derive from bone marrow, mature in the thymus, and are the primary effectors of cellular immune response; and the B lymphocytes, which arise in the bone marrow and are the antibody-producing cells of the body [214, 215]. Following their development, lymphocytes will migrate to secondary lymphoid organs such as the lymph nodes and spleen, which serve to capture circulating antigens. After exposure to these antigens, lymphocytes migrate to antigen-rich target sites throughout the body and employ their effector functions [214].

MSCs, including ASCs, have been shown to exhibit significant immunomodulatory effects on lymphocytes in vitro, via both contact-dependent and paracrine signaling. Several studies have compared the ability of ASCs and BMSCs to inhibit PBMC-derived lymphocyte proliferation [216, 217]. In lymphocytes, proliferation is most often assessed through the addition of carboxyfluorescein diacetate succinimidyl ester (CFSE) to the culture. CFSE is membrane permeable and is readily incorporated into cells. With each cell division, its signal is diluted twofold, allowing for generational tracking of populations via flow cytometry [218]. MSC-mediated antiproliferative effects were shown to be more robust when cells were directly cocultured as opposed to indirectly cocultured, illustrating the importance of cell–cell contact [216, 217]. These results highlight the ability of ASCs to modulate general lymphocyte populations and the potential to actively modulate the adaptive immune response (Fig. 4).

T cells

T cells are responsible for mediating cellular immunity. When T cells encounter activated APCs, such as DCs, macrophages, or B cells, they proliferate and differentiate into subtypes that coordinate to mediate the removal of the antigen [219, 220]. Depending on cytokines present in the environment, T cells are capable of differentiating into several different subtypes, the effects of which are defined by their cytokine expression profiles. T cells are also able to

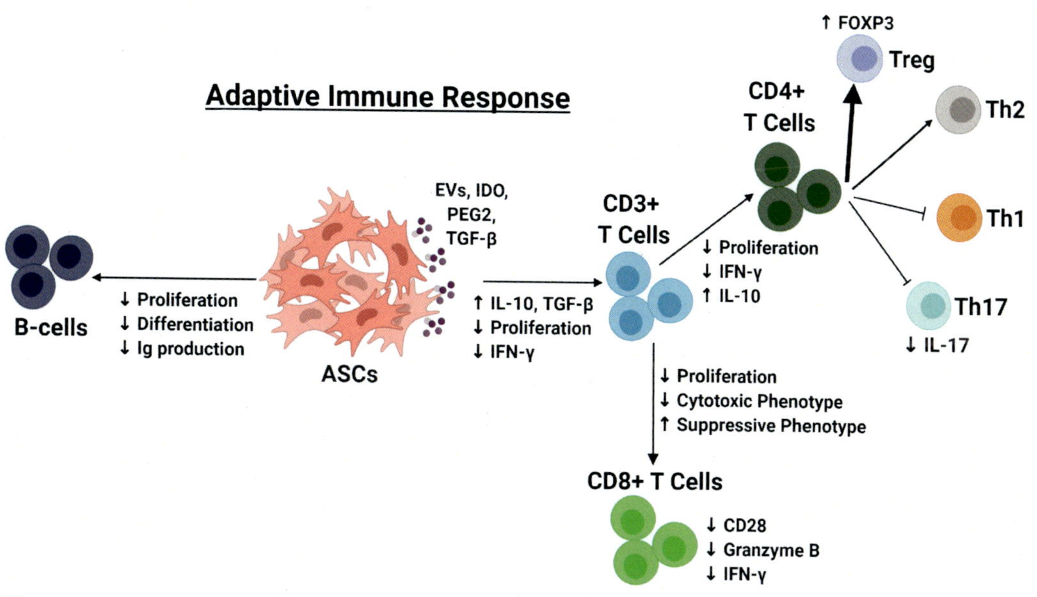

FIG. 4 In vitro effects of adipose stem cells (ASCs) on cells of the adaptive immune system. ASCs exhibit significant immunomodulatory effects on both B and T lymphocytes in vitro. This effect is mediated by both contact-dependent and paracrine signaling mechanisms. *ASCs*, adipose-derived stem cells; *CD28*, cluster of differentiation 28; *FOXP3*, forkhead box P3; *IDO*, indoleamine 2,3-dioxidase; *IFN-γ*, interferon-γ; *IL-10*, interleukin 10; *IL-17*, interleukin 17; *PEG2*, prostaglandin E2; *TGF-β*, transforming growth factor-beta; *Treg*, regulatory T cells; *Th1*, T-helper 1; *Th2*, T-helper 2; *Th17*, T-helper 17. *Created with Biorender.com*

maintain a sort of cellular memory for antigens, which allows them to respond more rapidly to a previously encountered antigen [221].

A defining feature of the T-cell lineage is the expression of CD3, which serves as a coreceptor in conjunction with the T-cell receptor (TCR). Isolation of T cells from PBMCs, spleen, or other lymphatic tissue can be accomplished via positive or negative selection. Selection for $CD3^+$ T cells selects for all subtypes and allows for whole-population interrogation. Following indirect coculture with ASCs, CD3+ T cells exhibited significantly decreased proliferation rates [222–224]. Additionally, this antiproliferative effect could be enhanced by direct coculture [223, 224]. Prestimulation of ASCs with proinflammatory cytokines such as IFN-γ or TNF-α significantly enhanced these antiproliferative effects [223]. ASC effects on T-cell proliferation are mediated by the secretion of IDO. Cocultured cells produced more IL-10 and PGE2 than isolated T-cell or ASC cultures, indicating communication-related alterations in their secretomes [222–224]. Another study by Chein et al. demonstrated an increase in the concentrations of the central antiinflammatory cytokines IL-10 and TGF-β following indirect coculture of activated T cells and ASCs [225]. Importantly, the protein concentration of proinflammatory cytokine IFN-γ was significantly reduced when ASCs were indirectly cocultured with activated $CD3^+$ T cells, in comparison with activated T-cells alone [225]. The results of these studies demonstrate the ability of the ASCs to modulate general T-cell differentiation and proliferation in vitro.

CD4$^+$ T cells

Also known as helper T cells, CD4$^+$ T cells carry TCRs that recognize antigens presented by MHC II molecules [215, 219, 220]. ASCs have been shown to be more robust in their suppression of helper T-cell proliferation than BMSCs and Wharton's jelly mesenchymal stem cells (WJMSCs) [224]. Though these results have been demonstrated to be dose-dependent, advanced age and chronic inflammatory disease states both have been shown to negatively impact ASCs ability to suppress T-cell proliferation [226, 227].

CD4$^+$ T cells can be further divided into subtypes defined by their cytokine expression and immune function. These include the proinflammatory, antitumorigenic Th1 and Th17 cells; the antiinflammatory, protumorigenic Th2 cells; and the antiinflammatory Treg cells [228]. Culturing CD4$^+$ T cells in ASC-conditioned medium elicits a shift towards to Treg subtype, as evidenced by a significant increase in forkhead box P3 (FoxP3) expression, the master regulator of Treg cell development [229]. This shift towards Tregs as well as the reported increase in IL-10 secretion suggests that ASCs can promote robust antiinflammatory responses in mixed T-cell populations.

CD8+ T cells

CD8$^+$ T cells, known as cytotoxic T cells, recognize antigens presented by MHC I molecules, which, in combination with CD80 or CD86 interaction, will result in activation and killing of infected or malignant cells [219, 220]. Similar to NK cells, CD8$^+$ T cells can secrete cytotoxic granules containing perforin and granzymes or employ Fas/Fas-L binding to induce caspase-mediated apoptosis in target cells [230, 231]. As previously seen with CD4$^+$ T cells, ASCs cultured directly with CD8$^+$ T cells suppress their proliferation in a dose-dependent manner [224]. These effects were found to be stronger than those of BMSCs or WJMSCs. Additionally, direct coculture of ASCs with CD8$^+$ T cells resulted in decreased CD8 expression [232]. Thus, ASCs push cytotoxic T cells towards a more suppressive, less cytotoxic phenotype. In these same cells, expression of CD28 was also significantly decreased, indicating a CD8$^+$ T-cell suppressor state [232]. Further, the shift towards a regulatory phenotype in CD8$^+$ T cells following ASC coculture was found to be functional, as ASCs were able to significantly reduce the production of granzyme B and IFN-γ [232]. Interestingly, unlike the effects observed with CD4$^+$ T cells, the immunosuppressive capabilities of ASCs on CD8$^+$ T cells were found to be unaffected by donor age [227].

B cells

B lymphocytes are responsible for mediating humoral immunity. Their primary function is to produce and secrete antibodies that recognize foreign antigens. Following antigen recognition via B-cell receptor (BCR), the B cells become activated, proliferate, and differentiate into antibody-secreting cells that may later become memory cells [233]. B cells are primarily identified by the expression of CD19 and can be isolated from PBMCs via negative selection for other cell types.

To date, the number of in vitro studies evaluating the immunomodulatory properties of ASCs on B cells has been minimal, but the overall consensus has been that ASCs are immunosuppressive. ASCs secrete factors that promote B-cell migration and chemoattraction to a greater extent than other mesenchymal cell types [234]. Although ASCs secrete distinct chemokines and produce other factors involved in cell motility and chemotaxis, the specific factor that promotes B-cell migration and chemoattraction is unknown at this point [234]. In comparison with BMSCs, direct coculture resulted in enhanced $CD19^{+}$ B-cell proliferation. This finding was enhanced when the cells were cocultured with ASCs [223]. Stimulation of ASCs with proinflammatory cytokines, which has previously been shown to increase their immunomodulatory capabilities, reversed the effect and inhibited B-cell proliferation to the same level as BMSC coculture [223].

Antibody production is one of the primary functions of B cells, thus, analysis of immunoglobulin (Ig) production following coculture is a simple heuristic used to assess the impact of MSCs on B-cell function [235, 236]. A study by Bochev et al. demonstrated that coculture of ASCs with stimulated PBMCs resulted in potent suppression of B-cell Ig production. The suppression was due to the ability of ASCs to inhibit the differentiation of B cells to Ig-producing cells and was found to be more robust than coculture with BMSCs [236]. Overall, these results suggest that ASCs are able to modulate the B-cell population in vitro and thus may possess significant immunomodulatory capacity in more complex systems.

Summary

While donor variability has not been thoroughly examined in the context of ASC-mediated immunomodulation of the innate or adaptive immune system, the ASC secretome is crucial to their influence over immune cells. It may, therefore, be hypothesized that impaired secretion of key factors such as PGE2, IDO, TGF-β, and IL-10 from donors with advanced age, elevated BMI, or underlying comorbidities would result in diminished immunomodulatory control over each of these specialized cells. Further studies need to be conducted to fully elucidate the impact of donor characteristics on immunomodulatory capabilities.

Conclusion and future directions

Minimal characterization of ASCs by their adherence to plastic, expression of well-defined surface markers, and the ability to differentiate into mesenchymal lineages ensures that scientists are examining related populations of cells. Further assessment of their secretome and immunomodulatory profiles may allow for the identification of ASC subpopulations, which can be tailored to specific translational applications. Already, researchers have identified several surface-marker expression patterns that are indicative of ASCs with unique characteristics. Future investigations into surface marker-defined subpopulations may lead to exciting new therapeutic approaches. Donor variation has also been shown to significantly affect the ability of ASCs to differentiate, grow, and produce immunomodulatory effects. Further research into the impact of donor characteristics such as age, BMI, and health status on therapeutic potential will allow for the selection of adipose tissue with optimal characteristics.

ASCs represent a valuable resource for use in regenerative medicine owing to their ease of acquisition, high cell yield, and potent immunomodulatory abilities. Thorough characterization of these cells is not only essential for defining them as adipose-derived MSCs but also necessary for collaborative efforts between labs. By fully characterizing ASCs, scientists will be able to share discoveries and reproduce results more effectively, leading to an accelerated rate of scientific breakthroughs.

References

[1] P.A. Zuk, et al., Human adipose tissue is a source of multipotent stem cells, Mol. Biol. Cell 13 (12) (2002) 4279–4295.
[2] G. Yu, et al., Isolation of human adipose-derived stem cells from lipoaspirates, Methods Mol. Biol. 702 (2011) 17–27.
[3] M. Dominici, et al., Minimal criteria for defining multipotent mesenchymal stromal cells. The International Society for Cellular Therapy position statement, Cytotherapy 8 (4) (2006) 315–317.
[4] P. Bourin, et al., Stromal cells from the adipose tissue-derived stromal vascular fraction and culture expanded adipose tissue-derived stromal/stem cells: a joint statement of the International Federation for Adipose Therapeutics and Science (IFATS) and the International Society for Cellular Therapy (ISCT), Cytotherapy 15 (6) (2013) 641.
[5] M.J. Gimble, J.A. Katz, A.B. Bunnell, Adipose-derived stem cells for regenerative medicine, Circul. Res. 100 (9) (2007) 1249–1260.
[6] P.C. Baer, H. Geiger, Adipose-derived mesenchymal stromal/stem cells: tissue localization, characterization, and heterogeneity, Stem Cells Int. 2012 (2012), 812693.
[7] Y. Kar Wey, et al., Phenotypic and functional characterization of long-term cryopreserved human adipose-derived stem cells, Sci. Rep. 5 (1) (2015).
[8] L.F. Zanata, et al., Effect of cryopreservation on human adipose tissue and isolated stromal vascular fraction cells: in vitro and in vsivo analyses, Plast. Reconstruct. Surg. 141 (2) (2018) 232e–243e.
[9] H. Gruber, et al., Human adipose-derived mesenchymal stem cells: serial passaging, doubling time and cell senescence, Biotech. Histochem. 87 (4) (2012) 303–311.
[10] T. Debnath, L.K. Chelluri, Standardization and quality assessment for clinical grade mesenchymal stem cells from human adipose tissue, Hematol. Transf. Cell Ther. 41 (1) (2019) 7–16.
[11] N.C. Truong, K.H.-T. Bui, P. Van Pham, Characterization of senescence of human adipose-derived stem cells after long-term expansion, Adv. Exp. Med. Biol. 1084 (2019) 109.
[12] R.I. Dmitrieva, et al., Bone marrow- and subcutaneous adipose tissue-derived mesenchymal stem cells: differences and similarities, Cell Cycle 11 (2) (2012) 377–383.
[13] R.H. Lee, et al., Characterization and expression analysis of mesenchymal stem cells from human bone marrow and adipose tissue, Cell. Physiol. Biochem. 14 (4-6) (2004) 311–324.
[14] W. Wagner, et al., Comparative characteristics of mesenchymal stem cells from human bone marrow, adipose tissue, and umbilical cord blood, Exp. Hematol. 33 (11) (2005) 1402–1416.
[15] Z. Yanxia, et al., Adipose-derived stem cell: a better stem cell than BMSC, Cell Res. 18 (S1) (2008) S165.
[16] M. Oedayrajsingh-Varma, et al., Adipose tissue-derived mesenchymal stem cell yield and growth characteristics are affected by the tissue-harvesting procedure, Cytotherapy 8 (2) (2006) 166–177.
[17] T.P. Frazier, et al., Body mass index affects proliferation and osteogenic differentiation of human subcutaneous adipose tissue-derived stem cells (Report), BMC Cell Biol. 14 (1) (2013).
[18] L. Kolaparthy, et al., Adipose tissue—adequate, accessible regenerative material, Int. J. Stem Cells 8 (2) (2015) 121–127.
[19] D.C. Ding, et al., Human adipose-derived stem cells cultured in keratinocyte serum free medium: donor's age does not affect the proliferation and differentiation capacities, J. Biomed. Sci. 20 (2013) 59.
[20] M. Kawagishi-Hotta, et al., Enhancement of individual differences in proliferation and differentiation potentials of aged human adipose-derived stem cells, Regen. Ther. 6 (2017) 29–40.
[21] J. Di Battista, et al., Proliferation and differentiation of human adipose-derived mesenchymal stem cells (ASCs) into osteoblastic lineage are passage dependent, Off. J. Int. Assoc. Inflamm. Soc. Eur. Hist. Res. Soc. 63 (11) (2014) 907–917.
[22] J. Li, H. Huang, X. Xu, Biological and genetic characteristics of mesenchymal stem cells in vitro derived from human adipose, umbilical cord and placenta, Int. J. Clin. Exp. Med. 10 (12) (2017) 16310–16318.
[23] J. Li, H. Huang, X. Xu, Biological characteristics and karyotiping of a new isolation method for human adipose mesenchymal stem cells in vitro, Tissue Cell 49 (3) (2017) 376–382.
[24] G. Liu, et al., Evaluation of the viability and osteogenic differentiation of cryopreserved human adipose-derived stem cells, Cryobiology 57 (1) (2008) 18–24.
[25] M.S. Choudhery, et al., Donor age negatively impacts adipose tissue-derived mesenchymal stem cell expansion and differentiation, J. Transl. Med. 12 (1) (2014).

[26] L. Badimon, J. Cubedo, Adipose tissue depots and inflammation: effects on plasticity and resident mesenchymal stem cell function, Cardiovasc. Res. 113 (9) (2017) 1064–1073.
[27] B. Gharibi, F.J. Hughes, Effects of medium supplements on proliferation, differentiation potential, and in vitro expansion of mesenchymal stem cells, Stem Cells Transl. Med. 1 (11) (2012) 771–782.
[28] S. Riis, et al., Comparative analysis of media and supplements on initiation and expansion of adipose-derived stem cells, Stem Cells Transl. Med. 5 (3) (2016) 314–324.
[29] J. Roxburgh, A. Metcalfe, Y. Martin, The effect of medium selection on adipose-derived stem cell expansion and differentiation: implications for application in regenerative medicine, Incorp. Methods Cell Sci. Int. J. Cell Cult. Biotechnol. 68 (4) (2016) 957–967.
[30] J. Czapla, et al., The effect of culture media on large-scale expansion and characteristic of adipose tissue-derived mesenchymal stromal cells, Stem Cell Res. Ther. 10 (1) (2019).
[31] B.M. Strem, et al., Multipotential differentiation of adipose tissue-derived stem cells, Keio J. Med. 54 (3) (2005) 132–141.
[32] S. Mohamed-Ahmed, et al., Adipose-derived and bone marrow mesenchymal stem cells: a donor-matched comparison, Stem Cell Res. Ther. 9 (1) (2018) 168.
[33] J.K. Fraser, et al., Differences in stem and progenitor cell yield in different subcutaneous adipose tissue depots, Cytotherapy 9 (5) (2007) 459–467.
[34] W. Jurgens, et al., Effect of tissue-harvesting site on yield of stem cells derived from adipose tissue: implications for cell-based therapies, Cell Tissue Res. 332 (3) (2008) 415–426.
[35] H.A. Tucker, B.A. Bunnell, Characterization of human adipose-derived stem cells using flow cytometry, Methods Mol. Biol. (Clifton, N.J.) 702 (2011) 121.
[36] D. Minteer, K.G. Marra, J.P. Rubin, Adipose-derived mesenchymal stem cells: biology and potential applications, Adv. Biochem. Eng. Biotechnol. 129 (2013) 59.
[37] P.C. Baer, et al., Comprehensive phenotypic characterization of human adipose-derived stromal/stem cells and their subsets by a high throughput technology (Report), Stem Cells Dev. 22 (2) (2013) 330.
[38] L. De Girolamo, et al., Stemness and osteogenic and adipogenic potential are differently impaired in subcutaneous and visceral adipose derived stem cells (ASCs) isolated from obese donors, Int. J. Immunopathol. Pharmacol. 26 (1_suppl) (2013) 11–21.
[39] A.C. Bowles, et al., Adipose stromal vascular fraction-mediated improvements at late-stage disease in a murine model of multiple sclerosis, Stem Cells 35 (2) (2017) 532–544.
[40] E.T. Camilleri, et al., Identification and validation of multiple cell surface markers of clinical-grade adipose-derived mesenchymal stromal cells as novel release criteria for good manufacturing practice-compliant production, Stem Cell Res. Ther. 7 (1) (2016) 107.
[41] L.Z. Poria Abdollah, Novel negative selection marker CD54 enhances differentiation of human adipose-derived mesenchymal stem cells, J. Clin. Cell. Immunol. (2013).
[42] T. Jiang, et al., Potent in vitro chondrogenesis of CD105 enriched human adipose-derived stem cells, Biomaterials 31 (13) (2010) 3564–3571.
[43] L.H. Pham, N.B. Vu, P. Van Pham, The subpopulation of CD105 negative mesenchymal stem cells show strong immunomodulation capacity compared to CD105 positive mesenchymal stem cells, Biomed. Res. Ther. 6 (4) (2019) 3131–3140.
[44] H. Suga, et al., Functional implications of CD34 expression in human adipose-derived stem/progenitor cells (ORIGINAL RESEARCH REPORT) (cluster of differentiation) (Report), Stem Cells Dev. 18 (8) (2009) 1201.
[45] Y.S. Choi, et al., Differentiation of human adipose-derived stem cells into beating cardiomyocytes, J. Cell Mol. Med. 14 (4) (2010) 878–889.
[46] C. Auxenfans, et al., Adipose-derived stem cells (ASCs) as a source of endothelial cells in the reconstruction of endothelialized skin equivalents, J. Tissue Eng. Regen. Med. 6 (7) (2012) 512–518.
[47] S. Jang, et al., Functional neural differentiation of human adipose tissue-derived stem cells using bFGF and forskolin, BMC Cell Biol. 11 (2010) 25.
[48] H. Aurich, et al., Hepatocyte differentiation of mesenchymal stem cells from human adipose tissue in vitro promotes hepatic integration in vivo, Gut 58 (4) (2009) 570–581.
[49] Y. Sakaguchi, et al., Comparison of human stem cells derived from various mesenchymal tissues: superiority of synovium as a cell source, Arth. Rheumat. 52 (8) (2005) 2521–2529.
[50] G. Pachon-Pena, et al., Stromal stem cells from adipose tissue and bone marrow of age-matched female donors display distinct immunophenotypic profiles, J. Cell. Physiol. 226 (3) (2011) 843–851.

[51] D.A. De Ugarte, et al., Comparison of multi-lineage cells from human adipose tissue and bone marrow, Cells Tissues Organs 174 (3) (2003) 101–109.
[52] D. Noël, et al., Cell specific differences between human adipose-derived and mesenchymal–stromal cells despite similar differentiation potentials, Exp. Cell Res. 314 (7) (2008) 1575–1584.
[53] V.V. Harmelen, et al., Effect of BMI and age on adipose tissue cellularity and differentiation capacity in women, Int. J. Obes. 27 (8) (2003) 889.
[54] C. Cramer, et al., Persistent high glucose concentrations alter the regenerative potential of mesenchymal stem cells (Report), Stem Cells Dev. 19 (12) (2010) 1875.
[55] S. Lopa, et al., Donor-matched mesenchymal stem cells from knee infrapatellar and subcutaneous adipose tissue of osteoarthritic donors display differential chondrogenic and osteogenic commitment, Eur. Cells Mater. 27 (2014) 298–311.
[56] K. Kornicka, et al., The effect of age on osteogenic and adipogenic differentiation potential of human adipose derived stromal stem cells (hASCs) and the impact of stress factors in the course of the differentiation process, Oxid. Med. Cell. Long. 2015 (2015).
[57] C. Siciliano, et al., The adipose tissue of origin influences the biological potential of human adipose stromal cells isolated from mediastinal and subcutaneous fat depots, Stem Cell Res. 17 (2) (2016) 342–351.
[58] T.C. Prestwich, O.A. Macdougald, Wnt/β-catenin signaling in adipogenesis and metabolism, Curr. Opin. Cell Biol. 19 (6) (2007) 612–617.
[59] C. Christodoulides, et al., Adipogenesis and WNT signalling, Trends Endocrinol. Metab. 20 (1) (2009) 16–24.
[60] Y. Tsurutani, et al., The roles of transforming growth factor-β and Smad3 signaling in adipocyte differentiation and obesity, Biochem. Biophys. Res. Commun. 407 (1) (2011) 68–73.
[61] K.G. Rana, et al., Transcriptional control of preadipocyte determination by Zfp423, Nature 464 (7288) (2010) 619.
[62] Z. Cao, R.M. Umek, S. McKnight, Regulated expression of 3 C/EBP isoforms during adipose conversion of 3T3-L1 cells, Genes Dev. 5 (9) (1991) 1538–1552.
[63] B.O. Park, R. Ahrends, M.N. Teruel, Consecutive positive feedback loops create a bistable switch that controls preadipocyte-to- adipocyte conversion, Cell Rep. 2 (4) (2012) 976–990.
[64] V.A. Payne, et al., C/EBP transcription factors regulate SREBP1c gene expression during adipogenesis, Biochem. J. 425 (1) (2009) 215.
[65] Z. Wu, et al., Conditional ectopic expression of C/EBP beta in NIH-3T3 cells induces PPAR gamma and stimulates adipogenesis, Genes Dev. 9 (19) (1995) 2350.
[66] Z. Wu, N.L. Bucher, S.R. Farmer, Induction of peroxisome proliferator-activated receptor gamma during the conversion of 3T3 fibroblasts into adipocytes is mediated by C/EBPbeta, C/EBPdelta, and glucocorticoids, Mol. Cell. Biol. 16 (8) (1996) 4128.
[67] Z. Wu, et al., Cross-regulation of C/EBPα and PPARγ controls the transcriptional pathway of adipogenesis and insulin sensitivity, Mol. Cell 3 (2) (1999) 151–158.
[68] S.-J. Yun, et al., Isoform-specific regulation of adipocyte differentiation by Akt/protein kinase Bα, Biochem. Biophys. Res. Commun. 371 (1) (2008) 138–143.
[69] H.H. Zhang, et al., Insulin stimulates adipogenesis through the Akt-TSC2-mTORC1 pathway (TSC2-mTOR control adipogenesis), PLoS One 4 (7) (2009), e6189.
[70] J.E. Kim, J. Chen, Regulation of peroxisome proliferator-activated receptor-[gamma] activity by mammalian target of rapamycin and amino acids in adipogenesis.(Signal Transduction), Diabetes 53 (11) (2004) 2748.
[71] J. Nakae, et al., The forkhead transcription factor Foxo1 regulates adipocyte differentiation, Dev. Cell 4 (1) (2003) 119.
[72] M. Armoni, et al., FOXO1 represses peroxisome proliferator-activated receptor-γ1 and -γ2 gene promoters in primary adipocytes: a novel paradigm to increase insulin sensitivity, J. Biol. Chem. 281 (29) (2006) 19881–19891.
[73] J.M. Gimble, F. Guilak, Adipose-derived adult stem cells: isolation, characterization, and differentiation potential, Cytotherapy 5 (5) (2003) 362–369.
[74] B.A. Bunnell, et al., Adipose-derived stem cells: Isolation, expansion and differentiation, Methods 45 (2) (2008) 115–120.
[75] J.M. Gimble, B.A. Bunnell, Adipose-Derived Stem Cells Methods and Protocols, Humana Press, New York, 2011.
[76] C. Vater, P. Kasten, M. Stiehler, Culture media for the differentiation of mesenchymal stromal cells, Acta Biomater 7 (2) (2011) 463–477.
[77] W.P. Cawthorn, E. Scheller, O. Macdougald, Adipose tissue stem cells meet preadipocyte commitment: going back to the future, J. Lipid Res. 53 (2) (2012) 227–246.

[78] T.K. Fam, A.S. Klymchenko, M. Collot, recent advances in fluorescent probes for lipid droplets, Mater. (Basel) 11 (9) (2018).
[79] T. Govender, et al., BODIPY staining, an alternative to the Nile Red fluorescence method for the evaluation of intracellular lipids in microalgae, Biores. Technol. 114 (2012) 507–511.
[80] L.-A.L.S. Harris, J.R. Skinner, N.E. Wolins, Imaging of neutral lipids and neutral lipid associated proteins, Methods in Cell Biol. 116 (2013) 213–226.
[81] B. Qiu, M.C. Simon, BODIPY 493/503 staining of neutral lipid droplets for microscopy and quantification by flow cytometry, Bio-Protocol 6 (17) (2016).
[82] M.E. Trujillo, P. Scherer, Adipose tissue-derived factors: Impact on health and disease, Endocr. Rev. 27 (7) (2006) 798.
[83] H. Skalnikova, et al., Mapping of the secretome of primary isolates of mammalian cells, stem cells and derived cell lines, Proteomics 11 (4) (2011) 691–708.
[84] I. Abishek, et al., Inflammatory lipid mediators in adipocyte function and obesity, Nat. Rev. Endocrinol. 6 (2) (2010) 71.
[85] S. Zvonic, et al., Secretome of primary cultures of human adipose-derived stem cells—modulation of serpins by adipogenesis, Mol. Cell. Proteom. 6 (1) (2007) 18–28.
[86] A. Strong, J. Gimble, B. Bunnell, Analysis of the pro- and anti-inflammatory cytokines secreted by adult stem cells during differentiation, Stem Cells Int. 2015 (2015) 412467.
[87] S. Gesta, et al., Evidence for a role of developmental genes in the origin of obesity and body fat distribution. (Author abstract), Proc. Natl. Acad. Sci. U. S. A. 103 (17) (2006) 6676.
[88] S.L. Hocking, et al., Intrinsic depot-specific differences in the secretome of adipose tissue, preadipocytes, and adipose tissue-derived microvascular endothelial cells, Diabetes 59 (12) (2010) 3008.
[89] E. Maury, et al., Adipokines oversecreted by omental adipose tissue in human obesity, Am. J. Physiol.-Endocrinol. Metab. 293 (3) (2007) E656–E665.
[90] O. Noriyuki, et al., Adipokines in inflammation and metabolic disease, Nat. Rev. Immunol. 11 (2) (2011) 85.
[91] G.I. Im, Y.W. Shin, K.B. Lee, Do adipose tissue-derived mesenchymal stem cells have the same osteogenic and chondrogenic potential as bone marrow-derived cells? Osteoarth. Cartil. 13 (10) (2005) 845–853.
[92] O. Hayashi, et al., Comparison of osteogenic ability of rat mesenchymal stem cells from bone marrow, periosteum, and adipose tissue, Calcif. Tissue Int. 82 (3) (2008) 238–247.
[93] P. Niemeyer, et al., Comparison of mesenchymal stem cells from bone marrow and adipose tissue for bone regeneration in a critical size defect of the sheep tibia and the influence of platelet-rich plasma, Biomaterials 31 (13) (2010) 3572–3579.
[94] R. Vishnubalaji, et al., Comparative investigation of the differentiation capability of bone-marrow- and adipose-derived mesenchymal stem cells by qualitative and quantitative analysis, Cell Tissue Res. 347 (2) (2012) 419–427.
[95] N.J. Panetta, D.M. Gupta, M.T. Longaker, Bone regeneration and repair, Curr. Stem Cell Res. Ther. 5 (2) (2010) 122–128.
[96] A.L. Strong, et al., Obesity inhibits the osteogenic differentiation of human adipose-derived stem cells, J. Transl. Med. 14 (2016) 27.
[97] W. Duan, M.J. Lopez, Effects of cryopreservation on canine multipotent stromal cells from subcutaneous and infrapatellar adipose tissue, Stem Cell Rev. Rep. 12 (2) (2016) 257–268.
[98] B.E. Grottkau, Y. Lin, Osteogenesis of adipose-derived stem cells, Bone Res. 1 (2) (2013) 133–145.
[99] F. Paduano, et al., Adipose tissue as a strategic source of mesenchymal stem cells in bone regeneration: a topical review on the most promising craniomaxillofacial applications, Int. J. Mol. Sci. 18 (10) (2017).
[100] S. Shaik, et al., Transcriptomic profiling of adipose derived stem cells undergoing osteogenesis by RNA-Seq, Sci. Rep. 9 (1) (2019) 11800.
[101] Q. Liu, et al., The role of the extracellular signal-related kinase signaling pathway in osteogenic differentiation of human adipose-derived stem cells and in adipogenic transition initiated by dexamethasone, Tissue Eng Part A 15 (11) (2009) 3487–3497.
[102] E.J. Tsang, B. Wu, P. Zuk, MAPK signaling has stage-dependent osteogenic effects on human adipose-derived stem cells in vitro, Connect Tissue Res 59 (2) (2018) 129–146.
[103] D.M. Lough, et al., Regulation of ADSC osteoinductive potential using notch pathway inhibition and gene rescue: a potential on/off switch for clinical applications in bone formation and reconstructive efforts, Plast. Reconstr. Surg. 138 (4) (2016) 642e-52e.

[104] Y. Liu, et al., Function of TGF-beta and p38 MAKP signaling pathway in osteoblast differentiation from rat adipose-derived stem cells, Eur. Rev. Med. Pharmacol. Sci. 17 (12) (2013) 1611–1619.
[105] J. Fan, et al., Enhanced osteogenesis of adipose-derived stem cells by regulating bone morphogenetic protein signaling antagonists and agonists, Stem Cells Transl. Med. 5 (4) (2016) 539–551.
[106] N.K. Dubey, et al., Revisiting the advances in isolation, characterization and secretome of adipose-derived stromal/stem cells, Int. J. Mol. Sci. 19 (8) (2018).
[107] S. Guidotti, et al., Enhanced osteoblastogenesis of adipose-derived stem cells on spermine delivery via beta-catenin activation, Stem Cells Dev. 22 (10) (2013) 1588–1601.
[108] R. Wu, et al., Long non-coding RNA HIF1A-AS2 facilitates adipose-derived stem cells (ASCs) osteogenic differentiation through miR-665/IL6 axis via PI3K/Akt signaling pathway, Stem Cell Res. Ther. 9 (1) (2018) 348.
[109] J. Xie, et al., Substrate elasticity regulates adipose-derived stromal cell differentiation towards osteogenesis and adipogenesis through beta-catenin transduction, Acta Biomater. 79 (2018) 83–95.
[110] J. Chang, et al., NF-κB inhibits osteogenic differentiation of mesenchymal stem cells by promoting β-catenin degradation, Proc. Natl. Acad. Sci. U. S. A. 110 (23) (2013) 9469–9474.
[111] X. Cao, et al., Naringin rescued the TNF-alpha-induced inhibition of osteogenesis of bone marrow-derived mesenchymal stem cells by depressing the activation of NF-small ka, CyrillicB signaling pathway, Immunol. Res. 62 (3) (2015) 357–367.
[112] Y.J. Wang, et al., Taxifolin enhances osteogenic differentiation of human bone marrow mesenchymal stem cells partially via NF-kappaB pathway, Biochem. Biophys. Res. Commun. 490 (1) (2017) 36–43.
[113] X. Chen, et al., Mechanical stretch-induced osteogenic differentiation of human jaw bone marrow mesenchymal stem cells (hJBMMSCs) via inhibition of the NF-κB pathway, Cell Death Dis. 9 (2) (2018) 207.
[114] Y. Wang, et al., Inhibition of PTGS1 promotes osteogenic differentiation of adipose-derived stem cells by suppressing NF-kB signaling, Stem Cell Res. Ther. 10 (1) (2019) 57.
[115] B. Levi, M.T. Longaker, Concise review: adipose-derived stromal cells for skeletal regenerative medicine, Stem Cells 29 (4) (2011) 576–582.
[116] A.T. Qureshi, et al., Human adipose-derived stromal/stem cell isolation, culture, and osteogenic differentiation, Methods Enzymol. 538 (2014) 67–88.
[117] A. Zolocinska, The expression of marker genes during the differentiation of mesenchymal stromal cells, Adv. Clin. Exp. Med. 27 (5) (2018) 717–723.
[118] K.S. Kang, et al., Regulation of osteogenic differentiation of human adipose-derived stem cells by controlling electromagnetic field conditions, Exp. Mol. Med. 45 (1) (2013), e6.
[119] B.E. Grottkau, et al., Comparison of effects of mechanical stretching on osteogenic potential of ASCs and BMSCs, Bone Res. 1 (3) (2013) 282–290.
[120] A.L. Strong, et al., Novel daidzein analogs enhance osteogenic activity of bone marrow-derived mesenchymal stem cells and adipose-derived stromal/stem cells through estrogen receptor dependent and independent mechanisms, Stem Cell Res. Ther. 5 (4) (2014) 105.
[121] C.D. Marshall, et al., In vitro and in vivo osteogenic differentiation of human adipose-derived stromal cells, Methods Mol. Biol. 1891 (2019) 9–18.
[122] O.A. Mohiuddin, et al., Decellularized adipose tissue hydrogel promotes bone regeneration in critical-sized mouse femoral defect model, Front. Bioeng. Biotechnol. 7 (2019) 211.
[123] Z. Hamidouche, et al., FHL2 mediates dexamethasone-induced mesenchymal cell differentiation into osteoblasts by activating Wnt/beta-catenin signaling-dependent Runx2 expression, FASEB J. 22 (11) (2008) 3813–3822.
[124] R.T. Franceschi, B.S. Iyer, Relationship between collagen synthesis and expression of the osteoblast phenotype in MC3T3-E1 cells, J. Bone Miner. Res. 7 (2) (1992) 235–246.
[125] F. Langenbach, J. Handschel, Effects of dexamethasone, ascorbic acid and beta-glycerophosphate on the osteogenic differentiation of stem cells in vitro, Stem Cell Res. Ther. 4 (5) (2013) 117.
[126] Q. Liu, et al., A comparative study of proliferation and osteogenic differentiation of adipose-derived stem cells on akermanite and beta-TCP ceramics, Biomaterials 29 (36) (2008) 4792–4799.
[127] F. Mussano, et al., Cytokine, chemokine, and growth factor profile characterization of undifferentiated and osteoinduced human adipose-derived stem cells, Stem Cells Int. 2017 (2017) 6202783.
[128] M.E. Bateman, et al., Osteoinductive effects of glyceollins on adult mesenchymal stromal/stem cells from adipose tissue and bone marrow, Phytomedicine 27 (2017) 39–51.
[129] X. Zhao, et al., Identification of key genes and pathways associated with osteogenic differentiation of adipose stem cells, J. Cell Physiol. 233 (12) (2018) 9777–9785.

[130] M. Cowper, et al., Human platelet lysate as a functional substitute for fetal bovine serum in the culture of human adipose derived stromal/stem cells, Cells 8 (7) (2019).

[131] F. Mussano, et al., Osteogenic differentiation modulates the cytokine, chemokine, and growth factor profile of ASCs and SHED, Int. J. Mol. Sci. 19 (5) (2018).

[132] S. Pagani, et al., Increased chondrogenic potential of mesenchymal cells from adipose tissue versus bone marrow-derived cells in osteoarthritic in vitro models, J. Cell Physiol. 232 (6) (2017) 1478–1488.

[133] L. Xu, et al., Tissue source determines the differentiation potentials of mesenchymal stem cells: a comparative study of human mesenchymal stem cells from bone marrow and adipose tissue, Stem Cell Res. Ther. 8 (1) (2017) 275.

[134] M.L. Gonzalez-Fernandez, et al., Study on viability and chondrogenic differentiation of cryopreserved adipose tissue-derived mesenchymal stromal cells for future use in regenerative medicine, Cryobiology 71 (2) (2015) 256–263.

[135] T.P. Hill, et al., Canonical Wnt/beta-catenin signaling prevents osteoblasts from differentiating into chondrocytes, Dev. Cell 8 (5) (2005) 727–738.

[136] H.J. Kim, G.I. Im, The effects of ERK1/2 inhibitor on the chondrogenesis of bone marrow- and adipose tissue-derived multipotent mesenchymal stromal cells, Tissue Eng. Part A 16 (3) (2010) 851–860.

[137] W. Bi, et al., Sox9 is required for cartilage formation, Nat. Genet. 22 (1) (1999) 85–89.

[138] T.M. Liu, et al., Zinc-finger protein 145, acting as an upstream regulator of SOX9, improves the differentiation potential of human mesenchymal stem cells for cartilage regeneration and repair, Arth. Rheum. 63 (9) (2011) 2711–2720.

[139] G. Coricor, R. Serra, TGF-beta regulates phosphorylation and stabilization of Sox9 protein in chondrocytes through p38 and Smad dependent mechanisms, Sci. Rep. 6 (2016) 38616.

[140] A. Karystinou, et al., Yes-associated protein (YAP) is a negative regulator of chondrogenesis in mesenchymal stem cells, Arth. Res. Ther. 17 (2015) 147.

[141] J. Ying, et al., Transforming growth factor-beta1 promotes articular cartilage repair through canonical Smad and Hippo pathways in bone mesenchymal stem cells, Life Sci. 192 (2018) 84–90.

[142] A.T. Mehlhorn, et al., Differential effects of BMP-2 and TGF-beta1 on chondrogenic differentiation of adipose derived stem cells, Cell Prolif. 40 (6) (2007) 809–823.

[143] H.J. Kim, G.I. Im, Combination of transforming growth factor-beta2 and bone morphogenetic protein 7 enhances chondrogenesis from adipose tissue-derived mesenchymal stem cells, Tissue Eng. Part A 15 (7) (2009) 1543–1551.

[144] L. Huang, et al., Synergistic effects of FGF-18 and TGF-beta3 on the chondrogenesis of human adipose-derived mesenchymal stem cells in the pellet culture, Stem Cells Int. 2018 (2018) 7139485.

[145] B. Johnstone, et al., In vitro chondrogenesis of bone marrow-derived mesenchymal progenitor cells, Exp. Cell Res. 238 (1) (1998) 265–272.

[146] J.R. Levick, Microvascular architecture and exchange in synovial joints, Microcirculation 2 (3) (1995) 217–233.

[147] M. Hirao, et al., Oxygen tension regulates chondrocyte differentiation and function during endochondral ossification, J. Biol. Chem. 281 (41) (2006) 31079–31092.

[148] R. Amarilio, et al., HIF1alpha regulation of Sox9 is necessary to maintain differentiation of hypoxic prechondrogenic cells during early skeletogenesis, Development 134 (21) (2007) 3917–3928.

[149] L. Yang, et al., Hypertrophic chondrocytes can become osteoblasts and osteocytes in endochondral bone formation, Proc. Natl. Acad. Sci. U. S. A. 111 (33) (2014) 12097–12102.

[150] X. Zhou, et al., Chondrocytes transdifferentiate into osteoblasts in endochondral bone during development, postnatal growth and fracture healing in mice, PLoS Genet. 10 (12) (2014), e1004820.

[151] G. Zhou, et al., Dominance of SOX9 function over RUNX2 during skeletogenesis, Proc. Natl. Acad. Sci. U. S. A. 103 (50) (2006) 19004–19009.

[152] M.C. Ronziere, et al., Chondrogenic potential of bone marrow- and adipose tissue-derived adult human mesenchymal stem cells, Biomed. Mater. Eng. 20 (3) (2010) 145–158.

[153] Y.J. Kim, H.J. Kim, G.I. Im, PTHrP promotes chondrogenesis and suppresses hypertrophy from both bone marrow-derived and adipose tissue-derived MSCs, Biochem. Biophys. Res. Commun. 373 (1) (2008) 104–108.

[154] K. Dzobo, et al., Fibroblast-derived extracellular matrix induces chondrogenic differentiation in human adipose-derived mesenchymal stromal/stem cells in vitro, Int. J. Mol. Sci. 17 (8) (2016).

[155] I. Sekiya, D.C. Colter, D.J. Prockop, BMP-6 enhances chondrogenesis in a subpopulation of human marrow stromal cells, Biochem. Biophys. Res. Commun. 284 (2) (2001) 411–418.

[156] S. Ansboro, et al., A chondromimetic microsphere for in situ spatially controlled chondrogenic differentiation of human mesenchymal stem cells, J. Control. Release 179 (1) (2014) 42–51.

[157] I.S. Yoon, et al., Proliferation and chondrogenic differentiation of human adipose-derived mesenchymal stem cells in porous hyaluronic acid scaffold, J. Biosci. Bioeng. 112 (4) (2011) 402–408.

[158] L. Calderon, et al., Type II collagen- hyaluronan hydrogel—a step towards a scaffold for intervertebral disc tissue engineering, Eur. Cells Mater. 20 (2010) 134–148.
[159] I. Gaudet, D. Shreiber, Characterization of methacrylated Type-I collagen as a dynamic, photoactive hydrogel, J. Biophys. Chem. 7 (1) (2012) 1–9.
[160] B. Sridhar, et al., A biosynthetic scaffold that facilitates chondrocyte-mediated degradation and promotes articular cartilage extracellular matrix deposition, Regen. Eng. Transl. Med. 1 (1) (2015) 11–21.
[161] A. Kabiri, et al., Effects of FGF-2 on human adipose tissue derived adult stem cells morphology and chondrogenesis enhancement in transwell culture, Biochem. Biophys. Res. Commun. 424 (2) (2012) 234–238.
[162] S.P. Grogan, et al., Visual histological grading system for the evaluation of in vitro-generated neocartilage, Tissue Eng. 12 (8) (2006) 2141–2149.
[163] R. Lattouf, et al., Picrosirius red staining: a useful tool to appraise collagen networks in normal and pathological tissues, J. Histochem. Cytochem. 62 (10) (2014) 751–758.
[164] D.D. Cissell, et al., A modified hydroxyproline assay based on hydrochloric acid in Ehrlich's solution accurately measures tissue collagen content, Tissue Eng. Part C Methods 23 (4) (2017) 243–250.
[165] S.W. Yi, et al., Gene expression profiling of chondrogenic differentiation by dexamethasone-conjugated polyethyleneimine with SOX trio genes in stem cells, Stem Cell Res. Ther. 9 (1) (2018) 341.
[166] I. Sekiya, et al., SOX9 enhances aggrecan gene promoter/enhancer activity and is up-regulated by retinoic acid in a cartilage-derived cell line, TC6, J. Biol. Chem. 275 (15) (2000) 10738–10744.
[167] L. Gao, et al., Effects of solid acellular type-I/III collagen biomaterials on in vitro and in vivo chondrogenesis of mesenchymal stem cells, Exp. Rev. Med. Dev. 14 (9) (2017) 717–732.
[168] T.M. Liu, et al., Identification of common pathways mediating differentiation of bone marrow- and adipose tissue-derived human mesenchymal stem cells into three mesenchymal lineages, Stem Cells 25 (3) (2007) 750–760.
[169] A. Bartholomew, et al., Mesenchymal stem cells suppress lymphocyte proliferation in vitro and prolong skin graft survival in vivo, Exp. Hematol. 30 (1) (2002) 42–48.
[170] R.S. Waterman, et al., A new mesenchymal stem cell (MSC) paradigm: polarization into a pro-inflammatory MSC1 or an Immunosuppressive MSC2 phenotype, PLoS One 5 (4) (2010), e10088.
[171] G.Y. Liu, et al., Short-term memory of danger signals or environmental stimuli in mesenchymal stem cells: implications for therapeutic potential, Cell Mol. Immunol. 13 (3) (2016) 369–378.
[172] J.M. Ryan, et al., Interferon-gamma does not break, but promotes the immunosuppressive capacity of adult human mesenchymal stem cells, Clin. Exp. Immunol. 149 (2) (2007) 353–363.
[173] V.G. Martinez, et al., Overexpression of hypoxia-inducible factor 1 alpha improves immunomodulation by dental mesenchymal stem cells, Stem Cell Res. Ther. 8 (1) (2017) 208.
[174] G. Hoeffel, F. Ginhoux, Ontogeny of tissue-resident macrophages, Front. Immunol. 6 (2015) 486.
[175] D.M. Mosser, J.P. Edwards, Exploring the full spectrum of macrophage activation, Nat. Rev. Immunol. 8 (12) (2008) 958–969.
[176] M. Daigneault, et al., The identification of markers of macrophage differentiation in PMA-stimulated THP-1 cells and monocyte-derived macrophages, PLoS One 5 (1) (2010), e8668.
[177] V. Rybalko, et al., Therapeutic potential of adipose-derived stem cells and macrophages for ischemic skeletal muscle repair, Regen. Med. 12 (2) (2017) 153–167.
[178] S. Adutler-Lieber, et al., Human macrophage regulation via interaction with cardiac adipose tissue-derived mesenchymal stromal cells, J. Cardiovasc. Pharmacol. Ther. 18 (1) (2013) 78–86.
[179] C. Manferdini, et al., Adipose stromal cells mediated switching of the pro-inflammatory profile of M1-like macrophages is facilitated by PGE2: in vitro evaluation, Osteoarth. Cartil. 25 (7) (2017) 1161–1171.
[180] H.J. Park, et al., Adipose-derived stem cells ameliorate colitis by suppression of inflammasome formation and regulation of M1-macrophage population through prostaglandin E2, Biochem. Biophys. Res. Commun. 498 (4) (2018) 988–995.
[181] R. Domenis, et al., Pro inflammatory stimuli enhance the immunosuppressive functions of adipose mesenchymal stem cells-derived exosomes, Sci. Rep. 8 (1) (2018) 13325.
[182] C. Li, et al., Macrophage polarization and meta-inflammation, Transl. Res. 191 (2018) 29–44.
[183] C.A. Roberts, A.K. Dickinson, L.S. Taams, The interplay between monocytes/macrophages and CD4(+) T cell subsets in rheumatoid arthritis, Front. Immunol. 6 (2015) 571.
[184] Z. Li, et al., Exosomes derived from mesenchymal stem cells attenuate inflammation and demyelination of the central nervous system in EAE rats by regulating the polarization of microglia, Int. Immunopharmacol. 67 (2019) 268–280.

[185] M.I. Guillen, et al., Paracrine anti-inflammatory effects of adipose tissue-derived mesenchymal stem cells in human monocytes, Front. Physiol. 9 (2018) 661.
[186] M. Sun, et al., Induction of macrophage M2b/c polarization by adipose tissue-derived mesenchymal stem cells, J. Immunol. Res. 2019 (2019) 7059680.
[187] P. Anderson, et al., Adipose-derived mesenchymal stromal cells induce immunomodulatory macrophages which protect from experimental colitis and sepsis, Gut 62 (8) (2013) 1131–1141.
[188] S. Stojanovic, S. Najman, The effect of conditioned media of stem cells derived from lipoma and adipose tissue on macrophages' response and wound healing in indirect co-culture system in vitro, Int. J. Mol. Sci. 20 (7) (2019).
[189] W.J. Song, et al., TSG-6 secreted by human adipose tissue-derived mesenchymal stem cells ameliorates dss-induced colitis by inducing M2 macrophage polarization in mice, Sci. Rep. 7 (1) (2017) 5187.
[190] K. Furuhashi, et al., Serum-starved adipose-derived stromal cells ameliorate crescentic GN by promoting immunoregulatory macrophages, J. Am. Soc. Nephrol. 24 (4) (2013) 587–603.
[191] H. Hattori, M. Ishihara, Altered protein secretions during interactions between adipose tissue- or bone marrow-derived stromal cells and inflammatory cells, Stem Cell Res. Ther. 6 (2015) 70.
[192] K. Kornicka, et al., Immunomodulatory properties of adipose-derived stem cells treated with 5-azacytydine and resveratrol on peripheral blood mononuclear cells and macrophages in metabolic syndrome animals, J. Clin. Med. 7 (11) (2018).
[193] M.A.A. Harrison, et al., Adipose-derived stem cells from obese donors polarize macrophages and microglia toward a pro-inflammatory phenotype, Cells 10 (1) (2020).
[194] J.Z. Li, et al., Comparison of adipose and bone marrowderived stem cells in protecting against oxLDLinduced inflammation in M1macrophagederived foam cells, Mol. Med. Rep. 19 (4) (2019) 2660–2670.
[195] D. Alvarez, E.H. Vollmann, U.H. von Andrian, Mechanisms and consequences of dendritic cell migration, Immunity 29 (3) (2008) 325–342.
[196] A. Castell-Rodríguez, et al., Dendritic cells: location, function, and clinical implications, Biol. Myelomonocyt. Cells (2017).
[197] N. Kadowaki, The divergence and interplay between pDC and mDC in humans, Front. Biosci. (Landmark Ed) 14 (2009) 808–817.
[198] R. Yanez, et al., Prostaglandin E2 plays a key role in the immunosuppressive properties of adipose and bone marrow tissue-derived mesenchymal stromal cells, Exp. Cell Res. 316 (19) (2010) 3109–3123.
[199] L.J. Zhou, T.F. Tedder, CD14(+) blood monocytes can differentiate into functionally mature CD83(+) dendritic cells, Proc. Natl. Acad. Sci. U. S. A. 93 (6) (1996) 2588–2592.
[200] M.K. Kim, J. Kim, Properties of immature and mature dendritic cells: phenotype, morphology, phagocytosis, and migration, Rsc Adv. 9 (20) (2019) 11230–11238.
[201] P. Blanco, et al., Dendritic cells and cytokines in human inflammatory and autoimmune diseases, Cytokine Growth Factor Rev. 19 (1) (2008) 41–52.
[202] E. Ivanova-Todorova, et al., Adipose tissue-derived mesenchymal stem cells are more potent suppressors of dendritic cells differentiation compared to bone marrow-derived mesenchymal stem cells, Immunol. Lett. 126 (1-2) (2009) 37–42.
[203] J. Valencia, et al., Comparative analysis of the immunomodulatory capacities of human bone marrow- and adipose tissue-derived mesenchymal stromal cells from the same donor, Cytotherapy 18 (10) (2016) 1297–1311.
[204] S.M. Melief, et al., Adipose tissue-derived multipotent stromal cells have a higher immunomodulatory capacity than their bone marrow-derived counterparts, Stem Cells Transl. Med. 2 (6) (2013) 455–463.
[205] Y.C. Wang, et al., The suppression effect of dendritic cells maturation by adipose-derived stem cells through TGF-beta1 related pathway, Exp. Cell Res. 370 (2) (2018) 708–717.
[206] H.J. Pegram, et al., Activating and inhibitory receptors of natural killer cells, Immunol. Cell Biol. 89 (2) (2011) 216–224.
[207] L.H. Glimcher, et al., Recent developments in the transcriptional regulation of cytolytic effector cells, Nat. Rev. Immunol. 4 (11) (2004) 900–911.
[208] G. Wang, et al., Comparison of the purity and vitality of natural killer cells with different isolation kits, Exp. Ther. Med. 13 (5) (2017) 1875–1883.
[209] G. Alter, J.M. Malenfant, M. Altfeld, CD107a as a functional marker for the identification of natural killer cell activity, J. Immunol. Methods 294 (1-2) (2004) 15–22.
[210] O. DelaRosa, et al., Human adipose-derived stem cells impair natural killer cell function and exhibit low susceptibility to natural killer-mediated lysis, Stem Cells Dev. 21 (8) (2012) 1333–1343.

[211] G.M. Spaggiari, et al., Mesenchymal stem cells inhibit natural killer-cell proliferation, cytotoxicity, and cytokine production: role of indoleamine 2,3-dioxygenase and prostaglandin E2, Blood 111 (3) (2008) 1327–1333.

[212] A. Ribeiro, et al., Mesenchymal stem cells from umbilical cord matrix, adipose tissue and bone marrow exhibit different capability to suppress peripheral blood B, natural killer and T cells, Stem Cell Res. Ther. 4 (5) (2013) 125.

[213] B. Blanco, et al., Immunomodulatory effects of bone marrow versus adipose tissue-derived mesenchymal stromal cells on NK cells: implications in the transplantation setting, Eur. J. Haematol. 97 (6) (2016) 528–537.

[214] F.A. Bonilla, H.C. Oettgen, Adaptive immunity, J. Allergy Clin. Immunol. 125 (2) (2010) S33–S40.

[215] E.J. Wherry, D. Masopust, Chapter 5 - Adaptive Immunity: Neutralizing, Eliminating, and Remembering for the Next Time, Elsevier Ltd, 2016, pp. 57–69.

[216] B. Puissant, et al., Immunomodulatory effect of human adipose tissue-derived adult stem cells: comparison with bone marrow mesenchymal stem cells, Brit. J. Haematol. 129 (1) (2005) 118–129.

[217] S. Wolbank, et al., Dose-dependent immunomodulatory effect of human stem cells from amniotic membrane: a comparison with human mesenchymal stem cells from adipose tissue, Tissue Eng. 13 (6) (2007) 1173.

[218] J. Hasbold, et al., Quantitative analysis of lymphocyte differentiation and proliferation in vitro using carboxyfluorescein diacetate succinimidyl ester, Immunol. Cell Biol. 77 (6) (1999) 516.

[219] P. Borgulya, et al., Development of the CD4 and CD8 lineage of T-cells - instruction versus selection, EMBO J. 10 (4) (1991) 913–918.

[220] J.K. Whitmire, R. Ahmed, Costimulation in antiviral immunity: differential requirements for CD4 + and CD8 + T cell responses, Curr. Opin. Immunol. (2000) 448–455.

[221] A. Lanzavecchia, F. Sallusto, Understanding the generation and function of memory T cell subsets, Curr. Opin. Immunol. 17 (3) (2005) 326–332.

[222] B. Kronsteiner, et al., Human mesenchymal stem cells from adipose tissue and amnion influence T-cells depending on stimulation method and presence of other immune cells, Stem cells Dev. 20 (12) (2011) 2115.

[223] C. Menard, et al., Clinical-grade mesenchymal stromal cells produced under various good manufacturing practice processes differ in their immunomodulatory properties: standardization of immune quality controls (report), Stem Cells Dev. 22 (12) (2013) 1789.

[224] M. Najar, et al., Mesenchymal stromal cells use PGE2 to modulate activation and proliferation of lymphocyte subsets: combined comparison of adipose tissue, Wharton's Jelly and bone marrow sources, Cell. Immunol. 264 (2) (2010) 171–179.

[225] C.-M. Chien, et al., Adipose-derived stem cell modulation of T-cell regulation correlates with heme oxgenase-1 pathway changes, Plast. Reconstruct. Surg. 138 (5) (2016) 1015–1023.

[226] O. Kizilay Mancini, et al., Age, atherosclerosis and type 2 diabetes reduce human mesenchymal stromal cell-mediated T-cell suppression, Stem Cell Res. Ther. 6 (1) (2015).

[227] L.W. Wu, et al., Donor age negatively affects the immunoregulatory properties of both adipose and bone marrow derived mesenchymal stem cells, Transplant Immunol. 30 (4) (2014) 122–127.

[228] W. Chen, Conversion of peripheral CD4+CD25- naive T cells to CD4+CD25+ regulatory T cells by TGF-beta induction of transcription factor Foxp3, J. Exp. Med. 198 (12) (2003) 1875–1887.

[229] E. Ivanova-Todorova, et al., Conditioned medium from adipose tissue-derived mesenchymal stem cells induces CD4+FOXP3+ cells and increases IL-10 secretion, J. Biomed. Biotechnol. 2012 (2012).

[230] J.A. Lopez, et al., Perforin forms transient pores on the target cell plasma membrane to facilitate rapid access of granzymes during killer cell attack, Blood 121 (14) (2013) 2659–2668.

[231] D. Hassin, et al., Cytotoxic T lymphocyte perforin and Fas ligand working in concert even when Fas ligand lytic action is still not detectable, Immunology 133 (2) (2011) 190–196.

[232] I. Hof-Nahor, et al., Human mesenchymal stem cells shift CD8+ T cells towards a suppressive phenotype by inducing tolerogenic monocytes, J. Cell Sci. 125 (19) (2012) 4640.

[233] J. Ollila, M. Vihinen, B cells, Int. J. Biochem. Cell Biol. 37 (3) (2005) 518–523.

[234] L. Barrio, et al., Human adipose tissue-derived mesenchymal stromal cells promote B-cell motility and chemoattraction, Cytotherapy 16 (12) (2014) 1692–1699.

[235] M. Samoylovich, et al., The influence of mesenchymal stromal cells on B-cell line growth and immunoglobulin synthesis, Cell Tissue Biol. 7 (3) (2013) 227–234.

[236] I. Bochev, et al., Mesenchymal stem cells from human bone marrow or adipose tissue differently modulate mitogen-stimulated B-cell immunoglobulin production in vitro, Cell Biol. Int. 32 (4) (2008) 384–393.

CHAPTER

4

Insights into the adipose stem cell niche in health and disease

Aaron C. Brown

Center for Molecular Medicine, Maine Medical Center Research Institute, Scarborough, ME, United States; School of Biomedical Sciences and Engineering, The University of Maine, Orono, ME, United States; Tufts University School of Medicine, Boston, MA, United States

Introduction

An understanding of adipose tissue regulation has broad clinical implications, as obesity is associated with an increased risk for diabetes, stroke, heart disease, and cancer, resulting in increased health care costs and decreased life expectancy [1]. Weight gain and obesity are triggered by chronic periods of food excess that result in a positive energy balance and excessive storage of lipids in adipose tissue, leading to inflammation, cellular stress, insulin resistance, and potentially diabetes [2]. The consumption of high-caloric food coupled with a sedentary lifestyle has triggered a global rise in obesity [3]. Diet and exercise alone are often not enough to sustain long-term weight loss as biological adaptations in the chronically obese can undermine these lifestyle modifications [4]. Thus, there is a need to develop new therapeutic strategies to address the public health risk of obesity.

Adipose tissue provides organs with structural support and protection from cold and, importantly, regulates energy balance and metabolic homeostasis [5]. Both humans and rodents possess white (WAT) and brown adipose tissue (BAT), which play distinct roles in metabolism. WAT is involved in energy storage, and its accumulation in overweight individuals correlates with metabolic syndrome, whereas BAT converts energy stored in lipids to heat, and its activity correlates positively with reduced risk of metabolic syndrome, making it an attractive therapeutic target tissue [1]. In addition to the "classical" BAT depot in the interscapular region, functionally similar "beige" adipose tissue can be induced in adult subcutaneous WAT (sWAT) in response to thermogenic activators of the sympathetic nervous system, such

https://doi.org/10.1016/B978-0-12-819376-1.00012-3

as cold-induced norepinephrine secretion [6]. Understanding the biological processes that regulate the development and proper maintenance of white, brown, and beige adipose tissues is key for developing therapies against metabolic disease.

A stem cell niche provides a specific microenvironment within a tissue in which direct cell–cell interactions and molecular signals either maintain stem cells in the undifferentiated state or promote their differentiation. The adipose niche holds a wide array of cell types including adipocytes, multipotent adipose-derived stem cells (ASCs), committed adipocyte progenitors (APs), endothelial cells, fibroblasts, immune cells, and vascular smooth muscle cells, which communicate via direct interactions or in a paracrine fashion through adipokine secretion [7]. In this chapter, we will provide an overview of the adipose stem cell niche and focus on the characterization of ASCs and APs, their interactions with other cell types in the niche, their molecular regulation, and the changes they undergo in response to increased adiposity that may play a role in the development of obesity-related disorders.

Defining adipocyte-derived stem cells and progenitors

The stromal vascular fraction (SVF) derived from the digestion of adipose tissue consists of a heterogeneous mixture of cells excluding mature adipocytes, but including ASCs, APs, fibroblasts, immune cells, epithelial cells, endothelial cells, and other cell types associated with the circulatory and nervous systems [8]. The terms stem cells and progenitor cells are often used interchangeably. However, their definition is controversial and still evolving [9]. Distinct traits of stem cells are pluripotency and their ability to replicate indefinitely. In contrast, progenitor cells can only replicate a limited number of times and are more differentiated than stem cells because they have committed to a specific lineage. A lack of suitable culture conditions for testing replicative capacity and the availability of unique surface markers has been a barrier for distinguishing between stem and progenitor cells derived from the adipose tissue niche. For the purposes of this chapter, ASCs will refer specifically to the population of adipose tissue resident mesenchymal stem cells (MSCs) with multilineage potential to form adipocytes, osteoblasts, chondrocytes, and other lineages. ASCs account for less than 0.1% of all SVF cells [10, 11]. APs (a.k.a. preadipocytes) will refer to cells committed to the adipocyte lineage, which normally represent 15%–35% of the SVF [10–13]. It is important to point out that APs often share the same cell surface phenotype as ASCs (unless otherwise noted) [11], but may express markers of adipocyte commitment such as peroxisome proliferator-activated receptor gamma (PPARγ). Lastly, "adipocyte precursors" in this chapter will generically refer to any cell in the niche with adipogenic potential, including ASCs, APs, and potentially other cell types.

Generation of adipocytes from adipocyte precursors

Adipose tissue possesses the unique ability to expand and contract under different physiological settings, such as overeating or diet and exercise. Generally, expansion of sWAT is associated with resistance to cardiometabolic syndrome during obesity, whereas expansion of visceral WAT (vWAT) is often associated with insulin resistance and diabetes [12, 13]. Expansion of adipose tissue mass can occur through hyperplasia, an increase in adipocyte cell number resulting from precursor proliferation, which is generally more metabolically favorable than through hypertrophy, an increase in adipocyte cell size via increased lipid storage [14]. In adults, the number of adipocytes stays relatively constant in both lean and obese individuals, where approximately 10% of mature adipocytes are replaced on a yearly basis through adipocyte differentiation from ASCs and APs [15, 16].

The first phase of adipocyte differentiation occurs when ASCs are induced by extracellular cues to form APs, which are specifically committed to an adipocyte fate. In the second phase, terminal differentiation of committed APs into mature, lipid-containing adipocytes is characterized by cell cycle arrest and involves a coordinated cascade of transcriptional events, mainly regulated by CCAAT-enhancer-binding proteins (C/EBPs), sterol regulatory element binding protein 1c (SREBP1c), and PPARγ [17–20]. PPARγ acts as the master regulator of adipogenesis, and its ablation blocks adipogenesis and the development of adipose tissue in mice [21, 22]. PPARγ expression is induced by C/EBPs and SREBPc1 and subsequently interacts directly with retinoid X receptor (RXR) to bind to PPARγ responsive regulatory elements and induce expression of genes involved in adipogenesis, lipid metabolism, inflammation, and maintenance of metabolic homeostasis [20, 23, 24]. Once formed, mature adipocytes acquire the molecular machinery necessary for lipid and glucose transport, insulin responsiveness, and the secretion of paracrine factors that act to control whole-body-mediated metabolic homeostasis [17]. Compared with white adipocytes, brown and beige adipocytes have additional transcription factors that regulate their differentiation and function and distinguish them from white adipocytes, including the expression of PR domain zinc finger protein 16 (PRDM16), early B cell factor 2 (EBF2), and peroxisome proliferator-activated receptor gamma coactivator 1-alpha (PGC1α) (reviewed in Ref. [18]). These transcription factors coordinate with PPARγ to increase expression of uncoupling protein-1 (UCP1) and activate UCP1-mediated thermogenic responses to β-adrenergic receptor stimuli.

Adipocyte precursors reside in the stromal vasculature fraction

The concept that adipocytes could be newly formed from a population of stem/progenitor cells was largely brought to light during the development of methods to separate adipose tissue into adipocytes and the SVF (reviewed in Ref. [8]). From these studies, it was found that adherent fibroblast-like cells within the SVF were competent for adipogenesis and subsequently referred to as preadipocytes [25–27]. However, because cell types other than adipocytes can be derived from the SVF (osteoblasts, chondrocytes, and myoblasts), these cultures likely reflect the presence of multipotent MSCs in addition to solely committed APs [8]. To gain further insight into adipocyte differentiation and cellular physiology and overcome limitations associated with variable differentiation and early senescence of SVF cells, a number of immortalized, clonal AP cell lines have been developed. For example, immortalized 3T3L1 cells were derived in the 1970s from mouse embryonic fibroblasts and are considered a committed AP cell line, which was used to determine that the transcription factor PPARγ was the master regulator of adipocyte differentiation and cooperates with C/ebpα to stimulate adipogenesis [28–31]. 3T3L1 APs are largely considered committed to formation of mature white adipocytes and appear to have a limited capacity for thermogenesis, but this has been the subject of some debate since the original derivation of these cells [32]. Other cell lines, including the mouse-derived HIB-1B and "Thermomouse" cell lines have been derived from interscapular BAT and express a robust thermogenic program to delineate mechanisms related to BAT thermogenesis [33, 34]. However, compared with primary isolated APs from specific adipose tissue depots, to what extent these immortalized cell lines represent any specific type of progenitor found in the native adipose tissue niche is unclear.

Immunophenotyping of ASCs and APs

The most widespread method used to distinguish adipocyte precursors from other cell types within the adipose tissue niche has been through cell surface marker analysis. The ability to identify and characterize adipocyte precursors within the SVF has been somewhat difficult for a number of reasons. First, distinct developmental origins are known to give rise to anatomically distinct adipocytes. This includes white, beige, and brown subtypes, where specific developmental programming may lead to distinct precursor lineages that give rise to a particular type of adipocyte. Also, as the SVF is cultured, surface marker expression may change, making it difficult to establish a definitive marker profile [8]. There are also differences seen between human and rodent studies, and many surface markers may not be specific to adipocyte precursors, but expressed on the surface of non-stem cell lineages [8]. In addition, surface marker profiles may differ or overlap between ASCs and more committed APs. For many of these reasons, it has been difficult to find a definitive adipocyte precursor cell surface marker profile. However, numerous markers have become useful for the study of these cells and their enrichment into populations with high adipogenic potential.

CD34 is a common MSC marker expressed on ASCs with high adipogenic potential, and in conjunction with depletion of common endothelial (CD31) and hematopoietic markers (CD45) can be used to enrich ASCs from the SVF [35, 36]. However, there may be inconsistent results with this marker, possibly due to its downregulation after cell culture [8, 37]. CD29 and Sca1 (mouse only) are also common MSC markers that can be used to identify and enrich ASCs. Along these lines, Friedman et al. described a subpopulation of CD29+:CD34+:Sca-1+:CD24+ undifferentiated ASCs that are capable of proliferating and differentiating into adipose tissue [10]. These precursors could also reconstitute a complete and normal WAT depot and rescue the diabetic phenotype that develops in lipodystrophic mice [10]. CD24 has been used to separate the existence of hierarchical populations of adipocyte precursors that exhibit different levels of adipocyte commitment and potential [38]. In adipose depots that exhibit a hyperplastic response, CD24+ ASCs experience a rapid and transient proliferation within days of high-fat diet (HFD) feeding [39]. Additionally, CD24+ ASCs represent a more primitive multipotent stem cell population, and CD24 expression is lost as cells begin to express adipocyte lineage-selective genes such as *PPARγ* and *C/ebpα* characteristic of more committed APs [38, 40].

Adipocyte precursors in the native niche

Adipocyte precursors located in the adipose extracellular matrix

Adipose tissue in the adult consists of distinct lobules containing clusters of adipocytes separated from each other by extracellular matrix (ECM) that forms concise separations known as septa [41] (Fig. 1A and B). Mathematical modeling in conjunction with

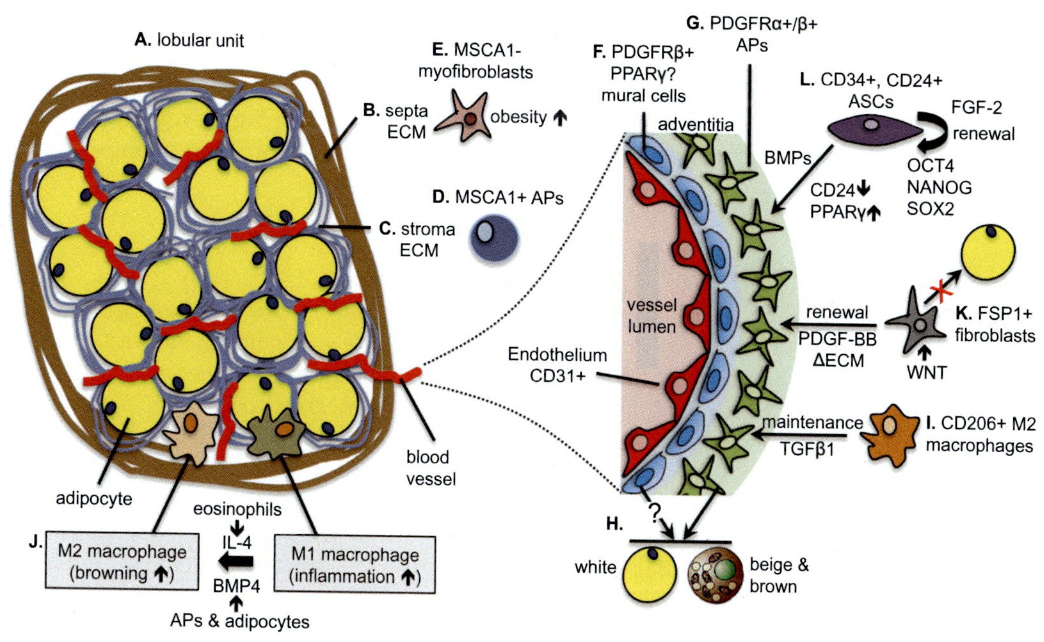

FIG. 1 Architecture and interactions within the adipose tissue niche. (A) Adipose tissue consists of clusters of adipocytes arranged in lobules (A), which are separated from each other by extracellular matrix (ECM) that forms the septa (B). In addition to the septa lining the outside of the lobule, the internal stroma also contains a distinct ECM compartment (C). (D) ECM within the stroma of the lobule contains CD34+/MSCA1+/CD271-adipocyte progenitors (APs) that are highly adipogenic and positioned between adipocytes or adjacent to the vasculature. (E) ECM within the septa contains CD34+/MSCA1−/CD271+ myofibroblasts that may aid in formation of the fibrous septa and contribute to fibrosis during obesity. (F) PDGFRβ+ mural cells located in the vessel walls may harbor adipogenic activity, but recent data contradict this and suggest PDGFRα+/β+APs (G) located within the vessel wall adventitia contribute to the majority of adipocytes found in white, beige, and brown adipocytes (H). (I) CD206+ M2 macrophages play a role in maintenance of APs through secretion of antiadipogenic TGFβ1, which prevents exhaustion of the AP pool by limiting excessive proliferation and premature senescence. (J) IL-4 and BMP4 represent two potential factors that may decrease the proportion of M1 macrophages in adipose tissue in favor of antiinflammatory M2 macrophages, which play a role in adipose tissue browning, protection from high-fat diet (HFD)-induced obesity, and improved metabolism. (K) FSP1+ fibroblasts found in the loose connective tissue outside the vessel wall rely on WNT signaling for proper function and lack adipogenic potential (red X). They contribute to the renewal and differentiation potential of APs through the secretion of PDGF-BB and remodeling of the ECM. (L) CD34+/CD24+ ASCs express markers of pluripotency (OCT4, NANOG, and SOX2) and undergo FGF2-mediated self-renewal. BMP ligands can induce multipotent adipose-derived stem cells (ASCs) to form committed APs (CD24−) that express adipocyte lineage-selective genes such as PPARγ.

experimental evidence suggests that these lobules emerge spontaneously through mechanical interactions between the adipocytes and fibers within the ECM [42]. Furthermore, recent evidence suggests that adipose tissue lobules consist of two structurally distinct ECM compartments, including the septa, which lines the outside of the lobule, and the internal stroma ECM (Fig. 1C) [43]. These two niches also contain distinct subsets of CD34+ precursor cells, with a highly adipogenic committed MSCA1+/CD271− progenitor subset enriched in the stroma, and a MSCA1−/CD271^{hi} subset found in the septa that marks myofibroblast precursors and may contribute to fibrous septa formation (Fig. 1D and E) [43]. In the stroma, APs are distinguished from pericytes and are not embedded in the capillary walls, but rather positioned between adipocytes or adjacent to the vasculature [43]. Interestingly, in humans there are intrinsic differences in these progenitor subsets in terms of their adipogenic and myofibroblastic capacities within the stroma and septa niches, with a higher myofibroblastic capacity in vWAT compared with sWAT during obesity that may contribute to fibrosis [43].

Perivascular cells as a source of adipocyte precursors

Early morphological studies demonstrated that adipocytes develop in close association with the vasculature, and it had been proposed years ago that pericytes, a subset of mural cells, which are known to modulate both endothelial cell proliferation and vessel contractility and possess multipotent MSC characteristics, may also represent an adipocyte precursor population (Fig. 1F) [41, 44–46]. In an attempt to find committed APs within the adipose tissue niche, Tang et al. hypothesized that they would express PPARγ, a central regulator of adipocyte formation [47, 48]. Through generation of a *PPARγ* reporter strain, they were able to show that *PPARγ*+ cells are found in PECAM+ (CD31) blood vessel walls of WAT. These cells also exhibited high adipogenic potential and expressed mural cell markers, including α-SMA, PDGFRβ, and NG2 [47]. The majority of the committed PPARγ+ APs also expressed Sca1 and CD34, but not CD105 (MSCs), CD45 (immune cells), TER-119 (erythrocytes), or Mac-1 (monocytes). Interestingly, *PPARγ*+ APs from the mural cell compartment were not found in the vessel walls of other tissues (skeletal and cardiac muscle, kidney, retina, pancreas, spleen, and lung), and mural cells from these additional tissues did not possess high adipogenic potential [47]. Furthermore, lineage tracing demonstrated that adipose tissue can derive from *Pdgfrβ*+mural cells, which have high adipogenic potential when transplanted into nude mice [47]. In a later study, using a doxycycline-inducible mural cell lineage tracking system based upon *Pdgfrβ* expression, it was shown that these perivascular APs also contribute toward adipocyte hyperplasia in vWAT during HFD feeding [49]. Overall, these results suggest that adipose tissue contains a unique population of APs that reside within the mural cell compartment.

Using vascular smooth muscle cell (*Myh11*, *α-Sma*) and mural cell Cre drivers (*Pdgfrβ*), recent studies have demonstrated that several subpopulations of beige adipocytes can also be lineage traced to the mural cell compartment [49–51]. Along these lines, studies in our laboratory have focused on the generation of beige adipocytes from induced pluripotent stem cells (iPSCs) [52]. These adipocytes derive from the splanchnic mesoderm, an embryonic tissue known to give rise to mural and vascular smooth muscle cells [53]. Splanchnic-derived mural-like MSCs are positive for α-SMA, PDGFRβ, and NG2 and form UCP1+ adipocytes that display the characteristic gene expression signature of beige adipocytes (CITED1+, CD137+,

TMEM26+), but not that of classical brown adipocytes, which normally express ZIC1 and a myogenic skeletal muscle signature [53].

Fibroblastic cells as a source of adipocyte precursors

Lineage tracing studies have also indicated that there are additional APs that do not reside within the mural cell compartment. PDGFRα is a pan-fibroblast marker on cells implicated as progenitors in many mesenchymal tissues [54]. For example, PDGFRα+ progenitors from WAT comprise a subpopulation within the CD34 and Sca-1 adipocyte precursor compartment that are bipotential [55]. These cells proliferate and can either give rise to white adipocytes during steady-state or high-fat feeding or beige adipocytes after pharmacological activation of the β3-adrenergic receptor (ADRB3) [55]. In this study, it was also found that PDGFRα+ progenitors reside close to the vasculature and exhibit long dendritic processes that contact multiple cells in the tissue microenvironment. However, PDGFRα+ APs in WAT were shown to be negative for PPARγ, α-SMA, and PDGFRβ, and therefore lie outside the mural cell compartment [55]. Interestingly, in a follow-up study, the researchers demonstrated that PDGFRα+ APs in sWAT could be induced to differentiate into beige adipocytes during acute (7 days) cold-induced "browning" [56]. On the other hand, 2 weeks of cold exposure is necessary to form beige adipocytes from PDGFRβ+ APs that reside in the mural compartment [49, 50]. Perhaps WAT-derived PDGFRα+ progenitors may be functionally distinguished from those in the mural cell compartment on the basis of their ability to rapidly respond to metabolic stress or adipogenic signals. Thus, beige adipogenesis may rely on different adipocyte precursor populations, each having its own distinct timing and activation characteristics [38].

Recent data is shedding more light on the discrepancies between mural (PDGFRβ+) and fibroblastic (PDGFRα+) origins of adipogenic precursors and their contributions to the formation of adipose tissue during HFD feeding, browning, and normal metabolic homoeostasis [57]. Using the simultaneous labeling of the adipose tissue vascular niche with three separate Cre drivers that label hematopoietic cells and endothelial cells (*Tie2-Cre*), mural cells, including both pericytes and vascular smooth muscle cells (*Tbx18-CreERT2*), and adipose tissue fibroblasts (*Pdgfra-MerCreMer*), it was shown that adipose tissue fibroblasts that reside within the blood vessel adventitia and adipose tissue capsule regions were the sole cell type of the vascular wall that contributed significantly to de novo adipogenesis [57]. Importantly, they demonstrated that PDGFRβ not only labels mural cells but adventitial PDGFRα+ APs as well, which may help to explain why earlier studies suggested a contribution from mural cells (Fig. 1G and H) [57]. Other studies also support this conclusion, whereby single-cell RNA sequencing has demonstrated that APs express both PDGFRα and PDGFRβ [58]. Interestingly, our data with multipotent, human iPSC-derived mural cells demonstrated that they acquired PDGFRα expression in conjunction with increased PPARγ expression upon transition to committed APs [52]. However, it remains to be clarified if PDGFRβ+ mural cells can give rise to double positive PDGFRα/PDGFRβ APs during adipocyte development in vivo, or if there are differences in the expression of these markers between mice and humans during de novo adipogenesis. Overall, the study of the cell surface marker profiles associated with distinct subpopulations of adipocyte precursors is continuously evolving as newly defined populations are being discovered.

Cell interactions within the adipose precursor niche

Adipocyte precursors are located in the perivascular region and stromal component of adipose tissue lobules and interact with both mature adipocytes and other cells of the precursor niche in vivo. We are only beginning to understand how these interactions influence adipocyte precursor fate and ultimately adipose tissue homeostasis. Mature adipocytes are activated in response to homeostatic and external signals such as weight gain, hypoxia, cold, exercise, and nutrition and secrete adipokines that can influence energy intake and expenditure [59]. Experiments with cultured adipocytes or adipose tissue explants have shown that mature adipocytes can secrete factors that either positively or negatively regulate adipocyte differentiation, with the total secretome resulting in the overall inhibition of AP differentiation [60].

Most evidence of cell–cell interactions that promote maintenance, proliferation, or commitment of adipocyte precursors toward differentiation of postmitotic, mature adipocytes has been focused on cells that reside in the SVF. However, it is unclear how the SVF and adipose tissue niche as a whole regulates the adipocyte precursor pool size and fate determination to control adipocyte size, number, and overall adipose homeostasis [61, 62]. Some limitations to this research include the overlap of markers between the different cell populations of the niche, making it difficult to distinguish the distinct cell types. In addition, the composition of the isolated SVF changes greatly during tissue culture expansion, even at low passages [63]. The main subpopulations of nucleated cells contained within the SVF include hematopoietic, endothelial, and stromal cells; however, a large degree of heterogeneity can exist between these populations based upon the anatomical location of the adipose tissue, tissue processing/culture methods, and the health or metabolic status of the individual [63, 64]. For example, adipocyte precursors isolated from the sWAT of obese patients exhibit increased expression of inflammatory genes, which correlates with a loss of stemness and increased commitment toward adipocyte differentiation [65]. Furthermore, APs from sWAT of obese patients lose the capacity to expand and form beige adipocytes when induced in cell culture [52, 66].

Adipocyte precursor interactions with endothelial cells

Since adipocyte precursors are in close contact with endothelial cells and adipose tissue development is closely associated with angiogenesis, these cell types may be regulated through direct cell–cell interactions or through paracrine signaling from endothelial cells [67]. Along these lines, it has been suggested that angiogenesis recruits adipocyte precursors and stimulates them to differentiate [68]. Furthermore, coculture with endothelial cells and mature adipocytes can promote immature preadipocyte development in vitro indirectly through adhesion of the endothelial cells with the mature adipocytes [69]. The relationship between adipocyte precursors and endothelial cells appears to be reciprocal. Adipocyte precursors promote endothelial cell proliferation and differentiation by secreting proangiogenic factors and stimulating blood vessel formation, suggesting that adipocyte differentiation and angiogenesis work in conjunction [63, 70]. Multipotent CD34+ ASCs that are within the α-SMA+/PDGFRβ+ pericyte subpopulation of the SVF can also stabilize the vasculature through a functional interaction with endothelial cells [71].

Adipocyte precursor interactions with immune cells

The expansion of adipose tissue during the development of obesity involves the accrual of immune cells that are linked to chronic inflammation and dysregulated metabolism; however, the role of immune cells in the adipocyte precursor niche is not well understood [72]. Immune cells of the SVF are CD34+, but can be distinguished from adipocyte precursors in their inability to adhere to culture dishes and are therefore removed during culture passage [63]. They can also be distinguished from endothelial cells by their absence of the endothelial cell surface marker CD31 [64]. Immune cells may provide an environment for neovascularization of adipose tissue undergoing ischemia through the secretion of cytokines and growth factors [63]. Additionally, numerous studies have shown that cells of the monocyte lineage (eosinophils, macrophages) play profound regulatory roles within the niche.

Macrophages represent the most abundant lineage of immune cells in adipose tissue and are involved in tissue repair, insulin sensitivity, fibrosis, and metabolic homeostasis [72]. These macrophages are categorized into (M1) proinflammatory macrophages, which worsen insulin resistance, and (M2) antiinflammatory macrophages, which improve insulin sensitivity [73–75]. It is likely that adipocyte precursors and macrophages interact, and it has been reported that M1 proinflammatory macrophages regulate the expression of angiogenic genes in preadipocytes [76, 77]. More recently, studies by Nawaz et al. demonstrated that CD206+ M2 macrophages can provide a niche for maintaining the pool of adipogenic precursors by preventing their exhaustion through overproliferation (Fig. 1I) [78]. By keeping adipogenic precursors in a state of hibernation, CD206+ M2 macrophages aid in unnecessary adipogenic precursor cell division, which would otherwise promote cell senescence. Furthermore, they showed that TGFβ1 expression by CD206+ M2 macrophages inhibited proliferation of PDGFRα+ adipogenic precursors. This is consistent with previous results demonstrating that TGFβ signaling is a known inhibitor of adipogenesis and suppressor of WAT browning [79, 80]. Along these lines, when CD206+ M2 macrophages are depleted in mice, enhanced browning of WAT occurred during cold exposure [78].

M2 macrophages have also been implicated in beige adipogenesis, suggesting different subpopulations of M2 macrophages may play distinct roles in adipose tissue biology. For example, during cold exposure in mice, activated eosinophils within the adipose tissue niche secrete IL-4 and IL-13 cytokines, which polarize macrophages toward an M2 fate (Fig. 1J) [81]. These M2 macrophages may, in turn, contribute to formation of beige adipocytes through the secretion of catecholamines [82]. However, whether M2 macrophages can secrete sufficient amounts of catecholamines to induce browning of WAT has recently been called into question [83]. Nevertheless, IL-4 secretion by immune cells may also act directly on APs isolated from sWAT of mice to promote beige adipogenesis [84, 85]. Our laboratory has demonstrated that IL-4 treatment of human mural-like adipocyte precursors (α-SMA+/PDGFRβ+/NG2+) significantly increased their ability to form beige adipocytes in vitro [52]. Overall, these studies demonstrate that M2 macrophages can influence the adipocyte precursor niche by promoting adipogenic precursor maintenance and commitment decisions.

Adipocyte precursor interactions with other cells in the niche

Recent studies in mice have shown that a population of fibroblasts marked by FSP1 are an important niche cell type that is necessary for the maintenance and adipogenic potential of

APs (Fig. 1K) [61]. FSP1+ fibroblasts express α-SMA and vimentin and do not harbor adipogenic potential, but are adjacent to PPARγ+ APs. WNT signaling regulates adipose tissue homeostasis by activating β-catenin and inhibiting differentiation of APs [86, 87]. Along these lines, WNT signaling in FSP1+ fibroblasts was shown to be essential for their function, as aberrant WNT signaling resulted in a loss of AP maintenance and adipogenic potential. Similarly, activation of WNT signaling in FSP1+ fibroblasts also resulted in a gradual loss of adipose tissue, but this was associated with resistance to diet-induced obesity. It was also found that alterations in FSP1+ fibroblasts resulted in decreased platelet-derived growth factor (PDGF-BB) signaling, which is necessary for maintenance of the AP pool. The authors also found that the decrease in PDGF-BB signaling altered the adipogenic potential of APs by changing the ability of the FSP1+ fibroblasts to regulate matrix metalloproteinase (MMP) expression and ECM remodeling in the microenvironment. Thus, FSP1+ fibroblasts regulate adipose tissue homeostasis by providing a niche for the regulation of AP maintenance and adipogenic potential [61].

Finally, cell types other than hematopoietic, endothelial, or stromal cells within the adipose tissue niche are also present and can have a profound effect in vivo. For example, sympathetic neurons release catecholamines in response to chronic cold exposure and can induce lipolysis in white adipocytes or differentiation of beige APs to form thermogenically active, mature beige adipocytes [49].

Fibroblast growth factor signaling and renewal of adipocyte precursors

Because ASCs hold promise in many regenerative applications, there is often a need for prolonged culture to obtain sufficient quantity for clinical applications. For many types of stem cells, FGF-2 is often used to supplement culture medium and aids in the preservation of self-renewal capacity and plasticity through multiple passages (Fig. 1L) [88, 89]. Adipocyte precursors grown for prolonged periods in culture exhibit a reduced capacity for proliferation, self-renewal, and differentiation potential. This is associated with a decrease in FGF-2 expression by adipocyte precursors, which can be restored by continuous treatment with recombinant FGF-2 [89]. Consistent with this, adipocyte precursors express fibroblast growth factor receptor 1 (FGFR1), which has a high affinity for FGF-2, and inhibition of this receptor results in reduced proliferation and inactivation of kinases including AKT, ERK, JNK, and p38 [90, 91]. Addition of FGF-2 to adipocyte precursors prior to treatment with an adipogenic differentiation cocktail increases PPARγ expression, which may help to explain its ability to increase adipocyte differentiation [88]. Interestingly, FGF-2 secreted by adipocyte precursors is exported to the cell surface, but not released into the culture medium, suggesting a functional autocrine loop [89]. Adipocyte precursors derived from subcutaneous and visceral adipose tissue isolated from bariatric surgery patients show reduced exportation of FGF-2 that is correlated with decreased proliferation, clonogenic potential, and poorer metabolic profiles [92]. It is worth noting that the positive effects of FGF-2 supplementation are limited to earlier passages of adipocyte precursors. Continuous FGF-2 supplementation exerts adverse effects at later passages due to a loss of FGFR1c expression and downstream STAT3 phosphorylation [93].

Bone morphogenetic proteins and commitment of adipogenic precursors

Probably no other growth factors have been studied more for their role in adipogenesis than that of the bone morphogenetic proteins (BMPs). BMP2, 4, 7, and others have been implicated in the direct regulation of ASCs and their commitment to APs, which can promote their differentiation toward white, beige, or brown adipocytes depending on the specific BMP ligand or in a context-dependent manner (Fig. 1L) [94–97].

BMP2 acts as a paracrine factor within human white adipose tissue

Cell culture studies with mouse cell lines have shown that addition of exogenous BMP2 can enhance adipogenesis in 3T3L1 preadipocytes and promote adipocyte lineage commitment in C3H10T1/2 MSCs [94, 98]. This occurs through SMAD1/5/8 signaling and increased expression of adipogenic transcription factors *PPAR*γ and *C/ebp*α [99, 100]. BMP2 is expressed in human abdominal and gluteal adipose tissue, as well as in preadipocytes isolated from these tissues [101]. In human abdominal APs, BMP2 signals through phosphorylation of SMAD1/5/8 and increases PPARγ expression and triacylglyceride accumulation [101]. Interestingly, several independent studies have shown that a polymorphism linked to BMP2 (rs979012) is associated with an increased waist-to-hip ratio and body mass index (BMI), further supporting a role of BMP signaling in adipose tissue biology [101–103].

BMP4 promotes a white or brown phenotype in a context-dependent manner

Earlier studies with BMP4 demonstrated that it commits C3H10T1/2 MSCs to an adipocyte lineage as measured by a high frequency of adipocyte formation following treatment with adipocyte differentiation inducers [94, 95, 104]. Transgenic overexpression of BMP4 in mouse adipocytes using the promoter of fatty acid binding protein 4 (*Fabp4*) leads to browning of inguinal WAT, increased energy expenditure, and protection from HFD-induced obesity with improved insulin sensitivity [105]. In contrast, *Bmp4*-deficient mice display white adipocyte hypertrophy and increased insulin resistance [105]. Interestingly, BMP4 expression in human WAT is inversely correlated with BMI, further suggesting a role for BMP4 in increased energy expenditure [105].

Whether BMP4 promotes a white or beige/brown phenotype may be cell type dependent. BMP4 can induce commitment of MSCs to preadipocytes that promote beige adipogenesis, but in mature adipocytes, BMP4 represses the beige phenotype in favor of a white phenotype [96]. BMP4 is expressed in both preadipocytes and mature adipocytes, but in WAT preadipocytes the effects of BMP4 to promote beige adipogenesis may be silenced owing to the expression of the BMP4/7 inhibitor gremlin 1 (GREM1) [106, 107]. BMP4 is also expressed in brown preadipocytes, and other cells found in the SVF, such as MSCs and endothelial cells, and is progressively decreased during terminal brown adipocyte differentiation [108]. This is the opposite to that observed in white adipocytes where BMP4 expression is higher in mature adipocytes than the SVF [109]. Thus, during the formation of beige and brown adipocytes, it has been proposed that BMP4 signaling decreases gradually from the stem cell niche to the mature adipocyte, likely because of local diffusion and the production of BMP4 antagonists [108].

Adipose-tissue-derived BMP4 regulation of beige and brown adipogenesis may also be due to its effects on other cells within the adipose tissue niche. In a recent study, it was found that FABP4-BMP4 transgenic mice activate M2 macrophage proliferation and inhibit M1 macrophages, leading to a large increase in M2 macrophages [109], which may explain the increased browning of WAT observed. In support of this, adoptive transfer of BMP4-induced M2 macrophages to sWAT induced expression of the brown adipocyte markers PRDM16, PGC1α, and UCP1 and increased whole-body oxygen consumption [109]. Thus, previous studies demonstrating that BMP4 promotes a browning effect on WAT, including increased energy expenditure and improved metabolism, could be in part explained by the role of BMP4 regulation of macrophages (Fig. 1J).

BMP7 promotes commitment of brown/beige adipocyte precursors

Studies with *Bmp7* null mice revealed that BMP7 is essential for both development and differentiation of brown adipocyte precursors to form interscapular BAT, as well as maintenance of its thermogenic program [97]. Interestingly, systemic administration of BMP7 to mice reverses obesity via an increase in energy expenditure as well as a reduction of appetite [110]. Genetic loss of the type 1A BMP receptor (*Bmpr1a*) in the MYF5+ lineage, the embryonic precursors to BAT [1], leads to a paucity of interscapular BAT and promotes beige adipogenesis within WAT through increased sympathetic nervous system input, suggesting crosstalk between the two adipose depots [111]. This compensation restores global BAT-mediated thermogenic capacity to maintain normal body temperature and resistance to diet-induced obesity [111]. Adipogenic precursors isolated from sWAT and treated with BMP7 show synergistic induction toward a BAT-like phenotype in the presence of β3-adrenergic agonists, suggesting a direct role for BMP signaling in beige adipogenesis [112, 113]. Lastly, BMP7 has been used for ex vivo generation of brown-like adipocytes from human embryonic stem cells and iPSCs, illustrating that this BMP is likely important during human BAT development [114].

The signaling and transcriptional mechanisms involved in BMP regulation of adipogenesis have not been well understood, but recent evidence suggests that BMP7 promotes brown and beige adipogenesis through SMAD-induced regulation of EBF2 and ZFP423 [115]. EBF2 is a transcription factor that is selectively expressed in brown and beige adipocyte precursors and regulates expression of brown adipose selective target genes, such as *Prdm16* [116–118]. The transcriptional regulator ZFP423 serves as a corepressor of EBF2 and, thus, is critical for maintaining white adipocyte identity through *Prdm16* suppression [119]. During activation of adipocyte precursors with BMP7, SMAD1/4 interacts with ZFP423 to disrupt the ZFP423–EBF2 protein complex, thus enabling EBF2 to drive expression of brown adipogenesis target genes such as *Prdm16* [119].

Adipose-derived stem cells express core regulators of self-renewal and pluripotency

OCT4, NANOG, and SOX2 expression in adipose-derived stem cells

Maintaining the self-renewal and multipotency capacities of the ASC pool is a key aspect of adipose tissue homeostasis [120]. OCT4, SOX2, and NANOG are transcription factors that suppress differentiation-associated genes and maintain pluripotency [121, 122]. They physically interact with each other to enhance or limit each other's expression and thus regulate target genes involved in self-renewal and pluripotency [121, 122]. Even though OCT4, NANOG, and SOX2 are expressed at lower levels in ASCs than in embryonic stem cells, a number of studies have demonstrated that these genes are associated with increased self-renewal and multipotency (Fig. 1L) [7, 123, 124]. Similar to CD34, the gold standard marker for ASCs that undergoes rapid downregulation during their ex vivo expansion, the expression of OCT4, NANOG, and SOX2 also decreases rapidly in cell culture, which may explain why earlier studies suggested that expression of these markers in ASCs was controversial [125]. Along these lines, methods that allow increased purity of primary CD34+ ASCs result in a two- to threefold higher expression of OCT4, NANOG, and SOX2 [126]. As ASCs undergo replicative senescence with increasing ex vivo expansion, increased reactive oxygen species (ROS) results in decreased proliferation, pluripotency, and expression of OCT4, NANOG, and SOX2 [127]. This senescence is accompanied by ROS-mediated downregulation of the transcription factor c-MAF, which binds directly to and regulates OCT4, NANOG, and SOX2 expression [127]. Lastly, factors that increase pluripotency of mouse embryonic stem cells, such as supplementation with LIF or overexpression of the mir-302 cluster, also support the ability of ASCs to maintain expression of pluripotency genes [125].

OCT4 and NANOG are required to maintain stemness in adipose-derived stem cells

Functional studies have demonstrated a role for pluripotency genes in the regulation of ASC function. Overexpression of OCT4 results in the demethylation of the regulatory regions of stemness genes including OCT4, NANOG, and SOX2 and improves ASC proliferation and multipotency [128]. Silencing of NANOG in ASCs causes downregulation of OCT4 and SOX2 gene expression [129], which is accompanied by a decrease in proliferation due to cell cycle arrest in G0/G1. In bone marrow-derived MSCs, OCT4 and NANOG have been shown to bind to the promoter of DNA methyltransferase 1 (DNMT1), which is responsible for the suppression of differentiation-associated genes by maintaining methylation status during DNA replication [130]. In agreement, NANOG repression in ASCs leads to a loss of pluripotency and differentiation potential via downregulation of DNMT1 [131–133].

Proper expression of *Nanog* and *Oct4* in ASCs may be in part controlled by programmed cell death 4 (*Pdcd4*), a selective protein translation inhibitor associated with diet-induced obesity, WAT inflammation, and insulin resistance [120, 134]. *Pdcd4* deficiency in mice is associated with an increase in *Oct4* and *Nanog* and results in increased stemness and proliferation of ASCs through enhanced AKT activation and cyclin D1 upregulation [120]. Interestingly, *Pdcd4* deficiency enhances the white to beige adipocyte transition, which results in increased energy expenditure and resistance to obesity in mice placed on a HFD [120]. These results suggest that increased stemness of ASCs may contribute favorably to the prevention of obesity and metabolic syndrome.

Genes specific to adipose tissue function may also regulate the expression of pluripotency genes in ASCs. For example, as ASC proliferation and differentiation potential decreases with continuing passage in culture, PPARγ and thyroid hormone receptor (TRβ) are significantly decreased in unison with a loss of OCT4 [135]. Treatment of ASCs with PPARγ agonists can increase OCT4 promoter activity and expression and restore differentiation capacity as ASCs age in culture [135]. During culture of ASCs, loss of proliferative capacity and stemness of ASCs is also associated with an age-related decline in proteasome complex and peptidase activities [136]. Specifically, proteosomal activation is associated with lower ROS, longer telomere length, and increased expression of OCT4, NANOG, and SOX2, concomitant with increased stemness [136]. Reciprocally, silencing of OCT4 or NANOG results in a significant reduction in proteosomal activity through decreased OCT4 binding of β2 and β5 proteosomal subunit promoters [136]. These data suggest that pluripotency genes may control stemness in part through regulation of proteostasis.

Changes in adipose-derived stem cells during obesity and metabolic disease

In response to HFD, adipocyte precursors contribute to adipocyte hyperplasia, although the degree *of* de novo adipogenesis varies among different adipose tissue depots [137]. In mice, adipogenesis at the onset of diet-induced obesity in the vWAT depot is initiated well before adipocyte hypertrophy takes place [39]. Similarly, expansion of human vWAT may also depend on a wave of adipocyte proliferation at the onset of obesity [138]. In humans and mice, fully developed obesity is associated with a reduction in adipocyte precursors and loss of their differentiation potential that can lead to adipocyte hypertrophy, adipose dysfunction, and metabolic syndrome [139, 140]. Thus, during energy excess and diabetes onset, adipocyte precursors appear to proliferate and differentiate into adipocytes efficiently, but are impaired at later stages, which may contribute to adipocyte hypertrophy and worsening metabolic abnormalities [7]. Furthermore, in obese individuals, a decline in ASC function may limit their overall multipotent and regenerative potential, which is critical for those patients that may benefit from autologous transplantation.

Compared with lean patients, ASCs isolated from sWAT and vWAT of obese patients show a reduction in cell proliferation, premature senescence, and lose their angiogenic potential and capacity for multilineage differentiation [65, 139, 141–143]. ASCs from obese patients show impairment in their ability to secrete proangiogenic factors including VEGF, HGF, FGF, and PDGF, which can impair angiogenesis, promote hypoxia, and result in cellular stress and adipocyte death [7] [132–135]. In the obese, the loss of ASC multipotency also correlates with a decrease in the expression of embryonic development and multilineage differentiation genes such as TBX15, HOXC10, and α-SMA [65, 142].

ASCs have been documented to possess potent immunosuppressive activity, which is important for controlling inflammation and immunopathologic reactions, as well as for implementation of immunomodulation therapies and allogeneic stem cell treatments [144, 145]. ASCs isolated from obese and type 2 diabetic patients display reduced immunosuppressive activities, including suppression of immune cell proliferation and polarization of macrophages to the M2 phenotype, compared with those from lean patients [146]. Interestingly, the inflammatory genes IL-1β, IL-8, and monocyte chemoattractant protein 1 (MCP1), previously correlated with an increase in BMI and associated with a risk of cardiovascular disease and type 2 diabetes are upregulated in ASCs from the obese [65]. These cytokines, along with increased expression of tumor necrosis factor-α (TNFα) and IL-6 in ASCs, may contribute to the onset or worsening of inflammation in adipose tissue and promote insulin resistance by recruiting and polarizing macrophages to the M1 subtype [7, 147–150].

Genes involved in stemness also appear dysregulated in obesity and type 2 diabetes. In ASCs from sWAT of obese patients, OCT4, SAL4, SOX15, and KLF4 showed decreased expression, whereas ASCs from omental adipose tissue showed increased expression [151]. In a separate study, increased expression of OCT4 and NANOG was observed in ASCs isolated from both vWAT and sWAT of diabetic compared with nondiabetic patients [152]. Other indicators of stemness, including decreased viability of ASCs in cell culture, lower telomerase activity, and decreased telomere length, further indicate dysregulation of self-renewal capacity in ASCs from obese patients [153]. Obesity-related inflammatory cytokines such as IL6 or TNFα shorten cilia in ASCs, thus affecting their ability to properly respond to stimuli [154]. Reversal of this phenotype with an inhibitor against Aurora A, a kinase involved in cilia disassembly, rescues cilia length and function in obese ASCs [155]. Interestingly, this reversal is associated with increased expression of self-renewal and stemness genes and suggests a potential strategy to combat obesity-related disorders [155].

Overall, disruption in networks of genes involved in stemness, inflammation, multilineage potential, and ASC trafficking and homing have all been attributed to playing a role in ASC dysfunction [65]. By interfering with adipose tissue remodeling, fueling inflammation, and promoting hypoxia, defective ASCs might contribute to the development of obesity and its related diseases [7, 156]. In support of this, supplementation of obese mice with ASCs from lean counterparts attenuates adipose inflammation, improves insulin action, and restores metabolic homeostasis [140]. Thus, effective proliferation, renewal, and differentiation of ASCs are of paramount importance for proper adipose tissue function, and restoration of ASCs may represent a promising strategy to combat obesity-related diseases [7, 120].

Changes in adipocyte progenitors during aging may contribute to metabolic dysfunction

Proliferative senescence in APs during aging may play a critical role in determining predisposition to metabolic disease. Deletion of telomerase reverse transcriptase (TERT) in either *Pdgfra* + or *Pdgfrb* + AP subpopulations in mice leads to telomere shortening and proliferative senescence. This is associated with adipocyte hypertrophy in WAT and systemic insulin resistance that is aggravated by a HFD [157]. Along these lines, telomere shortening in APs derived from subcutaneous adipose tissue in bariatric surgery patients is correlated with metabolic disease progression [157].

Beige adipocytes reside within WAT and expend energy to generate heat during cold exposure (termed cold-induced thermogenesis). It is well known that activated beige adipose tissue can stimulate weight loss and promote resistance to obesity, making it an appealing therapeutic target tissue [1, 158, 159]. Aging is a primary risk factor for obesity and is accompanied by a loss of beige adipose tissue, suggesting that loss of their energy-expending capacity might contribute to an obesity-prone phenotype with increased age [160]. In particular, several studies have demonstrated that cellular aging of beige APs is associated with decreased beige adipogenesis in older mice and humans [161, 162]. It has been demonstrated that premature induction of cellular senescence in SMA+ beige APs, either genetically through overexpression of p21, or pharmacologically via inhibition of CDK4/6 activity, increases expression of senescence-activating genes and blocks their ability to differentiate into active beige adipocytes upon cold exposure in mice [163]. Furthermore, reversal of senescence in beige SMA+ APs by targeting the p38/MAPK-p16^{Ink4a} pathway, either pharmacologically or genetically, reverses age-related inhibition of cold-induced beiging in mice [163]. Beige APs isolated from the SVF of human adipose tissue can also be rejuvenated through inhibition of the p38/MAPK pathway to restore beige adipocyte differentiation in cell culture [163]. In another study, a unique subset of CD81+/PDGFRa+ APs was discovered in sWAT of mice and humans that undergo de novo beige adipogenesis following cold exposure [164]. This highly proliferative population of APs shows decreased replicative capacity in aged mice, while those from young mice showed a repressed cellular senescence pathway. Furthermore, the CD81 surface receptor is required for de novo beige adipogenesis during cold exposure and results in diet-Induced obesity, glucose.

Intolerance, and adipose tissue inflammation when CD81 is ablated. In humans, there is a significant inverse correlation between the presence of CD81+ APs from sWAT and HOMA-IR, fasting blood glucose levels, diastolic blood pressure, and visceral adiposity. Overall, these results suggest that proliferative capacity and cellular senescence are important features required for APs to give rise to beige adipose tissue, and they may represent a therapeutic target to improve metabolic health in older individuals.

Conclusions and future perspectives

Information regarding the adipose tissue stem cell niche has been complicated by the inability to find specific markers for individual cell subpopulations, the lack of optimal cell culture conditions that maintain cells in their native state, and the absence of models for monitoring cells of the native niche in vivo. Advancements in new 3D culture techniques and discovery of new adipogenic growth factors may allow preservation and a more accurate representation of the native niche ex vivo. This becomes particularly important when determining the physiological abnormalities of adipogenic precursors from obese patients, since culture conditions may inaccurately reflect the native niche and lead to incorrect assumptions concerning their role in disease progression. In support of this, optimized 3D culture conditions that preserve the niche environment have already shown enhanced clinical effects [165].

Previous studies have suggested that a diverse repertoire of cell types contained within the niche (endothelial, hematopoietic, mural, fibroblastic) all have the potential to act as adipocyte precursors in vitro; however, lineage tracing studies in mice are beginning to shed light on the major populations that contribute to mature adipocytes during normal metabolic homeostasis, weight gain, and WAT browning. This includes a fibroblast stem cell population marked by PDGFRα and the mural cell marker PDGFRβ, which appears to contribute to the majority of mature adipocytes that are formed in adult mice [57]. Single-cell RNA sequencing and the discovery of non-cell surface markers for tracking and purification of adipogenic precursors will perhaps allow the further partition of these cells into a ordered hierarchy of adipogenic commitment and potential.

ASCs and progenitors isolated from subcutaneous and visceral adipose of obese patients show a reduction in cell proliferation, premature senescence, and lose their angiogenic potential and capacity for multilineage differentiation [65, 139–143]. Thus, at the later stages of obesity, loss of ASC and AP function may contribute to adipocyte hypertrophy associated with excess caloric intake and insulin resistance. ASCs derived from obese and type 2 diabetic patients display reduced immunosuppressive activity, including suppression of immune cell proliferation and the ability to polarize macrophages to the M2 phenotype [146]. Inflammatory cytokines associated with a risk of cardiovascular disease and type 2 diabetes are also upregulated in ASCs [65]. Taken together, the restoration of ASCs through transplantation might be an effective strategy to combat complications associated with obesity, as has been demonstrated in mice [140].

The molecular mechanisms that regulate the self-renewal and differentiation of ASCs or their downstream progenitors may also aid in the development of new therapies to combat obesity-related disorders. For example, mechanisms related to BMP7-induced differentiation of beige/brown adipocytes may provide clues toward development of therapies that increase thermogenesis and protect against metabolic syndrome. It is increasingly evident that macrophages have a significant impact on both renewal and the differentiation potential of adipogenic precursors, with M2 macrophage subsets serving to promote a healthy balance between renewal and differentiation [78]. The discovery of factors related to the polarization toward the M2 phenotype (e.g., BMP4/IL-4), or factors from these cells that act directly to promote beige adipocytes from adipocyte precursors, may be used to develop new therapies for weight loss [84, 85].

Conditions that allow functional coculture of different cell types found within the adipose precursor niche, including different subpopulations of ASCs, progenitors, macrophages, endothelial cells, neurons, mural cells, fibroblasts, and others, will be of increasing importance for streamlining the discovery of how these cells interact and promote or inhibit adipogenesis during normal, insulin-resistant, and inflammatory states. For example, the development of methods for the coculturing of sympathetic neurons and ASCs to study lipolysis and adipose tissue browning has not been successfully achieved, possibly owing to the secretion of factors in coculture that inhibit differentiation when in isolation outside of the native adipose tissue niche [166]. Finally, the research community is still attempting to unlock the therapeutic potential of ASCs for a broad range of diseases and regenerative applications, including reconstruction of soft tissue defects, wound healing and skin restoration, skeletal muscle/bone reconstruction, liver regeneration, and cardiac repair [167]. Future identification of new regulatory factors that control the adipocyte precursor niche will help to determine the optimal culturing methods for ASCs, which will be important not only for the study of metabolic diseases, but for the utilization of these cells in regenerative medicine.

Acknowledgments and funding support

This work was supported by NIH COBRE Award P20GM121301 (Aaron C. Brown, Lucy Liaw, and Clifford J. Rosen). Julieta Martino kindly provided assistance with illustrations and editing.

Competing interests

The author declares no competing interests.

References

[1] M. Harms, P. Seale, Brown and beige fat: development, function and therapeutic potential, Nat. Med. 19 (10) (2013) 1252–1263.

[2] C.N. Lumeng, A.R. Saltiel, Inflammatory links between obesity and metabolic disease, J. Clin. Invest. 121 (6) (2011) 2111–2117.

[3] V.S. Malik, W.C. Willett, F.B. Hu, Global obesity: trends, risk factors and policy implications, Nat. Rev. Endocrinol. 9 (1) (2013) 13–27.
[4] C.N. Ochner, A.G. Tsai, R.F. Kushner, T.A. Wadden, Treating obesity seriously: when recommendations for lifestyle change confront biological adaptations, Lancet Diabetes Endocrinol. 3 (4) (2015) 232–234.
[5] E.D. Rosen, B.M. Spiegelman, What we talk about when we talk about fat, Cell 156 (1–2) (2014) 20–44.
[6] B. Cannon, J. Nedergaard, Brown adipose tissue: function and physiological significance, Physiol. Rev. 84 (1) (2004) 277–359.
[7] F. Louwen, A. Ritter, N.N. Kreis, J. Yuan, Insight into the development of obesity: functional alterations of adipose-derived mesenchymal stem cells, Obes. Rev. 19 (7) (2018) 888–904.
[8] W.P. Cawthorn, E.L. Scheller, O.A. MacDougald, Adipose tissue stem cells meet preadipocyte commitment: going back to the future, J. Lipid Res. 53 (2) (2012) 227–246.
[9] R.M. Seaberg, D. van der Kooy, Stem and progenitor cells: the premature desertion of rigorous definitions, Trends Neurosci. 26 (3) (2003) 125–131.
[10] M.S. Rodeheffer, K. Birsoy, J.M. Friedman, Identification of white adipocyte progenitor cells in vivo, Cell 135 (2) (2008) 240–249.
[11] R. Berry, M.S. Rodeheffer, C.J. Rosen, M.C. Horowitz, Adipose tissue residing progenitors (adipocyte lineage progenitors and adipose derived stem cells (ADSC)), Curr. Mol. Biol. Rep. 1 (3) (2015) 101–109.
[12] M.J. Lee, Y. Wu, S.K. Fried, Adipose tissue heterogeneity: implication of depot differences in adipose tissue for obesity complications, Mol. Asp. Med. 34 (1) (2013) 1–11.
[13] F. Karpe, K.E. Pinnick, Biology of upper-body and lower-body adipose tissue—link to whole-body phenotypes, Nat. Rev. Endocrinol. 11 (2) (2015) 90–100.
[14] M.J. Lee, Y. Wu, S.K. Fried, Adipose tissue remodeling in pathophysiology of obesity, Curr. Opin. Clin. Nutr. Metab. Care 13 (4) (2010) 371–376.
[15] K.L. Spalding, E. Arner, P.O. Westermark, S. Bernard, B.A. Buchholz, O. Bergmann, et al., Dynamics of fat cell turnover in humans, Nature 453 (7196) (2008) 783–787.
[16] A. Rigamonti, K. Brennand, F. Lau, C.A. Cowan, Rapid cellular turnover in adipose tissue, PLoS One 6 (3) (2011), e17637.
[17] E.D. Rosen, O.A. MacDougald, Adipocyte differentiation from the inside out, Nat. Rev. Mol. Cell Biol. 7 (12) (2006) 885–896.
[18] C. Christodoulides, C. Lagathu, J.K. Sethi, A. Vidal-Puig, Adipogenesis and WNT signalling, Trends Endocrinol. Metab. 20 (1) (2009) 16–24.
[19] J.B. Kim, B.M. Spiegelman, ADD1/SREBP1 promotes adipocyte differentiation and gene expression linked to fatty acid metabolism, Genes Dev. 10 (9) (1996) 1096–1107.
[20] V.A. Payne, W.S. Au, C.E. Lowe, S.M. Rahman, J.E. Friedman, S. O'Rahilly, et al., C/EBP transcription factors regulate SREBP1c gene expression during adipogenesis, Biochem. J. 425 (1) (2009) 215–223.
[21] Y. Barak, M.C. Nelson, E.S. Ong, Y.Z. Jones, P. Ruiz-Lozano, K.R. Chien, et al., PPAR gamma is required for placental, cardiac, and adipose tissue development, Mol. Cell 4 (4) (1999) 585–595.
[22] E.D. Rosen, C.H. Hsu, X. Wang, S. Sakai, M.W. Freeman, F.J. Gonzalez, et al., C/EBPalpha induces adipogenesis through PPARgamma: a unified pathway, Genes Dev. 16 (1) (2002) 22–26.
[23] S.A. Kliewer, K. Umesono, D.J. Mangelsdorf, R.M. Evans, Retinoid X receptor interacts with nuclear receptors in retinoic acid, thyroid hormone and vitamin D3 signalling, Nature 355 (6359) (1992) 446–449.
[24] M. Ahmadian, J.M. Suh, N. Hah, C. Liddle, A.R. Atkins, M. Downes, et al., PPARgamma signaling and metabolism: the good, the bad and the future, Nat. Med. 19 (5) (2013) 557–566.
[25] C.W. Ng, W.J. Poznanski, M. Borowiecki, G. Reimer, Differences in growth in vitro of adipose cells from normal and obese patients, Nature 231 (5303) (1971) 445.
[26] W.J. Poznanski, I. Waheed, R. Van, Human fat cell precursors. Morphologic and metabolic differentiation in culture, Lab. Investig. 29 (5) (1973) 570–576.
[27] R.L. Van, C.E. Bayliss, D.A. Roncari, Cytological and enzymological characterization of adult human adipocyte precursors in culture, J. Clin. Invest. 58 (3) (1976) 699–704.
[28] H. Green, M. Meuth, An established pre-adipose cell line and its differentiation in culture, Cell 3 (2) (1974) 127–133.
[29] H. Green, O. Kehinde, An established preadipose cell line and its differentiation in culture. II. Factors affecting the adipose conversion, Cell 5 (1) (1975) 19–27.

[30] A. Chawla, M.A. Lazar, Peroxisome proliferator and retinoid signaling pathways co-regulate preadipocyte phenotype and survival, Proc. Natl. Acad. Sci. U. S. A. 91 (5) (1994) 1786–1790.
[31] P. Tontonoz, E. Hu, B.M. Spiegelman, Stimulation of adipogenesis in fibroblasts by PPAR gamma 2, a lipid-activated transcription factor, Cell 79 (7) (1994) 1147–1156.
[32] A.L. Olson, RalA signaling may reveal the true nature of 3T3-L1 adipocytes as a model for thermogenic adipocytes, Proc. Natl. Acad. Sci. U. S. A. 115 (30) (2018) 7651–7653.
[33] S. Klaus, L. Choy, O. Champigny, A.M. Cassard-Doulcier, S. Ross, B. Spiegelman, et al., Characterization of the novel brown adipocyte cell line HIB 1B. Adrenergic pathways involved in regulation of uncoupling protein gene expression, J. Cell Sci. 107 (Pt. 1) (1994) 313–319.
[34] A. Galmozzi, S.B. Sonne, S. Altshuler-Keylin, Y. Hasegawa, K. Shinoda, I.H.N. Luijten, et al., ThermoMouse: an in vivo model to identify modulators of UCP1 expression in brown adipose tissue, Cell Rep. 9 (5) (2014) 1584–1593.
[35] C. Sengenes, K. Lolmede, A. Zakaroff-Girard, R. Busse, A. Bouloumie, Preadipocytes in the human subcutaneous adipose tissue display distinct features from the adult mesenchymal and hematopoietic stem cells, J. Cell. Physiol. 205 (1) (2005) 114–122.
[36] H. Li, L. Zimmerlin, K.G. Marra, V.S. Donnenberg, A.D. Donnenberg, J.P. Rubin, Adipogenic potential of adipose stem cell subpopulations, Plast. Reconstr. Surg. 128 (3) (2011) 663–672.
[37] J.B. Mitchell, K. McIntosh, S. Zvonic, S. Garrett, Z.E. Floyd, A. Kloster, et al., Immunophenotype of human adipose-derived cells: temporal changes in stromal-associated and stem cell-associated markers, Stem Cells 24 (2) (2006) 376–385.
[38] C. Hepler, L. Vishvanath, R.K. Gupta, Sorting out adipocyte precursors and their role in physiology and disease, Genes Dev. 31 (2) (2017) 127–140.
[39] E. Jeffery, C.D. Church, B. Holtrup, L. Colman, M.S. Rodeheffer, Rapid depot-specific activation of adipocyte precursor cells at the onset of obesity, Nat. Cell Biol. 17 (4) (2015) 376–385.
[40] R. Berry, M.S. Rodeheffer, Characterization of the adipocyte cellular lineage in vivo, Nat. Cell Biol. 15 (3) (2013) 302–308.
[41] L. Napolitano, The differentiation of white adipose cells. An electron microscope study, J. Cell Biol. 18 (1963) 663–679.
[42] D. Peurichard, F. Delebecque, A. Lorsignol, C. Barreau, J. Rouquette, X. Descombes, et al., Simple mechanical cues could explain adipose tissue morphology, J. Theor. Biol. 429 (2017) 61–81.
[43] D. Esteve, N. Boulet, C. Belles, A. Zakaroff-Girard, P. Decaunes, A. Briot, et al., Lobular architecture of human adipose tissue defines the niche and fate of progenitor cells, Nat. Commun. 10 (1) (2019) 2549.
[44] K. Iyama, K. Ohzono, G. Usuku, Electron microscopical studies on the genesis of white adipocytes: differentiation of immature pericytes into adipocytes in transplanted preadipose tissue, Virchows Arch. B Cell Pathol. Incl. Mol. Pathol. 31 (2) (1979) 143–155.
[45] S. Cinti, M. Cigolini, O. Bosello, P. Bjorntorp, A morphological study of the adipocyte precursor, J. Submicrosc. Cytol. 16 (2) (1984) 243–251.
[46] M. Crisan, S. Yap, L. Casteilla, C.W. Chen, M. Corselli, T.S. Park, et al., A perivascular origin for mesenchymal stem cells in multiple human organs, Cell Stem Cell 3 (3) (2008) 301–313.
[47] W. Tang, D. Zeve, J.M. Suh, D. Bosnakovski, M. Kyba, R.E. Hammer, et al., White fat progenitor cells reside in the adipose vasculature, Science 322 (5901) (2008) 583–586.
[48] M.A. Lazar, PPAR gamma, 10 years later, Biochimie 87 (1) (2005) 9–13.
[49] L. Vishvanath, K.A. MacPherson, C. Hepler, Q.A. Wang, M. Shao, S.B. Spurgin, et al., Pdgfrbeta+ mural preadipocytes contribute to adipocyte hyperplasia induced by high-fat-diet feeding and prolonged cold exposure in adult mice, Cell Metab. 23 (2) (2016) 350–359.
[50] J.Z. Long, K.J. Svensson, L. Tsai, X. Zeng, H.C. Roh, X. Kong, et al., A smooth muscle-like origin for beige adipocytes, Cell Metab. 19 (5) (2014) 810–820.
[51] D.C. Berry, Y. Jiang, J.M. Graff, Mouse strains to study cold-inducible beige progenitors and beige adipocyte formation and function, Nat. Commun. 7 (2016) 10184.
[52] S. Su, A.R. Guntur, D.C. Nguyen, S.S. Fakory, C.C. Doucette, C. Leech, et al., A renewable source of human beige adipocytes for development of therapies to treat metabolic syndrome, Cell Rep. 25 (11) (2018) 3215–3228.e9.
[53] A.C. Brown, Brown adipocytes from induced pluripotent stem cells-how far have we come? Ann. N. Y. Acad. Sci. 1463 (2019) 9–22.

[54] R. Li, K. Bernau, N. Sandbo, J. Gu, S. Preissl, X. Sun, Pdgfra marks a cellular lineage with distinct contributions to myofibroblasts in lung maturation and injury response, Elife 7 (2018) e36865, https://doi.org/10.7554/eLife.36865.
[55] Y.H. Lee, A.P. Petkova, E.P. Mottillo, J.G. Granneman, In vivo identification of bipotential adipocyte progenitors recruited by beta3-adrenoceptor activation and high-fat feeding, Cell Metab. 15 (4) (2012) 480–491.
[56] Y.H. Lee, A.P. Petkova, A.A. Konkar, J.G. Granneman, Cellular origins of cold-induced brown adipocytes in adult mice, FASEB J. 29 (1) (2015) 286–299.
[57] P. Cattaneo, D. Mukherjee, S. Spinozzi, L. Zhang, V. Larcher, W.B. Stallcup, et al., Parallel lineage-tracing studies establish fibroblasts as the prevailing in vivo adipocyte progenitor, Cell Rep. 30 (2) (2020) 571–582.e2.
[58] R.B. Burl, V.D. Ramseyer, E.A. Rondini, R. Pique-Regi, Y.H. Lee, J.G. Granneman, Deconstructing adipogenesis induced by beta3-adrenergic receptor activation with single-cell expression profiling, Cell Metab. 28 (2) (2018) 300–309.e4.
[59] M. Coelho, T. Oliveira, R. Fernandes, Biochemistry of adipose tissue: an endocrine organ, Arch. Med. Sci. 9 (2) (2013) 191–200.
[60] T.D. Challa, L.G. Straub, M. Balaz, E. Kiehlmann, O. Donze, G. Rudofsky, et al., Regulation of de novo adipocyte differentiation through cross talk between adipocytes and preadipocytes, Diabetes 64 (12) (2015) 4075–4087.
[61] R. Zhang, Y. Gao, X. Zhao, M. Gao, Y. Wu, Y. Han, et al., FSP1-positive fibroblasts are adipogenic niche and regulate adipose homeostasis, PLoS Biol. 16 (8) (2018), e2001493.
[62] A. Nawaz, K. Tobe, M2-like macrophages serve as a niche for adipocyte progenitors in adipose tissue, J. Diabetes Investig. 10 (6) (2019) 1394–1400.
[63] Y. Sun, S. Chen, X. Zhang, M. Pei, Significance of cellular cross-talk in stromal vascular fraction of adipose tissue in neovascularization, Arterioscler. Thromb. Vasc. Biol. 39 (6) (2019) 1034–1044.
[64] P. Bourin, B.A. Bunnell, L. Casteilla, M. Dominici, A.J. Katz, K.L. March, et al., Stromal cells from the adipose tissue-derived stromal vascular fraction and culture expanded adipose tissue-derived stromal/stem cells: a joint statement of the International Federation for Adipose Therapeutics and Science (IFATS) and the International Society for Cellular Therapy (ISCT), Cytotherapy 15 (6) (2013) 641–648.
[65] B. Onate, G. Vilahur, S. Camino-Lopez, A. Diez-Caballero, C. Ballesta-Lopez, J. Ybarra, et al., Stem cells isolated from adipose tissue of obese patients show changes in their transcriptomic profile that indicate loss in stemcellness and increased commitment to an adipocyte-like phenotype, BMC Genomics 14 (2013) 625.
[66] A.L. Carey, C. Vorlander, M. Reddy-Luthmoodoo, A.K. Natoli, M.F. Formosa, D.A. Bertovic, et al., Reduced UCP-1 content in in vitro differentiated beige/brite adipocytes derived from preadipocytes of human subcutaneous white adipose tissues in obesity, PLoS One 9 (3) (2014), e91997.
[67] W.S. Kim, J. Han, S.J. Hwang, J.H. Sung, An update on niche composition, signaling and functional regulation of the adipose-derived stem cells, Expert. Opin. Biol. Ther. 14 (8) (2014) 1091–1102.
[68] D.L. Crandall, G.J. Hausman, J.G. Kral, A review of the microcirculation of adipose tissue: anatomic, metabolic, and angiogenic perspectives, Microcirculation 4 (2) (1997) 211–232.
[69] S. Aoki, S. Toda, T. Sakemi, H. Sugihara, Coculture of endothelial cells and mature adipocytes actively promotes immature preadipocyte development in vitro, Cell Struct. Funct. 28 (1) (2003) 55–60.
[70] D.O. Traktuev, D.N. Prater, S. Merfeld-Clauss, A.R. Sanjeevaiah, M.R. Saadatzadeh, M. Murphy, et al., Robust functional vascular network formation in vivo by cooperation of adipose progenitor and endothelial cells, Circ. Res. 104 (12) (2009) 1410–1420.
[71] D.O. Traktuev, S. Merfeld-Clauss, J. Li, M. Kolonin, W. Arap, R. Pasqualini, et al., A population of multipotent CD34-positive adipose stromal cells share pericyte and mesenchymal surface markers, reside in a periendothelial location, and stabilize endothelial networks, Circ. Res. 102 (1) (2008) 77–85.
[72] L. Russo, C.N. Lumeng, Properties and functions of adipose tissue macrophages in obesity, Immunology 155 (4) (2018) 407–417.
[73] C.N. Lumeng, J.L. Bodzin, A.R. Saltiel, Obesity induces a phenotypic switch in adipose tissue macrophage polarization, J. Clin. Invest. 117 (1) (2007) 175–184.
[74] S. Fujisaka, I. Usui, A. Bukhari, M. Ikutani, T. Oya, Y. Kanatani, et al., Regulatory mechanisms for adipose tissue M1 and M2 macrophages in diet-induced obese mice, Diabetes 58 (11) (2009) 2574–2582.
[75] J.I. Odegaard, A. Chawla, Pleiotropic actions of insulin resistance and inflammation in metabolic homeostasis, Science 339 (6116) (2013) 172–177.
[76] K. Sun, C.M. Kusminski, P.E. Scherer, Adipose tissue remodeling and obesity, J. Clin. Invest. 121 (6) (2011) 2094–2101.
[77] A. Takikawa, A. Mahmood, A. Nawaz, T. Kado, K. Okabe, S. Yamamoto, et al., HIF-1alpha in myeloid cells promotes adipose tissue remodeling toward insulin resistance, Diabetes 65 (12) (2016) 3649–3659.

[78] A. Nawaz, A. Aminuddin, T. Kado, A. Takikawa, S. Yamamoto, K. Tsuneyama, et al., CD206(+) M2-like macrophages regulate systemic glucose metabolism by inhibiting proliferation of adipocyte progenitors, Nat. Commun. 8 (1) (2017) 286.
[79] R.A. Ignotz, J. Massague, Type beta transforming growth factor controls the adipogenic differentiation of 3T3 fibroblasts, Proc. Natl. Acad. Sci. U. S. A. 82 (24) (1985) 8530–8534.
[80] H. Yadav, S.G. Rane, TGF-beta/Smad3 signaling regulates brown adipocyte induction in white adipose tissue, Front Endocrinol 3 (2012) 35.
[81] Y. Qiu, K.D. Nguyen, J.I. Odegaard, X. Cui, X. Tian, R.M. Locksley, et al., Eosinophils and type 2 cytokine signaling in macrophages orchestrate development of functional beige fat, Cell 157 (6) (2014) 1292–1308.
[82] S.M. van den Berg, A.D. van Dam, P.C. Rensen, M.P. de Winther, E. Lutgens, Immune modulation of brown(ing) adipose tissue in obesity, Endocr. Rev. 38 (1) (2017) 46–68.
[83] K. Fischer, H.H. Ruiz, K. Jhun, B. Finan, D.J. Oberlin, V. van der Heide, et al., Alternatively activated macrophages do not synthesize catecholamines or contribute to adipose tissue adaptive thermogenesis, Nat. Med. 23 (5) (2017) 623–630.
[84] M.W. Lee, J.I. Odegaard, L. Mukundan, Y. Qiu, A.B. Molofsky, J.C. Nussbaum, et al., Activated type 2 innate lymphoid cells regulate beige fat biogenesis, Cell 160 (1–2) (2015) 74–87.
[85] F. Lizcano, D. Vargas, A. Gomez, A. Torrado, Human ADMC-derived adipocyte thermogenic capacity is regulated by IL-4 receptor, Stem Cells Int. 2017 (2017) 2767916.
[86] D. Zeve, J. Seo, J.M. Suh, D. Stenesen, W. Tang, E.D. Berglund, et al., Wnt signaling activation in adipose progenitors promotes insulin-independent muscle glucose uptake, Cell Metab. 15 (4) (2012) 492–504.
[87] K.A. Longo, W.S. Wright, S. Kang, I. Gerin, S.H. Chiang, P.C. Lucas, et al., Wnt10b inhibits development of white and brown adipose tissues, J. Biol. Chem. 279 (34) (2004) 35503–35509.
[88] N. Kakudo, A. Shimotsuma, K. Kusumoto, Fibroblast growth factor-2 stimulates adipogenic differentiation of human adipose-derived stem cells, Biochem. Biophys. Res. Commun. 359 (2) (2007) 239–244.
[89] L.E. Zaragosi, G. Ailhaud, C. Dani, Autocrine fibroblast growth factor 2 signaling is critical for self-renewal of human multipotent adipose-derived stem cells, Stem Cells 24 (11) (2006) 2412–2419.
[90] D.A. Rider, C. Dombrowski, A.A. Sawyer, G.H. Ng, D. Leong, D.W. Hutmacher, et al., Autocrine fibroblast growth factor 2 increases the multipotentiality of human adipose-derived mesenchymal stem cells, Stem Cells 26 (6) (2008) 1598–1608.
[91] Y. Ma, N. Kakudo, N. Morimoto, F. Lai, S. Taketani, K. Kusumoto, Fibroblast growth factor-2 stimulates proliferation of human adipose-derived stem cells via Src activation, Stem Cell Res Ther 10 (1) (2019) 350.
[92] W. Oliva-Olivera, L. Coin-Araguez, S. Lhamyani, M. Clemente-Postigo, J.A. Torres, M.R. Bernal-Lopez, et al., Adipogenic impairment of adipose tissue-derived mesenchymal stem cells in subjects with metabolic syndrome: possible protective role of FGF2, J. Clin. Endocrinol. Metab. 102 (2) (2017) 478–487.
[93] Y. Cheng, K.H. Lin, T.H. Young, N.C. Cheng, The influence of fibroblast growth factor 2 on the senescence of human adipose-derived mesenchymal stem cells during long-term culture, Stem Cells Transl. Med. 9 (2019) 518–530.
[94] H. Huang, T.J. Song, X. Li, L. Hu, Q. He, M. Liu, et al., BMP signaling pathway is required for commitment of C3H10T1/2 pluripotent stem cells to the adipocyte lineage, Proc. Natl. Acad. Sci. U. S. A. 106 (31) (2009) 12670–12675.
[95] Q.Q. Tang, T.C. Otto, M.D. Lane, Commitment of C3H10T1/2 pluripotent stem cells to the adipocyte lineage, Proc. Natl. Acad. Sci. U. S. A. 101 (26) (2004) 9607–9611.
[96] S. Modica, L.G. Straub, M. Balaz, W. Sun, L. Varga, P. Stefanicka, et al., Bmp4 promotes a brown to white-like adipocyte shift, Cell Rep. 16 (8) (2016) 2243–2258.
[97] Y.H. Tseng, E. Kokkotou, T.J. Schulz, T.L. Huang, J.N. Winnay, C.M. Taniguchi, et al., New role of bone morphogenetic protein 7 in brown adipogenesis and energy expenditure, Nature 454 (7207) (2008) 1000–1004.
[98] V. Sottile, K. Seuwen, Bone morphogenetic protein-2 stimulates adipogenic differentiation of mesenchymal precursor cells in synergy with BRL 49653 (rosiglitazone), FEBS Lett. 475 (3) (2000) 201–204.
[99] K. Hata, R. Nishimura, F. Ikeda, K. Yamashita, T. Matsubara, T. Nokubi, et al., Differential roles of Smad1 and p38 kinase in regulation of peroxisome proliferator-activating receptor gamma during bone morphogenetic protein 2-induced adipogenesis, Mol. Biol. Cell 14 (2) (2003) 545–555.
[100] W. Jin, T. Takagi, S.N. Kanesashi, T. Kurahashi, T. Nomura, J. Harada, et al., Schnurri-2 controls BMP-dependent adipogenesis via interaction with Smad proteins, Dev. Cell 10 (4) (2006) 461–471.
[101] N.F. Denton, M. Eghleilib, S. Al-Sharifi, M. Todorcevic, M.J. Neville, N. Loh, et al., Bone morphogenetic protein 2 is a depot-specific regulator of human adipogenesis, Int. J. Obes. 43 (12) (2019) 2458–2468.

[102] D. Shungin, T.W. Winkler, D.C. Croteau-Chonka, T. Ferreira, A.E. Locke, R. Magi, et al., New genetic loci link adipose and insulin biology to body fat distribution, Nature 518 (7538) (2015) 187–196.
[103] E. Guiu-Jurado, M. Unthan, N. Bohler, M. Kern, K. Landgraf, A. Dietrich, et al., Bone morphogenetic protein 2 (BMP2) may contribute to partition of energy storage into visceral and subcutaneous fat depots, Obesity (Silver Spring) 24 (10) (2016) 2092–2100.
[104] R.R. Bowers, J.W. Kim, T.C. Otto, M.D. Lane, Stable stem cell commitment to the adipocyte lineage by inhibition of DNA methylation: role of the BMP-4 gene, Proc. Natl. Acad. Sci. U. S. A. 103 (35) (2006) 13022–13027.
[105] S.W. Qian, Y. Tang, X. Li, Y. Liu, Y.Y. Zhang, H.Y. Huang, et al., BMP4-mediated brown fat-like changes in white adipose tissue alter glucose and energy homeostasis, Proc. Natl. Acad. Sci. U. S. A. 110 (9) (2013) E798–E807.
[106] B. Gustafson, U. Smith, The WNT inhibitor Dickkopf 1 and bone morphogenetic protein 4 rescue adipogenesis in hypertrophic obesity in humans, Diabetes 61 (5) (2012) 1217–1224.
[107] B. Gustafson, A. Hammarstedt, S. Hedjazifar, J.M. Hoffmann, P.A. Svensson, J. Grimsby, et al., BMP4 and BMP antagonists regulate human white and beige adipogenesis, Diabetes 64 (5) (2015) 1670–1681.
[108] S. Modica, C. Wolfrum, The dual role of BMP4 in adipogenesis and metabolism, Adipocyte 6 (2) (2017) 141–146.
[109] S.W. Qian, M.Y. Wu, Y.N. Wang, Y.X. Zhao, Y. Zou, J.B. Pan, et al., BMP4 facilitates beige fat biogenesis via regulating adipose tissue macrophages, J. Mol. Cell Biol. 11 (1) (2019) 14–25.
[110] K.L. Townsend, R. Suzuki, T.L. Huang, E. Jing, T.J. Schulz, K. Lee, et al., Bone morphogenetic protein 7 (BMP7) reverses obesity and regulates appetite through a central mTOR pathway, FASEB J. 26 (5) (2012) 2187–2196.
[111] T.J. Schulz, P. Huang, T.L. Huang, R. Xue, L.E. McDougall, K.L. Townsend, et al., Brown-fat paucity due to impaired BMP signalling induces compensatory browning of white fat, Nature 495 (7441) (2013) 379–383.
[112] T.J. Schulz, T.L. Huang, T.T. Tran, H. Zhang, K.L. Townsend, J.L. Shadrach, et al., Identification of inducible brown adipocyte progenitors residing in skeletal muscle and white fat, Proc. Natl. Acad. Sci. U. S. A. 108 (1) (2011) 143–148.
[113] M. Elsen, S. Raschke, N. Tennagels, U. Schwahn, T. Jelenik, M. Roden, et al., BMP4 and BMP7 induce the white-to-brown transition of primary human adipose stem cells, Am. J. Physiol. Cell Physiol. 306 (5) (2014) C431–C440.
[114] M. Nishio, T. Yoneshiro, M. Nakahara, S. Suzuki, K. Saeki, M. Hasegawa, et al., Production of functional classical brown adipocytes from human pluripotent stem cells using specific hemopoietin cocktail without gene transfer, Cell Metab. 16 (3) (2012) 394–406.
[115] A.M. Blazquez-Medela, M. Jumabay, K.I. Bostrom, Beyond the bone: bone morphogenetic protein signaling in adipose tissue, Obes. Rev. 20 (5) (2019) 648–658.
[116] S. Rajakumari, J. Wu, J. Ishibashi, H.W. Lim, A.H. Giang, K.J. Won, et al., EBF2 determines and maintains brown adipocyte identity, Cell Metab. 17 (4) (2013) 562–574.
[117] W. Wang, M. Kissig, S. Rajakumari, L. Huang, H.W. Lim, K.J. Won, et al., Ebf2 is a selective marker of brown and beige adipogenic precursor cells, Proc. Natl. Acad. Sci. U. S. A. 111 (40) (2014) 14466–14471.
[118] R.R. Stine, S.N. Shapira, H.W. Lim, J. Ishibashi, M. Harms, K.J. Won, et al., EBF2 promotes the recruitment of beige adipocytes in white adipose tissue, Mol. Metab. 5 (1) (2016) 57–65.
[119] M. Shao, J. Ishibashi, C.M. Kusminski, Q.A. Wang, C. Hepler, L. Vishvanath, et al., Zfp423 maintains white adipocyte identity through suppression of the beige cell thermogenic gene program, Cell Metab. 23 (6) (2016) 1167–1184.
[120] Y. Bai, Q. Shang, H. Zhao, Z. Pan, C. Guo, L. Zhang, et al., Pdcd4 restrains the self-renewal and white-to-beige transdifferentiation of adipose-derived stem cells, Cell Death Dis. 7 (2016), e2169.
[121] L.A. Boyer, T.I. Lee, M.F. Cole, S.E. Johnstone, S.S. Levine, J.P. Zucker, et al., Core transcriptional regulatory circuitry in human embryonic stem cells, Cell 122 (6) (2005) 947–956.
[122] J. Wang, S. Rao, J. Chu, X. Shen, D.N. Levasseur, T.W. Theunissen, et al., A protein interaction network for pluripotency of embryonic stem cells, Nature 444 (7117) (2006) 364–368.
[123] P.C. Sachs, M.P. Francis, M. Zhao, J. Brumelle, R.R. Rao, L.W. Elmore, et al., Defining essential stem cell characteristics in adipose-derived stromal cells extracted from distinct anatomical sites, Cell Tissue Res. 349 (2) (2012) 505–515.
[124] P. Potdar, J. Sutar, Establishment and molecular characterization of mesenchymal stem cell lines derived from human visceral & subcutaneous adipose tissues, J. Stem Cells Regen. Med. 6 (1) (2010) 26–35.
[125] M.F. Taha, A. Javeri, S. Rohban, S.J. Mowla, Upregulation of pluripotency markers in adipose tissue-derived stem cells by miR-302 and leukemia inhibitory factor, Biomed. Res. Int. 2014 (2014) 941486.

[126] A. Higuchi, C.T. Wang, Q.D. Ling, H.H. Lee, S.S. Kumar, Y. Chang, et al., A hybrid-membrane migration method to isolate high-purity adipose-derived stem cells from fat tissues, Sci. Rep. 5 (2015) 10217.

[127] P.M. Chen, C.H. Lin, N.T. Li, Y.M. Wu, M.T. Lin, S.C. Hung, et al., c-Maf regulates pluripotency genes, proliferation/self-renewal, and lineage commitment in ROS-mediated senescence of human mesenchymal stem cells, Oncotarget 6 (34) (2015) 35404–35418.

[128] J.H. Kim, M.K. Jee, S.Y. Lee, T.H. Han, B.S. Kim, K.S. Kang, et al., Regulation of adipose tissue stromal cells behaviors by endogenic Oct4 expression control, PLoS One 4 (9) (2009), e7166.

[129] M. Pitrone, G. Pizzolanti, L. Tomasello, A. Coppola, L. Morini, G. Pantuso, et al., NANOG plays a hierarchical role in the transcription network regulating the pluripotency and plasticity of adipose tissue-derived stem cells, Int J Mol Sci. 18 (6) (2017) 1107.

[130] C.C. Tsai, P.F. Su, Y.F. Huang, T.L. Yew, S.C. Hung, Oct4 and nanog directly regulate Dnmt1 to maintain self-renewal and undifferentiated state in mesenchymal stem cells, Mol. Cell 47 (2) (2012) 169–182.

[131] M. Pitrone, G. Pizzolanti, A. Coppola, L. Tomasello, S. Martorana, G. Pantuso, et al., Knockdown of NANOG reduces cell proliferation and induces G0/G1 cell cycle arrest in human adipose stem cells, Int. J. Mol. Sci. 20 (10) (2019) 2580.

[132] Y. Wang, K.A. Kim, J.H. Kim, H.S. Sul, Pref-1, a preadipocyte secreted factor that inhibits adipogenesis, J. Nutr. 136 (12) (2006) 2953–2956.

[133] M.C. Mitterberger, S. Lechner, M. Mattesich, A. Kaiser, D. Probst, N. Wenger, et al., DLK1(PREF1) is a negative regulator of adipogenesis in CD105(+)/CD90(+)/CD34(+)/CD31(−)/FABP4(−) adipose-derived stromal cells from subcutaneous abdominal fat pats of adult women, Stem Cell Res. 9 (1) (2012) 35–48.

[134] Q. Wang, Z. Dong, X. Liu, X. Song, Q. Song, Q. Shang, et al., Programmed cell death-4 deficiency prevents diet-induced obesity, adipose tissue inflammation, and insulin resistance, Diabetes 62 (12) (2013) 4132–4143.

[135] L.T. Dao, E.Y. Park, O.K. Hwang, J.Y. Cha, H.S. Jun, Differentiation potential and profile of nuclear receptor expression during expanded culture of human adipose tissue-derived stem cells reveals PPARgamma as an important regulator of Oct4 expression, Stem Cells Dev. 23 (1) (2014) 24–33.

[136] M. Kapetanou, N. Chondrogianni, S. Petrakis, G. Koliakos, E.S. Gonos, Proteasome activation enhances stemness and lifespan of human mesenchymal stem cells, Free Radic. Biol. Med. 103 (2017) 226–235.

[137] Q.A. Wang, C. Tao, R.K. Gupta, P.E. Scherer, Tracking adipogenesis during white adipose tissue development, expansion and regeneration, Nat. Med. 19 (10) (2013) 1338–1344.

[138] P. Arner, D.P. Andersson, A. Thorne, M. Wiren, J. Hoffstedt, E. Naslund, et al., Variations in the size of the major omentum are primarily determined by fat cell number, J. Clin. Endocrinol. Metab. 98 (5) (2013) E897–E901.

[139] B. Onate, G. Vilahur, R. Ferrer-Lorente, J. Ybarra, A. Diez-Caballero, C. Ballesta-Lopez, et al., The subcutaneous adipose tissue reservoir of functionally active stem cells is reduced in obese patients, FASEB J. 26 (10) (2012) 4327–4336.

[140] Q. Shang, Y. Bai, G. Wang, Q. Song, C. Guo, L. Zhang, et al., Delivery of adipose-derived stem cells attenuates adipose tissue inflammation and insulin resistance in obese mice through remodeling macrophage phenotypes, Stem Cells Dev. 24 (17) (2015) 2052–2064.

[141] T.P. Frazier, J.M. Gimble, J.W. Devay, H.A. Tucker, E.S. Chiu, B.G. Rowan, Body mass index affects proliferation and osteogenic differentiation of human subcutaneous adipose tissue-derived stem cells, BMC Cell Biol. 14 (2013) 34.

[142] M. Roldan, M. Macias-Gonzalez, R. Garcia, F.J. Tinahones, M. Martin, Obesity short-circuits stemness gene network in human adipose multipotent stem cells, FASEB J. 25 (12) (2011) 4111–4126.

[143] L.M. Perez, A. Bernal, N. San Martin, B.G. Galvez, Obese-derived ASCs show impaired migration and angiogenesis properties, Arch. Physiol. Biochem. 119 (5) (2013) 195–201.

[144] G. Zheng, G. Qiu, M. Ge, J. He, L. Huang, P. Chen, et al., Human adipose-derived mesenchymal stem cells alleviate obliterative bronchiolitis in a murine model via IDO, Respir. Res. 18 (1) (2017) 119.

[145] G.M. Spaggiari, L. Moretta, Cellular and molecular interactions of mesenchymal stem cells in innate immunity, Immunol. Cell Biol. 91 (1) (2013) 27–31.

[146] C. Serena, N. Keiran, V. Ceperuelo-Mallafre, M. Ejarque, R. Fradera, K. Roche, et al., Obesity and type 2 diabetes alters the immune properties of human adipose derived stem cells, Stem Cells 34 (10) (2016) 2559–2573.

[147] K.R. Silva, S. Liechocki, J.R. Carneiro, C. Claudio-da-Silva, C.M. Maya-Monteiro, R. Borojevic, et al., Stromal-vascular fraction content and adipose stem cell behavior are altered in morbid obese and post bariatric surgery ex-obese women, Stem Cell Res Ther 6 (2015) 72.

[148] L.M. Perez, B. de Lucas, V.V. Lunyak, B.G. Galvez, Adipose stem cells from obese patients show specific differences in the metabolic regulators vitamin D and Gas5, Mol. Genet. Metab. Rep. 12 (2017) 51–56.
[149] M.J. Lee, J. Kim, M.Y. Kim, Y.S. Bae, S.H. Ryu, T.G. Lee, et al., Proteomic analysis of tumor necrosis factor-alpha-induced secretome of human adipose tissue-derived mesenchymal stem cells, J. Proteome Res. 9 (4) (2010) 1754–1762.
[150] G.E. Kilroy, S.J. Foster, X. Wu, J. Ruiz, S. Sherwood, A. Heifetz, et al., Cytokine profile of human adipose-derived stem cells: expression of angiogenic, hematopoietic, and pro-inflammatory factors, J. Cell. Physiol. 212 (3) (2007) 702–709.
[151] R.S. Patel, G. Carter, G. El Bassit, A.A. Patel, D.R. Cooper, M. Murr, et al., Adipose-derived stem cells from lean and obese humans show depot specific differences in their stem cell markers, exosome contents and senescence: role of protein kinase C delta (PKCdelta) in adipose stem cell niche, Stem Cell Investig. 3 (2016) 2.
[152] P. Dentelli, C. Barale, G. Togliatto, A. Trombetta, C. Olgasi, M. Gili, et al., A diabetic milieu promotes OCT4 and NANOG production in human visceral-derived adipose stem cells, Diabetologia 56 (1) (2013) 173–184.
[153] L.M. Perez, A. Bernal, B. de Lucas, N. San Martin, A. Mastrangelo, A. Garcia, et al., Altered metabolic and stemness capacity of adipose tissue-derived stem cells from obese mouse and human, PLoS One 10 (4) (2015), e0123397.
[154] A. Ritter, A. Friemel, N.N. Kreis, S.C. Hoock, S. Roth, U. Kielland-Kaisen, et al., Primary cilia are dysfunctional in obese adipose-derived mesenchymal stem cells, Stem Cell Rep. 10 (2) (2018) 583–599.
[155] A. Ritter, N.N. Kreis, S. Roth, A. Friemel, L. Jennewein, C. Eichbaum, et al., Restoration of primary cilia in obese adipose-derived mesenchymal stem cells by inhibiting Aurora A or extracellular signal-regulated kinase, Stem Cell Res Ther 10 (1) (2019) 255.
[156] L. Badimon, J. Cubedo, Adipose tissue depots and inflammation: effects on plasticity and resident mesenchymal stem cell function, Cardiovasc. Res. 113 (9) (2017) 1064–1073.
[157] Z. Gao, A.C. Daquinag, C. Fussell, Z. Zhao, Y. Dai, A. Rivera, et al., Age-associated telomere attrition in adipocyte progenitors predisposes to metabolic disease, Nat. Metab. 2 (12) (2020) 1482–1497.
[158] A.M. Cypess, S. Lehman, G. Williams, I. Tal, D. Rodman, A.B. Goldfine, et al., Identification and importance of brown adipose tissue in adult humans, N. Engl. J. Med. 360 (15) (2009) 1509–1517.
[159] J. Wu, P. Cohen, B.M. Spiegelman, Adaptive thermogenesis in adipocytes: is beige the new brown? Genes Dev. 27 (3) (2013) 234–250.
[160] E. Zoico, S. Rubele, A. De Caro, N. Nori, G. Mazzali, F. Fantin, et al., Brown and Beige adipose tissue and aging, Front. Endocrinol. 10 (2019) 368.
[161] N.H. Rogers, A. Landa, S. Park, R.G. Smith, Aging leads to a programmed loss of brown adipocytes in murine subcutaneous white adipose tissue, Aging Cell 11 (6) (2012) 1074–1083.
[162] T. Yoneshiro, S. Aita, M. Matsushita, Y. Okamatsu-Ogura, T. Kameya, Y. Kawai, et al., Age-related decrease in cold-activated brown adipose tissue and accumulation of body fat in healthy humans, Obesity (Silver Spring) 19 (9) (2011) 1755–1760.
[163] D.C. Berry, Y. Jiang, R.W. Arpke, E.L. Close, A. Uchida, D. Reading, et al., Cellular aging contributes to failure of cold-induced beige adipocyte formation in old mice and humans, Cell Metab. 25 (2) (2017) 481.
[164] Y. Oguri, K. Shinoda, H. Kim, D.L. Alba, W.R. Bolus, Q. Wang, et al., CD81 controls beige fat progenitor cell growth and energy balance via FAK signaling, Cell 182 (3) (2020) 563–577.e20.
[165] R. Dai, Z. Wang, R. Samanipour, K.I. Koo, K. Kim, Adipose-derived stem cells for tissue engineering and regenerative medicine applications, Stem Cells Int. 2016 (2016) 6737345.
[166] L.C. Turtzo, R. Marx, M.D. Lane, Cross-talk between sympathetic neurons and adipocytes in coculture, Proc. Natl. Acad. Sci. U. S. A. 98 (22) (2001) 12385–12390.
[167] U.D. Wankhade, M. Shen, R. Kolhe, S. Fulzele, Advances in adipose-derived stem cells isolation, characterization, and application in regenerative tissue engineering, Stem Cells Int. 2016 (2016) 3206807.

CHAPTER

5

Mechanisms of adipose-derived stem cell aging and the impact on therapeutic potential

Xiaoyin Shan and Ivona Percec

Department of Surgery, Perelman School of Medicine, University of Pennsylvania, Philadelphia, PA, United States

Introduction

With the improvement in health care and living conditions, the general population has enjoyed extended lifespan. Along with this, the population of persons aged 65 years and older is estimated to double between 2012 and 2050, from 43.1 to 83.7 million in the United States [1]. This pattern has led to a significant paradigm shift in modern medicine, whereby the focus of treating terminal diseases has shifted to earlier interventions for the prevention and management of aging as well as total healthy lifespan optimization. As individuals age, the capacities of tissue to maintain homeostasis diminish. In response, the nascent field of regenerative medicine has developed to address the general decline in homeostatic mechanisms resulting in age-related disorders such as diabetes, cardiovascular disease, neurodegenerative diseases, and cancer [2]. Tissue-resident stem cells play a pivotal role in the maintenance of organismal homeostasis through their supportive paracrine functions, differentiation, and self-renewal. These vital functions are thought to be regulated in large part by adult stem cell populations and their associated niches. Age-related decline in the function of the stem cells results in a reduced capacity to maintain tissue homeostasis and biologic function. Stem cell rejuvenation thus presents an attractive strategy to combat aging and associated diseases.

ASCs have been explored extensively for regenerative medicine applications owing to abundant donor tissue, ease of harvest, multilineage differentiation potential, trophic factor secretion, and low immunogenicity and oncogenic capacity [3–8]. Given the importance of

Scientific Principles of Adipose Stem Cells
https://doi.org/10.1016/B978-0-12-819376-1.00017-2

ASCs in regenerative medicine, extensive research has been devoted to the understanding of key cellular processes affecting their clinical efficacy and basic biology. In this chapter, we focus on the mechanisms of ASC aging and the impact on their therapeutic potential.

General aging background

Biological aging is characterized by a system-wide decline in the physiological functions of tissues and organs over time [9, 10]. At the cellular level, it is manifested as multiple complex phenotypes including genomic instability, telomere attrition, epigenetic alterations, loss of protein homeostasis, deregulated nutrient sensing, mitochondrial dysfunction, cellular senescence, stem cell exhaustion, and altered intercellular communication [11]. Emerging evidence suggests that conditions due to the loss of cellular homeostasis, including oxidative stress, inflammation, protein aggregation, endoplasmic reticulum (ER) stress, metabolic stress, and perturbation of mitochondrial function promote many aging-associated pathological disorders [9, 10, 12] (Fig. 1). Therefore, aging can be considered the predominant risk factor for most diseases and conditions impacting healthy lifespan, and extensive research is devoted to the understanding of pathogenic mechanism of diseases during aging [13].

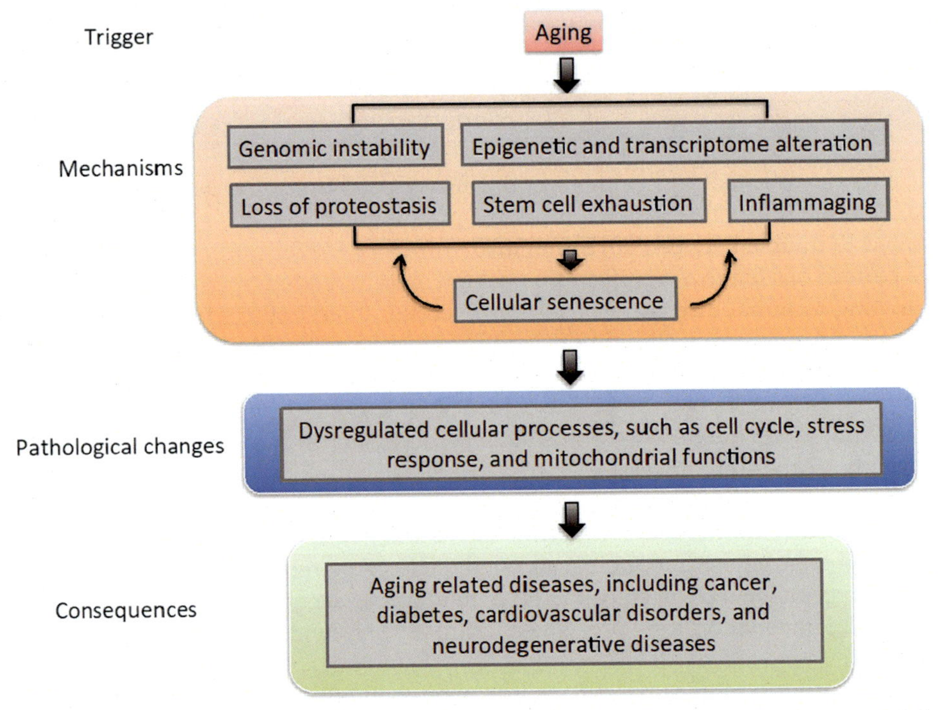

FIG. 1 Aging as a major risk factor for human pathologies. Aging leads to dysregulation of many cellular pathways, contributing to the pathogenesis of aging-associated diseases.

The function of adult stem cells contributes significantly to healthy aging. Resident stem cells are crucial in replenishing tissue specific cells through their lineage differentiation to maintain tissue and organ homeostasis. Environmental stress accumulated over time results in a wide variety of molecular and cellular damage characteristics that significantly deplete adult stem cell function.

Adipose-derived stem cell aging mechanisms

Global cellular aging has been the subject of broad investigation; however, relatively little is known about the unique features of primary human adult stem cell aging. Evidence suggests that impact of aging from stem cells differs from that of terminally differentiated cells owing to their unique functions in lineage differentiation and self-renewal [14]. Coordination between self-renewal and differentiation is critical for adult stem cells to fulfill their role in tissue repair. These two processes are closely intertwined with cell cycle progression [15–17]. Further, lineage-restricted differentiation of stem cells is important for tissue regeneration and homeostasis [18]. Prior studies, using human pluripotent stem cells, have shown that progression of the G1 phase of the cell cycle could determine cell fate propensity of human pluripotent stem cells [19, 20]. The process was regulated by cyclin D1–3, which controls differentiation signals such as the TGF-β-Smad2/3 pathway [19, 20]. Although the features of pluripotent stem cells and adult stem cells are profoundly different, the fundamental principles in the control of the cell cycle processes are likely to be similar. Thus, modulating cell cycle progression in adult stem cells could be interesting if the same results are obtained.

Stem cell niches provide the microenvironment that critically influences stem cell regeneration [21]. Consequently, aging-related changes in the niche, such as matrix composition, stiffness, and cell type, could play a role in adult stem cell maintenance and differentiation, resulting in an age-dependent phenomenon. Dynamic environmental stimuli such as proinflammatory cytokines, such as TNF-a and IL-6, that increase with age could further influence stem cell cycle regulation by interacting with members of the cell cycle regulatory network [15, 22, 23]. A thorough understanding of the molecular mechanisms governing aging and differentiation potential of adult stem cells is critically important, but currently incomplete. To decipher the molecular pathways governing the aging process in stem cells, adult stem cells from different tissues have been used. Specifically, adult stem cells derived from primary human adipose tissue have been utilized in our laboratory to study aging mechanisms.

Adipose tissue plays a key role in a variety of organismal functions, including energy storage, nutrient sensing, and temperature regulation. In response to specific stimuli, a role for adipose tissue is further evoked in endocrine, immune-modulatory, and tissue repair activities [4]. During aging, adipose tissue exhibits a global decline in function, such as differentiation potential, in combination with pathological changes in lineage differentiations resulting in chronic inflammation and the presence of senescent cells within the tissue [4]. Significantly, resident adipose tissue macrophages have been shown to play an important role in maintaining the inflammatory state in adipose tissue. Macrophages are known to secrete a vast range of biomolecules, including cytokines, to sustain the chronic inflammatory state, frequently termed as inflammaging [24, 25]. Consequently, the macroenvironment in aging adipose tissue with elevated proinflammatory cytokines is distinctively malleable, and its alteration is believed to impact the microenvironment niche of resident stem cells [4].

ASCs, similar to bone marrow-derived stem cells (BSCs), are capable of multilineage differentiation, including adipogenic, osteogenic, and chondrogenic, among others [26]. Emerging evidence suggests that the ASC niche is the vasculature within adipose tissue [27, 28], and that ASCs can be further classified as supra-adventitial [29, 30]. Despite the important roles played by these cells, the molecular mechanisms responsible for their aging are poorly understood. We have recently shown that aging ASCs manifest a series of characteristic changes, including alterations in cell cycle progression, protein synthesis, and small ubiquitin-related modifier (SUMO) protein expression after heat shock [31, 32]. Specifically, we examined genome-wide transcriptional networks during chronological aging in primary human ASCs and demonstrated that the transcriptome of aging ASCs is distinctly more stable than that of age-matched fibroblasts and, further, that age-dependent modifications in cell cycle progression and translation initiation specifically characterize aging ASCs in conjunction with increased nascent protein synthesis and a distinctly shortened G1 phase. Our results reveal novel chronological aging mechanisms in ASCs that are inherently different from differentiated cells and that may reflect an organismal attempt to meet the increased demands of tissue and organ homeostasis during aging. We then investigated age-dependent, genome-wide alterations in the chromatin accessibility of primary human ASCs in comparison with age-matched fibroblasts via assay for transposase-accessible chromatin with sequencing (ATAC-seq) technology. Our results demonstrate that aging ASCs globally possess more stable chromatin accessibility profiles as compared with aging fibroblasts, suggesting that robust regulatory mechanisms maintain adult stem cell chromatin structure against aging. Furthermore, we observed age-dependent subtle changes in promoter nucleosome positioning in selective pathways during aging, concurrent with altered SUMO protein expression under stress conditions. SUMO proteins are involved in protein sumoylation, a type of protein posttranslational modification. Stress response pathways are regulated at multiple levels, including by sumoylation. Together, our data suggest a significant role for nucleosome positioning in sumoylation pathway regulation in the stress response during adult stem cell aging. The differences described here between the chromatin structure of human ASCs and fibroblasts will further elucidate the mechanisms regulating gene expression during aging in both stem cells and differentiated cells. Our ongoing studies, as well as previous studies [33–35], suggest that aging ASCs exhibit altered lineage differentiation potential, resulting in compromised linage differentiation and reduced self-renewal, negatively impacting stem cell pools. Given the previous observation with human pluripotent stem cells [19, 20], where progression of G1 phase of cell cycle impacts cell fate propensity of human pluripotent stem cells, our discovery offers a link between altered cell cycle control and lineage differentiation potential of ASCs. Thus, the modulation of cell cycle progression could contribute to a strategy for the rejuvenation of aging ASCs.

During chronological aging, adult stem cells are increasingly subject to stress from the accumulation of toxic metabolites, DNA damage, epigenetic alterations, aggregation of damaged proteins, and mitochondrial dysfunction [14]. Aging ASCs subject to such stressors exhibit decreased differentiation potentials and proliferative capacity, negatively impacting their clinical efficacy [36, 37]. Induction of stress response pathways, such as proteostasis, DNA damage repair, and mitochondrial respiratory metabolism, has been a conserved feature across organisms during aging to adapt to various forms of cellular damage [38]. However, these mechanisms become compromised during aging [39]. For example, the regulatory molecules involved in the heat shock response are known to become impaired during aging owing to destructive processes that act on cells and organs over the lifetime [40]. Interestingly, alterations in global protein sumoylation, a form of protein translational modification, have been observed in cultured human cells after heat shock, where cells are subjected to elevated temperature, an acute proteotoxic stressor [41, 42]. Our studies demonstrated a significant change in SUMO-1 protein expression in old ASCs under stress conditions, representing increased sensitivity to stress-regulated SUMO-1 protein expression in old as compared with young ASCs. Concurrently, we have also observed age-dependent changes in nucleosome positioning in the promoter region of SUMO proteins. Our data thus suggest a significant role for nucleosome positioning, a mechanism of transcription regulation, in sumoylation pathway regulation in stress response during adult stem cell aging. Since age-related genomic and epigenetic changes influence cellular pathways during ASC aging, modulating nucleosome positioning could provide another strategy for combating aging of ASCs.

Studies using stem cells from other tissues, such as the hematopoietic system, intestine, muscle, brain, skin, and germline, have also contributed to the identification of aging phenotypes and potential mechanisms in aging [43]. Although each specific population of stem cells is functionally different, common age-related cellular processes including telomere attrition, DNA damage and mutation, cellular senescence, nutrient sensing, and metabolism are likely to play a role in ASC aging.

How adipose-derived stem cell aging impacts current clinical application of adipose-derived stem cells

ASCs are considered among the most promising cell sources for regenerative medicine applications [3–5, 7, 8, 44]. Their successful isolation and characterization have led to an increasing number of clinical studies, in total comprising over 1400 patients and encompassing almost all organ systems, between 2005 and 2020 [6]. However, a recent study showed that the age of a stem cell donor, secondary only to human leukocyte antigen (HLA) matching, is the most important characteristic influencing the outcome of a stem cell transplant [45, 46]. Furthermore, accumulating evidence has demonstrated that, in aging and related conditions, modulating systemic parameters and the tissue microenvironment can significantly impact therapeutic success [47–51].

A large number of clinical trial studies have been carried out to examine the role of ASCs in different conditions and fields, such as ischemia, sclerosis, fistula and osteoarthritis, and clinical dermatology [52, 53]. Anecdotally, we have observed positive results in sensory return after fat grafting during female genital mutilation reconstruction surgeries. In clinical applications, ASCs contribute to tissue regeneration through multiple functions, including modulating the immune response, providing signals for differentiation, migration, enzymatic reactions, and promoting balanced tissue homeostasis [52]. The impact of ASC aging on any of these functions is likely to compromise the outcome of ASC-based therapies. Emerging evidence suggests that ASCs harvested from old donors demonstrate senescent features, with a higher content of β-galactosidase positive cells, resulting in decreased proliferation and skewed lineage differentiation potential [33, 36, 37]. In addition, ASC migration and wound healing ability are compromised during aging [36, 54]. At the molecular level, aging leads to the expression of genes involved in senescence, e.g., CHEK1 required for checkpoint-mediated cell cycle arrest in response to DNA damage, and p16, an inhibitor of CDK4 kinase, and altered expression of DNA damage repair genes as well as micro-RNA production [37]. Results from a study using canines examining ASCs isolated from defined age groups similarly demonstrated aging-related decline in ASC proliferation, cell survival genes, and differentiation potentials [55].

Together, these observations confirm that aging impacts ASCs at the molecular and functional level, thereby impairing tissue regeneration and homeostasis. These findings additionally suggest a negative influence on the outcome of clinical applications when aging ASCs are employed, particularly when differentiation potential, migration, and wound healing functions are involved.

Current molecular attempts at rejuvenating stem cells for clinical applications

Strategies for stem cell rejuvenation are increasingly being investigated to overcome age-related changes that negatively impact the efficacy of ASCs in clinical applications for tissue regeneration [45, 46, 48, 50, 51, 56]. Insights into stem cell rejuvenation from clinical and model organism studies suggest that specifically targeting microenvironment-independent intrinsic stem cell properties negatively affected by aging is critically important for improving stem cell function and their therapeutic application in regenerative medicine [47, 50, 51, 56].

The successful rejuvenation of aging stem cells via synthetic or natural agents is directly dependent on the understanding of aging mechanisms in these cells. Extensive research efforts have been devoted to deciphering the complex regulatory networks in these aging cells. Early experiments using animal models led to the conclusion that the aging process can indeed be modulated by extrinsic factors [44, 57–64]. The identification of cellular pathways and key factors involved in this regulatory network has provided novel insight into the rejuvenation of aging stem cells. Specifically, the modulation of separate signaling pathways, including Notch, TGF-b, JAK/STAT, p38 MAPK, sirtuins, oxytocin, and mTOR, has resulted in altered aging phenotypes in both animal models and human stem cells [62, 65–77]. In addition to the regulation of these signaling pathways, plasma factors, such as chemokine CCL1 and GDF1, a secreted ligand of the TGF-beta, have also been shown to modulate aging phenotypes [64, 78, 79]. The complexity of the regulatory network secondary to the crosstalk among pathways has posed a significant challenge for the identification of effective molecular approaches to rejuvenate aging stem cells, in part because multiple factors may be required at specific concentrations to obtain the desired effect. Although several reagents from the phytochemical and flavonoid family have been proposed for stem cell rejuvenation, their potential carcinogenic risks for clinical applications have yet to be investigated [51].

Summary

ASCs have been extensively explored for clinical applications secondary to their many positive characteristics, as well as immune-modulatory, secretory, migratory, proliferation, and differentiation functions. A large number of clinical trials have been carried out to evaluate their potential in treating a wide range of diseases [80]. Accumulating evidence suggests that ASCs are effective in the treatment of inflammatory and autoimmune diseases [80]. Significantly, the migratory function of ASCs allows them to play a paracrine role in promoting tissue regeneration [80].

However, the functions of ASCs decline during aging. Hence, their efficacy in clinical applications is compromised in the aged population, the group most likely to benefit from their therapeutic characteristics. One strategy that has been actively explored to address this concern is stem cell rejuvenation through the modulation of cellular pathways that regulate aging in these cells. A thorough understanding of the molecular mechanisms responsible specifically for stem cell aging is critical for the development of effective strategies to rejuvenate these cells. Newly developed reagents and protocols will need extensive safety tests prior to clinical application. Tapping into the pool of currently available anti-aging small molecules, e.g., Sirt1-activating compounds (STACs) [81], should be considered first in order to accelerate the drug development process, because these compounds have been studied extensively, leading to a good understanding of the mechanism.

ASCs are a heterogeneous cell population by nature, a characteristic that contributes to their biological variability. As such, it would be important to establish whether the specific population components change with aging. To obtain predictable therapeutic results, cell populations must be carefully characterized. Further, the functions of each subpopulation must be evaluated to permit selection of specific subpopulations that are most effective for a given clinical application. It is conceivable that the refinement of ASC populations may significantly improve treatment outcomes.

With increased knowledge of stem cell aging mechanisms, in particular, ASC-specific aging mechanisms, it will be feasible to develop strategies to combat ASC aging and to improve the outcomes of their clinical application in the aged population, thereby prolonging healthy lifespan in patients.

References

[1] J.M. Ortman, V.A. Velkoff, H. Hogan, An Aging Nation: The Older Population in the United States, United States Census Bureau, Economics and Statistics Administration, US, 2014.

[2] G.C. Gurtner, M.J. Callaghan, M.T. Longaker, Progress and potential for regenerative medicine, Annu. Rev. Med. 58 (2007) 299–312, https://doi.org/10.1146/annurev.med.58.082405.095329.

[3] L. Frese, P.E. Dijkman, S.P. Hoerstrup, Adipose tissue-derived stem cells in regenerative medicine, Transfus. Med. Hemother. 43 (2016) 268–274, https://doi.org/10.1159/000448180.

[4] A.K. Palmer, J.L. Kirkland, Aging and adipose tissue: potential interventions for diabetes and regenerative medicine, Exp. Gerontol. 86 (2016) 97–105, https://doi.org/10.1016/j.exger.2016.02.013.

[5] V.V. Miana, E.A.P. Gonzalez, Adipose tissue stem cells in regenerative medicine, Ecancermedicalscience 12 (2018) 822, https://doi.org/10.3332/ecancer.2018.822.

[6] N.M. Toyserkani, et al., Concise review: a safety assessment of adipose-derived cell therapy in clinical trials: a systematic review of reported adverse events, Stem Cells Transl. Med. 6 (2017) 1786–1794, https://doi.org/10.1002/sctm.17-0031.

[7] K. Waked, J. Colle, M. Doornaert, V. Cocquyt, P. Blondeel, Systematic review: the oncological safety of adipose fat transfer after breast cancer surgery, Breast 31 (2017) 128–136, https://doi.org/10.1016/j.breast.2016.11.001.

[8] M. De Decker, et al., Breast cancer and fat grafting: efficacy, safety and complications-a systematic review, Eur. J. Obstet. Gynecol. Reprod. Biol. 207 (2016) 100–108, https://doi.org/10.1016/j.ejogrb.2016.10.032.

[9] R. Ren, A. Ocampo, G.H. Liu, J.C. Izpisua Belmonte, Regulation of stem cell aging by metabolism and epigenetics, Cell Metab. 26 (2017) 460–474, https://doi.org/10.1016/j.cmet.2017.07.019.

[10] C.E. Riera, C. Merkwirth, C.D. De Magalhaes Filho, A. Dillin, Signaling networks determining life span, Annu. Rev. Biochem. 85 (2016) 35–64, https://doi.org/10.1146/annurev-biochem-060815-014451.
[11] C. Lopez-Otin, M.A. Blasco, L. Partridge, M. Serrano, G. Kroemer, The hallmarks of aging, Cell 153 (2013) 1194–1217, https://doi.org/10.1016/j.cell.2013.05.039.
[12] B. Liu, et al., Depleting the methyltransferase Suv39h1 improves DNA repair and extends lifespan in a progeria mouse model, Nat. Commun. 4 (2013) 1868, https://doi.org/10.1038/ncomms2885.
[13] C. Franceschi, et al., The continuum of aging and age-related diseases: common mechanisms but different rates, Front. Med. 5 (2018) 61, https://doi.org/10.3389/fmed.2018.00061.
[14] J. Oh, Y.D. Lee, A.J. Wagers, Stem cell aging: mechanisms, regulators and therapeutic opportunities, Nat. Med. 20 (2014) 870–880, https://doi.org/10.1038/nm.3651.
[15] T. Burdon, A. Smith, P. Savatier, Signalling, cell cycle and pluripotency in embryonic stem cells, Trends Cell Biol. 12 (2002) 432–438.
[16] M.G. Scioli, et al., The biomolecular basis of adipogenic differentiation of adipose-derived stem cells, Int. J. Mol. Sci. 15 (2014) 6517–6526, https://doi.org/10.3390/ijms15046517.
[17] K.W. Orford, D.T. Scadden, Deconstructing stem cell self-renewal: genetic insights into cell-cycle regulation, Nat. Rev. Genet. 9 (2008) 115–128, https://doi.org/10.1038/nrg2269.
[18] E. Fuchs, J.A. Segre, Stem cells: a new lease on life, Cell 100 (2000) 143–155, https://doi.org/10.1016/s0092-8674(00)81691-8.
[19] S. Pauklin, L. Vallier, The cell-cycle state of stem cells determines cell fate propensity, Cell 155 (2013) 135–147, https://doi.org/10.1016/j.cell.2013.08.031.
[20] S. Pauklin, P. Madrigal, A. Bertero, L. Vallier, Initiation of stem cell differentiation involves cell cycle-dependent regulation of developmental genes by Cyclin D, Genes Dev. 30 (2016) 421–433, https://doi.org/10.1101/gad.271452.115.
[21] D.L. Jones, A.J. Wagers, No place like home: anatomy and function of the stem cell niche, Nat. Rev. Mol. Cell Biol. 9 (2008) 11–21, https://doi.org/10.1038/nrm2319.
[22] H. Bruunsgaard, M. Pedersen, B.K. Pedersen, Aging and proinflammatory cytokines, Curr. Opin. Hematol. 8 (2001) 131–136.
[23] H. Harashima, N. Dissmeyer, A. Schnittger, Cell cycle control across the eukaryotic kingdom, Trends Cell Biol. 23 (2013) 345–356, https://doi.org/10.1016/j.tcb.2013.03.002.
[24] C. Nathan, Secretory products of macrophages: twenty-five years on, J. Clin. Invest. 122 (2012) 1189–1190, https://doi.org/10.1172/jci62930.
[25] S. Gordon, The macrophage: past, present and future, Eur. J. Immunol. 37 (Suppl. 1) (2007) S9–17, https://doi.org/10.1002/eji.200737638.
[26] P.A. Zuk, The adipose-derived stem cell: looking back and looking ahead, Mol. Biol. Cell 21 (2010) 1783–1787, https://doi.org/10.1091/mbc.E09-07-0589.
[27] G. Lin, et al., Defining stem and progenitor cells within adipose tissue, Stem Cells Dev. 17 (2008) 1053–1063, https://doi.org/10.1089/scd.2008.0117.
[28] D.O. Traktuev, et al., A population of multipotent CD34-positive adipose stromal cells share pericyte and mesenchymal surface markers, reside in a periendothelial location, and stabilize endothelial networks, Circ. Res. 102 (2008) 77–85, https://doi.org/10.1161/CIRCRESAHA.107.159475.
[29] L. Zimmerlin, et al., Stromal vascular progenitors in adult human adipose tissue, Cytometry A 77 (2010) 22–30, https://doi.org/10.1002/cyto.a.20813.
[30] X. Cai, Y. Lin, P.V. Hauschka, B.E. Grottkau, Adipose stem cells originate from perivascular cells, Biol. Cell. 103 (2011) 435–447, https://doi.org/10.1042/BC20110033.
[31] X. Shan, et al., Transcriptional and cell cycle alterations mark aging of primary human adipose-derived stem cells, Stem Cells 35 (2017) 1392–1401, https://doi.org/10.1002/stem.2592.
[32] X. Shan, C. Roberts, Y. Lan, I. Percec, Age alters chromatin structure and expression of SUMO proteins under stress conditions in human adipose-derived stem cells, Sci. Rep. 8 (2018) 11502, https://doi.org/10.1038/s41598-018-29775-y.
[33] M.S. Choudhery, M. Badowski, A. Muise, J. Pierce, D.T. Harris, Donor age negatively impacts adipose tissue-derived mesenchymal stem cell expansion and differentiation, J. Transl. Med. 12 (2014) 8, https://doi.org/10.1186/1479-5876-12-8.
[34] W. Wu, L. Niklason, D.M. Steinbacher, The effect of age on human adipose-derived stem cells, Plast. Reconstr. Surg. 131 (2013) 27–37, https://doi.org/10.1097/PRS.0b013e3182729cfc.

[35] X. Ye, et al., Age-related changes in the regenerative potential of adipose-derived stem cells isolated from the prominent fat pads in human lower eyelids, PLoS One 11 (2016), https://doi.org/10.1371/journal.pone.0166590, e0166590.

[36] M. Liu, et al., Adipose-derived Mesenchymal stem cells from the elderly exhibit decreased migration and differentiation abilities with senescent properties, Cell Transplant. 26 (2017) 1505–1519, https://doi.org/10.1177/0963689717721221.

[37] E.U. Alt, et al., Aging alters tissue resident mesenchymal stem cell properties, Stem Cell Res. 8 (2012) 215–225, https://doi.org/10.1016/j.scr.2011.11.002.

[38] M.C. Haigis, B.A. Yankner, The aging stress response, Mol. Cell 40 (2010) 333–344, https://doi.org/10.1016/j.molcel.2010.10.002.

[39] D.J. Dues, et al., Aging causes decreased resistance to multiple stresses and a failure to activate specific stress response pathways, Aging 8 (2016) 777–795, https://doi.org/10.18632/aging.100939.

[40] S.K. Calderwood, A. Murshid, T. Prince, The shock of aging: molecular chaperones and the heat shock response in longevity and aging—a mini-review, Gerontology 55 (2009) 550–558, https://doi.org/10.1159/000225957.

[41] E.A. Niskanen, et al., Global SUMOylation on active chromatin is an acute heat stress response restricting transcription, Genome Biol. 16 (2015) 153, https://doi.org/10.1186/s13059-015-0717-y.

[42] F. Golebiowski, et al., System-wide changes to SUMO modifications in response to heat shock, Sci. Signal. 2 (2009) ra24, https://doi.org/10.1126/scisignal.2000282.

[43] M.B. Schultz, D.A. Sinclair, When stem cells grow old: phenotypes and mechanisms of stem cell aging, Development 143 (2016) 3–14, https://doi.org/10.1242/dev.130633.

[44] K. Takahashi, S. Yamanaka, Induction of pluripotent stem cells from mouse embryonic and adult fibroblast cultures by defined factors, Cell 126 (2006) 663–676, https://doi.org/10.1016/j.cell.2006.07.024.

[45] B.E. Shaw, et al., Development of an unrelated donor selection score predictive of survival after HCT: donor age matters Most, Biol. Blood Marrow Transplant. 24 (2018) 1049–1056, https://doi.org/10.1016/j.bbmt.2018.02.006.

[46] J.M. Bastida, et al., Influence of donor age in allogeneic stem cell transplant outcome in acute myeloid leukemia and myelodisplastic syndrome, Leuk. Res. 39 (2015) 828–834, https://doi.org/10.1016/j.leukres.2015.05.003.

[47] J. Neves, P. Sousa-Victor, H. Jasper, Rejuvenating strategies for stem cell-based therapies in aging, Cell Stem Cell 20 (2017) 161–175, https://doi.org/10.1016/j.stem.2017.01.008.

[48] S. Ikehara, M. Li, Stem cell transplantation improves aging-related diseases, Front. Cell Dev. Biol. 2 (2014) 16, https://doi.org/10.3389/fcell.2014.00016.

[49] S.H. Kao, et al., GSK3beta controls epithelial-mesenchymal transition and tumor metastasis by CHIP-mediated degradation of Slug, Oncogene 33 (2014) 3172–3182, https://doi.org/10.1038/onc.2013.279.

[50] S. Zhang, Z. Dong, Z. Peng, F. Lu, Anti-aging effect of adipose-derived stem cells in a mouse model of skin aging induced by D-galactose, PLoS One 9 (2014), https://doi.org/10.1371/journal.pone.0097573, e97573.

[51] K. Honoki, Preventing aging with stem cell rejuvenation: feasible or infeasible? World J. Stem Cells 9 (2017) 1–8, https://doi.org/10.4252/wjsc.v9.i1.1.

[52] M. Gaur, M. Dobke, V.V. Lunyak, Mesenchymal stem cells from adipose tissue in clinical applications for dermatological indications and skin aging, Int. J. Mol. Sci. 18 (2017), https://doi.org/10.3390/ijms18010208.

[53] M.E. Bateman, A.L. Strong, J.M. Gimble, B.A. Bunnell, Concise review: using fat to fight disease: a systematic review of nonhomologous adipose-derived stromal/stem cell therapies, Stem Cells 36 (2018) 1311–1328, https://doi.org/10.1002/stem.2847.

[54] M. Zhang, et al., The effect of age on the regenerative potential of human eyelid adipose-derived stem cells, Stem Cells Int. 2018 (2018) 5654917, https://doi.org/10.1155/2018/5654917.

[55] J. Lee, et al., Effect of donor age on the proliferation and multipotency of canine adipose-derived mesenchymal stem cells, J. Vet. Sci. 18 (2017) 141–148, https://doi.org/10.4142/jvs.2017.18.2.141.

[56] G. Kaur, C. Cai, Current progress in the rejuvenation of aging stem/progenitor cells for improving the therapeutic effectiveness of myocardial repair, Stem Cells Int. 2018 (2018) 9308301, https://doi.org/10.1155/2018/9308301.

[57] B.M. Carlson, J.A. Faulkner, Muscle transplantation between young and old rats: age of host determines recovery, Am. J. Phys. 256 (1989) C1262–C1266, https://doi.org/10.1152/ajpcell.1989.256.6.C1262.

[58] C.M. McCay, F. Pope, W. Lunsford, G. Sperling, P. Sambhavaphol, Parabiosis between old and young rats, Gerontologia 1 (1957) 7–17.

[59] F.C. Ludwig, R.M. Elashoff, Mortality in syngeneic rat parabionts of different chronological age, Trans. N. Y. Acad. Sci. 34 (1972) 582–587.

[60] Z. Song, et al., Alterations of the systemic environment are the primary cause of impaired B and T lymphopoiesis in telomere-dysfunctional mice, Blood 115 (2010) 1481–1489, https://doi.org/10.1182/blood-2009-08-237230.
[61] R.P. Lanza, et al., Extension of cell life-span and telomere length in animals cloned from senescent somatic cells, Science 288 (2000) 665–669, https://doi.org/10.1126/science.288.5466.665.
[62] I.M. Conboy, et al., Rejuvenation of aged progenitor cells by exposure to a young systemic environment, Nature 433 (2005) 760–764, https://doi.org/10.1038/nature03260.
[63] L. Lapasset, et al., Rejuvenating senescent and centenarian human cells by reprogramming through the pluripotent state, Genes Dev. 25 (2011) 2248–2253, https://doi.org/10.1101/gad.173922.111.
[64] S.A. Villeda, et al., The ageing systemic milieu negatively regulates neurogenesis and cognitive function, Nature 477 (2011) 90–94, https://doi.org/10.1038/nature10357.
[65] M.E. Carlson, M. Hsu, I.M. Conboy, Imbalance between pSmad3 and notch induces CDK inhibitors in old muscle stem cells, Nature 454 (2008) 528–532, https://doi.org/10.1038/nature07034.
[66] A.S. Brack, et al., Increased Wnt signaling during aging alters muscle stem cell fate and increases fibrosis, Science 317 (2007) 807–810, https://doi.org/10.1126/science.1144090.
[67] M.E. Carlson, H.S. Silva, I.M. Conboy, Aging of signal transduction pathways, and pathology, Exp. Cell Res. 314 (2008) 1951–1961, https://doi.org/10.1016/j.yexcr.2008.03.017.
[68] Y. Wang, J. Kellner, L. Liu, D. Zhou, Inhibition of p38 mitogen-activated protein kinase promotes ex vivo hematopoietic stem cell expansion, Stem Cells Dev. 20 (2011) 1143–1152, https://doi.org/10.1089/scd.2010.0413.
[69] F.D. Price, et al., Inhibition of JAK-STAT signaling stimulates adult satellite cell function, Nat. Med. 20 (2014) 1174–1181, https://doi.org/10.1038/nm.3655.
[70] X.Y. Zhai, et al., Knockdown of SIRT6 enables human bone marrow mesenchymal stem cell senescence, Rejuvenation Res. 19 (2016) 373–384, https://doi.org/10.1089/rej.2015.1770.
[71] M.E. Carlson, et al., Molecular aging and rejuvenation of human muscle stem cells, EMBO Mol. Med. 1 (2009) 381–391, https://doi.org/10.1002/emmm.200900045.
[72] C. Elabd, et al., Oxytocin controls differentiation of human mesenchymal stem cells and reverses osteoporosis, Stem Cells 26 (2008) 2399–2407, https://doi.org/10.1634/stemcells.2008-0127.
[73] C. Elabd, et al., Oxytocin is an age-specific circulating hormone that is necessary for muscle maintenance and regeneration, Nat. Commun. 5 (2014) 4082, https://doi.org/10.1038/ncomms5082.
[74] M. Petersson, Opposite effects of oxytocin on proliferation of osteosarcoma cell lines, Regul. Pept. 150 (2008) 50–54, https://doi.org/10.1016/j.regpep.2008.02.007.
[75] X. Guo, X.F. Wang, Signaling cross-talk between TGF-beta/BMP and other pathways, Cell Res. 19 (2009) 71–88, https://doi.org/10.1038/cr.2008.302.
[76] A.S. Yamashita, et al., Notch pathway is activated by MAPK signaling and influences papillary thyroid cancer proliferation, Transl. Oncol. 6 (2013) 197–205, https://doi.org/10.1593/tlo.12442.
[77] C. Chen, Y. Liu, Y. Liu, P. Zheng, mTOR regulation and therapeutic rejuvenation of aging hematopoietic stem cells, Sci. Signal. 2 (2009) ra75, https://doi.org/10.1126/scisignal.2000559.
[78] M. Sinha, et al., Restoring systemic GDF11 levels reverses age-related dysfunction in mouse skeletal muscle, Science 344 (2014) 649–652, https://doi.org/10.1126/science.1251152.
[79] M.A. Egerman, et al., GDF11 increases with age and inhibits skeletal muscle regeneration, Cell Metab. 22 (2015) 164–174, https://doi.org/10.1016/j.cmet.2015.05.010.
[80] M. Patrikoski, B. Mannerstrom, S. Miettinen, Perspectives for clinical translation of adipose stromal/stem cells, Stem Cells Int. 2019 (2019) 5858247, https://doi.org/10.1155/2019/5858247.
[81] B.P. Hubbard, D.A. Sinclair, Small molecule SIRT1 activators for the treatment of aging and age-related diseases, Trends Pharmacol. Sci. 35 (2014) 146–154, https://doi.org/10.1016/j.tips.2013.12.004.

CHAPTER

6

Human pluripotent nontumorigenic multilineage differentiating stress enduring (Muse) cells isolated from adipose tissue: A new paradigm in regenerative medicine and cell therapy

Karen L. Leung and Gregorio D. Chazenbalk

Department of Obstetrics and Gynecology, University of California Los Angeles (UCLA), Los Angeles, CA, United States

Discovery of Muse and adipose-derived multilineage-differentiating stress-enduring cells

Multilineage-differentiating stress-enduring (Muse) cells were first discovered by serendipity in bone marrow (BM) by Dr. Mari Dezawa's team in 2010 [1]. While rushing to get to a wine gathering in the evening, the researcher unintentionally cultured skeletal muscle cells generated from mesenchymal stem cells (MSCs) in a protease trypsin solution without any nutrients overnight. To her surprise, a small number of cells were not only discovered to have survived the harsh environment of the trypsin solution, but also proliferated and formed clusters upon culture for 7–10 days in single-cell suspension, similar to characteristics of embryonic stem (ES) cells. Later, Muse cells were also shown to exhibit multiple pluripotent markers, including NANOG, Oct3/4, Par-4, and SSEA3 [2], becoming a novel adult pluripotent stem cell population for study in the field of regenerative medicine.

In 2013, a group from UCLA led by Dr. Gregorio Chazenbalk discovered a population of pluripotent stem cells from adipose tissue lipoaspirate with similar characteristics as Muse cells, termed adipose-derived Muse (Muse-AT) cells [3]. Muse-AT cells were also discovered

Scientific Principles of Adipose Stem Cells
https://doi.org/10.1016/B978-0-12-819376-1.00004-4

by serendipity, when an unexpected failure of a laboratory equipment exposed all cells from the adipose tissue to severe cellular stress including long-term exposure to the proteolytic enzyme collagenase, serum deprivation, low temperatures, and hypoxia [3, 4]. Normally, such severe cellular stress conditions would cause all cells from the adipose tissue to die; however, Muse-AT cells were found to survive such conditions.

While Muse cells isolated from fibroblasts and BM aspirates are now defined as 100% SSEA3+ (stage-specific embryonic antigen) cells [2], with SSEA3 being a cellular marker of pluripotency typically expressed by pluripotent stem cells such as embryonic and iPS cells, Muse-AT cells are only 60% SSEA3+ cells [4]. Therefore, it is important to distinguish between Muse cells being isolated via cell sorting using an SSEA3+ antibody [2], and Muse-AT cells isolated by severe cellular stress. For Muse-AT cell isolation, lipoaspirate is first incubated with collagenase at 37°C for 30 min to release adipocytes and the SVF, followed by an additional incubation for 16 h in severe cellular stress conditions as described previously [3, 4] (Fig. 1A). Muse-AT cells in adipose tissue lipoaspirate are defined by their ability to survive such high levels of cellular stress; therefore, while they are a heterogenous cell population, further purification for Muse-AT cells is unnecessary, as more than 95% of the isolated Muse-AT cells already exhibit pluripotent qualities and are capable of triploblastic differentiation [3, 4].

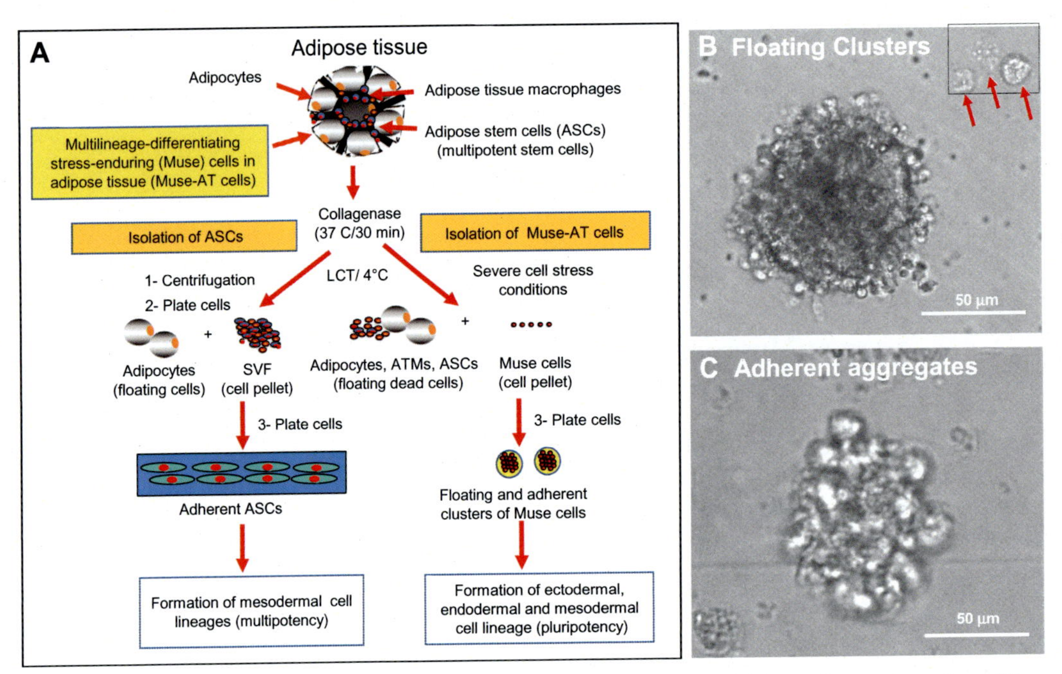

FIG. 1 Isolation and morphology of adipose-derived muse (Muse-AT) cells. (A) Schematic representation of Muse-AT isolation under severe cellular stress. (B) Muse-AT cells can grow in suspension, forming cell clusters and individual cells, and (C) Muse-AT cells can adhere to the dish, forming cell aggregates. *Reproduced from Fig. 1A, C, and D in Heneidi, S., Simerman, A.A., Keller, E., Singh, P., Li, X., Dumesic, D.A., Chazenbalk, G., Awakened by cellular stress: isolation and characterization of a novel population of pluripotent stem cells derived from human adipose tissue, PLoS One 8 (2013) e64752. https://doi.org/10.1371/journal.pone.0064752 under CC-BY license.*

The singular survival of Muse-AT cells under severe cellular stress conditions suggests that Muse-AT cells exist in a quiescent state in normal physiological conditions but become activated when the physiological cellular environment is disrupted and undergoes stress. Quiescence is a characteristic of other adult stem cells such as hematopoietic [5–7] and epithelial stem cells [8, 9], whereby quiescence preserves the capacity of cells to self-renew and survive at a critical time when other cell populations in the physiological niche fail to do so.

Components of adipose tissue

Adipose tissue is a master regulator of systemic lipid storage in the body, acting as an energy depot and endocrine organ with a major role in maintaining metabolic homeostasis [10]. When adipose tissue does not function sufficiently, metabolic and cardiovascular diseases related to excess adiposity arise, including type-2 diabetes, dyslipidemia, fatty liver disease, hypertension, and coronary heart disease [11]. Adipose tissue dysfunction in the development of such diseases is regulated in part by hormones and adipokines secreted by adipose tissue, via paracrine and endocrine secretion regulating appetite control, fat distribution, insulin sensitivity and secretion, energy expenditure, and inflammation [11, 12].

Adipose tissue is comprised of adipocytes, multipotent ASCs, adipose stem progenitors (preadipocytes), adipose tissue macrophages (ATMs), vascular endothelial cells, and other cells such as nerve cells, myeloid dendritic cells, stromal cells, fibroblasts, and extracellular matrix (ECM) components that provide structural and biochemical support [13]. This heterogeneity of cells within adipose tissue creates a dynamic environment allowing for crosstalk between different cell types to govern tissue homeostasis [14–17]. ASCs have been shown to differentiate not only along adipocyte, osteoblast, chondrocyte, and other mesenchymal cell lineages, but also toward cardiomyocytes and endothelial cells [15]. Preadipocytes have the capacity to differentiate to ATMs [18]. Similarly, ATMs also have the capacity to differentiate into preadipocytes that eventually increase the population of adipocytes [14]. Thus, cell populations within adipose tissue demonstrate high levels of cell plasticity and are continually shifting from one cell type to another, perhaps in response to environmental cues and the metabolic or tissue repair needs of the body.

Adipose stem cells

Adipose tissue can be easily obtained from the body and is an abundant source of adult stem cells, 100–500-fold more than that of BM [19]. Due to their abundance and ease of acquisition, ASCs have traditionally been studied by researchers worldwide on their biology and clinical applicability since their first isolation and classification in 2001 [20]. ASCs can be readily isolated from the stromal vascular fraction (SVF) of lipoaspirate or from surgically removed adipose tissue after enzymatic digestion [21–23]. Cultured ASCs are defined by their potential to differentiate down adipogenic, osteogenic, and chondrogenic pathways as well as staining positive for CD73, CD90, and CD105 while staining negative for CD34 (although typically positively expressed by ASCs from freshly isolated SVF, ASCs that have been cultured through multiple passages lack CD34 expression [24]) and CD31 [25].

The morphology, gene expression, and differentiation characteristics of ASCs are similar to MSCs derived from BM and umbilical cord [26, 27].

Because of their ability to differentiate into a variety of different cell lineages and their capacity to secrete a broad selection of cytokines, chemokines, and growth factors to elicit autocrine and paracrine effects, ASCs have become a popular field of study in regenerative medicine. ASCs have been investigated for use in sport injuries [28, 29]; implanted in soft tissues to promote the healing process in cutaneous wounds [30]; studied for antiinflammatory potential in treating multiple sclerosis, rheumatoid arthritis, and diabetes mellitus [31]; and used in investigation of neurodegenerative disease treatment including Alzheimer's disease, Parkinson's disease [32], intervertebral disc, amyotrophic lateral sclerosis [33, 34], multiple system atrophy, and traumatic brain injury [35, 36].

Adipose-derived Muse cells vs adipose stem cells

Differences between isolation methodology

While sharing the same biological niche in the SVF as ASCs, Muse-AT cells are a different stem cell population in adipose tissue and can be isolated separate from ASCs. ASCs are isolated via filtering digested adipose tissue through a 150-μm nylon mesh following digestion with collagenase at 37°C for 45–60 min, while Muse-AT cells are isolated following an additional incubation with collagenase at 4°C for 16 h such that severe cellular stress conditions are introduced [3]. Muse-AT cells are the only cell component of adipose tissue that survive such harsh environment, as Muse-AT cells are in quiescence under normal physiological conditions but are activated under those stressful cell conditions.

Differences in gene expression

Gene expression studies have shown over 800 differentially expressed genes between Muse-AT cells and ASCs isolated from the same adipose tissue involved in functions of cell death and survival, embryonic development, tissue development, cellular assembly and organization, and cellular function and maintenance [3]. The expression patterns of many of these genes provide plausible explanations to the fundamental characteristics of Muse-AT cells. For example, genes associated with DNA repair in cell death and survival are upregulated in Muse-AT cells, suggesting the ability of Muse-AT cells to resist DNA damage and maintain function during events of cellular stress [3].

Another notable upregulated gene is CXCL2, a chemokine regularly overexpressed in cancerous tissue and shown to play a role in the survival and pervasiveness of some cancers through invasive cancer treatment [3, 37]. CXCL2 has also been shown to be involved with stem cell homing, whereby quiescent cells are mobilized from their dormant state [37]. Increased expression of CXCL2 in Muse-AT cells, therefore, may at least partially explain the ability of Muse-AT cells to survive extreme cellular stress and become activated from quiescence, whereas ASCs do not survive in such conditions.

Remarkably, Muse-AT cells also exhibit higher levels of Delta-like protein (DLK), a marker for preadipocytes and newly formed adipocytes, than do ASCs [14]. This is consistent with a

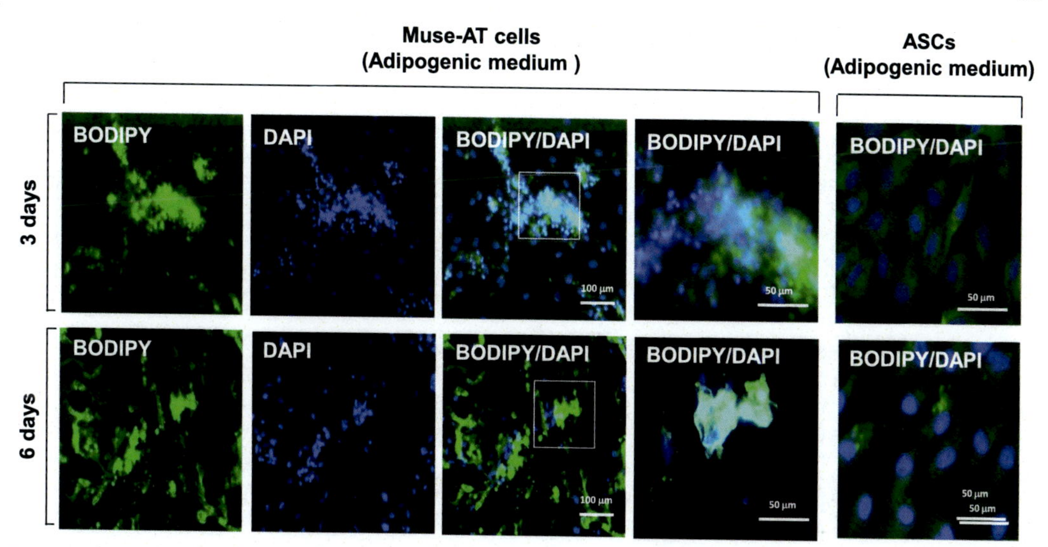

FIG. 2 Adipose-derived Muse (Muse-AT) cell differentiation to adipocytes. Isolated Muse-AT cells and adipose stem cells (ASCs) derived from the same lipoaspirate were grown as adherent cells in the presence of adipocyte differentiation medium. At 3–6 days, Muse-AT cells differentiate into adipocytes much faster than ASCs under the same culture conditions, indicated by the amount of BODIPY-C16 lipid drops accumulated inside the cells. *Reproduced from Fig. 3B in Heneidi, S., Simerman, A.A., Keller, E., Singh, P., Li, X., Dumesic, D.A., Chazenbalk, G., Awakened by cellular stress: isolation and characterization of a novel population of pluripotent stem cells derived from human adipose tissue, PLoS One 8 (2013) e64752. https://doi.org/10.1371/journal.pone.0064752 under CC-BY license.*

much faster adipocyte differentiation process in Muse-AT cells as compared with ASCs [3]. Normally, ASCs take around 2.5 weeks to fully differentiate into mature adipocytes, but the same process only takes 3–6 days with Muse-AT cells, as measured via immunocytochemistry by fluorescent dye BODIPY-C_{16}, which detects lipid droplet formation [3] (Fig. 2). This phenomenon may be related to the stronger preference Muse-AT cells have to commit to the adipogenic lineage, which may be potentially linked to an epigenetic memory of its tissue of origin.

Most importantly, Muse-AT cells overexpress pluripotent stem cell markers relative to ASCs, specifically SOX2, Oct3/4, and REX1 by three- to fourfold [3]. This differential expression of pluripotent stem cell markers indicates the difference in differentiation capacity between Muse-AT cells and ASCs and highlights the intrinsic pluripotent property of Muse-AT cells.

Pluripotency vs multipotency

By definition, Muse-AT cells are pluripotent stem cells as they express many pluripotent stem cell markers and are able to differentiate into cells of all three germ layers [3, 4]. On the other hand, ASCs are a type of MSC, whereby they are only able to differentiate into cells of mesodermal origin [38]. ASCs differentiate into adipocytes, osteoblasts, and myocytes, although they may also be able to differentiate into other mesodermal cells such as endothelial cells [39]. Due to their limited differentiation capacity, ASCs are considered to be multipotent, only able to differentiate into a limited range of cell types.

Properties of adipose-derived Muse cells

Pluripotency of adipose-derived Muse cells

A key characteristic of Muse-AT cells is their pluripotency, based in their ability to differentiate into any of the three germ layers [3]. Muse-AT cells uniquely form clusters when growing in suspension in nonadherent dishes (Fig. 1B) and form aggregates in adherent dishes (Fig. 1C), [3, 4, 40], similar to the traditionally gold-standard ES and iPS cells. These cell clusters stain double positive for CD105 and SSEA3, markers for MSCs and ES cells, respectively [3, 25, 41]. Moreover, Muse-AT cells express pluripotent stem cell markers, including SSEA4, NANOG, Oct 3/4, Sox2, and TRA1-60 (Fig. 3) [3, 4], through at least eight passages in vitro. Under medium-specific induction, Muse-AT cells also express NK2–5, GATA3 and α-fetoprotein, and MAP2, characteristic of mesodermal, endodermal, and ectodermal cell lines, respectively [3, 4]. Furthermore, Muse-AT cells seem to display an epigenetic memory similar to iPS cells, as Muse-AT cells preferentially differentiate spontaneously into adipocytes despite their pluripotency and capacity to differentiate into any cell type, as if they have an epigenetic memory of their tissue of origin [3].

Nontumorigenic properties of adipose-derived Muse cells

While having similar pluripotent qualities to ES and iPS cells, a difference between the cell types is that Muse-AT cells are nontumorigenic. Historically, ES and iPS cell research has been struggling to circumvent technical issues of teratoma formation for ES and iPS cell clinical application. Clinical trials using ES cells to treat spinal cord injuries were complicated when human ES cell transplantation into immune deficient mice resulted in teratoma formation, leading to discontinuation of the trials in the fifth enrolled patient [42]. Similarly, in 2014, the RIKEN institute in Japan stopped treating their second patient with iPS cells in their clinical trials on macular degeneration following discovery of mutations in the patient's iPS cells [43].

Muse-AT cells, on the other hand, display nontumorigenic activity in vivo, as well as a stable karyotype in culture (Fig. 4A and B) [4, 40]. This phenotype could be explained by the low expression levels of the RNA binding protein gene involved with tumorigenesis, Lin28, and high levels of expression of Let7, a micro-RNA regulating embryonic development, tumor suppression, and phenotypic differentiation, detected in Muse-AT cells [40]. Lin28 and Let7 regulate opposing processes, thus, with a high Let7/Lin28 ratio, Muse-AT cells may be nontumorigenic, unlike ES and iPS cells, which have a high Lin28/Let7 ratio [44]. By this theory, altering the Let7/Lin28 ratio of Muse-AT cells, possible through genetic modification, may likely reverse its nontumorigenic properties, and if accomplished through genetic modification techniques, may also potentially introduce mutations to the cells and cause tumorigenesis if injected into tissue in vivo.

Additional gene expression studies by microarray have also shown Muse-AT cells to have a lower expression of genes involved with tissue development, cellular assembly and organization, cellular function and maintenance, DNA replication, repair, and cell cycling, indicating their low rate of growth in contrast with ES and iPS cells [3]. Certain pluripotent stem cell markers in Muse cells, specifically NANOG and Oct3/4, also exhibit increased methylation patterns [45], which may further contribute to the nontumorigenic properties of Muse-AT cells.

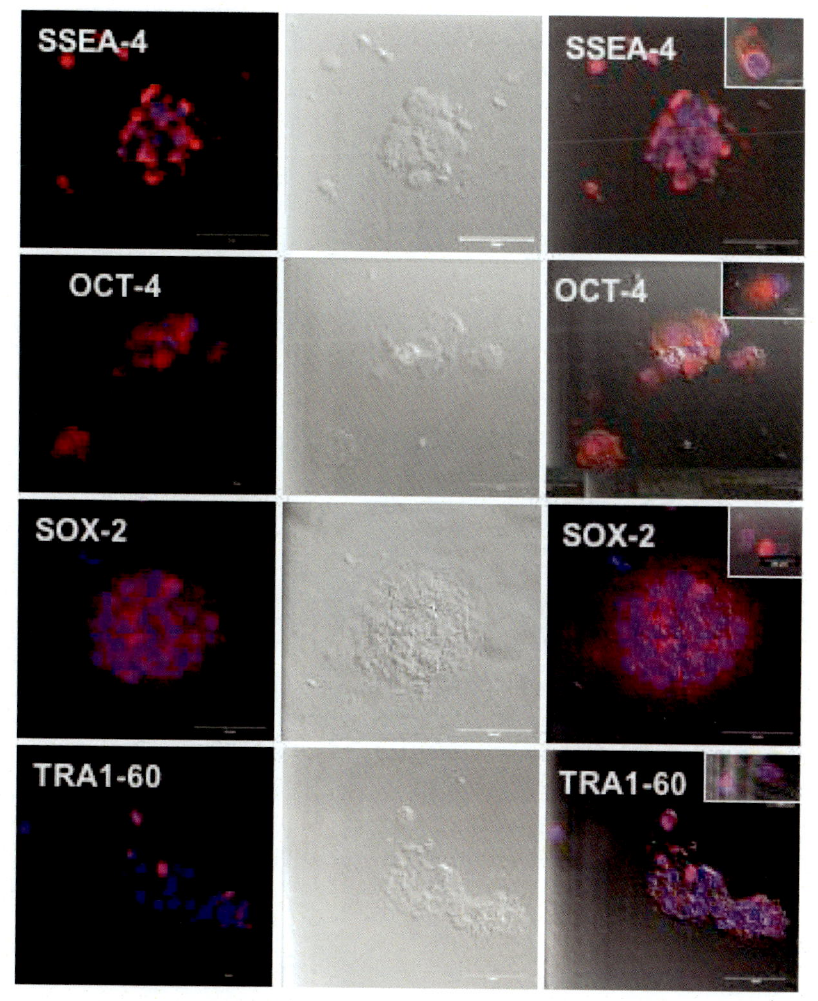

FIG. 3 Expression of pluripotent stem cell markers SSEA4, Oct-4, Sox-2, and TRA1-60 in adipose-derived Muse (Muse-AT) cells. Left panel: immunofluorescence microscopy of Muse-AT cells stained for SSEA-4, Oct-4, Sox-2, and TRA1-60. *Center panel*: micrograph of all cells resistant to severe cellular stress. *Right panel*: overlay of immunofluorescence microscopy of Muse-AT cells (*left panel*) and all cells resistant to severe cellular stress (*middle panel*) indicating that over 95% of all cells resistant to severe cellular stress are Muse-AT cells. *Reproduced from Fig. 1B in Fisch, S.C., Gimeno, M.L., Phan, J.D., Simerman, A.A., Dumesic, D.A., Perone, M.J., Chazenbalk, G.D., Pluripotent nontumorigenic multilineage differentiating stress enduring cells (muse cells): a seven-year retrospective, Stem Cell Res. Ther. 8 (2017) 227. https://doi.org/10.1186/s13287-017-0674-3 under CC-BY license.*

Immunomodulatory properties of adipose-derived Muse cells

As components of mesenchymal tissue, Muse-AT cells exhibit similar immunomodulatory properties as previously reported in MSCs [46]. When two different murine macrophage-related cell lines were incubated with Muse-AT conditioned culture media in vitro, there was a decrease in proinflammatory TNF-α levels secreted by these macrophages, demonstrating the immunomodulatory activity of Muse-AT cells through their

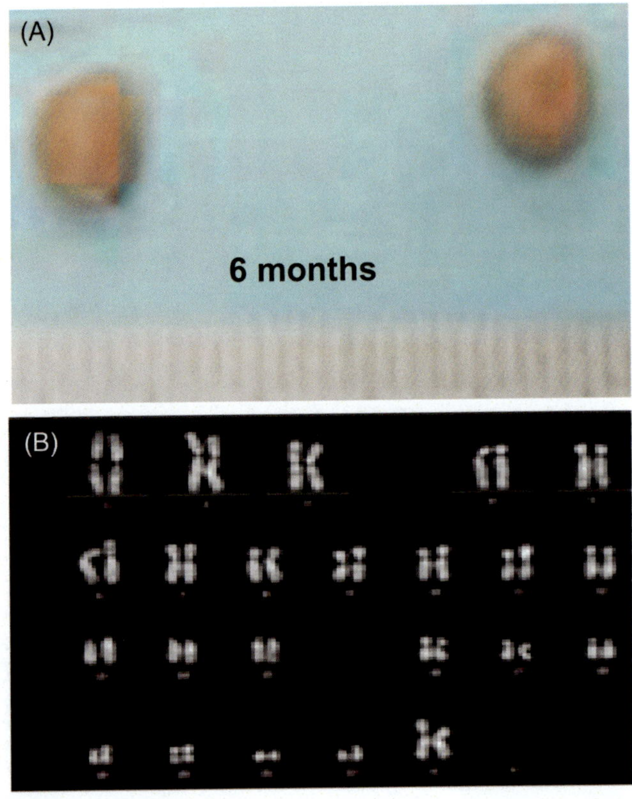

FIG. 4 (A) Adipose-derived Muse (Muse-AT) cells transplanted to testes of severe combined immunodeficiency (SCID) mice did not generate teratomas similar to untreated testes. (B) Muse-AT cells grown in culture showed normal karyotype. *Reproduced from Fig. 4 in Gimeno, M.L., Fuertes, F., Barcala Tabarrozzi, A.E., Attorressi, A.I., Cucchiani, R., Corrales, L., Oliveira, T.C., Sogayar, M.C., Labriola, L., Dewey, R.A., Perone, M.J., Pluripotent nontumorigenic adipose tissue-derived muse cells have immunomodulatory capacity mediated by transforming growth factor-β1, Stem Cells Transl. Med. 6 (2017) 161–173. https://doi.org/10.5966/sctm.2016-0014 under CC-BY license.*

secreted products [4]. Muse-AT conditioned medium also diminished antigen-specific stimulation of Th1-type cytokines and favored the secretion of IL-10 in both T cells and macrophages [4].

The immunomodulatory capacity of Muse-AT cells has been shown to be mainly based on TGF-β action, a key immunosuppressive cytokine under specific circumstances [4]. When Muse-AT cells were cultured in normal Dulbecco's Modified Eagle Medium (DMEM), there was an increase in TGF-β secretion into the culture medium [4]. The TGF-β signaling pathways involve TGF-β ligands binding to its respective receptors and phosphorylating a family of SMAD proteins, specifically SMAD2 in host cells [4, 47]. In Muse-AT cell conditioned medium, antigen-specific stimulated T lymphocytes exhibited high intracellular levels of phosphorylated SMAD2, suggesting that Muse-AT cells have immunomodulatory properties by secreting TGF-β [4]. When an inhibitor of type I TGF-β receptor was added to the Muse-AT conditioned medium, the immune regulatory activity of Muse-AT conditioned medium was almost completely reversed, but neutralizing anti-TGF-β1 in the culture medium of antigen-specific stimulation of T lymphocytes reestablished IFN-γ secretion [4], further showing that Muse-AT cells may control T-cell and macrophage functions through regulating the TGF-β signaling pathway.

Evolutionary bases of adipose-derived Muse cells

The Muse-AT cell gene profile shares many similarities with other stem cells described in very primitive species, suggesting involvement with a highly conserved cellular mechanism related to cell survival and regeneration in response to cellular stress and acute injury. The expression of genes involved in cell death and survival, DNA replication and repair, the cell cycle, and embryonic development in Muse-AT cells offers insight into the evolutionary significance of Muse-AT cells, and suggests that Muse-AT cells may be the missing link between mammals and less complex organisms in the evolutionary chain of tissue regeneration, since genes related to regeneration ability have been lost through time in many species according to the evolutionary standpoint [3, 40, 48].

Muse-AT cells share many characteristics present in autonomous regeneration processes of more-primitive species, most notably planarians. Adult stem cells in planarians, also known as neoblasts, are comprised of totipotent, pluripotent, and multipotent mitotically active stem cells and make up 20%–35% of all planarian cells [49]. Similar to how Muse-AT cells home to damaged tissue, when expression of signaling proteins becomes activated, neoblasts migrate to the injury site and initiate tissue regeneration and repair [50, 51]. At the site of injury, undifferentiated cells form the regenerative blastema and after 3–4 weeks, the amputated or injured area is restored to original morphology and function [52]. Furthermore, much like Muse-AT cells, neoblasts also do not form teratomas due to a high Let7/Lin28 ratio [49]. Interestingly, when planarians are put in stressful situations such as starvation, leading to a shrinking of body size via apoptotic mechanisms, the mitotic index of neoblasts increases to replace lost cells [53, 54]. This response to acute injury or stress in planarians resembles the manner in which Muse-AT cells are activated by cellular stress, suggesting the presence of a possible highly conserved mechanism and indicating a critical role of Muse-AT cells in both the evolutionary conservation of pluripotency and adaptations to cellular stress, spanning the course of 500 million years, from the first appearance of planarians to the present.

Besides having similar components as neoblasts, many differentially expressed genes in Muse-AT cells also have homologs present in organisms such as *Saccharomyces cerevisiae*, *Caenorhabditis elegans*, *Chlamydomonas*, *orpedo californica*, and *Drosophila* [3, 48]. Thus, it is likely that Muse cells exhibit a highly conserved cellular mechanism for survival in response to severe cell stress that may date back to the age when more-primitive organisms, especially planarians, were prominent [55].

Clinical perspective of adipose-derived Muse cells

For decades, the field of regenerative medicine has been searching for a population of pluripotent stem cells that can be used safely for clinical treatment of tissue damage and disease that may benefit from stem cell therapy. ES and iPS cells gained the initial spotlight for such potential, but concerns regarding teratoma formation when transplanted into animal models set back their clinical applicability. Additionally, the use of embryos to derive ES cells brought into discussion whether it is ethical to destroy embryos for research purposes and pushed for a need to find pluripotent stem cell populations that can be obtained from adult tissues without teratoma formation upon transplantation.

In 2010, Dr. Mari Dezawa's team discovered Muse cells from BM aspirates under long-term trypsin incubation, currently redefined to 100% of the cell population having SSEA3+ expression [2], with similar pluripotent and nontumorigenic properties as Muse-AT cells [3, 4]. Since then, many animal model studies using human Muse cells isolated from various mesenchymal tissues, including BM, skin, and adipose tissue have demonstrated the efficacy of Muse cells to repair damaged liver, kidney, neural and cerebrovascular tissues, lung, and cardiac tissue, owing to their high capacity of migration and integration into damaged tissue to replenish cells and repair tissue function [56–65].

In the liver, Muse cells have demonstrated the capacity to reverse widespread liver damage and improve chronic liver disease. BM-derived Muse cells differentiate into hepatocytes in vivo, and when intravenously injected into a damaged liver mouse model, the injected Muse cells were able to successfully implant and remain integrated in the liver [58, 59]. The injected Muse cells were also found restricted in the local inflammatory sites of the liver without integrating into other tissues except the lungs 2 weeks after intravenous induction [58]. After injection of Muse cells into the damaged liver, there was a consequent increase in albumin levels and decrease in bilirubin production, signifying functional improvement following Muse cell treatment [58].

Muse cell treatment has also been shown to be applicable in treating dysfunctional kidneys. Intravenously injected Muse cells into focal segmental glomerulosclerosis (FSGS) mouse models, a model for the precursor of chronic kidney disease, preferentially integrate into damaged glomeruli and improve the function of the kidneys as seen through improved urine protein and plasma creatinine levels as well as creatinine clearance [63]. Muse cells also significantly reduce glomerular sclerosis and interstitial fibrosis [63].

Furthermore, Muse cells have also demonstrated in vivo applicability in rodent models of cerebrovascular and neural tissue damage. When Muse cells are implanted in vivo after cerebral hemorrhage, mice implanted with Muse cells recover motor skills at accelerated rates [60]. Similarly, mice with Muse cells integrated into the motor and sensory cortex demonstrate heightened neural circuit function, potentially through electrophysiological improvement as suggested by hind-limb somatosensory evoked potentials with higher amplitudes [62]. Moreover, Muse cells have been found mobilized from BM into the blood stream circulation instantly after an ischemic stroke and integrated into damaged neural tissue to potentially regenerate neural cells through spontaneous differentiation, as suggested by their expression of neuronal factors Tuj-1 and NeuN [56, 66]. Supporting these results, Muse-AT cells have the capacity to differentiate in vitro into neuron-like cells detectable by morphology and by axon- and dendrite-specific markers (Fig. 5A and B).

Recently, a study investigating stem cell treatments for lung ischemia found that human Muse cells were able to migrate to the injured lung of a rat model and suppress apoptosis as well as stimulate proliferation of host alveolar cells better than MSC treatment [64]. Corresponding in vitro studies also found that Muse cells migrated to serum from lung-

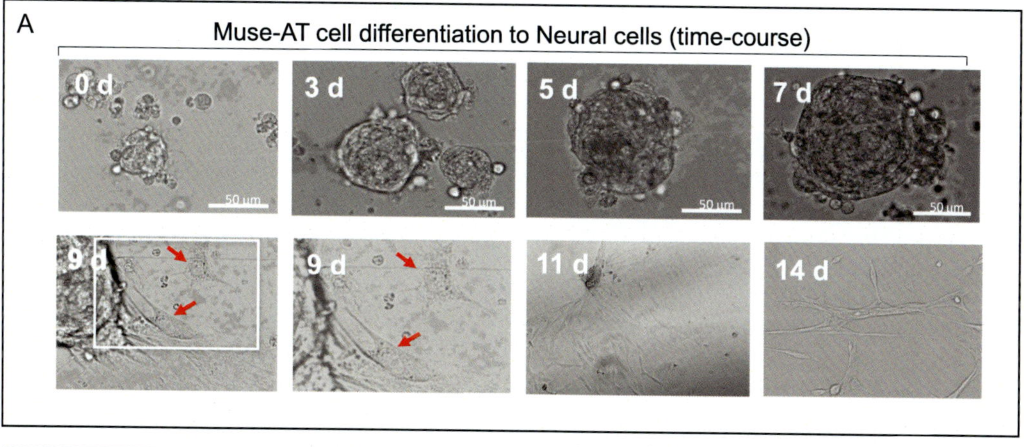

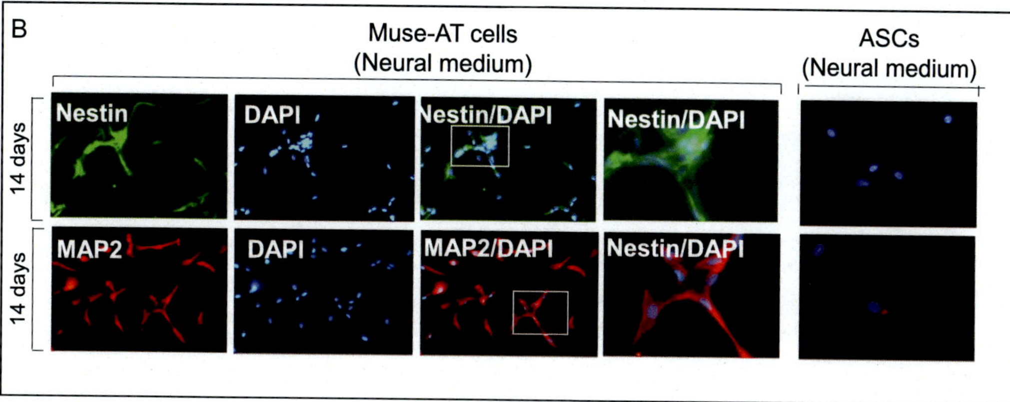

FIG. 5 Adipose-derived Muse (Muse-AT) cell differentiation to neural-like cells. (A) Isolated Muse-AT cells were grown as nonadherent cells in the presence of Neurobasal medium/B-27 supplement serum-free/kanamycin/glutamine/bFGF and EGF for 7 days. Formation of neural cell spheres was detected. These neural cell spheres were then grown as adherent cells for an additional 7 days in DMEM 2% FCS/bFGF/BDNF. Newly formed neural-like cells are indicated by red arrows in micrographs. (B) Newly formed neural-like cells at 14 days stained positive for Nestin and MAP2 immunofluorescent markers. Comparatively, adipose stem cells (ASCs) stained negative for both markers. *Reproduced from Fig. 5B and C in Heneidi, S., Simerman, A.A., Keller, E., Singh, P., Li, X., Dumesic, D.A., Chazenbalk, G., Awakened by cellular stress: isolation and characterization of a novel population of pluripotent stem cells derived from human adipose tissue, PLoS One 8 (2013) e64752. https://doi.org/10.1371/journal.pone.0064752 under CC-BY license.*

injured rats and produced beneficial substances such as keratinocyte growth factor (KGF), hepatocyte growth factor, angiopoietin-1, and prostaglandin E2 [64]. These studies confirm that Muse cells are able to successfully ameliorate lung ischemia in animal models via pleiotropic effects and suggest the potential of Muse cells to treat lung injuries in the future.

Increased Muse cell numbers in peripheral blood of acute myocardial infarction patients has also been shown to be associated with improved cardiac function [61]. Subsequently, the therapeutic efficacy of Muse cells to repair aortic aneurysms was evaluated by intravenous injection of Muse cells into mice and was found to attenuate aneurysm dilation [57].

Further studies investigating the mechanisms of this phenomenon in rabbits have revealed the role of the sphingosine monophosphate system, specifically the S1P–S1PR2 axis, with S1P being one of the most potent signaling sphingolipids that govern the formation and development of multiple organ systems, in mediating the homing of Muse cells into damaged cardiac tissue as opposed to the SDF-1-CXCR3 system as the major player in mediating MSC migration [65]. In conjunction with previous mechanistic studies, this mechanism may be occurring alongside upregulation of CXCL2 expression [3, 40], a chemokine important in stem cell homing, and increased TGF-β action over immune cells to exhibit immunomodulatory properties [4].

Successful animal models led to further studies in the application of Muse cells in human clinical therapy and extension into other clinical areas. In January 2019, the Life Science Institute, headed by one of Japan's largest enterprises, Mitsubishi Corporation, officially began an exploratory clinical trial on the "Muse cell-based product" CL2020, a product claimed to be "Muse cell rich," to treat acute myocardiac infarction and ischemic stroke, and expects to obtain approval for commercial manufacturing by the year 2021 [67]. Concurrently, research areas in musculoskeletal regeneration, a field that has thus far not yet evaluated the potential of Muse cells in developing therapy, have also been considering the possible use of Muse cells as a novel and safe cell source to treat musculoskeletal disorders, although there remains the need for additional studies to evaluate its efficacy [68].

Most studies on Muse cells thus far have used Muse cells derived from MSCs, most commonly from BM and fibroblasts, but Muse-AT cells are easily accessible from lipoaspirate material and can be obtained at a low cost after a simple procedure. Most importantly, unlike other well-established pluripotent stem cells, both Muse and Muse-AT cells have the capacity to differentiate into all three germ layers without teratoma formation, making them a promising candidate for use in stem cell therapy and tissue regeneration.

Skepticism behind adipose-derived Muse cells

The existence of Muse and Muse-AT cells and its considerable promise has been demonstrated in numerous studies over the several years since their discovery. Yet, there exist others who remain skeptical about these cells today. One reason may be because Muse-AT cells challenge the existing paradigm of ES and iPS cells as gold-standard pluripotent cell populations by introducing a new pluripotent cell population that exists in adult tissues. Another reason may be confusion of Muse-AT cells with other proclaimed pluripotent adult stem cells, such as multipotent adult progenitor cells (MAPCs) [69], very small embryonic-like stem cells (VSELs) [70], or stimulus-triggered acquisition of pluripotency (STAP) cells. While the similar claims of pluripotent qualities among these cells may convolute the differences between Muse-AT cells and these cell types, Muse-AT cells differ in terms of isolation methodology, functional characteristics, and potency, as well as having substantially more supporting evidence for their existence and capacity.

MAPCs are among one of the cell types critics have compared Muse and Muse-AT cells to. MAPCs were discovered in 2001 from BM and found to differentiate into ectodermal, mesodermal, and endodermal cell types [69, 71]. When MAPCs derived from rodents were injected into mouse embryos, they contributed to multiple somatic tissues in the mouse, but these findings were not replicated in studies completed with human MAPCs [71]. In fact, when culturing MAPCs from humans in vitro, the expansion potential was found to decrease as compared with studies done with MAPCs derived from mice [71]. Despite the original authors' mention that the results could be affected by different culture conditions between the two MAPC types, it seems that further studies need to be done on MAPCs to confirm their levels of pluripotency in humans, whereas pluripotent qualities in Muse cells have been heavily demonstrated in both animal and human studies.

In 2008, another population of pluripotent cells was found in 0.01% of BM, named VSELs [70]. These cells were isolated solely by cell size and immunophenotype of few markers, without evaluating the formal pluripotency criteria, leading some researchers to question their existence owing to possibilities of artifactual expression as well as challenge their true pluripotency [72]. While subsequent studies also found these cells in other tissues, other groups have had difficulty isolating these cells even after performing numerous experiments, and instead demonstrated that these cells may be multipotent rather than pluripotent [72]. In addition, there has been evidence of aneuploidy in VSEL lines [73], whereas there has been no evidence showing mutation or chromosomal damage in Muse cells to date.

Presumably the most controversial population of adult stem cells to date has been STAP cells, which made global headlines in 2014 for its discovery followed by its irreproducibility, ultimately leading to retraction of its original publication in Nature. The isolation method of Muse cells by severe cellular stress has been compared with that assumed for STAP cells [74, 75] by some scientists, but this comparison ultimately deems invalid as it was later reported that there may have been embryonic cell contamination in the original STAP cell study and that STAP cells in fact do not exist. In contrast, the existence, properties, and reproducibility of Muse cells have been confirmed by numerous studies performed by seven independent groups [48], drawing evidence from both in vitro and in vivo studies.

Many critics of Muse and Muse-AT cells base their skepticism on their misleading relationship to other cell populations that are unclear in their existence and properties, but Muse and Muse-AT cells in actuality are a separate entity from these cells.

Conclusions

Over the last decades, regenerative medicine has been a growing trend in the scientific research arena. With unbelievable potential to fully heal damaged tissues and organs, as well as offer solutions to diseases and conditions that are currently beyond repair, many scientists around the world are investigating the elusive mechanisms underlying stem cell regulation of growth and regrowth, the heart of regenerative medicine. Without doubt, many challenges and debates have risen throughout the course of research using ES and iPS cells as the gold-standard pluripotent cell populations in the field, including bioethical issues with using ES cells and artificial genetic manipulation with iPS cells. Moreover, both cell types exhibit tumorigenesis in vivo, further restraining the feasibility of ES and iPS cells in clinical use.

Muse and Muse-AT cells serve as novel adult pluripotent stem cell populations that counter these issues while having a regenerative capacity in vivo, rendering the cells an ideal candidate for use in disease treatment. While pluripotency of ES and iPS cells is partially defined by their ability to form teratomas in vivo, there is an overwhelming amount of evidence supporting the pluripotent differentiation capacity of Muse cells. Furthermore, Muse cells uniquely have an innate capacity to home to damaged tissue, something not present in any other pluripotent stem cell populations. Since their discovery, Muse cells have repeatedly exhibited promising qualities and translational relevance for disease treatment not evident in other pluripotent stem cells, making Muse cells a very promising candidate in the field of regenerative medicine that will promote innovation, growth, and patient safety.

Acknowledgments

Supported in part by the Eunice Kennedy Shriver National Institute of Child Health & Human Development and the National Institutes of Health through the cooperative agreement U54 HD071836.

References

[1] M. Dezawa, The muse cell discovery, thanks to wine and science, in: M. Dezawa (Ed.), Muse Cells: Endogenous Reparative Pluripotent Stem Cells, Advances in Experimental Medicine and Biology, Springer Japan, Tokyo, 2018, pp. 1–11. https://doi.org/10.1007/978-4-431-56847-6_1.

[2] Y. Kuroda, S. Wakao, M. Kitada, T. Murakami, M. Nojima, M. Dezawa, Isolation, culture and evaluation of multilineage-differentiating stress-enduring (muse) cells, Nat. Protoc. 8 (2013) 1391–1415. https://doi.org/10.1038/nprot.2013.076.

[3] S. Heneidi, A.A. Simerman, E. Keller, P. Singh, X. Li, D.A. Dumesic, G. Chazenbalk, Awakened by cellular stress: isolation and characterization of a novel population of pluripotent stem cells derived from human adipose tissue, PLoS One 8 (2013), e64752. https://doi.org/10.1371/journal.pone.0064752.

[4] M.L. Gimeno, F. Fuertes, A.E. Barcala Tabarrozzi, A.I. Attorressi, R. Cucchiani, L. Corrales, T.C. Oliveira, M.C. Sogayar, L. Labriola, R.A. Dewey, M.J. Perone, Pluripotent nontumorigenic adipose tissue-derived muse cells have immunomodulatory capacity mediated by transforming growth factor-β1, Stem Cells Transl. Med. 6 (2017) 161–173. https://doi.org/10.5966/sctm.2016-0014.

[5] F. Arai, A. Hirao, M. Ohmura, H. Sato, S. Matsuoka, K. Takubo, K. Ito, G.Y. Koh, T. Suda, Tie_2/angiopoietin-1 signaling regulates hematopoietic stem cell quiescence in the bone marrow niche, Cell 118 (2004) 149–161. https://doi.org/10.1016/j.cell.2004.07.004.

[6] T. Cheng, N. Rodrigues, H. Shen, Y. Yang, D. Dombkowski, M. Sykes, D.T. Scadden, Hematopoietic stem cell quiescence maintained by p21cip1/waf1, Science 287 (2000) 1804–1808. https://doi.org/10.1126/science.287.5459.1804.

[7] Y. Kunisaki, I. Bruns, C. Scheiermann, J. Ahmed, S. Pinho, D. Zhang, T. Mizoguchi, Q. Wei, D. Lucas, K. Ito, J.C. Mar, A. Bergman, P.S. Frenette, Arteriolar niches maintain haematopoietic stem cell quiescence, Nature 502 (2013) 637–643. https://doi.org/10.1038/nature12612.

[8] L. Li, H. Clevers, Coexistence of quiescent and active adult stem cells in mammals, Science 327 (2010) 542–545. https://doi.org/10.1126/science.1180794.

[9] T. Umemoto, M. Yamato, K. Nishida, J. Yang, Y. Tano, T. Okano, Limbal epithelial side-population cells have stem cell-like properties, including quiescent state, Stem Cells Dayt. Ohio 24 (2006) 86–94. https://doi.org/10.1634/stemcells.2005-0064.

[10] P.E. Scherer, Adipose tissue: from lipid storage compartment to endocrine organ, Diabetes 55 (2006) 1537–1545. https://doi.org/10.2337/db06-0263.

[11] M. Blüher, Adipose tissue dysfunction contributes to obesity related metabolic diseases, Best Pract. Res. Clin. Endocrinol. Metab. Comp. Obes. 27 (2013) 163–177. https://doi.org/10.1016/j.beem.2013.02.005.

[12] C.R. Balistreri, C. Caruso, G. Candore, The role of adipose tissue and Adipokines in obesity-related inflammatory diseases [WWW document], Mediat. Inflamm. (2010). https://doi.org/10.1155/2010/802078.

[13] S. Gesta, Y.-H. Tseng, C.R. Kahn, Developmental origin of fat: tracking obesity to its source, Cell 131 (2007) 242–256. https://doi.org/10.1016/j.cell.2007.10.004.

[14] G. Chazenbalk, C. Bertolotto, S. Heneidi, M. Jumabay, B. Trivax, J. Aronowitz, K. Yoshimura, C.F. Simmons, D.A. Dumesic, R. Azziz, Novel pathway of adipogenesis through cross-talk between adipose tissue macrophages,

adipose stem cells and adipocytes: evidence of cell plasticity, PLoS One 6 (2011). https://doi.org/10.1371/journal.pone.0017834.

[15] J.K. Fraser, R. Schreiber, B. Strem, M. Zhu, Z. Alfonso, I. Wulur, M.H. Hedrick, Plasticity of human adipose stem cells toward endothelial cells and cardiomyocytes, Nat. Clin. Pract. Cardiovasc. Med. 3 (Suppl 1) (2006) S33–S37. https://doi.org/10.1038/ncpcardio0444.

[16] V. Pellegrinelli, S. Carobbio, A. Vidal-Puig, Adipose tissue plasticity: how fat depots respond differently to pathophysiological cues, Diabetologia 59 (2016) 1075–1088. https://doi.org/10.1007/s00125-016-3933-4.

[17] P. Schling, G. Löffler, Cross talk between adipose tissue cells: impact on pathophysiology, Physiology 17 (2002) 99–104. https://doi.org/10.1152/nips.01349.2001.

[18] G. Charrière, B. Cousin, E. Arnaud, M. André, F. Bacou, L. Penicaud, L. Casteilla, Preadipocyte conversion to macrophage. Evidence of plasticity, J. Biol. Chem. 278 (2003) 9850–9855. https://doi.org/10.1074/jbc.M210811200.

[19] D.-T. Chu, T. Nguyen Thi Phuong, N.L.B. Tien, D.K. Tran, L.B. Minh, V.V. Thanh, P. Gia Anh, V.H. Pham, V. Thi Nga, Adipose tissue stem cells for therapy: an update on the progress of isolation, culture, storage, and clinical application, J. Clin. Med. 8 (2019) 917. https://doi.org/10.3390/jcm8070917.

[20] P.A. Zuk, M. Zhu, H. Mizuno, J. Huang, J.W. Futrell, A.J. Katz, P. Benhaim, H.P. Lorenz, M.H. Hedrick, Multilineage cells from human adipose tissue: implications for cell-based therapies, Tissue Eng. 7 (2001) 211–228. https://doi.org/10.1089/107632701300062859.

[21] A.C. Boquest, A. Shahdadfar, J.E. Brinchmann, P. Collas, Isolation of stromal stem cells from human adipose tissue, Methods Mol. Biol. Clifton NJ 325 (2006) 35–46. https://doi.org/10.1385/1-59745-005-7:35.

[22] J. Gimble, F. Guilak, Adipose-derived adult stem cells: isolation, characterization, and differentiation potential, Cytotherapy 5 (2003) 362–369. https://doi.org/10.1080/14653240310003026.

[23] K.C. Hicok, M.H. Hedrick, Automated isolation and processing of adipose-derived stem and regenerative cells, Methods Mol. Biol. Clifton NJ 702 (2011) 87–105. https://doi.org/10.1007/978-1-61737-960-4_8.

[24] A. Scherberich, N.D. Di Maggio, K.M. McNagny, A familiar stranger: CD34 expression and putative functions in SVF cells of adipose tissue, World J. Stem Cells 5 (2013) 1–8. https://doi.org/10.4252/wjsc.v5.i1.1.

[25] M. Dominici, K. Le Blanc, I. Mueller, I. Slaper-Cortenbach, F. Marini, D. Krause, R. Deans, A. Keating, D. Prockop, E. Horwitz, Minimal criteria for defining multipotent mesenchymal stromal cells. The International Society for Cellular Therapy position statement, Cytotherapy 8 (2006) 315–317. https://doi.org/10.1080/14653240600855905.

[26] D. Legzdina, A. Romanauska, S. Nikulshin, T. Kozlovska, U. Berzins, Characterization of senescence of culture-expanded human adipose-derived mesenchymal stem cells, Int. J. Stem Cells 9 (2016) 124–136. https://doi.org/10.15283/ijsc.2016.9.1.124.

[27] W. Wagner, F. Wein, A. Seckinger, M. Frankhauser, U. Wirkner, U. Krause, J. Blake, C. Schwager, V. Eckstein, W. Ansorge, A.D. Ho, Comparative characteristics of mesenchymal stem cells from human bone marrow, adipose tissue, and umbilical cord blood, Exp. Hematol. 33 (2005) 1402–1416. https://doi.org/10.1016/j.exphem.2005.07.003.

[28] M.J. Eagan, P.A. Zuk, K.-W. Zhao, B.E. Bluth, E.J. Brinkmann, B.M. Wu, D.R. McAllister, The suitability of human adipose-derived stem cells for the engineering of ligament tissue, J. Tissue Eng. Regen. Med. 6 (2012) 702–709. https://doi.org/10.1002/term.474.

[29] K.-M. Jang, H.C. Lim, J.H. Bae, Mesenchymal stem cells for enhancing biologic healing after anterior cruciate ligament injuries, Curr. Stem Cell Res. Ther. 10 (2015) 535–547. https://doi.org/10.2174/1574888x10666150528153025.

[30] S.-K. Choi, J.-K. Park, J.-H. Kim, K.-M. Lee, E. Kim, K.-S. Jeong, W.B. Jeon, Integrin-binding elastin-like polypeptide as an in situ gelling delivery matrix enhances the therapeutic efficacy of adipose stem cells in healing full-thickness cutaneous wounds, J. Control. Release Off. J. Control. Release Soc. 237 (2016) 89–100. https://doi.org/10.1016/j.jconrel.2016.07.006.

[31] R.A. Sabol, A.C. Bowles, A. Côté, R. Wise, N. Pashos, B.A. Bunnell, Therapeutic potential of adipose stem cells, Adv. Exp. Med. Biol. (2018). https://doi.org/10.1007/5584_2018_248.

[32] H.S. Choi, H.J. Kim, J.-H. Oh, H.-G. Park, J.C. Ra, K.-A. Chang, Y.-H. Suh, Therapeutic potentials of human adipose-derived stem cells on the mouse model of Parkinson's disease, Neurobiol. Aging 36 (2015) 2885–2892. https://doi.org/10.1016/j.neurobiolaging.2015.06.022.

[33] A. Gugliandolo, P. Bramanti, E. Mazzon, Mesenchymal stem cells: a potential therapeutic approach for amyotrophic lateral sclerosis? Stem Cells Int. 2019 (2019) 3675627. https://doi.org/10.1155/2019/3675627.

[34] N.P. Staff, N.N. Madigan, J. Morris, M. Jentoft, E.J. Sorenson, G. Butler, D. Gastineau, A. Dietz, A.J. Windebank, Safety of intrathecal autologous adipose-derived mesenchymal stromal cells in patients with ALS, Neurology 87 (2016) 2230–2234. https://doi.org/10.1212/WNL.0000000000003359.
[35] N.S. Kappy, S. Chang, W.M. Harris, M. Plastini, T. Ortiz, P. Zhang, J.P. Hazelton, J.P. Carpenter, S.A. Brown, Human adipose-derived stem cell treatment modulates cellular protection in both in vitro and in vivo traumatic brain injury models, J. Trauma Acute Care Surg. 84 (2018) 745–751. https://doi.org/10.1097/TA.0000000000001770.
[36] N. Tajiri, S.A. Acosta, M. Shahaduzzaman, H. Ishikawa, K. Shinozuka, M. Pabon, D. Hernandez-Ontiveros, D.W. Kim, C. Metcalf, M. Staples, T. Dailey, J. Vasconcellos, G. Franyuti, L. Gould, N. Patel, D. Cooper, Y. Kaneko, C.V. Borlongan, P.C. Bickford, Intravenous transplants of human adipose-derived stem cell protect the brain from traumatic brain injury-induced neurodegeneration and motor and cognitive impairments: cell graft biodistribution and soluble factors in young and aged rats, J. Neurosci. 34 (2014) 313–326. https://doi.org/10.1523/JNEUROSCI.2425-13.2014.
[37] L. Li, R. Bhatia, Stem cell quiescence, Clin. Cancer Res. Off. J. Am. Assoc. Cancer Res. 17 (2011) 4936–4941. https://doi.org/10.1158/1078-0432.CCR-10-1499.
[38] G. Chamberlain, J. Fox, B. Ashton, J. Middleton, Concise review: mesenchymal stem cells: their phenotype, differentiation capacity, immunological features, and potential for homing, Stem Cells Dayt. Ohio 25 (2007) 2739–2749. https://doi.org/10.1634/stemcells.2007-0197.
[39] F. Paino, M. La Noce, D. Di Nucci, G.F. Nicoletti, R. Salzillo, A. De Rosa, G.A. Ferraro, G. Papaccio, V. Desiderio, V. Tirino, Human adipose stem cell differentiation is highly affected by cancer cells both in vitro and in vivo: implication for autologous fat grafting, Cell Death Dis. 8 (2017), e2568. https://doi.org/10.1038/cddis.2016.308.
[40] A.A. Simerman, D.A. Dumesic, G.D. Chazenbalk, Pluripotent muse cells derived from human adipose tissue: a new perspective on regenerative medicine and cell therapy, Clin. Transl. Med. 3 (2014) 12. https://doi.org/10.1186/2001-1326-3-12.
[41] J.K. Henderson, J.S. Draper, H.S. Baillie, S. Fishel, J.A. Thomson, H. Moore, P.W. Andrews, Preimplantation human embryos and embryonic stem cells show comparable expression of stage-specific embryonic antigens, Stem Cells Dayt. Ohio 20 (2002) 329–337. https://doi.org/10.1634/stemcells.20-4-329.
[42] C.T. Scott, D. Magnus, Wrongful termination: lessons from the Geron clinical trial, Stem Cells Transl. Med. 3 (2014) 1398–1401. https://doi.org/10.5966/sctm.2014-0147.
[43] K. Garber, RIKEN suspends first clinical trial involving induced pluripotent stem cells, Nat. Biotechnol. 33 (2015) 890–891. https://doi.org/10.1038/nbt0915-890.
[44] J.E. Thornton, R.I. Gregory, How does Lin28 let-7 control development and disease? Trends Cell Biol. 22 (2012) 474–482. https://doi.org/10.1016/j.tcb.2012.06.001.
[45] S. Wakao, M. Kitada, Y. Kuroda, T. Shigemoto, D. Matsuse, H. Akashi, Y. Tanimura, K. Tsuchiyama, T. Kikuchi, M. Goda, T. Nakahata, Y. Fujiyoshi, M. Dezawa, Multilineage-differentiating stress-enduring (muse) cells are a primary source of induced pluripotent stem cells in human fibroblasts, Proc. Natl. Acad. Sci. U. S. A. 108 (2011) 9875–9880. https://doi.org/10.1073/pnas.1100816108.
[46] S.M. Melief, S.B. Geutskens, W.E. Fibbe, H. Roelofs, Multipotent stromal cells skew monocytes towards an anti-inflammatory interleukin-10-producing phenotype by production of interleukin-6, Haematologica 98 (2013) 888–895. https://doi.org/10.3324/haematol.2012.078055.
[47] H.-H. Hu, D.-Q. Chen, Y.-N. Wang, Y.-L. Feng, G. Cao, N.D. Vaziri, Y.-Y. Zhao, New insights into TGF-β/Smad signaling in tissue fibrosis, Chem. Biol. Interact. 292 (2018) 76–83. https://doi.org/10.1016/j.cbi.2018.07.008.
[48] S.C. Fisch, M.L. Gimeno, J.D. Phan, A.A. Simerman, D.A. Dumesic, M.J. Perone, G.D. Chazenbalk, Pluripotent nontumorigenic multilineage differentiating stress enduring cells (muse cells): a seven-year retrospective, Stem Cell Res Ther 8 (2017) 227. https://doi.org/10.1186/s13287-017-0674-3.
[49] D.E. Wagner, I.E. Wang, P.W. Reddien, Clonogenic neoblasts are pluripotent adult stem cells that underlie planarian regeneration, Science 332 (2011) 811–816. https://doi.org/10.1126/science.1203983.
[50] S. Chera, L. Ghila, K. Dobretz, Y. Wenger, C. Bauer, W. Buzgariu, J.-C. Martinou, B. Galliot, Apoptotic cells provide an unexpected source of Wnt3 signaling to drive hydra head regeneration, Dev. Cell 17 (2009) 279–289. https://doi.org/10.1016/j.devcel.2009.07.014.
[51] D. Wenemoser, P.W. Reddien, Planarian regeneration involves distinct stem cell responses to wounds and tissue absence, Dev. Biol. 344 (2010) 979–991. https://doi.org/10.1016/j.ydbio.2010.06.017.

[52] S.J. Odelberg, Cellular plasticity in vertebrate regeneration, Anat. Rec. B. New Anat. 287 (2005) 25–35. https://doi.org/10.1002/ar.b.20080.

[53] J. Baguñà, Mitosis in the intact and regenerating planarian Dugesia mediterranea n.sp. I. Mitotic studies during growth, feeding and starvation, J. Exp. Zool. 195 (1976) 53–64. https://doi.org/10.1002/jez.1401950106.

[54] N.J. Oviedo, P.A. Newmark, A. Sánchez Alvarado, Allometric scaling and proportion regulation in the freshwater planarian Schmidtea mediterranea, Dev. Dyn. Off. Publ. Am. Assoc. Anat. 226 (2003) 326–333. https://doi.org/10.1002/dvdy.10228.

[55] D. Kültz, Molecular and evolutionary basis of the cellular stress response, Annu. Rev. Physiol. 67 (2005) 225–257. https://doi.org/10.1146/annurev.physiol.67.040403.103635.

[56] E. Hori, Y. Hayakawa, T. Hayashi, S. Hori, S. Okamoto, T. Shibata, M. Kubo, Y. Horie, M. Sasahara, S. Kuroda, Mobilization of pluripotent multilineage-differentiating stress-enduring cells in ischemic stroke, J. Stroke Cerebrovasc. Dis. Off. J. Natl. Stroke Assoc. 25 (2016) 1473–1481. https://doi.org/10.1016/j.jstrokecerebrovasdis.2015.12.033.

[57] K. Hosoyama, S. Wakao, Y. Kushida, F. Ogura, K. Maeda, O. Adachi, S. Kawamoto, M. Dezawa, Y. Saiki, Intravenously injected human multilineage-differentiating stress-enduring cells selectively engraft into mouse aortic aneurysms and attenuate dilatation by differentiating into multiple cell types, J. Thorac. Cardiovasc. Surg. 155 (2018) 2301–2313.e4. https://doi.org/10.1016/j.jtcvs.2018.01.098.

[58] M. Iseki, Y. Kushida, S. Wakao, T. Akimoto, M. Mizuma, F. Motoi, R. Asada, S. Shimizu, M. Unno, G. Chazenbalk, M. Dezawa, Muse cells, nontumorigenic pluripotent-like stem cells, have liver regeneration capacity through specific homing and cell replacement in a mouse model of liver fibrosis, Cell Transplant. 26 (2017) 821–840. https://doi.org/10.3727/096368916X693662.

[59] H. Katagiri, Y. Kushida, M. Nojima, Y. Kuroda, S. Wakao, K. Ishida, F. Endo, K. Kume, T. Takahara, H. Nitta, H. Tsuda, M. Dezawa, S.S. Nishizuka, A distinct subpopulation of bone marrow mesenchymal stem cells, muse cells, directly commit to the replacement of liver components, Am. J. Transplant. Off. J. Am. Soc. Transplant. Am. Soc. Transpl. Surg. 16 (2016) 468–483. https://doi.org/10.1111/ajt.13537.

[60] N. Shimamura, K. Kakuta, L. Wang, M. Naraoka, H. Uchida, S. Wakao, M. Dezawa, H. Ohkuma, Neuro-regeneration therapy using human muse cells is highly effective in a mouse intracerebral hemorrhage model, Exp. Brain Res. 235 (2017) 565–572. https://doi.org/10.1007/s00221-016-4818-y.

[61] T. Tanaka, K. Nishigaki, S. Minatoguchi, T. Nawa, Y. Yamada, H. Kanamori, A. Mikami, H. Ushikoshi, M. Kawasaki, M. Dezawa, S. Minatoguchi, Mobilized muse cells after acute myocardial infarction predict cardiac function and remodeling in the chronic phase, Circ. J. Off. J. Jpn. Circ. Soc. 82 (2018) 561–571. https://doi.org/10.1253/circj.CJ-17-0552.

[62] H. Uchida, T. Morita, K. Niizuma, Y. Kushida, Y. Kuroda, S. Wakao, H. Sakata, Y. Matsuzaka, H. Mushiake, T. Tominaga, C.V. Borlongan, M. Dezawa, Transplantation of unique subpopulation of fibroblasts, muse cells, ameliorates experimental stroke possibly via robust neuronal differentiation, Stem Cells Dayt. Ohio 34 (2016) 160–173. https://doi.org/10.1002/stem.2206.

[63] N. Uchida, Y. Kushida, M. Kitada, S. Wakao, N. Kumagai, Y. Kuroda, Y. Kondo, Y. Hirohara, S. Kure, G. Chazenbalk, M. Dezawa, Beneficial effects of systemically administered human muse cells in adriamycin nephropathy, J. Am. Soc. Nephrol. 28 (2017) 2946–2960. https://doi.org/10.1681/ASN.2016070775.

[64] H. Yabuki, T. Watanabe, H. Oishi, M. Katahira, M. Kanehira, Y. Okada, Muse cells and ischemia-reperfusion lung injury, Adv. Exp. Med. Biol. 1103 (2018) 293–303. https://doi.org/10.1007/978-4-431-56847-6_16.

[65] Y. Yamada, S. Wakao, Y. Kushida, S. Minatoguchi, A. Mikami, K. Higashi, S. Baba, T. Shigemoto, Y. Kuroda, H. Kanamori, M. Amin, M. Kawasaki, K. Nishigaki, M. Taoka, T. Isobe, C. Muramatsu, M. Dezawa, S. Minatoguchi, S1P-S1PR2 Axis mediates homing of muse cells into damaged heart for long-lasting tissue repair and functional recovery after acute myocardial infarction, Circ. Res. 122 (2018) 1069–1083. https://doi.org/10.1161/CIRCRESAHA.117.311648.

[66] T. Yamauchi, Y. Kuroda, T. Morita, H. Shichinohe, K. Houkin, M. Dezawa, S. Kuroda, Therapeutic effects of human multilineage-differentiating stress enduring (MUSE) cell transplantation into infarct brain of mice, PLoS One 10 (2015), e0116009. https://doi.org/10.1371/journal.pone.0116009.

[67] M. Dezawa, Clinical trials of muse cells, Adv. Exp. Med. Biol. 1103 (2018) 305–307. https://doi.org/10.1007/978-4-431-56847-6_17.

[68] X. Pang, N. Burdekin, C. Li, Z. Zheng, An urgent demand for novel, safe cell sources for musculoskeletal regeneration, Med. One 3 (2018). https://doi.org/10.20900/mo.20180010.

[69] M. Reyes, C.M. Verfaillie, Characterization of multipotent adult progenitor cells, a subpopulation of mesenchymal stem cells, Ann. N. Y. Acad. Sci. 938 (2001) 231–233. discussion 233-235 https://doi.org/10.1111/j.1749-6632.2001.tb03593.x.

[70] M.Z. Ratajczak, E.K. Zuba-Surma, M. Wysoczynski, J. Ratajczak, M. Kucia, Very small embryonic-like stem cells: characterization, developmental origin, and biological significance, Exp. Hematol. 36 (2008) 742–751. https://doi.org/10.1016/j.exphem.2008.03.010.

[71] R. Sambathkumar, M. Kumar, C.M. Verfaillie, Chapter 12—multipotent adult progenitor cells, in: A. Atala, R. Lanza, A.G. Mikos, R. Nerem (Eds.), Principles of Regenerative Medicine, third ed., Academic Press, Boston, 2019, pp. 181–190. https://doi.org/10.1016/B978-0-12-809880-6.00012-6.

[72] M. Miyanishi, Y. Mori, J. Seita, J.Y. Chen, S. Karten, C.K.F. Chan, H. Nakauchi, I.L. Weissman, Do pluripotent stem cells exist in adult mice as very small embryonic stem cells? Stem Cell Rep. 1 (2013) 198–208. https://doi.org/10.1016/j.stemcr.2013.07.001.

[73] A. Heider, R. Danova-Alt, D. Egger, M. Cross, R. Alt, Murine and human very small embryonic-like cells: a perspective, Cytom. Part J. Int. Soc. Anal. Cytol. 83 (2013) 72–75. https://doi.org/10.1002/cyto.a.22229.

[74] P. Knoepfler, Nature Yanks Article that Was Actually Advertisement on Controversial Stem Cells [WWW Document], The Niche, 2019. https://ipscell.com/2019/02/nature-yanks-article-that-was-actually-advertisement-on-controversial-stem-cells/. accessed 5.15.20.

[75] P. Knoepfler, Dubious MUSE Cells Are in 4 Japanese Stem Cell Trials [WWW Document], The Niche, 2019. https://ipscell.com/2019/11/dubious-muse-cells-are-in-4-japanese-stem-cell-trials/. accessed 5.15.20.

CHAPTER

7

Adipose stem cell homing and routes of delivery

Ganesh Swaminathan[a], *Yang Qiao*[a,b], *Bhavesh D. Kevadiya*[a], *Lucille A. Bresette*[a], *Daniel D. Liu*[a], *and Avnesh S. Thakor*[a]

[a]Interventional Regenerative Medicine and Imaging Laboratory, Department of Radiology, Stanford University School of Medicine, Palo Alto, CA, United States [b]Department of Interventional Radiology, The University of Texas MD Anderson Cancer Center, Houston, TX, United States

Introduction

This chapter will provide an overview of the current strategies for adipose stem cell (ASC) delivery and homing in regenerative medicine. ASCs, like other types of mesenchymal stem cells (MSC), possess the ability to migrate toward regions of inflammation and function as agents for promoting tissue regeneration and wound healing once they arrive at these regions of inflammation. In response to tissue damage, the local microenvironment creates a chemokinetic signaling gradient that facilitates ASC homing while sometimes also providing signals for their differentiation [1–3]. ASCs can also be modified to migrate toward exogenously produced signals; for example, ASCs labeled with superparamagnetic iron oxide (SPIO) have demonstrated the ability to migrate preferentially toward magnetic arrays in vitro, independently of endogenously secreted chemokinetic signals [4]. Although the use of exogenous signals for ASC targeting is still being developed, this approach may eventually lead to the development of a user-controllable technique to target ASC in vivo such that they can more efficiently reach target sites at sufficient concentrations. To increase the efficacy of delivering ASCs to specific target areas, different routes of administration have also been investigated. Although most preclinical and clinical studies have administered ASCs via an intravenous (IV) injection, alternative routes to increase the efficiency of delivering these cells closer to target areas include intraarterial (IA) [5, 6] and intraparenchymal (i.e., intracerebral/intracerebroventricular [7, 8]), intramyocardial [9], and intraarticular [10, 11] injections with

Scientific Principles of Adipose Stem Cells
https://doi.org/10.1016/B978-0-12-819376-1.00016-0

approaches such as intranasal administration used to bypass issues related to the blood–brain barrier [12]. Additionally, standardization and the institution of good manufacturing practices regarding ASC isolation and culturing will be vital to the establishment of ASC therapy as a legitimate medical therapy.

Homing steps

After ASCs are administered into the body, they travel to damaged tissues along a gradient of chemotactic proteins [13]. Once in the systemic circulation, ASCs migrate to the target site via a systematic process in five distinct steps: (i) tethering and rolling, (ii) activation, (iii) arrest, (iv) transmigration or diapedesis, and (v) migration [14, 15] (Fig. 1). A thorough understanding of the homing system employed by ASCs is essential to best optimize and harness their therapeutic potential.

The first step of homing is *tethering*, which involves binding of CD44 ligands on ASCs to selectins on endothelial cells. Tethering allows ASCs to then "*roll*" along the vessel wall [14]. However, the specific selectin(s) to which ASC ligands bind remain unknown. Given that hematopoietic cell *E*- and L-selectin ligand (HCELL) and the P-selectin glycoprotein ligand-1 (PSGL-1) are not expressed on ASCs, it was initially thought that their corresponding selectins were not involved in ASC tethering and rolling [14]. Interestingly, anti-P-selectin antibodies have been shown to suppress ASC tethering, therefore suggesting that P-selectin may be a viable candidate used to tether ASCs, although the tethering must take place via a ligand

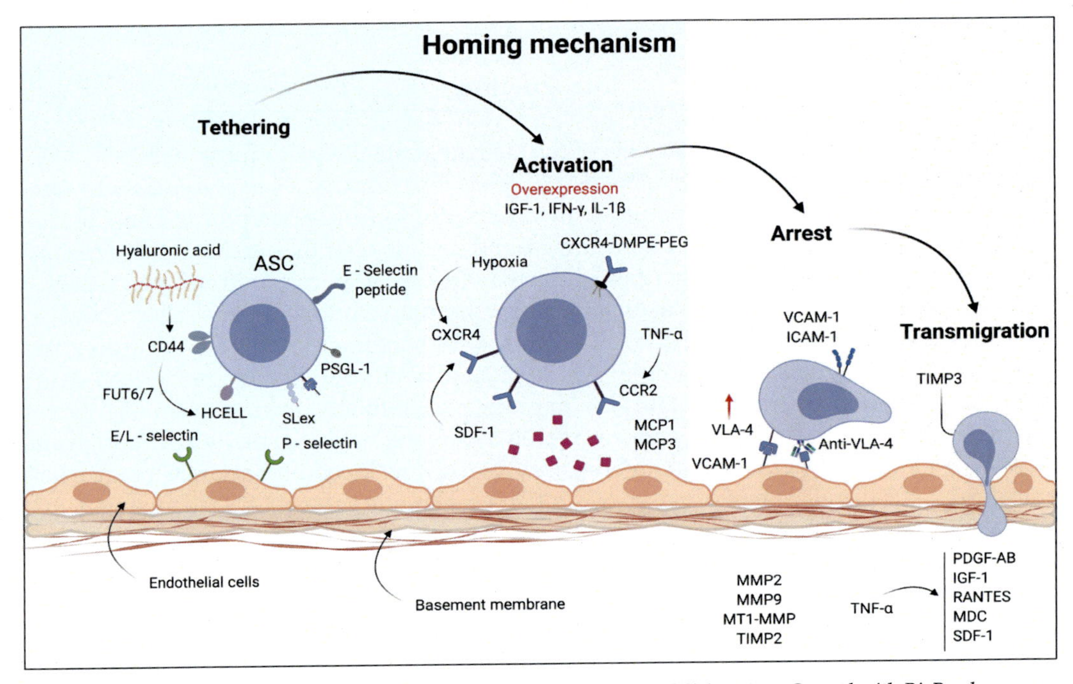

FIG. 1 Schematic showing the different steps in adipose stem cell (ASC) homing. *Created with BioRender.*

other than PSGL-1. Furthermore, ASCs have been shown to employ CD24 as a P-selectin ligand [16]. Recently, Galectin-1, a member of the lectin family, was found to be a ligand for P-selectin. Although Galectin-1 is known to be involved in immune regulation, its role in homing of ASCs needs further investigation [17].

Tethering is followed by *activation* (second step) and *cell arrest* (third step). Both activation and cell arrest processes are coregulated by inflammatory signals. *Activation* is mainly driven by the chemokines and their binding to their receptors. Chemokine receptors are G-protein-coupled receptors present on the cell surface and are capable of stimulating cell migration and homeostasis when activated by binding of chemokines, cytokines that are upregulated for recruitment of immune cells, or stem cells in response to inflammatory or immune signals [18]. Stromal cell-derived factor (SDF)-1, a key chemokine upregulated in inflammation, and vascular cell adhesion molecule (VCAM)-1, a glycoprotein expressed in endothelial cells upon activation by inflammatory cytokines, are key mediators for the activation and migration of ASCs [19–22]. Thus, inflammation triggers increased expression of VCAM-1 and SDF-1. VCAM-1 upregulates the expression of chemokine receptor CXCR4 in ASCs and thereby facilitates the binding of its ligand, SDF-1, while SDF-1 induces adhesion of ASCs onto VCAM-1 via activation of very late antigen 4 (VLA; integrin $\alpha4\beta1$) by mechanisms not completely understood [23–25]. Although overexpression of CXCR4 on ASCs has been shown to increase homing, the presence of CXCR4 on ASCs is contested, and there is a possibility that other receptors such as C-X-C chemokine receptor type 7 (CXCR7), monocyte chemoattractant protein 1 (MCP-1), monocyte chemoattractant protein 3 (MCP-3), C-C motif chemokine receptor 1 (CCR1), C-C motif chemokine receptor 4 (CCR4), C-C motif chemokine receptor 7 (CCR7), C-C motif chemokine receptor 9 (CCR9), C-C motif chemokine receptor 10 (CCR10), C-X-C chemokine receptor type 5 (CXCR5), and C-X-C chemokine receptor type 6 (CXCR6) may also play a role in the activation step [26–28].

The fourth step, *transmigration* or *diapedesis*, is the transcellular migration of ASCs through the endothelial cell layer and underlying basement membrane [25]. This occurs by the release of matrix metalloproteinases (MMPs) by ASCs, which result in the breakdown of the basement membrane. Under homeostatic conditions, the expression and activity of MMPs are regulated by tissue inhibitors of metalloproteinases (TIMPs) that are expressed by various cells, including endothelial cells, platelets, and smooth muscle cells. During inflammation, the surge in inflammatory cytokines promotes increased activity of MMPs by downregulating TIMPs, thus enabling localized tissue remodeling [29]. Thus, studies have demonstrated the role of MMP2, MT1-MMP, and TIMP1 in the migration of MSCs. Selective knockdown of MMP2 and MT1-MMP adversely impacted migration of MSCs in vitro, while knockdown of TIMP1 increased MSC invasion Interestingly, TIMP2 suppression also results in decreased migration, but this observation is due to the role of TIMP2 in the maturation and activation of MMP2 [30, 31]. While we know that the transmigration step is largely facilitated by MMPs, it is likely that this process involves several other proteins.

The fifth and final step of homing is *cell migration* through the interstitium to the site of injury or inflammation. Such migration of ASCs is driven by the chemotactic signals that are released at the site of inflammation. Chemotactic signals are a gradient of cytokines and growth factors such as platelet-derived growth factor (PDGF)-AB, insulin-like growth factor (IGF)-1, hepatocyte growth factor (HGF), and epidermal growth factor. In vitro MSC migration studied by both Ponte et al. [32] and Mishima et al. [33] showed significant influence in migration due to PDGF followed by IGF. Notably, other factors such as vascular endothelial growth factor (VEGF), interleukin-8 (IL8), and bone morphogenetic proteins (BMPs)- 4 and 7 was also found to promote MSC migration [33]. Therefore, this suggests that MSC migration and engraftment at the injury site is largely driven by a myriad of growth factors, which are likely secreted by the tissue-resident cells.

Strategies to enhance homing of ASCs

One of the major challenges in ASC cell therapy is that only a limited number of cells reach the target site. Since most of the cells delivered by IV injection are retained in the lungs, the regenerative outcome is dependent on the few cells (roughly 1%) that localize to the target tissue [34]. This may account for some of the limited efficacy seen when ASCs have been used in clinical trials [35]. Since increasing the overall dose of ASCs for therapy is expensive and requires time for culture expansion, with increased potential for contamination and exposure to xenofactors in serum, studies have aimed at enhancing both the therapeutic and homing potential of ASCs by priming them using certain biomolecules or different culture conditions, or even through genetic modifications of the cell.

Priming

Use of inflammatory cytokines

Priming ASCs using biomolecules is a widely employed approach to improve the efficacy of ASC therapy. By exposing MSCs to inflammatory cytokines, MSCs become activated and exhibit better immunomodulatory and regenerative properties [36]. Several studies have employed interferon gamma (IFNγ) for upregulating the expression of anti-inflammatory factors such as cyclooxygenase (COX-2), transforming growth factor beta (TGFβ), and HGF [36–38], as well as that of immunomodulatory genes such as intracellular adhesion molecule 1 (ICAM1), vascular cell adhesion molecule 1 (VCAM1), human leukocyte antigen DR isotype (HLA-DR), C—C motif chemokine ligand 8 (CCL8), C-X-C motif chemokine ligand 9 (CXCL9), and C-X-C motif chemokine ligand 10 (CXCL10) [39, 40]; many of these biomolecules are also involved in MSC homing. For example, an improved therapeutic response was shown when ASCs were pretreated with IFNγ prior to being administered for treatment in a mouse model of obliterative bronchiolitis [41]. Another inflammatory cytokine used to prime MSCs is TNFα. ASCs preconditioned with TNFα promoted endothelial progenitor cell homing and angiogenesis via an increased release of IL6 and IL8 in a murine ischemic hindlimb model [42]. Mechanistic studies have shown that preconditioning of ASCs with TNFα increased cell proliferation, mobilization, and osteogenic differentiation by the activation of signaling pathways, such as extracellular signal-regulated kinase (ERK) and p38 mitogen-activated protein kinase (MAPK) to increase bone regeneration [43]. The ERK–MAPK signaling pathway is critical for driving differentiation of ASCs and is, thus, involved in MSC fate determination [44, 45]. Similarly, TNFα preconditioning upregulated immunomodulatory factors such as indoleamine 2.3 dioxygenase (IDO), prostaglandin E2 (PGE2), and HGF [46], although the expression levels were much lower compared with after stimulation by IFNγ. Hence, a combinatorial approach has been used by several studies to maximize the effects of both IFNγ and TNFα. This approach has resulted in activated ASCs secreting several key immunomodulatory factors, including PGE2, IL-10, and MCP-1 [47]. Studies have also investigated the effects of combination of other proinflammatory cytokines, such as IL17, TNFα, and IFNγ, and showed that IL17 enhanced the immunomodulatory effects of TNFα and IFNγ in a mouse model [48], whereas treatment with IL1β, IL6, and IL23 was shown to maintain the immunomodulatory effects of ASCs and upregulated the release of TGFβ and IL10, thus promoting regenerative effects [49]. Therefore, preconditioning using cytokines improved the efficacy of ASCs, though more studies are necessary to elucidate the optimal combinations for specific effects.

Hypoxia

Optimal oxygen levels are critical for maintaining cellular homeostasis both in vitro and in vivo. While ASCs are predominantly cultured under 21% O_2 in vitro, upon in vivo administration of ASCs, the cells often engraft at sites that are relatively hypoxic (i.e., around 5% O_2 or less) [50]. Hence, culture conditions that recapitulate the in vivo conditions, which ASCs will end up in, may be important to maintain and leverage their regenerative properties [51, 52]. Several studies have investigated the impact of hypoxia on ASCs and have shown that hypoxia promotes the stemness of ASCs via an increase in the expression of markers of stemness such as Oct4, Sox2, and Nanog [53, 54]. Additionally, ASCs cultured under hypoxia had improved immunomodulatory [55], angiogenic [56, 57], and migratory capacities [58].

When hypoxia-treated ASCs were delivered in vivo, they demonstrated improved viability and engraftment into target tissues [59]. Additionally, ASCs promoted survival of neighboring cells [60], thereby suggesting their promise for therapeutic applications. For instance, hypoxia-treated ASCs enhanced angiogenesis via increased release of growth factors such as vascular endothelial growth factor (VEGF) and HGF in studies involving ischemic lesions [59, 61]. Although hypoxia pretreatment of ASCs can be used as a way of stimulating MSCs, to apply hypoxia as a standard mode of stimulating MSCs requires further research to determine the appropriate oxygen levels relative to which target tissue/condition MSCs are being used to treat. Furthermore, maintenance of quality control during and after hypoxia treatment using advanced culture systems is also needed to ensure their reliability and consistency.

Genetic modification

Genetic modification offers the potential to enhance genes responsible for antiapoptosis, retention, and homing. To prolong survival and stemness of ASCs, targeted overexpression of both Sox2 and Oct4 has been performed, which demonstrated enhanced proliferation and stemness of ASCs [62]. Furthermore, using a vector-independent approach, by transfection of stem cell-specific miR-302, a type of microRNA that functions in posttranscriptional gene silencing, and treatment of leukemia inhibitory factor (LIF), improved proliferation and reduced oxidative stress was achieved [63, 64]. Since excess reactive oxygen species (ROS) can also affect the survival of transplanted cells, studies have also investigated ways to overexpress superoxide dismutase 2 (SOD2), which can eliminate ROS, thereby enabling cells to resist hypoxic stress. Hence, ASCs overexpressing SOD2 have also been shown to have a significant improvement in survival compared with control ASCs both in vitro and in vivo [65]. Since SDF-1 is also upregulated in inflammatory and ischemic microenvironments, stem cells overexpressing its ligand, CXCR4, showed enhanced migration and homing to initiate repair [66, 67]. Similarly, transplantation of ASCs overexpressing granulocyte chemotactic protein 2 (GCP2) in experimental myocardial infarction mouse models was shown to result in reduction in infarct size and improvement in cardiac function [68, 69]. Thus, targeted overexpression of key chemotactic molecules can improve the functioning and migration of MSCs, though additional research will be required to fine-tune the exact molecules and the amount of their overexpression.

Routes of MSC delivery

The route of administration of MSCs for therapy requires several considerations such as the type of organ, pathology, and desired mechanism of action [35, 70, 71]. While preclinical and clinical studies have used different routes of MSC delivery, the most common method is systemic IV injection, though other more directed delivery approaches such as IA or intraparenchymal injections have also been employed (Fig. 2).

Intravenous delivery

IV injection remains the most common choice of delivery because of its ease and limited risk. A meta-analysis of over 750 clinical trials involving MSCs showed IV injections were

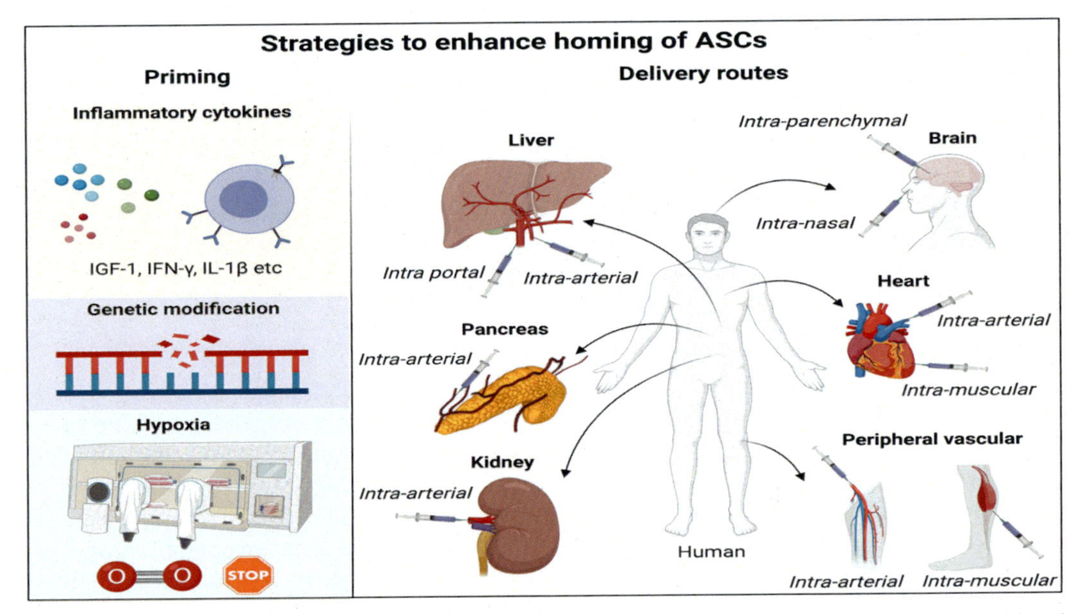

FIG. 2 Schematic showing different strategies for enhancing adipose stem cell (ASC) homing. *Created with BioRender.*

employed in 43% of trials [72]. While IV injections are minimally invasive and allow for multiple injections, biodistribution studies have shown that there is a significant retention of MSCs in lungs in the first 24–48 h following injection [73]. It is believed that the size of MSCs contributes to their entrapment in lungs, as administration of vasodilators can decrease their entrapment [74]. This pulmonary "first-pass" effect can be a major problem in the homing and engraftment of ASCs at the target organ, as seen in studies using these cells to treat patients with osteogenesis imperfecta [75, 76] and graft versus host disease [77]. While many studies have shown a beneficial therapeutic effect following IV injection, this could be attributed to them using a larger dose to compensate for the pulmonary "first-pass" effect. Kean et al. analyzed a total of 276 clinical trials involving MSCs and found 21% of studies had a 30-fold difference in the mean dose per IV injection per person [78], thus highlighting the increased MSC dosage requirements for IV delivery.

Intraarterial delivery

IA injection can overcome some of the limitations associated with IV delivery, notably circumvention of the pulmonary "first-pass" effect which, in turn, will increase their distribution at the target organ. For example, Toupet et al. compared the efficacy of IA versus IV injection of ASCs in mouse models of osteoarthritis [79], and found that 15% of MSCs infused via IA engrafted in inflamed knee joints, while poor engraftment was observed following IV infusion. Similarly, other studies have reported improved efficacy of IA-administered MSCs in the context of kidney injury [80, 81], ischemic stroke [82, 83], cirrhosis [84], and renovascular disease [85]. Another advantage of IA delivery has been the ability to use smaller doses, which, in turn, reduces the risk of cell embolus formation [82, 86, 87] given that cells are administered in vessels with higher flow rates, which results in less stasis and therefore less cell clumping. Despite these advantages, Argibay et al. reported adverse effects such as cerebral lesions induced by MSCs occluding the brain vessels upon IA delivery in rat model of ischemia, which was not observed upon IV delivery [88].

Intraparenchymal delivery

One major advantage of delivering MSCs directly at, or next to, the site of damage is that they can immediately engraft and initiate repair. Indeed, MSCs used in clinical studies for the treatment of cardiac disease, such as acute myocardial infarction (AMI) [89] and ischemic heart failure [90], have been injected using an open transepicardial approach. While the overall efficacy of direct intraparenchymal approaches are still being investigated, a meta-analysis of preclinical and clinical studies in the setting of AMI has shown that transendocardial injections (i.e., IA delivery) led to more favorable outcomes than transepicardial [91]. Challenges associated with this approach include potential tissue injury as well as the promotion of MSC differentiation due to direct injection into the tissue microenvironment [92, 93]. Intramuscular (IM) delivery of MSCs has also been considered as an alternative to IV administration for amyotrophic lateral sclerosis (ALS) [94] and peripheral artery disease including critical limb ischemia [95, 96]; in the latter case, MSCs have been shown to promote angiogenesis and revascularization of ischemic tissue [97]. Importantly, IM delivery of MSCs within skeletal muscle allowed for their retention to over 6 weeks, thereby allowing for sustained therapeutic effects on the target tissue. For instance, Bura et al. reported significantly improved leg pain and reduction in ulcer size in limb ischemia patients following IM delivery of autologous stromal vascular fraction cells derived from abdominal wall fat, which resulted in pain-free walking [96]. Similarly, several other preclinical studies reported improved engraftment of MSCs after IM delivery in animal models of ALS [94], diabetic polyneuropathy [98], and ischemia [99]. As IM delivery has a good safety profile, it could be a safe alternative method for ASC delivery compared with intravascular injection, especially given that it can mitigate any prothrombotic side effects [100].

Current challenges in adipose stem cell homing therapy

Despite using well-established protocols for isolation, culture, maintenance, cryopreservation, and banking, there is a lack of standardization across all centers using ASC therapies. As a result, there is likely to be a large amount of heterogeneity in ASCs used across studies, which will also create inherent variability in their therapeutic and homing potential. To address this issue, assays that directly assess ASC migration toward exogenously produced signals are in the early phases of development. An example of this is demonstrated in the ability of ASCs labeled with SPIO nanoparticles to migrate preferentially toward magnetic arrays in vitro, independently of endogenously secreted chemokinetic signals, which when administered clinically would allow for evaluation of real-time image tracking of targetable ASCs [4]. Such approaches will also allow us to ensure effective targeting of transplanted ASCs, thus helping us to understand the therapeutic efficacy of ASCs in different disease tissues.

References

[1] Z. Lu, S.I. Roohani-Esfahani, H. Zreiqat, Mimicking bone microenvironment for directing adipose tissue-derived mesenchymal stem cells into osteogenic differentiation, Methods Mol. Biol. 1202 (2014) 161–171.

[2] L.S. Litvinova, et al., Secretion of niche signal molecules in conditions of osteogenic differentiation of multipotent mesenchymal stromal cells induced by textured calcium phosphate coating, Biomed. Khim. 65 (4) (2019) 339–346.

[3] Y. Chen, et al., Adiposederived mesenchymal stem cells exhibit tumor tropism and promote tumorsphere formation of breast cancer cells, Oncol. Rep. 41 (4) (2019) 2126–2136.

[4] A.J. El Haj, et al., An in vitro model of mesenchymal stem cell targeting using magnetic particle labelling, J. Tissue Eng. Regen. Med. 9 (6) (2015) 724–733.

[5] Y. Qiao, et al., Magnetic resonance and photoacoustic imaging of brain tumor mediated by mesenchymal stem cell labeled with multifunctional nanoparticle introduced via carotid artery injection, Nanotechnology 29 (16) (2018) 165101.

[6] S.H. Oh, et al., Early neuroprotective effect with lack of long-term cell replacement effect on experimental stroke after intra-arterial transplantation of adipose-derived mesenchymal stromal cells, Cytotherapy 17 (8) (2015) 1090–1103.

[7] K.A. Chang, et al., The therapeutic effects of human adipose-derived stem cells in Alzheimer's disease mouse models, Neurodegener. Dis. 13 (2–3) (2014) 99–102.

[8] F. Ezquer, et al., Intravenous administration of anti-inflammatory mesenchymal stem cell spheroids reduces chronic alcohol intake and abolishes binge-drinking, Sci. Rep. 8 (1) (2018) 4325.

[9] J. Otto Beitnes, et al., Intramyocardial injections of human mesenchymal stem cells following acute myocardial infarction modulate scar formation and improve left ventricular function, Cell Transplant. 21 (8) (2012) 1697–1709.
[10] C.H. Jo, et al., Intra-articular injection of mesenchymal stem cells for the treatment of osteoarthritis of the knee: a proof-of-concept clinical trial, Stem Cells 32 (5) (2014) 1254–1266.
[11] W. Wang, W. Cao, Treatment of osteoarthritis with mesenchymal stem cells, Sci. China Life Sci. 57 (6) (2014) 586–595.
[12] A. Mangraviti, et al., Non-virally engineered human adipose mesenchymal stem cells produce BMP4, target brain tumors, and extend survival, Biomaterials 100 (2016) 53–66.
[13] F. Nitzsche, et al., Concise review: MSC adhesion cascade-insights into homing and transendothelial migration, Stem Cells 35 (6) (2017) 1446–1460.
[14] R. Sackstein, et al., Ex vivo glycan engineering of CD44 programs human multipotent mesenchymal stromal cell trafficking to bone, Nat. Med. 14 (2) (2008) 181–187.
[15] M. Ullah, D.D. Liu, A.S. Thakor, Mesenchymal stromal cell homing: mechanisms and strategies for improvement, iScience 15 (2019) 421–438.
[16] A.M. Bailey, et al., Agent-based model of therapeutic adipose-derived stromal cell trafficking during ischemia predicts ability to roll on P-selectin, PLoS Comput. Biol. 5 (2) (2009), e1000294.
[17] H. Suila, et al., Human umbilical cord blood-derived mesenchymal stromal cells display a novel interaction between P-selectin and galectin-1, Scand. J. Immunol. 80 (1) (2014) 12–21.
[18] S. Francois, et al., Local irradiation not only induces homing of human mesenchymal stem cells at exposed sites but promotes their widespread engraftment to multiple organs: a study of their quantitative distribution after irradiation damage, Stem Cells 24 (4) (2006) 1020–1029.
[19] T.T. Lau, D.A. Wang, Stromal cell-derived factor-1 (SDF-1): homing factor for engineered regenerative medicine, Expert. Opin. Biol. Ther. 11 (2) (2011) 189–197.
[20] R.F. Wynn, et al., A small proportion of mesenchymal stem cells strongly expresses functionally active CXCR4 receptor capable of promoting migration to bone marrow, Blood 104 (9) (2004) 2643–2645.
[21] H. Gao, et al., Activation of signal transducers and activators of transcription 3 and focal adhesion kinase by stromal cell-derived factor 1 is required for migration of human mesenchymal stem cells in response to tumor cell-conditioned medium, Stem Cells 27 (4) (2009) 857–865.
[22] M. Honczarenko, et al., Human bone marrow stromal cells express a distinct set of biologically functional chemokine receptors, Stem Cells 24 (4) (2006) 1030–1041.
[23] V.F. Segers, et al., Mesenchymal stem cell adhesion to cardiac microvascular endothelium: activators and mechanisms, Am. J. Physiol. Heart Circ. Physiol. 290 (4) (2006) H1370–H1377.
[24] B. Ruster, et al., Mesenchymal stem cells display coordinated rolling and adhesion behavior on endothelial cells, Blood 108 (12) (2006) 3938–3944.
[25] C. Steingen, et al., Characterization of key mechanisms in transmigration and invasion of mesenchymal stem cells, J. Mol. Cell. Cardiol. 44 (6) (2008) 1072–1084.
[26] S. Bobis-Wozowicz, et al., Genetically modified adipose tissue-derived mesenchymal stem cells overexpressing CXCR4 display increased motility, invasiveness, and homing to bone marrow of NOD/SCID mice, Exp. Hematol. 39 (6) (2011) 686–696.e4.
[27] Y. Wang, et al., CXCR-7 receptor promotes SDF-1alpha-induced migration of bone marrow mesenchymal stem cells in the transient cerebral ischemia/reperfusion rat hippocampus, Brain Res. 1575 (2014) 78–86.
[28] Y. Shao, et al., Overexpression of CXCR7 promotes mesenchymal stem cells to repair phosgene-induced acute lung injury in rats, Biomed. Pharmacother. 109 (2019) 1233–1239.
[29] C. Ries, et al., MMP-2, MT1-MMP, and TIMP-2 are essential for the invasive capacity of human mesenchymal stem cells: differential regulation by inflammatory cytokines, Blood 109 (9) (2007) 4055–4063.
[30] A. De Becker, et al., Migration of culture-expanded human mesenchymal stem cells through bone marrow endothelium is regulated by matrix metalloproteinase-2 and tissue inhibitor of metalloproteinase-3, Haematologica 92 (4) (2007) 440–449.
[31] H. Will, et al., The soluble catalytic domain of membrane type 1 matrix metalloproteinase cleaves the propeptide of progelatinase A and initiates autoproteolytic activation. Regulation by TIMP-2 and TIMP-3, J. Biol. Chem. 271 (29) (1996) 17119–17123.
[32] A.L. Ponte, et al., The in vitro migration capacity of human bone marrow mesenchymal stem cells: comparison of chemokine and growth factor chemotactic activities, Stem Cells 25 (7) (2007) 1737–1745.

[33] Y. Mishima, M. Lotz, Chemotaxis of human articular chondrocytes and mesenchymal stem cells, J. Orthop. Res. 26 (10) (2008) 1407–1412.
[34] I.M. Barbash, et al., Systemic delivery of bone marrow-derived mesenchymal stem cells to the infarcted myocardium: feasibility, cell migration, and body distribution, Circulation 108 (7) (2003) 863–868.
[35] J. Galipeau, L. Sensebe, Mesenchymal stromal cells: clinical challenges and therapeutic opportunities, Cell Stem Cell 22 (6) (2018) 824–833.
[36] J.M. Ryan, et al., Interferon-gamma does not break, but promotes the immunosuppressive capacity of adult human mesenchymal stem cells, Clin. Exp. Immunol. 149 (2) (2007) 353–363.
[37] J. Croitoru-Lamoury, et al., Interferon-gamma regulates the proliferation and differentiation of mesenchymal stem cells via activation of indoleamine 2,3 dioxygenase (IDO), PLoS One 6 (2) (2011), e14698.
[38] C. Noone, et al., IFN-gamma stimulated human umbilical-tissue-derived cells potently suppress NK activation and resist NK-mediated cytotoxicity in vitro, Stem Cells Dev. 22 (22) (2013) 3003–3014.
[39] M. Yamamoto, et al., Early expression of plasma CCL8 closely correlates with survival rate of acute graft-vs.-host disease in mice, Exp. Hematol. 39 (11) (2011) 1101–1112.
[40] M. Muller, et al., Review: the chemokine receptor CXCR3 and its ligands CXCL9, CXCL10 and CXCL11 in neuroimmunity—a tale of conflict and conundrum, Neuropathol. Appl. Neurobiol. 36 (5) (2010) 368–387.
[41] G. Zheng, et al., Human adipose-derived mesenchymal stem cells alleviate obliterative bronchiolitis in a murine model via IDO, Respir. Res. 18 (1) (2017) 119.
[42] Y.W. Kwon, et al., Tumor necrosis factor-alpha-activated mesenchymal stem cells promote endothelial progenitor cell homing and angiogenesis, Biochim. Biophys. Acta 1832 (12) (2013) 2136–2144.
[43] Z. Lu, et al., Activation and promotion of adipose stem cells by tumour necrosis factor-alpha preconditioning for bone regeneration, J. Cell. Physiol. 228 (8) (2013) 1737–1744.
[44] Y. Mei, et al., miR-21 modulates the ERK–MAPK signaling pathway by regulating SPRY2 expression during human mesenchymal stem cell differentiation, J. Cell. Biochem. 114 (6) (2013) 1374–1384.
[45] F. Ng, et al., PDGF, TGF-beta, and FGF signaling is important for differentiation and growth of mesenchymal stem cells (MSCs): transcriptional profiling can identify markers and signaling pathways important in differentiation of MSCs into adipogenic, chondrogenic, and osteogenic lineages, Blood 112 (2) (2008) 295–307.
[46] S.J. Prasanna, et al., Pro-inflammatory cytokines, IFNgamma and TNFalpha, influence immune properties of human bone marrow and Wharton jelly mesenchymal stem cells differentially, PLoS One 5 (2) (2010), e9016.
[47] R. Domenis, et al., Pro inflammatory stimuli enhance the immunosuppressive functions of adipose mesenchymal stem cells-derived exosomes, Sci. Rep. 8 (1) (2018) 13325.
[48] X. Han, et al., Interleukin-17 enhances immunosuppression by mesenchymal stem cells, Cell Death Differ. 21 (11) (2014) 1758–1768.
[49] A. Pourgholaminejad, et al., The effect of pro-inflammatory cytokines on immunophenotype, differentiation capacity and immunomodulatory functions of human mesenchymal stem cells, Cytokine 85 (2016) 51–60.
[50] G.H. Goossens, E.E. Blaak, Adipose tissue oxygen tension: implications for chronic metabolic and inflammatory diseases, Curr. Opin. Clin. Nutr. Metab. Care 15 (6) (2012) 539–546.
[51] D.M. Panchision, The role of oxygen in regulating neural stem cells in development and disease, J. Cell. Physiol. 220 (3) (2009) 562–568.
[52] A. Mohyeldin, T. Garzon-Muvdi, A. Quinones-Hinojosa, Oxygen in stem cell biology: a critical component of the stem cell niche, Cell Stem Cell 7 (2) (2010) 150–161.
[53] J.R. Choi, et al., Impact of low oxygen tension on stemness, proliferation and differentiation potential of human adipose-derived stem cells, Biochem. Biophys. Res. Commun. 448 (2) (2014) 218–224.
[54] Y. Yamamoto, et al., Low oxygen tension enhances proliferation and maintains stemness of adipose tissue-derived stromal cells, Biores. Open Access 2 (3) (2013) 199–205.
[55] M. Roemeling-van Rhijn, et al., Effects of hypoxia on the immunomodulatory properties of adipose tissue-derived mesenchymal stem cells, Front. Immunol. 4 (2013) 203.
[56] A. Efimenko, et al., Angiogenic properties of aged adipose derived mesenchymal stem cells after hypoxic conditioning, J. Transl. Med. 9 (2011) 10.
[57] L. Liu, et al., Hypoxia preconditioned human adipose derived mesenchymal stem cells enhance angiogenic potential via secretion of increased VEGF and bFGF, Cell Biol. Int. 37 (6) (2013) 551–560.
[58] J.H. Kim, et al., The pivotal role of reactive oxygen species generation in the hypoxia-induced stimulation of adipose-derived stem cells, Stem Cells Dev. 20 (10) (2011) 1753–1761.

[59] X. Hu, et al., Transplantation of hypoxia-preconditioned mesenchymal stem cells improves infarcted heart function via enhanced survival of implanted cells and angiogenesis, J. Thorac. Cardiovasc. Surg. 135 (4) (2008) 799–808.
[60] S.T. Hollenbeck, et al., Tissue engraftment of hypoxic-preconditioned adipose-derived stem cells improves flap viability, Wound Repair Regen. 20 (6) (2012) 872–878.
[61] L. Leroux, et al., Hypoxia preconditioned mesenchymal stem cells improve vascular and skeletal muscle fiber regeneration after ischemia through a Wnt4-dependent pathway, Mol. Ther. 18 (8) (2010) 1545–1552.
[62] S.M. Han, et al., Enhanced proliferation and differentiation of Oct4- and Sox2-overexpressing human adipose tissue mesenchymal stem cells, Exp. Mol. Med. 46 (2014), e101.
[63] J.Y. Kim, et al., MicroRNA-302 induces proliferation and inhibits oxidant-induced cell death in human adipose tissue-derived mesenchymal stem cells, Cell Death Dis. 5 (2014), e1385.
[64] M.F. Taha, et al., Upregulation of pluripotency markers in adipose tissue-derived stem cells by miR-302 and leukemia inhibitory factor, Biomed. Res. Int. 2014 (2014) 941486.
[65] A. Parniczky, et al., Genetic analysis of human chymotrypsin-like elastases 3A and 3B (CELA3A and CELA3B) to assess the role of complex formation between proelastases and procarboxypeptidases in chronic pancreatitis, Int. J. Mol. Sci. 17 (12) (2016) 2148.
[66] M. Janowski, Functional diversity of SDF-1 splicing variants, Cell Adhes. Migr. 3 (3) (2009) 243–249.
[67] H.H. Cho, et al., Overexpression of CXCR4 increases migration and proliferation of human adipose tissue stromal cells, Stem Cells Dev. 15 (6) (2006) 853–864.
[68] M. Kim, et al., CXCR4 overexpression in human adipose tissue-derived stem cells improves homing and engraftment in an animal limb ischemia model, Cell Transplant. 26 (2) (2017) 191–204.
[69] S.W. Kim, et al., Mesenchymal stem cells overexpressing GCP-2 improve heart function through enhanced angiogenic properties in a myocardial infarction model, Cardiovasc. Res. 95 (4) (2012) 495–506.
[70] C.H. Nijboer, et al., Intranasal stem cell treatment as a novel therapy for subarachnoid hemorrhage, Stem Cells Dev. 27 (5) (2018) 313–325.
[71] J.Q. Yin, J. Zhu, J.A. Ankrum, Manufacturing of primed mesenchymal stromal cells for therapy, Nat. Biomed. Eng. 3 (2) (2019) 90–104.
[72] M. Kabat, et al., Trends in mesenchymal stem cell clinical trials 2004-2018: is efficacy optimal in a narrow dose range? Stem Cells Transl. Med. 9 (1) (2020) 17–27.
[73] U.M. Fischer, et al., Pulmonary passage is a major obstacle for intravenous stem cell delivery: the pulmonary first-pass effect, Stem Cells Dev. 18 (5) (2009) 683–692.
[74] J. Gao, et al., The dynamic in vivo distribution of bone marrow-derived mesenchymal stem cells after infusion, Cells Tissues Organs 169 (1) (2001) 12–20.
[75] E.M. Horwitz, et al., Isolated allogeneic bone marrow-derived mesenchymal cells engraft and stimulate growth in children with osteogenesis imperfecta: implications for cell therapy of bone, Proc. Natl. Acad. Sci. U. S. A. 99 (13) (2002) 8932–8937.
[76] E.M. Horwitz, et al., Transplantability and therapeutic effects of bone marrow-derived mesenchymal cells in children with osteogenesis imperfecta, Nat. Med. 5 (3) (1999) 309–313.
[77] L. Fouillard, et al., Infusion of allogeneic-related HLA mismatched mesenchymal stem cells for the treatment of incomplete engraftment following autologous haematopoietic stem cell transplantation, Leukemia 21 (3) (2007) 568–570.
[78] T.J. Kean, et al., MSCs: delivery routes and engraftment, cell-targeting strategies, and immune modulation, Stem Cells Int. 2013 (2013) 13.
[79] K. Toupet, et al., Survival and biodistribution of xenogenic adipose mesenchymal stem cells is not affected by the degree of inflammation in arthritis, PLoS One 10 (1) (2015), e0114962.
[80] F. Togel, et al., Bioluminescence imaging to monitor the in vivo distribution of administered mesenchymal stem cells in acute kidney injury, Am. J. Physiol. Renal Physiol. 295 (1) (2008) F315–F321.
[81] M. Morigi, P. De Coppi, Cell therapy for kidney injury: different options and mechanisms—mesenchymal and amniotic fluid stem cells, Nephron Exp. Nephrol. 126 (2) (2014) 59.
[82] P. Walczak, et al., Dual-modality monitoring of targeted intraarterial delivery of mesenchymal stem cells after transient ischemia, Stroke 39 (5) (2008) 1569–1574.
[83] M. Watanabe, D.R. Yavagal, Intra-arterial delivery of mesenchymal stem cells, Brain Circ. 2 (3) (2016) 114–117.
[84] K.T. Suk, et al., Transplantation with autologous bone marrow-derived mesenchymal stem cells for alcoholic cirrhosis: phase 2 trial, Hepatology 64 (6) (2016) 2185–2197.

[85] A. Saad, et al., Autologous mesenchymal stem cells increase cortical perfusion in renovascular disease, J. Am. Soc. Nephrol. 28 (9) (2017) 2777–2785.
[86] M. Janowski, et al., Cell size and velocity of injection are major determinants of the safety of intracarotid stem cell transplantation, J. Cereb. Blood Flow Metab. 33 (6) (2013) 921–927.
[87] L. Li, et al., Effects of administration route on migration and distribution of neural progenitor cells transplanted into rats with focal cerebral ischemia, an MRI study, J. Cereb. Blood Flow Metab. 30 (3) (2010) 653–662.
[88] B. Argibay, et al., Intraarterial route increases the risk of cerebral lesions after mesenchymal cell administration in animal model of ischemia, Sci. Rep. 7 (2017) 40758.
[89] V. Karantalis, et al., Autologous mesenchymal stem cells produce concordant improvements in regional function, tissue perfusion, and fibrotic burden when administered to patients undergoing coronary artery bypass grafting: the prospective randomized study of mesenchymal stem cell therapy in patients undergoing cardiac surgery (PROMETHEUS) trial, Circ. Res. 114 (8) (2014) 1302–1310.
[90] T.M. Yau, et al., Intramyocardial injection of mesenchymal precursor cells and successful temporary weaning from left ventricular assist device support in patients with advanced heart failure: a randomized clinical trial, JAMA 321 (12) (2019) 1176–1186.
[91] A.J. Kanelidis, et al., Route of delivery modulates the efficacy of mesenchymal stem cell therapy for myocardial infarction: a meta-analysis of preclinical studies and clinical trials, Circ. Res. 120 (7) (2017) 1139–1150.
[92] C. Kan, et al., Microenvironmental factors that regulate mesenchymal stem cells: lessons learned from the study of heterotopic ossification, Histol. Histopathol. 32 (10) (2017) 977–985.
[93] G. Bauer, M. Elsallab, M. Abou-El-Enein, Concise review: a comprehensive analysis of reported adverse events in patients receiving unproven stem cell-based interventions, Stem Cells Transl. Med. 7 (9) (2018) 676–685.
[94] P. Petrou, et al., Safety and clinical effects of mesenchymal stem cells secreting neurotrophic factor transplantation in patients with amyotrophic lateral sclerosis: results of phase 1/2 and 2a clinical trials, JAMA Neurol. 73 (3) (2016) 337–344.
[95] P.K. Gupta, et al., A double blind randomized placebo controlled phase I/II study assessing the safety and efficacy of allogeneic bone marrow derived mesenchymal stem cell in critical limb ischemia, J. Transl. Med. 11 (2013) 143.
[96] A. Bura, et al., Phase I trial: the use of autologous cultured adipose-derived stroma/stem cells to treat patients with non-revascularizable critical limb ischemia, Cytotherapy 16 (2) (2014) 245–257.
[97] S. Hamidian Jahromi, J.E. Davies, Concise review: skeletal muscle as a delivery route for mesenchymal stromal cells, Stem Cells Transl. Med. 8 (5) (2019) 456–465.
[98] J.W. Han, et al., Bone marrow-derived mesenchymal stem cells improve diabetic neuropathy by direct modulation of both angiogenesis and myelination in peripheral nerves, Cell Transplant. 25 (2) (2016) 313–326.
[99] J.R. Beegle, et al., Preclinical evaluation of mesenchymal stem cells overexpressing VEGF to treat critical limb ischemia, Mol. Ther. Methods Clin. Dev. 3 (2016) 16053.
[100] B. Soria-Juan, et al., Cost-effective, safe, and personalized cell therapy for critical limb ischemia in type 2 diabetes mellitus, Front. Immunol. 10 (2019) 1151.

CHAPTER

8

Bioreactors and microphysiological systems for adipose-based pharmacologic screening

Mallory D. Griffin and Rosalyn D. Abbott

Biomedical Engineering, Carnegie Mellon University, Pittsburgh, PA, United States

Introduction

In the pharmaceutical industry, in vivo animal and clinical testing is a critical step to registration of a pharmaceutical with the Food and Drug Administration (FDA). Pharmaceuticals undergo two application processes with the FDA. The first, the investigational new drug (IND) application, occurs after preclinical animal testing [1]. Of the 5000 pharmaceuticals that undergo preclinical evaluation, only 5 will be approved as an IND [2]. This is primarily because of low in vitro to in vivo extrapolation. False positives from in vitro screening results in time-consuming and costly animal studies. In addition, dose–response curves from in vitro data do not successfully predict in vivo dose–response curves. Inaccurate predictions of dose response increase the number and size of trials needed to optimize dosage prior to clinical trials. Should a drug be approved as an IND, it undergoes clinical trials, where the approval rating again remains low. Differences between human and animal physiology cause pharmaceuticals successful in preclinical animal testing to fail in clinical trials. Only one in five drugs tested clinically will have their New Drug Application (NDA) approved [1–3]. The extensive testing combined with high failure rates puts the cost of development of one approved drug over $2.6 billion dollars [4]. Because of this, recent efforts have focused on improving in vitro testing of drugs prior to animal studies.

Improved in vitro models of tissues and organs can reduce time and costs associated with animal studies. In addition, in vitro models reduce ethical concerns surrounding animal studies by improving adherence to the 3Rs, replace, reduce, and refine, developed by Russell and Burch in 1959 [5]. More physiologically complex in vitro models can reduce false positive responses, reduce the size of animal studies by providing more accurate toxicity and dose–response data, or increase the likelihood of success in vivo.

Scientific Principles of Adipose Stem Cells
https://doi.org/10.1016/B978-0-12-819376-1.00011-1

MPSs refer to in vitro models of tissues or organs constructed of single cell-type culture systems [6]. They bridge the gap between standard cell culture techniques and in vivo models by integrating more sophisticated aspects such as three-dimensional (3D) matrices, perfusion culture, vascularization, and organ-specific cellular architecture [6–8]. Bioreactors, usually systems of a larger scale than MPS, refer to tissue culture devices that closely monitor and control environmental and operating conditions (e.g., pH, temperature, pressure, nutrient supply, and waste removal) [9]. As sustainable MPS typically require close maintenance of these variables. The difference between these systems is ambiguous, and we will discuss MPS and bioreactors simultaneously.

Adipose tissue is a highly dynamic endocrine organ with significant influence on health and disease pathology. Its prevalence throughout the body enables close contact with other tissues allowing for local signaling effects on major organs [10–14]. It also has systemic effects via the secretion of hundreds of adipokines [15, 16]. Thus, it is a key pathological organ and promising therapeutic target in a variety of disorders including obesity [17, 18], metabolic syndrome [19, 20], type 2 diabetes [21], cardiovascular disease [22–28], nonalcoholic fatty liver disease [29–32], cancer [33–36], polycystic ovarian syndrome [37, 38], and lipodystrophy disorders [18, 39], among others. In addition, adipose tissue can accumulate hydrophobic compounds, which influences drug and chemical biodistribution and toxicokinetic profiles [40, 41]. Thus, adipose tissue is a crucial organ in understanding absorption/distribution/metabolism/excretion/toxicology (ADMET) data.

MPS and bioreactors provide 3D and dynamic environments that enable mechanotransduction, perfusion for nutrient delivery and waste removal, coculture systems, and spatiotemporal chemical gradients [42]. However, adipose tissue poses unique complications for in vitro culture. Mature adipocytes are terminally differentiated, preventing proliferation, and are nonadherent. Large lipid depots within mature adipocytes also make the cells fragile and prone to rupture. Adipocytes can be differentiated from multiple stem cell sources including induced pluripotent stem cells (iPSCs) [43], embryonic stem cells (ESCs) [44], bone marrow-derived stem cells (BMDSs) [45], and adipose-derived stem cells (ASCs) [46]. However, adipocytes require lengthy differentiation times, which means long-term culture models are needed to achieve a mature phenotype.

Adipose tissue also has specific attributes that are advantageous over other tissue types, including minimally invasive harvesting and the ability to capture discarded samples from voluntary procedures (i.e., liposuction). This also allows for the creation of patient-specific models to capture population differences in lipolysis, glucose uptake, and lipid content in white adipose tissue (WAT) [47]. There are also differences in metabolism and in disease risk based on individual fat depots [48] and patient demographics such as race/ethnicity [49]. Moreover, adipose tissue contains a variety of cell types, including fibroblasts, endothelial cells, immune cells, etc., that can be individually isolated, allowing for coculture models that maintain patient specificity.

Components of adipose-based microphysiological systems and bioreactors

A variety of techniques have been employed in recent years to create adipose-based MPS/bioreactors including organs-on-a-chip, cell sheets, and organoids (Table 1). Here, we define the minimum components of MPS/bioreactors for adipose-based pharmacologic screening:

TABLE 1 Summary of adipose-based microphysiological systems/bioreactors.

System	Type	Cell type	Culture time	Flow rate	References
Membrane mature adipocyte aggregate cultures (MAAC)	Transwell	Primary murine and human adipocytes	14 days	N/A	[140]
Three-dimensional adipose tissue model	Scaffold	Primary hASCs and hESC-mesenchymal stem cells	21 days	N/A	[141]
Sandwiched white adipose tissue (SWAT)	Cell sheets	Primary hWAT and hASCs	53 days	N/A	[142]
Uniform adipose spheroids	Spheroid	Primary and immortalized hASCs and murine ASCs	49 days	N/A	[138]
Selective differentiation of hiPSCs to brown or white adipocytes	Spheroid	hiPSCs	30 days	N/A	[143]
Embryoid body adipogenic differentiation of hiPSCs and hESCs	Spheroid	hiPSCs and hESCs	24 days	N/A	[144]
Adipogenic potential of hiPSC and hESC embryoid bodies	Spheroid	hiPSCs and hESCs	14 days	N/A	[145]
Differentiation of human pluripotent stem cells to white or brown adipocytes	Spheroid	hiPSCs and hESCs	21 days	N/A	[43]
High-efficiency differentiation of human pluripotent stem cells to brown adipocytes	Spheroid	hiPSCs and hESCs	10 days	N/A	[146]
	Spheroid	Primary hASCs			[127]
Two-compartment system simulating insulin-dependent glucose uptake in adipose tissues	Spheroids/ microfluidic device	3T3-L1 preadipocytes	10 days (static) <1 day (perfusion)	3.6 mL/h	[147]
Primary adipose tissue culture in microfluidic interface	Microfluidic device	Primary murine adipocytes	1 day	40 μL/h	[148]
3D-printed template for macro-to-micro interfacing	Microfluidic device	Primary murine adipose tissue explant	1 day	120 μL/h	[149]
μMUX	Microfluidic device	Primary murine adipose tissue explant	7 days (static), 1 day (perfusion)	7.2 mL/h	[150]
Dual-chip microfluidic device	Microfluidic device	3T3-L1 preadipocytes	<1 day	4.8 mL/h	[151]
Reversibly sealed multilayer microfluidic device	Microfluidic device	3T3-L1 preadipocytes	<1 day	480 μL/h	[152]

Continued

TABLE 1 Summary of adipose-based microphysiological systems/bioreactors—cont'd

System	Type	Cell type	Culture time	Flow rate	References
WAT-on-a-chip	Microfluidic device	3T3-L1 preadipocytes	9 days	20 μL/h	[153]
Human WAT-on-a-chip	Microfluidic device	Primary human adipocytes	36 days	40 μL/h	[154]
Microfluidic device for on-line solid-phase extraction (SPE) and mass spectrometry (MS) analysis of secreted metabolites	Microfluidic device	3T3-L1 preadipocytes	2–3 days	45 μL/h	[155]
Reversibly sealed microfluidic cell chamber	Microfluidic device	3T3-L1 preadipocytes	<1 day	480 μL/h to 4.8 mL/h	[156]
Human ex vivo adipose tissue on-chip	Microfluidic device	Primary human adipose tissue	7 days	1.5–36.0 μL/h	[157]
Automated microfluidic large-scale integration platform for differentiating human adult stem cells	Microfluidic device	Primary hASCs	14 days	27.6 μL/h	[158]
Perfusion bioreactor for using 3DKUBE	Microfluidic device	Primary hASCs	14 days	20.4 mL/h	[159]
Perfusion-based cell culture device	Microfluidic device	Primary hASCs	14 days	18–54 mL/h	[96]
Primary murine adipose tissue chip	Microfluidic device	Primary murine ASCs	19 days	5 μL/h	[130]

Abbreviations: ASC, *adipose-derived stem cell;* hASC, *human adipose-derived stem cell;* hESC, *human embryonic stem cell;* hiPSC, *human induced pluripotent stem cell;* hWAT, *human white adipose tissue.*

adipocytes, extracellular matrix (ECM), and static or perfusion culture (Fig. 1). In addition, we will discuss materials for organ-on-a-chip technology and other considerations for increasing sophistication.

Adipocytes

Brown *versus* white adipose tissue

WAT is the most abundant adipose tissue type in adults and is characterized by unilocular adipocytes. Brown adipose tissue (BAT), on the other hand, consists of multilocular adipocytes (Fig. 2) that burn energy through nonshivering thermogenesis [50]. Because of BAT's high metabolic activity, it is linked to leanness and weight loss [51, 52], whereas high levels of WAT are linked to various health disorders. Because of its close link to a disease state, WAT is a more obvious model for adipose-based pharmacologic screening. However, BAT has been shown to lose its thermogenic capacity and move toward a WAT-like phenotype in overweight and insulin-resistant states [53]. Thus, BAT models are relevant in analyzing potentially harmful WAT-inducing effects of pharmacologics [53]. Additionally, because of its

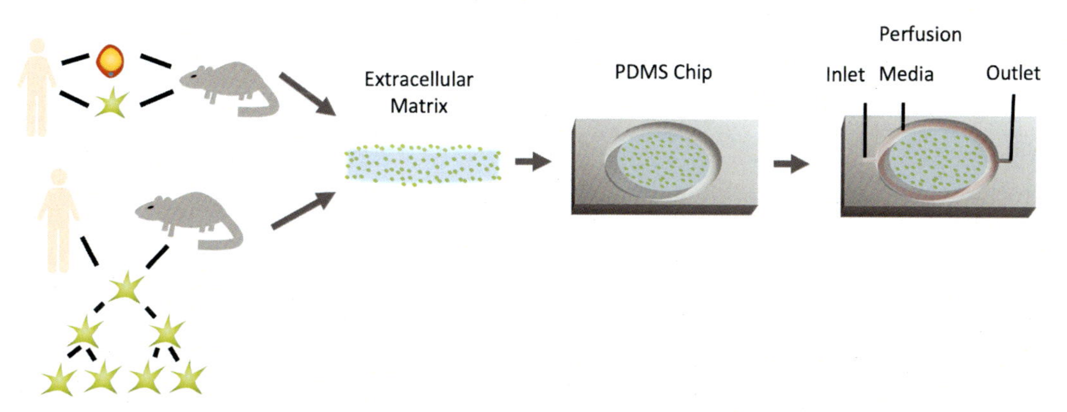

FIG. 1 Human adipose-derived stem cells (hASCS) differentiated in silk scaffolds showing small multilocular lipid depots (left) versus mature human adipocytes with large unilocular lipid depots (right).

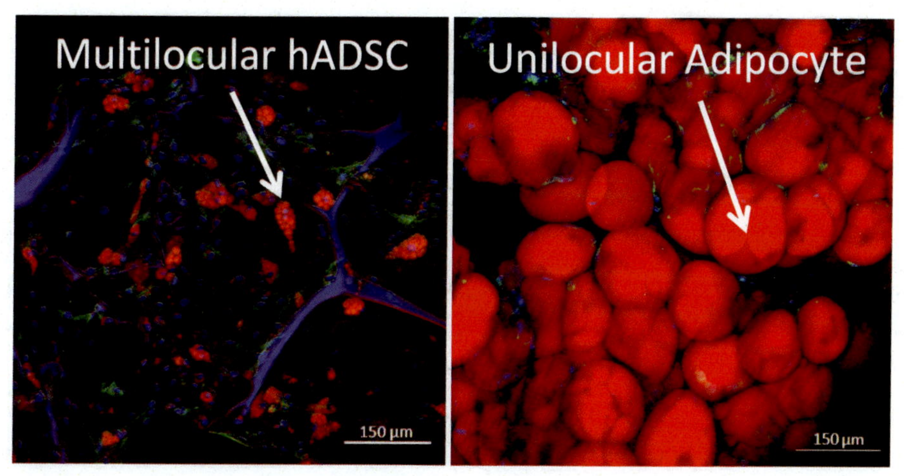

FIG. 2 Minimum components necessary for organ-on-a-chip platform. Primary or immortalized cells may be derived from human or animal adipose tissue. Mature or differentiated adipocytes are embedded in ECM, which is integrated into a microfluidic device.

high metabolic activity, induction of a BAT-like phenotype in WAT is one potential mechanism for antiobesity therapeutics [54, 55]. Thus, both WAT and BAT-based in vitro models are relevant for pharmacologic screening.

Subcutaneous *versus* visceral adipose tissue

WAT exists in intramuscular regions as well as in the form of subcutaneous and visceral adipose tissue, which are defined based on their anatomical position beneath the skin and lining internal organs in the body, respectively. Though visceral adipose tissue has a greater pathological effect toward type 2 diabetes and metabolic syndrome [56–58], cardiovascular disease [56, 59, 60], and cancer [61, 62], it is more difficult to obtain, and thus subcutaneous adipose tissue is used far more in research. Metabolic differences exist between subcutaneous and visceral adipose tissue as well as changes in adipokine secretion that vary with body mass index (BMI) and increased visceral adipose accumulation [63, 64]. Thus, subcutaneous and visceral adipose tissue-derived cells also produce varying physiological responses in vitro. As the exact mechanisms of adipose tissue pathology in obesity, cardiovascular disease, cancer, etc. are not understood, more research is needed to determine the optimal source of adipose derivatives for pharmacologic screening. The molecular signaling pathways between adipose tissue and neighboring organs are complex, and it is likely that different models will require cells sourced from one or more different depots.

Primary cells

Two sources exist for integration of primary cells in an MPS/bioreactor: primary mature adipocytes and differentiated ASCs. Differentiated ASCs are commonly used because they are plastic-adherent, are easily expanded in vitro, and can be cryopreserved. They are readily differentiated to adipocytes using well-established differentiation media usually containing insulin, indomethacin, dexamethasone, and 3-isobutyl-1-methylxanthine [65]. Differentiated ASCs first exhibit a multilocular lipid phenotype, indicative of immature adipocytes. Over an extended culture period, the lipid droplets begin to fuse to form a single, large unilocular lipid droplet characteristic of mature white adipocytes (Fig. 2).

Human ASCs (hASCs) are easily isolated from human adipose tissue obtained as tissue particles from liposuction procedures or excised en bloc [66, 67]. They offer the obvious and significant advantage of physiological accuracy. Rodents have anatomical and metabolic differences in their subcutaneous adipose tissue when compared with humans [68], and target genes in obesity-related disorders are not necessarily conserved across species [69]. However, primary animal cells have the advantage of allowing for selection from animals with specific diets, sex, and transgenic disease models, and for collection from various adipose tissue depots. Deveaud et al. compared epididymal and inguinal fat pads in rats and found differences in mitochondrial activity [70]. Palou et al. found the retroperitoneal fat depot in rats had higher expression of lipogenic and lipolytic genes compared with mesenteric and inguinal fat depots [71]. Kadowaki, et al. found sex- and depot-specific expression of PPARγ in rats [72]. Such studies comparing sex and fat depots would be significantly more challenging in humans. Moreover, animal-based MPS/bioreactors allow for in vitro modeling with cells from knock-in/knock-out tissues or coculture systems with tissues where it is difficult to obtain primary human samples (i.e., pancreas, liver, neural). They are also relevant in the development of veterinary drugs. Thus, while hASCs may be easily accessible, there is a role for animal cell models in in vitro testing.

Mature adipocytes can also be isolated from primary human or animal adipose tissue. Unlike young, differentiated ASCs, they offer the advantage of being fully differentiated and of a WAT phenotype. This minimizes the time in culture required to reach a mature, physiologically relevant tissue. However, mature adipocytes are terminally differentiated and so do not proliferate in culture [73] and are not cryopreserved well [74–76]. Because of this, they require large volumes of primary tissue to obtain enough cells for high-throughput screening. Mature adipocytes also offer the same advantages and disadvantages for human- and animal-derived cells as outlined above.

Embryonic stem cells and induced pluripotent stem cells

Adipocytes can be derived from human ESCs (hESCs) and/or human iPSCs (hiPSCs) [77]. Because adipose tissue can generally be obtained from patients voluntarily and noninvasively, use of hESCs and hiPSCs for adipose cultures has been limited. In addition, hESCs pose ethical concerns and can be difficult and costly to obtain owing to lack of donors [78]. A disadvantage of hiPSCs compared with primary ASCs is the need for additional steps and time to induce a mesenchymal stem cell phenotype [79]. However, unlike ASCs, hiPSCs offer the advantage of unlimited proliferation and differentiation capacity [78]. This means hiPSCs can be used to obtain large volumes of cells for high-throughput screening or to create coculture models with all cell types derived from the same cell line [42]. Additionally, hiPSCs are patient-specific and could potentially create multiorgan autologous models [42].

Immortalized cells

While ASCs proliferate in vitro, they may lose proliferation and differentiation capacity with extensive passaging, creating limitations for high-throughput experiments where large quantities of identical cells are required. Immortalized cell lines offer the advantage of being able to obtain large quantities for high-throughput experiments [42]. The most common cell line used in in vitro adipose tissue cultures is the immortalized 3T3-L1 murine preadipocyte line. They have been cultured in 2D extensively [80, 81] and can be differentiated to adipocytes [82]. These cells have the advantage of extensive literature data to compare to but may lack relevance to human physiology. The major advantage of immortalized hASCs is the ability to obtain larger quantities of cells for high-throughput experiments. They also have a longer shelf life for repeat studies, and are more cost-effective. Some researchers have also successfully immortalized their own line of hASCs isolated from primary human adipose tissue [83–85]. However, while these studies demonstrate the successful application of immortalized ASCs to in vitro cultures, immortalized cells have drawbacks that should be considered. Cell lines express a differential phenotype to primary cells, and the risk of contamination with another cell type can be high. Additionally, while they may proliferate indefinitely, genotypic and phenotypic changes can occur with higher passage numbers [86].

Extracellular matrix

Considerations

The environment surrounding the cells in an MPS/bioreactor should mimic the in vivo ECM as closely as possible. Adipocytes are buoyant, fragile, and nonadherent. Therefore, they require a 3D matrix for encapsulation. The ECM should promote adhesion, proliferation, and differentiation.

One major consideration for adipose-based systems is ECM substrate stiffness. ASCs can differentiate to other mesenchymal lineages, and stiffer substrates have been shown to encourage osteogenic differentiation of ASCs over adipogenic [87–89]. Additionally, the material should be able to withstand relevant shear stresses that occur during perfusion without degradation unless they are protected by semipermeable membranes. Matrix pore size should also be a major consideration for an adipose tissue culture as adipocytes themselves are large (up to 150 μm [90]) and ASCs expand greatly in size during differentiation [77]. Substrate pores should be able to accommodate large cells and changes in cell size during differentiation [77].

Materials for microphysiological systems/bioreactors

Most organ-on-a-chip systems use a biomaterial to encapsulate the adipocytes or ASCs within the chip; however, the chip's material itself is important. Pharmacologic testing requires high-throughput screening systems. This means large quantities of chips will be used, and data will need to be repeatable over time and across laboratories. Thus, a material for manufacturing in vitro therapeutic testing systems needs to be low-cost, easy to obtain, easy to process, and processed with replicability. Polydimethylsiloxane (PDMS) is often used because it has tunable elastic properties, low cost, low autofluorescence, biocompatibility, optical clarity, and easy moldability [8]. However, it can also absorb small hydrophobic drugs, which may limit its potential for pharmacologic screening [8]. To avoid hydrophobic drug absorption, Domansky et al. developed a polyurethane with similar properties to PDMS but with less absorption of hydrophobic compounds [91]. Polystyrene has also been used for organ-on-a-chip technology, though it is rigid and may not be suitable for every design [92]. Finally, another alternative to PDMS is a PDMS coating that prevents absorption of hydrophobic compounds [92–94]. This may be the most easily integrated technique as PDMS is widely used.

Static versus perfusion

Culture system perfusion provides a continuous supply of nutrients and waste removal not present in static cultures. Although it increases the complexity of the system, perfusion cultures provide a physiological advantage over static cultures by mimicking physiologic shear stresses and mechanical cues for mechanotransduction. This is particularly important for adipose-based cultures, as shear stress has been shown to influence differentiation of adipocytes. Abbott et al. found that perfusion increased lipid accumulation in differentiating hASCs [95]. However, adipocytes experience low shear stress in vivo (<0.01 Pa [96–98]), and excessive shear stresses can lead to rupture. Additionally, there may be different responses to shear stress over time. A shear stress of 1 Pa at early stages of differentiation in 3T3-L1 preadipocytes was not shown to affect lipid accumulation, but equivalent shear stresses applied at late stages of differentiation decreased lipid accumulation and adipogenic gene expression [99]. hASCs can also differentiate to other mesenchymal lineages such as osteogenic, chondrogenic, skeletal myogenic, cardiac, neurogenic, and endothelial lineages [100]. Mechanical cues including shear stress have been shown to influence differentiation toward these other lineages [101, 102]. Organ-on-a-chip systems have a unique advantage in this regard over other MPS/bioreactors, as the microchannels provide laminar flow at their small dimensions (<1 mm), which mimics in vivo blood flow [8]. The shear stress can also be altered by changing flow rates and channel sizes. Most adipose organ-on-a-chip models also integrate a porous membrane that protects cells from high shear stresses, enabling increased medium flow rates without compromising viability. These membranes can also protect ECMs from degradation, especially in long-term cultures. Additionally, faster flow rates allow for more immediate delivery of molecular cues to the cells and dilute analytes for analysis. Therefore, shear stress needs to be optimized for each MPS/bioreactor to maximize nutrient delivery and waste removal while maintaining viability of the cells and supporting differentiation to adipocytes.

While the physiological relevance of perfusion systems is a major justification for flow systems, the increase in complexity they require should be considered. More sophisticated training and equipment is necessary over static cultures, which is impactful for drug development experiments with large sample numbers. Additionally, air bubbles may be generated during perfusion, which poses a risk of cell injury, and ECM degradation may be increased with perfusion [8]. These factors should be considered in adipose-based screening, as the majority of adipose-related disorders are chronic and will require long-term cultures, which increases the probability of these complications.

Other considerations

Chemical gradients

Uptake and release of glycerol and nonesterified fatty acids between adipocytes and the interstitial space is concentration gradient dependent and changes with fasting versus fed states [103]. Transport of glycerol and glucose from the adipose interstitial space to systemic circulation also occurs along a concentration gradient [104–106]. Organ-on-a-chip technology allows for the creation of chemical gradients across samples, making these devices well suited for studying these metabolic events. Cimetta et al. created a bioreactor that produced chemical gradients in media across wells [107]. This idea could be applied when testing dosage of pharmaceuticals.

Coculture components of microphysiological systems/bioreactors

A major component of MPS/bioreactors is their ability to enable coculture models of tissues and organs. As discussed, inflammation and vascularization play a crucial role in adipose tissue pathology and are thus highly regarded coculture components.

Inflammation

Inflammation is a key component for adipose-based screening systems because of its crucial role in the pathophysiology of multiple adipose-related disorders, including obesity and type 2 diabetes. Macrophages have been linked extensively to insulin resistance and are a promising method for inducing a diabetic phenotype in vitro. Macrophage accumulation in subcutaneous adipose tissue is associated with increased adiposity [108]. Additionally, estimates suggest an increase from 10% macrophages in lean patients to 40% in obese patients [109]. Interestingly, macrophage accumulation is also associated with lipolysis during weight loss. This subsides over time in normal-weight patients but does not lessen in obese patients [110]. High-fat diets are also associated with increased M1 macrophage (proinflammatory) accumulation and a higher M1-to-M2 (antiinflammatory) ratio [111, 112]. Because of this, macrophages may serve as a physiologically relevant model for inducing inflammation in adipocytes in vitro.

Macrophages have been cocultured with 3T3-L1 preadipocytes [82, 113] and adipocytes [114] and were shown to induce insulin resistance and an inflammatory state [82, 113]. Additionally, Liu et al. incorporated a human monocyte cell line with differentiated hASCs. The addition of monocytes did not result in a change in adiponectin or IL-6 expression but did move the adipocytes to a more insulin-resistant phenotype [98]. Thus, macrophage cocultures are a promising mechanism for inducing a diabetic phenotype and should be considered in adipose-based MPS/bioreactors for diabetic drug development.

Other immune cells have also been able to induce an inflammatory phenotype in vitro. Nitta et al. [115] cocultured splenocytes with 3T3-L1 preadipocytes and found increased expression of monocyte chemoattractant protein-1 (MCP-1) and interleukin-6 (IL-6). Interestingly, tumor necrosis factor-α (TNF-α) expression was decreased in cultures with direct contact between cell types but increased in cultures with paracrine signaling alone. This fits with other data showing certain inflammatory cytokines such as TNF-α may only be secreted locally, whereas other cytokines such as IL-6 may be secreted systemically. The latter can therefore contribute to pathology of other organs directly, whereas local inflammatory cytokines may exert a pathological effect by affecting other systemic adipokines such as leptin or adiponectin [116]. Paracrine-signaling cytokines may therefore be more applicable to isolated adipose tissue models and systemic cytokines relevant to multiorgan models with a shared medium. More research is needed to discern precise mechanisms of inflammation so that a physiologically accurate method can be used to create an inflammatory phenotype in vitro.

Vascularization

Adipose tissue is a highly vascularized connective tissue. During human fetal development, adipogenesis is closely associated with vasculogenesis and angiogenesis [117–119]. To improve physiological accuracy, some MPS/bioreactors incorporate vascularization. This may be in the form of a coculture with endothelial cells [120, 121] or a vascular network where the endothelial cells are housed in a compartment separate from the adipocytes [122]. Integration of endothelial cells allows for paracrine signaling with adipocytes. While vascularization is an essential feature of healthy adipose tissue, it has also been linked to insulin resistance [118, 123], and so crosstalk between endothelial cells and adipocytes is crucial in modeling both healthy and diseased adipose tissue. Additionally, in the case of perfusion systems, confining the convective transport to an endothelial-lined compartment protects adipocytes from high shear stresses and recreates the endothelial cell barrier. It should also be noted that perfusion may be more relevant in vascularized adipose tissue models, as perfusion may play a role in inducing vascularization [124].

Coculture models with adipocytes and endothelial cells have been successful in vitro [125, 126], and the more sophisticated models cite formation of a lumen [121, 127]. More work is needed to create an in vitro system that has an entire vascular network with a functioning endothelial cell barrier. A major barrier to multiorgan models is the requirement for one universal medium. Vascularization could be a solution to this requirement, by acting as a barrier to selectively regulate nutrient delivery [8].

Current adipose-based microphysiological systems/bioreactors

Adipose-tissue-on-a-chip systems

Compared with other organ systems, the field of adipose organ-on-a-chip systems is relatively young. Nakayama et al. created a disposable three-compartment micro-cell culture device out of PDMS. Magnetic stirrers were used to create an internal pumping system that allowed for two setups: (1) compartment-specific perfusion allows for separate culture conditions for each cell type; while this prevents cross-talk, it allows for different media and perfusion rate conditions for each compartment, which allows for distinct differentiation of cell types; (2) after differentiation, connecting valves can be opened to allow for perfusion throughout the entire device, allowing for cross-talk and mimicking the in vivo system [41]. However, such a pumping system would not provide laminar flow as a standard microfluidic pumping system does. While this may not affect the phenotype of mature adipocytes, recreating laminar flow is relevant in modeling vasculature.

Liu et al. created an adipose-on-a-chip system with hASCs and human monocytes (U937 mononuclear cells). In this system, a porous barrier containing microchannels separated differentiated hASCs from monocytes while allowing for paracrine signaling [98]. Importantly, the authors were able to induce varying levels of shear stress by injecting medium volumes into one, two, or three inlets. This allowed them to measure differential effects of shear stress on cell differentiation. Additionally, functionalized magnetic beads were used to capture secretion of cytokines for noninvasive assessment of analytes over time. The device also contained a third chamber not used in this study but that allowed for integration of additional cell types [98]. Results from this device showed that adipocytes in coculture with monocytes had decreased insulin response, indicating that the model may be able to induce insulin resistance. Moreover, images showed induction of a "macrophage-like" morphology from the coculture, indicating a possible role for adipocytes in macrophage differentiation [98].

Kongsuphol et al. cultured differentiated primary hASCs with human peripheral blood mononuclear cells (PBMC) in a microfluidic device. The device contained two cell culture chambers, each of which contained three compartments: (1) adipocytes, (2) PBMC, and (3) immunoassay compartment used for noninvasive sample collection. The compartments were separated by a polyethylene terephthalate (PET) membrane [128]. Fluid was perfused through three syringe pumps per cell culture chamber (six syringe pumps per chip). Cocultures with PBMCs showed increased total lipid area with or without inflammation via lysophosphatidic acid (LPA). PBMC cell distribution was similar on-chip and in static cultures [128]. They were also able to show decreases in insulin sensitivity in differentiated hASCs with LPA-induced inflammation. This was ameliorated with addition of the type 2 diabetes drug metformin. They also tested docosapentaenoic acid, an omega-3 polyunsaturated fatty acid shown to improve insulin sensitivity in mice, and found a decrease in IL-6 secretion [128]. These findings validate the model as a platform for testing pharmaceuticals.

A PDMS device containing a cell culture chamber and two microchannels was used to model vascularized human adipose tissue. A hydrogel solution with vascular endothelial cells and fibroblasts was injected into a PDMS cell culture chamber. Microchannels were lined with vascular endothelial cells to form a lumen. Perfusable 3D microvascular beds were created with endothelial cells and fibroblasts cocultured with hASCs in a fibrin hydrogel to model vascularized adipose tissue in vitro [129]. Authors used a flow rate of 36 μL/h for shear stress in the compartments of 1×10^{-5} Pa [129].

Tanataweethum et al. cultured primary murine ASCs in a chip made up of PDMS and silicone chambers separated by a polycarbonate membrane. They found increased cell viability in a perfusion system (5 μL/h) over static culture [130]. They also demonstrated fusion of lipids into unilocular lipid droplets, indicative of a mature phenotype. They found a 1000-fold decrease in shear stress using a dual-layer device where the cells are separated from perfusion by a polycarbonate membrane versus a single-layer device where the cells are exposed directly to fluid flow. The single-layer device also demonstrated detachment of cells and decreased viability. The single-layer and dual-layer devices had a shear stress of 4.3×10^{-5} Pa and 6.88×10^{-7} Pa, respectively [130]. This was calculated based on the Loskill et al. model for peak flow rate in the bottom chamber.

Human-on-a-chip systems

To assess therapeutics for multiorgan disorders such as metabolic syndrome, an ideal platform would incorporate multiple organs with pathological effects in obesity and diabetes, such as pancreas, gut, etc., allowing for crosstalk between numerous cell types (Table 2). It would also allow for integration of organs involved in metabolic clearing, such as liver and kidney. A major challenge with multiorgan models is the need for a common medium between all cell types. Zhang et al. developed a multiorgan microfluidic model with lung, liver, kidney, and adipose tissue, using a common medium with growth factors [131].

TABLE 2 Summary of adipose-based coculture microphysiological systems/bioreactors.

System type	Adipose-derived cells	Other cell types	References
Microfluidic device	Primary murine adipocytes	Hepatocytes	[160]
Microfluidic device	Primary rat adipocytes	Primary rat hepatocytes and human hepatocarcinoma Hep G2	[41, 161]
Microfluidic device	Primary hASCs	Immortalized human monocytes	[98]
Microfluidic device	Primary hASCs	Peripheral blood mononuclear cells (PBMCs)	[128]
Microfluidic device	Primary hASCs	primary human adipose microvascular endothelial cells (hAMECs)	[129]
Microfluidic system	Primary human adipocytes	HUVECs	[122]
Spheroids	Primary hASCs	HUVECs	[120]
Layered coculture construct	Primary hASCs	Human dermal fibroblasts, Human bone marrow mesenchymal stem cells, and HUVECs	[162]
Scaffold	Primary hASCs	HUVECs	[121]

Abbreviations: hASC, *human adipose-derived stem cell;* HUVEC, *human umbilical vein endothelial cell.*

They specialized the common medium to lung cells by using controlled-release gelatin microspheres to distribute a lung-specific growth factor to that compartment. This concept can be integrated with other growth factors to help maintain individual cell-type function in a human-on-a-chip system.

Microcarriers

Microcarriers are microbeads capable of supporting cell adhesion and growth. They are usually on the order of 90–350 μm and suspended in medium in a bioreactor for culture. They have a high surface-area-to-volume ratio, which increases cell adhesion. This allows for high-density cell cultures in a more confined space than when using standard techniques. Cell detachment can be an issue in microcarrier cultures because of mechanical forces in the bioreactors. Different surface coatings have been utilized to enhance cell attachment [132]. Because microcarriers require adhesive cell lines, adipocyte culture on microcarriers is unrealistic. However, for high-throughput drug testing, large quantities of cells are required, and microcarriers can be used for in vitro expansion of ASCs in a simplified setup [133–136].

Organoids

Organoids are small self-assembled 3D tissue constructs derived from stem cells. By differentiating the stem cells in culture during organoid formation, cell-directed organ-specific 3D structures can be self-assembled [137]. Adipose-based organoids are often made by a hanging drop method where cells are cultured in low-adherence plates and inverted to promote detachment and self-assembly into spheroids [138]. Klingelhutz et al. also used ultra-low-attachment plates with a round bottom to form organoids [138], and Daquinag et al. used magnetic nanoparticles to draw cells to the meniscus of medium, where they self-assembled into organoids. Organoids allow for the study of organ development with fewer resources required for high-throughput screening because of their small culture size. However, creation of multiorgan models with organoids or cultures allowing for crosstalk remains complex. Layered organoids with multiple cell types have been created, but this may be an unrealistic mechanism for recreating an entire organ system in vitro. Additionally, because of their spherical structure and poor diffusivity, cells in the core can become nutrient-deprived and accumulate waste. To combat this, organoids have been integrated into microfluidic chips to develop larger-scale organoids and with specific structures including a lumen that allows for perfusion. This eliminates the poorly nourished core of the organoid and improves nutrient delivery and waste removal [139]. However, this has not been done with adipose-based cultures, and so the concepts need to be translated to this tissue type.

Assessments of microphysiological systems and bioreactors

MPS/bioreactors need to allow for cell function assessments through destructive or nondestructive mechanisms. While destructive endpoints compromise the integrity of the sample, requiring increased sample number, they can be informative. In nondestructive analyses, the sample remains viable and multiple assessments can be run on the same sample without compromising the system. The ideal MPS/bioreactor will allow for multiple noninvasive assessments of the system so that cell phenotype can be continuously monitored over time. This allows for development of longitudinal data from a single sample, reducing variability due to sample-to-sample variation. As drug testing aims for high-throughput screening, conducting all assessments on one sample rather than one sample per assessment reduces trial size.

Adipokines

Adipose-secreted cytokines (adipokines) are important markers of cell function. Adipose tissue has been shown to secrete hundreds of adipokines [15, 16] that can have local autocrine and paracrine effects as well as systemic effects [163, 164]. Adipose tissue secretomes vary between depots (subcutaneous and visceral WAT [165, 166]) and between WAT and BAT [167]. Changes to the adipose tissue secretome are also associated with various disease states of adipose tissue [164–166] and other tissues, including neurovascular disease [168], cardiovascular disease [169], arthritis [170], adipose tissue browning [171], cirrhosis [172], and cancer [173], and so their crosstalk will be important in future disease-in-a-dish models. Thus, adipose-based MPS/bioreactors require assessments of adipokine secretion. ELISAs can be used for expression of adipokines such as leptin [174, 175], adiponectin [98], TNF-α [128, 174, 115], IL-6 [98, 128, 174, 115], and interleukin-1β (IL-1β) [128, 115]. Leptin can also be measured from medium samples using a radioimmunoassay [176].

Lipogenesis

The key marker of successful adipogenic differentiation is the accumulation of intracellular lipids (lipogenesis), usually visualized with fluorescent lipid staining using Oil Red O [81, 175, 177, 178], AdipoRed [95, 130, 179], or LipidTox [130]. Staining of intracellular lipids also displays lipogenesis in mature adipocytes, as occurs with obesity. Staining can be done for imaging or fluorescently quantified, which are both destructive analyses. Intracellular triglyceride levels can also be quantified using a colorimetric enzymatic assay [95], although this is also a destructive endpoint. Additionally, Li et al. used a fluorescently labeled fatty acid analog to monitor real-time fatty acid uptake in adipocytes [150].

Lipolysis

Another key marker of adipocyte function is lipolysis. Adipocytes continually fluctuate between states of lipogenesis and lipolysis. Measurement of lipolysis is an important marker of a healthy adipocyte phenotype. During lipolysis, triglycerides are hydrolyzed to form glycerol and free fatty acids, which are secreted for use by other organs. Lipolysis can be measured via expression of lipolytic enzymes or via quantification of lipolytic end products (glycerol and free fatty acids). Colorimetric enzymatic assay kits (i.e., Abcam Free Glycerol Assay Kit, Cayman Glycerol Colorimetric Kit, BioAssay Systems EnzyChrom Glycerol Assay

Kit) can be used on medium samples for nondestructive measurement of both glycerol [95, 175, 179, 180] and free fatty acids [180]. It should be noted that glycerol secretions can result from breakdown of mono- and diglycerides as well as triglycerides, but these molecules account for very small amounts of glycerol secretion. Moreover, glycerol is not further metabolized at all, whereas free fatty acids can be reutilized by adipocytes [181, 182]. Thus, glycerol secretion is a more accurate measurement of lipolysis than free fatty acids. Lipolysis can also be stimulated using isoproterenol or forskolin to allow for a functional readout of adipocytes and can be done repeatedly without affecting adipocyte viability. Lipolysis stimulation with isoproterenol or forskolin can also be used as a marker of differentiation [130, 180]. A concentration of 10 μM isoproterenol or forskolin is usually used for this [180, 183, 184], though 0.1 μM isoproterenol has also been used [130].

Insulin response

Functional response tests of adipocytes also include insulin-stimulated glucose uptake. Glucose uptake can be measured noninvasively from medium samples using glucose oxidase assays [147]. However, traditional nondestructive assays for glucose tend to have low accuracy and low sensitivity. Lee et al. measured glucose uptake in differentiated 3T3-L1 preadipocytes using radioactive glucose and measuring radioactivity with a liquid scintillation analyzer [185]. However, this method is destructive. The fluorescent D-glucose analog 2-[*N*-(7-nitrobenz-2-oxa-1,3-diazol-4-yl) amino]-2-deoxy-D-glucose (2-NBDG) has been used to image glucose uptake in living HepG2 human hepatocarcinoma cells and L6 rat skeletal muscle cells. Liu et al. used a fluorescence glucose analog on their adipose-on-a-chip device to visualize glucose uptake in real time but could not quantify results via image analysis [98]. However, they and Kongsuphol et al. were able to quantify fluorescence readouts from lysed cell solutions with a plate reader [128]. While the latter is a destructive assessment, this technique has been used noninvasively in hepatocarcinoma cells [186] and skeletal muscle cells [186] and could be refined for real-time glucose-uptake imaging and quantification. Another technique is the use of a clinical glucose/lactate analyzer that can analyze media for glucose concentrations and lactic acid dehydrogenase (LDH) secretions [187]. However, this technique requires access to specialized machinery and so may not be applicable to wide-scale assessments for drug development.

Nondestructive imaging techniques

Optical imaging modalities can track adipose constructs over time, providing concurrent spatial information as well as information on matrix organization, cell morphology, and functionality. A common imaging technique is two-photon excited fluorescence (TPEF), which has been used to noninvasively quantify endogenous markers of adipogenic differentiation in a perfusion bioreactor system [159] as well as in hASCs in silk scaffolds [188]. Third-harmonic generation (THG) is another noninvasive imaging technique that has been used in conjunction with TPEF to monitor cell metabolism and lipogenesis in hASC silk scaffolds [189]. In addition, Abbott et al. used coherent anti-Stokes Raman scattering (CARS) and second-harmonic generation imaging to noninvasively image lipids and collagen, respectively, in primary human adipose tissue cultured in silk scaffolds [47]. This provided nondestructive visualization of intracellular lipid accumulation, allowing for time-course analysis of lipogenesis. Finally, Kongsuphol et al. quantified lipid content directly from 2D brightfield images which, though less precise than other techniques, is a simple and easy way to nondestructively track lipid accumulation over time [128].

Oxygen tension

Oxygen tension has been shown to play a role in various disease states, including obesity and inflammation. Lack of capillary formation during adipogenesis can result in hypoxia and subsequent inflammation. This is particularly relevant in obesity, where adipogenesis occurs rapidly on a large scale [118, 190–195]. Thus, methods of inducing hypoxia in vitro would add this key element to create physiologically relevant adipose disease models. Additionally, thermogenesis in BAT results in cell catabolism and high oxygen consumption, so measurement of oxygen levels is used to assess BAT phenotype [182]. In vitro culture systems mimicking hypoxia and normoxia have been established [196, 197] but, to our knowledge, not in the context of adipose tissue. Still, establishment of varying oxygen tensions in an MPS/bioreactor warrants attention.

Than et al. used a phosphorescent oxygen-sensitive probe that emits signal inversely proportional to oxygen levels. However, its use requires limiting oxygen exchange by coating cells in mineral oil, which may prevent performing of other assessments on the same sample [198]. Weyand et al. used phosphorescent microbeads to noninvasively monitor cellular oxygen levels using a commercially available system [199]. Dib et al. used another commercially available system, Seahorse XF24 Extracellular Flux Analyzer, to monitor oxygen consumption rate in mouse WAT [200]. Thomas et al. developed a noninvasive oxygen-sensing thin film using HUVECs by incorporating the phosphorescent dye, Pt(II) meso-tetrakis(pentafluorophenyl)porphine (PtTFPP), into PDMS thin films [201]. While oxygen tension is an important outcome marker, specialized equipment is required, and thus, limited research has focused on this area.

Future directions

The Defense Advanced Research Projects Agency (DARPA) and the National Institutes of Health (NIH) have made substantial investments in organ-on-a-chip development, and DARPA has requested development of a 10-organ chip model for drug development testing [202]. However, ongoing challenges must be overcome before entire human-on-a-chip systems become dependable. Multiorgan systems that mimic the in vivo environment must use one universal medium for all cell types to allow for proper interorgan communication [8]. Defining a medium that can maintain viability and phenotype of multiple cell types is a significant hurdle in current attempts at multiorgan models. Moreover, the use of stem cells in MPS/bioreactors makes the issue more challenging, as differentiation to a specific cell type typically requires highly defined culture media [203]. Additionally, most cell culture media use xenogeneic serum containing undefined concentrations of hormones, growth factors, amino acids, etc. To allow for reproducibility, a universal medium will have to be chemically defined [203]. In addition to a universal medium, specific cell types require specific cell–ECM interactions to maintain phenotype, and organ-specific ECMs will be required [8, 203]. There also remain hurdles in scaling of multiorgan models [147]. In vitro organ system models will need to mimic relative sizes and metabolic activity of organs in vivo. This is so that ADMET and PK/PD data are representative of the in vivo response [6, 204]. Organ size and metabolic activity will also need to be in proportion to the universal medium so that secreted metabolites are not diluted or overly concentrated [203]. These scaling hurdles add complexity to adipose-based systems because of the highly variable body fat percentages in humans. Adipose-based MPS/bioreactors may need to be scalable in size and metabolic activity to properly mimic a range of BMIs [205]. Additionally, MPS/bioreactors need to be optimized for high-throughput screening. This means reducing sample numbers, cell counts, costs of materials, and ease of production. This will allow for the large-scale testing needed to bring a pharmacologic to clinical trial. The ultimate objective of in vitro models for drug testing is to create systems that are physiologically accurate enough that in vivo testing can be avoided.

References

[1] The Drug Development Process, 2018. https://www.fda.gov/patients/learn-about-drug-and-device-approvals/drug-development-process.

[2] Angiogenesis Models, Modulators, and Clinical Applications, first ed., Springer US, New York, NY, 1998.

[3] I. Kola, J. Landis, Can the pharmaceutical industry reduce attrition rates? Nat. Rev. Drug Discov. 3 (2004) 711.

[4] J.A. DiMasi, H.G. Grabowski, R.W. Hansen, Innovation in the pharmaceutical industry: new estimates of R&D costs, J. Health Econ. 47 (2016) 20–33.

[5] W.M.S. Russell, R.L. Burch, The Principles of Humane Experimental Technique, Methuen & Co ltd, London, 1959.

[6] J.P. Wikswo, The relevance and potential roles of microphysiological systems in biology and medicine, Exp. Biol. Med. (Maywood) 239 (9) (2014) 1061–1072.

[7] G.A. Truskey, Human microphysiological systems and organoids as in vitro models for toxicological studies, Front. Public Health 6 (2018) 185.

[8] S.N. Bhatia, D.E. Ingber, Microfluidic organs-on-chips, Nat. Biotechnol. 32 (8) (2014) 760–772.

[9] I. Martin, D. Wendt, M. Heberer, The role of bioreactors in tissue engineering, Trends Biotechnol. 22 (2) (2004) 80–86.

[10] V. Mohamed-Ali, J.H. Pinkney, S.W. Coppack, Adipose tissue as an endocrine and paracrine organ, Int. J. Obes. Relat. Metab. Disord.: J. Int. Assoc. Study Obes. 22 (12) (1998) 1145–1158.

[11] L. Badimon, B. Oñate, G. Vilahur, Adipose-derived mesenchymal stem cells and their reparative potential in ischemic heart disease, Revista espanola de cardiologia (English ed.) 68 (7) (2015) 599–611.

[12] T. Ronti, G. Lupattelli, E. Mannarino, The endocrine function of adipose tissue: an update, Clin. Endocrinol. 64 (4) (2006) 355–365.

[13] X. Li, T. Ma, J. Sun, M. Shen, X. Xue, Y. Chen, Z. Zhang, Harnessing the secretome of adipose-derived stem cells in the treatment of ischemic heart diseases, Stem Cell Res Ther 10 (1) (2019) 196.

[14] C. Antoniades, 'Dysfunctional' adipose tissue in cardiovascular disease: a reprogrammable target or an innocent bystander? Cardiovasc. Res. 113 (9) (2017) 997–998.

[15] S. Lehr, S. Hartwig, D. Lamers, S. Famulla, S. Müller, F.-G. Hanisch, C. Cuvelier, J. Ruige, K. Eckardt, D.M. Ouwens, H. Sell, J. Eckel, Identification and validation of novel adipokines released from primary human adipocytes, Mol. Cell. Proteomics 11 (1) (2012) (M111.010504).

[16] J. Zhong, S.A. Krawczyk, R. Chaerkady, H. Huang, R. Goel, J.S. Bader, G.W. Wong, B.E. Corkey, A. Pandey, Temporal profiling of the secretome during adipogenesis in humans, J. Proteome Res. 9 (10) (2010) 5228–5238.

[17] M. Lafontan, Adipose tissue and adipocyte dysregulation, Diabetes Metab. 40 (1) (2014) 16–28.

[18] A. Garg, Adipose tissue dysfunction in obesity and lipodystrophy, Clin. Cornerstone 8 (2006) S7–S13.

[19] R.H. Unger, G.O. Clark, P.E. Scherer, L. Orci, Lipid homeostasis, lipotoxicity and the metabolic syndrome, Biochim. Biophys. Acta Mol. Cell Biol. Lipids 1801 (3) (2010) 209–214.
[20] S. Virtue, A. Vidal-Puig, Adipose tissue expandability, lipotoxicity and the metabolic syndrome — an allostatic perspective, Biochim. Biophys. Acta Mol. Cell Biol. Lipids 1801 (3) (2010) 338–349.
[21] K. Cusi, The role of adipose tissue and lipotoxicity in the pathogenesis of type 2 diabetes, Curr. Diab. Rep. 10 (4) (2010) 306–315.
[22] E. Ha Elizabeth, C. Bauer Robert, Emerging roles for adipose tissue in cardiovascular disease, Arterioscler. Thromb. Vasc. Biol. 38 (8) (2018) e137–e144.
[23] E.K. Oikonomou, C. Antoniades, The role of adipose tissue in cardiovascular health and disease, Nat. Rev. Cardiol. 16 (2) (2019) 83–99.
[24] A. Smekal, J. Vaclavik, Adipokines and cardiovascular disease: a comprehensive review, Biomed. Pap. 161 (1) (2017) 31–40.
[25] H. Berg Anders, E. Scherer Philipp, Adipose tissue, inflammation, and cardiovascular disease, Circ. Res. 96 (9) (2005) 939–949.
[26] W.B. Lau, K. Ohashi, Y. Wang, H. Ogawa, T. Murohara, X.-L. Ma, N. Ouchi, Role of Adipokines in cardiovascular disease, Circ. J. 81 (7) (2017) 920–928.
[27] V.B. Patel, S. Shah, S. Verma, G.Y. Oudit, Epicardial adipose tissue as a metabolic transducer: role in heart failure and coronary artery disease, Heart Fail. Rev. 22 (6) (2017) 889–902.
[28] A.D. Ogorodnikova, M. Kim, A.P. McGinn, P. Muntner, U. Khan, R.P. Wildman, Incident cardiovascular disease events in metabolically benign obese individuals, Obesity (Silver Spring, Md) 20 (3) (2012) 651–659.
[29] R. Parker, The role of adipose tissue in fatty liver diseases, Liver Res. 2 (1) (2018) 35–42.
[30] E. Buzzetti, M. Pinzani, E.A. Tsochatzis, The multiple-hit pathogenesis of non-alcoholic fatty liver disease (NAFLD), Metabolism 65 (8) (2016) 1038–1048.
[31] S. Milić, D. Lulić, D. Štimac, Non-alcoholic fatty liver disease and obesity: biochemical, metabolic and clinical presentations, World J. Gastroenterol. 20 (28) (2014) 9330–9337.
[32] K. Qureshi, G.A. Abrams, Metabolic liver disease of obesity and role of adipose tissue in the pathogenesis of nonalcoholic fatty liver disease, World J. Gastroenterol. 13 (26) (2007) 3540–3553.
[33] A.J. Cozzo, A.M. Fuller, L. Makowski, Contribution of adipose tissue to development of cancer, Compr. Physiol. 8 (1) (2017) 237–282.
[34] C. Himbert, M. Delphan, D. Scherer, L.W. Bowers, S. Hursting, C.M. Ulrich, Signals from the adipose microenvironment and the obesity-cancer link-A systematic review, Cancer Prev. Res. (Phila.) 10 (9) (2017) 494–506.
[35] D.F. Quail, A.J. Dannenberg, The obese adipose tissue microenvironment in cancer development and progression, Nat. Rev. Endocrinol. 15 (3) (2019) 139–154.
[36] H. Salaün, J. Thariat, M. Vignot, Y. Merrouche, S. Vignot, Obesity and cancer, Bull. Cancer 104 (1) (2017) 30–41.
[37] S. Poli Mara, B.L. Sheila, S. Fabíola, M.M. Debora, Adipose tissue dysfunction, adipokines, and low-grade chronic inflammation in polycystic ovary syndrome, Reproduction 149 (5) (2015) R219–R227.
[38] H.F. Escobar-Morreale, Polycystic ovary syndrome: definition, aetiology, diagnosis and treatment, Nat. Rev. Endocrinol. 14 (5) (2018) 270–284.
[39] A. Garg, A.K. Agarwal, Lipodystrophies: disorders of adipose tissue biology, Biochim. Biophys. Acta 1791 (6) (2009) 507–513.
[40] E. Jackson, R. Shoemaker, N. Larian, L. Cassis, Adipose tissue as a site of toxin accumulation, Compr. Physiol. 7 (4) (2017) 1085–1135.
[41] H. Nakayama, H. Kimura, M. Nishikawa, K. Komori, T. Fujii, Y. Sakai, Development of a Disposable Multi-Compartment Micro-Cell Culture Device, IEEE, 2007.
[42] J. Rogal, A. Zbinden, K. Schenke-Layland, P. Loskill, Stem-cell based organ-on-a-chip models for diabetes research, Adv. Drug Deliv. Rev. 140 (2019) 101–128.
[43] T. Ahfeldt, R.T. Schinzel, Y.-K. Lee, D. Hendrickson, A. Kaplan, D.H. Lum, R. Camahort, F. Xia, J. Shay, E.P. Rhee, C.B. Clish, R.C. Deo, T. Shen, F.H. Lau, A. Cowley, G. Mowrer, H. Al-Siddiqi, M. Nahrendorf, K. Musunuru, R.E. Gerszten, J.L. Rinn, C.A. Cowan, Programming human pluripotent stem cells into white and brown adipocytes, Nat. Cell Biol. 14 (2) (2012) 209–219.
[44] C. Dani, A.G. Smith, S. Dessolin, P. Leroy, L. Staccini, P. Villageois, C. Darimont, G. Ailhaud, Differentiation of embryonic stem cells into adipocytes in vitro, J. Cell Sci. 110 (11) (1997) 1279.

[45] M.F. Pittenger, A.M. Mackay, S.C. Beck, R.K. Jaiswal, R. Douglas, J.D. Mosca, M.A. Moorman, D.W. Simonetti, S. Craig, D.R. Marshak, Multilineage potential of adult human mesenchymal stem cells, Science 284 (5411) (1999) 143–147.
[46] O. Huttala, R. Mysore, J.R. Sarkanen, T. Heinonen, V.M. Olkkonen, T. Ylikomi, Differentiation of human adipose stromal cells in vitro into insulin-sensitive adipocytes, Cell Tissue Res. 366 (1) (2016) 63–74.
[47] R.D. Abbott, F.E. Borowsky, C.A. Alonzo, A. Zieba, I. Georgakoudi, D.L. Kaplan, Variability in responses observed in human white adipose tissue models, J. Tissue Eng. Regen. Med. (2017).
[48] I.J. Neeland, P. Poirier, J.-P. Després, Cardiovascular and metabolic heterogeneity of obesity: clinical challenges and implications for management, Circulation 137 (13) (2018) 1391–1406.
[49] Q. Qi, A.M. Stilp, T. Sofer, J.-Y. Moon, B. Hidalgo, A.A. Szpiro, T. Wang, M.C.Y. Ng, X. Guo, MEta-analysis of type 2 DIabetes in African Americans (MEDIA) Consortium, Y.-D.I. Chen, K.D. Taylor, M.L. Aviles-Santa, G. Papanicolaou, J.S. Pankow, N. Schneiderman, C.C. Laurie, J.I. Rotter, R.C. Kaplan, Genetics of type 2 diabetes in U.S. Hispanic/Latino individuals: results from the hispanic community health study/study of latinos (HCHS/SOL), Diabetes 66 (5) (2017) 1419–1425.
[50] T.C.L. Bargut, M.B. Aguila, C.A. Mandarim-de-Lacerda, Brown adipose tissue: updates in cellular and molecular biology, Tissue Cell 48 (5) (2016) 452–460.
[51] P. Seale, M.A. Lazar, Brown fat in humans: turning up the heat on obesity, Diabetes 58 (7) (2009) 1482.
[52] A.M. Cypess, C.R. Kahn, Brown fat as a therapy for obesity and diabetes, Curr. Opin. Endocrinol. Diabetes Obes. 17 (2) (2010) 143–149.
[53] T.C.L. Bargut, A.C.A.G. Silva-e-Silva, V. Souza-Mello, C.A. Mandarim-de-Lacerda, M.B. Aguila, Mice fed fish oil diet and upregulation of brown adipose tissue thermogenic markers, Eur. J. Nutr. 55 (1) (2016) 159–169.
[54] T.H. Carsten, W.K. Florian, Adipose tissue browning in mice and humans, J. Endocrinol. 241 (3) (2019) R97–R109.
[55] B. Cannon, J.A.N. Nedergaard, Brown adipose tissue: function and physiological significance, Physiol. Rev. 84 (1) (2004) 277–359.
[56] S.A. Ritchie, J.M.C. Connell, The link between abdominal obesity, metabolic syndrome and cardiovascular disease, Nutr. Metab. Cardiovasc. Dis. 17 (4) (2007) 319–326.
[57] B.L.o. Wajchenberg, Subcutaneous and visceral adipose tissue: their relation to the metabolic syndrome, Endocr. Rev. 21 (6) (2000) 697–738.
[58] P. Mathieu, Abdominal obesity and the metabolic syndrome: a surgeon's perspective, Can. J. Cardiol. 24 (Suppl D) (2008) 19D–23D.
[59] N. González, Z. Moreno-Villegas, A. González-Bris, J. Egido, Ó. Lorenzo, Regulation of visceral and epicardial adipose tissue for preventing cardiovascular injuries associated to obesity and diabetes, Cardiovasc. Diabetol. 16 (1) (2017) 44.
[60] Y. Wu, A. Zhang, D.J. Hamilton, T. Deng, Epicardial fat in the maintenance of cardiovascular health, Methodist Debakey Cardiovasc. J. 13 (1) (2017) 20–24.
[61] T.-H. Oh, J.-S. Byeon, S.-J. Myung, S.-K. Yang, K.-S. Choi, J.-W. Chung, B. Kim, D. Lee, J.H. Byun, S.J. Jang, J.-H. Kim, Visceral obesity as a risk factor for colorectal neoplasm, J. Gastroenterol. Hepatol. 23 (3) (2008) 411–417.
[62] D.V. Schapira, R.A. Clark, P.A. Wolff, A.R. Jarrett, N.B. Kumar, N.M. Aziz, Visceral obesity and breast cancer risk, Cancer 74 (2) (1994) 632–639.
[63] V. Van Harmelen, S. Reynisdottir, P. Eriksson, A. Thörne, J. Hoffstedt, F. Lönnqvist, P. Arner, Leptin secretion from subcutaneous and visceral adipose tissue in women, Diabetes 47 (6) (1998) 913.
[64] J.N. Fain, A.K. Madan, M.L. Hiler, P. Cheema, S.W. Bahouth, Comparison of the release of Adipokines by adipose tissue, adipose tissue matrix, and adipocytes from visceral and subcutaneous abdominal adipose tissues of obese humans, Endocrinology 145 (5) (2004) 2273–2282.
[65] Z. Mohammadi, J.T. Afshari, M.R. Keramati, D.H. Alamdari, M. Ganjibakhsh, A.M. Zarmehri, A. Jangjoo, M.H. Sadeghian, M.A. Ameri, L. Moinzadeh, Differentiation of adipocytes and osteocytes from human adipose and placental mesenchymal stem cells, Iran. J. Basic Med. Sci. 18 (3) (2015) 259–266.
[66] T. Alstrup, M. Eijken, A.B. Bohn, B. Møller, T.E. Damsgaard, Isolation of adipose tissue–derived stem cells: enzymatic digestion in combination with mechanical distortion to increase adipose tissue–derived stem cell yield from human aspirated fat, Curr. Protoc. Stem Cell Biol. 48 (1) (2019) e68.
[67] M. Locke, J. Windsor, P.R. Dunbar, Human adipose-derived stem cells: isolation, characterization and applications in surgery, ANZ J. Surg. 79 (4) (2009) 235–244.

[68] D.E. Chusyd, D. Wang, D.M. Huffman, T.R. Nagy, Relationships between rodent white adipose fat pads and human white adipose fat depots, Front. Nutr. 3 (2016) 10.
[69] K.J. Gaulton, Mechanisms of type 2 diabetes risk loci, Curr. Diab. Rep. 17 (9) (2017) 72.
[70] C. Deveaud, B. Beauvoit, B. Salin, J. Schaeffer, M. Rigoulet, Regional differences in oxidative capacity of rat white adipose tissue are linked to the mitochondrial content of mature adipocytes, Mol. Cell. Biochem. 267 (1–2) (2004) 157–166.
[71] M. Palou, T. Priego, J. Sánchez, A.M. Rodríguez, A. Palou, C. Picó, Gene expression patterns in visceral and subcutaneous adipose depots in rats are linked to their morphologic features, Cell. Physiol. Biochem. 24 (5–6) (2009) 547–556.
[72] K. Kadowaki, K. Fukino, E. Negishi, K. Ueno, Sex differences in PPARγ expressions in rat adipose tissues, Biol. Pharm. Bull. 30 (4) (2007) 818–820.
[73] J.M. Ntambi, K. Young-Cheul, Adipocyte differentiation and gene expression, J. Nutr. 130 (12) (2000) 3122S–3126S.
[74] L.L. Pu, Cryopreservation of adipose tissue, Organ 5 (3) (2009) 138–142.
[75] L.L.Q. Pu, X. Cui, J. Li, B.F. Fink, M.L. Cibull, D. Gao, The fate of cryopreserved adipose aspirates after in vivo transplantation, Aesthet. Surg. J. 26 (6) (2006) 653–661.
[76] S.-M. Hwang, J.-S. Lee, H.-D. Kim, Y.-H. Jung, H.-I. Kim, Comparison of the viability of cryopreserved fat tissue in accordance with the thawing temperature, Arch. Plast. Surg. 42 (2) (2015) 143–149.
[77] C.T. Gomillion, K.J.L. Burg, Stem cells and adipose tissue engineering, Biomaterials 27 (36) (2006) 6052–6063.
[78] K.H. Narsinh, J. Plews, J.C. Wu, Comparison of human induced pluripotent and embryonic stem cells: fraternal or identical twins? Mol. Ther. 19 (4) (2011) 635–638.
[79] J. Yu, K. Hu, K. Smuga-Otto, S. Tian, R. Stewart, I.I. Slukvin, J.A. Thomson, Human induced pluripotent stem cells free of vector and transgene sequences, Science 324 (5928) (2009) 797.
[80] Y. Mizunoe, Y. Sudo, N. Okita, H. Hiraoka, K. Mikami, T. Narahara, A. Negishi, M. Yoshida, R. Higashibata, S. Watanabe, H. Kaneko, D. Natori, T. Furuichi, H. Yasukawa, M. Kobayashi, Y. Higami, Involvement of lysosomal dysfunction in autophagosome accumulation and early pathologies in adipose tissue of obese mice, Autophagy 13 (4) (2017) 642–653.
[81] D. Papineau, A. Gagnon, A. Sorisky, Apoptosis of human abdominal preadipocytes before and after differentiation into adipocytes in culture, Metabolism 52 (8) (2003) 987–992.
[82] Y. Watanabe, Y. Nagai, H. Honda, N. Okamoto, S. Yamamoto, T. Hamashima, Y. Ishii, M. Tanaka, T. Suganami, M. Sasahara, K. Miyake, K. Takatsu, Isoliquiritigenin attenuates adipose tissue inflammation in vitro and adipose tissue fibrosis through inhibition of innate immune responses in mice, Sci. Rep. 6 (2016) 23097.
[83] C. Darimont, I. Zbinden, O. Avanti, P. Leone-Vautravers, V. Giusti, P. Burckhardt, A.M.A. Pfeifer, K. Macé, Reconstitution of telomerase activity combined with HPV-E7 expression allow human preadipocytes to preserve their differentiation capacity after immortalization, Cell Death Differ. 10 (9) (2003) 1025–1031.
[84] V. Zilberfarb, F. Pietri-Rouxel, R. Jockers, S. Krief, C. Delouis, T. Issad, A.D. Strosberg, Human immortalized brown adipocytes express functional beta3-adrenoceptor coupled to lipolysis, J. Cell Sci. 110 (7) (1997) 801.
[85] B.G. Vu, F.A. Gourronc, D.A. Bernlohr, P.M. Schlievert, A.J. Klingelhutz, Staphylococcal Superantigens stimulate immortalized human adipocytes to produce chemokines, PLoS One 8 (10) (2013) e77988.
[86] G. Kaur, J.M. Dufour, Cell lines, Spermatogenesis 2 (1) (2012) 1–5.
[87] J. Xie, D. Zhang, C. Zhou, Q. Yuan, L. Ye, X. Zhou, Substrate elasticity regulates adipose-derived stromal cell differentiation towards osteogenesis and adipogenesis through β-catenin transduction, Acta Biomater. 79 (2018) 83–95.
[88] B. Trappmann, J.E. Gautrot, J.T. Connelly, D.G.T. Strange, Y. Li, M.L. Oyen, M.A. Cohen Stuart, H. Boehm, B. Li, V. Vogel, J.P. Spatz, F.M. Watt, W.T.S. Huck, Extracellular-matrix tethering regulates stem-cell fate, Nat. Mater. 11 (7) (2012) 642–649.
[89] H. Lv, H. Wang, Z. Zhang, W. Yang, W. Liu, Y. Li, L. Li, Biomaterial stiffness determines stem cell fate, Life Sci. 178 (2017) 42–48.
[90] T. McLaughlin, C. Lamendola, N. Coghlan, T.C. Liu, K. Lerner, A. Sherman, S.W. Cushman, Subcutaneous adipose cell size and distribution: relationship to insulin resistance and body fat, Obesity 22 (3) (2014) 673–680.
[91] K. Domansky, D.C. Leslie, J. McKinney, J.P. Fraser, J.D. Sliz, T. Hamkins-Indik, G.A. Hamilton, A. Bahinski, D.E. Ingber, Clear castable polyurethane elastomer for fabrication of microfluidic devices, Lab Chip 13 (19) (2013) 3956–3964.

[92] B. Zhang, A. Korolj, L.B.F. Lun, M. Radisic, Advances in organ-on-a-chip engineering, Nat. Rev. Mater. 3 (8) (2018) 257–278.
[93] H. Sasaki, H. Onoe, T. Osaki, R. Kawano, S. Takeuchi, Parylene-coating in PDMS microfluidic channels prevents the absorption of fluorescent dyes, Sensors Actuators B Chem. 150 (1) (2010) 478–482.
[94] K. Ren, Y. Zhao, J. Su, D. Ryan, H. Wu, Convenient method for modifying poly(dimethylsiloxane) to be airtight and resistive against absorption of small molecules, Anal. Chem. 82 (14) (2010) 5965–5971.
[95] R.D. Abbott, W.K. Raja, R.Y. Wang, J.A. Stinson, D.L. Glettig, K.A. Burke, D.L. Kaplan, Long term perfusion system supporting adipogenesis, Methods (San Diego, Calif.) 84 (2015) 84–89.
[96] Y. Liu, P. Kongsuphol, S.B.N. Gourikutty, Q. Ramadan, Human adipocyte differentiation and characterization in a perfusion-based cell culture device, Biomed. Microdevices 19 (3) (2017) 18.
[97] K.M. Kim, Y.J. Choi, J.-H. Hwang, A.R. Kim, H.J. Cho, E.S. Hwang, J.Y. Park, S.-H. Lee, J.-H. Hong, Shear stress induced by an interstitial level of slow flow increases the osteogenic differentiation of mesenchymal stem cells through TAZ activation, PLoS One 9 (3) (2014) e92427.
[98] Y. Liu, P. Kongsuphol, S.Y. Chiam, Q.X. Zhang, S.B.N. Gourikutty, S. Saha, S.K. Biswas, Q. Ramadan, Adipose-on-a-chip: a dynamic microphysiological in vitro model of the human adipose for immune-metabolic analysis in type II diabetes, Lab Chip 19 (2) (2019) 241–253.
[99] J. Choi, S.Y. Lee, Y.-M. Yoo, C.H. Kim, Maturation of adipocytes is suppressed by fluid shear stress, Cell Biochem. Biophys. 75 (1) (2017) 87–94.
[100] C. Argentati, F. Morena, M. Bazzucchi, I. Armentano, C. Emiliani, S. Martino, Adipose stem cell translational applications: from bench-to-bedside, Int. J. Mol. Sci. 19 (11) (2018) 3475.
[101] Y.-X. Qin, M. Hu, Mechanotransduction in musculoskeletal tissue regeneration: effects of fluid flow, loading, and cellular-molecular pathways, Biomed. Res. Int. 2014 (2014) 863421.
[102] S.-H. Park, W.Y. Sim, B.-H. Min, S.S. Yang, A. Khademhosseini, D.L. Kaplan, Chip-based comparison of the osteogenesis of human bone marrow- and adipose tissue-derived Mesenchymal stem cells under mechanical stimulation, PLoS One 7 (9) (2012) e46689.
[103] S.D. O'Donovan, M. Lenz, R.G. Vink, N.J.T. Roumans, T.M.C.M. de Kok, E.C.M. Mariman, R.L.M. Peeters, N.-A.W. van Riel, M.A. van Baak, I.C.W. Arts, A computational model of postprandial adipose tissue lipid metabolism derived using human arteriovenous stable isotope tracer data, PLoS Comput. Biol. 15 (10) (2019) e1007400.
[104] S.W. Coppack, D.L. Chinkes, J.M. Miles, B.W. Patterson, S. Klein, A multicompartmental model of in vivo adipose tissue glycerol kinetics and capillary permeability in lean and obese humans, Diabetes 54 (7) (2005) 1934.
[105] R.G. Tiessen, M.M. Rhemrev-Boom, J. Korf, Glucose gradient differences in subcutaneous tissue of healthy volunteers assessed with ultraslow microdialysis and a nanolitre glucose sensor, Life Sci. 70 (21) (2002) 2457–2466.
[106] W. Regittnig, M. Ellmerer, G. Fauler, G. Sendlhofer, Z. Trajanoski, H.-J. Leis, L. Schaupp, P. Wach, T.R. Pieber, Assessment of transcapillary glucose exchange in human skeletal muscle and adipose tissue, Am. J. Physiol. Endocrinol. Metab. 285 (2) (2003) E241–E251.
[107] E. Cimetta, C. Cannizzaro, R. James, T. Biechele, R.T. Moon, N. Elvassore, G. Vunjak-Novakovic, Microfluidic device generating stable concentration gradients for long term cell culture: application to Wnt3a regulation of β-catenin signaling, Lab Chip 10 (23) (2010) 3277–3283.
[108] E.O.M. de Victoria, X. Xu, J. Koska, A.M. Francisco, M. Scalise, A.W. Ferrante, J. Krakoff, Macrophage content in subcutaneous adipose tissue, Diabetes 58 (2) (2009) 385.
[109] S.P. Weisberg, D. McCann, M. Desai, M. Rosenbaum, R.L. Leibel, A.W. Ferrante Jr., Obesity is associated with macrophage accumulation in adipose tissue, J. Clin. Invest. 112 (12) (2003) 1796–1808.
[110] A. Kosteli, E. Sugaru, G. Haemmerle, J.F. Martin, J. Lei, R. Zechner, A.W. Ferrante Jr., Weight loss and lipolysis promote a dynamic immune response in murine adipose tissue, J. Clin. Invest. 120 (10) (2010) 3466–3479.
[111] S. Fujisaka, I. Usui, A. Bukhari, M. Ikutani, T. Oya, Y. Kanatani, K. Tsuneyama, Y. Nagai, K. Takatsu, M. Urakaze, M. Kobayashi, K. Tobe, Regulatory mechanisms for adipose tissue M1 and M2 macrophages in diet-induced obese mice, Diabetes 58 (11) (2009) 2574–2582.
[112] C.N. Lumeng, J.L. Bodzin, A.R. Saltiel, Obesity induces a phenotypic switch in adipose tissue macrophage polarization, J. Clin. Invest. 117 (1) (2007) 175–184.
[113] Y. Sakamoto, J. Kanatsu, M. Toh, A. Naka, K. Kondo, K. Iida, The dietary Isoflavone Daidzein reduces expression of pro-inflammatory genes through PPARα/γ and JNK pathways in adipocyte and macrophage co-cultures, PLoS One 11 (2) (2016) e0149676.

[114] E. Karkeni, J. Marcotorchino, F. Tourniaire, J. Astier, F. Peiretti, P. Darmon, J.-F. Landrier, Vitamin D limits chemokine expression in adipocytes and macrophage migration in vitro and in male mice, Endocrinology 156 (5) (2015) 1782–1793.
[115] C.F. Nitta, R.A. Orlando, Crosstalk between immune cells and adipocytes requires both paracrine factors and cell contact to modify cytokine secretion, PLoS One 8 (10) (2013) e77306.
[116] M.S. Desruisseaux, M.E.T. Nagajyothi, H.B. Tanowitz, P.E. Scherer, Adipocyte, adipose tissue, and infectious disease, Infect. Immun. 75 (3) (2007) 1066.
[117] C.M. Poissonnet, A.R. Burdi, F.L. Bookstein, Growth and development of human adipose tissue during early gestation, Early Hum. Dev. 8 (1) (1983) 1–11.
[118] S. Corvera, O. Gealekman, Adipose tissue angiogenesis: impact on obesity and type-2 diabetes, Biochim. Biophys. Acta 1842 (3) (2014) 463–472.
[119] Y. Cao, Angiogenesis modulates adipogenesis and obesity, J. Clin. Invest. 117 (9) (2007) 2362–2368.
[120] R. Yao, Y. Du, R. Zhang, F. Lin, J. Luan, A biomimetic physiological model for human adipose tissue by adipocytes and endothelial cell cocultures with spatially controlled distribution, Biomed. Mater. 8 (4) (2013) 045005.
[121] J.H. Kang, J.M. Gimble, D.L. Kaplan, In vitro 3D model for human vascularized adipose tissue, Tissue Eng. A 15 (8) (2009) 2227–2236.
[122] E. Iori, B. Vinci, E. Murphy, M.C. Marescotti, A. Avogaro, A. Ahluwalia, Glucose and fatty acid metabolism in a 3 tissue in-vitro model challenged with Normo- and Hyperglycaemia, PLoS One 7 (4) (2012) e34704.
[123] H.-K. Sung, K.-O. Doh, J.E. Son, J.G. Park, Y. Bae, S. Choi, S.M.L. Nelson, R. Cowling, K. Nagy, I.P. Michael, G.Y. Koh, S.L. Adamson, T. Pawson, A. Nagy, Adipose vascular endothelial growth factor regulates metabolic homeostasis through angiogenesis, Cell Metab. 17 (1) (2013) 61–72.
[124] V. Bassaneze, V.G. Barauna, C. Lavini-Ramos, J. Kalil, I.T. Schettert, A.A. Miyakawa, J.E. Krieger, Shear stress induces nitric oxide–mediated vascular endothelial growth factor production in human adipose tissue Mesenchymal stem cells, Stem Cells Dev. 19 (3) (2009) 371–378.
[125] S. Reggio, C. Rouault, C. Poitou, J.-C. Bichet, E. Prifti, J.-L. Bouillot, S. Rizkalla, D. Lacasa, J. Tordjman, K. Clément, Increased basement membrane components in adipose tissue during obesity: links with TGFβ and metabolic phenotypes, J. Clin. Endocrinol. Metab. 101 (6) (2016) 2578–2587.
[126] V. Pellegrinelli, C. Rouault, N. Veyrie, K. Clément, D. Lacasa, Endothelial cells from visceral adipose tissue disrupt adipocyte functions in a three-dimensional setting: partial rescue by Angiopoietin-1, Diabetes 63 (2) (2014) 535.
[127] S. Muller, I. Ader, J. Creff, H. Leménager, P. Achard, L. Casteilla, L. Sensebé, A. Carrière, F. Deschaseaux, Human adipose stromal-vascular fraction self-organizes to form vascularized adipose tissue in 3D cultures, Sci. Rep. 9 (1) (2019) 7250.
[128] P. Kongsuphol, S. Gupta, Y. Liu, S.B.N. Gourikutty, S.K. Biswas, Q. Ramadan, In vitro micro-physiological model of the inflamed human adipose tissue for immune-metabolic analysis in type II diabetes, Sci. Rep. 9 (1) (2019) 4887.
[129] J. Paek, S.E. Park, Q. Lu, K.-T. Park, M. Cho, J.M. Oh, K.W. Kwon, Y.-s. Yi, J.W. Song, H.I. Edelstein, J. Ishibashi, W. Yang, J.W. Myerson, R.Y. Kiseleva, P. Aprelev, E.D. Hood, D. Stambolian, P. Seale, V.R. Muzykantov, D. Huh, Microphysiological engineering of self-assembled and Perfusable microvascular beds for the production of vascularized three-dimensional human microtissues, ACS Nano 13 (7) (2019) 7627–7643.
[130] N. Tanataweethum, A. Zelaya, F. Yang, R.N. Cohen, E.M. Brey, A. Bhushan, Establishment and characterization of a primary murine adipose tissue-chip, Biotechnol. Bioeng. 115 (8) (2018) 1979–1987.
[131] C. Zhang, Z. Zhao, N.A. Abdul Rahim, D. van Noort, H. Yu, Towards a human-on-chip: culturing multiple cell types on a chip with compartmentalized microenvironments, Lab Chip 9 (22) (2009) 3185–3192.
[132] S. Derakhti, S.H. Safiabadi-Tali, G. Amoabediny, M. Sheikhpour, Attachment and detachment strategies in microcarrier-based cell culture technology: a comprehensive review, Mater. Sci. Eng. C 103 (2019) 109782.
[133] C.-Y. Lin, C.-H. Huang, Y.-K. Wu, N.-C. Cheng, J. Yu, Maintenance of human adipose derived stem cell (hASC) differentiation capabilities using a 3D culture, Biotechnol. Lett. 36 (7) (2014) 1529–1537.
[134] M. Gadelorge, M. Bourdens, N. Espagnolle, C. Bardiaux, J. Murrell, L. Savary, S. Ribaud, B. Chaput, L. Sensebé, Clinical-scale expansion of adipose-derived stromal cells starting from stromal vascular fraction in a single-use bioreactor: proof of concept for autologous applications, J. Tissue Eng. Regen. Med. 12 (1) (2018) 129–141.
[135] C. Frye, C. Patrick, Three-dimensional adipose tissue model using low shear bioreactors, In Vitro Cell. Dev. Biol. Anim. 42 (5–6) (2006) 109–114.

[136] B. Cunha, T. Aguiar, S.B. Carvalho, M.M. Silva, R.A. Gomes, M.J.T. Carrondo, P. Gomes-Alves, C. Peixoto, M. Serra, P.M. Alves, Bioprocess integration for human mesenchymal stem cells: from up to downstream processing scale-up to cell proteome characterization, J. Biotechnol. 248 (2017) 87–98.
[137] H. Clevers, Modeling development and disease with Organoids, Cell 165 (7) (2016) 1586–1597.
[138] A.J. Klingelhutz, F.A. Gourronc, A. Chaly, D.A. Wadkins, A.J. Burand, K.R. Markan, S.O. Idiga, M. Wu, M.J. Potthoff, J.A. Ankrum, Scaffold-free generation of uniform adipose spheroids for metabolism research and drug discovery, Sci. Rep. 8 (1) (2018) 523.
[139] F. Yu, W. Hunziker, D. Choudhury, Engineering microfluidic organoid-on-a-chip platforms, Micromachines 10 (3) (2019) 165.
[140] M.J. Harms, Q. Li, S. Lee, C. Zhang, B. Kull, S. Hallen, A. Thorell, I. Alexandersson, C.E. Hagberg, X.-R. Peng, A. Mardinoglu, K.L. Spalding, J. Boucher, Mature human white adipocytes cultured under membranes maintain identity, function, and can transdifferentiate into brown-like adipocytes, Cell Rep. 27 (1) (2019) 213–225 (e5).
[141] R.Y. Wang, R.D. Abbott, A. Zieba, F.E. Borowsky, D.L. Kaplan, Development of a three-dimensional adipose tissue model for studying embryonic exposures to obesogenic chemicals, Ann. Biomed. Eng. 45 (7) (2017) 1807–1818.
[142] F.H. Lau, K. Vogel, J.P. Luckett, M. Hunt, A. Meyer, C.L. Rogers, O. Tessler, C.L. Dupin, H.S. Hilaire, K.N. Islam, T. Frazier, J.M. Gimble, S. Scahill, Sandwiched white adipose tissue: a microphysiological system of primary human adipose tissue, Tissue Eng. Part C Methods 24 (3) (2017) 135–145.
[143] T. Mohsen-Kanson, A.-L. Hafner, B. Wdziekonski, Y. Takashima, P. Villageois, A. Carrière, M. Svensson, C. Bagnis, B. Chignon-Sicard, P.-A. Svensson, L. Casteilla, A. Smith, C. Dani, Differentiation of human induced pluripotent stem cells into brown and white adipocytes: role of Pax3, Stem Cells 32 (6) (2014) 1459–1467.
[144] D. Taura, M. Noguchi, M. Sone, K. Hosoda, E. Mori, Y. Okada, K. Takahashi, K. Homma, N. Oyamada, M. Inuzuka, T. Sonoyama, K. Ebihara, N. Tamura, H. Itoh, H. Suemori, N. Nakatsuji, H. Okano, S. Yamanaka, K. Nakao, Adipogenic differentiation of human induced pluripotent stem cells: comparison with that of human embryonic stem cells, FEBS Lett. 583 (6) (2009) 1029–1033.
[145] M. Noguchi, K. Hosoda, M. Nakane, E. Mori, K. Nakao, D. Taura, Y. Yamamoto, T. Kusakabe, M. Sone, H. Sakurai, J. Fujikura, K. Ebihara, K. Nakao, In vitro characterization and engraftment of adipocytes derived from human induced pluripotent stem cells and embryonic stem cells, Stem Cells Dev. 22 (21) (2013) 2895–2905.
[146] M. Nishio, T. Yoneshiro, M. Nakahara, S. Suzuki, K. Saeki, M. Hasegawa, Y. Kawai, H. Akutsu, A. Umezawa, K. Yasuda, K. Tobe, A. Yuo, K. Kubota, M. Saito, K. Saeki, Production of functional classical Brown adipocytes from human pluripotent stem cells using specific Hemopoietin cocktail without gene transfer, Cell Metab. 16 (3) (2012) 394–406.
[147] C. Moraes, J.M. Labuz, B.M. Leung, M. Inoue, T.-H. Chun, S. Takayama, On being the right size: scaling effects in designing a human-on-a-chip, Integr Biol (Camb) 5 (9) (2013) 1149–1161.
[148] L.A. Godwin, J.C. Brooks, L.D. Hoepfner, D. Wanders, R.L. Judd, C.J. Easley, A microfluidic interface for the culture and sampling of adiponectin from primary adipocytes, Analyst 140 (4) (2015) 1019–1025.
[149] J.C. Brooks, K.I. Ford, D.H. Holder, M.D. Holtan, C.J. Easley, Macro-to-micro interfacing to microfluidic channels using 3D-printed templates: application to time-resolved secretion sampling of endocrine tissue, Analyst 141 (20) (2016) 5714–5721.
[150] X. Li, J.C. Brooks, J. Hu, K.I. Ford, C.J. Easley, 3D-templated, fully automated microfluidic input/output multiplexer for endocrine tissue culture and secretion sampling, Lab Chip 17 (2) (2017) 341–349.
[151] A.M. Clark, K.M. Sousa, C. Jennings, O.A. MacDougald, R.T. Kennedy, Continuous-flow enzyme assay on a microfluidic chip for monitoring glycerol secretion from cultured adipocytes, Anal. Chem. 81 (6) (2009) 2350–2356.
[152] A.M. Clark, K.M. Sousa, C.N. Chisolm, O.A. MacDougald, R.T. Kennedy, Reversibly sealed multilayer microfluidic device for integrated cell perfusion and on-line chemical analysis of cultured adipocyte secretions, Anal. Bioanal. Chem. 397 (7) (2010) 2939–2947.
[153] P. Loskill, T. Sezhian, K.M. Tharp, F.T. Lee-Montiel, S. Jeeawoody, W.M. Reese, P.-J.H. Zushin, A. Stahl, K.E. Healy, WAT-on-a-chip: a physiologically relevant microfluidic system incorporating white adipose tissue, Lab Chip 17 (9) (2017) 1645–1654.
[154] J. Rogal, C. Binder, E. Kromidas, C. Probst, S. Schneider, K. Schenke-Layland, P. Loskill, WAT's up!? – organ-on-a-chip integrating human mature white adipose tissues for mechanistic research and pharmaceutical applications, bioRxiv (2019) 585141.

[155] C.E. Dugan, J.P. Grinias, S.D. Parlee, M. El-Azzouny, C.R. Evans, R.T. Kennedy, Monitoring cell secretions on microfluidic chips using solid-phase extraction with mass spectrometry, Anal. Bioanal. Chem. 409 (1) (2017) 169–178.

[156] C.E. Dugan, R.T. Kennedy, Measurement of lipolysis products secreted by 3T3-L1 adipocytes using microfluidics, Methods Enzymol. 538 (2014) 195–209.

[157] A. Zambon, A. Zoso, O. Gagliano, E. Magrofuoco, G.P. Fadini, A. Avogaro, M. Foletto, S. Quake, N. Elvassore, High temporal resolution detection of patient-specific glucose uptake from human ex vivo adipose tissue on-chip, Anal. Chem. 87 (13) (2015) 6535–6543.

[158] X. Wu, N. Schneider, A. Platen, I. Mitra, M. Blazek, R. Zengerle, R. Schüle, M. Meier, In situ characterization of the mTORC1 during adipogenesis of human adult stem cells on chip, Proc. Natl. Acad. Sci. U. S. A. 113 (29) (2016) E4143–E4150.

[159] A. Ward, K.P. Quinn, E. Bellas, I. Georgakoudi, D.L. Kaplan, Noninvasive metabolic imaging of engineered 3D human adipose tissue in a perfusion bioreactor, PLoS One 8 (2) (2013) e55696.

[160] M.A. Guzzardi, C. Domenici, A. Ahluwalia, Metabolic control through hepatocyte and adipose tissue cross-talk in a multicompartmental modular bioreactor, Tissue Eng. A 17 (11 – 12) (2011) 1635–1642.

[161] H. Nakayama, H. Kimura, K. Komori, T. Fujii, Y. Sakai, Development of a disposable three-compartment micro-cell culture device for toxicokinetic study in humans and its preliminary evaluation, 2020. http://citeseerx.ist.psu.edu/viewdoc/download?doi=10.1.1.531.7941&rep=rep1&type=pdf.

[162] J. Michael Sorrell, M.A. Baber, D.O. Traktuev, K.L. March, A.I. Caplan, The creation of an in vitro adipose tissue that contains a vascular–adipocyte complex, Biomaterials 32 (36) (2011) 9667–9676.

[163] J. Eckel, Chapter 2 - Adipose tissue: a major secretory organ, in: J. Eckel (Ed.), The Cellular Secretome and Organ Crosstalk, Academic Press, 2018, pp. 9–63.

[164] N. Ouchi, J.L. Parker, J.J. Lugus, K. Walsh, Adipokines in inflammation and metabolic disease, Nat. Rev. Immunol. 11 (2) (2011) 85–97.

[165] S.K. Fried, D.A. Bunkin, A.S. Greenberg, Omental and subcutaneous adipose tissues of obese subjects release interleukin-6: depot difference and regulation by Glucocorticoid1, J. Clin. Endocrinol. Metab. 83 (3) (1998) 847–850.

[166] K. Samaras, N.K. Botelho, D.J. Chisholm, R.V. Lord, Subcutaneous and visceral adipose tissue gene expression of serum adipokines that predict type 2 diabetes, Obesity (Silver Spring, Md.) 18 (5) (2010) 884–889.

[167] A. Ali Khan, J. Hansson, P. Weber, S. Foehr, J. Krijgsveld, S. Herzig, M. Scheideler, Comparative secretome analyses of primary murine white and brown adipocytes reveal novel Adipokines, Mol. Cell. Probes 17 (12) (2018) 2358–2370.

[168] R. Opatrilova, M. Caprnda, P. Kubatka, V. Valentova, S. Uramova, V. Nosal, L. Gaspar, L. Zachar, I. Mozos, D. Petrovic, J. Dragasek, S. Filipova, D. Büsselberg, A. Zulli, L. Rodrigo, P. Kruzliak, V. Krasnik, Adipokines in neurovascular diseases, Biomed. Pharmacother. 98 (2018) 424–432.

[169] R. Shibata, N. Ouchi, K. Ohashi, T. Murohara, The role of adipokines in cardiovascular disease, J. Cardiol. 70 (4) (2017) 329–334.

[170] E.C.d.S. Fatel, F.T. Rosa, A.N.C. Simão, I. Dichi, Adipokines in rheumatoid arthritis, Adv. Rheumatol. (Lond. Engl.) 58 (1) (2018) 25.

[171] A. Rodríguez, S. Becerril, S. Ezquerro, L. Méndez-Giménez, G. Frühbeck, Crosstalk between adipokines and myokines in fat browning, Acta Physiol. (Oxf. Engl.) 219 (2) (2017) 362–381.

[172] C. Buechler, E.M. Haberl, L. Rein-Fischboeck, C. Aslanidis, Adipokines in liver cirrhosis, Int. J. Mol. Sci. 18 (7) (2017) 1392.

[173] A. Booth, A. Magnuson, J. Fouts, M. Foster, Adipose tissue, obesity and adipokines: role in cancer promotion, Horm. Mol. Biol. Clin. Invest. 21 (1) (2015) 57–74.

[174] J.-P. Bastard, M. Maachi, J.T. van Nhieu, C. Jardel, E. Bruckert, A. Grimaldi, J.-J. Robert, J. Capeau, B. Hainque, Adipose tissue IL-6 content correlates with resistance to insulin activation of glucose uptake both in vivo and in vitro, J. Clin. Endocrinol. Metab. 87 (5) (2002) 2084–2089.

[175] E. Bellas, K. Marra, D. Kaplan, Sustainable three-dimensional tissue model of human adipose tissue, Tissue Eng. Part C Methods 19 (10) (2013) 745–754.

[176] M.E. Trujillo, S. Sullivan, I. Harten, S.H. Schneider, A.S. Greenberg, S.K. Fried, Interleukin-6 regulates human adipose tissue lipid metabolism and Leptin production in vitro, J. Clin. Endocrinol. Metab. 89 (11) (2004) 5577–5582.

[177] J. Prawitt, A. Niemeier, M. Kassem, U. Beisiegel, J. Heeren, Characterization of lipid metabolism in insulin-sensitive adipocytes differentiated from immortalized human mesenchymal stem cells, Exp. Cell Res. 314 (4) (2008) 814–824.

[178] L.F. Liu, C.M. Craig, L.L. Tolentino, O. Choi, J. Morton, H. Rivas, S.W. Cushman, E.G. Engleman, T. McLaughlin, Adipose tissue macrophages impair preadipocyte differentiation in humans, PLoS One 12 (2) (2017) e0170728.

[179] R.D. Abbott, R.Y. Wang, M.R. Reagan, Y. Chen, F.E. Borowsky, A. Zieba, K.G. Marra, J.P. Rubin, I.M. Ghobrial, D.L. Kaplan, The use of silk as a scaffold for mature, sustainable unilocular adipose 3D tissue engineered systems, Adv. Healthc. Mater. 5 (2016) 1667–1677.

[180] M. Schweiger, T.O. Eichmann, U. Taschler, R. Zimmermann, R. Zechner, A. Lass, Measurement of lipolysis, Methods Enzymol. 538 (2014) 171–193.

[181] M. Vaughan, D. Steinberg, Effect of hormones on lipolysis and esterification of free fatty acids during incubation of adipose tissue in vitro, J. Lipid Res. 4 (2) (1963) 193–199.

[182] R.D. Abbott, F.E. Borowsky, K.P. Quinn, D.L. Bernstein, I. Georgakoudi, D.L. Kaplan, Non-invasive assessments of adipose tissue metabolism in vitro, Ann. Biomed. Eng. 44 (3) (2016) 725–732.

[183] D.L. Brasaemle, D.M. Levin, D.C. Adler-Wailes, C. Londos, The lipolytic stimulation of 3T3-L1 adipocytes promotes the translocation of hormone-sensitive lipase to the surfaces of lipid storage droplets, Biochim. Biophys. Acta Mol. Cell Biol. Lipids 1483 (2) (2000) 251–262.

[184] P. Baskaran, B. Thyagarajan, Measurement of Basal and Forskolin-stimulated Lipolysis in Inguinal Adipose Fat Pads, J. Vis. Exp. (2017) 55625 (125).

[185] A. Lee, K.-M. Choi, W.-B. Jung, H. Jeong, G.-Y. Kim, J.H. Lee, M.K. Lee, J.T. Hong, Y.-S. Roh, S.-H. Sung, H.-S. Yoo, Enhancement of glucose uptake by meso-dihydroguaiaretic acid through GLUT4 up-regulation in 3T3-L1 adipocytes, Molecules (Basel, Switzerland) 22 (9) (2017) 1423.

[186] C. Zou, Y. Wang, Z. Shen, 2-NBDG as a fluorescent indicator for direct glucose uptake measurement, J. Biochem. Biophys. Methods 64 (3) (2005) 207–215.

[187] D.M. Minteer, M.T. Young, Y.-C. Lin, P.J. Over, J.P. Rubin, J.C. Gerlach, K.G. Marra, Analysis of type II diabetes mellitus adipose-derived stem cells for tissue engineering applications, J. Tissue Eng. 6 (2015) (2041731415579215–2041731415579215).

[188] K.P. Quinn, E. Bellas, N. Fourligas, K. Lee, D.L. Kaplan, I. Georgakoudi, Characterization of metabolic changes associated with the functional development of 3D engineered tissues by non-invasive, dynamic measurement of individual cell redox ratios, Biomaterials 33 (21) (2012) 5341–5348.

[189] T. Chang, M.S. Zimmerley, K.P. Quinn, I. Lamarre-Jouenne, D.L. Kaplan, E. Beaurepaire, I. Georgakoudi, Non-invasive monitoring of cell metabolism and lipid production in 3D engineered human adipose tissues using label-free multiphoton microscopy, Biomaterials 34 (34) (2013) 8607–8616.

[190] M.E. Rausch, S. Weisberg, P. Vardhana, D.V. Tortoriello, Obesity in C57BL/6J mice is characterized by adipose tissue hypoxia and cytotoxic T-cell infiltration, Int. J. Obes. 32 (3) (2008) 451–463.

[191] N. Hosogai, A. Fukuhara, K. Oshima, Y. Miyata, S. Tanaka, K. Segawa, S. Furukawa, Y. Tochino, R. Komuro, M. Matsuda, I. Shimomura, Adipose tissue hypoxia in obesity and its impact on Adipocytokine Dysregulation, Diabetes 56 (4) (2007) 901.

[192] M. Pasarica, O.R. Sereda, L.M. Redman, D.C. Albarado, D.T. Hymel, L.E. Roan, J.C. Rood, D.H. Burk, S.R. Smith, Reduced adipose tissue oxygenation in human obesity: evidence for rarefaction, macrophage chemotaxis, and inflammation without an angiogenic response, Diabetes 58 (3) (2009) 718–725.

[193] Z. Michailidou, S. Turban, E. Miller, X. Zou, J. Schrader, P.J. Ratcliffe, P.W.F. Hadoke, B.R. Walker, J.P. Iredale, N.M. Morton, J.R. Seckl, Increased angiogenesis protects against adipose hypoxia and fibrosis in metabolic disease-resistant 11β-hydroxysteroid dehydrogenase type 1 (HSD1)-deficient mice, J. Biol. Chem. 287 (6) (2012) 4188–4197.

[194] L. Hodson, S.M. Humphreys, F. Karpe, K.N. Frayn, Metabolic signatures of human adipose tissue hypoxia in obesity, Diabetes 62 (5) (2013) 1417–1425.

[195] J. Ye, Z. Gao, J. Yin, Q. He, Hypoxia is a potential risk factor for chronic inflammation and adiponectin reduction in adipose tissue of Ob/Ob and dietary obese mice, Am. J. Physiol. Endocrinol. Metab. 293 (4) (2007) E1118–E1128.
[196] C.-C. Peng, W.-H. Liao, Y.-H. Chen, C.-Y. Wu, Y.-C. Tung, A microfluidic cell culture array with various oxygen tensions, Lab Chip 13 (16) (2013) 3239–3245.
[197] G. Khanal, K. Chung, X. Solis-Wever, B. Johnson, D. Pappas, Ischemia/reperfusion injury of primary porcine cardiomyocytes in a low-shear microfluidic culture and analysis device, Analyst 136 (17) (2011) 3519–3526.
[198] A. Than, H.L. He, S.H. Chua, D. Xu, L. Sun, M.K.-S. Leow, P. Chen, Apelin enhances brown adipogenesis and browning of white adipocytes, J. Biol. Chem. 290 (23) (2015) 14679–14691.
[199] B. Weyand, M. Nöhre, E. Schmälzlin, M. Stolz, M. Israelowitz, C. Gille, H.P. von Schroeder, K. Reimers, P.M. Vogt, Noninvasive oxygen monitoring in three-dimensional tissue cultures under static and dynamic culture conditions, BioResearch Open Access 4 (1) (2015) 266–277.
[200] L. Dib, A. Bugge, S. Collins, LXRα fuels fatty acid-stimulated oxygen consumption in white adipocytes, J. Lipid Res. 55 (2) (2014) 247–257.
[201] P.C. Thomas, M. Halter, A. Tona, S.R. Raghavan, A.L. Plant, S.P. Forry, A noninvasive thin film sensor for monitoring oxygen tension during in vitro cell culture, Anal. Chem. 81 (22) (2009) 9239–9246.
[202] D.E. Watson, R. Hunziker, J.P. Wikswo, Fitting tissue chips and microphysiological systems into the grand scheme of medicine, biology, pharmacology, and toxicology, Exp. Biol. Med. (Maywood) 242 (16) (2017) 1559–1572.
[203] Y.I. Wang, C. Carmona, J.J. Hickman, M.L. Shuler, Multiorgan microphysiological Systems for Drug Development: strategies, advances, and challenges, Adv. Healthc. Mater. 7 (2) (2018), https://doi.org/10.1002/adhm.201701000.
[204] U. Marx, T.B. Andersson, A. Bahinski, M. Beilmann, S. Beken, F.R. Cassee, M. Cirit, M. Daneshian, S. Fitzpatrick, O. Frey, C. Gaertner, C. Giese, L. Griffith, T. Hartung, M.B. Heringa, J. Hoeng, W.H. de Jong, H. Kojima, J. Kuehnl, M. Leist, A. Luch, I. Maschmeyer, D. Sakharov, A.J.A.M. Sips, T. Steger-Hartmann, D.A. Tagle, A. Tonevitsky, T. Tralau, S. Tsyb, A. van de Stolpe, R. Vandebriel, P. Vulto, J. Wang, J. Wiest, M. Rodenburg, A. Roth, Biology-inspired microphysiological system approaches to solve the prediction dilemma of substance testing, ALTEX 33 (3) (2016) 272–321.
[205] K.M. Gadde, C.K. Martin, H.-R. Berthoud, S.B. Heymsfield, Obesity: pathophysiology and management, J. Am. Coll. Cardiol. 71 (1) (2018) 69–84.

PART 3

Adipose cell therapy and regenerative medicine

CHAPTER

9

Adipose stem cells and donor demographics: Impact of anatomic location, donor sex, race, BMI, and health

Adam Cottrill[a], *Yasamin Samadi*[a], *and Kacey Marra*[b,c]

[a]Department of Plastic Surgery, University of Pittsburgh, Pittsburgh, PA, United States
[b]Departments of Plastic Surgery and Bioengineering, University of Pittsburgh, Pittsburgh, PA, United States [c]McGowan Institute for Regenerative Medicine, University of Pittsburgh, Pittsburgh, PA, United States

Introduction

Sources of adipose stem cells

Similar to bone marrow, adipose tissue derives from the mesoderm and contains a supportive multipotent mesenchymal stem cell (MSC) population, termed adipose stem cells (ASCs). However, collection of ASCs is considerably simpler than bone marrow MSCs as ASCs can be extracted from tissues discarded from minimally invasive, low-risk procedures such as liposuction. Therefore, due to its high abundance and simplistic harvesting, adipose tissue is a popular source for stem cells, with important roles in homeostasis, metabolism regulation, and aging. Clinically, ASCs have many therapeutic applications, including soft tissue replacement and wound repair.

Adipose tissues are broadly categorized into three primary groups: white adipose tissue (WAT), bone marrow adipose tissue (BMAT), and brown adipose tissue (BAT). WAT is found subcutaneously and in visceral depots that contain numerous adipocytes with high turnover. Both BAT and WAT have lipolytic and lipogenic functions that are involved in energy dissipation and accumulation, respectively. BAT is termed as such owing to its abundance of mitochondrial content, cytochromes, and vascularization [1]. BAT usually maintains a role in

Scientific Principles of Adipose Stem Cells
https://doi.org/10.1016/B978-0-12-819376-1.00014-7

utility (such as providing ways to heat the body) compared with the other types of fat. It has numerous unique properties such as expression of uncoupling protein 1 (UCP-1). It is also used for heat production and contains a lower number of ASCs than does WAT, which is rich in ASCs and more abundant in various areas of the body [1, 2].

For a cell to be considered an ASC, it must meet minimal criteria set by the International Society for Cellular Therapy (ISCT): (1) adherence to plastic; (2) expression of CD73, CD90, and CD105 but lack of expression of CD45, CD34, CD14 or CD11b, CD79a or CD19, and HLA-Dr surface markers; and (3) potential to differentiate into preadipocytes, chondrocytes, and osteoblasts [3]. The frequency and distribution of ASCs in unique anatomic WAT depots is dependent on a variety of factors, including age, location of harvest, and harvesting technique. Location of harvest becomes an important factor; BAT is found primarily in periadrenal regions, perirenal, axillary, and cervical in fetus and neonate, though these regions become WAT in adult humans. WAT has a wider area of distribution and is deposited in larger amounts per area. Sites that WAT can be derived from include subcutaneous regions of the abdomen, thighs and buttocks, intestines (visceral fat), perirenal space, omentum, retroorbital space, and bone marrow [1, 4]. Many subcutaneous depots are located superficially and in the deep region of the abdomen. Subcutaneous depots are usually the best source of stem cells in comparison with other fat depots, exhibiting increased multipotency and stemness. Studies have discerned that superficial tissue and abdominal tissue are a better source of stromal vascular fraction (SVF), and another recently released study found that ASC and SVF count was increased in WAT from inner and outer thigh rather than abdomen [5]. The undisputed best source of ASCs and SVF is still controversial, with study data presenting conflicting evidence. Further research that controls for confounding variables such as sex, age, harvesting and processing technique utilized, race and ethnicity, BMI, and smoking status would be beneficial to isolate the most efficient source of ASCs while removing extraneous factors (Fig. 1).

ASCs that are harvested from different sites of the body have demonstrated different characteristics. ASC yields are the greatest in subcutaneous depots, with the highest concentrations found in arm depots and the greatest plasticity found in inguinal tissues [6]. ASCs from BAT more readily undergo skeletal myogenic differentiation than ASCs from WAT. Much work has been done on the optimal source of ASCs, but the preferred clinical harvest site may differ based on ease of access, frequency, yield, and even the patient.

Clinically relevant harvest sites

Harvest sites for ASCs are important to the properties of the ASCs derived and have differences in plasticity and primary function depending on location and type of fat used. This renders the identification of clinical harvest sites an important objective. Primary sites of extraction include subcutaneous regions of abdomens, thighs and buttocks, perirenal space, omentum, and bone marrow. Additional sites may include superficial abdominal and mesenteric sites that show increased multipotency and stemness in addition to a potentially improved source of SVF [1]. It is important to note, however, that each of these fat depots have fundamental differences from one another. Omental fat has been supported to be physiologically unique from subcutaneous fat, whereas before they were considered one source. Researchers also provided evidence that fat progenitors from two of the primary visceral deposits, omental and mesenteric, exhibit distinct gene expression profiles [7].

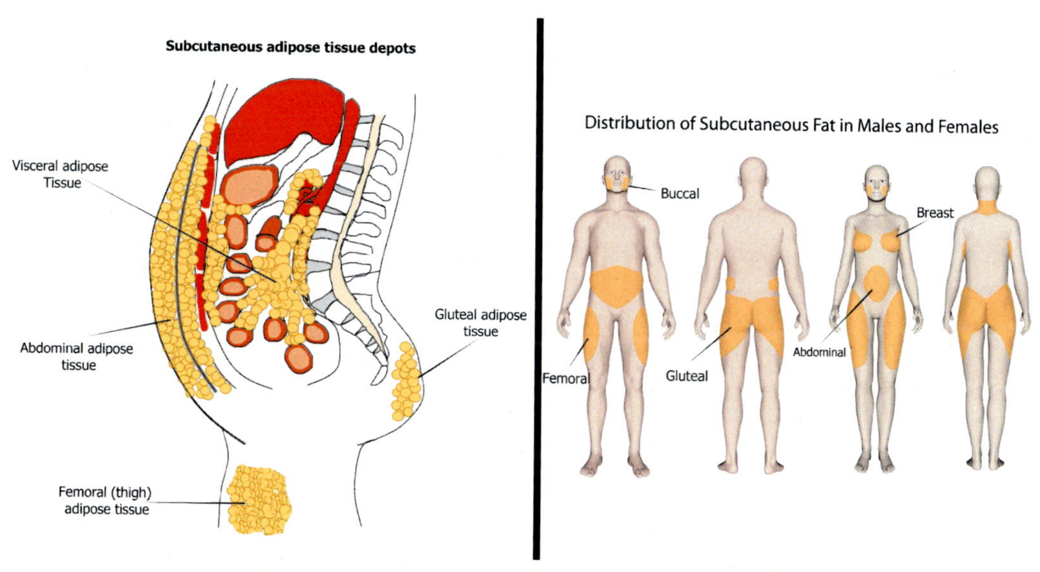

FIG. 1 Anatomical location of fat deposits. *Adapted from: https://www.mdpi.com/2079-7737/8/2/23/xml, and https://successiblelife.com/8-simple-fat-burning-and-toning-exercises-for-women/ – stock image of human outlines with fat from Google – public domain*

Primary methods of retrieval include en bloc resection and lipoaspiration, two of the most common methods for harvest, with the latter being the easiest and with fewer complications. Abdominally sourced fat and fat from the inner and outer thigh both produce viable amounts of ASCs and SVF to be utilized as clinically relevant harvest sites. Another relevant site of harvested fat in male and female patients is the buccal fat pad, which was shown to be a promising source of ASCs for osteogenic differentiation [1, 8]. Buccal fat is extremely easy to harvest along with abdominal and thigh fat, and liposuction, resection, and Coleman's technique are the three techniques generally utilized to extract tissues, with liposuction resulting in the best cell yield and viability compared with resection [1, 8]. Both methods (liposuction and resection) tend to utilize the Coleman technique where possible for more efficient processing.

Each of these areas can act as relevant clinical harvest sites without significant difference in viability of ASCs, as observed in a study comparing extraction from the abdominal, buccal, and hip regions and showing all ASCs to be multipotent and differentiated toward the osteogenic, adipogenic, and chondrogenic lineages in the presence of respective medium with similar morphology [5, 8, 9]. However, currently there is no definitive evidence to suggest an optimal harvesting area in terms of yield, proliferation rate, differentiation capacity, and viability between cells harvested from different sites; there are multiple clinically relevant harvest sites in terms of isolation for ASCs that may be dependent on a number of factors [8].

Effects of harvesting technique

In addition to harvest site, technique is equally as important in terms of cell yield and viability. Multiple harvesting techniques can be utilized for extraction. Studies were conducted comparing blood–oil waste and number of ASCs harvested. Multiple methods such as direct excision, liposuction, and power-assisted liposuction (PAL) with and without Coleman's processing technique were examined. Blood–oil discards from fat centrifugation were also compared, and it was found that liposuction and blood–oil waste yielded fewer ASCs than a method utilizing Coleman's without centrifugation [10]. Numbers of SVF cells and ASCs with direct excisions were typically much higher than those with the Coleman technique [10–12]. The number of total SVF cells from the Coleman technique with centrifugation was higher than without centrifugation; however, the number of ASCs obtained through utilization of centrifugation did not differ significantly [10]. The Coleman technique was the best processing technique following direct excision and liposuction. The techniques did not differ significantly in terms of blood–oil fraction (the ratio of blood to oil) [10]. Another study found that the number of viable cells was not affected by the type of mechanical technique utilized, with cultured ASCs expressing the appropriate phenotype [13]. A separate study investigated the effects of local or tumescent anesthesia during liposuction on ASCs. Number of viable ASCs was significantly lower in the local anesthetic compared with the tumescent group [14]. Survival was also much lower in a dose-dependent manner in the lidocaine group compared with the phosphate-buffered saline (PBS) controls [14, 15].

Bajek et al. studied harvesting techniques such as surgical resection, PAL, and laser-assisted liposuction (LAL) and compared them with one another [11]. The number of ASCs obtained was, on average, the highest in resection, followed by LAL and PAL, which were not significantly different from one another in cell count. Colonies formed by resection and PAL were significantly higher when compared with average colonies formed by LAL [11]. The study concluded that PAL is the preferred method, owing to high proliferation and slow senescence, but there are conflicting reports indicating that, though wet liposuction produced fewer ASCs than dry liposuction, the properties of fat extracted through either method remain similar [16].

Fontesa et al. tested direct excision against syringe hand aspiration and suction-assisted liposuction, i.e., PAL, using various pressures [12]. A wet versus dry aspiration and tumescent technique test using these methods was performed. It was found that PAL had a 47% higher adipocyte count when aspirated at low compared with high pressure, and cell viability was significantly higher at day 7 with low pressure. PAL maintained similar quality and quantity of ASCs as manual, but PAL cells had higher levels of differentiation marker expression [12].

The best cannula size for harvesting in these techniques is undecided, but bigger is better in that the optimally sized cannula is one that will be large enough to avoid shear stress and preserve integrity of SVF cells upon extraction [12]. Dry aspiration consists of direct aspiration without any preparation solution and results in 20%–50% of the aspirated volume comprising blood, while wet aspiration injects the donor site with a solution (i.e., saline, anesthetics) prior to aspiration and results in 4%–30% of aspirated volume comprising blood [12]. A super-wet technique was developed and reported by Fodor et al., with a reduced blood loss of 1%–2% of aspirated volume composed of blood [17, 18]. Another technique that can be used with the above methods is the tumescent technique that presents a large volume of infiltrate to total aspirate volume and majorly reduces blood loss to approximately 1% of total aspirated volume while also not requiring general anesthesia [11, 12]. It is generally considered a safer procedure for larger aspiration and has improved aesthetics. PAL was shown

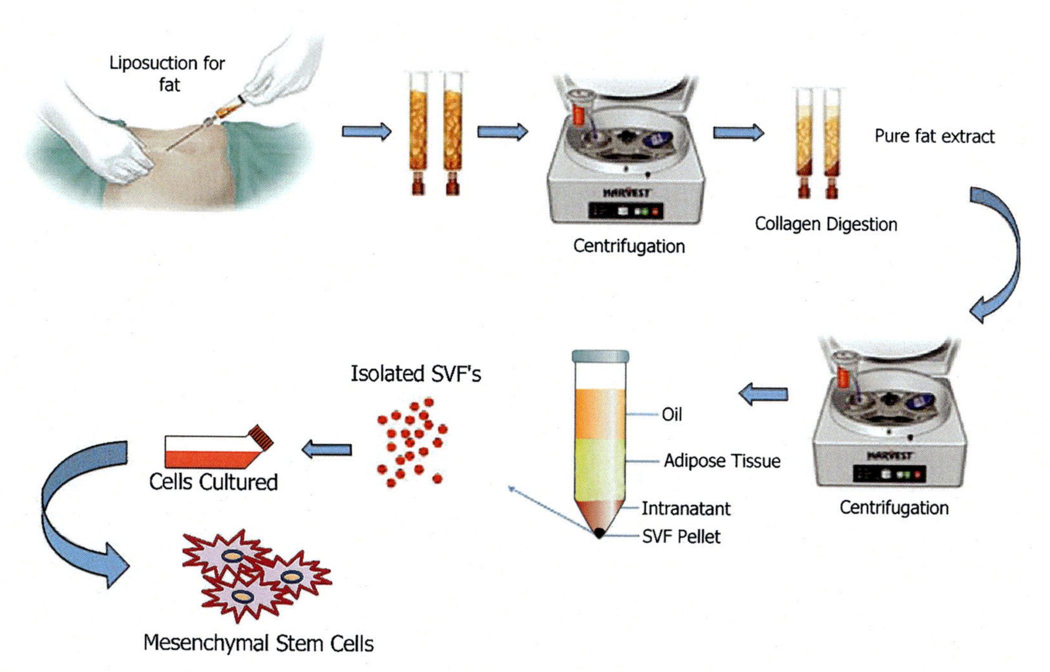

FIG. 2 ASC harvesting technique. *Adapted from: http://www.jcasonline.com/article.asp?issn=0974-2077;year=2016;volume=9;issue=3;spage=152;epage=156;aulast=Tan;type=3*

to have lower ASC yield in SVF than tissue resection, and manual liposuctions showed a higher potential of viable ASCs than waterjet assisted. PAL was better than LAL and was generally better with lower pressure (350 mmHg) compared with higher pressures (700 mmHg) and obtained a higher cell yield. PAL and manual liposuction had similar quality and quantity of ASCs, while PAL demonstrated higher expression of differentiation markers [12] (Fig. 2).

Harvesting techniques vary widely in implementation and effect. Tissue resection and direct excision yield optimal results that are closely followed by the Coleman technique and PAL. It is difficult to compare and definitively say that one method is better than others owing to donor site differences, and the differences between gender, age, and other demographics will be explored later in this chapter.

Patient demographics

Source, technique, and harvest location have significant amounts of research to support preferred methods. As adipose tissue use becomes utilized more often in regenerative medicine, it is important to understand the differences in patient demographics, such as sex and differences in composition or stemness that occur through race and ethnicity differences (Figs. 3 and 4).

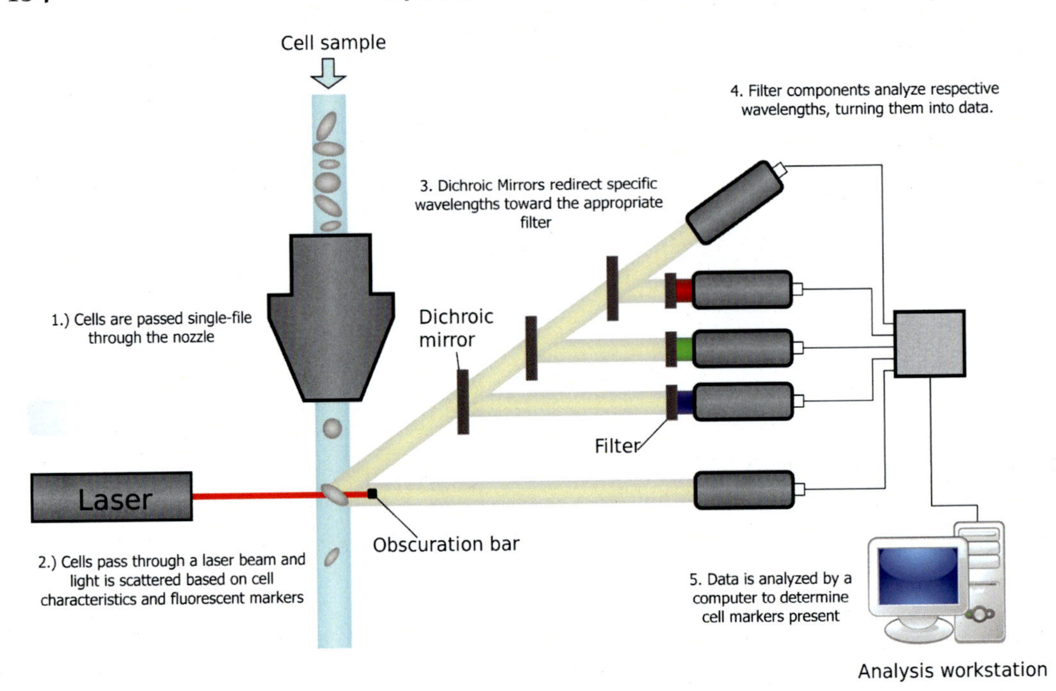

FIG. 3 Schematic diagram of a flow cytometry (FC) device used to measure physical and chemical characteristics of cells. *Adapted from:* *https://www.researchgate.net/publication/259450329_Flow_Cytometry_Bioinformatics*

Male versus female

It is widely accepted that there are anatomical differences in many tissues, systems, and cellular processes between the sexes. Much of this can be attributed to sex steroid differences, which can influence growth, metabolism, and processes. Adipose tissue is no different in that sex steroid regulation in men and women affects certain aspects of location, amount, and, to a lesser extent, the function. On average, females have greater fat stores than men, and before puberty, females have greater subcutaneous fat deposits than men of comparable age [19]. Women tend to have more fat in gluteal and peripheral regions of the body (and generally higher body fat percentage), resulting in the generalized "pear"-shaped body from deposition occurring more heavily in the hips and thighs [19, 20]. Males, on the other hand, are more likely to have significant amounts of abdominal fat and a higher percentage of body fat attributed to visceral adipose tissues (VAT), and to be more susceptible to abdominal adiposity than women [19, 20].

Sex steroids can influence properties of adipose tissue in various ways, including production and function of adipose tissues and utilization of fat stores. Estrogens reduce inflammatory signaling while improving the action of certain hormones such as insulin. Estrogens have a protective attribute that inhibits fat deposition in subcutaneous fat depots, which is why females prepuberty usually have higher subcutaneous fat levels than males pre- and postpuberty [19]. Male steroid hormones, androgens, have an opposite effect and are deleterious, causing an increase in the size of abdominal adipocytes while having an adipogenic

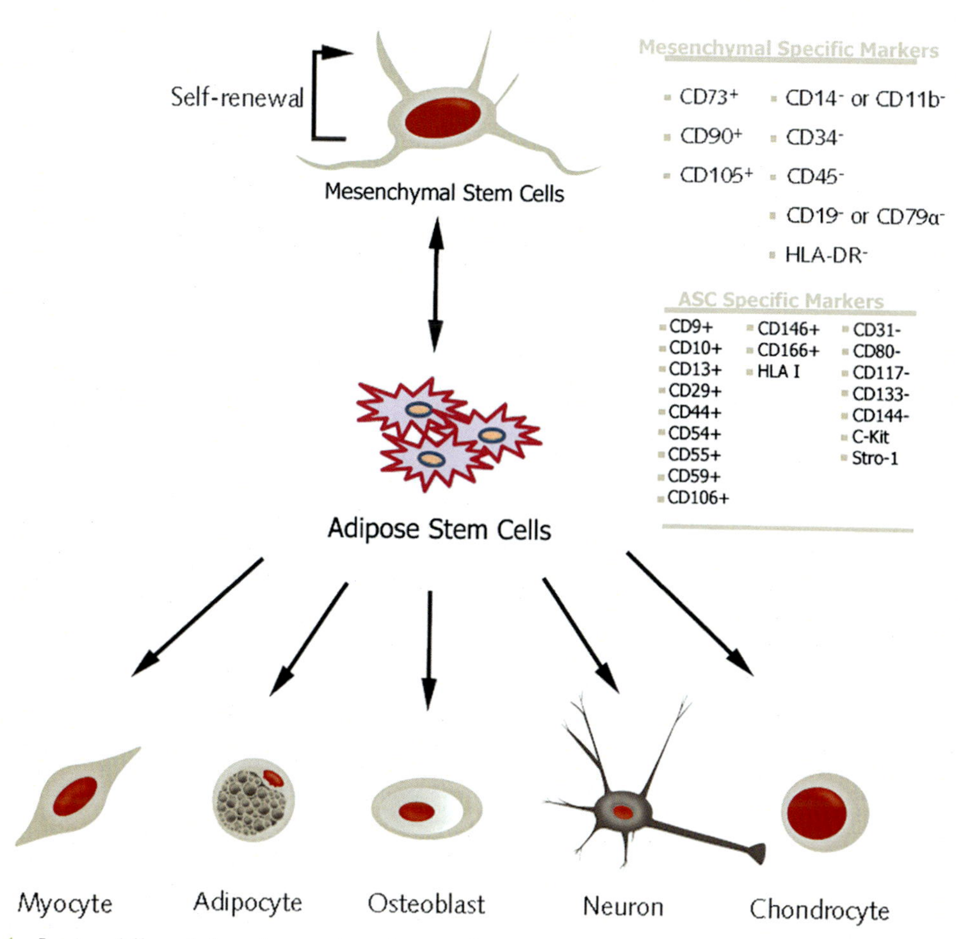

FIG. 4 In vitro differentiation of mesenchymal stem cells into adipose stem cells (ASCs), various progenitor cells, and ASC specific cellular markers. *Adapted from Sigma Aldrich protocol page – https://www.sigmaaldrich.com/technical-documents/protocols/biology/cell-culture/mesenchymal-stem-cell-differentiation.html*

capability that increases insulin sensitivity and improves insulin resistance. The effects of testosterone have been observed to increase subcutaneous adipose tissue (SAT) and VAT [19]. It becomes that differences in adiposity and fat distribution between the sexes are very closely associated with whole-body metabolism and health, along with effects of hormonal differences determined through the use of twin studies [20]. Twin studies have demonstrated that genetic factors greatly contribute to BMI variance (up to 70%), and the study of a few rare genetic syndromes that have differential effects on adiposity between males and females has allowed this effect to be examined more closely. It was found that a number of polymorphisms in the estrogen receptor alpha gene are associated with fat distribution, a relationship that is potentially restricted to females [20].

In vitro studies show differences in lipoprotein lipase (LPL) activity, insulin-stimulated glucose uptake, and lipid synthesis between the sexes, with higher activity usually seen in females compared with males [20]. In vitro studies are generally an oversimplification, with the actual processes and mechanisms in vivo being more complicated. In vivo differences include storage of meal-derived fatty acids in SAT being higher in women than in men, and fatty acid storage in VAT being higher in men than in women [20].

Sex differences affect the release of energy from fatty tissues—women generally have higher levels of fat; because the primary function of adipose tissue is to store energy as intracellular triglycerides, and once again mobilize these stores as free fatty acids, it may be presumed that women have lower levels of lipolysis because they tend to store more fat. The opposite was found to be true—women had significantly higher lipolysis in comparison with resting expenditure than men did by about 40%. When accounting for a higher dependence of fat oxidation for women, fatty acid oxidation was comparable between the two, but women still demonstrated higher rates of lipolysis and higher levels of circulating free fatty acids than men [20]. A summary of the differences can be found in Table 1.

In relation to sex differences in ASC capabilities, Fitzgerald et al. found that 17-beta-estradiol was able to influence adipogenesis by increasing preadipocyte replication, and researchers found that high concentrations of dehydroepiandrosterone (DHEA) and rogen-related steroids blocked adipose conversion processes without a change in adipogenesis [19]. Adiponectin is a hormone that is responsible for regulating glucose levels and causing fatty acid breakdown and is produced in adipose tissues. An increasing amount of evidence has discerned that sex steroids such as testosterone and dihydrotestosterone (DHT) have inhibitive effects on 3T3-L1 preadipocyte differentiation [19, 21]. PPAR-gamma-2 expression was much greater (2.89-fold) in ASCs from female mice compared with male, but other studies did not see the same sex difference in yield and proliferation, a disparity that is potentially introduced by the age of the donor [21, 22]. A separate team found that ASCs sourced from younger patients proliferated at a faster rate than those from older patients, and apoptosis of ASCs was observed to be lower in younger individuals than in older. Researchers discerned that superficial fat depots of all ranges exhibited lower apoptosis, and this may also contribute to perceived differences in problems with replication above [23].

Sex and its influence on adipose tissue and ASCs is a topic that has been investigated thoroughly to account for the factors and differences that can occur. It is potentially difficult to isolate novel differences between sexes when discussing adipose tissue and stem cells owing to the many differences in technique, harvest location, age, and other miscellaneous factors between studies that are typically not controlled for. Uncovering novel sex differences has the potential to greatly progress the utilization and application of harvested cells between the sexes.

TABLE 1 Sex differences in fat deposition.

Sex	Sex hormone	Effect of hormone	Fat deposit site	Primary storage of fatty acids
Male	Androgens	Increased size of abdominal adipocytes	Abdominal, visceral	Higher in visceral adipose tissues
Female	Estrogens	Inhibits fat deposition in subcutaneous fat depots	Gluteal, thighs	Higher in subcutaneous adipose tissues

Race and ethnicity

The effect of race and ethnicity of patient donor cells combined with the characteristic differences due to sex can have a significant impact on ASCs. Torriani et al. selectively quantified intermuscular adipose tissue (IMAT), focusing on a number of factors including sex, ethnicity, and overall adiposity of subjects [24]. The study of African American, Asian, and Caucasian sedentary volunteers produced results as follows: at low levels of adiposity, all three groups studied were observed to have comparable levels of IMAT. However, with increasing levels of adiposity, African Americans demonstrated the greatest increase in IMAT per kilogram of fat as total level of adiposity rose among all groups evaluated [24]. An increase in overall adiposity resulted in the ratio of VAT to IMAT deposit to increase in Caucasians and Asians when compared with African Americans. Another study by Bacha et al. compared obese African Americans to obese Caucasian adolescents, and found that VAT was 30% lower in African Americans despite the factors BMI and total percentage body fat being held constant between the two [24].

Staiano et al. studied the differences in age-adjusted total body fat (TBF) between four groups: males, females, African Americans, and Caucasians [25]. It was observed after age adjustment that TBF was greater for African Americans and females compared with Caucasians and males who, when compared with one another, saw no significant differences in TBF. VAT of whites and males was greater than African Americans and females, respectively, as well with similar patterns occurring in SAT and VAT.

Carroll et al. studied differences in adipose tissue characteristics in African Americans, Hispanics, and white men and women over the age of 45. Researchers found that, for men and women in all groups, waist circumference and BMI were not significantly different [26]. Among the men, both white and Hispanic men had greater total VAT than the African American men, and among the women, Hispanics and whites had greater VAT than African Americans [26]. This study demonstrated that it may be beneficial to have limitation ranges for waist circumference or BMI to control for risk factors in different ethnic and racial groups in clinical studies.

Cumulatively, racial and ethnic adipose studies demonstrate significant differences between varying backgrounds in relation to total body fat, VAT, SAT, and BMI or waist circumference. The differences in these traits are imperative to understand to improve technique for adipose harvesting and to understand location and difference of fat depots between sex and ethnicity. Differences in adipose tissue deposition and volume in relation to racial differences potentially indicate that ASCs will have differing stemness potentials. Further studies should be performed that isolate ASCs from donors with differing ethnic backgrounds and compare their stemness and ability to differentiate after various passages. A prospective study could potentially produce significant data that will contribute to understanding the effect of race and ethnicity on adipose-derived MSCs.

Effects of patient health

There has been extensive research performed on donor site differences, and the most optimal harvesting technique, but there are still many factors of patient health that may affect ASCs that should be studied in controlled experimental settings, including the differences of stem cells received from donors that have varying BMI, immune status, and smoking status, which will be explored further in the chapter.

Body mass index

Varghese et al. performed a systematic review on factors affecting adipocyte function and viability, including the effect of BMI [21]. Eight studies were assessed that demonstrated the effects that an increasing BMI had on adipocyte function and viability. In the largest study to date ($n=189$), 30 obese women demonstrated a reduction in the viability of mature adipocytes per gram of adipose tissue, and there were distinct differences in differentiation capacity of ASCs when compared with increasing BMI [21]. These findings have been fortified by many others that found similar conclusions regarding a decrease in differentiation and proliferative capacity in relation with increasing BMI [27–29].

Frazier et al. examined ASCs in obese individuals and found that compromised osteogenic and early adipogenic potential correlated with the ability to form colonies in vitro, which was also found to be inversely proportional to the individual BMI [27]. Proliferative ability, differentiation potential, relative cell volume and complexity, and colony-forming potential of ASCs was also examined. Researchers observed that BMI did not affect late-time-point adipogenic differentiation in vivo and that it negatively correlated with extracellular matrix mineralization [27]. BMI did not affect ASC relative cell size and complexity in vitro as seen in some other studies [27].

BMI studies support that stem cells' differentiation ability, migration ability, and angiogenic and proliferative abilities diminish as BMI increases. Studies provide significant evidence that BMI is an important factor to consider for donors and fat storage. Perez et al. studied changes in telomerase activity, with a focus on DNA telomere length [30]. The data determined that ASCs had a decreased life cycle and reduced self-renewal abilities, which resulted in earlier apoptosis; this led to an overall decreased viability of adipose cells for long-term storage and has important ramifications for donor recovery and use [30].

The BMI studies mentioned above support that the cells show less viability and ability to differentiate as BMI increases, but mechanisms were not explored. Tang et al. studied the reduction in differentiation of enlarged ASCs in relation to mitogen-activated protein-4-kinase-4 (MAP4K4) expression [31]. The results concluded that the reduction of differentiation potential could be linked to MAP4K4 expression of adipogenesis by inhibition of PPAR-gamma activation, which is responsible for processes of adipogenesis.

It is important to note, however, that the negative effects of high BMI on adipose tissue and stem cells are somewhat reversible. Researchers determined that, after significant weight loss, subcutaneous tissues returned to a noninflammatory status [21]. Bariatric surgery and diet-induced long-term caloric restriction naturally reprogrammed ASCs, making cells similar to naïve ASCs with an improved viability and lifespan as compared with before [21]. ASCs that were harvested from formerly obese individuals underwent adipogenic differentiation more readily than those from nonobese individuals. As a counterpoint, there have also been multiple studies (six) that reported no significant association of BMI with ASC yield or function. The topic of BMI effect on ASCs needs to be further studied before a definite decision can be made; however, the consensus appears to be that BMI has a negative impact on many implications of ASC application, potential, and lifespan.

Immune status

In patients with severe trauma or that are immunocompromised. ASCs may often be used as an ongoing therapy, potentially being given in conjunction with chemotherapy and other radiation treatments that inhibit the immune system. It is important that situations involving a compromised immune status are thoroughly studied to prevent adverse side effects occurring in cell therapy patients. Varghese et al. studied the effects of immunosuppressive medications such as alemtuzumab and tacrolimus following composite tissue transplantation [21]. The drugs were shown to have deleterious effects on the ASCs, reducing their proliferative capacity and viability based on the dosage administered. Exposure to tamoxifen, a breast cancer treatment that acts as an estrogen modulator, resulted in apoptosis and inhibition of proliferation and differentiation in a dose- and time-dependent manner [21]. This finding is relevant for soft-tissue reconstruction, as ASCs may be used in conjunction with breast cancer treatments to maintain the form and natural contours of the human body. Liang et al. demonstrated that there was no reported difference in the ASCs when exposed to three common therapeutic agents [32].

Characteristics of ASCs such as immunomodulatory effects and the ability to reduce chronic immunosuppression are dependent on dose and cell passage number [33]. ASCs exhibited an efficient immunomodulatory effect that was able to suppress the proliferation of responder cells, an interaction that was shown to be contact-independent, with cells showing specific cytokine secretion [33].

Understanding the ability of ASCs to act in an immunomodulatory capacity and their interaction with medicines used to treat cancers and commonly used chemotherapeutic agents will further broaden the application of use for ASCs. The higher number of suitable situations for ASCs produces more information in support of ASCs as a therapeutic. Information can then be utilized in a manner that allows the implementation of more complicated clinical application of ASCs in the future that will reach a wider audience.

Smoking

Smoking is known to have detrimental effects on the bioactivity of many cells and processes throughout the body, but the effect of smoking on the bioactivity of MSCs has not been extensively studied. Barwinska et al. studied the effects of smoking on adipose stromal cells and found that it dramatically decreased the ability to promote recovery of blood perfusion in the ischemic tissue and support endothelial vasculogenic activity [34]. This was determined in response to many studies showing that MSCs increase the rate of tissue and organ recovery after acute injuries. Barwinska et al. examined cigarette-smoke effects on ASC morphology and proliferation rate, and found that proliferation, adipogenic differentiation, and cell surface markers were the same for the smoke and no-smoke groups [34]. In therapeutic testing of smoking on ASCs, the data revealed that no-smoke ASCs improved blood flow in chronic limb ischemia, while the cigarette-smoke group ASCs were ineffective at doing so. Barwinska et al. also tested the vasculogenic activity of both groups and observed that the cigarette-smoke groups were only able to produce vascular networks that were much simpler and less dense than the no-smoke ASCs [34].

Results indicate that there is no effect of cigarette smoke on the morphology and proliferation rate of ASCs but reveal significant differences in the therapeutic effects of cells that are exposed. Further evidence to corroborate existing data will need to be produced before the true effects of smoke exposure can be ascertained.

Clinical translation

Studies of adipose tissue retention based on donor demographics

The process of autologous fat grafting was first described by Dr. Neuber in Germany in 1893, during which time he used adipose tissue harvested from his patient's arm to repair a defect in an infraorbital scar. Fat grafting, also known as lipotransfer, is a commonly used procedure in regenerative and reconstructive surgery, with uses ranging from mitigation of volume deficits and improvement of contour to reduction of fibrosis and hypertrophic scarring. Although possessing great therapeutic promise, outcomes are variable, with fat survival rate varying from 15% to 83% within 6 months to 3.7 years of postsurgery follow-up [35, 36]. In addition to complications including fibrosis and scarring, complications corresponding to inadequate vascularization of the graft can occur [37].

Diabetes is known to impair wound healing, causing apprehension among surgeons about performing autologous fat grafting in diabetic patients. Concerns regarding the low rate of survivability of fat grafts in diabetic patients may be related to microvascular problems that are characteristic of the disease, in addition to a defect in the normally expected hypoxia-driven neovascularization. To address this gap in knowledge, Choi et al. investigated the survival rate and quality of autologous fat transfer in the Sprague–Dawley (SD) and Otsuka Long-Evans Tokushima Fatty (OLETF) rat, a rat model of diabetes. Through ultrasound analysis, autologous fat transfer survival rate was shown to be lower in SD than in OLETF rats (40.52% versus 53.38%), with a corresponding higher graft failure rate in OLETF rats (31.25% versus 6.25%).

In addition to metabolic derangements such as diabetes, lifestyle factors such as tobacco smoking may influence fat graft retention. Researchers studied fat graft retention in a SD-type rat model of cigarette smoking through a passive smoke exposure system, finding cigarette smoking to have a detrimental effect on fat graft survival. Researchers observed significant differences in graft retention in addition to serum levels of cotinine, tissue malondialdehyde (MDA), adipose tissue/fibrosis ratio, stem cell counts, perilipin positive cell density, and inflammation density between nonexposed versus cigarette smoke exposed treatment groups [38].

Clinical studies

Few studies investigating the characteristics that contribute to fat graft retention and survival in humans have been conducted. Patient factors including age, sex, BMI, donor site, and presence of concurrent patient pathologies and metabolic derangements, in addition to lifestyle factors such as tobacco smoking, are not yet fully discerned and remain a subject of interest. In a study enrolling 24 female patients, divided into age groups based on younger patients ≤45 years and older patients ≥46 years, and BMI defined as normal weight with a BMI <25 or overweight with a BMI of ≥25, fat viability was compared between donor sites including the lower abdomen, inner thigh, and flank. Younger patients demonstrated greater viability of lower abdominal fat than older patients, while older patients demonstrated higher adipocyte viability of the flank region in comparison with younger patients. Younger patients demonstrated greater adipocyte viability of the lower abdominal adipose tissue than that of the flank, while no differences were seen in older patients. No differences in fat viability were observed from inner thigh fat, or among the different BMI groups. Differences in patient age in relation to harvest site of fat graft may be an important factor in determining optimal patient harvest site for optimal graft retention.

Clinical studies demonstrate varying effects of patient age on fat graft retention. In a prospective study of 66 patients undergoing autologous fat transfer of the midface region, no significant differences in volume retention or fat resorption were observed between three different age groups including age ranges of 39–49 years, 50–59 years, and 60–70 years in the span of 1 year follow-up postsurgery [39]. In another prospective study, analysis of 142 patients with unilateral craniofacial contour deformities who underwent autologous free fat grafting, patient age, Parry–Romberg syndrome, previous craniofacial bone surgery, graft volume, and forehead unit were found to be independent negative predictors of fat graft retention (all $P < 0.05$), while the cheek unit was found to be an independent positive (all $P < 0.05$) predictor of fat retention [40].

Factors such as patient BMI have also been demonstrated to have an impact on graft viability, as mentioned previously. Adipose tissue harvested from high-BMI donors has been shown to demonstrate greater resistance to hypoxia-induced apoptosis as assessed via perilipin staining. Under hypoxic conditions, high patient BMI is also associated with a significant positive correlation with ANGPTL4, a gene that plays a role in cell survival. Molecular mechanisms such as an adipocyte's innate resistance to hypoxic conditions in addition to macrophage activation have been found to influence fat graft retention. Interestingly, induced differentiation of M1 macrophages, which are proinflammatory in nature, is found to be negatively correlated with BMI under hypoxic conditions [41]. Patient factors, such as fluctuations in body weight, may affect the volume of adipose tissue retained. The effects of cigarette smoking in a retrospective study of 18 cigarette-smoking patients undergoing facial lipotransfer with follow-up at a mean of 19.3 months demonstrated 40% mean fat survival rate, with the authors concluding that patients who smoke have decreased fat viability in comparison with nonsmokers, in addition to decreased impairment in skin quality [42]. However, the patients in this study were not compared with nonsmoking patients receiving autologous fat grafting.

It is important to remember that many of the methods for application of adipose tissues in the clinical setting originated in research. The fat graft studies on BMI, age, demographic, technique, harvest site, etc. that have been discussed in this paper are the groundwork for the clinical applications in use today. When an improvement or a breakthrough occurs in a laboratory owing to a better understanding of one of these aspects, the field advances together, clinical and experimental. The intrinsic versatility of ASCs (multipotency, ability to be differentiated) goes hand in hand with fat grafting and the work involved in this process. Fat grafting is one the major uses of adipose tissues in the clinical setting and is able to effectively make use of the ASC biology because of the preceding research studies and work that has contributed to this knowledge. This allows fat grafting to have diverse and effective clinical applications that involve greater retention, overall viability, and ease of access while still having an overall effective ability to work as multipotent stem cells. This procedure highlights the importance of groundwork research that contributes to the application and effective use of clinical techniques and how both processes loop back and support one another, ensuring a constantly evolving and expanding field of knowledge for application and study.

References

[1] N.K. Dubey, et al., Revisiting the advances in isolation, characterization and secretome of adipose-derived stromal/stem cells, Int. J. Mol. Sci. 19 (8) (2018).

[2] Z. Si, et al., Adipose-derived stem cells: Sources, potency, and implications for regenerative therapies, Biomed. Pharmacother. 114 (June 2019) (2019).

[3] P. Palumbo, et al., Methods of Isolation, characterization and expansion of human adipose-derived stem cells (ASCs): An overview, Int. J. Mol. Sci. 19 (7) (2018).

[4] K. Kishi, et al., Distribution of adipose-derived stem cells in adipose tissues from human cadavers, J. Plast. Reconstr. Aesthet. Surg. 63 (10) (2010) 1717–1722.

[5] A. Tsekouras, et al., Comparison of the viability and yield of adipose-derived stem cells (ASCs) from different donor areas, In Vivo 31 (6) (2017) 1229–1234.

[6] P. Palumbo, et al., In vitro evaluation of different methods of handling human liposuction aspirate and their effect on adipocytes and adipose derived stem cells, J. Cell. Physiol. 230 (8) (2015) 1974–1981.

[7] T. Tchkonia, et al., Identification of depot-specific human fat cell progenitors through distinct expression profiles and developmental gene patterns, Am. J. Physiol. Endocrinol. Metab. 292 (1) (2007) E298–E307.

[8] W.J. Jurgens, et al., Effect of tissue-harvesting site on yield of stem cells derived from adipose tissue: implications for cell-based therapies, Cell Tissue Res. 332 (3) (2008) 415–426.

[9] M. Rezai Rad, et al., Impact of tissue harvesting sites on the cellular behaviors of adipose-derived stem cells: implication for bone tissue engineering, Stem Cells Int. 2017 (2017) 9.

[10] T. Iyyanki, et al., Harvesting technique affects adipose-derived stem cell yield, Aesthet. Surg. J. 35 (4) (2015) 10.

[11] A. Bajek, et al., Does the harvesting technique affect the properties of adipose-derived stem cells?-the comparative biological characterization, J. Cell. Biochem. 118 (5) (2017) 1097–1107.

[12] T. Fontes, et al., Autologous fat grafting: harvesting techniques, Ann. Med. Surg. (Lond.) 36 (December 2018) 7.

[13] M.J. Oedayrajsingh-Varma, et al., Adipose tissue-derived mesenchymal stem cell yield and growth characteristics are affected by the tissue-harvesting procedure, Cytotherapy 8 (2) (2006) 166–177.

[14] W.Z. Wang, et al., Lidocaine-induced ASC apoptosis (tumescent vs. local Anesthesia), Aesthetic Plast. Surg. 38 (5) (2014) 1017–1023.

[15] H. Nie, et al., Effect of lidocaine on viability and gene expression of human adipose–derived mesenchymal stem cells: an in vitro study, PM&R 11 (11) (2019) 1218–1227.

[16] C. Muscari, et al., Comparison between stem cells harvested from wet and dry Lipoaspirates, Connect. Tissue Res. 54 (1) (2013) 34–40.

[17] R.J. Rohrich, S.J. Beran, P.B. Fodor, The role of subcutaneous infiltration in suction-assisted lipoplasty: a review, Plast. Reconstr. Surg. 99 (2) (1997) 514–519 (discussion 520-6).

[18] Y. Ullmann, et al., Searching for the favorable donor site for fat injection: in vivo study using the nude mice model, Dermatol. Surg. 31 (10) (2005) 1304–1307.

[19] S.J. Fitzgerald, et al., A new approach to study the sex differences in adipose tissue, J. Biomed. Sci. 25 (1) (2018) 89.

[20] K. Karastergiou, et al., Sex differences in human adipose tissues—the biology of pear shape, Biol. Sex Differ. 3 (1) (2012) 13.

[21] J. Varghese, et al., Systematic review of patient factors affecting adipose stem cell viability and function: implications for regenerative therapy, Stem Cell Res. Ther. 8 (1) (2017) 45.

[22] Z. Tao, et al., Estradiol signaling mediates gender difference in visceral adiposity via autophagy, Cell Death Dis. 9 (3) (2018) 309.

[23] D. Minteer, K.G. Marra, J.P. Rubin, Adipose-derived mesenchymal stem cells: biology and potential applications, Adv. Biochem. Eng. Biotechnol. 129 (2013) 59–71.

[24] M. Torriani, S. Grinspoon, Racial differences in fat distribution: the importance of intermuscular fat, Am. J. Clin. Nutr. 81 (4) (2005) 731–732.
[25] A.E. Staiano, et al., Ethnic and sex differences in visceral, subcutaneous, and total body fat in children and adolescents, Obesity (Silver Spring) 21 (6) (2013) 1251–1255.
[26] J.F. Carroll, et al., Visceral fat, waist circumference, and BMI: impact of race/ethnicity, Obesity (Silver Spring) 16 (3) (2008) 600–607.
[27] T.P. Frazier, et al., Body mass index affects proliferation and osteogenic differentiation of human subcutaneous adipose tissue-derived stem cells, BMC Cell Biol. 14 (2013) 34.
[28] P. Isakson, et al., Impaired preadipocyte differentiation in human abdominal obesity: role of Wnt, tumor necrosis factor-alpha, and inflammation, Diabetes 58 (7) (2009) 1550–1557.
[29] L.M. Perez, et al., Altered metabolic and stemness capacity of adipose tissue-derived stem cells from obese mouse and human, PLoS One 10 (4) (2015), e0123397.
[30] L.M. Pérez, et al., Obese-derived ASCs show impaired migration and angiogenesis properties, Arch. Physiol. Biochem. 119 (5) (2013) 195–201.
[31] X. Tang, et al., An RNA interference-based screen identifies MAP4K4/NIK as a negative regulator of PPARgamma, adipogenesis, and insulin-responsive hexose transport, Proc. Natl. Acad. Sci. U. S. A. 103 (7) (2006) 2087–2092.
[32] W. Liang, et al., Human adipose tissue derived mesenchymal stem cells are resistant to several chemotherapeutic agents, Cytotechnology 63 (5) (2011) 523–530.
[33] M. Waldner, et al., Characteristics and Immunomodulating functions of adipose-derived and bone marrow-derived mesenchymal stem cells across defined human leukocyte antigen barriers, Front. Immunol. 9 (2018) 1642.
[34] D. Barwinska, et al., Cigarette smoking impairs adipose stromal cell vasculogenic activity and abrogates potency to ameliorate ischemia, Stem Cells 36 (6) (2018) 856–867.
[35] N.-Z. Yu, et al., A systemic review of autologous fat grafting survival rate and related severe complications, Chin Med J (Engl) 128 (9) (2015) 1245–1251.
[36] V. Hromadkova, et al., The CD34+ cell number alone predicts retention of the human fat-graft volume in a nude mouse model, Folia Biol. (Praha) 65 (2) (2019) 64–69.
[37] S. Mou, et al., Extracellular vesicles from human adipose derived stem cells for the improvement of angiogenesis and fat grafting application, Plast. Reconstr. Surg. 144 (2019) 869–880.
[38] A. Ercan, et al., Effects of cigarette smoke on fat graft survival in an experimental rat model, Aesthetic Plast. Surg. 43 (3) (2019) 815–825.
[39] J.D. Meier, R.A. Glasgold, M.J. Glasgold, Autologous fat grafting: long-term evidence of its efficacy in midfacial rejuvenation, Arch. Facial Plast. Surg. 11 (1) (2009) 24–28.
[40] R. Denadai, et al., Predictors of autologous free fat graft retention in the management of craniofacial contour deformities, Plast. Reconstr. Surg. 140 (1) (2017) 50e–61e.
[41] S. Wang, et al., Molecular mechanisms of adipose tissue survival during severe hypoxia: implications for autologous fat graft performance, Plast. Reconstr. Surg. Glob. Open 7 (6) (2019), e2275.
[42] B. Ozalp, C. Cakmakoglu, The effect of smoking on facial fat grafting surgery, J. Craniofac. Surg. 28 (2) (2017) 449–453.

CHAPTER

10

Immunomodulatory properties of adipose stem cells in vivo: Preclinical and clinical applications

Matthias Waldner[a], *Fuat Baris Bengur*[b], *and Lauren Kokai*[c,d]

[a]Department of Plastic Surgery and Hand Surgery, University Hospital Zurich, Zurich, Switzerland [b]Department of Plastic Surgery, University of Pittsburgh, Pittsburgh, PA, United States [c]Departments of Plastic Surgery and Bioengineering, University of Pittsburgh, Pittsburgh, PA, United States [d]McGowan Institute for Regenerative Medicine, University of Pittsburgh, Pittsburgh, PA, United States

Introduction

The interaction between mesenchymal stem cells (MSCs) and the human immune system has been intensively studied over the last three decades. Since the initial discovery of their suppressive effects on lymphocyte proliferation in 2002 [1], multiple additional studies have described low immunogenicity and beneficial immunomodulatory properties of bone-marrow-derived MSCs (BMSCs) in vivo and in vitro. Because allogenic MSCs (cells from two individuals within the same species that are genetically dissimilar and thus immunologically incompatible) have showed little or imperceptible immune activation while maintaining beneficial modulation of the immunologic response [2], they are considered a gold standard cell-based therapy for immunomodulation. As such, there is substantial evidence of BMSC efficacy from preclinical and clinical applications treating a wide range of immune dysfunctions, from autoimmune disorders to allograft antigenicity in solid organ transplantation.

Similar properties as in BMSC were described for ASCs in 2005, and since then, ASCs have also been fervently investigated for multiple applications in immunomodulation [3]. Furthermore, because of their ease of harvest, proliferative capability, potent antiinflammatory and immunosuppressive effects, and genetic stability, ASCs are a highly appealing source for clinical use compared with other stem cell types. In 2006, ASCs were applied for the first time as a

https://doi.org/10.1016/B978-0-12-819376-1.00019-6

treatment in a graft vs host disease (GvHD) model associated with allogeneic hematopoietic transplantation. Results showed improved survival via inhibition of T-cell proliferation and significant reduction in proinflammatory cytokines such as tumor necrosis factor (TNF-α), interferon gamma (IFN-γ), and interleukin-12 (IL-12) [4]. Following this study, numerous additional applications were investigated that suggest ASCs are a therapeutic solution for a wide range of previously untreatable clinical problems.

The aim of this chapter is to describe the immunomodulatory effects of ASCs, with focus toward induction of immune cell differentiation and activation in vivo. The preclinical studies on this topic will be highlighted based on the disease models being studied. Finally, results of ongoing clinical trials and other clinical studies will be discussed.

Immunomodulatory effects

The immune system can be categorized as innate or adaptive protections from foreign pathogens. Innate immunity refers to nonspecific and rapid defense mechanisms such as epithelial barriers, neutrophils, macrophages, natural killer (NK) cells, or dendritic cells (DCs). Adaptive immunity, on the other hand, is a complex response that involves B cells and T cells, is highly specific, develops over long periods, and includes a memory function that makes future responses to antigens more efficient. Major histocompatibility complex (MHC) is an integral part of the adaptive immune response that comprises cell surface proteins responsible for antigen presentation to T cells and binding to T-cell receptors. These complexes are encoded by human leukocyte antigen (HLA) genes in humans.

Immunomodulatory properties of both BMSCs and ASCs are considered to be similar as they both lack HLA-DR, a MHC class II receptor, and they both express low levels of MHC class I, which allows them to be less immunogenic than other cell types and evade immune responses [3, 5]. Furthermore, both cell types similarly decrease T-cell proliferation and inflammatory cytokine production [6], and suppress lymphocyte reactivity in mixed lymphoid reaction. Furthermore, BMSCs and ASCs have both demonstrated efficacy in reducing inflammation in vivo. This immune-privileged status allows the use of allogeneic MSCs across different MHC-barriers (i.e., from one individual to another) without the need for immunosuppression. In addition, ASCs have been demonstrated to be more potent immunomodulatory effectors than BMSCs derived from age-matched donors, and ASCs have higher levels of cytokine secretion such as interleukin-6 (IL-6) and transforming growth factor beta (TGF-β1) [7]. Finally, ASCs are also more potent inhibitors of DC differentiation as they functionally inhibit the expression of potent costimulatory molecules CD80, CD83, and CD86 and upregulate antiinflammatory cytokine interleukin-10 (IL-10) secretion [8].

The umbrella term "immunomodulatory effect" of ASCs can be divided into two distinct actions: immunosuppressive or immunostimulatory. The latter is a dynamic effect that has been shown to be influenced by ASC culture expansion or "passaging." While primary stromal vascular cells (SVF) or low passage (P0) ASCs tend to demonstrate alloreactivity and promote allogeneic T-cell proliferation in mixed lymphocyte reactions (MLRs), ASCs at subsequent passages increasingly suppress lymphocyte proliferative; thus, ASCs are considered to have lost their immunostimulatory properties [5]. This can be partially explained by the heterogeneity of primary SVF obtained immediately following adipose enzymatic digestion, which includes some cell populations of antigen-presenting cell (APC) with MHC class I and II surface markers that trigger immune response in earlier passages [5, 9, 10].

Mechanisms

Immunomodulatory ASC effects are mainly used for their immunosuppressive properties, as there is abundant evidence from both in vitro and in vivo studies. ASCs regulate the immune system by two main mechanisms: direct cell-to-cell contact and paracrine-mediated, indirect communication via secretion of growth factors, extracellular vesicles, and other soluble factors [11].

T cells

ASCs modulate the immune response to alloantigens or other stimuli by suppressing or modifying T-cell proliferation and/or by directing T-cell differentiation. The primary laboratory technique used to monitor the effect of ASCs on lymphocytes after stimulation is the MLR, which uses alternative approaches to induce T-cell activation, including use of mitogens such as the NF-κB activator phorbol 12-myristate 13-acetate or CD3/CD28 stimulation. ASCs are then added either directly or indirectly to activated T cells to observe modulations of proliferative responses in lymphocyte subsets. The presence of ASCs in MLRs has been shown to inhibit proliferation of naïve $CD4^+$ T-helper (Th) cells. These effects appear to impede the cell cycle in the G0/G1 phase, reducing the number of activated T cells. In addition to their suppressive effects, ASCs direct differentiation of Th0 toward Th1, Th2, Th17, and $CD4^+$ forkhead box protein P3 $(FoxP3)^+$ regulatory T cells (Tregs). Significantly, MSC induction of $CD4^+$, $FoxP3^+$ Tregs has been hypothesized to be a central point in their immunomodulatory effects. On this axis, MSCs not only result in an inhibitory effect of the immune response to allogeneic antigens, but they also impart a tolerogenic influence. ASC induction of tolerance is therefore amplified by the influence of Tregs on DCs and cytotoxic T lymphocytes (CTLs), subsequentially altering their activation and differentiation. The balance between Treg activation and self-reacting T cells is of utmost importance to prevent autoimmunity and is therefore an important immune checkpoint.

Soluble factor indoleamine 2,3-dioxygenase (IDO) is an enzyme that catalyzes tryptophan to kynurenine, another key pathway in T-cell activation that can be inhibited by ASCs. Production of IDO by ASC is stimulated by interferons from activated immune cells and both suppresses immune cells and promotes generation of Tregs. Other important factors secreted by MSC are TGF-β and hepatocyte growth factor (HGF), as evident by significant loss of immunosuppressive effect when blocked. Further factors such as prostaglandin E2 (PGE2) have a role in dampening the immune response by induction of macrophages to release antiinflammatory IL-10, and HLA-G5 is crucial to suppress T-cell function and develop immunogenic tolerance [12].

Th1 cells are a primary cell type involved in cell-mediated immune responses, activating effector cells such as macrophages and/or cytotoxic $CD8^+$ T cells. ASCs modulate the function of Th1 cells and reduce secretion of proinflammatory cytokines such as IFN-γ and IL-2 and their susceptibility to inflammatory signals. The effects of MSCs are dependent on the humoral environment, establishing a complex balance between the proinflammatory and antiinflammatory cytokines. In the presence of ASCs, production of antiinflammatory IL-4 by Th2 cells is increased. This mechanism is important in autoimmune diseases, where MSCs reduce IL-4 and IL-5 production and increase Th1 cell activity. For example, in an allergic rhinitis model, ASCs inhibited IL-4 and IL-5 production and enhanced IFN-γ production, therefore inhibiting eosinophilic inflammation in the nasal mucosa partly via downregulation of Th2 and shifting to a Th1 immune response to allergens [13]. Similar effects have been shown on the activity of Th17 $CD4^+$ cells, which are involved in autoimmune diseases. Treatment of activated T cells with ASCs results in a specific plasticity of Th17 cells toward IL-10-secreting Th17 cells with increased mobility [14].

$CD8^+$ CTLs are the main effector cell of the immune system for targeting cells with intracellular pathogens via MHC class I, which is present on almost all cell types. CTLs play an important role in autoimmunity, acute rejection, and GvHD. Activation of CTLs is triggered following the interaction of the T-cell receptor with the MHC class I complex presenting processed peptide fragments. A second signal conducted via coreceptors CD80 or CD86 on APCs, monocytes, DCs, or B cells is necessary to trigger the activation of CTLs. Presence of ASCs during T-cell activation inhibits $CD8^+$ cell proliferation and downregulates the expression of the CD8 receptor, reducing cytotoxicity and increasing CTLs' regulatory phenotype. The downregulation of costimulatory signals (CD80, CD86) on monocytes by MSC results in a shift of DCs toward a suppressive phenotype and, therefore, in an inhibition of CTL activation [15].

B cells

B cells are central lymphocytes involved in various immunological diseases through cytokine secretion, antigen presentation, and antibody production. After solid organ- or vascularized composite allotransplantation (VCA), B cells are instrumental in the development of humoral rejection, namely, through production of donor-specific HLA antibodies (DSAs). In addition, B cells provide costimulatory signals that activate T cells. Furthermore, certain B-cell subsets called B regs ($CD19^+$ $CD38^{high}$ $CD24^{high}$) have regulatory functions and maintain immunological hemostasis via secretion of IL10.

ASC have been shown to influence the proliferation and differentiation of B cells, depending on the inflammatory conditions. These findings suggest that under inflammatory conditions, ASCs can suppress B-cell proliferation, while under conditions of immunological quiescence they support the formation of B regs [16]. For example, in a mouse model of systemic lupus erythematosus, ASCs decreased the size and number of germinal centers and effector B cells and expanded the population of B regs in spleen, ameliorating the autoimmune response [17].

One important mechanism in the interaction between ASC and B cells is the IDO-mediated tryptophan catabolism. By reducing tryptophan concentrations, B-cell proliferation is inhibited as described in T cells. The interaction between MSCs and B cells seems to be both cell-contact dependent and via soluble factors. IFN-γ is a key regulator of the effects of MSC on B cells, due to the induction of IDO activity. High levels of IFN-γ, present in an inflammatory microenvironment, lead to MSC-mediated inhibition of B-cell proliferation and reduced antibody production, while low levels support proliferation and differentiation into B regs.

Macrophages, natural killer cells, and dendritic cells

Macrophages are highly plastic myeloid cells that regulate cell proliferation in response to inflammatory conditions. While activated macrophages exist in a phenotype spectrum, they can be broadly classified into two main categories: proinflammatory M1 classically activated macrophages and M2 alternatively activated antiinflammatory and proangiogenic macrophages. ASCs crosstalk with macrophages and induce monocytes toward an M2 phenotype with immunomodulatory and antiinflammatory functions independent of cell-to-cell contact [18]. This effect is thought to be facilitated by the secretion of TGF-β and via tumor necrosis factor-inducible gene 6 protein (TSG-6) [19].

NK cells are important for recognition and elimination of cells infected by intracellular pathogens and tumor cells by interacting with MHC I molecules on the cell surface. ASCs have inhibitory potential on cytokine-induced NK cells, resulting in a G0–G1 cell cycle arrest, similar as in T cells. In the setting of cancer, strong inhibition of NK cells is not helpful as it would hinder graft versus tumor effects, important for neoplastic diseases such as leukemia.

DCs are APCs involved in the adaptive immune response and perform a key role in the interaction between innate and adaptive immune response. After capturing, processing, and presenting antigens, DCs interact with T cells, resulting in either recognition of self-antigens or activation. In addition to their antigen processing and presenting capabilities, DCs migrate to secondary lymphoid organs to prime lymphatic cells. The presence of ASC has been shown to reduce costimulatory signals such as CD80 and CD86 on DCs. Production of IDO by DCs is associated with higher levels of TGF-β secretion by ASC and therefore exerts an immunomodulating function on DC. ASC paracrine activity demonstrates robust and reliable suppression of DC maturation and T-cell-activating function, an additional mechanism for combating inflammation and autoimmune conditions.

Purine metabolism

One mechanism through which inflammation is initiated following tissue injury or insult is through release of extracellular adenosine triphosphate (ATP) and various other nucleotides, which act as danger signals and chemotactically recruit and activate neutrophils, macrophages, DCs, and memory T cells [20]. Extracellular ATP activates ATP-gated ion channels known as P2X receptors and G-protein-coupled P2Y receptors. P2X or P2Y receptors control, among many things, calcium entry into cells, membrane pore formation, macrophage chemotaxis, microglial activation, and platelet activation [21–23]. As a counterbalance to purine-induced inflammation, the ectoenzymes CD39 and CD73 rapidly convert extracellular ATP to adenosine, which promotes angiogenesis and is antiinflammatory, activating A_{2A} and A_{2B} adenosine receptors on most immune cells. Adenosine receptor coupling to intracellular pathways is highly cell type dependent and temporally regulated. For example, in macrophages, A_{2A} receptors are potently upregulated after exposure to NF-κB activators TNF-α and IL-1α [24]. Following A_{2A} ligation, adenosine can either compete with the NF-κB pathway by activating adenylate cyclase to generate intracellular cAMP and activate protein kinase A (PKA), or induce hypoxia inducible factor (HIF) through activation of MAPKs and PKC [25].

Effects of ASCs on the metabolism of purines are another integral function demonstrating immunomodulatory properties. It has been demonstrated that purinergic receptors and ectoenzymes are present on the surface of MSCs, and purine signaling likely has an important role in differentiation, proliferation, cell death, and successful engraftment of ASCs in the extracellular environment [26]. Ectonucleotidases including CD39 (ectonucleoside triphosphate diphosphohydrolase) and CD73 (ecto-5′-nucleotidase) are the most prominent cell-surface enzymes that regulate purinergic signaling by converting ATP to ADP and to the nucleoside adenosine. Expression levels of CD73 have been linked with pronounced antiinflammatory activity by attenuating infiltration of $CCR2^+$ macrophages and upregulating antiinflammatory genes [27]. This, in fact, has been proposed as a possible explanation for the inconsistencies in regenerative and immunomodulatory results with MSCs.

Migration and homing: In vivo tracking

To establish their immunomodulating potential in vivo, ASC must be transferred to the organism after being processed ex situ. Whether applied intravascularly or locally, the distribution of ASCs within the organism is an important component in the immunomodulation mechanism. After the cells are administered, in addition to migrating to the target area, they can migrate to various anatomical locations including lymphoid organs. To study these mechanisms, in vivo cell tracking has provided new insights into homing, migration, and long-term fate of applied cells. These techniques are also helpful for revealing the complex roles that other cell types play in the biological process. New techniques allowing a combination of optical and magnetic resonance imaging allow real-time tracking and high-resolution 3D imaging of cell distribution [28]. These studies have demonstrated that the fate of applied ASCs is dependent on the method of application and that cell migration occurs after initial inoculation. When applied intravenously, a significant number of cells become trapped in filtering organs, such as the small arterioles in the lung and liver [29]. This effect is decreased when applied intraarterially.

The capability of MSCs to migrate toward the sites of inflammation through chemotaxis is one of the major advantages of MSC-based therapies. ASCs have been shown to home to multiple tissues such as myocardium, bone, and muscle, and they may even pass the blood–brain barrier when inflammatory stimuli are present. The expression of integrin-α4 by ASCs enables cells to migrate via the VCAM-1 pathway [30], rendering them more mobile than bone-marrow-derived cells. Multiple chemokine signals such as IFN-γ and TNF-α are also known to be involved in the regulation of cell migration, leading ASCs to the location of demand, although conflicting diseases such as infections may interfere with the correct homing of cells.

Transplantation

The main drawback of organ or tissue transplantation is the chronic need for immunosuppression to prevent graft rejection. Various immunosuppressive agents have been developed over the years, but almost all of them can lead to severe side effects after long-term use, such as higher incidence of malignancies, diabetes, cardiovascular diseases, renal failure, or GvHD. One way to overcome the sequelae of chronic immunosuppression is development of tolerance, meaning that no rejection against the graft occurs while maintaining an acceptable graft function. Use of MSCs as immunosuppressive therapies suggests that durable tolerance may be achievable, enabling immunosuppressants to be reduced or eliminated. Due to their profound effects on the immune system, ASCs have become strong candidates for immunomodulation in transplantation patients [31]. In addition to the suppressive effects on the immune response, ASCs have tolerogenic properties, supporting donor cell engraftment and promoting chimerism [32]. In solid organ transplantation, the use of ASC in the peritransplant period has been increasingly investigated over the last decade. ASCs are also actively investigated in the fields of hematopoietic and vascularized composite allotransplantation [9, 33]. The cell source for the use of ASCs is also another area of research where studies have shown no significant difference in the immunosuppressive effect of ASCs from donor-derived, recipient-derived, or third-party cells [34]. In general, when MSCs are used in the setting of transplantation, initial concomitant immunosuppressive therapy is necessary to overcome the acute rejection phase [35]. Toxicity and functional studies have demonstrated that the presence of immunosuppressive agents and lymphocyte-depleting agents did not have detrimental effects on ASC at clinically relevant doses. However, the interaction of ASC and the presence of immunosuppressive drugs in vivo is still not fully understood. In addition, GvHD needs to be prevented while maintaining the fine balance between the donor and recipient cells.

Solid organ transplantation

Kidney transplantation

Kidney transplantation is a treatment option for end-stage kidney disease, as successful transplant reduces mortality and improves quality of life of patients compared with long-term dialysis. BMSCs have been investigated in multiple preclinical studies and clinical trials and show promise for autologous and allogeneic (donor-derived) cells, as the infusion of cells is well tolerated by the patients [36]. Their main effects are through promotion of donor-specific tolerance via Tregs and APCs. Similarly, ASCs have been shown to significantly prolong graft survival in animal experimental models by reducing the severity of acute rejection and downregulating activated T-cell response. Applications can vary between systemic and local (e.g., intraarterial injection to the renal artery) administrations. In a rat model, autologous ASCs injected through the renal artery reduced the severity of acute rejection and led to prolonged graft survival by reducing the number of infiltrated $CD4^+$ and $CD8^+$ T cells [37]. In addition, increased levels of the antiinflammatory TSG-6 was found to be effective for suppressing alloreactive T cells through downregulating CD44. Synergistic immunosuppressive effect of autologous ASC and OX40-OX40L, a costimulatory receptor expressed primarily on activated $CD4^+$ and $CD8^+$ T cells, blockade attenuated the acute rejection in a rat model [38]. This prolonged graft survival, reduced rejection, downregulated IFN-γ, and upregulated IL-10, TGF-β, and $FoxP3^+$.

Heart transplantation

The main causes for end-stage heart failure are coronary artery disease, nonischemic cardiomyopathies, and restrictive or congenital heart diseases. In patients with symptomatic end-stage heart failure despite elaborated medical therapy, cardiac transplantation is the therapy of choice. However, heart transplantation is associated with a substantial mortality during the first year. Pretreatment with MSCs has been shown to prolong graft survival via differential secretion of IFN-γ and improves graft function if combined with certain immunosuppressive agents [35]. Due to the unique regenerative capacities of MSC, the dual effect of improved graft repair and immunomodulation is of great interest in the field of allogeneic heart transplantation. ASCs have been applied after preconditioning with Toll-like receptor (TLR) 3 activator in a fully mismatched MHC heterotopic heart transplantation in mice [39]. This resulted in increased graft survival by inhibition of lymphocyte proliferation and upregulation of an effector molecule of Tregs, fibrinogen-like protein 2 (FGL2).

Vascularized composite allotransplantation

Vascularized composite allotransplantation describes transfer of a composite graft containing multiple tissues, such as skin, muscle, bone, fat, nerves, and lymph nodes, mainly for the life-enhancing purpose to achieve functional recovery and patient satisfaction. It differs from solid organ transplantation in that composite tissue grafts have more diverse and extremely antigenic properties that necessitate more potent immunosuppressive treatments. Clinical applications include hand and forearm, face, uterus, penis, abdominal wall transplantation, and more. Skin can be considered as the most immunogenic component among the tissues being transferred in a VCA and is the first tissue type to demonstrate signs of rejection. Therefore, skin allograft models are used to test outcomes in VCA trials. The use of ASC in combination with immunodepleting and immunosuppressive agents has been studied in multiple preclinical models with prolonged survival [9]. In addition, ASCs were also found to ameliorate chronic graft vasculopathy, which is a long-term sequela of VCA, after initial acute rejection.

Systemic applications of ASC results in a prolonged graft survival in a rat hindlimb transplantation model [32]. In these models, different groups have demonstrated that a single dose of ASC can induce transient peripheral chimerism and an increased level of Tregs. In addition, inhibition of Th17 response was also identified with increased skin allograft survival [40]. Although both ASCs and BMSCs extend allograft survival, they utilize different mechanistic and molecular pathways, and their mechanisms are not yet well defined [41]. Combination of multiple injections of ASCs with antilymphocyte serum and transient cyclosporin A also resulted in prolonged allograft survival [42]. The underlying mechanisms were identified as suppressing T-cell proliferation, increasing levels of $CD4^+$ $CD25^+$ $FoxP3^+$ Treg, and upregulating TGF-β and IL-10 levels. Similar effects were observed when recipient-derived autologous ASCs were used [43]. It appears that the increased levels of $CD4^+$ $CD25^+$ $FoxP3^+$ Tregs are also important in tolerance induction and maintenance in recipients of MHC-complex mismatched VCAs [44]. The effects of prolonged survival, reduced inflammatory cell infiltration, and induced Tregs were maintained even without the addition of immunosuppressant to the ASCs [45]. Accumulation of immunosuppressive M2 macrophages, suppression of B cells, prevention of DC differentiation, and downregulation of costimulatory molecules have also been described as potential mechanisms for the immunomodulatory role of ASCs in VCA [31]. Chemokine receptor 7 (CCR7) was also found to be effective in targeted migration of ASCs to secondary lymphoid organs to intensify their in vivo immunomodulatory effect [46], with future promise regarding targeting strategies to improve therapeutic effects of ASCs.

To maximize the therapeutic potential of ASCs, single dose versus repetitive applications of ASCs have been tested [47]. When compared with single injections, repetitive, posttransplant injections of donor-derived ASCs were found to be more effective in improving graft survival, increasing Treg levels, and reducing infiltrative leukocytes. Therefore, it was concluded that the timing and dosing of ASC administration are crucial for VCA survival and development of peripheral chimerism.

Graft versus host disease

Allogeneic hematopoietic stem cell transplantation (HCT) is an established therapy for different hematological or immune disorders. This is performed by transplantation of hematopoietic stem cells and progenitor cells from different sources (such as bone marrow, peripheral blood, cord blood), as either autologous or allogeneic from related or nonrelated donors. In hematological malignancies, the donor T cells play an important role to effectively recognize and attack host tumor cells. Unfortunately, after HCT, donor cells can recognize host tissues as the target cells owing to their inflammatory triggers and cause GvHD. Gastrointestinal system, respiratory tract, skin, and liver are the most affected tissues in the recipients. It is the most common complication after HCT and a main cause for non-relapse mortality in these patients. Current preventative and treatment options have the disadvantages of infectious complications, delayed immune reconstitution, and high risk of relapse.

MSCs hold great potential to treat GvHD with their capacity to exert similar immunomodulating effects on recipient and donor T cells by suppressing CD4, CD8, and NK cells, inducing Tregs, and inhibiting the proliferation of B cells. Multiple studies including clinical trials using BMSC have been performed, demonstrating various results to prevent therapy-refractive GvHD. ASCs have also been applied, resulting in survival improvement via inhibition of T-cell proliferation and significant reduction in proinflammatory cytokines [4]. A possible explanation for the decreased severity of GvHD and improved survival can be the influence of ASCs on the balance of IL-4 and INF-γ and the ability to promote long-term hematopoiesis [48]. This probably causes a change in the cytokine milieu from a proinflammatory environment to an antiinflammatory environment. Nonetheless, in a more recent humanized mouse model, although MSCs from various origins had different effects on immune cells both in vitro and in vivo, none of them were able to significantly prevent death from GvHD [49]. Additionally, in vitro procoagulant activities of ASCs were highlighted with the suggestion of close monitoring while they are being applied.

Clinical applications of ASCs for GvHD are currently under investigation for their safety and efficacy. Initially, patients with severe refractory acute GvHD were successfully treated with ASCs from HLA-mismatched unrelated donors as a last-resort treatment option [50, 51]. In a clinical trial, patients with chronic GvHD underwent ASC treatment with cyclosporine and prednisone, and relapse and mortality due to infection was not observed [52]. The trial demonstrated that this treatment can be safe and is likely to have an impact on the disease course in the study, with an increase in CD19, CD4, and TNF-α levels and a temporary decrease in NK cells. In the various other clinical studies investigating the effects of MSC after systemic application, there exists a great variation of protocols regarding dose, timepoints, cell passaging, and isolation strategies. Therefore, the potential of stem cell treatment for GvHD deserves further exploration through future structured clinical trials.

Autoimmune disorders

Inflammatory bowel disease

IBD is determined by the two major disorders ulcerative colitis and Crohn's disease. While ulcerative colitis mainly involves the colon, Crohn's disease can involve inflammation of any part of the gastrointestinal system. Both disorders are developed by a pathological interaction between the intestinal immune system and microbial factors. The main underlying mechanism is considered as dysfunction of mucosal T cells, resulting in a prolonged and uncontrolled cytokine production and inflammation. The ongoing inflammation can lead to disruption of the intestinal epithelial barrier, causing bacterial colonization or fistulas. The mainstay of treatment for the Crohn's fistulas is focused toward surgical interventions with immunosuppressants and antibiotics. However, these approaches can lead to high recurrence rates, and postoperative complications such as incontinence.

ASCs have been investigated for the treatment of fistulas because of their immunomodulatory properties, mainly inhibition of T-cell proliferation [53]. In a mouse model of colitis, systemic ASC infusion has been shown to strongly reduce the mucosal inflammation by downregulating Th1-driven response, decreasing inflammatory cytokines (such as TNF-α, IFN-γ), increasing the regulatory cytokine IL-10, and inducing $CD4^+$ $CD25^+$ $FoxP3^+$ Tregs [54]. This, in fact, ameliorated both clinical and histopathological signs of the disease, protected against severe sepsis, and reduced the mortality [55]. The mechanism of Tregs in ASC therapy has been evaluated, and it was revealed ASCs induce Tregs by activating latent TGF-β via thrombospondin-1 (TSP-1), in a process that is independent of $CD103^+$ DC induction [56]. In addition, PGE2 secreted from the ASCs was shown to reduce inflammation by increasing $FoxP3^+$ Tregs [57].

In multiple clinical trials, ASCs have been applied to treat fistulas, resulting in an increased rate of fistula healing and closure [58–60]. These effects on clinical outcomes were likely due to the capability of ASCs to reduce inflammation, which is established through an increase in antiinflammatory, a decrease in proinflammatory cytokines, and conversion of macrophages to a regulatory phenotype [61]. In addition, induction of IDO and subsequent inhibition of T-lymphocyte function and proliferation of Tregs is also a proposed mechanism [62]. Multiple meta-analyses revealed that there is a significant improvement of fistula closure and reduction of recurrence when ASCs have been used, especially in patients with complex fistulas that cannot be healed by conventional procedures [63, 64]. Although the current data highly support their beneficial role, further research and large-scale randomized controlled trails are needed to confirm their clinical effects.

Rheumatoid arthritis and osteoarthritis

RA is an autoimmune disease characterized by chronic inflammation of the synovium and progressive bone and cartilage destruction. With progression, RA leads to deformity through weakness of tendons and destruction of joints with ongoing erosion of the bone and cartilage. If not treated or responsive to therapy, RA leads to a loss of function with inability to carry out daily activities. Mechanisms underlying RA are not fully understood, but genetic variations (e.g., MHC I), environmental stimuli, and dysregulation of the immune response are held responsible. The inflammatory environment in joints affected by RA is the result of proinflammatory Th1 and Th17 cell activation, macrophages secreting proinflammatory cytokines, and production of autoantibodies by B cells. Despite the novel developments in the treatment including antiinflammatory drugs and biological agents, there still is a need for a more effective therapy with fewer side effects. MSCs are a promising source to address the ongoing inflammation involved in cartilage and bone damage.

Systemic administration of human ASCs in mice resulted in a decreased severity of arthritis through reduced number of pathogenic GM-CSF$^+$ CD4$^+$ T cells in the spleen and peripheral blood, and increased number of FoxP3$^+$ CD4$^+$ and IL10$^+$ IL17$^-$ CD4$^+$ Tregs, and Th17 cells expressing antiinflammatory IL10 response in the lymph nodes [14]. In addition, downregulation of proinflammatory cytokines TNF-α, IL-1β, and IL-6 was observed [65]. Homing of MSC to lymphoid tissues is a major step after systemic applications to exert their immunomodulatory effects. Therefore, intralymphatic administration has been proposed to improve efficacy of the therapy [66]. Similar mechanisms of the antiinflammatory response were observed leading to reduced severity and progression of arthritis. To localize the effects of the treatment to the area of the disease, intraarticular injections directly to the targeted joint have been applied. In a mouse model, this resulted in reduced intraarticular inflammation and increased regeneration of the damaged cartilage through inhibition of synovial fibroblasts and activated macrophages, and upregulation of TSG-6 and TGF-β1 [67]. Clinically, the safety and preliminary evidence of efficiency have been described. Intravenous infusion of allogeneic ASCs in a small cohort was well tolerated without the presence of adverse events and had promising results in terms of possible efficacy [68].

Osteoarthritis (OA) is a degenerative joint disease characterized by irreversible damage to the cartilage, subchondral bone, and the surrounding soft tissue. As the cartilage within the joint space is avascular, it has limited ability of self-renewal, leading to progressive loss and ultimately degeneration of the joint itself. Effective treatment options are currently limited, and applications of ASCs, especially in the form of SVF, have yielded promising clinical results with minimal complications [69]. Intraarticular injection of human ASCs in a mouse model of arthritis has demonstrated their safety, and tracking of the biodistribution of the injected cells revealed that the majority of the ASCs remained at the joint and a number of cells migrated to the bone marrow, adipose tissue, and muscle [70]. Clinically, various trials have given intraarticular injections of autologous ASCs and SVF to patients with OA of the knee [61]. Intraarticular injection of ASCs has been shown to improve function and reduce pain of the knee joint without causing adverse events [71–73]. In addition, ASCs also reduced cartilage defects by regeneration of the articular cartilage [71].

Type 1 diabetes mellitus

Type 1 diabetes mellitus (DM) is an immune-mediated disease characterized by destruction of the insulin-producing pancreatic β-cells and severe hypoinsulinemia, hyperglycemia, and abnormal glucose metabolism. The treatment is inevitably exogenous insulin replacement, which makes it difficult for the patients to maintain ideal blood insulin levels. MSCs, especially ASCs, have been promising resources in the treatment, with their regenerative and immunomodulatory potential by upregulation of Tregs and downregulation of cytotoxic T cells [74]. In addition to their function as a source of insulin-producing cells for transplantation [75, 76] and their protective properties from graft rejection, their immunomodulatory role to support and improve the function and proliferation of resident pancreatic islets has been an area of interest [77, 78]. The first evidence on the effect of ASCs on regulation of the immune response demonstrated recovery in a mouse model through downregulation of the $CD4^+$ Th1-biased immune response and enhancement of Tregs in the pancreatic lymph nodes [79]. In addition, inflammatory cell infiltration and IFN-γ levels were reduced within the pancreas itself, while insulin, pancreatic duodenal homeobox-1, and active TGF-β1 expressions were improved. The roles of immunomodulatory cytokines were elucidated in a streptozotocin-induced diabetic mouse model, where vascular endothelial growth factor and tissue inhibitor of metalloproteinase 1 (TIMP-1) were found to be upregulated and contribute to the prevention of pancreatic beta cell death [80].

Multiple sclerosis

Multiple sclerosis (MS), is a neurological autoimmune disease of the central nervous system, characterized by chronic inflammation, demyelination, and axonal damage. Different potential mechanisms for the underlying pathophysiology exist, of which the most common is the inflammatory reaction involving autoreactive T- and B-cells. In later stages of the disease, microglial activation and chronic neurodegeneration are predominant, where lymphocyte infiltration and production of inflammatory cytokine such as IFN-γ cause loss of oligodendrocytes. The use of ASCs or SVF has been shown to significantly delay the onset of disease, reduce signs of motor impairment, and slow disease progression in animal models of MS (experimental autoimmune encephalomyelitis) [81]. Systemic administration of ASCs has been shown to both induce neuroregeneration by endogenous oligodendrocyte progenitors and suppress the inflammation and subsequent demyelination [30]. The latter, immunomodulatory effects on the autoimmune response, have been orchestrated by inducing a Th2-type cytokine shift of antigen-specific CD4 T cells in lymph nodes. In addition to primarily homing into lymphoid organs, ASCs have been also identified in the central nervous system possibly via an expressed activity of a4-integrin-dependent migration that adheres to inflamed parts. The role of induced CD4$^+$ CD25$^+$ FoxP3$^+$ Tregs and increased IL-4 secretion have also been demonstrated in mouse models where intraperitoneal injections of ASCs were applied with resulting immunomodulatory and neuroprotective effects [82, 83]. In addition, splenocyte proliferation and IL-17 secretion were downregulated. SVF has also been demonstrated to be effective in ameliorating the neuroinflammation via increased levels of IL-10 and TGF-β, and induced Tregs [84, 85]. SVF treatment correlated with diminished activity of the Th type 1 cells, and during early stages, ASCs were found to enhance the expansion of T cells through IL-6 signaling [85]. In general, application of ASC or SVF can reduce the number of lesions, decrease demyelination, and even improve remyelination in the CNS via secretion of a broad range of cytokines (such as TGF-β, IL-10, IL-4, IDO) and inhibition of T-cell activation and infiltration [61].

Other conditions

In addition to the above-mentioned uses, there have been numerous applications of ASCs for their immunomodulatory properties in various other fields.

Ischemic conditions

In ischemic conditions such as myocardial infarction (MI) or stroke, ASCs have shown value for improved recovery. Their effects are not only exerted by their role in increased angiogenesis [86, 87], or differentiation and remodeling [88], but also through immunomodulatory effects on the ongoing inflammation after the ischemia. In models of myocardial ischemia, ASCs have shown improved function by reducing inflammation and fibrosis and increasing angiogenesis.

Immune mechanisms are activated after organ ischemic injuries and followed by generation of oxidative stress and reactive oxygen species. ASCs have been found to reduce the inflammation induced by oxidative stress and reactive oxygen species in a rat model of ischemic stroke [89]. In a murine cerebral ischemia model, systemic administration of ASCs has shown their neuroprotective effects; and the expression of numerous growth factors, cytokines, and chemokines related to inflammation, immune response, and cell death were found to be significantly changed [90]. Similar effects have been demonstrated with intraarterial transplantation of ASCs into carotid artery in a mouse model of cerebral ischemia, where inflammation during the early stages was attenuated [91]. In terms of the allogeneic and autogenic cell source in treatment of cerebral ischemia, allogenic ASCs have demonstrated significantly stronger expressions of IL-2 and IFN-γ, and local accumulation of CD4$^+$, CD8$^+$ T-lymphocytes, and microglial cells [92].

Neurodegenerative disorders

The pathogenesis and progressions of neurodegenerative diseases are modulated up to some level by the degree of ongoing neuroinflammation including proinflammatory cytokines such as TNF-α, IL-1, IL-6, and IL-12 [93]. Therefore, this enabled the development of a new therapeutic area for ASCs in neurodegenerative disorders, in which they have been already applied for their neurogenic and neuroprotective potential, such as for Parkinson's disease [94]. In a mouse model of Alzheimer's disease, ASCs have been shown to reduce oxidative stress and improve cognitive impairment [95]. In a mouse model of amyotrophic lateral sclerosis model, ASCs have demonstrated reduced levels of p38 mitogen activated protein kinase, which is involved in inflammation and neuronal death [96]. In an early clinical trial on patients with amyotrophic lateral sclerosis, intrathecal administration of ASCs appeared to be safe, but without any significant improvements [97].

Chronic obstructive pulmonary disease and idiopathic pulmonary fibrosis

ASCs have demonstrated their immunomodulatory capacity by their ability to reduce inflammation, neutrophils, and oxidative damage in the airways in models of chronic obstructive pulmonary disease, a progressive lung disease associated with chronic bronchitis and lung parenchyma destruction [98], and idiopathic pulmonary fibrosis, a progressive fibrotic lung disorder [99, 100].

Vasculitis

In a mouse model of Kawasaki disease, systemic administration of ASCs lowered the inflammation in coronary arteries by downregulation of the levels of proinflammatory cytokines, such as IL-1b, IL-12, IL-17, IFN- γ, and TNF-α [101].

Conclusions and future directions

The immunomodulatory properties of ASCs make them a valuable treatment option for numerous immune-mediated diseases. In addition to their ability of self-renewal and uses in regenerative medicine, the number of preclinical studies and clinical trials focusing on the in vivo immunomodulatory effects of these cells and outcomes of their use has been increasing over the past years. However, they are still far from becoming standardized methods owing to complexity in isolation, culture, differentiation, and characterization, and regulatory concerns. Although it might seem like they are an easily accessible source of stem cells, patient factors such as age, sex, body mass index, and smoking status have major effects on their function. In addition, altering the immune system by means of affecting the ability of the protective cells to recognize foreign cells or exacerbating the immune response could potentially lead to unpredicted outcomes. Therefore, expansion of their current applications and discovery of novel applications require a thorough and in-depth analysis considering microenvironmental and patient-related factors. Understanding the basic mechanisms of how the cells react within an in vivo system will aid in standardization, which in turn will facilitate clinical translation via creation of experimental protocols, and less variability within the results for clinical trials with larger cohorts. Once the barriers that prevent convenient clinical translation have been lifted, the future is bright in terms of applying recently developing

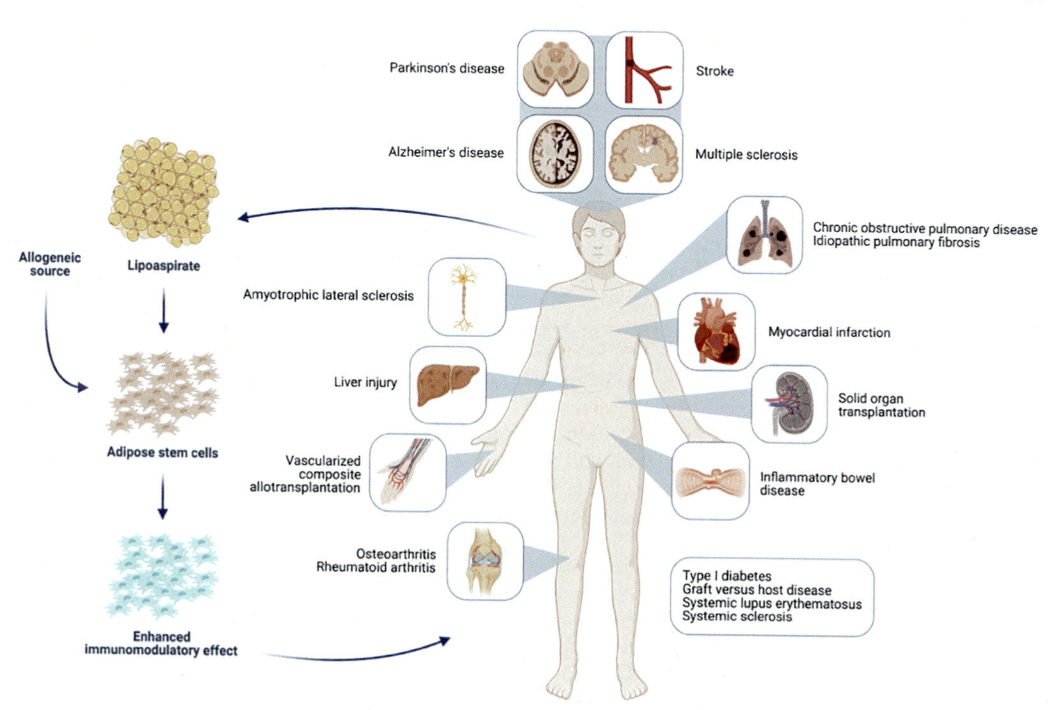

FIG. 1 Potential future clinical applications of adipose stem cell-based immunomodulatory therapies, based on currently available in vivo studies. Created with Biorender.com.

technologies of genetical engineering, to effectuate the ideal cell with enhanced immunomodulatory functions related to the target disease [12] (Fig 1).

References

[1] A. Bartholomew, C. Sturgeon, M. Siatskas, K. Ferrer, K. McIntosh, S. Patil, et al., Mesenchymal stem cells suppress lymphocyte proliferation in vitro and prolong skin graft survival in vivo, Exp. Hematol. 30 (1) (2002) 42–48.

[2] W.T. Tse, J.D. Pendleton, W.M. Beyer, M.C. Egalka, E.C. Guinan, Suppression of allogeneic T-cell proliferation by human marrow stromal cells: implications in transplantation, Transplantation 75 (3) (2003) 389–397.

[3] B. Puissant, C. Barreau, P. Bourin, C. Clavel, J. Corre, C. Bousquet, et al., Immunomodulatory effect of human adipose tissue-derived adult stem cells: comparison with bone marrow mesenchymal stem cells, Br. J. Haematol. 129 (1) (2005) 118–129.

[4] R. Yanez, M.L. Lamana, J. Garcia-Castro, I. Colmenero, M. Ramirez, J.A. Bueren, Adipose tissue-derived mesenchymal stem cells have in vivo immunosuppressive properties applicable for the control of the graft-versus-host disease, Stem Cells 24 (11) (2006) 2582–2591.

[5] K. McIntosh, S. Zvonic, S. Garrett, J.B. Mitchell, Z.E. Floyd, L. Hammill, et al., The immunogenicity of human adipose-derived cells: temporal changes in vitro, Stem Cells 24 (5) (2006) 1246–1253.

[6] J. Valencia, B. Blanco, R. Yanez, M. Vazquez, C. Herrero Sanchez, M. Fernandez-Garcia, et al., Comparative analysis of the immunomodulatory capacities of human bone marrow- and adipose tissue-derived mesenchymal stromal cells from the same donor, Cytotherapy 18 (10) (2016) 1297–1311.

[7] S.M. Melief, J.J. Zwaginga, W.E. Fibbe, H. Roelofs, Adipose tissue-derived multipotent stromal cells have a higher immunomodulatory capacity than their bone marrow-derived counterparts, Stem Cells Transl. Med. 2 (6) (2013) 455–463.

[8] E. Ivanova-Todorova, I. Bochev, M. Mourdjeva, R. Dimitrov, D. Bukarev, S. Kyurkchiev, et al., Adipose tissue-derived mesenchymal stem cells are more potent suppressors of dendritic cells differentiation compared to bone marrow-derived mesenchymal stem cells, Immunol. Lett. 126 (1–2) (2009) 37–42.
[9] M. Pappalardo, L. Montesano, F. Toia, A. Russo, S. Di Lorenzo, F. Dieli, et al., Immunomodulation in vascularized composite allotransplantation: what is the role for adipose-derived stem cells? Ann. Plast. Surg. 82 (2) (2019) 245–251.
[10] A.A. Leto Barone, S. Khalifian, W.P. Lee, G. Brandacher, Immunomodulatory effects of adipose-derived stem cells: fact or fiction? Biomed. Res. Int. 2013 (2013) 383685.
[11] S. Al-Ghadban, B.A. Bunnell, Adipose tissue-derived stem cells: Immunomodulatory effects and therapeutic - potential, Physiology (Bethesda) 35 (2) (2020) 125–133.
[12] S. Ceccarelli, P. Pontecorvi, E. Anastasiadou, C. Napoli, C. Marchese, Immunomodulatory effect of adipose-derived stem cells: the cutting edge of clinical application, Front. Cell Dev. Biol. 8 (2020) 236.
[13] K.S. Cho, H.K. Park, H.Y. Park, J.S. Jung, S.G. Jeon, Y.K. Kim, et al., IFATS collection: Immunomodulatory effects of adipose tissue-derived stem cells in an allergic rhinitis mouse model, Stem Cells 27 (1) (2009) 259–265.
[14] M. Lopez-Santalla, P. Mancheno-Corvo, R. Menta, J. Lopez-Belmonte, O. DelaRosa, J.A. Bueren, et al., Human adipose-derived mesenchymal stem cells modulate experimental autoimmune arthritis by modifying early adaptive T cell responses, Stem Cells 33 (12) (2015) 3493–3503.
[15] I. Hof-Nahor, L. Leshansky, S. Shivtiel, L. Eldor, D. Aberdam, J. Itskovitz-Eldor, et al., Human mesenchymal stem cells shift CD8+ T cells towards a suppressive phenotype by inducing tolerogenic monocytes, J. Cell Sci. 125 (Pt 19) (2012) 4640–4650.
[16] F. Luk, L. Carreras-Planella, S.S. Korevaar, S.F.H. de Witte, F.E. Borras, M.G.H. Betjes, et al., Inflammatory conditions dictate the effect of mesenchymal stem or stromal cells on B cell function, Front. Immunol. 8 (2017) 1042.
[17] M.J. Park, S.K. Kwok, S.H. Lee, E.K. Kim, S.H. Park, M.L. Cho, Adipose tissue-derived mesenchymal stem cells induce expansion of interleukin-10-producing regulatory B cells and ameliorate autoimmunity in a murine model of systemic lupus erythematosus, Cell Transplant. 24 (11) (2015) 2367–2377.
[18] J. Xie, T.J. Jones, D. Feng, T.G. Cook, A.A. Jester, R. Yi, et al., Human adipose-derived stem cells suppress elastase-induced murine abdominal aortic inflammation and aneurysm expansion through paracrine factors, Cell Transplant. 26 (2) (2017) 173–189.
[19] J.S. Heo, Y. Choi, H.O. Kim, Adipose-derived mesenchymal stem cells promote M2 macrophage phenotype through exosomes, Stem Cells Int. 2019 (2019) 7921760.
[20] H.B. Da Silva, L.K. Beura, H. Wang, E.A. Hanse, R. Gore, M.C. Scott, et al., The purinergic receptor P2RX7 directs metabolic fitness of long-lived memory CD8+ T cells, 559 (7713) (2018) 264–268.
[21] R.D. Fields, B. Stevens, ATP: an extracellular signaling molecule between neurons and glia, Trends Nurosci. 23 (12) (2000) 625–633.
[22] S. Mason, A. Paradiso, R. Boucher, Regulation of transepithelial ion transport and intracellular calcium by extracellular ATP in human normal and cystic fibrosis airway epithelium, Br. J. Pharmacol. 103 (3) (1991) 1649–1656.
[23] G. Ulate, S.R. Scott, J. González, J.A. Gilabert, A.R. Artalejo, Extracellular ATP regulates exocytosis by inhibiting multiple Ca 2+ channel types in bovine chromaffin cells, Pflugers Arch. 439 (3) (2000) 304–314.
[24] G. Haskó, P. Pacher, A2A receptors in inflammation and injury: lessons learned from transgenic animals, J. Leukoc. Biol. 83 (3) (2008) 447–455.
[25] B.B. Fredholm, Y. Chern, R. Franco, M. Sitkovsky, Aspects of the general biology of adenosine A2A signaling, Prog. Neurobiol. 83 (5) (2007) 263–276.
[26] T. Glaser, A.R. Cappellari, M.M. Pillat, I.C. Iser, M.R. Wink, A.M. Battastini, et al., Perspectives of purinergic signaling in stem cell differentiation and tissue regeneration, Purinergic Signal 8 (3) (2012) 523–537.
[27] K. Tan, H. Zhu, J. Zhang, W. Ouyang, J. Tang, Y. Zhang, et al., CD73 expression on mesenchymal stem cells dictates the reparative properties via its anti-inflammatory activity, Stem Cells Int. 2019 (2019) 8717694.
[28] M. Allard, D. Cote, L. Davidson, J. Dazai, R.M. Henkelman, Combined magnetic resonance and bioluminescence imaging of live mice, J. Biomed. Opt. 12 (3) (2007), 034018.
[29] C. Toma, W.R. Wagner, S. Bowry, A. Schwartz, F. Villanueva, Fate of culture-expanded mesenchymal stem cells in the microvasculature: in vivo observations of cell kinetics, Circ. Res. 104 (3) (2009) 398–402.
[30] G. Constantin, S. Marconi, B. Rossi, S. Angiari, L. Calderan, E. Anghileri, et al., Adipose-derived mesenchymal stem cells ameliorate chronic experimental autoimmune encephalomyelitis, Stem Cells 27 (10) (2009) 2624–2635.

[31] K.B. Stivers, J.E. Beare, P.M. Chilton, S.K. Williams, C.L. Kaufman, J.B. Hoying, Adipose-derived cellular therapies in solid organ and vascularized-composite allotransplantation, Curr. Opin. Organ Transplant. 22 (5) (2017) 490–498.

[32] J.A. Plock, J.T. Schnider, W. Zhang, R. Schweizer, W. Tsuji, N. Kostereva, et al., Adipose- and bone marrow-derived mesenchymal stem cells prolong graft survival in vascularized composite Allotransplantation, Transplantation 99 (9) (2015) 1765–1773.

[33] R. Heyes, A. Iarocci, Y. Tchoukalova, D.G. Lott, Immunomodulatory role of mesenchymal stem cell therapy in vascularized composite allotransplantation, J. Transp. Secur. 2016 (2016) 6951693.

[34] M. Waldner, W. Zhang, I.B. James, K. Allbright, E. Havis, J.M. Bliley, et al., Characteristics and Immunomodulating functions of adipose-derived and bone marrow-derived mesenchymal stem cells across defined human leukocyte antigen barriers, Front. Immunol. 9 (2018) 1642.

[35] E. Eggenhofer, P. Renner, Y. Soeder, F.C. Popp, M.J. Hoogduijn, E.K. Geissler, et al., Features of synergism between mesenchymal stem cells and immunosuppressive drugs in a murine heart transplantation model, Transpl. Immunol. 25 (2–3) (2011) 141–147.

[36] F. Casiraghi, N. Perico, M. Cortinovis, G. Remuzzi, Mesenchymal stromal cells in renal transplantation: opportunities and challenges, Nat. Rev. Nephrol. 12 (4) (2016) 241–253.

[37] T. Kato, M. Okumi, M. Tanemura, K. Yazawa, Y. Kakuta, K. Yamanaka, et al., Adipose tissue-derived stem cells suppress acute cellular rejection by TSG-6 and CD44 interaction in rat kidney transplantation, Transplantation 98 (3) (2014) 277–284.

[38] T. Liu, Y. Zhang, Z. Shen, X. Zou, X. Chen, L. Chen, et al., Immunomodulatory effects of OX40Ig gene-modified adipose tissue-derived mesenchymal stem cells on rat kidney transplantation, Int. J. Mol. Med. 39 (1) (2017) 144–152.

[39] Z. Bao, J. Li, P. Zhang, Q. Pan, B. Liu, J. Zhu, et al., Toll-like receptor 3 activator preconditioning enhances modulatory function of AdiposeDerived mesenchymal stem cells in a fully MHC-mismatched murine model of heterotopic heart transplantation, Ann. Transplant. 25 (2020), e921287.

[40] R.A. Larocca, P.M. Moraes-Vieira, E.J. Bassi, P. Semedo, D.C. de Almeida, M.B. da Silva, et al., Adipose tissue-derived mesenchymal stem cells increase skin allograft survival and inhibit Th-17 immune response, PLoS One 8 (10) (2013), e76396.

[41] R. Zamora, S.K. Ravuri, J.A. Plock, Y. Vodovotz, V.S. Gorantla, Differential inflammatory networks distinguish responses to bone marrow-derived versus adipose-derived mesenchymal stem cell therapies in vascularized composite allotransplantation, J. Trauma Acute Care Surg. 83 (1 Suppl 1) (2017), S50-S8.

[42] Y.R. Kuo, C.C. Chen, S. Goto, I.T. Lee, C.W. Huang, C.C. Tsai, et al., Modulation of immune response and T-cell regulation by donor adipose-derived stem cells in a rodent hind-limb allotransplant model, Plast. Reconstr. Surg. 128 (6) (2011), 661e-72e.

[43] Y.R. Kuo, C.C. Chen, Y.C. Chen, C.M. Chien, Recipient adipose-derived stem cells enhance recipient cell engraftment and prolong allotransplant survival in a miniature swine hind-limb model, Cell Transplant. 26 (8) (2017) 1418–1427.

[44] H.Y. Cheng, N. Ghetu, W.C. Huang, Y.L. Wang, C.G. Wallace, C.J. Wen, et al., Syngeneic adipose-derived stem cells with short-term immunosuppression induce vascularized composite allotransplantation tolerance in rats, Cytotherapy 16 (3) (2014) 369–380.

[45] S.H. Jeong, Y.H. Ji, E.S. Yoon, Immunosuppressive activity of adipose tissue-derived mesenchymal stem cells in a rat model of hind limb allotransplantation, Transplant. Proc. 46 (5) (2014) 1606–1614.

[46] T. Ma, S. Luan, R. Tao, D. Lu, L. Guo, J. Liu, et al., Targeted migration of human adipose-derived stem cells to secondary lymphoid organs enhances their Immunomodulatory effect and prolongs the survival of Allografted vascularized composites, Stem Cells 37 (12) (2019) 1581–1594.

[47] J.A. Plock, J.T. Schnider, R. Schweizer, W. Zhang, W. Tsuji, M. Waldner, et al., The influence of timing and frequency of adipose-derived mesenchymal stem cell therapy on immunomodulation outcomes after vascularized composite Allotransplantation, Transplantation 101 (1) (2017) e1–e11.

[48] M. Jiang, X. Bi, X. Duan, N. Pang, H. Wang, H. Yuan, et al., Adipose tissue-derived stem cells modulate immune function in vivo and promote long-term hematopoiesis in vitro using the aGVHD model, Exp. Ther. Med. 19 (3) (2020) 1725–1732.

[49] C. Gregoire, C. Ritacco, M. Hannon, L. Seidel, L. Delens, L. Belle, et al., Comparison of mesenchymal stromal cells from different origins for the treatment of graft-vs.-host-disease in a humanized mouse model, Front. Immunol. 10 (2019) 619.

[50] B. Fang, Y. Song, Q. Lin, Y. Zhang, Y. Cao, R.C. Zhao, et al., Human adipose tissue-derived mesenchymal stromal cells as salvage therapy for treatment of severe refractory acute graft-vs.-host disease in two children, Pediatr. Transplant. 11 (7) (2007) 814–817.
[51] B. Fang, Y. Song, L. Liao, Y. Zhang, R.C. Zhao, Favorable response to human adipose tissue-derived mesenchymal stem cells in steroid-refractory acute graft-versus-host disease, Transplant. Proc. 39 (10) (2007) 3358–3362.
[52] M. Jurado, C. De La Mata, A. Ruiz-Garcia, E. Lopez-Fernandez, O. Espinosa, M.J. Remigia, et al., Adipose tissue-derived mesenchymal stromal cells as part of therapy for chronic graft-versus-host disease: a phase I/II study, Cytotherapy 19 (8) (2017) 927–936.
[53] F. De Francesco, M. Romano, L. Zarantonello, C. Ruffolo, D. Neri, N. Bassi, et al., The role of adipose stem cells in inflammatory bowel disease: from biology to novel therapeutic strategies, Cancer Biol. Ther. 17 (9) (2016) 889–898.
[54] M.A. Gonzalez, E. Gonzalez-Rey, L. Rico, D. Buscher, M. Delgado, Adipose-derived mesenchymal stem cells alleviate experimental colitis by inhibiting inflammatory and autoimmune responses, Gastroenterology 136 (3) (2009) 978–989.
[55] E. Gonzalez-Rey, P. Anderson, M.A. Gonzalez, L. Rico, D. Buscher, M. Delgado, Human adult stem cells derived from adipose tissue protect against experimental colitis and sepsis, Gut 58 (7) (2009) 929–939.
[56] H. Takeyama, T. Mizushima, M. Uemura, N. Haraguchi, J. Nishimura, T. Hata, et al., Adipose-derived stem cells ameliorate experimental murine colitis via TSP-1-dependent activation of latent TGF-beta, Dig. Dis. Sci. 62 (8) (2017) 1963–1974.
[57] J.H. An, W.J. Song, Q. Li, S.M. Kim, J.I. Yang, M.O. Ryu, et al., Prostaglandin E2 secreted from feline adipose tissue-derived mesenchymal stem cells alleviate DSS-induced colitis by increasing regulatory T cells in mice, BMC Vet. Res. 14 (1) (2018) 354.
[58] D. Garcia-Olmo, D. Herreros, I. Pascual, J.A. Pascual, E. Del-Valle, J. Zorrilla, et al., Expanded adipose-derived stem cells for the treatment of complex perianal fistula: a phase II clinical trial, Dis. Colon Rectum 52 (1) (2009) 79–86.
[59] W.Y. Lee, K.J. Park, Y.B. Cho, S.N. Yoon, K.H. Song, D.S. Kim, et al., Autologous adipose tissue-derived stem cells treatment demonstrated favorable and sustainable therapeutic effect for Crohn's fistula, Stem Cells 31 (11) (2013) 2575–2581.
[60] M. Garcia-Arranz, M.D. Herreros, C. Gonzalez-Gomez, P. de la Quintana, H. Guadalajara, T. Georgiev-Hristov, et al., Treatment of Crohn's-related Rectovaginal fistula with allogeneic expanded-adipose derived stem cells: a phase I-IIa clinical trial, Stem Cells Transl. Med. 5 (11) (2016) 1441–1446.
[61] M.E. Bateman, A.L. Strong, J.M. Gimble, B.A. Bunnell, Concise review: using fat to fight disease: a systematic review of nonhomologous adipose-derived stromal/stem cell therapies, Stem Cells 36 (9) (2018) 1311–1328.
[62] J. Panés, D. García-Olmo, G. Van Assche, J.F. Colombel, W. Reinisch, D.C. Baumgart, et al., Expanded allogeneic adipose-derived mesenchymal stem cells (Cx601) for complex perianal fistulas in Crohn's disease: a phase 3 randomised, double-blind controlled trial, Lancet 388 (10051) (2016) 1281–1290.
[63] S. Choi, B.G. Jeon, G. Chae, S.J. Lee, The clinical efficacy of stem cell therapy for complex perianal fistulas: a meta-analysis, Tech. Coloproctol. 23 (5) (2019) 411–427.
[64] F. Cheng, Z. Huang, Z. Li, Efficacy and safety of mesenchymal stem cells in treatment of complex perianal fistulas: a meta-analysis, Stem Cells Int. 2020 (2020) 8816737.
[65] L. Zhang, X.Y. Wang, P.J. Zhou, Z. He, H.Z. Yan, D.D. Xu, et al., Use of immune modulation by human adipose-derived mesenchymal stem cells to treat experimental arthritis in mice, Am. J. Transl. Res. 9 (5) (2017) 2595–2607.
[66] P. Mancheno-Corvo, M. Lopez-Santalla, R. Menta, O. DelaRosa, F. Mulero, B. Del Rio, et al., Intralymphatic Administration of Adipose Mesenchymal Stem Cells Reduces the severity of collagen-induced experimental arthritis, Front. Immunol. 8 (2017) 462.
[67] H. Ueyama, T. Okano, K. Orita, K. Mamoto, S. Sobajima, H. Iwaguro, et al., Local transplantation of adipose-derived stem cells has a significant therapeutic effect in a mouse model of rheumatoid arthritis, Sci. Rep. 10 (1) (2020).
[68] J.M. Alvaro-Gracia, J.A. Jover, R. Garcia-Vicuna, L. Carreno, A. Alonso, S. Marsal, et al., Intravenous administration of expanded allogeneic adipose-derived mesenchymal stem cells in refractory rheumatoid arthritis (Cx611): results of a multicentre, dose escalation, randomised, single-blind, placebo-controlled phase Ib/IIa clinical trial, Ann. Rheum. Dis. 76 (1) (2017) 196–202.
[69] E.T. Hurley, Y. Yasui, A.L. Gianakos, D. Seow, Y. Shimozono, G. Kerkhoffs, et al., Limited evidence for adipose-derived stem cell therapy on the treatment of osteoarthritis, Knee Surg. Sports Traumatol. Arthrosc. 26 (11) (2018) 3499–3507.

[70] K. Toupet, M. Maumus, J.A. Peyrafitte, P. Bourin, P.L. van Lent, R. Ferreira, et al., Long-term detection of human adipose-derived mesenchymal stem cells after intraarticular injection in SCID mice, Arthritis Rheum. 65 (7) (2013) 1786–1794.
[71] C.H. Jo, Y.G. Lee, W.H. Shin, H. Kim, J.W. Chai, E.C. Jeong, et al., Intra-articular injection of mesenchymal stem cells for the treatment of osteoarthritis of the knee: a proof-of-concept clinical trial, Stem Cells 32 (5) (2014) 1254–1266.
[72] Y.M. Pers, L. Rackwitz, R. Ferreira, O. Pullig, C. Delfour, F. Barry, et al., Adipose mesenchymal stromal cell-based therapy for severe osteoarthritis of the knee: a phase I dose-escalation trial, Stem Cells Transl. Med. 5 (7) (2016) 847–856.
[73] N. Yokota, M. Yamakawa, T. Shirata, T. Kimura, H. Kaneshima, Clinical results following intra-articular injection of adipose-derived stromal vascular fraction cells in patients with osteoarthritis of the knee, Regen Ther. 6 (2017) 108–112.
[74] H. Takahashi, N. Sakata, G. Yoshimatsu, S. Hasegawa, S. Kodama, Regenerative and transplantation medicine: Cellular therapy using adipose tissue-derived mesenchymal stromal cells for type 1 diabetes mellitus, J. Clin. Med. 8 (2) (2019).
[75] S.S. Roy, M. Mukherjee, S. Bhattacharya, C.N. Mandal, L.R. Kumar, S. Dasgupta, et al., A new cell secreting insulin, Endocrinology 144 (4) (2003) 1585–1593.
[76] M.G. Amer, A.S. Embaby, R.A. Karam, M.G. Amer, Role of adipose tissue derived stem cells differentiated into insulin producing cells in the treatment of type I diabetes mellitus, Gene 654 (2018) 87–94.
[77] R.R. Bhonde, P. Sheshadri, S. Sharma, A. Kumar, Making surrogate beta-cells from mesenchymal stromal cells: perspectives and future endeavors, Int. J. Biochem. Cell Biol. 46 (2014) 90–102.
[78] H.P. Lin, T.M. Chan, R.H. Fu, C.P. Chuu, S.C. Chiu, Y.H. Tseng, et al., Applicability of adipose-derived stem cells in type 1 diabetes mellitus, Cell Transplant. 24 (3) (2015) 521–532.
[79] E.J. Bassi, P.M. Moraes-Vieira, C.S. Moreira-Sa, D.C. Almeida, L.M. Vieira, C.S. Cunha, et al., Immune regulatory properties of allogeneic adipose-derived mesenchymal stem cells in the treatment of experimental autoimmune diabetes, Diabetes 61 (10) (2012) 2534–2545.
[80] T.M. Kono, E.K. Sims, D.R. Moss, W. Yamamoto, G. Ahn, J. Diamond, et al., Human adipose-derived stromal/stem cells protect against STZ-induced hyperglycemia: analysis of hASC-derived paracrine effectors, Stem Cells 32 (7) (2014) 1831–1842.
[81] J.A. Semon, C. Maness, X. Zhang, S.A. Sharkey, M.M. Beuttler, F.S. Shah, et al., Comparison of human adult stem cells from adipose tissue and bone marrow in the treatment of experimental autoimmune encephalomyelitis, Stem Cell Res. Ther. 5 (1) (2014) 2.
[82] F. Yousefi, M. Ebtekar, M. Soleimani, S. Soudi, S.M. Hashemi, Comparison of in vivo immunomodulatory effects of intravenous and intraperitoneal administration of adipose-tissue mesenchymal stem cells in experimental autoimmune encephalomyelitis (EAE), Int. Immunopharmacol. 17 (3) (2013) 608–616.
[83] F. Yousefi, M. Ebtekar, S. Soudi, M. Soleimani, S.M. Hashemi, In vivo immunomodulatory effects of adipose-derived mesenchymal stem cells conditioned medium in experimental autoimmune encephalomyelitis, Immunol. Lett. 172 (2016) 94–105.
[84] A.C. Bowles, A.L. Strong, R.M. Wise, R.C. Thomas, B.Y. Gerstein, M.F. Dutreil, et al., Adipose stromal vascular fraction-mediated improvements at late-stage disease in a murine model of multiple sclerosis, Stem Cells 35 (2) (2017) 532–544.
[85] A.C. Bowles, R.M. Wise, B.Y. Gerstein, R.C. Thomas, R. Ogelman, R.C. Manayan, et al., Adipose stromal vascular fraction attenuates TH1 cell-mediated pathology in a model of multiple sclerosis, J. Neuroinflammation 15 (1) (2018) 77.
[86] C. Valina, K. Pinkernell, Y.H. Song, X. Bai, S. Sadat, R.J. Campeau, et al., Intracoronary administration of autologous adipose tissue-derived stem cells improves left ventricular function, perfusion, and remodelling after acute myocardial infarction, Eur. Heart J. 28 (21) (2007) 2667–2677.
[87] K. Schenke-Layland, B.M. Strem, M.C. Jordan, M.T. DeEmedio, M.H. Hedrick, K.P. Roos, et al., Adipose tissue-derived cells improve cardiac function following myocardial infarction, J. Surg. Res. 153 (2) (2009) 217–223.
[88] Y. Zhu, T. Liu, K. Song, R. Ning, X. Ma, Z. Cui, ADSCs differentiated into cardiomyocytes in cardiac microenvironment, Mol. Cell. Biochem. 324 (1–2) (2009) 117–129.
[89] K.H. Chen, C.H. Chen, C.G. Wallace, C.M. Yuen, G.S. Kao, Y.L. Chen, et al., Intravenous administration of xenogenic adipose-derived mesenchymal stem cells (ADMSC) and ADMSC-derived exosomes markedly reduced brain infarct volume and preserved neurological function in rat after acute ischemic stroke, Oncotarget 7 (46) (2016) 74537–74556.

[90] D. Jeon, K. Chu, S.T. Lee, K.H. Jung, J.J. Ban, D.K. Park, et al., Neuroprotective effect of a cell-free extract derived from human adipose stem cells in experimental stroke models, Neurobiol. Dis. 54 (2013) 414–420.
[91] S.H. Oh, C. Choi, D.J. Chang, D.A. Shin, N. Lee, I. Jeon, et al., Early neuroprotective effect with lack of long-term cell replacement effect on experimental stroke after intra-arterial transplantation of adipose-derived mesenchymal stromal cells, Cytotherapy 17 (8) (2015) 1090–1103.
[92] Z. Yu, T. Wenyan, S. Xuewen, D. Baixiang, W. Qian, W. Zhaoyan, et al., Immunological effects of the intraparenchymal administration of allogeneic and autologous adipose-derived mesenchymal stem cells after the acute phase of middle cerebral artery occlusion in rats, J. Transl. Med. 16 (1) (2018) 339.
[93] L. Guzman-Martinez, R.B. Maccioni, V. Andrade, L.P. Navarrete, M.G. Pastor, N. Ramos-Escobar, Neuroinflammation as a common feature of neurodegenerative disorders, Front. Pharmacol. 10 (2019) 1008.
[94] M.K. McCoy, T.N. Martinez, K.A. Ruhn, P.C. Wrage, E.W. Keefer, B.R. Botterman, et al., Autologous transplants of adipose-derived adult stromal (ADAS) cells afford dopaminergic neuroprotection in a model of Parkinson's disease, Exp. Neurol. 210 (1) (2008) 14–29.
[95] Y. Yan, T. Ma, K. Gong, Q. Ao, X. Zhang, Y. Gong, Adipose-derived mesenchymal stem cell transplantation promotes adult neurogenesis in the brains of Alzheimer's disease mice, Neural Regen. Res. 9 (8) (2014) 798–805.
[96] C.V. Fontanilla, H. Gu, Q. Liu, T.Z. Zhu, C. Zhou, B.H. Johnstone, et al., Adipose-derived stem cell conditioned media extends survival time of a mouse model of amyotrophic lateral sclerosis, Sci. Rep. 5 (1) (2015).
[97] Staff NP, N.N. Madigan, J. Morris, M. Jentoft, E.J. Sorenson, G. Butler, et al., Safety of intrathecal autologous adipose-derived mesenchymal stromal cells in patients with ALS, Neurology 87 (21) (2016) 2230–2234.
[98] A. Ghorbani, A. Feizpour, M. Hashemzahi, L. Gholami, M. Hosseini, M. Soukhtanloo, et al., The effect of adipose derived stromal cells on oxidative stress level, lung emphysema and white blood cells of Guinea pigs model of chronic obstructive pulmonary disease, Daru 22 (1) (2014) 26.
[99] P. Llontop, D. Lopez-Fernandez, B. Clavo, J.L. Afonso Martín, M.D. Fiuza-Pérez, M. García Arranz, et al., Airway transplantation of adipose stem cells protects against bleomycin-induced pulmonary fibrosis, J. Invest. Med. 66 (4) (2018) 739–746.
[100] H. Jiang, J. Zhang, Z. Zhang, S. Ren, C. Zhang, Effect of transplanted adiposederived stem cells in mice exhibiting idiopathic pulmonary fibrosis, Mol. Med. Rep. 12 (4) (2015) 5933–5938.
[101] R. Uchimura, T. Ueda, R. Fukazawa, J. Hayakawa, R. Ohashi, N. Nagi-Miura, et al., Adipose tissue-derived stem cells suppress coronary arteritis of Kawasaki disease in vivo, Pediatr. Int. 62 (1) (2020) 14–21.

CHAPTER

11

Clinical experience with adipose tissue enriched with adipose stem cells

Shawn Loder[a], *Danielle Minteer*[b], *and J. Peter Rubin*[c,d]

[a]Department of Plastic Surgery, University of Pittsburgh, Pittsburgh, PA, United States [b]Plastic & Reconstructive Surgery, University of Pittsburgh Medical Center, Pittsburgh, PA, United States [c]Departments of Plastic Surgery and Bioengineering, University of Pittsburgh, Pittsburgh, PA, United States [d]McGowan Institute for Regenerative Medicine, University of Pittsburgh, Pittsburgh, PA, United States

A brief history

The autologous fat graft has existed as a reconstructive adjunct as far back as 1893 when it was first proposed by Dr. Gustav Neuber to correct atrophic scarring [1]. Autologous fat grafting remained within the realm of surgical free fat transfer until 1974 when Drs. Arpad and Giorgio Fischer developed what was to be the predecessor of the modern liposuction cannula [2]. Advances in technique and equipment by French surgeon Dr. Pierre Fournier and introduction of "wet" techniques by Dr. Yves-Gerard Illouz ultimately led to the development of modern tumescent liposuction by Dr. Jeffrey Klein in 1985 [2]. These works were expanded upon by Dr. Sydney Coleman in 1987 when he introduced what has become the gold standard in liposuction—the Coleman technique—a process of low-pressure liposuction, brief centrifugation, and careful lipofilling with respect to the architecture of the native tissue [3]. The evolution in technique over this period coalesced throughout the modern era as introduction of new technology and advances in our understanding of adipose biology have expanded the indications for fat transfer well beyond its initial description as a filler for atrophic scars. As fat grafting has become safer and more reliable, its popularity has grown and, with it, interest in the future of the science of lipotransfer.

Here, we will discuss the current state of fat grafting and ASCs in the clinical sphere, with a focus on current technologies, indications, and new developments likely to reach our patients in the near future.

Scientific Principles of Adipose Stem Cells
https://doi.org/10.1016/B978-0-12-819376-1.00001-9

The value of fat

Adipose tissue has been described as the "ideal" filler for adding or replacing tissue volume [4]. In most patients, fat is easily available with minimal donor site morbidity and is a comparatively inexpensive and lasting option for patients looking for filler or tissue sculpting. Aesthetically, adipose is valued for the natural appearance provided by carefully placed aliquots. Furthermore, as an autologous filler, adipose is biocompatible with its recipient tissue and triggers a very limited immunologic response [4]. Additionally, fat is valued not only as a volume filler but also for its ability to improve tissue quality. This effect is driven, in no small part, by the range of useful cellular and acellular components that make up each fat graft.

Adipose composition

Adipose tissue is a heterogenous mix broadly composed of mature adipocytes, extracellular matrix (ECM), and the stromal vascular fraction (SVF) [5]. SVF was first described in 1964 as a byproduct of collagenase digestion of adipose tissue in rats. It has since been further characterized and subdivided into several cellular components, including ASCs, endothelial cells, hematopoietic cells, pericytes, and vascular smooth muscle cells, fibroblasts, and a mix of immunologic cells, erythrocytes, and platelets [5, 6]. While the exact effect and science behind each component in this niche is reserved for other areas of this text, it is critical for the clinician to understand the relationship between the composition and density of each cellular component and the viability, proliferation, and effect of the graft. The clinical manipulation of this composition, a process called "enrichment," is discussed later in this chapter.

Adipose-derived stem cells

The ASC is an adult mesenchymal stem cell similar in cytometric profile to both dermal fibroblasts and bone-marrow derived stem cells (BMSCs) [7]. Initially described as an adipocyte precursor, these cells demonstrate multipotency with the ability to reliably differentiate into chondrogenic, osteogenic, myogenic, and adipogenic lineages in vivo [8]. Additionally, these cells have potent angiogenic induction, immunoregulatory, and paracrine signaling roles that remain incompletely understood [9]. In contrast to other mesenchymal populations that are either limited in number or inaccessible for therapeutic manipulation, ASCs are readily available in clinical medicine as a byproduct of suction-assisted lipectomy (SAL) also known as liposuction. Adipose harvest and lipotransfer thus represents the most accessible mesenchymal population currently available in the clinical sphere, and it is critical for physicians interested in ASC-based therapies to understand the basic techniques, technologies, and skills required to optimize autologous fat transfer.

Tools and techniques

Principles of adipose harvest

Early fat grafting was fraught with significant variability in graft survival, and improving stability and viability of grafts has been a prime force guiding clinical advancement. As described by Dr. Coleman in 1987, the eponymous Coleman technique was centered on three key principles: (1) use of a blunt cannula and low suction to minimize shear trauma, (2) refinement of the fat to remove the nonviable components, and (3) careful arrangement of the adipose aliquots within the recipient tissue to sequester each aliquot within the recipient architecture. These principles are carefully designed to minimize injury to the adipose graft and maximize the vascular bed available after transfer [7]. Work by Dr. Dennis Orgill and his team proposed further that the initial limiting factor on adipose transfer is oxygen diffusion and thus described both a "microribbon model" in which fat injections greater than 2.5–2.6 mm in radius exhibit central necrosis (Fig. 1) and a "fluid accommodation model" in which grafted tissue reaches a critical interstitial fluid pressure at approximately 9 mmHg [10]. In the fluid accommodation model, increases in interstitial pressure beyond 9 mmHg will progressively compress surrounding capillaries, resulting in decreased perfusion and, consequently, decreased oxygen delivery [10]. Taken together, these principles suggest that controlled placement of slender aliquots of graft and meticulous avoidance of overfilling will provide the clinician with the most viable and reliable result.

Clinical perspective

In the setting of autologous fat grafting, liposuction refers to the first stage of the overall procedure, i.e., tissue harvest. It is the disaggregation and collection of adipose tissue from a preselected donor site (Fig. 2) for processing and transfer to the recipient field for which the operating physician seeks to make a therapeutic or aesthetic change. Liposuction, however, is a complex clinical domain in its own right, with a broad range of technologies and techniques that allow the surgeon to sculpt or create a balanced physique and aesthetic body contour for their patients. Liposuction, or suction lipectomy, is one of the most commonly performed aesthetic operations today and is both adjunct and cornerstone to many other facial and body contouring procedures.

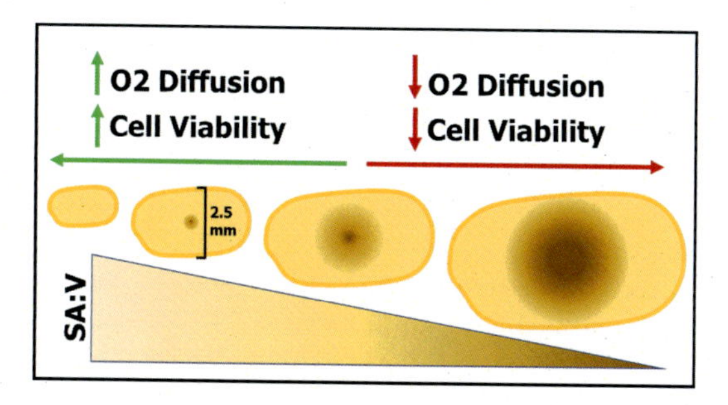

FIG. 1 O_2 diffusion decreases as the diameter of adipose aliquots grows. Significant mismatch between surface area and volume (SA:V) leads to central hypoxia and loss of cell viability.

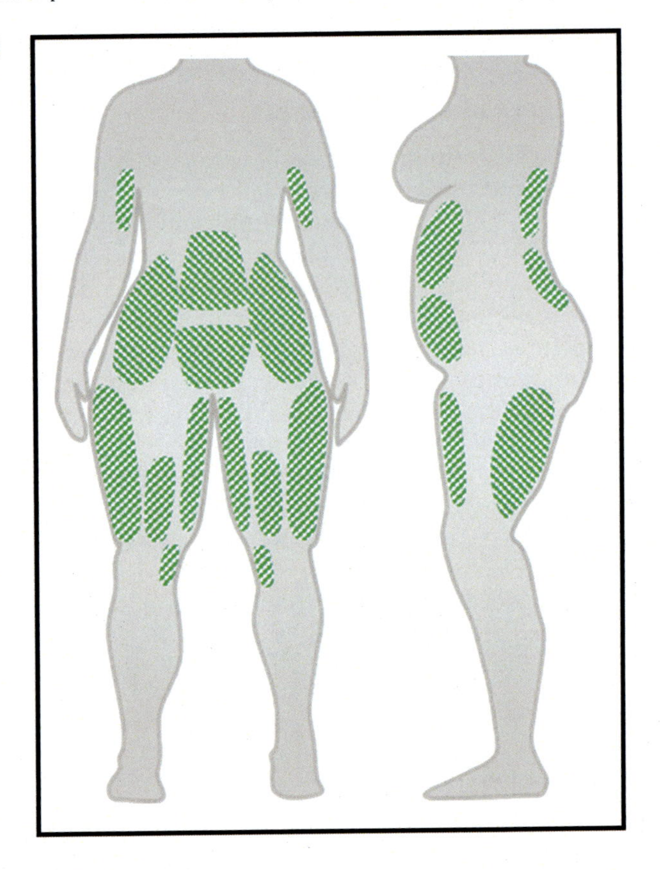

FIG. 2 Common adipose donor sites (green) utilized in modern liposuction.

The principles of adipose harvest, as discussed previously, are critical to ensuring a viable graft. Collection of adipose for grafting requires (1) meticulous preparation of the donor site; (2) careful selection of both liposuction modality and cannula class; and (3) preparation, purification, and/or augmentation of the graft prior to transfer. When liposuction is utilized for both its inherent aesthetic contouring properties as well as for collection of fat intended for secondary grafting, then the complexity of the operation increases as the surgeon must carefully balance the desire to maximize donor harvest with a need to ensure a pleasing visual outcome.

Donor site preparation

Tumescence

The introduction of wet, super-wet, and later tumescent techniques has revolutionized the safety and efficiency of liposuction. These techniques refer to the infiltration of a "wetting solution" commonly containing dilute lidocaine and epinephrine. This caries several benefits in that it provides pain control and by vasoconstriction minimizes the risk of injury to vessels and, subsequently, bleeding. Additionally, the tumescent technique temporarily alters adipose tissue turgor and increases rigidity and deformation resistance, improving its ability to be cleaved by a liposuction cannula. A comparison of the four currently accepted classes of liposuction pretreatment has shown that (1) dry technique involving no infiltration results in 20%–45% of the aspirated volume being blood loss, (2) wet technique including infiltration of 200–300mL of wetting solution per site with blood loss approximately 4%–30% of the total aspirate volume, (3) super-wet technique applying 1mL of wetting solution per 1mL of expected aspirate with approximately 1% of the total volume expected to be blood, and (4) tumescent technique in which 3–4mL of wetting solution per 1mL of expected aspirate is infiltrated with less than 1% blood loss expected [11]. Clinically, use of tumescence allows for more extensive liposuction in more anatomically sensitive areas, with decreased risk of pain, bleeding, or injury to unwanted structures.

At a cellular level, lidocaine is associated in vitro with suppression of adipocyte proliferation and lipolytic function. The effects of concurrent lidocaine and epinephrine are transient, however, and have not been demonstrated to permanently affect the metabolic activity of adipocytes [11–13]. When cells are collected via dry versus tumescent approaches, however, adipocyte viability is improved with the tumescent technique. This is believed to be secondary to a reduction in mechanical stresses that affect the viability of the graft. Currently, the general consensus is that there is no significant risk to adipose graft viability with modern wetting solution [12].

Introduction of buffers and antioxidant compounds such as ascorbic acid *N*-acetylcysteine has been described in the literature [13, 14]. Buffers function to raise the pH of wetting solution and thus neutralize the acidic pH of amide-type anesthetics such as lidocaine. Additionally, antioxidant compounds have been described as a method to reduce oxidative stress on the graft. Antioxidants in particular have been described in vitro to improve the viability of adipocytes and, thus, graft retention. Currently, supplementation of the wetting solution with antioxidants has not been sufficiently demonstrated to either benefit or adversely affect the viability or function of the SVF in vivo.

Pretunneling

Prior to removal of adipose for grafting, several techniques have been described to pretunnel or redistribute adipose in the donor tissue bed. This preliminary step may be utilized for several different purposes. Passing a blunt or spatulated tip cannula through an adipose deposit disrupts the tissue architecture and ECM [15]. This leads to tissue disaggregation, for which benefits have been described of loosening tissue, decreasing resistance to the cannula proper, and assisting the operating surgeon in developing a safe deep or superficial plane of approach for further liposuction (Fig. 3). These techniques are based around the theory that adipose tissue that has been disaggregated will provide less resistance, less friction, and, thus, less shear, ultimately improving yield and reducing operator fatigue.

Liposuction modalities

A key determinant of lipotransfer outcomes is harvest technique, which varies in the degree and mechanism of negative pressure utilized to disaggregate and then withdraw native

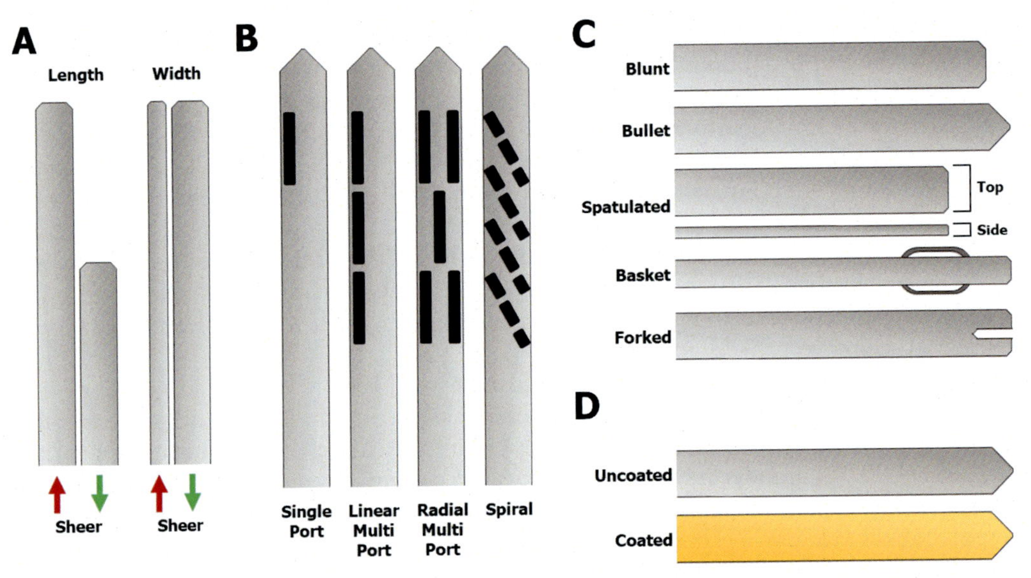

FIG. 3 Cannula variability. (A) Shear and, thus, barotrauma are directly correlated with cannula length and indirectly related to cannula width. (B) Common patterns of cannula pores. (C) Common patterns of cannula tips. (D) Use of a coating agent to minimize friction/resistance.

adipose tissue into and through the cannula. Additionally, some modalities provide mechanical aid to either more safely or thoroughly disrupt tissue to aid with surgeon fatigue.

Suction-assisted lipectomy

Suction pressure is the most direct determinant of barotrauma on a given adipose donor. Higher vacuum pressures are more capable of pulling tissue into the cannula and thus facilitate a more rapid harvest; however, as originally proposed by Dr. Coleman, lower suction and consequently lower shear stress result in greater graft survival [4, 16]. In this vein, the most basic liposuction modality is the manual aspiration technique now more commonly known as suction-assisted lipectomy (SAL). This is the process by which the liposuction cannula mechanically morcellates native adipose and retrieves the aliquots directly under negative pressure. Results are variable across the literature; however, initial studies demonstrated superior adipose yield and viability with manual aspiration techniques [16–18]. These differences were not evident in later studies using tumescent conditions [19]. There is, however, an overall trend towards greater cell viability with pressures below 700 mmHg. An evolution of this technique to overcome the limitations of surgeon fatigue is power-assisted liposuction (PAL) by which a power source such as compressed air or electricity is utilized to either vibrate or piston the liposuction cannula in such a way as to augment the operator's ability to disaggregate native tissues. As with variations in SAL pressures and techniques, there is no consistent relationship in the literature between adipocyte or ASC viability in PAL versus SAL options at similar negative pressure parameters [20].

There now exist several liposuction modalities that, in addition to the traditional SAL techniques, provide a separate nonmechanical mechanism of tissue disruption. Commonly described mechanisms include water jet-assisted liposuction (WAL), ultrasound-assisted liposuction (UAL), and the technologically related vibration amplification of sound energy at resonance (VASER)-assisted liposuction (VAL), radiofrequency-assisted liposuction (RFAL), and the family of laser-assisted liposuction (LAL) techniques. Many of these technologies have been further developed and specialized through various proprietary methods; however, we will briefly discuss the technology as a whole herein.

Water-jet assisted liposuction

WAL utilizes differential resistance of adipose and other internal structures to a blade of pressurized water, saline, or tumescent wetting solution to preferentially separate fat from vascular and lymphatic tissue [21]. This technique is described for both aesthetic and functional liposuction and has found a niche in lymphedema debulking, often marketed as sparing residual lymph channels. While ultimately not conclusive, there is evidence that WAL techniques result in only scant collection of lymphatic tissue in the harvested aspirate [22]. Additionally, there is evidence to suggest that the usage of WAL is equivalent to SAL techniques in regard to adipocyte and ASC viability [1].

Ultrasound-assisted liposuction

UAL represents a technique by which the initial disaggregation of tissue is performed by ultrasonic energy as opposed to mechanical means. Modern UAL is typically described as a three-stage technique beginning with (1) infiltration of wetting material into the intermediate space between the deep to superficial adipose planes, (2) application of an ultrasonic probe at frequencies greater than 16 kHz to selectively lyse the underlying adipose in a deep to superficial manner, and (3) evacuation of lysed tissues via traditional suction cannula [22]. UAL is a popular, though somewhat controversial, technique within the field of liposuction. Mechanistically, the application of ultrasonic energy leads to mechanical fragmentation, tumescent cavitation, and thermal energy generation [23]. Because of this, UAL techniques have been described as having a greater capacity to emulsify adipose trapped in fibrotic areas not typically amenable to traditional SAL techniques [24–26]. UAL techniques are thus more often recommended in cases of secondary liposuction in a scarred field. These benefits are contrasted, however, by an increased risk of thermal injury, which is most commonly described when an internal ultrasound probe is utilized. When used for autologous adipose harvest, external UAL has demonstrated a histologic and cell viability profile similar to that of SAL [27]. Internal UAL, however, has been associated with thermal liquefaction of tissues, with increased damage to adipocytes [28]. Innovation in UAL technology has led to the development of third-generation devices such as the VASER probe. VASER is a proprietary technology that utilizes a small, selectively grooved probe with either pulsed or continuous ultrasonic energy to emulsify adipose tissue [29–32]. Currently, third-generation UAL devices such as VASER have demonstrated a greater safety profile versus their older counterparts. In regard to lipotransfer, VASER devices have demonstrated similar cell yield, viability, and differentiation versus SAL and PAL modalities [27, 29–32].

Laser-assisted liposuction

The final modality we will discuss in this chapter is LAL. In this modality, laser light of different wavelengths is used to target specific chromophores within the adipose layer to induce selective photothermolysis and thus preferentially lyse and liberate adipose tissue. Depending on the chromophore, targeted LAL techniques may additionally be utilized for photocoagulation of microvasculature and contracture of dermal collagen [22, 33]. Because of this, LAL is described as resulting in improvements in postoperative blood loss versus other liposuction modalities. LAL may additionally be utilized as an adjunctive technique prior to SAL for improved aspiration of lysed tissues [22].

As discussed, there are several wavelengths of light used to affect specific chromophores. The 1064nm Nd:YAG laser is most well described with an appropriate safety profile [34, 35]. Because of the wide range of laser modalities, wavelengths, and orientation, there are several proprietary LAL technologies, each with different recommended indications and benefits [33]. Disadvantages of LAL include risk of thermal injury, and as with WAL and UAL modalities, there is an increased startup cost and procedural time associated with the technique. Unfortunately, as a modality of adipose harvest, LAL techniques have been associated with decreased ASC viability, differentiation, and proliferation [36]. As such, further optimization remains necessary to maximize yield versus other liposuction techniques.

Cannula selection

Fine-tuning of the adipose harvest technique requires an understanding of how equipment selection determines grafting outcomes and survival. As before, the degree of mechanical shear and barotrauma plays a significant role in adipose particle injury. There are several characteristics associated with a given cannula's efficacy and its effect on graft viability (Fig. 4).

Cannula dimensions and porosity

A cannula's width, length, and porosity define the degree of barotrauma introduced to the graft. Shear rate of a fluid within a cylinder is directly proportional to velocity and inversely related to the diameter of the cylinder. This correlates with greater shearing deformation and, thus, reduced cell viability of fat harvested with progressively thinner cannulas [37, 38].

FIG. 4 Representation of the trilaminar skin demonstrating the anatomic planes of the deep and superficial adipose tissues.

The exact dimensions of an ideal cannula remain undetermined, with extensive research examining 2–6-mm cannulas with variable results [37–40]. While wider cannulas are associated with greater cellular survival and retention of native architecture, they are mechanically more likely to leave furrows and contour irregularities in the tissue and often must be directed more deeply during tissue harvest to avoid suboptimal aesthetic results [20, 37–40]. The number and diameter of each aperture within the cannula similarly affects cell viability as the total surface area is inversely related to the pressure at each hole [20, 37–40]. Additionally, arrangement of apertures determines the direction of harvest in three-dimensional (3D) space. A linear arrangement of perforations thus allows the operator to focus on a single problem area or to protect a delicate structure such as the underside of the dermis. Radially oriented perforations are not suitable for delicate work in this manner, as they are designed for maximum harvest efficiency.

Microcannulas

Cannulas with a diameter of less than 2 mm are regarded as microcannulas. These cannulas generally are regarded as having a higher degree of barotrauma versus those with more traditional dimensions; however, their finer dimensions allow for more superficial liposuction and sculpting. Clinically, microcannulas can be utilized in superficial areas of the face where more traditional cannula sizes would leave significant contour irregularities. These cannulas have additionally been utilized for injection of filler in fine features and delicate structures, given their narrower profile, increased mobility, and tip control.

Cannula variation

Other variations in cannula design include curvature of the cannula shaft and variation in the cannula head. Liposuction initially started with the use of modified uterine curettes or "sharp" cannulas but has evolved to primarily use any of a variety of modern blunt cannulas. This design of blunt cannula is intended to minimize cannula passage into surrounding tissues, thus localizing the zone of injury to the adipose compartment where the clinician is working. Bullet tips, having a more pointed tip than the standard blunt cannula, are designed to improve the penetration of the cannula and are useful in pretunneling and dissection-based liposuction techniques. Flattened or spatulated tips are designed to be used in PAL as a means to better guide the cannula in orientation with fibrous tissues to assist the surgeon to find a less traumatic dissection plane. Finally, dissector tips are designed specifically for the breakdown of non-adipose tissue such as scar or adhesions. V-shaped tips or forked cannulas are classes of dissector tip [40–42]. In regard to cannula shaft, curved cannulas are designed to more easily control the plane of harvest by changing between a superficial and deep angle of penetration.

A final note on cannula selection is in regard to surface material. A standard cannula is composed of stainless steel; however, there are several commercial brands that utilize coated cannulas. The coated surfaces such as those treated with polytetrafluoroethylene (i.e., Teflon) or zirconium nitride are utilized in liposuction cannulas. Coating decreases friction on the outer surface of the cannula. This has not been extensively studied in the literature regarding graft viability.

Graft processing

After collection, the adipose autograft is provided in an impure form, a mixture of nonviable oils, tumescence, blood, and other debris that is a byproduct of liposuction. Several of these "impurities" are highly inflammatory and/or immunogenic and can be detrimental to viability of the graft [2]. Additionally, inadequate separation leads to overestimation of the volume of material and can confound the operator's ability to judge the final result. Separation of this mixture into architecturally preserved adipose as well as SVF and ASCs is necessary to generate a consistent result. There are several methods by which the solid fraction of the harvested material is separated, including gravity separation, centrifugation as described in Dr. Coleman's original work, filtration and washing, and use of a sterile gauze mesh to facilitate selective absorption of unwanted impurities.

Gravity and centrifugal separation

Historically, the core techniques utilized for lipoaspirate processing utilized gravity to separate distinct layers on the basis of density. Through technological advances and reduced equipment costs, the most common practice now involves centrifugation to save time. Utilizing a standard speed of rotation approximately 1200 *g* (3000 rpm) for 3–5 min, lipoaspirate is separated into three phases, with a distinct solid pellet at the bottom of the syringe (Fig. 5) [4].

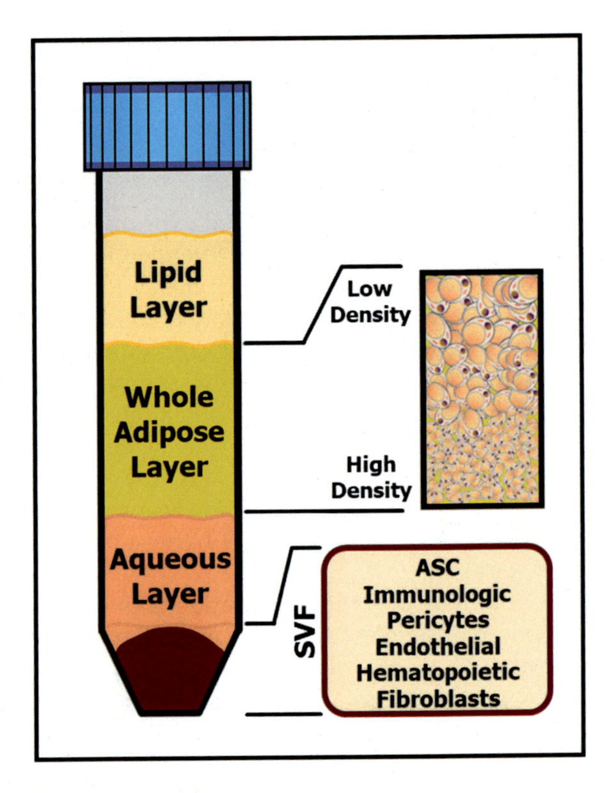

FIG. 5 Layering of adipose components after centrifugation. The adipose layer is further subdivided into higher- and lower-density components.

The top oil layer consists of liberated lipids and fat-soluble compounds and is typically decanted from the container. The central layer is typically preserved for lipotransfer and is composed of architecturally intact adipose tissue. While the central layer is homogenous, on visible inspection it is more accurately a heterogenous gradient with increasing density of viable cells as one moves lower in the column. High-density adipose has been described as containing a greater number of intact endothelial cells, decreased histologic evidence of fibrosis, and a greater percent of successful engraftment [2, 17]. The final liquid layer is composed of aqueous components, including blood, wetting solution, and other aqueous cellular fluids. This layer is typically discarded.

The solid pellet at the bottom of this column is SVF. This heterogenous mixture contains all intact nucleated cells that are either not mature adipocytes or not trapped within the adipose tissue layer. Transfer of SVF or of isolated ASCs is dependent on isolation of this pellet, and thus, centrifugation is the most common mechanism in the clinical setting by which the SVF is collected.

Clinically, centrifugation is one of the more efficient techniques for large-volume adipose separation; however, there is significant evidence that higher speeds of rotation and prolonged centrifugation lead to cell lysis and decreased cell viability, and can compromise the integrity of adipose tissues. The combination of gravity and centrifugation techniques is the most common processing practice described in the clinical setting.

Tissue filtration and washing

Mesh filtration with washing of intact adipose tissue is a second method by which intact autograft is separated from both the lipid and aqueous phases. A metal or nylon mesh is typically utilized to trap the adipose tissue, while an isotonic solution such as saline is passed through the sieve. This technique is highly efficient for reduction of erythrocyte and leukocyte contaminants [7, 43]. It is typically regarded as equivalent to other modalities for graft processing in terms of adipose survival and can be utilized as part of an automated process for high-volume lipotransfer with effective collection and concentration of adipose graft.

There are reports in the literature of adipose tissue necrosis on the use of metal filters and the formation of nodules with filtration techniques; however, these are not widely described in clinical practice [44]. This technique in isolation does not allow for density-based separation of tissue and is not appropriate for collection of the SVF pellet given the need for continuous washing. Filtration techniques are, however, utilized in intraoperative isolation and purification of the ASCs and will be further discussed in that context below.

Absorptive purification

Absorptive purification refers to the use of a tightly woven gauze or mesh such as Telfa to selectively absorb the aqueous and lipid components of the lipoaspirate by capillary action. This technique is typically described as minimizing shear stress on the tissue similarly to the previously described gravity separation. When compared with centrifugation and filtration, this technique can be more time intensive and can be difficult to scale up to high-volume lipotransfer. Gauze rolling, however, has been described as providing efficient reduction in liquid components while atraumatically preserving adipose architecture, cell viability, and percent engraftment [45–47]. When utilized as a preprocessing step prior to isolation of ASCs from the lipoaspirate, this technique is described as providing the greatest percent fraction of SVF cells [48–50].

Millifat, microfat, and nanofat

As the indications for autologous fat transfer have evolved so has the need for greater refinement of the adipose graft. The adipose parcel harvested from a standard 3-mm cannula (9 gauge), termed macrofat, typically provides a parcel size of 1.0–2.5 mm in diameter. These parcels are well suited for the replacement of volume deficits and, as described above, can be harvested with minimal graft morbidity.

In the context of very superficial grafts or grafts that necessitate very fine alteration in contour, additional processing has been described to create smaller adipose parcel sizes for a more homogenous product. In current clinical usage, adipose tissue that has been processed to parcel sizes greater than 1.0 mm but less than 2.5 mm are denoted as millifat [51]. Parcels between 1.0 and 0.5 mm are described as microfat [51]. Both millifat and microfat can be generated either by direct harvest with cannula size as small as 0.7 mm in diameter or by postprocessing techniques such as mincing. The parcels collected in macrofat, millifat, and microfat, while variable in diameter, all maintain native adipose architecture with viable mature adipocytes and a near-native composition of SVF and ASCs. These grafts are particularly valuable in the aesthetic and reconstructive approach to features such as the lip and eyelid where the persistent contour of macrofat globules provides a suboptimal cosmetic result.

Further advancement in this direction has led to the development of nanofat. Nanofat is an autologous adipose product with parcel sizes less than 0.5 mm [52]. These are generated utilizing shear-based techniques such as continuous passage through a closed Leur-to-Leur system. Given the degree of postprocessing and shear required to generate nanofat, these grafts lose their native architecture, and a majority of the mature adipocytes are nonviable. As ASCs and the SVF niche are more resistant to shear compared with adipocytes, these grafts retain a highly enriched population of mesenchymal cells and bioactive cellular components [53]. Nanofat is thus regarded less as an option for filler, and rather as a technique for the transfer of ASCs and other adipose-derived components for soft-tissue rejuvenation and improvement in skin quality. In the clinical setting, nanofat processing techniques are valuable, as they are typically mechanical in nature and thus are less subject to regulatory scrutiny.

Stromal vascular fraction purification and enrichment

Isolation of the SVF and, more specifically, ASCs is a more recent innovation in lipotransfer as fat-cell transplantation strategies have been developed. One example of this is cell-assisted lipotransfer (CAL). Isolation of ASCs for clinical usage typically occurs in one of two ways: (1) isolation of nanofat as described above and (2) collagenase-assisted enzymatic degradation of morselized adipose. Enzymatic methods differ from the nanofat processing in that the shear forces are significantly reduced; however, the introduction of collagenase requires multiple filtration and dilution steps to make the autograft safe for reintroduction to the patient. Both methods have been utilized in clinical practice, and the ideal technique for ASC isolation in the clinical sphere remains to be determined.

After isolation, SVF or ASCs can either be utilized as a distinct graft or can be used to "enrich" a sample of autologous fat graft (Fig. 6). This process of CAL is currently

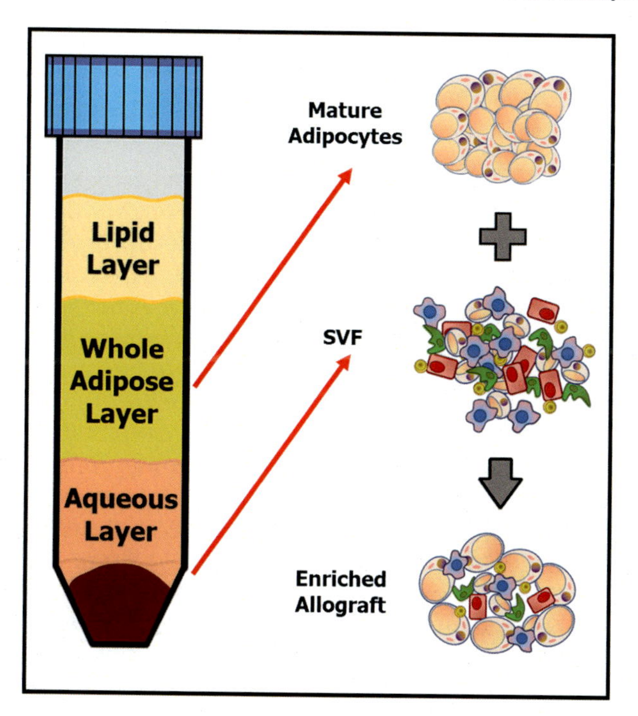

FIG. 6 Schematic representing the generation of an enriched autograft sample.

incompletely understood, and its value remains an active subject of study and debate [54, 55]. There is evidence of improved graft viability independent of recipient site when used for lipofilling; however, CAL has not been demonstrated to actually decrease the number of procedures necessary to achieve an aesthetic result [54, 55]. Additionally, it remains to be determined how exactly CAL affects the complication profile of adipose autograft.

Principles of lipotransfer

Recipient site variables

Recipient site selection is typically guided by clinical need as opposed to anatomic convenience; however, there are several variables that are associated with differences in autograft retention. Vascularity of the donor site represents the key factor associated with adipose survival. Age, irradiation, and scar, while indications for adipose transfer in their own right, are associated with decreased vascularity. As described previously, adipose survival is dependent on maximizing oxygen diffusion and revascularization; thus, sites that demonstrate poor vascular beds may not be adequate to sustain moderate-to-large volume grafts and may require serial fat grafting to achieve a lasting result.

Mobility of the recipient site is additionally described as a factor leading to poor engraftment. Muscle, given its robust vascular supply, was initially identified as a viable bed for adipose transfer; however, later studies seemed to demonstrate that repetitive motion and contraction of muscle did not provide the graft with the stability necessary for vascular integration [4]. Labial and glabellar zones of the face are similarly mobile and, despite significant vascularity, have historically had variable long-term results after lipotransfer [56].

Preparation of the recipient site

Pretreatment of the recipient site is a series of maneuvers designed to improve either the vascularity or tissue quality to better host and maintain the autograft upon transfer. Multiple classes of pretreatment have been described in the literature, including external suction/expansion technique, implantation of alloplastic materials and/or internal expanders, microneedling, preconditioning via controlled ischemia, and administration of either SVF or exogenous growth factors [57]. These techniques can be roughly divided into those that are intended to reshape the soft tissue envelope, such as microneedling and volume expansion, and the remainder, which are intended to improve the vascular bed. In preclinical studies, each of these systems has demonstrated improved angiogenesis and graft survival [57]. The majority of these techniques, however, have not yet made the transition fully into clinical practice. External volume expansion is currently the most clinically studied of these techniques via the Brava system [44, 45]. This system, which is specialized for the breast, utilizes a negative-pressure external volume expander to both increase the available space for lipotransfer and increase soft-tissue vascularity of the breast [44, 45]. From a clinical perspective, it is important to consider that each of these techniques requires one if not more interventions prior to, during, or after lipotransfer. As this involves a greater time, financial burden, and risk, these techniques must still be adequately assessed for their value in autologous fat grafts. Given their nature as adjuncts designed to improve the graft site, it remains possible that the relevance of recipient site pretreatment will be in those compromised sites, such as burns, scars, and irradiated tissue, where current lipotransfer techniques are limited.

Reinjection technique

According to the guidelines put forth by Drs. Coleman and Orgill, the goals of reinjection are to transfer each individual adipose aliquot fully surrounded by vascularized tissue to minimize central necrosis and promote rapid engraftment. This is often performed in a criss-cross pattern to avoid focal overgrafting at or around the injection site (Fig. 7). Avoidance of overgrafting and, thus, pressure necrosis is commonly achieved by the formation of "microribbons" of adipose. The "microribbon" model of lipotransfer is best achieved with the following technical points: (1) injection occurs only as the cannula is withdrawn, (2) resistance to injection is incrementally overcome to prevent pressure buildup and uncontrolled tissue deposition, and (3) injection occurs at a slow and steady rate previously described between 0.5 and 1 mL/s. These three rules are necessary to minimize shear stress, maximize the homogeneity of the adipose aliquot "microribbon," and to ensure that the autograft remains within the desired tissue plane as generated by the cannula [10].

Additionally, the ratio of cannula width to length discussed above for cannula selection in liposuction applies to both liposuction and lipofilling cannulas; however, there are additional considerations for lipofilling. Specifically, if a cannula used during lipofilling is too wide, the aliquot of adipose collected may have a surface-area-to-volume ratio inadequate to survive off diffusion in the immediate postengraftment period [46]. This may lead to central necrosis and, thus, an increase in complications and a less aesthetic result [46].

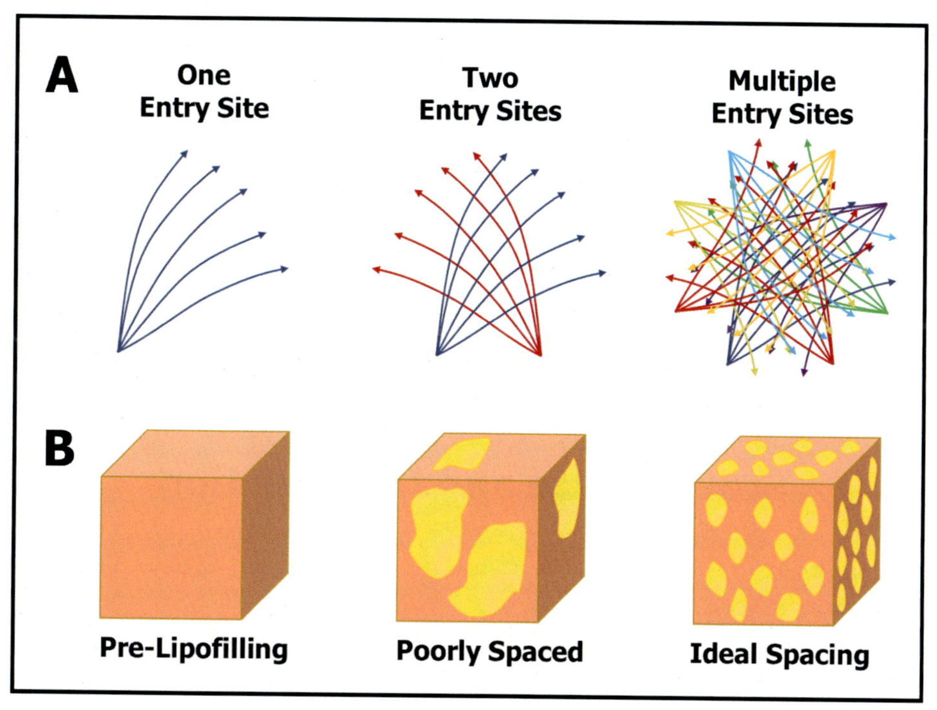

FIG. 7 (A) Schematic representing the possible options for criss-cross pattern lipofilling with multiple entry sites. (B) Representation of idea spacing and intercalation of adipose aliquots in tissue.

Serial lipotransfer

Overgrafting, as described above, is the point at which a grafted tissue reaches critical interstitial pressure where any further filling decreases capillary perfusion and, thus, oxygen delivery. In Dr. Orgill's "fluid accommodation model," this point is reached at approximately 9 mmHg. This model is based on native tissue elasticity and vascularity and must be seen as a general guideline, especially in poorly vascularized and contracted tissues such as those seen in scar or irradiation. To avoid overgrafting, serial lipotransfer is often utilized (Fig. 8). This technique relies on each successive graft to both expand the soft tissue envelope and bring in new and more vascularized soft tissue [10]. Serial lipotransfer has been demonstrated to be safe, effective, and generally well tolerated in clinical practice. There is, however, always an increased cost and risk to the patient with multiple procedures, which must be weighed against the benefit gained from this technical modification.

High-volume (megavolume) lipotransfer

Mention must be made of the concept of high-volume lipotransfer. Notably in the context of the reinjection techniques described above, successful engraftment appears to favor lower volumes of injection [47, 58]. Anatomically, this appears to be true within the setting of any given pass of the cannula; however, significant volume can be transferred, assuming a healthy and well-vascularized bed of sufficient volume [1, 59]. To promote effective engraftment of

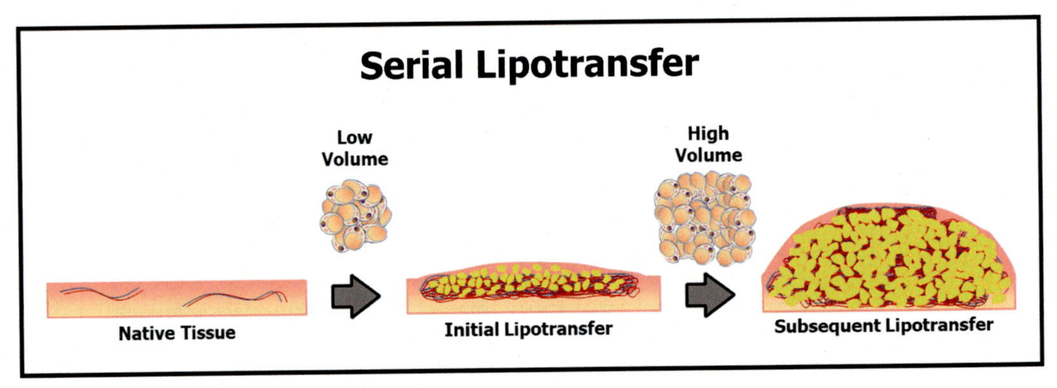

FIG. 8 Tissue expansion and improvement of vascularity with lipotransfer.

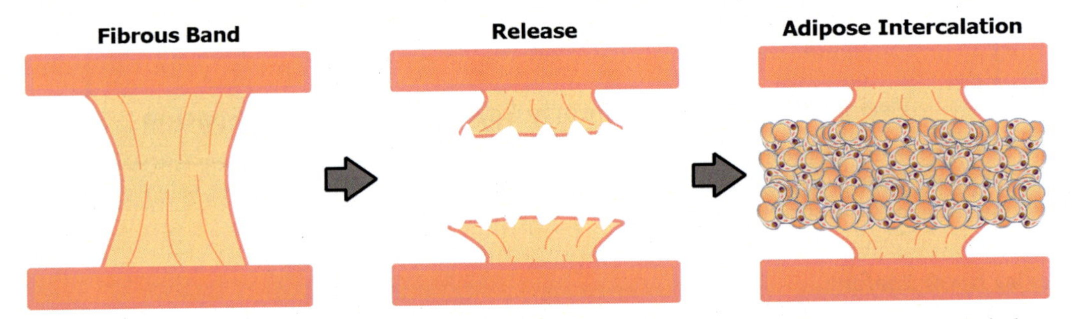

FIG. 9 Schematic representing a simplified scar release with intercalation of healthy adipose tissue at site of release.

larger volumes of autograft, meticulous cannula technique is critical. Overtunneling as with overfilling can lead to formation of negative space where adipose can aggregate and become hypoxic [58]. A latticework pattern utilizing multiple points of entry has been described in 3D space to maximize the amount of volume that remains vascularized upon transfer [1, 59]. This technique is most commonly utilized for aesthetic and reconstructive lipotransfer where space filling and volume expansion are necessary to the final result.

Scar release and intercalation

A final adjunctive technique to be described here is the release of fibrotic tissue and intercalation of adipose between two previously adherent layers (Fig. 9). This technique may be facilitated during tunneling by utilization of forked cannulas as described above or may be achieved by microneedling or needle perforation [60]. The theory behind scar release during lipotransfer is that intercalation of healthy adipose will prevent fibrotic scarring and reformation of the adhesion, thus allowing for smooth gliding and expansion of the tissue [60]. In the literature, this is described as percutaneous aponeurotomy and lipofilling and is a technique utilized for management of several different types of fibrotic pathology that will be touched upon later in this chapter.

Risks and complications

Lipotransfer, while generally well tolerated, is not without risks, and it is important that both clinician and patient be aware of the possible complications of these procedures. Complications of the process are easily divided into those related to the donor site, the graft itself, and the recipient site. Liposuction, while significantly safer since the introduction of tumescent anesthesia, can lead to pain, paresthesia, ecchymosis, seroma and/or hematoma formation, contour irregularities and hypodermal fibrosis, scarring and/or dyspigmentation of the cannula entry site, infection, and thermal, vascular, or mechanical injury to either the overlying dermis or, more critically, to deep structures with muscle, peritoneum, or large caliber vessels [41]. These injuries are fortunately quite rare, and the more common complications, such as bruising, discomfort, and seroma, are typically minimal and easily managed. An additional risk associated with tumescence anesthesia is lidocaine toxicity. Doses in tumescent anesthesia reach between 35 and 55 mg/kg, well above the limit of 7 mm/kg typically associated with concurrent infiltration of lidocaine and epinephrine [61]. Because of the delay in systemic absorption seen in tumescent, this is typically well tolerated; however, there are case reports of cardiac arrest secondary to lidocaine toxicity from tumescent anesthesia, and it is critical that the surgeon take careful assessment of risk factors and minimize excess lidocaine administration where possible [62].

We have previously discussed several factors that may contribute to loss or compromise of the autograft. As a general rule, the more gently the adipose tissue is handled and the less time it spends outside the body, the greater the take. Excellent sterile technique is critical to minimizing risk of graft infection. Fat that has not been optimally handled has a higher risk of fibrosis, calcification, and oil cyst formation [63, 64]. Calcifications are of particular concern as they can present clinically as new, palpable mass. The aesthetic presence of a new mass in areas such as the breast puts patients through significant anxiety and unnecessary procedures and testing. Given the central need for effective angiogenesis and capillary blood flow in the early postengraftment period, avoidance of vasoconstrictors such as nicotine is also necessary to avoid diminished graft survival. Lipofilling additionally carries many similar risks to the recipient site as described above for liposuction.

Additional risks include the possibility of thrombotic or fat emboli. Large-volume liposuction and/or lipotransfer can take a prolonged period of time while the patient is sedated. This carries the thrombotic risks of similarly extensive surgical procedures, and thus, patients must be evaluated preoperatively for thrombotic risk factors. In the perioperative setting, appropriate prophylaxis is necessary. Thromboembolic events including deep vein thrombosis (DVT) and pulmonary embolism are a serious complication and must be accounted for as part of the surgical plan. Fat emboli also carry a serious risk of morbidity and mortality depending on the location of injection. In the face and particularly on the nasal dorsum, fat transfer carries a documented risk of intravascular penetration and migration of adipose tissue to deposit in the ophthalmic artery, leading to transient or permanent blindness [65]. When fat is inappropriately deposited into larger vessels, such as the gluteal vasculature, a clinically significant embolus can migrate centrally, leading to pulmonary embolism and death [66]. Respect for anatomic planes and native vasculature is, thus, critical to avoid the most devastating consequences of autologous fat grafting.

One final point that bears mentioning is the historical concern that adipose autograft may be associated with oncologic risk. This was a serious initial concern with the advent of adipose autograft to the breast for reconstruction after mastectomy defect in the setting of breast cancer. Extensive clinical and preclinical studies over the last 30 years have demonstrated no increased oncologic risk with lipotransfer to the breast or any other recipient site [67–71]. Similarly, given the risk of fat necrosis and calcification described above, there is concern that lipotransfer to the breast may affect breast cancer detection on mammography. Fortunately, this has also not borne out in clinical practice [67–71].

Indications

Form versus *function*

The clinical indications for lipotransfer are broad and have only continued to grow with the advent of ASC-based therapies. Fat transfer has now been described and/or utilized in almost every anatomic site and for purposes across numerous surgical specialties. Broadly speaking, adipose-derived autografts are utilized for one of two distinct but interrelated purposes: (1) fat may be used anatomically either as a filler to provide shape, bulk, and contour to a tissue or architecturally to change tissue quality via release and intercalation of adipose tissue within a typically fibrotic or constricted tissue bed, or (2) adipose, SVF, or ASCs can be utilized functionally for their cellular and/or biochemical benefits.

The original and still predominant usage of autologous tissue is to fill space. As described above, adipose is a readily available, inexpensive, and lasting option to fill, expand, or contour tissue. Lipotransfer is widely utilized for both aesthetic and reconstructive purposes and has been leveraged to reshape and restructure anatomic deficiencies for almost all parts of the body. Because of its ability to provide bulk and to break up painful and tethered scar tissue, adipose has additionally found a role in the treatment of painful scars and amputation site dysesthesia. Additionally, adipose is now being utilized for a more direct mechanical role in the maintenance, contouring, and closure of surgical and anatomic sphincters, fistulas, and ostomies.

Functional transfer may be utilized to directly influence tissue quality or indirectly via immunomodulatory, antiinflammatory, or proangiogenic signaling. These effects are thought to be directed by resident ASCs and other cellular portions of the SVF resident within adipose deposits. These cells have demonstrated capacity for induction of dermal hyperplasia, de novo collagen synthesis with restoration of a more native alignment of the ECM, and neoangiogenesis [72]. In damaged and attenuated tissues, this translates to a thicker and more resilient dermis and a more robust vascular network. Functional transfer in this manner has seen use in scar rejuvenation, burn and chronic wound management, and the treatment and prophylaxis of radiation dermatitis [7, 73]. Additionally, ASCs present as a multipotent and immunomodulatory cell, which has opened up the use of functional fat grafting to a broad range of inflammatory and degenerative musculoskeletal disorders that we will describe below.

As our understanding of lipotransfer improves, modern approaches to fat grafting often harness fat's ability both as a filler and as a functional tissue enhancer. As the field expands, the number of indications and treatments utilizing adipose-based techniques will continue to grow. Given this, we have sought to summarize some of the more common indications below.

Aesthetic and reconstructive fat grafting

When utilized for aesthetic purpose, adipose tissue typically is used to contour, augment, and volumize or rejuvenate. Adipose is widely used to volumize and recontour the face, augment the breast and buttocks, and rejuvenate aging tissues such as the hands and forearms. Here, we will describe areas where fat grafting and, more specifically, ASC-enriched grafting has been utilized to improve tissue aesthetics, contour, and quality.

Facial fat grafting

Facial fat grafting is an aesthetic procedure that straddles the boundary between use of fat as a filler and its effect on tissue rejuvenation. Modern facial rejuvenation utilizes the anatomic concept of facial fat compartments. With aging, there is a well-characterized deflation of these compartments particularly in the periorbital region of the face, and surgical intervention thus focuses on selective lipofilling within these adipose compartments to generate a youthful facial contour. For patients who are looking for augmentation as opposed to rejuvenation, adipose tissue is regularly used in the lips, chin, jawline, and neck to a more masculine or feminine contour as desired [7, 74]. Because of the exceedingly thin tissues found in areas such as the lips, nasal dorsum, and eyelids, more refined grafts—millifat and microfat as previously described—are utilized in a deep plane to minimize contour irregularities [75, 76]. In addition to the aforementioned facial compartments, lipotransfer is invaluable in correction of acne scars, deep rhytids, and facial folds such as the nasolabial fold, tear troughs, and marionette lines of the face [77, 78]. When utilized in this manner, adipose has an advantage over most other fillers in that there is limited risk of rejection or allergic reaction and that the adipose tissue will provide a long-lasting outcome. The face is highly susceptible to asymmetry, with obvious deformity generated by even small degrees of unilateral atrophic change. This is of particular importance with congenital diseases such as hemifacial microsomia and scleroderma or acquired lipodystrophies [79–81]. Here, adipose is again valuable as a relatively noninvasive means to correct unilateral soft-tissue defects.

ASC-directed therapies and nanofat are being developed for facial rejuvenation to capitalize on the proposed tissue quality benefit ascribed to these techniques. There are data to suggest that adipose tissue has the capacity to interfere with dermal elastosis and thus provide direct improvement to the collagen and elastin networks that are disrupted with photoaged skin [82]. This has led to utilization of nanofat in the treatment of superficial skin pathology such as rhytids where traditional adipose transfer would lead to contour irregularity. Nanofat has additionally been described as a potential treatment to address attenuation of already-thin skin seen in disorders such as blepharochalasis [7, 52].

Fat grafting to the breast

The breast is a common site for both aesthetic and reconstructive fat grafting. During breast reconstruction after a mastectomy, defect lipofilling is traditionally utilized as an adjunctive procedure either at time of initial reconstruction or to correct defects that occur with time.

Liposuction as a primary reconstructive modality is more commonly associated with treatment of contour irregularity occurring after biopsy, lumpectomy, or abscess drainage [83, 84]. Additionally, while not the primary treatment used in clinical practice, adipose autograft has been described for the softening of capsular contracture and may represent a treatment option for lower-grade capsules to delay or avoid breast implant removal [85]. Fat, nanofat, and isolated ASCs are additionally used in the treatment of radiation dermatitis and in rejuvenation of irradiated and constricted tissue, the benefits and biology of which we will discuss later in this text.

Aesthetic fat grafting to the breast plays a role similar to reconstructive lipotransfer. Adipose tissue is commonly utilized to improve areas of deflation, contour abnormalities, breast asymmetry, and congenital breast deformities. Fat transfer breast augmentation is a useful technique utilizing serial lipotransfer for women that want to avoid implant-based augmentation and are willing to accept more modest short-term results for a more natural appearance and feel. Because of the risk of graft resorption, adipose-based breast augmentation typically relies on overfilling as compensation for the final result. Because of the risk of overgrafting and fat necrosis as described earlier, this often means that the target volume fill for each individual lipotransfer is often quite less than might be expected for an implant-based augmentation for the same breast footprint [86–88].

Rejuvenation of the hand and upper extremity

After the face, the hands represent one of the most highly visible anatomic structures in daily life. Hands are frequently exposed to environmental and traumatic insults, ultimately resulting in attenuated skin, scar, and atrophic tissues. In addition to the cutaneous sequalae of photoaging, the hand loses subcutaneous volume, leading to increased prominence of bones, tendons, and veins. The hand is thus one of the more vulnerable sites for poor cosmetic outcomes with aging [89, 90]. Adipose transfer represents one of the safer and more lasting options for volumization and recontouring of the hand and, in particular, of the dorsal hand and webspace. The goal of hand rejuvenation in this manner is to improve subcutaneous fullness, soften the appearance of exposed tendons and veins, and improve overall skin quality—an effect potentially enhanced by ASC activity. Hand rejuvenation is a well-tolerated procedure and typically with high patient satisfaction, and is rapidly growing as a cosmetic option for aesthetic treatment of aging skin [89, 90].

Functional fat grafting

Trauma, oncologic resection, and iatrogenic injury may result in a spectrum of deformity that can include absent or deficient tissues, amputation, hypotrophic scars, and/or tethered scars. Any of these can lead to significant asymmetry and contour defect for which adipose autograft is well suited for reconstruction. As filler, fat provides a soft and ultimately well-vascularized tissue bed; however, many of these defects present with additional functional, architectural, and mechanical defects that must ultimately be addressed. In the next two sections, we will discuss uses of adipose tissue that extend beyond its use as a space-filling graft.

Architectural and mechanical fat grafting

Spanning the divide between fat for filler and for function, there are several techniques that utilize intercalation of adipose tissue to change the mechanical properties of the recipient tissue. Scarred and fibrous tissues are inherently nonpliable, and this technique is useful in fibrotic and contractile pathology, such as burn and scar contractures across the joint, to soften tissue in cases of microsomia, and has been described for Dupuytren's contracture of the hand [91]. This technique shares similarities with the use of adipose flaps for prevention of tendon adhesions and perineural scar, and unsurprisingly, lipotransfer has been explored as adjunct treatment after tenolysis and/or neurolysis for that very indication [92–94]. The anatomic basis for architectural fat grafting in this manner relies on the physical barrier provided by adipose tissue. Healthy adipose is well vascularized and highly compliant, allowing surrounding tissues to glide, stretch, and move though the intercalated adipose more freely. Adipose thus serves a buffer role, preventing either reformation of a contractile band or serving as an anchor for the fibrotic scar tissue and thus sparing the underlying structure.

Fat grafting to change the mechanical properties of an anatomic site additionally spans the boundary between functional fat grafting and its use as a filler. Adipose may be used for obliteration of dead space to augment sinus or fistula closure, to improve vocalization and air flow in cases of velopharyngeal incompetence or injury to the vocal cords, and for laryngoplasty and myringoplasty [95–98]. Hypotrophic adipose may also be augmented, as in the case of pressure ulcers, as an adjunctive therapy to prevent recurrent injury by improving soft-tissue bulk. Adipose tissue can be utilized as a physical barrier for treatment of urinary or fecal incontinence via periureteric, perianal, or sphincteric engraftment or to enhance the shape and fit of an ostomy appliance [99–102]. The range of indications by which fat grafting may be used to alter the structure and, thus, mechanical function of a site is extensive and varied, represented mostly by small-scale studies and case reports.

Tissue rejuvenation

Restoration of a well-vascularized, pliable, and aesthetic integument after injury represents one of the most sought-after goals in both aesthetic and reconstructive surgery. Cutaneous attenuation and loss of integrity can occur with aging, trauma and scar formation, burns, radiation, and a host of dermatologic pathologies. The skin's native function depends on its trilaminar structure consisting of an epidermal barrier, dense dermal support, and a hypodermal layer through which vascular, lymphatic, and neural elements invest the dermis. In native tissue, the hypodermis is the source of ASCs and is the fat from which both adipose autograft and SVF are collected. After severe and/or chronic injury, there is typically disruption of this trilaminar structure with disordered and dysmorphic healing of dermal and hypodermal elements. While mesenchymal cells are present within the dermis as fibroblasts and the hair bulb in hair-bearing portions of the skin, a major supporting role for dermal and epidermal health stems from ASCs resident within the hypodermal niche. Subdermal fat graft, direct transfer of ASCs, and/or SVF all function to introduce healthy mesenchymal cells into the hypodermal space with the goal of restoring the support network necessary to maintain healthy skin.

Adipose transfer to a severely scarred or attenuated tissue bed works by several different mechanisms. Fat can act as a lipofilling or intercalating agent as previously described, restoring contour, bulk, pliability, and elasticity to the tissue. As will be discussed later in this chapter, adipose autograft can act as a release for entrapped dermal–hypodermal nerve fibers to help address the pain and dysesthesia associated with scarred tissue. There are, currently poorly understood, antimelanogenic effects associated with fat grafting that can potentially minimize scar dyspigmentation [103].

At the cellular level, ASCs act a potent induction agent for neovascularization of the hypodermal capillary network improving blood flow to the overlying dermis. There is a poorly understood interplay between dermal resident fibroblasts and ASCS such that adipose autograft can induce collagen neosynthesis and ameliorate elastosis, allowing for the formation of a more natural and structurally appropriate dermal ECM [82, 103]. Additionally, ASCs along with other cellular components of the SVF have a potent immunomodulatory effect capable of suppressing local immune response [104].

The restoration of a healthy biochemical niche is critical for rejuvenation of tissues that have been systemically injured, such as those in irradiated beds or in the setting of poorly controlled diabetes. Radiation injury in particular leads to progressive obliteration of the microvasculature, cutaneous fragility and permeability, and fibrosis [73]. With time, the tissue becomes progressively more hypoxic, leading to a cascade of worsening tissue injury. This same process is described in adipose transplant to diabetic wounds of the lower extremity where the microvasculature is similarly compromised. Adipose and, specifically, ASC transplantation is presumed to break this cycle by introducing healthy, proangiogenic tissue into the wound bed that can, with time, improve tissue oxygenation and restore cutaneous integrity [73]. Because of the extent of injury and the need to overcome damaged tissue with new healthy cells, serial lipotransfer is often necessary to address these defects.

Fat grafting for pain

Neuropathic pain is a difficult problem to address, with few readily apparent solutions. Pain is a multifactorial process and is often resistant and recalcitrant to therapy. Fat grafting has been proposed as one potential solution. Lipotransfer addresses the problem of neuropathic pain and, in particular, pain after trauma in several ways. In the setting of amputation and neuroma, fat grafting superficial to the nerve body can provide significant relief as a mechanical buffer against direct contact and, thus, repetitive irritation. As described above, autograft can be used to release entrapped dermal–hypodermal nerve fibers. The adipose tissue then provides a physical barrier to prevent reentrapment and scarring. This is a mechanism similar to the use of adipose tissue to prevent perineural scar and as an adjunct to neurolysis to prophylax against scar in the future. In the case of perineural fat grafting in treatment of a nonreconstructable neuroma, the neuroma is resected and the autograft is placed around and adjacent to the stump. This has been demonstrated in clinical practice to reduce neuropathic pain and improve function of the extremity [94, 105]. This has additionally been utilized successfully in treatment of painful pedal fat pad atrophy utilizing perforating adipose injections into the substance of the pedal fat pad [106–108]. There are also reports of fat grafting being successfully utilized as adjunct after surgical decompression for focal migraines [109]. Whether this technique utilizes a similar mechanism to perineural fat grafting as described above remains an open area of research at this time.

Fat grafting for Raynaud's phenomenon

Similar to the treatment of scleroderma described previously, autologous fat grafting has been described in small-scale studies as a successful therapeutic option for relief from symptoms of Raynaud's phenomenon. Raynaud's phenomenon refers to a painful and progressive decrease in blood flow to the extremities that can, in part, be traced to aberrant sympathetic tone leading to pathologic vasoconstriction. In a limited clinical trial perivascular fat grafting was found to lead to improved blood flow and decreased pain to the distal digits. The exact mechanism for this effect is currently unclear [110, 111].

On the horizon: Clinical trials

Fat grafting has emerged as a promising application for clinical use and remains a rapidly growing area of research. In 2020, there are nearly 900 clinical trials actively registered though the US National Library of Medicine (NLM) at the National Institutes of Health utilizing adipose tissues. A portion of these trials are exploring other avenues for adipose technology, namely, brown adipose tissue activation, adipose tissue's role in metabolism, and the influence of exercise on adipose tissue. Here we will evaluate those applications that examine direct therapeutic interventions and the usage of ASCs for treatment of human disease. Specifically, we will discuss ASCs in regenerative medicine for the following: osteoarthritis (OA) in the knee, spinal cord injury (SCI), amyotrophic lateral sclerosis (ALS), multiple sclerosis (MS), myocardial infarction, fistulas, lung diseases, and others.

Osteoarthritis and musculoskeletal disease

Adipose transfer has been utilized in the last decade for treatment of musculoskeletal disorders such as OA and degenerative disk disease. In the small joints of the hand, the basal thumb, and the temporomandibular joint (TMJ), adipose has been described as a filler, providing a barrier to prevent painful rubbing between two articular surfaces [112, 113]. In larger joints and in the case of degenerative disk disease, application of ASC has been described; however, the mechanism of effect has not yet been elucidated [114, 115]. Proposed theories include the application of mesenchymal cells to injured surfaces and introduction of prochondrogenic factors.

Currently, it is uncertain whether ASCs are able to survive the compressive load and relatively avascular environment of the synovial space; however, several clinical trials are currently in progress to identify what, if any, value adipose-based stem cell therapies may have for degenerative joint diseases. Andia et al. summarized and reviewed many of these studies to evaluate OA treatment options including BMSCs, platelet-rich plasma (PRP), and ASCs [116]. Currently there is significant variability between the approaches used, including combination of ASCs with PRP, the direct inoculation of ASCs to the knee [NCT03399630], and utilization of variable adipose donors including autologous infrapatellar fat pads, autologous adipose via lipotransfer, and industrial sources such as K-STEM CELL, which had been harvested from abdominal fat via liposuction, prepared under good manufacturing practices, and tested prior to shipping for use [117–123].

Recently, Lee et al. assessed the efficacy and safety of a single intraarticular injection of ASCs for patients with knee OA [124]. ASCs were injected into the knees of 12 participants and were compared with those in a control group of normal saline and followed for 6 months. Outcomes included the Western Ontario and McMaster Universities Osteoarthritis Index (WOMAC) score, clinical and radiologic examination, change of cartilage defect via magnetic resonance imaging (MRI), and adverse events. Those injected with ASCs had a significant improvement in their WOMAC scores at 6 months, whereas there was no change in WOMAC scores of those in the control group. MRI of those receiving ASCs showed no significant change in cartilage defect, but those in the control group experienced an increase in cartilage defect at 6 months. No serious adverse events were observed.

Traumatic spinal cord injury

Injuries of the spinal cord represent a devastating and difficult-to-treat clinical scenario and are an active area of research. Mesenchymal cells, from both bone marrow and adipose-derived sources are actively being evaluated for the treatment of SCI [NCT02981576, NCT01624779, NCT01769872, NCT04064957] [125].

The CELLTOP study [NCT03308565] investigated the safety and efficacy of intrathecal autologous ASCs in patients with blunt traumatic SCI. In their initial report, investigators describe one participant, age 53 years, a survivor of a surfing accident who sustained a high cervical grade A SCI with subsequent neurologic improvement [126]. The patient plateaued within 6 months after injury, then, upon enrolling in the study, received an intrathecal injection of 100 million autologous ASCs 11 months after injury. At 3, 6, 12, and 18 months postinjection, both motor and sensory scores suggested meaningful improvement. Separately, [NCT01274975] eight male participants who had survived a SCI more than 12 months prior received an intravenous, single dosage of 4×10^8 autologous ASCs [127]. Throughout the 3-month study period, no serious adverse events were observed.

Amyotrophic lateral sclerosis

Investigators at the Mayo Clinic conducted an ASC dose-escalation safety study assessing intrathecal treatment for ALS in 27 participants [128]. ASCs were isolated, then expanded in culture and cryopreserved for 8 weeks prior to injection. Autologous ASCs were infused intrathecally after a standard lumbar puncture in either one single dose of 1×10^7 or two monthly doses of 1×10^8 cells. Weekly follow-up visits occurred in the first 4 weeks after treatment, then every 3 months until 2 years, death, or loss to follow-up. Measured outcomes included clinical evaluations, an ALS questionnaire, blood and cerebrospinal fluid (CSF) sampling, and MRI of the neuroaxis. ASC intrathecal treatment was found to be safe for ALS. Dose-dependent changes were observed in MRI, CSF, and temporary lumbosacral-radicular pain. The most commonly reported events were found to be temporary low-back and radicular leg pain in subjects treated with the highest dose level. Longitudinal questionnaires confirmed continued progression of disease in all participants. Four participants died during the study period, and autopsies were performed and found no sign of tumor formation.

Stroke therapy

As with SCI above, there is growing interest in the use of mesenchymal cell as treatment for recalcitrant neurologic injury, and ASCs have been studied in both China and Spain as therapeutic adjuncts after stroke. In one study [NCT02813512], sponsored by Gwo Xi Stem Cell Applied Technology, three participants with chronic stroke received autologous ASC injection into the brain and were assessed by MRI and eight standardized stroke indices at 1, 3, and 6 months posttreatment. Results have not been disseminated. In Spain, a separate study group [NCT01678534] enrolled 19 participants with ischemic stroke; participants were blinded and randomized into either a placebo or treatment arm. Treatment was an intravenous dose of autologous ASCs administered into the brain within the first 2 weeks of stroke symptom onset. Safety was the primary outcome of the study, measured by adverse events, complications, and tumor development; and efficacy was the secondary outcome, as determined by stroke volume measured by MRI and two stroke indices. Participants were followed for 2 years. The ASC treatment was well tolerated with no complications, and the infusion of autologous ASCs was feasible and safe. Study participants receiving ASC treatment demonstrated improved neurologic recovery [129].

Multiple sclerosis

There are multiple reports of autologous ASC usage for MS [130–132]. Fernandez et al. in 2018 reports a multicenter study [NCT01056471] where 30 participants were randomized into either a placebo group ($N=11$), ASC low-dose treatment group ($N=10$), or ASC high-dose treatment group ($N=9$). ASCs were harvested from lipectomy at a good manufacturing practices facility within 12 h of tissue collection, cultured for days until 80% confluence, resuspended in Ringer's lactate, and then administered intravenously through a peripheral venous catheter over 2 h using an infusion pump. Placebo injections consisted of Ringer's lactate in identical packages. Outcome measures included safety as determined by adverse events, clinical variables, immunologic assessments, analysis of CSF, MRI, evoked potentials (EPs), optical coherence tomography (OCT) of the retina, cognition, and quality of life measured by several questionnaires over 12 months postinjection. Two participants in the placebo group died during the study from events unrelated to the intervention (choking/bronchial aspiration and respiratory infection). No serious AEs were considered to be related to the treatment. No clinically significant changes in clinical variables, CSF, OCT, cognition, or quality of life were observed. A decrease in MRI nonnormalized cerebral volume in the low-dose ASC was observed at 12 months compared with baseline. EPs showed no significant differences between groups or over time, though there were promising trends towards efficacy that lead the investigators to suggest larger future studies.

Erectile dysfunction

ASCs have been shown to safely and effectively improve outcomes in patients with erectile dysfunction (ED) [133]. Seventeen male patients with ED were enrolled in a prospective, open-label, single-arm, single-center study [NCT02240823]. Adipose tissue was harvested with WAL, then ASCs isolated using the automated processing system, Celution®

800/CRS (Cytori Therapeutics, San Diego, CA USA). Participants received an injection of autologous ASCs into the corpus cavernosum and were followed for 6 months postinjection. No serious adverse events were observed. During the study period, 8 of 17 participants recovered erectile function following ASC injection with the ability to complete intercourse.

Vocal fold paralysis

To evaluate the safety of ASCs in the glottal gap in the unilateral paralysis of the vocal cord, 14 patients were enrolled and randomized into either an experimental arm ($N=7$) to receive a fat graft enriched with autologous ASCs or a control arm ($N=7$) that was treated with centrifuged autologous fat (CAF) alone [NCT02904824] [134]. Autologous ASCs were harvested via liposuction and processed through the Celution® System. Two adverse events were observed, septic shock and glottis oedema, which occurred in the experimental arm and were related to the intubation procedure, not to the experimental intervention. Both participants were withdrawn from the study. Voice quality was evaluated preinjection and at 180 days postinjection. Both groups exhibited an objective improvement in voice quality, and both groups were efficient in restoring the phonation function when examined using distortion, biomechanical, and gap features.

Alopecia

Eight patients with androgenetic alopecia were enrolled into an interventional study [NCT02729415] to evaluate effect of autologous ASCs on hair density and thickness up to 6 months postprocedure. While results have not yet been published, the study has been reported as recently completed (January 2019) on clinicaltrials.gov.

Lung diseases

Investigators have turned to ASCs for providing a regenerative medicine solution to address various respiratory conditions, including acute respiratory distress syndrome, chronic obstructive pulmonary disease (COPD), combined pulmonary fibrosis and emphysema (CPFE), and idiopathic pulmonary fibrosis [135–138]. Tzilas et al. in 2015 enrolled eight COPD and five CPFE patients to receive a series of three infusions of ASCs, which were harvested via lipoaspiration. All participants were smokers and were followed for 12 months postinfusion. No serious adverse events and no significant change in pulmonary function tests were observed.

Dry macular degeneration

Oner et al. describes a clinical case series investigating the safety and efficacy of ASC implantation in four patients with optic nerve diseases [139]. Through 6 months of follow-up, no subject had any systemic ocular complications. All experienced visual activity improvement, visual field improvement, and improvement in multifocal electroretinography recordings. Additionally, all four patients had choroidal thickening.

Wound healing

ASCs have demonstrated promising results to improve wound healing in preclinical models; however, no clinical research study has been published yet [140]. There is one study registered on clinicaltrials.gov [NCT02092870]. The study enrolled 25 participants and harvested autologous ASCs from lipoaspirate, which were then delivered to the wound in the form of multiple injections in a single treatment. Percent change in wound size was evaluated at 12 weeks. Results are pending at publication of this text.

Rotator cuff repair

ASCs have been used in combination with fibrin glue to repair rotator cuff tears [141]. In total, 161 patients with full-thickness rotator cuff tear were randomized into either a conventional group ($N=85$) to undergo arthroscopic rotator cuff repair alone, or an injection group ($N=76$) to undergo arthroscopic rotator cuff repair with ASC injection. Autologous ASCs were isolated from lipoaspirate, cultured, and then loaded into a scaffold of fibrin glue and applied with a spinal needle into the site of the tendon–bone approximation and over the repaired tendon after arthroscopic fluid was extracted. Measurements of pain, range of motion, function, and MRI, to assess repaired tendon structural integrity, were collected for 28 months postprocedure. Pain, range of motion, and functional measures improved significantly in both groups, but with no significant difference between groups. MRI indicated a lower retear rate in the group that received ASCs compared with the conventional group.

Another study is registered on clinicaltrials.gov as completed as of December 2018, but no results are yet posted [NCT02918136]. This study enrolled 18 patients with partial-thickness rotator cuff tear into either a group that received ASC injection or a group that received cortisone injection.

Degenerative disc disease

To address chronic discogenic low-back pain, Kumar et al. assessed the safety and tolerability of an intradiscal implantation of ASCs plus hyaluronic acid over 12 months [NCT02338271] [142]. Ten patients were enrolled in the study and received a single dose (either high or low) of ASCs, which had been isolated from liposuction and then cultured until 80%–90% confluence was reached prior to injection. No serious adverse events occurred during the study period. Pain and satisfaction scores significantly improved for six of the ten participants (three in each dosage group), three of which exhibited increased water content based on increased apparent diffusion coefficient on diffusion MRI.

In a separate study, investigators evaluated the safety and efficacy of autologous ASCs combined with PRP intradiscally into patients with degenerative disc disease [NCT02097862] [143]. Fifteen participants received a 1 mL injection of ASC and PRP suspension and were followed for 12 months postinjection. No serious adverse events were observed. Flexion, pain, and short-form questionnaires reflected significantly improved outcomes.

Heart disease

Because of the multipotency of ASCs and their demonstrated ability to differentiate into endothelial cells and cardiomyocytes in vivo, they have been used in clinical research trials for myocardial infarction and chronic myocardial ischemia [144–147]. In the MyStromalCell trial [NCT01449032] [147], ASCs were harvested via liposuction, isolated, and then plated in culture. ASCs were differentiated toward endothelial cell by stimulation with VEGF-A165 culture medium. ASCs were delivered into viable myocardium (identified via 3D electromechanical mapping) in 10–15 injections of 0.2 [2] mL each. Participants were followed for 3 years, and safety and exercise outcomes were compared with a placebo treatment of phosphate-buffered saline (PBS). Treatment with ASCs was found to be safe and trended toward increased exercise capacity versus placebo.

Additionally, one group has successfully utilized a new surgical technique for reducing myocardial scar after a heart attack, the adipose graft transposition procedure (AGTP), involving the existing fat surrounding the heart [148]. Ten study participants were evaluated through 1 year of follow-up; five received AGTP on the nonrevascularizable area, and five were left untouched. AGTP was found to be safe, and no differences in adverse, clinical, or arrhythmic events were observed between groups. No significant difference in cardiac MRI was observed between groups at 1 year; however, those treated with AGTP had a slightly smaller left ventricular end systolic volume and necrosis ratio. The authors caution interpretation of these results and indicate that further studies are necessary.

Fistula

While past investigators have observed promising results treating perianal fistulas with cultured ASCs or BMSCs, the process of cell expansion and culture is timely, expensive, and requires special facilities [149–156]. Because of the recalcitrant nature of Crohn's disease, there are now several efforts to identify a safe, efficient, and effective ASC-based therapeutic for closure of Crohn's [NCT03555773, NCT01011244, NCT00992485, NCT03803917, NCT01372969, NCT01314079, NCT00999115, NCT01803347] and non-Crohn's fistulae [NCT00475410, NCT01020825]. One study was conducted in Spain to determine the safety and feasibility of expanded ASCs to treat Crohn's-related rectovaginal fistula in 10 patients [149,158,150-157,159]. Twenty million ASCs were injected into the vaginal walls and fistula tract. If unhealed at 12 weeks postinjection, a second dose of 40 million ASCs was delivered, and participants were followed for 1 year after the last injection. ASC injection proved to be safe for treatment, and 60% of participants achieved complete healing. One group evaluated the use of freshly harvested fat tissue injection for anovaginal fistulas and complex fistulas caused by Crohn's disease in 21 patients [157, 159]. Injection of freshly collected autologous fat tissue was found to be safe and resulted in complete fistula healing in 57% of study participants after 6 months.

Fat grafting

As ASCs are being utilized for their regenerative properties, fat grafting is being utilized for its mechanical properties in conjunction with the regenerative potential within the tissue.

Ruane et al. found that fat tissue, injected under the metatarsals of 17 patients, significantly reduced pain outcomes in those with pedal fat pad atrophy [108]. Six months postinjection, there was an increase in tissue thickness and volume, as well as improvements in activities of daily living. In a separate study, the same investigators randomized participants into either an interventional group ($N=18$), which received fat injections under the metatarsals with 2 years of follow-up, or a control group ($N=13$), which was provided standard of care for 1 year, then foot fat injections with 1 year of follow-up [106-108, 160]. Investigators found that fat injections into the foot provide long-lasting improvements in pain and function, and may prevent against worsening if administered earlier following onset. Additionally, the same group found that pedal fat grafting improved skin quality, which could contribute to improved clinical outcomes [160]. This group has also evaluated fat injections for fat pad atrophy of the heel, and has observed promising results through 12 months of follow-up [161].

In addition to pedal applications, fat grafting has been studied in other weight-bearing applications, such as at lower-extremity amputation sites [101]. When injected at lower-extremity amputation sites, fat tissue was injected to provide padding over the bony structures of five injured military personnel suffering from pain and limited function over 2 years. There was a nonsignificant trend toward improvement in pain scores. As a continuation of this pilot, a second study to assess fat grafting at lower-extremity amputation sites of ten patients was conducted by the same group (NCT02076022) [162]. Over 2 years, participants exhibited improvement in pain scores, hypersensitivity, and prosthetic fit. Six of the 10 enrolled participants discontinued pain and/or antianxiety medication throughout the study.

Autologous fat grafting has been studied in non-weight bearing applications, such as craniofacial deformities [163]. Twenty participants received fat grafting to a craniofacial deformity, and graft volume was measured over 9 months via computed tomography. No serious adverse events were observed. Satisfaction with appearance, social relationships, and social functioning improved at 9 months. Graft volume retention at 3 months strongly predicted 9-month retention, and those who underwent a second procedure exhibited similar volume retention as the first time.

Romberg's disease

Koh et al. evaluated the effect of ASCs in microfat grafting of patients with Parry-Romberg disease on fat graft volume (NCT01309061) [164]. Ten participants were enrolled and randomized into either a microfat graft + ASC treatment pathway or a microfat graft only pathway. ASCs were isolated from abdominal fat via liposuction and cultured for 2 weeks, then injected with secondary fat grafts. Analyses included gross appearance, patient satisfaction, and measurement of hemifacial volume via a 3D camera (Vectra, Canfield Scientific) and 3D CT scan over 15 months of follow-up. Fat graft survival and fat graft retention was better in those supplemented with ASCs than microfat grafts alone.

Summary

Adipose autograft and ASC therapeutics represent one of the fastest growing and most broadly applicable techniques in plastic surgery. Here, we discussed the basics of the technique for adipose harvest, processing, and transfer as well as briefly discussed current and proposed therapeutic avenues that utilize adipose tissue. As we develop a greater understanding of the tools, techniques, and the tissue itself, adipose-based therapy is expected to continue to grow and is an important part of the modern plastic surgeon's armamentarium.

References

[1] E.R. Zielins, E.A. Brett, M.T. Longaker, D.C. Wan, Autologous fat grafting: the science behind the surgery, Aesthet. Surg. J. 36 (4) (2016) 488–496, https://doi.org/10.1093/asj/sjw004.

[2] E. Bellini, M.P. Grieco, E. Raposio, A journey through liposuction and liposculture: review, Ann. Med. Surg. (Lond.) 24 (2017) 53–60, https://doi.org/10.1016/j.amsu.2017.10.024.

[3] I.B. James, S.R. Coleman, J.P. Rubin, Fat, stem cells, and platelet-rich plasma, Clin. Plast. Surg. 43 (3) (2016) 473–488, https://doi.org/10.1016/j.cps.2016.03.017.

[4] A.L. Strong, P.S. Cederna, J.P. Rubin, S.R. Coleman, B. Levi, The current state of fat grafting: a review of harvesting, processing, and injection techniques, Plast. Reconstr. Surg. 136 (4) (2015) 897–912, https://doi.org/10.1097/PRS.0000000000001590.

[5] M. Rodbell, Metabolism of isolated fat cells. I. effects of hormones on glucose metabolism and lipolysis, J. Biol. Chem. 239 (1964) 375–380.

[6] A. Bajek, N. Gurtowska, J. Olkowska, L. Kazmierski, M. Maj, T. Drewa, Adipose-derived stem cells as a tool in cell-based therapies, Arch. Immunol. Ther. Exp. 64 (6) (2016) 443–454.

[7] F. Simonacci, N. Bertozzi, M.P. Grieco, E. Grignaffini, E. Raposio, Procedure, applications, and outcomes of autologous fat grafting, Ann. Med. Surg. (Lond). 20 (2017) 49–60, https://doi.org/10.1016/j.amsu.2017.06.059.

[8] R.L. Van, C.E. Bayliss, D.A. Roncari, Cytological and enzymological characterization of adult human adipocyte precursors in culture, J. Clin. Invest. 58 (3) (1976) 699–704.

[9] C.M. Rivera-Cruz, J.J. Shearer, M. Figueiredo Neto, M.L. Figueiredo, The immunomodulatory effects of mesenchymal stem cell polarization within the tumor microenvironment niche, Stem Cells Int. 2017 (2017) 4015039, https://doi.org/10.1155/2017/4015039.

[10] R.K. Khouri Jr., R.E. Khouri, J.R. Lujan-Hernandez, K.R. Khouri, L. Lancerotto, D.P. Orgill, Diffusion and perfusion: the keys to fat grafting, Plast. Reconstr. Surg. Glob. Open 2 (9) (2014) e220, https://doi.org/10.1097/GOX.0000000000000183.

[11] J. Sood, L. Jayaraman, N. Sethi, Liposuction: anaesthesia challenges, Indian J. Anaesth. 55 (3) (2011) 220–227, https://doi.org/10.4103/0019-5049.82652.

[12] J.H. Moore Jr., J.W. Kolaczynski, L.M. Morales, R.V. Considine, Z. Pietrzkowski, P.F. Noto, J.F. Caro, Viability of fat obtained by syringe suction lipectomy: effects of local anesthesia with lidocaine, Aesthet. Plast. Surg. 19 (4) (1995) 335–339.

[13] A. Lunger, T. Ismail, A. Todorov, J. Buergin, F. Lunger, I. Oberhauser, M. Haug, D.F. Kalbermatten, R.D. Largo, I. Martin, A. Scherberich, D.J. Schaefer, Improved adipocyte viability in autologous fat grafting with ascorbic acid-supplemented tumescent solution, Ann. Plast. Surg. (2019), https://doi.org/10.1097/SAP.0000000000001857 (Epub ahead of print).

[14] J. Gillis, S. Gebremeskel, K.D. Phipps, L.A. MacNeil, C.J. Sinal, B. Johnston, P. Hong, M. Bezuhly, Effect of N-acetylcysteine on adipose-derived stem cell and autologous fat graft survival in a mouse model, Plast. Reconstr. Surg. 136 (2) (2015) 179e–188e, https://doi.org/10.1097/PRS.0000000000001443.

[15] A. Khanna, G. Filobbos, Avoiding unfavourable outcomes in liposuction, Indian J. Plast. Surg. 46 (2) (2013) 393–400, https://doi.org/10.4103/0970-0358.118618.

[16] L.L. Pu, S.R. Coleman, X. Cui, R.E. Ferguson Jr., H.C. Vasconez, Autologous fat grafts harvested and refined by the Coleman technique: a comparative study, Plast. Reconstr. Surg. 122 (3) (2008) 932–937, https://doi.org/10.1097/PRS.0b013e3181811ff0.

[17] J.L. Crawford, B.A. Hubbard, S.H. Colbert, C.L. Puckett, Fine tuning lipoaspirate viability for fat grafting, Plast. Reconstr. Surg. 126 (4) (2010) 1342–1348, https://doi.org/10.1097/PRS.0b013e3181ea44a9.

[18] L. Charles-de-Sá, N.F. Gontijo de Amorim, D. Dantas, J.V. Han, P. Amable, M.V. Teixeira, P.L. de Araújo, W. Link, R. Borojevich, G. Rigotti, Influence of negative pressure on the viability of adipocytes and mesenchymal

stem cell, considering the device method used to harvest fat tissue, Aesthet. Surg. J. 35 (3) (2015) 334–344, https://doi.org/10.1093/asj/sju047.

[19] M. Keck, J. Kober, O. Riedl, H.B. Kitzinger, S. Wolf, T.M. Stulnig, M. Zeyda, A. Gugerell, Power assisted liposuction to obtain adipose-derived stem cells: impact on viability and differentiation to adipocytes in comparison to manual aspiration, J. Plast. Reconstr. Aesthet. Surg. 67 (1) (2014) e1–e8, https://doi.org/10.1016/j.bjps.2013.08.019.

[20] T. Fontes, I. Brandão, R. Negrão, M.J. Martins, R. Monteiro, Autologous fat grafting: harvesting techniques, Ann. Med. Surg. (Lond). 36 (2018) 212–218, https://doi.org/10.1016/j.amsu.2018.11.005.

[21] J.J. Stutz, D. Krahl, Water jet-assisted liposuction for patients with lipoedema: histologic and immunohistologic analysis of the aspirates of 30 lipoedema patients, Aesthet. Plast. Surg. 33 (2) (2009) 153–162, https://doi.org/10.1007/s00266-008-9214-y.

[22] S.M. Shridharani, J.M. Broyles, A. Matarasso, Liposuction devices: technology update, Med. Devices (Auckl.) 7 (2014) 241–251, https://doi.org/10.2147/MDER.S47322.

[23] D.L. Miller, N.B. Smith, M.R. Bailey, G.J. Czarnota, K. Hynynen, I.R. Makin, Bioeffects Committee of the American Institute of Ultrasound in Medicine, Overview of therapeutic ultrasound applications and safety considerations, J. Ultrasound Med. 31 (4) (2012) 623–634.

[24] R.J. Rohrich, S.J. Beran, J.M. Kenkel, W.P. Adams Jr., F. DiSpaltro, Extending the role of liposuction in body contouring with ultrasound-assisted liposuction, Plast. Reconstr. Surg. 101 (4) (1998) 1090–1102 (discussion 1117-9).

[25] P.B. Fodor, J. Watson, Personal experience with ultrasound-assisted lipoplasty: a pilot study comparing ultrasound-assisted lipoplasty with traditional lipoplasty, Plast. Reconstr. Surg. 101 (4) (1998) 1103–1116 (discussion 1117-9).

[26] M.S. Beckenstein, J.C. Grotting, Ultrasound-assisted lipectomy using the solid probe: a retrospective review of 100 consecutive cases, Plast. Reconstr. Surg. 105 (6) (2000) 2161–2174 (discussion 2175-9).

[27] D. Duscher, Z.N. Maan, A. Luan, M.M. Aitzetmüller, E.A. Brett, D. Atashroo, A.J. Whittam, M.S. Hu, G.G. Walmsley, K.S. Houschyar, A.F. Schilling, H.G. Machens, G.C. Gurtner, M.T. Longaker, D.C. Wan, Ultrasound-assisted liposuction provides a source for functional adipose-derived stromal cells, Cytotherapy 19 (12) (2017) 1491–1500, https://doi.org/10.1016/j.jcyt.2017.07.013.

[28] R.J. Rohrich, D.E. Morales, J.E. Krueger, M. Ansari, O. Ochoa, J. Robinson Jr., S.J. Beran, Comparative lipoplasty analysis of in vivo-treated adipose tissue, Plast. Reconstr. Surg. 105 (6) (2000) 2152–2158 (discussion 2159-60).

[29] M.L. Jewell, P.B. Fodor, E.B. de Souza Pinto, M.A. Al Shammari, Clinical application of VASER- -assisted lipoplasty: a pilot clinical study, Aesthet. Surg. J. 22 (2) (2002) 131–146, https://doi.org/10.1067/maj.2002.123377.

[30] M.E. Schafer, K.C. Hicok, D.C. Mills, S.R. Cohen, J.J. Chao, Acute adipocyte viability after third-generation ultrasound-assisted liposuction, Aesthet. Surg. J. 33 (5) (2013) 698–704, https://doi.org/10.1177/1090820X13485239.

[31] E.B. de Souza Pinto, P.C. Abdala, C.M. Maciel, F.P. dos Santos, R.P. de Souza, Liposuction and VASER, Clin. Plast. Surg. 33 (1) (2006) 107–115 (vii).

[32] M.W. Nagy, P.F. Vanek Jr., A multicenter, prospective, randomized, single-blind, controlled clinical trial comparing VASER-assisted Lipoplasty and suction-assisted Lipoplasty, Plast. Reconstr. Surg. 129 (4) (2012) 681e–689e, https://doi.org/10.1097/PRS.0b013e3182442274.

[33] J.C. McBean, B.E. Katz, Laser lipolysis: an update, J. Clin. Aesthet. Dermatol. 4 (7) (2011) 25–34.

[34] A. Goldman, U. Wollina, E.C. de Mundstock, Evaluation of tissue tightening by the subdermal Nd: YAG laser-assisted liposuction versus liposuction alone, J Cutan Aesthet Surg 4 (2) (2011) 122–128, https://doi.org/10.4103/0974-2077.85035.

[35] S. Mordon, A.F. Eymard-Maurin, B. Wassmer, J. Ringot, Histologic evaluation of laser lipolysis: pulsed 1064-nm Nd:YAG laser versus cw 980-nm diode laser, Aesthet. Surg. J. 27 (3) (2007) 263–268, https://doi.org/10.1016/j.asj.2007.03.005.

[36] M.T. Chung, A.S. Zimmermann, K.J. Paik, S.D. Morrison, J.S. Hyun, D.D. Lo, A. McArdle, D.T. Montoro, G.G. Walmsley, K. Senarath-Yapa, M. Sorkin, R. Rennert, H.H. Chen, A.S. Chung, D. Vistnes, G.C. Gurtner, M.T. Longaker, D.C. Wan, Isolation of human adipose-derived stromal cells using laser-assisted liposuction and their therapeutic potential in regenerative medicine, Stem Cells Transl. Med. 2 (10) (2013) 808–817, https://doi.org/10.5966/sctm.2012-0183.

[37] D.O. Beck, K. Davis, R.J. Rohrich, Enhancing Lipoaspirate efficiency by altering liposuction cannula design, Plast. Reconstr. Surg. Glob. Open 2 (10) (2014) e222, https://doi.org/10.1097/GOX.0000000000000101.
[38] V.L. Young, H.J. Brandon, The physics of suction-assisted lipoplasty, Aesthet. Surg. J. 24 (3) (2004) 206–210, https://doi.org/10.1016/j.asj.2004.03.001.
[39] X. Yang, F.M. Egro, T. Jones, W.V. Nerone, M. Yousefpour, J.A. Gusenoff, J.P. Rubin, L.E. Kokai, Comparison of adipose particle size on autologous fat graft retention in a rodent model, Plast. Aesthet. Res. 7 (2020) 8. https://doi.org/10.20517/2347-9264.2019.63.
[40] J.C. Kirkham, J.H. Lee, M.A. Medina 3rd, M.C. McCormack, M.A. Randolph, W.G. Austen Jr., The impact of liposuction cannula size on adipocyte viability, Ann. Plast. Surg. 69 (4) (2012) 479–481, https://doi.org/10.1097/SAP.0b013e31824a459f.
[41] V.V. Dixit, M.S. Wagh, Unfavourable outcomes of liposuction and their management, Indian J. Plast. Surg. 46 (2) (2013) 377–392, https://doi.org/10.4103/0970-0358.118617.
[42] A.O. Momoh, S. Colakoglu, C. de Blacam, M.S. Curtis, B.T. Lee, The forked liposuction cannula: a novel approach to the correction of cicatricial contracture deformities in breast reconstruction, Ann. Plast. Surg. 69 (3) (2012) 256–259, https://doi.org/10.1097/SAP.0b013e3182275d8f.
[43] K.W. Minn, K.H. Min, H. Chang, S. Kim, E.J. Heo, Effects of fat preparation methods on the viabilities of autologous fat grafts, Aesthet. Plast. Surg. 34 (5) (2010) 626–631, https://doi.org/10.1007/s00266-010-9525-7.
[44] M.N. Mirzabeigi, M. Lanni, C.S. Chang, R.Y. Stark, S.J. Kovach, L.C. Wu, J.M. Serletti, L.P. Bucky, Treating breast conservation therapy defects with Brava and fat grafting: technique, outcomes, and safety profile, Plast. Reconstr. Surg. 140 (3) (2017) 372e–381e, https://doi.org/10.1097/PRS.0000000000003626.
[45] R.K. Khouri, M. Eisenmann-Klein, E. Cardoso, B.C. Cooley, D. Kacher, E. Gombos, T.J. Baker, Brava and autologous fat transfer is a safe and effective breast augmentation alternative: results of a 6-year, 81-patient, prospective multicenter study, Plast. Reconstr. Surg. 129 (5) (2012) 1173–1187, https://doi.org/10.1097/PRS.0b013e31824a2db6.
[46] I.B. James, D.A. Bourne, G. DiBernardo, S.S. Wang, J.A. Gusenoff, K. Marra, J.P. Rubin, The architecture of fat grafting II: impact of cannula diameter, Plast. Reconstr. Surg. 142 (5) (2018) 1219–1225, https://doi.org/10.1097/PRS.0000000000004837.
[47] D.A. Bourne, I.B. James, S.S. Wang, K.G. Marra, J.P. Rubin, The architecture of fat grafting: what lies beneath the surface, Plast. Reconstr. Surg. 137 (3) (2016) 1072–1079, https://doi.org/10.1097/01.prs.0000479992.10986.ad.
[48] E.C. Cleveland, N.J. Albano, A. Hazen, Roll, spin, wash, or filter? Processing of Lipoaspirate for autologous fat grafting: an updated, evidence-based review of the literature, Plast. Reconstr. Surg. 136 (4) (2015) 706–713, https://doi.org/10.1097/PRS.0000000000001581.
[49] O. Canizares Jr., J.E. Thomson, R.J. Allen Jr., E.H. Davidson, J.P. Tutela, P.B. Saadeh, S.M. Warren, A. Hazen, The effect of processing technique on fat graft survival, Plast. Reconstr. Surg. 140 (5) (2017) 933–943, https://doi.org/10.1097/PRS.0000000000003812.
[50] M. Pfaff, W. Wu, E. Zellner, D.M. Steinbacher, Processing technique for lipofilling influences adipose-derived stem cell concentration and cell viability in lipoaspirate, Aesthet. Plast. Surg. 38 (1) (2014) 224–229, https://doi.org/10.1007/s00266-013-0261-7.
[51] S. Cohen, H. Womack, A. Ghanem, Fat grafting for facial rejuvenation through injectable tissue replacement and regeneration: a differential, standardized, anatomic approach, Clin. Plast. Surg. 47 (1) (2020) 31–41, https://doi.org/10.1016/j.cps.2019.08.005.
[52] P. Tonnard, A. Verpaele, G. Peeters, M. Hamdi, M. Cornelissen, H. Declercq, Nanofat grafting: basic research and clinical applications, Plast. Reconstr. Surg. 132 (4) (2013) 1017–1026, https://doi.org/10.1097/PRS.0b013e31829fe1b0.
[53] T.M. Gause, R.E. Kling, W.N. Sivak, K.G. Marra, J.P. Rubin, L.E. Kokai, Particle size in fat graft retention: a review on the impact of harvesting technique in lipofilling surgical outcomes, Adipocyte 3 (4) (2014) 273–279, https://doi.org/10.4161/21623945.2014.957987.
[54] J. Laloze, A. Varin, J. Gilhodes, N. Bertheuil, J.L. Grolleau, J. Brie, J. Usseglio, L. Sensebe, T. Filleron, B. Chaput, Cell-assisted lipotransfer: friend or foe in fat grafting? Systematic review and meta-analysis, J. Tissue Eng. Regen. Med. 12 (2) (2018) e1237–e1250, https://doi.org/10.1002/term.2524.
[55] Y. Zhou, J. Wang, H. Li, X. Liang, J. Bae, X. Huang, Q. Li, Efficacy and safety of cell-assisted Lipotransfer: a systematic review and meta-analysis, Plast. Reconstr. Surg. 137 (1) (2016) 44e–57e, https://doi.org/10.1097/PRS.0000000000001981.

[56] A. Mojallal, C. Shipkov, F. Braye, P. Breton, J.L. Foyatier, Influence of the recipient site on the outcomes of fat grafting in facial reconstructive surgery, Plast. Reconstr. Surg. 124 (2) (2009) 471–483, https://doi.org/10.1097/PRS.0b013e3181af023a.

[57] C.M. Oranges, J. Striebel, M. Tremp, S. Madduri, D.F. Kalbermatten, Y. Harder, D.J. Schaefer, The preparation of the recipient site in fat grafting: a comprehensive review of the preclinical evidence, Plast. Reconstr. Surg. 143 (4) (2019) 1099–1107, https://doi.org/10.1097/PRS.0000000000005403.

[58] R.K. Khouri, G. Rigotti, E. Cardoso, R.K. Khouri Jr., T.M. Biggs, Megavolume autologous fat transfer: part II. Practice and techniques, Plast. Reconstr. Surg. 133 (6) (2014) 1369–1377, https://doi.org/10.1097/PRS.0000000000000179.

[59] M.T. Chung, K.J. Paik, D.A. Atashroo, J.S. Hyun, A. McArdle, K. Senarath-Yapa, E.R. Zielins, R. Tevlin, C. Duldulao, M.S. Hu, G.G. Walmsley, A. Parisi-Amon, A. Momeni, J.R. Rimsa, G.W. Commons, G.C. Gurtner, D.C. Wan, M.T. Longaker, Studies in fat grafting: part I. effects of injection technique on in vitro fat viability and in vivo volume retention, Plast. Reconstr. Surg. 134 (1) (2014) 29–38, https://doi.org/10.1097/PRS.0000000000000290.

[60] R.K. Khouri, J.M. Smit, E. Cardoso, N. Pallua, L. Lantieri, I.M. Mathijssen, R.K. Khouri Jr., G. Rigotti, Percutaneous aponeurotomy and lipofilling: a regenerative alternative to flap reconstruction? Plast. Reconstr. Surg. 132 (5) (2013) 1280–1290, https://doi.org/10.1097/PRS.0b013e3182a4c3a9.

[61] N.F. Holt, Tumescent anaesthesia: its applications and well tolerated use in the out-of-operating room setting, Curr. Opin. Anaesthesiol. 30 (4) (2017) 518–524, https://doi.org/10.1097/ACO.0000000000000486.

[62] S. Mrad, C. El Tawil, W.A. Sukaiti, R. Bou Chebl, G. Abou Dagher, Z. Kazzi, Cardiac arrest following liposuction: a case report of lidocaine toxicity, Oman Med. J. 34 (4) (2019) 341–344, https://doi.org/10.5001/omj.2019.66.

[63] A. Hamza, V. Lohsiriwat, M. Rietjens, Lipofilling in breast cancer surgery, Gland Surg. 2 (1) (2013) 7–14, https://doi.org/10.3978/j.issn.2227-684X.2013.02.03.

[64] S.R. Coleman, A.P. Saboeiro, Fat grafting to the breast revisited: safety and efficacy, Plast. Reconstr. Surg. 119 (3) (2007) 775–785 (discussion 786-7).

[65] K.T. Loh, J.J. Chua, H.M. Lee, J.T. Lim, G. Chuah, B. Yim, B.K. Puah, Prevention and management of vision loss relating to facial filler injections, Singap. Med. J. 57 (8) (2016) 438–443, https://doi.org/10.11622/smedj.2016134.

[66] M.M. Mofid, S. Teitelbaum, D. Suissa, A. Ramirez-Montañana, D.C. Astarita, C. Mendieta, R. Singer, Report on mortality from gluteal fat grafting: recommendations from the ASERF task force, Aesthet. Surg. J. 37 (7) (2017) 796–806, https://doi.org/10.1093/asj/sjx004.

[67] T.K. Krastev, S.J. Schop, J. Hommes, A.A. Piatkowski, E.M. Heuts, R.R.W.J. van der Hulst, Meta-analysis of the oncological safety of autologous fat transfer after breast cancer, Br. J. Surg. 105 (9) (2018) 1082–1097, https://doi.org/10.1002/bjs.10887.

[68] S.L. Spear, C.N. Coles, B.K. Leung, M. Gitlin, M. Parekh, D. Macarios, The safety, effectiveness, and efficiency of autologous fat grafting in breast surgery, Plast. Reconstr. Surg. Glob. Open 4 (8) (2016) e827, https://doi.org/10.1097/GOX.0000000000000842.

[69] K.A. Gutowski, ASPS Fat Graft Task Force, Current applications and safety of autologous fat grafts: a report of the ASPS fat graft task force, Plast. Reconstr. Surg. 124 (1) (2009) 272–280, https://doi.org/10.1097/PRS.0b013e3181a09506.

[70] M.M.A. Silva, L.E. Kokai, V.S. Donnenberg, J.L. Fine, K.G. Marra, A.D. Donnenberg, M.S. Neto, J.P. Rubin, Oncologic safety of fat grafting for autologous breast reconstruction in an animal model of residual breast cancer, Plast. Reconstr. Surg. 143 (1) (2019) 103–112, https://doi.org/10.1097/PRS.0000000000005085.

[71] R. Schweizer, W. Tsuji, V.S. Gorantla, K.G. Marra, J.P. Rubin, J.A. Plock, The role of adipose-derived stem cells in breast cancer progression and metastasis, Stem Cells Int. 2015 (2015) 120949, https://doi.org/10.1155/2015/120949.

[72] M. Gaur, M. Dobke, V.V. Lunyak, Mesenchymal stem cells from adipose tissue in clinical applications for dermatological indications and skin aging, Int. J. Mol. Sci 18 (1) (2017), https://doi.org/10.3390/ijms18010208 (Pii: E208).

[73] G. Rigotti, A. Marchi, M. Galiè, G. Baroni, D. Benati, M. Krampera, A. Pasini, A. Sbarbati, Clinical treatment of radiotherapy tissue damage by lipoaspirate transplant: a healing process mediated by adipose-derived adult stem cells, Plast. Reconstr. Surg. 119 (5) (2007) 1409–1422 (discussion 1423-4).

[74] S.R. Coleman, Facial recontouring with lipostructure, Clin. Plast. Surg. 24 (2) (1997) 347–367.
[75] J.M. Serra-Renom, J.M. Serra-Mestre, Periorbital rejuvenation to improve the negative vector with blepharoplasty and fat grafting in the malar area, Ophthalmic Plast. Reconstr. Surg. 27 (6) (2011) 442–446, https://doi.org/10.1097/IOP.0b013e318224b0d5.
[76] W.P. Kao, Y.N. Lin, T.Y. Lin, Y.H. Huang, C.K. Chou, H. Takahashi, T.Y. Shieh, K.P. Chang, S.S. Lee, C.S. Lai, S.-D. Lin, T.M. Lin, Microautologous fat transplantation for primary augmentation rhinoplasty: long-term monitoring of 198 Asian patients, Aesthet. Surg. J. 36 (6) (2016) 648–656, https://doi.org/10.1093/asj/sjv253.
[77] J.W. Groen, T.K. Krastev, J. Hommes, J.A. Wilschut, M.J.P.F. Ritt, R.R.J.W. van der Hulst, Autologous fat transfer for facial rejuvenation: a systematic review on technique, efficacy, and satisfaction, Plast. Reconstr. Surg. Glob. Open 5 (12) (2017) e1606, https://doi.org/10.1097/GOX.0000000000001606.
[78] S. Tenna, A. Cogliandro, M. Barone, V. Panasiti, M. Tirindelli, C. Nobile, P. Persichetti, Comparative study using autologous fat grafts plus platelet-rich plasma with or without fractional CO2 laser resurfacing in treatment of acne scars: analysis of outcomes and satisfaction with FACE-Q, Aesthet. Plast. Surg. 41 (3) (2017) 661–666, https://doi.org/10.1007/s00266-017-0777-3.
[79] G.C. Slack, C.J. Tabit, K.A. Allam, H.K. Kawamoto, J.P. Bradley, Parry-Romberg reconstruction: beneficial results despite poorer fat take, Ann. Plast. Surg. 73 (3) (2014) 307–310, https://doi.org/10.1097/SAP.0b013e31827aeb0d.
[80] M. Gheisari, A. Ahmadzadeh, N. Nobari, B. Iranmanesh, N. Mozafari, Autologous fat grafting in the treatment of facial scleroderma, Dermatol. Res. Pract. 2018 (2018) 6568016, https://doi.org/10.1155/2018/6568016.
[81] N. Del Papa, F. Caviggioli, D. Sambataro, E. Zaccara, V. Vinci, G. Di Luca, A. Parafioriti, E. Armiraglio, W. Maglione, R. Polosa, F. Klinger, M. Klinger, Autologous fat grafting in the treatment of fibrotic perioral changes in patients with systemic sclerosis, Cell Transplant. 24 (1) (2015) 63–72, https://doi.org/10.3727/096368914X674062.
[82] L. Charles-de-Sá, N.F. Gontijo-de-Amorim, C. Maeda Takiya, R. Borojevic, D. Benati, P. Bernardi, A. Sbarbati, G. Rigotti, Antiaging treatment of the facial skin by fat graft and adipose-derived stem cells, Plast. Reconstr. Surg. 135 (4) (2015) 999–1009, https://doi.org/10.1097/PRS.0000000000001123.
[83] R.A. Mann, T.N.S. Ballard, D.L. Brown, A.O. Momoh, E.G. Wilkins, J.H. Kozlow, Autologous fat grafting to lumpectomy defects: complications, imaging, and biopsy rates, J. Surg. Res. 231 (2018) 316–322, https://doi.org/10.1016/j.jss.2018.05.023.
[84] S.S. Ergün, E.G. Baygöl, R.B. Kayan, İ.M. Kuzu, O. Akman, Correction of brassiere strap grooves with fat injections, Aesthet. Surg. J. 35 (5) (2015) 561–564, https://doi.org/10.1093/asj/sjv007.
[85] S. Papadopoulos, G. Vidovic, M. Neid, A. Abdallah, Using fat grafting to treat breast implant capsular contracture, Plast. Reconstr. Surg. Glob. Open 6 (11) (2018) e1969, https://doi.org/10.1097/GOX.0000000000001969.
[86] S.R. Coleman, A.P. Saboeiro, Primary breast augmentation with fat grafting, Clin. Plast. Surg. 42 (3) (2015) 301–306 (vii) https://doi.org/10.1016/j.cps.2015.03.010.
[87] E. Delay, S. Guerid, The role of fat grafting in breast reconstruction, Clin. Plast. Surg. 42 (3) (2015) 315–323. vii https://doi.org/10.1016/j.cps.2015.03.003.
[88] V. Pinsolle, A. Chichery, J.L. Grolleau, J.P. Chavoin, Autologous fat injection in Poland's syndrome, J. Plast. Reconstr. Aesthet. Surg. 61 (7) (2008) 784–791, https://doi.org/10.1016/j.bjps.2007.11.033.
[89] S.R. Coleman, Hand rejuvenation with structural fat grafting, Plast. Reconstr. Surg. 110 (7) (2002) 1731–1744 (discussion 1745-7).
[90] D. Hoang, M.I. Orgel, D.A. Kulber, Hand rejuvenation: a comprehensive review of fat grafting, J. Hand Surg. Am. 41 (5) (2016) 639–644, https://doi.org/10.1016/j.jhsa.2016.03.006.
[91] U. Tuncel, A. Kurt, M. Gumus, O. Aydogdu, N. Güzel, O. Demir, Preliminary results with non-centrifuged autologous fat graft and percutaneous aponeurotomy for treating Dupuytren's disease, Hand Surg. Rehabil. 36 (5) (2017) 350–354, https://doi.org/10.1016/j.hansur.2017.05.004.
[92] O.E. Damgaard, P.A. Siemssen, Lipografted tenolysis, J. Plast. Reconstr. Aesthet. Surg. 63 (8) (2010) e637–e638, https://doi.org/10.1016/j.bjps.2010.02.004.
[93] M.R. Colonna, M.C. Scarcella, S. d'Alcontres Fd, G. Delia, F. Lupo, Should fat graft be recommended in tendon scar treatment? Considerations on three cases (two feet and a severe burned hand), Eur. Rev. Med. Pharmacol. Sci. 18 (5) (2014) 753–759.
[94] L. Vaienti, R. Gazzola, F. Villani, P.C. Parodi, Perineural fat grafting in the treatment of painful neuromas, Tech. Hand Up. Extrem. Surg. 16 (1) (2012) 52–55, https://doi.org/10.1097/BTH.0b013e31823cd218.

[95] D.Y. Lee, Y.H. Kim, Can fat-plug myringoplasty be a good alternative to formal myringoplasty? A systematic review and meta-analysis, Otol. Neurotol. 39 (4) (2018) 403–409, https://doi.org/10.1097/MAO.0000000000001732.

[96] J.M. Lasso, D. Poletti, B. Scola, P. Gómez-Vilda, A.I. García-Martín, M.E. Fernández-Santos, Injection Laryngoplasty using autologous fat enriched with adipose-derived regenerative stem cells: a safe therapeutic option for the functional reconstruction of the glottal gap after unilateral vocal fold paralysis, Stem Cells Int. 2018 (2018) 8917913, https://doi.org/10.1155/2018/8917913.

[97] G. Cantarella, R.F. Mazzola, M. Gaffuri, E. Iofrida, P. Biondetti, L.V. Forzenigo, L. Pignataro, S. Torretta, Structural fat grafting to improve outcomes of vocal Folds' fat augmentation: long-term results, Otolaryngol. Head Neck Surg. 158 (1) (2018) 135–143, https://doi.org/10.1177/0194599817739256.

[98] E. Nigh, G.A. Rubio, J. Hillam, M. Armstrong, L. Debs, S.R. Thaller, Autologous fat injection for treatment of velopharyngeal insufficiency, J. Craniofac. Surg. 28 (5) (2017) 1248–1254, https://doi.org/10.1097/SCS.0000000000003702.

[99] A. Shafik, Perianal injection of autologous fat for treatment of sphincteric incontinence, Dis. *Colon rectum* 38 (6) (1995) 583–587.

[100] V. Kirchin, T. Page, P.E. Keegan, K.O. Atiemo, J.D. Cody, S. McClinton, P. Aluko, Urethral injection therapy for urinary incontinence in women, Cochrane Database Syst. Rev. 7 (2017) CD003881, https://doi.org/10.1002/14651858.CD003881.pub4.

[101] D.A. Bourne, R.D. Thomas, J. Bliley, G. Haas, A. Wyse, A. Donnenberg, V.S. Donnenberg, I. Chow, R. Cooper, S. Coleman, K. Marra, P.F. Pasquina, J.P. Rubin, Amputation-site soft-tissue restoration using adipose stem cell therapy, Plast. Reconstr. Surg. 142 (5) (2018) 1349–1352, https://doi.org/10.1097/PRS.0000000000004889.

[102] S. Czerniak, J.A. Gusenoff, Z.M. MacIsaac, R.P. CBram, D. Amar, C. Seynnaeve, D. Medich, S. Coleman, J.P. Rubin, Fat grafting for improved ileostomy ostomy device fit: a case report, Wound Manag. Prev. 65 (3) (2019) 38–44.

[103] A. Mojallal, C. Lequeux, C. Shipkov, et al., Improvement of skin quality after fat grafting: clinical observation and an animal study, Plast. Reconstr. Surg. (2009), https://doi.org/10.1097/PRS.0b013e3181b17b8f.

[104] M.S. Choudhery, S.N. Jan, M.M. Bashir, F.A. Khan, Z. Hidayat, H.H. Ansari, M. Sohail, A.B. Bajwa, H.B. Shami, A. Hanif, F. Aziz, Unfiltered Nanofat injections rejuvenate Postburn scars of Face, Ann. Plast. Surg. 82 (1) (2019) 28–33.

[105] M. Alessandri-Bonetti, F.M. Egro, P. Persichetti, S.R. Coleman, R.J. Peter, The role of fat grafting in alleviating neuropathic pain: a critical review of the literature, Plast. Reconstr. Surg. Glob. Open 7 (5) (2019) e2216, https://doi.org/10.1097/GOX.0000000000002216.

[106] J.A. Gusenoff, R.T. Mitchell, K. Jeong, D.K. Wukich, B.R. Gusenoff, Autologous fat grafting for pedal fat pad atrophy: a prospective randomized clinical trial, Plast. Reconstr. Surg. 138 (5) (2016) 1099–1108.

[107] D.M. Minteer, B.R. Gusenoff, J.A. Gusenoff, Fat grafting for pedal fat pad atrophy in a 2-year, prospective, randomized, crossover, single-center clinical trial, Plast. Reconstr. Surg. 142 (6) (2018) 862e–871e, https://doi.org/10.1097/PRS.0000000000005006.

[108] E.J. Ruane, D.M. Minteer, A.J. Wyse, B.R. Gusenoff, J.A. Gusenoff, Volumetric analysis in autologous fat grafting to the foot, Plast. Reconstr. Surg. 144 (3) (2019) 463e–470e, https://doi.org/10.1097/PRS.0000000000005956.

[109] B. Guyuron, N. Pourtaheri, Therapeutic role of fat injection in the treatment of recalcitrant migraine headaches, Plast. Reconstr. Surg. 143 (3) (2019) 877–885, https://doi.org/10.1097/PRS.0000000000005353.

[110] Bank J, S.M. Fuller, G.I. Henry, L.S. Zachary, Fat grafting to the hand in patients with Raynaud phenomenon: a novel therapeutic modality, Plast. Reconstr. Surg. 133 (5) (2014) 1109–1118, https://doi.org/10.1097/PRS.0000000000000104.

[111] N. Del Papa, G. Di Luca, R. Andracco, et al., Regional grafting of autologous adipose tissue is effective in inducing prompt healing of indolent digital ulcers in patients with systemic sclerosis: results of a monocentric randomized controlled study, Arthritis Res. Ther. 21 (1) (2019) 7 (Published 2019 Jan 7) https://doi.org/10.1186/s13075-018-1792-8.

[112] C. Herold, H.O. Rennekampff, R. Groddeck, S. Allert, Autologous fat transfer for thumb carpometacarpal joint osteoarthritis: a prospective study, Plast. Reconstr. Surg. 140 (2) (2017) 327–335, https://doi.org/10.1097/PRS.0000000000003510.

[113] W. Van Bogaert, N. De Meurechy, M.Y. Mommaerts, Autologous fat grafting in Total temporomandibular joint replacement surgery, Ann. Maxillofac. Surg. 8 (2) (2018) 299–302, https://doi.org/10.4103/ams.ams_165_18.

[114] D. Spasovski, V. Spasovski, Z. Baščarević, M. Stojiljković, M. Vreća, M. Anđelković, S. Pavlović, Intra-articular injection of autologous adipose-derived mesenchymal stem cells in the treatment of knee osteoarthritis, J. Gene Med. 20 (1) (2018), https://doi.org/10.1002/jgm.3002.

[115] J.M. Lamo-Espinosa, G. Mora, J.F. Blanco, F. Granero-Moltó, J.M. Núñez-Córdoba, S. López-Elío, E. Andreu, F. Sánchez-Guijo, J.D. Aquerreta, J.M. Bondía, A. Valentí-Azcárate, M. Del Consuelo Del Cañizo, E.M. Villarón, J.R. Valentí-Nin, F. Prósper, Intra-articular injection of two different doses of autologous bone marrow mesenchymal stem cells versus hyaluronic acid in the treatment of knee osteoarthritis: long-term follow up of a multicenter randomized controlled clinical trial (phase I/II), J. Transl. Med. 16 (1) (2018) 213, https://doi.org/10.1186/s12967-018-1591-7.

[116] I. Andia, N. Maffulli, Biological therapies in regenerative sports medicine, Sports Med. 47 (5) (2017) 807–828, https://doi.org/10.1007/s40279-016-0620-z.

[117] K.H.T. Bui, T.D. Duong, N.T. Nguyen, T.D. Nguyen, V.T. Le, V.T. Mai, N.L.C. Phan, D.M. Le, N.K. Phan, P.V. Pham, Symptomatic knee osteoarthritis treatment using autologous adipose derived stem cells and platelet-rich plasma: a clinical study, Biomed. Res. Ther. 1 (01) (2014) 02–08.

[118] N. Gibbs, R. Diamond, E.O. Sekyere, W.D. Thomas, Management of knee osteoarthritis by combined stromal vascular fraction cell therapy, platelet-rich plasma, and musculoskeletal exercises: a case series, J. Pain Res. 8 (2015) 799–806, https://doi.org/10.2147/JPR.S92090.

[119] Y.S. Kim, O.R. Kwon, Y.J. Choi, D.S. Suh, D.B. Heo, Y.G. Koh, Comparative matched-pair analysis of the injection versus implantation of mesenchymal stem cells for knee osteoarthritis, Am. J. Sports Med. 43 (11) (2015) 2738–2746, https://doi.org/10.1177/0363546515599632.

[120] J. Pak, J.J. Chang, J.H. Lee, S.H. Lee, Safety reporting on implantation of autologous adipose tissue-derived stem cells with platelet-rich plasma into human articular joints, BMC Musculoskelet. Disord. 14 (2013) 337, https://doi.org/10.1186/1471-2474-14-337.

[121] P.B. Fodor, S.G. Paulseth, Adipose derived stromal cell (ADSC) injections for pain Management of Osteoarthritis in the human knee joint, Aesthet. Surg. J. 36 (2) (2016) 229–236, https://doi.org/10.1093/asj/sjv135.

[122] Y.G. Koh, S.B. Jo, O.R. Kwon, D.S. Suh, S.W. Lee, S.H. Park, Y.J. Choi, Mesenchymal stem cell injections improve symptoms of knee osteoarthritis, Arthroscopy 29 (4) (2013) 748–755, https://doi.org/10.1016/j.arthro.2012.11.017.

[123] C.H. Jo, Y.G. Lee, W.H. Shin, H. Kim, J.W. Chai, E.C. Jeong, J.E. Kim, H. Shim, J.S. Shin, I.S. Shin, J.C. Ra, S. Oh, K.S. Yoon, Intra-articular injection of mesenchymal stem cells for the treatment of osteoarthritis of the knee: a proof-of-concept clinical trial, Stem Cells 32 (5) (2014) 1254–1266, https://doi.org/10.1002/stem.1634.

[124] W.S. Lee, H.J. Kim, K.I. Kim, G.B. Kim, W. Jin, Intra-articular injection of autologous adipose tissue-derived mesenchymal stem cells for the treatment of knee osteoarthritis: a phase IIb, randomized, placebo-controlled clinical trial, Stem Cells Transl. Med. 8 (6) (2019) 504–511, https://doi.org/10.1002/sctm.18-0122.

[125] N.K. Venkataramanaa, P. Rakhi, Mesenchymal Stem Cells in Spinal Cord Injury. Topics in Paraplegia, IntechOpen, 2014.

[126] M. Bydon, A.B. Dietz, S. Goncalves, F.M. Moinuddin, M.A. Alvi, A. Goyal, Y. Yolcu, C.L. Hunt, K.L. Garlanger, A.S. Del Fabro, R.K. Reeves, A. Terzic, A.J. Windebank, W. Qu, CELLTOP clinical trial: first report from a phase 1 trial of autologous adipose tissue–derived mesenchymal stem cells in the treatment of paralysis due to traumatic spinal cord injury, Mayo Clin. Proc. (2019), https://doi.org/10.1016/j.mayocp.2019.10.008 (pii:S0025-6196(19)30871-7). (Epub ahead of print).

[127] J.C. Ra, I.S. Shin, S.H. Kim, S.K. Kang, B.C. Kang, H.Y. Lee, Y.J. Kim, J.Y. Jo, E.J. Yoon, H.J. Choi, E. Kwon, Safety of intravenous infusion of human adipose tissue-derived mesenchymal stem cells in animals and humans, Stem Cells Dev. 20 (8) (2011) 1297–1308, https://doi.org/10.1089/scd.2010.0466.

[128] Staff NP, N.N. Madigan, J. Morris, M. Jentoft, E.J. Sorenson, G. Butler, D. Gastineau, A. Dietz, A.J. Windebank, Safety of intrathecal autologous adipose-derived mesenchymal stromal cells in patients with ALS, Neurology 87 (21) (2016) 2230–2234.

[129] E. Díez-Tejedor, M. Gutiérrez-Fernández, P. Martínez-Sánchez, B. Rodríguez-Frutos, G. Ruiz-Ares, M.L. Lara, B.F. Gimeno, Reparative therapy for acute ischemic stroke with allogeneic mesenchymal stem cells from adipose tissue: a safety assessment: a phase II randomized, double-blind, placebo-controlled, single-center, pilot clinical trial, J. Stroke Cerebrovasc. Dis. 23 (10) (2014) 2694–2700, https://doi.org/10.1016/j.jstrokecerebrovasdis.2014.06.011.

[130] O. Fernández, G. Izquierdo, V. Fernández, L. Leyva, V. Reyes, M. Guerrero, A. León, C. Arnaiz, G. Navarro, M.D. Páramo, A. Cuesta, B. Soria, A. Hmadcha, D. Pozo, R. Fernandez-Montesinos, M. Leal, I. Ochotorena,

P. Gálvez, M.A. Geniz, F.J. Barón, R. Mata, C. Medina, C. Caparrós-Escudero, A. Cardesa, N. Cuende, Research Group Study EudraCT 2008-004015-35, Adipose-derived mesenchymal stem cells (AdMSC) for the treatment of secondary-progressive multiple sclerosis: a triple blinded, placebo controlled, randomized phase I/II safety and feasibility study, PLoS One. 13 (5) (2018) e0195891, https://doi.org/10.1371/journal.pone.0195891.

[131] N.H. Riordan, T.E. Ichim, W.P. Min, H. Wang, F. Solano, F. Lara, M. Alfaro, J.P. Rodriguez, R.J. Harman, A.N. Patel, M.P. Murphy, R.R. Lee, B. Minev, Non-expanded adipose stromal vascular fraction cell therapy for multiple sclerosis, J. Transl. Med. 7 (2009) 29, https://doi.org/10.1186/1479-5876-7-29.

[132] A. Stepien, N.L. Dabrowska, M. Maciagowska, R.P. Macoch, A. Zolocinska, S. Mazur, K. Siennicka, E. Frankowska, R. Kidzinski, M. Chalimoniuk, Z. Pojda, Clinical application of autologous adipose stem cells in patients with multiple sclerosis: preliminary results, Mediat. Inflamm. 2016 (2016) 5302120.

[133] M.K. Haahr, C.H. Jensen, N.M. Toyserkani, D.C. Andersen, P. Damkier, J.A. Sørensen, L. Lund, S.P. Sheikh, Safety and potential effect of a single intracavernous injection of autologous adipose-derived regenerative cells in patients with erectile dysfunction following radical prostatectomy: an open-label phase I clinical trial, EBioMedicine 5 (2016) 204–210, https://doi.org/10.1016/j.ebiom.2016.01.024.

[134] J.M. Lasso, D. Poletti, B. Scola, P. Gómez-Vilda, A.I. García-Martín, M.E. Fernández-Santos, Injection Laryngoplasty using autologous fat enriched with adipose-derived regenerative stem cells: a safe therapeutic option for the functional reconstruction of the glottal gap after unilateral vocal fold paralysis, Stem Cells Int. 2018 (2018) 8917913, https://doi.org/10.1155/2018/8917913.

[135] G. Zheng, L. Huang, H. Tong, Q. Shu, Y. Hu, M. Ge, K. Deng, L. Zhang, B. Zou, B. Cheng, J. Xu, Treatment of acute respiratory distress syndrome with allogeneic adipose-derived mesenchymal stem cells: a randomized, placebo-controlled pilot study, Respir. Res. 15 (2014) 39, https://doi.org/10.1186/1465-9921-15-39.

[136] V. Tzilas, E. Bouros, D. Fourla, A. Zimopoulos, N. Papdopoulos, G. Koliakos, G. Kolios, I. Pneumatikos, V. Aidinis, A. Tzouvelekis, D. Bouros, Prospective phase 1 open clinical trial to study the safety of adipose derived mesenchymal stem cells (ADMSCs) in COPD and combined pulmonary fibrosis and emphysema (CPFE), Eur. Respir. J. 46 (2015) OA1970, https://doi.org/10.1183/13993003.congress-2015.OA1970.

[137] D.J. Weiss, R. Casaburi, R. Flannery, M. LeRoux-Williams, D.P. Tashkin, A placebo-controlled, randomized trial of mesenchymal stem cells in COPD, Chest 143 (6) (2013) 1590–1598, https://doi.org/10.1378/chest.12-2094.

[138] A. Tzouvelekis, V. Paspaliaris, G. Koliakos, P. Ntolios, E. Bouros, A. Oikonomou, A. Zissimopoulos, N. Boussios, B. Dardzinski, D. Gritzalis, A. Antoniadis, M. Froudarakis, G. Kolios, D. Bouros, A prospective, non-randomized, no placebo-controlled, phase Ib clinical trial to study the safety of the adipose derived stromal cells-stromal vascular fraction in idiopathic pulmonary fibrosis, J. Transl. Med. 11 (2013) 171, https://doi.org/10.1186/1479-5876-11-171.

[139] A. Oner, Z.B. Gonen, D.G. Sevim, N. Sinim Kahraman, M. Unlu, Six-month results of suprachoroidal adipose tissue-derived mesenchymal stem cell implantation in patients with optic atrophy: a phase 1/2 study, Int. Ophthalmol. 39 (12) (2019) 2913–2922, https://doi.org/10.1007/s10792-019-01141-5.

[140] I. James, D. Bourne, M. Silva, E. Havis, K. Albright, L. Zhang, N. Kostereva, S. Wang, G. DiBernardo, R. Guest, J. Lei, A. Almadori, L. Satish, K. Marra, J.P. Rubin, Adipose stem cells enhance excisional wound healing in a porcine model, J. Surg. Res. 229 (2018) 243–253, https://doi.org/10.1016/j.jss.2018.03.068.

[141] Y.S. Kim, C.H. Sung, S.H. Chung, S.J. Kwak, Y.G. Koh, Does an injection of adipose-derived mesenchymal stem cells loaded in fibrin glue influence rotator cuff repair outcomes? A clinical and magnetic resonance imaging study, Am. J. Sports Med. 45 (9) (2017) 2010–2018, https://doi.org/10.1177/0363546517702863.

[142] H. Kumar, D.H. Ha, E.J. Lee, J.H. Park, J.H. Shim, T.K. Ahn, K.T. Kim, A.E. Ropper, S. Sohn, C.H. Kim, D.K. Thakor, S.H. Lee, I.B. Han, Safety and tolerability of intradiscal implantation of combined autologous adipose-derived mesenchymal stem cells and hyaluronic acid in patients with chronic discogenic low back pain: 1-year follow-up of a phase I study, Stem Cell Res Ther 8 (1) (2017) 262, https://doi.org/10.1186/s13287-017-0710-3.

[143] K. Comella, R. Silbert, M. Parlo, Effects of the intradiscal implantation of stromal vascular fraction plus platelet rich plasma in patients with degenerative disc disease, J. Transl. Med. 15 (1) (2017) 12, https://doi.org/10.1186/s12967-016-1109-0.

[144] J.H. Houtgraaf, W.K. den Dekker, B.M. van Dalen, T. Springeling, R. de Jong, R.J. van Geuns, M.L. Geleijnse, F. Fernandez-Aviles, F. Zijlsta, P.W. Serruys, H.J. Duckers, First experience in humans using adipose tissue–derived regenerative cells in the treatment of patients with ST-segment elevation myocardial infarction, J. Am. Coll. Cardiol. 59 (5) (2012) 539–540, https://doi.org/10.1016/j.jacc.2011.09.065.

[145] I.A. Panfilov, R. de Jong, S. Takashima, H.J. Duckers, Clinical study using adipose-derived mesenchymal-like stem cells in acute myocardial infarction and heart failure, Methods Mol. Biol. 1036 (2013) 207–212, https://doi.org/10.1007/978-1-62703-511-8_16.

[146] E.C. Perin, R. Sanz-Ruiz, P.L. Sánchez, J. Lasso, R. Pérez-Cano, J.C. Alonso-Farto, E. Pérez-David, M.E. Fernández-Santos, P.W. Serruys, H.J. Duckers, J. Kastrup, S. Chamuleau, Y. Zheng, G.V. Silva, J.T. Willerson, F. Fernández-Avilés, Adipose-derived regenerative cells in patients with ischemic cardiomyopathy: the PRECISE trial, Am. Heart J. 168 (1) (2014) 88–95.e2, https://doi.org/10.1016/j.ahj.2014.03.022.

[147] A.A. Qayyum, M. Haack-Sørensen, A.B. Mathiasen, E. Jørgensen, A. Ekblond, J. Kastrup, Adipose-derived mesenchymal stromal cells for chronic myocardial ischemia (MyStromalCell trial): study design, Regen. Med. 7 (3) (2012) 421–428, https://doi.org/10.2217/rme.12.17.

[148] A. Bayes-Genis, P. Gastelurrutia, M.L. Cámara, A. Teis, J. Lupón, C. Llibre, E. Zamora, X. Alomar, X. Ruyra, S. Roura, A. Revilla, J.A. San Román, C. Gálvez-Montón, First-in-man safety and efficacy of the adipose graft transposition procedure (AGTP) in patients with a myocardial scar, EBioMedicine 7 (2016) 248–254, https://doi.org/10.1016/j.ebiom.2016.03.027.

[149] M.D. Herreros, M. Garcia-Arranz, H. Guadalajara, P. De-La-Quintana, D. Garcia-Olmo, FATT Collaborative Group, Autologous expanded adipose-derived stem cells for the treatment of complex cryptoglandular perianal fistulas: a phase III randomized clinical trial (FATT 1 fistula advanced therapy trial 1) and long-term evaluation, Dis. Colon Rectum 55 (7) (2012) 762–772, https://doi.org/10.1097/DCR.0b013e318255364a.

[150] F. de la Portilla, F. Alba, D. García-Olmo, J.M. Herrerías, F.X. González, A. Galindo, Expanded allogeneic adipose-derived stem cells (eASCs) for the treatment of complex perianal fistula in Crohn's disease: results from a multicenter phase I/IIa clinical trial, Int. J. Color. Dis. 28 (3) (2013) 313–323, https://doi.org/10.1007/s00384-012-1581-9.

[151] A.B. Dietz, E.J. Dozois, J.G. Fletcher, G.W. Butler, D. Radel, A.L. Lightner, M. Dave, J. Friton, A. Nair, E.T. Camilleri, A. Dudakovic, A.J. van Wijnen, W.A. Faubion, Autologous mesenchymal stem cells, applied in a bioabsorbable matrix, for treatment of perianal fistulas in patients with Crohn's disease, Gastroenterology 153 (1) (2017) 59–62 (e2) https://doi.org/10.1053/j.gastro.2017.04.001.

[152] D. García-Olmo, M. García-Arranz, D. Herreros, I. Pascual, C. Peiro, J.A. Rodríguez-Montes, A phase I clinical trial of the treatment of Crohn's fistula by adipose mesenchymal stem cell transplantation, Dis. *Colon rectum* 48 (7) (2005) 1416–1423.

[153] D. Garcia-Olmo, D. Herreros, I. Pascual, J.A. Pascual, E. Del-Valle, J. Zorrilla, P. De-La-Quintana, M. Garcia-Arranz, M. Pascual, Expanded adipose-derived stem cells for the treatment of complex perianal fistula: a phase II clinical trial, Dis. *Colon rectum* 52 (1) (2009) 79–86, https://doi.org/10.1007/DCR.0b013e3181973487.

[154] W.Y. Lee, K.J. Park, Y.B. Cho, S.N. Yoon, K.H. Song, D.S. Kim, S.H. Jung, M. Kim, H.W. Yoo, I. Kim, H. Ha, C.S. Yu, Autologous adipose tissue-derived stem cells treatment demonstrated favorable and sustainable therapeutic effect for Crohn's fistula, Stem Cells 31 (11) (2013) 2575–2581, https://doi.org/10.1002/stem.1357.

[155] I. Molendijk, B.A. Bonsing, H. Roelofs, K.C. Peeters, M.N. Wasser, G. Dijkstra, C.J. van der Woude, M. Duijvestein, R.A. Veenendaal, J.J. Zwaginga, H.W. Verspaget, W.E. Fibbe, A.E. van der Meulen-de Jong, D.W. Hommes, Allogeneic bone marrow–derived mesenchymal stromal cells promote healing of refractory perianal fistulas in patients with Crohn's disease, Gastroenterology 149 (4) (2015) 918–927 (e6) https://doi.org/10.1053/j.gastro.2015.06.014.

[156] J. Panés, D. García-Olmo, G. Van Assche, J.F. Colombel, W. Reinisch, D.C. Baumgart, A. Dignass, M. Nachury, M. Ferrante, L. Kazemi-Shirazi, J.C. Grimaud, F. de la Portilla, E. Goldin, M.P. Richard, A. Leselbaum, S. Danese, ADMIRE CD Study Group Collaborators, Expanded allogeneic adipose-derived mesenchymal stem cells (Cx601) for complex perianal fistulas in Crohn's disease: a phase 3 randomised, double-blind controlled trial, Lancet 388 (10051) (2016) 1281–1290, https://doi.org/10.1016/S0140-6736(16)31203-X.

[157] G. Naldini, A. Sturiale, B. Fabiani, I. Giani, C. Menconi, Micro-fragmented adipose tissue injection for the treatment of complex anal fistula: a pilot study accessing safety and feasibility. Techniques in coloproctology, Tech. Coloproctol. 22 (2) (2018) 107–113, https://doi.org/10.1007/s10151-018-1755-8.

[158] M. García-Arranz, M.D. Herreros, C. González-Gómez, P. de la Quintana, H. Guadalajara, T. Georgiev-Hristov, J. Trébol, D. Garcia-Olmo, Treatment of Crohn's-related rectovaginal fistula with allogeneic expanded-adipose derived stem cells: a phase I–IIa clinical trial, Stem Cells Transl. Med. 5 (11) (2016) 1441–1446, https://doi.org/10.5966/sctm.2015-0356.

[159] A. Dige, H.T. Hougaard, J. Agnholt, B.G. Pedersen, M. Tencerova, M. Kassem, K. Krogh, L. Lundby, Efficacy of injection of freshly collected autologous adipose tissue into perianal fistulas in patients with Crohn's disease, Gastroenterology 156 (8) (2019) 2208–2216 (e1) https://doi.org/10.1053/j.gastro.2019.02.005.

[160] I. James, B. Gusenoff, S.S. Wang, G. Dibernardo, D.T. Minteer, J. Gusenoff, Fat grafting improves foot pain associated with fat pad atrophy of the heel: early findings from a randomized controlled clinical trial, Plast. Reconstr. Surg. Glob. Open 6 (4 Suppl) (2018) 107–108, https://doi.org/10.1097/01.GOX.0000534002.53419.d4.

[161] S.E. Farber, D. Minteer, B.R. Gusenoff, J.A. Gusenoff, The influence of fat grafting on skin quality in cosmetic foot grafting: a randomized, cross-over clinical trial, Aesthet. Surg. J. 39 (4) (2019) 405–412, https://doi.org/10.1093/asj/sjy168.

[162] F.M. Egro, D.A. Bourne, D.M. Minteer, et al., Adipose stem cell therapy for amputation site soft tissue restoration: a prospective, randomized controlled clinical trial, Plast. Reconstr. Surg. Glob. Open 7 (8 Suppl) (2019) 84, https://doi.org/10.1097/01.GOX.0000584696.77367.cb.

[163] D.A. Bourne, J. Bliley, I. James, A.D. Donnenberg, V.S. Donnenberg, B.F. Branstetter, G.L. Haas, E. Radomsky, E.M. Meyer, M.E. Pfeifer, S.A. Brown, K.G. Marra, S. Coleman, J.P. Rubin, Changing the paradigm of craniofacial reconstruction: a prospective clinical trial of autologous fat transfer for craniofacial deformities, Ann. Surg. (2019), https://doi.org/10.1097/SLA.0000000000003318 (Epub ahead of print).

[164] K.S. Koh, T.S. Oh, H. Kim, I.W. Chung, K.W. Lee, H.B. Lee, E.J. Park, J.S. Jung, I.S. Shin, J.C. Ra, J.W. Choi, Clinical application of human adipose tissue–derived mesenchymal stem cells in progressive hemifacial atrophy (parry-romberg disease) with microfat grafting techniques using 3-dimensional computed tomography and 3-dimensional camera, Ann. Plast. Surg. 69 (3) (2012) 331–337, https://doi.org/10.1097/SAP.0b013e31826239f0.

CHAPTER

12

Adipose-derived stem cells for wound healing and fibrosis

Yasamin Samadi[a], *Francesco M. Egro*[a], *Ricardo Rodriguez*[b], *and Asim Ejaz*[a]

[a]Department of Plastic Surgery, University of Pittsburgh Medical Center, Pittsburgh, PA, United States [b]CosmeticSurg, LLC, Luthersville, MD, United States

Introduction

The new millennium has brought forth a newly discovered type of adult stem cell, known as the adipose-derived stem cell (ASC), a novel therapy of interest in the treatment of complicated and chronic wounds. ASCs, first isolated in 2001 from human adipose tissue, are pluripotent stem cells that have the ability to differentiate into cells originating from all three germinal layers: the endoderm, mesoderm, and ectoderm [1, 2]. They have much greater abundance and relative ease of harvest in comparison with bone-marrow-derived mesenchymal stem cells (BM-MSCs), and have the ability to secrete several cytokines and growth factors necessary in promotion of wound healing [3]. In addition, ASCs are capable of directly differentiating into cells that may serve as building blocks for tissues in wound repair, making them an innovative and very promising therapeutic tool in the management of complicated and chronic wounds. ASCs possess significant potential in the promotion of wound healing and mitigation of fibrosis, as demonstrated not only through in vitro and in vivo animal studies but also through recent clinical trials. This chapter explores the role of ASCs in wound healing, and the treatment and mitigation of radiation-induced fibrosis, hypertrophic scarring, diabetic wounds, and burn injuries. In addition, clinical implications and prospects of this relatively new, yet promising aspect of regenerative research will be discussed.

Skin: The largest organ of the body

The skin, the largest organ of the body, is the first to come into contact with the external environment, and thereby serves as our first line of defense against environmental insults. It

https://doi.org/10.1016/B978-0-12-819376-1.00005-6

comprises three layers: the epidermis, dermis, and hypodermis [4]. The epidermis is the outermost layer of the skin, responsible for protection against pathogens and water loss, and consists of a stratum corneum, stratum granulosum, stratum lucidum, stratum spinosum, and stratum basalis. Immediately below the epidermis is the dermis, comprising an extracellular matrix composed of collagen fibrils, elastic fibers, and microfibrils interspersed in a hyaluronan- and proteoglycan-rich matrix [4]. The dermis provides tensile strength and elasticity to the skin. It is the home of mechanoreceptors, nociceptors, thermoreceptors, hair follicles, sweat glands, apocrine glands, sebaceous glands, and blood and lymphatic vessels. The hypodermis, also known as subcutaneous tissue, is not technically part of skin itself, but consists of adipose tissue and blood vessels that play a pertinent role in the mechanical attachment of the integument to underlying muscles and bones, in addition to providing insulation and maintaining body temperature.

The skin not only serves as a physical barrier but also as an immunological barrier to environmental insults [5]. Langerhans cells in the skin serve a key immunological role by providing surveillance for the skin and, along with the dermal dendritic cells, perform an antigen-presenting role [6]. In addition, the dermis harbors mast cells and T cells with important roles in radiation-induced immune responses [7–11]. The skin is not only crucial to maintaining a primary defense against external physical threats; it is also fundamental to maintaining homeostasis. To demonstrate this fact, loss of 15% of the integument may result in serious consequences such as water loss and an inability to maintain a homeostatic body temperature, proving life-threatening to the patient affected. Further implications of serious injury to the skin manifest as social and psychological challenges to the victim, in addition to economic impacts to the healthcare system.

Wound healing

Wound healing is a complex and dynamic biological process by which the body regenerates and repairs injuries, with the ultimate objective of restoring damaged tissue to its original form and function. Regeneration refers to the process of differentiation of preexisting stem or precursor cells into cells that replace damaged tissue to its original form. Repair mechanisms pertain to the replacement of damaged tissue through deposition of proteins and components different from those comprising the original tissue structure, forming replacement tissue of a fibrous nature consisting mainly of collagen. This complex and dynamic process is facilitated by a variety of cytokines, chemokines, and growth factors. Tissue repair is mediated by growth factors that contribute to cellular proliferation of parenchymal cells and the production of an extracellular matrix. A wound may undergo both regeneration and repair; however, if the damage to the tissue is significant, then displacement and destruction of preexisting stem and precursor cells may leave only repair mechanisms in place to mitigate the damage [12]. Normal wound repair is initiated upon injury to cells and their components and may progress through distinct stages that occur on a time continuum separately or simultaneously: hemostasis, inflammation, proliferation, and remodeling. The adequate progression of each stage is pertinent to successful wound healing, and deviation from the normal course may lead to delayed or pathological healing.

Hemostasis

Initial tissue injury leads to disruption of the vascular endothelium, exposure of the basal lamina, extravasation of blood constituents, and activation of platelets. Vascular injury induces vasoconstriction of the injured blood vessels within 10–15 min, by effectively concentrating clotting factors within a smaller space. This promotes the coagulation cascade in which thrombocytes and platelets ultimately contribute to the formation of a fibrin network that not only blocks further blood loss and invasion of pathogens, but also provides a preliminary matrix for platelet aggregation, cellular migration, and fibroblast proliferation and differentiation. Aggregation and activation of blood constituents also results in the subsequent release of cytokines and growth factors involved in the deposition of a provisional extracellular matrix (transforming growth factor-β), chemotaxis (platelet-derived growth factor), epithelialization (fibroblast and epidermal growth factor), and angiogenesis (vascular endothelial growth factor) [12–14]. Platelets and polymorphonuclear leukocytes entrapped in a blood clot also release cytokines that further amplify this aggregation process and initiate the coagulation cascade. Furthermore, polymorphonuclear cells secrete chemokines that attract immune cells involved in the inflammatory phase of wound healing [15, 16].

Inflammation

The inflammatory phase is characterized by a platelet response and subsequent release of chemokines, cytokines, and an influx of inflammatory polymorphonuclear leukocytes and other immune cells within 1–2 days. Inflammatory cells play a pertinent role in the clearance of necrotic tissue and debris from the area of injury through the process of phagocytosis. Cells such as mastocytes, gamma delta cells, and Langerhans cells are responsible for secreting chemokines and cytokines that contribute to the inflammatory phase, physically characterized by erythema, edema, and warmth [11, 17]. Skin-resident Langerhans and dendritic cells phagocytose foreign antigens and present them on the cell surface to T cells [17–20]. Within a few hours after injury, the release of cytokines such as IL-1β, tumor necrosis factor-α (TNF-α), and IFN-γ at the site of injury results in an expression of cell adhesion molecules, P- and *E*-selectins, as well as ICAM-1, and ICAM-2, which are pertinent to neutrophil adhesion and diapedesis. Neutrophils, attracted by chemokines, including IL-8, growth-related oncogene-α, and monocyte chemoattractant protein-1 (MCP-1), in turn transmigrate across the endothelial cell wall of blood capillaries to the site of injury, where they function in phagocytosis and clearance of dead cells and cellular debris. This response may last a few days, before neutrophils at the wound site are replaced by macrophages, the majority of which are recruited from the blood in the form of monocytes [16].

Within 48 h of an injury causing a lesion, monocytes from the circulation are recruited to the site of damage via chemokine signaling, where they differentiate into macrophages, and along with resident macrophages already present in the damaged tissue, work as antigen-presenting cells, thereby aiding neutrophils in phagocytosing cells containing potentially harmful foreign antigens. Macrophages can be activated via the classical or alternative pathway and play a role in transitioning from the exudative stages of wound healing to the proliferative stage. M1 macrophages, activated by the classical pathway, are proinflammatory. In contrast, M2 macrophages, activated through the alternative pathway, are antiinflammatory and proangiogenic. Both types of macrophages release growth factors including platelet-derived growth factor (PDGF) and vascular endothelial growth factor (VEGF), responsible for new tissue growth in the area of defect [21–25]. The presence of macrophages is essential for wound healing, arguably more so than neutrophils, as demonstrated through experiments in which wounds treated with antimacrophage serum displayed decreased clearance of fibrin, neutrophils, erythrocytes, and debris [26,27]. A summary of cytokines and their role in wound healing is presented in Table 1 [14].

Proliferation

The proliferative phase of wound healing occurs 3–21 days postinjury and is characterized by angiogenesis, epithelialization, collagen deposition, and the formation of granulation tissue, which replaces the defective volume lost during the wound resorption. The proliferative phase of wound healing is characterized by collagen and fibronectin synthesis and deposition by fibroblasts [28]. Fibroblasts migrate to the site of damage, where they differentiate into myofibroblasts and synthesize proteins such as collagen and elastin that are key components of a new extracellular matrix [29]. Type III collagen is the main collagen type deposited as part of the provisional extracellular matrix, before being subsequently replaced with type I collagen during the remodeling phase of healing [30]. Severing of blood vessels and loss of blood flow to the area of damage is followed by ensuing hypoxia, resulting in the release of nitric oxide (NO) by endothelial cells, thereby stimulating the release of VEGF by a variety

TABLE 1 Summary of inflammatory cytokines involved in wound healing.

Cytokine	Cells of origin	Role in inflammatory response	Clinical relevance
EGF	Platelets, macrophages, fibroblasts	Angiogenesis, reepithelialization	Decreased in chronic wounds
FGF	Macrophages, mast cells, T lymphocytes, endothelial cells	Epithelialization	Decreased in chronic wounds
IFNα	Monocytes, macrophages	Decreases collagen production	Used in treatment of hypertrophic/keloid scars
PDGF	Platelets, macrophages, fibroblasts	Fibroblast recruitment, myofibroblast stimulation	Recombinant form used in treatment of diabetic ulcers
TNFα	Macrophages, T lymphocytes, keratinocytes	Leukocyte chemoattraction	Elevated levels linked to deficient healing
TGFβ	Platelets, fibroblasts, macrophages	Reepithelialization, wound fibroplasia	Elevated in hypertrophic and keloid scars
IL-1	Keratinocytes, neutrophils, macrophages	Leukocyte chemoattraction, wound fibroplasia	Increased in chronic wounds
IL-8	Macrophages, endothelial cells	Reepithelialization, angiogenesis	Increased in delayed healing
IL-10	Monocytes, lymphocytes	Downregulation of other cytokines, limits fibroblast proliferation	Necessary for scarless fetal wound healing

EGF, epidermal growth factor; *FGF*, fibroblast growth factor; *IFNα*, interferon alpha; *PDGF*, platelet-derived growth factor; *TNFα*, tumor necrosis factor alpha; *TGFβ*, transforming growth factor beta; *IL*, interleukin.
Reproduced from Janis, J.E. and B. Harrison, Wound healing: part I. Basic science. *Plastic and reconstructive surgery, 2016.* ***138****(3S): p. 9S-17S.*

of cells, a prominent mediator of angiogenesis. Mediators such as fibroblast growth factor (FGF) and PDGF also promote formation of new blood vessels.

Angiogenesis occurs through the proliferation of vascular endothelial cells present and lining older intact blood vessels, resulting in new blood vessels branching out into the newly formed extracellular matrix, and the formation of anastomosis between vessels [31]. As blood flow is returned to the wound bed, return of oxygen to the tissue participates in a negative feedback loop, controlling the extent of angiogenesis and extracellular matrix deposition. This autoregulatory mechanism, when functioning in an unremarkable manner, is crucial in preventing abnormal scar formation. The end result is the formation of granulation tissue, consisting of a highly vascularized extracellular matrix replacing the defective region caused by injury. The proliferative phase is also characterized by epithelialization of the epidermis, during which epithelial cells at the stratum basalis, or the innermost layer of the epidermis, proliferate and migrate upwards toward the surface of the epidermis [32]. The size of the wound is minimized through the action of myofibroblasts, through a process known as wound contraction. Myofibroblasts maintain a tight grip on the edges of the wound and bring the wound edges closer together through contractions that are similar to those observed in smooth muscle cells. After the proliferative phase, cells that originally played a role in regeneration and repair may undergo apoptotic mechanisms when no longer needed [32].

Remodeling

During the remodeling phase, excess collagen and extracellular matrix components are degraded by matrix metalloproteases (MMPs), resulting in reorganization and restoration of an organized skin architecture based on glycosaminoglycans and proteoglycans [17]. The remodeling phase is characterized by apoptosis of cells that are no longer of value in the healing response. During the remodeling phase, collagen fibers self-assemble into microfibrils in both longitudinal and lateral directions. In addition, collagen fibers realign along the tension lines of the tissue, further stabilizing the healing tissue by increasing its tensile strength [33]. Approximately 3 months postinjury, the wound reaches 80% of its original tensile strength but is ultimately unable to reach the full capacity of its original uninjured counterpart [34].

Modern therapeutic approaches

Wound healing is a highly regulated and relatively fragile process. Certain pathologies can lead to aberrant functioning in the wound healing mechanism. Pathophysiological processes resulting in a delay or malfunction of the normal phases of wound healing may prevent progression of healing in an orderly and timely fashion. This may occur in cases involving infected wounds, radiation-induced fibrosis, vascular pathologies, decubitus ulcers, or metabolic derangements such as diabetes and obesity, resulting in the presence of complicated and chronic wounds [35].

A variety of solutions have been attempted in the treatment of wounds, ranging from the use of dressings to engineered skin substitutes, but all demonstrate limitations and low efficacies. Creams, ointments, and dressings are the mainstay treatments for many wounds, but are characterized by variable efficacies and a treatment course that is painful, costly, and long-drawn. Surgical management includes debridement and excision of necrotic tissue, as well as coverage of the open wound. These coverage modalities may be temporary, such as xenografts, allografts, or other skin substitutes. Although skin substitutes are initially able to promote granulation tissue formation, thereby providing a protective barrier against dehydration and infection, they are subsequently removed owing to immunological rejection. Definitive coverage modalities include autograft, cultured epidermal autograft, or tissue rearrangement. The use of autografts is limited by the availability of donor skin, which can be a major issue in large burn injuries and is associated with donor site morbidity. Furthermore, the use of split-thickness skin grafts or cultured epidermal autografts does not replace the loss of subcutaneous tissue and adnexal structures, and can lead to contractures and functional impairment. Local flaps and adjacent tissue transfer are useful surgical techniques employed for coverage of wounds but can only be considered in selected patients who have stable wounds, are nutritionally and medically optimized, and have sufficient adjacent tissue to be transferred.

Mesenchymal stem cells

The increasing burden of chronic wounds on quality of life and longevity, in addition to limitations in current treatment modalities, has led to the pursuit of alternative treatment options. The pluripotent nature of stem cells has made them a desirable candidate to overcome at least some of the limitations of current wound care options. Mesenchymal stem cells (MSCs) have long been a subject of study in the mitigation of chronic wounds [36, 37]. These adult stem cells have been shown to have paracrine functions through the secretion of cytokines and growth factors that play a role in the migration and proliferation of cells involved in tissue repair at the site of injury and are known to have both antiinflammatory and angiogenic properties [38, 39].

Recently, however, a proinflammatory phenotype of MSCs has been described [40]. The proinflammatory MSC phenotype nomenclature is similar to that of the macrophage (M1 = inflammatory, M2 = regenerative); thus, the MSC1 phenotype is proinflammatory, whereas the MSC2 is regenerative. Expression of the MSC1 phenotype occurs when lipopolysaccharide (LPS, i.e., degradation products of bacterial cell walls) stimulates TLR4, or when in the presence of low levels of inflammatory cytokines IFN-γ and TNF-α. This is important for early injury responses, where the MSC1 secretes cytokines to summon an inflammatory response to deal with injury or pathogens. As IFN-γ and TNF-α levels rise beyond a certain threshold indicating an inflammatory state, or TLR3 is activated, the MSC adopts an MSC2 immunomodulatory phenotype [41]. The MSC1 phenotype could have a significant role in combating cancer, whereas the MSC2 phenotype can have a tumor progression role [42]. The exact role of these MSC phenotypes in wound healing is still unclear.

MSCs isolated from bone marrow and placental tissue have great potential in wound healing applications. BM-MSCs, isolated from bone marrow aspirate, have been the subject of interest in chronic wound repair [43]. However, there are limitations in the availability and procurement of bone marrow and placental samples for isolation of MSCs. BM-MSCs tend to display impaired proliferation, earlier senescence, and chondrogenic response with increasing patient age. Although adipose stem cells display senescence similar to that of BM-MSCs, as demonstrated with decreasing telomere length and beta-galactosidase activity with age [44], they do not demonstrate any significant changes in cell proliferation or quality with increasing patient age [45] and demonstrate much stronger proliferation ability than BM-MSCs. Studies have showed a

$40\times$ higher yield of stem cells from adipose tissue than that derived from the bone marrow [46]. For example, stem cells isolated from 1 gram of adipose tissue through liposuction may yield approximately 3.5×10^5 to 1×10^6 ASCs. In comparison, the process of MSC isolation from 1 g of bone marrow yields about 500 to 5×10^4 cells [47]. ASCs can be cultured to yield up to $1000\times$ the initial harvest number at isolation [48]. Furthermore, ASCs have been shown to retain their normal diploid karyotype after 100 generations of culturing. A 1-mL volume of lipoaspirate isolated from human adipose tissue yields about $(2.5–3.75)\times10^5$ ASCs within a 4- to 6-day period of culturing cells in culture with medium containing 10% fetal bovine serum [12]. Liposuction, although known to damage mature adipocytes, does not affect the functionality of ASCs [49]. As opposed to other MSCs, which are derived from invasive and painful procedures including harvest from the bone marrow and have a limited capacity in terms of quantity, ASCs may be easily and repeatedly harvested under general anesthesia with minimal discomfort and risk to the patient, and are easily isolated from subcutaneous tissue of the abdomen, thighs, or arms using methods such as resection or liposuction. Due to its relatively abundant quantity in the body, subcutaneous adipose tissue proves to be the most feasible source for isolation of high quantities of multipotent stem cells [50].

Adipose tissue removed through surgical resection or liposuction, which is often discarded as surgical waste, is not only a readily available source of stem cells in the field of regenerative medicine but is also an aesthetically beneficial procedure and a financially feasible source of valuable cells utilized in both bench and clinical regenerative research. A comparison of BM-MSCs and ASCs is summarized in Table 2.

TABLE 2 Advantages and disadvantages of bone-marrow-derived stem mesenchymal stem cells versus adipose-derived stem cells used for wound healing.

Stem cell type	Advantages	Disadvantages
BM-MSCs	• Multipotent • Availability • Immunosuppressive • Both autograft and allograft are possible • No ethical issues • Limited replicative lifespan (reduced risk of malignant formation) • No risk of immune rejection (particularly in autograft approach)	• Limited self-renewal ability • Propensity to form osseous tissue • Invasive harvesting procedure
ASCs	• Multipotent • Availability • Easy to isolate and obtain • Immunosuppressive • Both autograft and allograft are possible • No ethical issues • No risk of immune rejection (particularly in autograft approach)	• Limited self-renewal ability • Propensity to form osseous tissue • Higher replicative lifespan than BM-MSCs (risk of malignant formation)

BM-MSCs, bone-marrow-derived stem mesenchymal stem cells; *ASCs*, adipose-derived stem cells.
Adapted from A. Hassanshahi, et al., Adipose-derived stem cells for wound healing. J. Cell. Physiol., 234(6) (2019) 7903–7914.

Adipose-derived stem cells

ASCs are a type of adult stem cell, also known as postnatal stem cells. Their relatively recent discovery at the turn of the century has led to an advent of research that appears promising in the field of regenerative medicine. ASCs have been isolated from both subcutaneous and visceral adipose tissues [51], but given the ease of access and low morbidity, subcutaneous tissue is the most common source of lipoaspirate and ASCs in the body. ASCs were first isolated from human adipose tissue in 2001[1]. Phenotypically, ASCs are defined as $CD34^+CD90^+CD29^+CD45^-CD31^-$ and are described as cells of mesenchymal origin, exhibiting trilineage differentiation capabilities into bone, cartilage, or fat [52–57]. Further research has demonstrated the ability of ASCs to differentiate into cells consistent with neurons [58], oligodendrocytes [59], functional Schwann cells [60, 61], and cells of epidermal lineage [62]. Furthermore, studies showing endodermal differentiation of ASCs into hepatocytes and pancreatic islet cells [63, 64] add pertinent evidence to the most current theory that ASCs may be of pluripotent, rather than multipotent, nature, meaning that they may be capable of forming cells derived from all three embryonic germ layers.

Isolation of adipose-derived stem cells

There is no standardized method for the isolation of ASCs [65], but the original technique described by Zuk et al. [1] remains the most widely used. This is characterized by the chopping of adipose tissue into small pieces approximately 1 mm in size using forceps, enzymatic digestion with collagenase type II, centrifugation, isolation, and culture. Digestion of adipose tissue produces two main fractions: the adipocyte fraction, and what is known as the stromal vascular fraction (SVF) [66]. The SVF comprises ASCs, lymphocytes, endothelial cells, pericytes, and fibroblasts [67]. After isolation, the SVF is cultured overnight. ASCs adhere to the plastic culture flasks and are therefore easily identifiable under a microscope. Cells may have doubling times ranging from about 40–120 h during the logarithmic phase of growth. Doubling times may be influenced by the type and location of adipose tissue, by harvesting procedure, and by age of the donor from which tissue was harvested [68]. The timing of isolation and propagation method may have an impact on the immunophenotype and expression kinetics of ASCs. For example, researchers have demonstrated freshly isolated ASCs to be highly positive for CD34, CD117, and HLA-DR, with lower expression of the cell differentiation markers CD105 and especially CD166 in comparison with cultured ASCs.

Wound healing properties of adipose-derived stem cells (Fig. 1)

Regenerative properties of the adipose-derived stem cell

ASCs have multiple properties that render them ideal candidates for regenerative medicine and wound healing applications. ASCs not only have regenerative properties, but they also possess antiinflammatory, immunomodulatory, antiapoptotic, and antifibrotic qualities that may play a role in enhancing and modulating the various phases of wound healing, making them an ideal potential form of therapeutic management in the mitigation of various wounds and disfiguring tissue pathologies, including radiation-induced fibrosis, diabetic

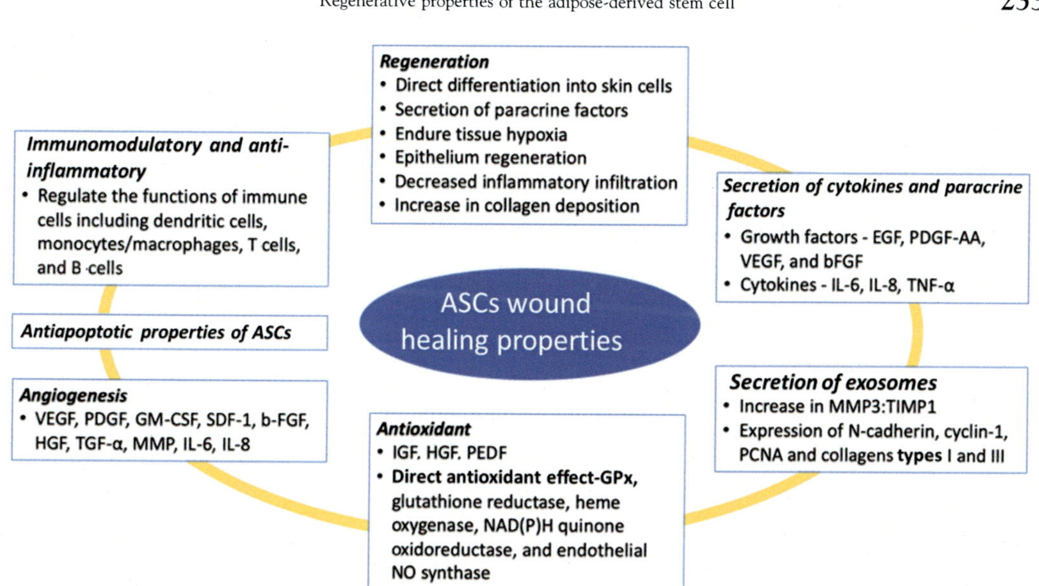

FIG. 1 Wound healing properties and mechanisms of ASCs. ASCs may improve the process of wound healing through several mechanisms: regeneration, secretion of cytokines and paracrine factors, secretion of exosomes, antioxidant action, angiogenesis, antiapoptotic action, and immune modulatory and antiinflammatory action. *ASC*, adipose-derived stem cell; *EGF*, epidermal growth factor; *PDGF*, platelet-derived growth factor; *VEGF*, vascular endothelial growth factor; *bFGF*, basic fibroblast growth factor; *IL*, interleukin; *TNF-α*, tumor necrosis factor-α; *MMP*, matrix metalloproteinase; *TIMP1*, tissue inhibitor of matrix metalloproteinases-1; *PCNA*, proliferating cell nuclear antigen; *IGF*, insulin-like growth factor; *HGF*, hepatocyte growth factor; *PEDF*, pigment epithelium-derived factor; *GPx*, glutathione peroxidase; *NO*, nitric oxide; *GM-CSF*, granulocyte-macrophage colony-stimulating factor; *SDF-1*, stromal cell-derived factor 1.

ulcers, and partial- and full-thickness burns (Fig. 1) [69]. ASCs have been demonstrated to play a pertinent role in the cutaneous wound healing process. ASCs have a direct impact on fibroblast activation, proliferation, migration, and collagen synthesis and promote the proliferation of human dermal fibroblasts through both direct cell-to-cell contact and paracrine activation via the expression of cytokines. They play a role in regulating the transcription of extracellular matrix proteins, resulting in an upregulation of collagen types I and III, and fibronectin, while downregulating matrix metalloprotease 1 (MMP-1) [70]. ASCs have an activating role in the migration of fibroblastic cells in both in vitro and in vivo models of healing.

Although ASCs have been shown to be effective in the treatment of acute and chronic wounds in preclinical and clinical settings, their exact mechanism of action remains under study. ASCs have the ability to promote tissue regeneration through direct differentiation into skin cells, and most prominently via the secretion of paracrine factors that function in the downregulation of the inflammatory response and contribute to differentiation of stem cells into keratinocytes, endothelial cells, and fibroblasts [68]. The regenerative ability of ASCs may be partially attributed to their ability to endure tissue hypoxia. Most of the mature adipocytes resorb after autologous fat transfer, but ASCs survive the hypoxic environment, proliferate, and differentiate into mature adipocytes [71]. Owing to the heterogeneity of the ASC population, the multilineage differentiation of ASC into mesoderm-derived cells observed by researchers was initially thought to have been possibly due to the presence of multiple precursor populations in the SVF. However, early in the discovery of ASCs, Zuk et al. demonstrated their clonogenic properties, observing multilineage differentiation from single ASC clones [1]. Since the discovery of their pluripotency [2], ASCs have been shown to differentiate into adipogenic, osteogenic, chondrogenic, myogenic, and cardiomyogenic cells, in addition to pancreatic β-cells, hepatocytes, and neurogenic-type cells, marking their significant potential for clinical utility in regenerative medicine [50, 72, 73].

ASCs are able to differentiate into keratinocytes in vitro through culturing of autologous ASCs in an imitation eco-environment containing fibrin complexes [74], and can differentiate into K5- and K14-positive cells, which produce high levels of keratinocyte growth factor (KGF), also known as FGF7, which plays a pertinent role in epithelialization during the proliferative phase of wound healing. This property has also been demonstrated in an in vivo murine model by Ebrahimian et al., during which injection of green fluorescent protein (GFP)-transfected ASCs into the dorsal surface of mice that had been subjected to skin punched wounds, combined with either a one-time dose of 20 Gy irradiation or no irradiation, were subsequently visualized in the epidermis and dermis. At day 14 post induction of punch wounds and irradiation, researchers observed improved dermal wound healing and closure, collagen production, and viscoelasticity in all treatment groups treated with ASCs. ASCs localized to the epidermis and dermis, where they expressed cytokeratins CK5 and CK14, known protein markers of epidermal keratinocytes, demonstrating their ability to differentiate into keratinocytes, thereby enhancing regenerative mechanisms in wound healing [75].

In other research involving animal models, such as a rabbit skin injury model, topical administration of ASCs, especially autologous ASCs, to the wound site was associated with improved epithelial regeneration, decreased inflammatory infiltration, an increased rate of wound healing, and an increase in collagen deposition in damaged tissue compared with control treatment groups and tissues treated with BM-MSCs [76]. Another fascinating ability of ASCs is their ability to differentiate into fibroblasts, which are crucial in the process of wound repair. Fibroblasts secrete various bioactive factors that are necessary in wound healing. ASCs activate fibroblasts, in addition to increasing the recruitment of endothelial cells and immune cells such as macrophages. Through these mechanisms, ASCs promote granulation tissue formation, as has been demonstrated in a rabbit model of full-thickness cutaneous injury, to a greater extent than do BM-MSCs [77].

Cells of regenerative potential, including ASCs, are a pertinent topic of interest in tissue engineering, an emerging area of scientific research in the repair of chronic and complicated wounds. For example, in a canine model of acute vocal cord injury, scientists were able to induce differentiation of ASCs into fibroblast-like cells, which they injected into the area of injury at the site of the vocal cord of dogs, assessing characteristics of extracellular protein components such as collagen, elastin, fibronectin, decorin (DCN), and hyaluronic acid from 15 days to 6 months following implantation, observing significantly enhanced wound healing in the ASC-treated group [78]. Vocal cord tissues treated with fibroblast-like cells showed greater collagen and elastin content in comparison with groups treated with vocal cord fibroblasts, suggesting that ASCs may have greater potential than vocal cord fibroblasts in the recovery of vocal cord vibration post injury. The development of additional ways of engineering tissues that will be of clinical use is underway [79].

Secretion of cytokines and paracrine factors

In addition to their ability to directly differentiate into many types of cells, ASCs stimulate the repair and regeneration of injured tissues through secretion of a variety of factors, including cytokines, chemokines, and growth factors. These factors play a role in promotion of healing through paracrine mechanisms, promoting both the immune response to injury, neoangiogenesis, and the proliferation and migration of endogenous cells [80, 81]. Notably, the paracrine mechanisms by which ASCs function is arguably their most important characteristic contributing to their regenerative potential.

The role of cytokines secreted by ASCs have been investigated in in vitro and in vivo settings. Adipose stem cell-conditioned medium (ASC-CM), demonstrated to enhance the cutaneous wound healing process, was found to be high in EGF, PDGF-AA, VEGF, and bFGF content. VEGF, bFGF, and PDGF-AA significantly stimulate the migration of skin fibroblasts. In addition, bFGF and EGF found in adipose-conditioned medium significantly induce proliferation of fibroblasts. ASC-conditioned medium promotes a greater enhancement of fibroblast proliferation and migration than any single cell alone, suggesting a synergistic role of these cytokines in the cutaneous wound healing process [82].

ASCs secrete cytokines, including IL-6 and IL-8, which play a pertinent role in epithelial regeneration during the proliferative phase of wound healing [80], and may be induced to secrete IL-6 and IL-8 through the action of TNF-α, a key inflammatory cytokine, in an in vitro setting. The TNF-α-conditioned medium was tested in an in vivo rat excisional cutaneous wound model, resulting in acceleration of wound closure, proliferation, angiogenesis, reepithelization, and infiltration of immune cells into the wound bed, thereby enhancing the healing process. In contrast, depletion of IL-6 and IL-8 through immunoprecipitation stunted wound healing despite supplementation of the conditioned ASC medium with TNF-α, further highlighting the paracrine mechanism by which ASCs promote cutaneous wound healing [83].

Secretion of exosomes

ASCs have the ability to secrete exosomes, or extracellular vesicles typically 40–100 nm in size that contain proteins, mRNA, miRNA, and other molecules that have been implicated to play a role in paracrine signaling [84]. Recent research has demonstrated the potential of ASC exosomes to promote wound healing in in vivo murine models. Rodents injected intravenously with ASC exosomes have demonstrated decreased scar formation size and an increased ratio of collagen III, a more stable form of collagen with greater tensile strength, to collagen I, which is prominent in the provisional extracellular matrix during the proliferative phase of wound healing. ASC exosomes prevent the differentiation of fibroblasts into myofibroblasts and increase the ratio of transforming growth factor-β3 (TGF-β3) to TGF-β 1 in vivo. They promote the remodeling phase of wound healing through increasing the expression of metalloproteinase-3 (MMP3) by skin dermal fibroblasts via activation of the extracellular signal-regulated kinase (ERK)/mitogen-activated protein kinase (MAPK) pathway, leading to an increase in the ratio of MMP3 to tissue inhibitor of matrix metalloproteinase-1 (TIMP1). Thereby, ASC exosomes regulate wound healing and mitigate scar formation by promoting a balance between factors involved in the proliferative and remodeling phases of wound healing [85].

Exosomes released by ASCs are thought to promote and facilitate soft tissue and cutaneous wound healing by enhancing characteristics of fibroblasts. ASC exosomes are taken up and internalized by fibroblasts, and induce gene expression of N-cadherin, cyclin-1, PCNA, and collagen types I and III, thereby stimulating cell migration, proliferation, and collagen synthesis in a dose-dependent manner. They are able to localize to the area of cutaneous injury, as has been demonstrated in in vivo tracing experiments. Mice treated with systemic administration of exosomes display an increase in collagen I and III production during the early stages of wound healing. A decrease in collagen expression during later stages of healing points to the role of the ASC exosomes in regulation of the remodeling phase of wound healing and prevention of hypertrophic scar formation [86]. Furthermore, cotransplantation of ASC-derived exosomes with autologous fat in a murine model has been shown to improve volume retention by enhancing vascularization and regulating the inflammatory response [87].

Promotion of angiogenesis

Compromise of blood flow to the site of injury can lead to the formation of chronic wounds, especially in the presence of peripheral vascular disease and diabetes. Fat grafting, a procedure involving the transfer of fat, which naturally contains ASCs, from one region of the body to another region of interest, has been shown to promote vascularization in fibrotic skin [88]. This phenomenon is due to the presence of ASCs in adipose tissue, which enhance neoangiogenesis in ischemic tissue and can thus be a potentially powerful therapeutic tool for the treatment of wounds in ischemic diseases [80].

ASCs secrete various cytokines, including VEGF, PDGF, granulocyte-macrophage colony-stimulating factor, stromal cell-derived factor 1 (SDF-1), basic fibroblast growth factor (b-FGF), hepatocyte growth factor (HGF), transforming growth factor alpha (TGF-α), MMPs, and IL-6 and IL-8, which play pertinent roles in neoangiogenesis [80, 81, 89]. At the site of tissue injury, the ensuing inflammatory state induces tissue hypoxia, not only increasing the proliferation rate of ASCs, but also enhancing wound healing though upregulation of VEGF and bFGF [90]. VEGF plays the most pertinent role in angiogenesis and vasculogenesis through the recruitment and mobilization of endothelial progenitor cells [91]. The VEGF family of molecules, which are secreted by macrophages, fibroblasts, keratinocytes, and even platelets [92], exists in many forms, of which VEGF-A is the most potent with regard to angiogenesis. VEGF-A binds to tyrosine kinase receptors, known as VEGF receptor-1 (VEGFR-1) and VEGF receptor-2 (VEGFR-2). VEGF-A binding to its receptor will stimulate a signaling cascade involving the activation of the MAPK pathway, which induces proliferation; protein kinase B, which inhibits apoptosis; and Src kinase, focal adhesion kinase, and p38 MAPK, which leads to cell migration [93].

In an in vivo murine model of skin wounds subjected to either no irradiation or a one-time dose of 20 Gy irradiation, Ebrahimian et al. demonstrated improved VEGF plasma levels and improved capillary density and skin perfusion following ASC treatment compared with controls. They also observed the incorporation of ASCs into capillary structures using immunohistochemical staining techniques for CD31, a marker of endothelial cells [75]. ASCs contain a subpopulation of stem cells that can form capillary-like tubes dependent on the PDGF and bFGF signaling pathway. Migration of human ASCs has been shown to increase in response to increased concentrations of PDGF-BB [94].

When cultured with VEGF in an in vitro setting, ASCs have been demonstrated to express endothelial cell markers. They have been shown to differentiate into endothelial cells in both in vivo and in vitro studies, resulting in a subsequent improvement of blood flow and neoangiogenesis to ischemic lower extremities of mice [95]. In addition, local injection of autologous ASCs isolated from the inguinal fat pads of ICR mice into random pattern skin flaps either into the pedicle base or 1.5 cm distal to the pedicle has been demonstrated to mitigate flap necrosis and increase survival area. Treatment groups that were administered ASCs showed a statistically significant increase in flap viability as assessed via measurements during postoperative day 7 in comparison with groups treated with either medium containing no ASCs, with mature adipocytes, or bFGF only [96].

HGF, secreted by ASCs, also plays an important role in angiogenesis; when inhibited, ASCs are unable to promote the migration and proliferation of endothelial cells [97]. In addition to demonstrating great potential for stimulating and promoting angiogenesis, ASCs have the ability to function in situ as pericytes, which are multifunctional mural cells of the microcirculation that wrap around endothelial cells that line the capillaries and venules, thereby providing vascular stability to vascular structures, and functioning in communication with endothelial cells in response to stimuli [98]. The proangiogenic characteristics of ASCs are not only due to their ability to secrete growth factors such as VEGF and HGF, but also their capacity to differentiate into vascular endothelium [72, 81, 94, 99–104].

Antioxidant properties of adipose-derived stem cells

Decrease in vascularization of fibrotic tissue results in hypoxia and increased reactive oxygen species (ROS) production. In addition, influx of macrophages at the site of radiation injury further enhances ROS production and accumulation. Increase in vascularity upon ASC treatment exerts antioxidative effects. Several secreted factors such as insulin-like growth factor (IGF), HGF, pigment epithelium-derived factor (PEDF), and PDGF are believed to contribute to the antioxidant properties of ASCs [90, 105–107]. In addition, ASCs confer a direct antioxidant effect through upregulation of antioxidant enzymes such as glutathione peroxidase (GPx), glutathione reductase, heme oxygenase, NAD(*P*)H quinone oxidoreductase, and endothelial NO synthase [88].

Immunomodulatory and antiinflammatory effects

A prominent contributing factor to chronic complicated wounds is a prolongation of the inflammatory response. ASCs modulate and regulate the immune and inflammatory response mainly through the secretion of paracrine factors [107–110]. ASCs regulate the functions of immune cells, including dendritic cells, monocytes/macrophages, T cells, and B cells, and through other mechanisms that remain to be elucidated [68]. ASCs induce apoptosis of activated macrophages. They participate in cell–cell binding and paracrine signaling in an in vitro setting, which ultimately suppresses lymphocyte proliferation [68]. They interact with monocytes and granulocytes via the cell adhesion molecule Thy-1. ASCs have potent antiinflammatory properties, interacting with innate and adaptive immune responses, and they have the capacity to downregulate T-cell proliferation [111–113] They are also able to exert immunosuppressive effects via the release of factors such as prostaglandin E2, known to suppress acute inflammatory mediators, and indoleamine 2,3-dioxygenase, the rate-limiting step in tryptophan metabolism that inhibits natural killer cells and the differentiation and proliferation of Th2-type CD4+ T cells, but promotes the differentiation and proliferation of regulatory T cells and myeloid-derived suppressor cells [111, 114, 115]. In response to oxidative stress, ASCs secrete TGF-β1, which promotes the differentiation of premature helper T cells to regulatory T cells. Furthermore, ASCs mediate a phenotypic switch from proinflammatory M1-type macrophages to antiinflammatory M2-type macrophages, thereby contributing to a mitigation of the inflammatory response [116]. The activation of the M2 phenotype of macrophages by ASCs in turn results in the release of VEGF and IL-10 by these cells, in addition to the release of HGF and FGF-b [9, 117], which ameliorate inflammation and promote angiogenesis and lymphangiogenesis at the site of the injury [9, 10]. The promotion of the M2 phenotype has been shown to improve autologous fat graft survival in a murine model through promotion of angiogenesis [118].

ASCs also secrete galectin-1 and galactin-3, a family of beta-galactoside-binding proteins that play a role modulating cell–cell and cell–matrix interactions. Galectin-1 and galectin-3 play a pertinent role in suppression of the proliferation of T cells through the metabolism of L-arginine, a necessary factor in T-cell proliferation and function [68]. At the site of tissue injury, monocytes and leukocytes promote activation of ASCs, causing the release of various molecules including IL-6, NO, indoleamine 2,3- dioxygenase, and prostaglandin E2, ultimately resulting in a prevention of differentiation and maturation of dendritic cells, B-lymphocyte differentiation into plasma cells, and T-cell proliferation and development [68].

In an in vivo mouse model for interstitial lung fibrosis injury, Kotani et al. demonstrated the ability of intravenous ASC injection to inhibit both inflammation and fibrosis in the lungs through inhibition of macrophage production of proinflammatory cytokines, including TNF-α and IL-12 [119]. Furthermore, ASCs have demonstrated immunosuppressive properties in in vivo murine models of colitis, arthritis, and graft-versus-host disease [120].

Evidence of adipose-derived stem cells in wound healing and fibrosis

Burns

In 2015, fire and heat resulted in 67 million injuries worldwide [121], resulting in about 2.9 million hospitalizations and 176,000 deaths. Every year, 1.1 million people suffer from burn injuries that require medical attention in the United States alone [81]. While large burns can be fatal, advances in critical care have improved survival outcomes [122, 123]. Burn injuries have different possible etiologies, including thermal, chemical, electrical, radiation, or friction. Burns can impact the skin and subcutaneous tissue to various degrees. Superficial or first-degree burns result from damage to the epidermis and usually heal within 3 days without sequelae and thus can be managed conservatively. Partial or second-degree burns extend to the underlying dermis and, depending on the amount of dermis involved, can lead to prolonged healing times, scarring, and contractures. Full-thickness or third-degree burns extend through the entire dermis, whereas fourth-degree burns extend into the subcutaneous tissue, muscle, and/or bone. The mainstay for deep second-, third-, and fourth-degree burns is surgical excision and grafting owing to the expected prolonged healing time, which can otherwise lead to severe sequelae including infection, hypertrophic scarring, and contractures.

Burn injuries have a significant impact on the quality and longevity of patients' lives. Long-term outcomes following burn injuries are dependent on the surface area of the burn injury, patient age, sex, and comorbidities. Burns covering large surface areas of the body may result in significant heat and water loss, affecting the thermoregulation and homeostatic mechanism of the body. Superficial and partial thickness burns may be managed conservatively. Full-thickness burns, which do not heal by themselves, require invasive surgical procedures such as skin grafting. Skin grafts may be autologous or may consist of a skin substitute mainly derived from porcine skin. Although substitute skin grafts are initially able to provide a protective barrier against dehydration and infection, they are subsequently removed because of immunological rejection. More invasive surgical methods may be utilized, depending on the severity of the injury. For example, in the case of circumferential burns of the limbs or thorax, a patient will undergo an urgent surgical procedure known as escharotomy to prevent ensuing circumferential fibrosis, which may in turn result in a stifling of the vascular supply, and tissue necrosis.

The survival of patients with burns has improved during the past few decades owing to the use of skin grafts. In the United States, an estimated 96% of patients admitted to a burn center survive [124]. However, healing results in the replacement of full-thickness burns with fibrosis and, in turn, contractures, not only affecting the appearance of the tissue but also resulting in limitations in range of movement. These injuries present a major challenge to clinicians in that there are significant limitations to using partial-thickness and full-thickness skin grafts in mitigating the damage. There are also limitations regarding harvest of skin from a donor site, which may not be of an adequate quantity to fully treat the full surface area of tissue damage [125]. Limitations include not only the sourcing of the skin from another donor site, which may lead to morbidity at the donor site, but also the extent to which the graft mitigates the damage. A skin graft may cover the surface of burned tissue but is not able to replace the lost subcutaneous tissue and structures involved. In addition, the resulting scar tissue may lead to a redistribution of nerve fibers, resulting in neuroinflammation and pain. Patients may be left with chronic high-intensity pain symptoms that may be characterized as a type of neuropathic pain, resulting in significant disability. In 2013, an estimated 1.2 million years lived with disability and 12.3 million disability-adjusted life years were due to burn-related injuries [126].

ASCs have shown significant potential in alleviating the damage induced by burns, predominantly through their antiinflammatory and antineuroinflammatory properties. Several in vitro and in vivo studies have demonstrated the antiinflammatory effects of ASCs, which are thought to be mediated by increased IL-10 expression, in turn resulting in a suppression of the NFκB signaling pathway. In a study using a rat model of burn injury, researchers tested the effect of autologous transplantation of ASCs derived from the inguinal region of rats on neuroinflammation associated with third-degree burn injuries. Autologous transplantation of ASCs to the area of injury at the paw of the rat decreased local inflammation at the site of burn injury, the paw, and central neuroinflammation, as analyzed via a sectioning of the spinal cord at the lumbar level, in addition to reducing apoptosis and autophagy in the spinal cord in comparison with rats treated with silver sulfadiazine cream. Burn injuries result in a reduction in the mechanical threshold, or the mechanical stimulus, that rats can withstand at the site of the injury. With the treatment of ASCs, mechanical threshold was restored [127].

In a rat model of deep partial-thickness burns, Feng et al. demonstrated improved wound healing as analyzed by measuring mean wound area as well as performing histological and immunohistochemical analysis, in animals treated with intradermal injection of 5×10^5 ASCs. Statistically significant improvement in wound healing, which was prominent in the first 2 weeks of treatment and characterized by increased angiogenesis and skin appendage regeneration, was demonstrated in comparison with the control group. In another study of a left lateral abdominal region thermal burn injury model of Wistar rats, treatment of rats with intradermal ASC transplantation in two stages, the first on the same day of the burn injury, and the second equivalent dose 4 days after induction of burn injury, resulted in a reduction in scar tissue area, an increase in collagen type III deposition, and a reduction in lymphatic vessels [128].

The use of ASCs in the healing of burn-related injuries has also been conducted in porcine models. In a Göttingen minipig model of full-thickness burns, the method of administration by which ASCs would be most effective in the healing of burns was examined. Autologous ASCs were administered either through multiple injections in the perimeter of the burn wound or sprayed onto the wound surface (0.25×10^6 viable cells/cm^2). There was no significant loss of ASC viability or number using the spray method. Both treatment groups treated with injection or spray displayed increased wound reepithelization in comparison with the control group. In addition, both modes of treatment had similar effects on epithelization, immunomodulation, and angiogenesis [129].

Clinical studies assessing ASCs in acute burns and burn-related scars are lacking, likely owing to regulation challenges. The closest attempt was performed by Gentile et al., who compared patients who had burn sequelae and posttraumatic scars treated with SVF-enhanced autologous fat grafting ($n=10$) with those who were treated with traditional centrifuged autologous fat grafting ($n=10$). Patients treated with SVF-enhanced autologous fat grafting demonstrated a 63% maintenance of contour restoration after 1 year compared with only 39% of the control group ($P<0.0001$). Other clinical studies have assessed the impact of autologous fat grafting on acute burns and burn-related scars. The largest case series ($n=240$) was published by Piccolo et al., who reported overall improvement of scar quality except for three wounds. However, conclusions are hard to draw because of the inclusion of nonburn-related traumatic wounds as well as a lack of objective measures [130]. There exists only one published randomized controlled trial assessing the outcomes of eight 10×5 cm burn scars of eight pediatric patients followed up for 12 months. Four of the scars were injected with autologous fat grafts and the other four were injected with normal saline, with observers and patients both blinded to the treatment groups. No significant improvement in mature scars was identified in terms of clinical examination and Vancouver Scar Scale scores [131]. Given the underpowered study and the fact that fat grafting and not ASCs or SVF were used, it is difficult to derive conclusions regarding the direct impact of ASCs on wound healing of burn-related injuries from this study.

Hypertrophic scarring

Hypertrophic scars are an undesirable variant of wound healing, characterized by the excessive proliferation and production of fibroblasts, myofibroblasts, and extracellular matrix deposition, restricted to the area within the original wound, usually occurring 1–2 months after injury. Adolescents and pregnant women are more likely to experience hypertrophic scarring, whereas the lowest incidence is found in people with albinism. Furthermore, it is estimated that 70% of deep partial and full-thickness burns resulting in hypertrophic scarring are likely due to the stimulation of dermal fibroblasts to produce collagen and inflammatory mediators such as TGF-β1, which in turn stimulates fibroblasts to deposit elastin and collagen [132].

ASCs are thought to play a bimodal role in the regulation of scar formation. ASCs promote the proliferation and differentiation of fibroblasts during the proliferative phase of wound healing, yet inhibit the excessive proliferation and migration of hypertrophic scar fibroblasts and reduce the expression of related cytokines. Few studies have investigated the effects of ASCs on mitigation of hypertrophic scarring. In one such study, Deng et al. isolated hypertrophic scar-derived fibroblasts (HSFs) and ASCs from patients, which were cocultured with ASCs after treatment with exogenous TGF-β1, a known modulator of fibrosis. After 5 days of ASC coculture treatment, decreased expression of collagen I, collagen III, fibronectin (FN), TGF-β1, IL-6, IL-8, CTGF, and alpha-smooth muscle actin (α-SMA), in addition to an increase in DCN, MMP-1, and TIMP-1, was observed in HSFs. The increase in expression levels of Col 1, FN, TGF-β1, IL-6, CTGF, and α-SMA after initial treatment with exogenous TGF-β1 was reversed by 5 days of ASC coculture treatment. After ASC coculture treatment, the protein expression of TGF-β1 and its related intracellular signal pathway molecules, including p-smad2, p-smad3, p-Stat3, and p-ERK, were downregulated. ASCs inhibited proliferation, migration, and contractility of HSFs [133].

Diabetic wounds

Diabetes mellitus is the most prevalent metabolic illness in the world, resulting in an estimated 4%–10% prevalence of diabetic foot ulcers [134], and an estimated cost of $9–$13 billion annually in the United States [135]. Diabetic ulcers are characterized by aberrant wound healing due to various factors. Hyperglycemia leads to dysfunction of proteins and enzymes. The basement membrane permeability is altered, which in turn affects the delivery of oxygen and nutrients. Tissues are prone to microvascular and macrovascular compromise, which can greatly impact the delivery of oxygen and nutrients. These factors lead to an impairment in prompt closure of wounds, resulting in a prolongation of the inflammatory phase characterized by MMP-2 and MMP-9 overexpression, thereby interfering with appropriate deposition of an extracellular matrix, which is pertinent to granulation tissue formation [136, 137]. The compromise of granulation tissue formation results in a decrease in the tensile strength of the resulting healed wound [138]. The hyperglycemic, ischemic, and impaired immune system state also predisposes the wound to infections and further compromises wound healing. These ulcers are often intractable; they require a prolonged therapeutic period with rehabilitation training after complete recovery, often resulting in amputations, thereby contributing to a significant decrease in quality of life for affected patients [139].The use of artificial skin has proved beneficial in the treatment of diabetic wounds in clinical settings. However, such therapeutic approaches may be limited by patient factors such as a relatively large wound area and/or prolonged ischemia [139]. Although not yet demonstrated to be fully efficacious in the treatment of chronic diabetic wounds, ASCs have been shown to accelerate wound healing in both porcine and murine models of diabetes. In a porcine diabetic model, researchers isolated ASCs from adipose tissue of the flank region of diabetic pigs, which they injected into surgically created full-thickness wounds with low- or high-dose ASCs or endothelial differentiated ASCs (EC/ASCs) on day 0 and day 15. In another treatment group, they administered 2 mL of serum-free m199 culture medium primed with either ASCs or human umbilical vein endothelial cells every 3 days, assessing wound healing at days 0, 10, 15, 20, and 28. Wounds treated with either ASCs or EC/ASCs showed a decreased inflammatory response and increased angiogenesis when compared with the control group. Both cellular therapy with ASCs or EC/ASCs and topical therapy were shown to increase the rate of wound healing in the diabetic porcine model [140]. Similar outcomes of healing using ASCs have been demonstrated in mice. Freshly isolated autologous ASCs containing an atelocollagen matrix with silicone membrane (ACMS) topically applied to full-thickness skin defects of diabetic (*db/db*) mice resulted in significantly increased granulation tissue formation, capillary formation, and epithelialization in comparison with mice treated with ACMS alone [141]. In a murine diabetic model, injection of autologous ASCs into the area surrounding the induced wound results in decreased expression of the genes uPA and MMP-9. While a upregulation in the expression of the TIMP1 and MMP2 was observed [142].

Unfortunately, to the best of our knowledge, no clinical studies have directly examined the impact of ASCs on diabetic wounds. However, clinical trials that examined the impact of ASCs on limb ischemia included patients who had concomitant diabetic wounds. In 2012, Lee et al. conducted a pilot study using ASCs in 15 patients with critical limb ischemia, three of which had diabetic foot ulcers. A total of 3×10^8 ASCs/mL was injected at 60 points into the muscles of the lower extremities. No complications were observed, and clinical improvement occurred in 66.7% of patients. Of note, two of the three patients with diabetic foot ulcers experienced improvement in wound healing, while one patient required an amputation [143]. In 2014, the ACellDREAM clinical trial used ASCs on seven patients with lower limb ischemia who were deemed unfit for revascularization. Three of these patients had concomitant diabetes. A total of 200 million cells was injected intramuscularly into the ischemic leg of patients with no complications. Overall, the transcutaneous oxygen pressure tended to increase in most patients with enhanced wound healing. Among the three diabetic patients treated, two patients experienced a decrease in the number of ulcers by 50%, but one patient required an amputation [144]. In 2013, Marino et al. assessed 20 patients with peripheral arterial disease and leg ulcers (including 18 with diabetes) who were either treated with ASCs injected at the edges of the ulcer ($n=10$) or left untreated (control, $n=10$). ASC-treated wounds showed marked reductions in the ulcers' diameter (complete healing occurred in 6 of the 10 ASC-treated patients), depth, and associated pain, with no complications [145]. The clinical results appear promising thus far, but the small sample size and ischemic etiology, a confounding factor, prevents definitive conclusions regarding the efficacy of ASCs in the treatment of diabetic wounds.

Radiation-induced fibrosis

In 2016, there were an estimated 15.5 million people with a history of cancer living in the United States [146]. About half of patients diagnosed with cancer undergo radiotherapy as a primary or adjuvant treatment [147], with an estimated half-million cancer patients requiring radiotherapy annually [148, 149]. The golden standard of treatment in modern radiotherapy, known as external beam therapy, is characterized by the use of high-energy packets consisting of photons beams and particles of electrons, protons, and neutrons to a submillimeter precision [150]. Despite advancements in precision and localization of the modern radiotherapeutic approach, radiotherapy is associated with acute and chronic side effects [151, 152]. Almost 95% of the patients receiving radiation therapy develop acute or late skin reactions [17, 153, 154]. Acute side effects may occur during or within weeks of the procedure and may include irritation, erythema, epilation, and edema of the skin, as well as mucositis, alopecia of targeted glabrous areas, possibly progressing to blistering, dry desquamation, moist desquamation, and postinflammatory hyperpigmentation [17]. Chronic side effects are observed within 6 months to years post irradiation and may include fibrosis, chronic ulceration, pain at the area of treatment, atrophy, neural damage, vascular damage, and telangiectasias [151, 152]. Fibrosis results in a reduction in tissue elasticity, thus decreasing limb range of motion. As the fibrosis of skin and subcutaneous tissue develops, the range of motion decreases with contractures, and pain, skin necrosis, ulceration, and lymphedema ensue [155, 156]. Systemically, cardiac effects, endocrine disorders, and infertility may occur owing to irradiation, depending on the target area treated [17]. These chronic side effects have a significant impact on the longevity and quality of life of patients. Radiation dose, area and volume of tissue irradiated, time of exposure, radiosensitivity of exposed tissue, and time gap between fractioned delivery of radiations are directly correlated to the probability of developing side effects of radiation therapy [151, 152, 157, 158]. There are currently few treatment modalities available in the mitigation of fibrosis, but autologous fat grafting and ASCs have shown potential in alleviating chronic fibrosis [17, 69] given their net antifibrotic, angiogenic, antioxidant, regenerative, and antiapoptotic properties.

The various etiologies presented resulting in the dysregulation of one or more phases of wound healing contribute to the development of chronic wounds, quantitatively and arbitrarily defined as wounds that do not heal adeptly within a 1- to 3-month time span. Although differing in etiologies, chronic wounds are characterized by the presence of high levels of proinflammatory cytokines, ROS, proteases, and persistent microbial infection. In the case of radiation-induced tissue injury, the dysregulation of the healing process is characterized by a prolonged proliferative phase and a defective remodeling phase, resulting in excess deposition of collagen and extracellular matrix components, ultimately leading to fibrosis at the site of injury [17, 159].

Radiation to the skin results in damage to keratinocytes, fibroblasts, and endothelial cells, and leads to cell damage and apoptosis, resulting in a breach of the skin barrier. Skin-resident Langerhans and dendritic cells are responsible for the uptake of invading antigens and presentation of antigens to T cells [18, 19, 160]. Acute radiation damage to keratinocytes and ROS result in the subsequent release of cytokines and chemokines such as IL-1α, IL-1β, IL-6, IL-8, CCL4, CXCL10, and CCL2 [152], which play a role in attracting an influx of inflammatory cells to the site of damage. Radiation-induced tissue damage is characterized by an upregulation of TGF-β1 and collagen types 1–6. TGF-β, a multifunctional cytokine belonging to the transforming growth factor superfamily, is thought to play a pertinent role in the development of chronic fibrosis and has been shown to be activated by ionizing radiation [161–167]. TGF comprises three isoforms, TGF-β1, TGF-β2, and TGF-β3. TGF-β1 signaling via the canonical Smad pathway leads to transcription of specific target genes with profibrotic effects, resulting in increased proliferation and differentiation of fibroblasts to myofibroblasts, which in turn leads to increased collagen type I and III synthesis and deposition [168–170]. TGF-β1 also activates myofibroblasts through MAP kinases (p38, JNK, and ERK) independent of the Smad pathway [171]. Following radiation injury, cytokines and growth factors such as CXCR4 and SDF-1 mediate the migration and proliferation of endothelial cells, which promote angiogenesis and vasculogenesis, even further supported by influx of bone marrow origin cells to the radiation injury site [172].

ASCs have been demonstrated to have antifibrotic properties, making them an ideal candidate in the treatment of radiation-induced fibrosis. ASCs release bFGF, epidermal growth factor (EGF), PDGF, VEGF, and HGF in response to mechanical stress [173]. ASCs downregulate TGF-β, a multifunctional cytokine implicated in radiation-induced fibrosis [17]. TGF-β1 binds to its receptor and triggers a transcriptional response that facilitates proliferation and differentiation of fibroblasts to myofibroblasts [4]. Hypoxia and production of ROS and reactive nitrogen species by fibrotic tissues also contribute to induction and persistence of fibrosis [69]. Studies have shown a decrease in TGF-β gene expression in fibrotic tissue upon ASC/lipoaspirate administration. Injection of ASCs in a Göttingen minipig model of cutaneous radiation syndrome, not only downregulated TGF-β expression but also decreased scar size, as exhibited by a sustained epidermal recovery with a multilayered appearance similar to healthy skin in the ASC-treated pigs, and an increase in MMP expression [174]. In a rabbit model of radiation-induced muscular fibrosis, Sun et al. observed a decrease in collagenous fibrosis area and TGF-β1 expression upon ASC administration [175]. In another study, Wei and colleagues demonstrated a decrease in fibrosis in a rabbit model of radiation-induced fibrosis. Rabbits treated with ASCs had a decreased proportion of collagen fibers to the total wound area than those of the control group, at 4 weeks, 8 weeks, and 26 weeks post irradiation, as well as a decrease in protein expression and integrated optimal density of TGF-β1 [176]. More recent studies point to the ability of ASCs to facilitate the influx of bone marrow stem cells to fibrotic regions [177, 178]. In an in vivo murine mouse model of radiation-induced fibrosis using a one-time dose of 35 Gy, Ejaz et al. observed an increase in TGF-β1 expression and decrease in limb range of motion upon initial radiation treatment. Significant improvement in limb excursion, as assessed by measuring the degrees of limb motion and assessing restoration of normal epithelium architecture, was demonstrated in mice following ASC injection. The study demonstrated that transwell coculture of ASCs with irradiated human foreskin fibroblasts resulted in downregulation of profibrotic genes, including TGF-β, CTGF, IL1, NF-κB, and TNF, in addition to collagens 1–6 in the irradiated cells. The phenomena observed is currently thought to be mediated through the paracrine effects of ASCs. Further support for a possible paracrine mechanism of ASCs was demonstrated when ASC-conditioned medium reduced the expression of collagen proteins in TGF-β-treated human dermal fibroblasts [179].

The major paracrine factor of interest that plays a role in the downregulation of TGF-β1 expression observed in the ASC transwell coculture model is thought to be HGF, a single inactive polypeptide secreted by mesenchymal cells and cleaved into its active form in the extracellular space by serine proteases [69]. Like TGF-β, HGF is widely distributed in the extracellular matrix of most tissues. Its receptor, hepatocyte growth factor receptor (HGFR), is expressed by mesenchymal, endothelial, and epithelial cell lines. The receptor has a complex structure and triggers many intracellular pathways. HGF is known to play a pertinent role in morphogenesis, cell proliferation, and cell migration [17]. It displays antiinflammatory effects through reduction of NF-κB, which results in the downregulation of expression of inflammatory cytokines, including TNF-α, INF-γ, MCP-1, and IL-12 [180, 181]. In addition, HGF plays a direct role in the prevention of translocation of TGF-β1 to the nucleus, thereby preventing TGF-β1 from exerting its profibrotic effects [180]. Antifibrotic effects of HGF have been demonstrated in skin, myocardial, pulmonary, hepatic, and renal fibrosis [69, 182–185].

HGF produced by ASCs may also play an indirect role in the mitigation of radiation-induced injury by promoting the migration and homing of bone marrow cells involved in tissue repair to the site or tissue irradiation, supporting its possible role as a chemokine involved in stimulating migration of bone marrow cells [69]. bFGF is known to promote HGF release by ASCs via the c-Jun N-terminal kinase (JNK) pathway [186]. bFGF mediates antifibrotic effects of ASC-based therapies through induction of myofibroblast apoptosis [187], collagen remodeling, MMP-1 upregulation, TIMP-1 downregulation, and inhibition of TGF-β1 signaling [188]. Bone-marrow-derived cells play a pertinent role in the repair of irradiated tissue. Irradiation is known to induce the migration of endothelial progenitor cells, mesenchymal cells, and melanocytic cells residing in the bone marrow to the area of tissue irradiation. The majority of these cells are CD11b-expressing myelomonocytic cells, which release angiogenic factors, promoting angiogenesis and vessel repair [172, 189]. Despite their antifibrotic properties, [174], ASCs may also secrete profibrotic factors that are responsible for stimulating the proliferation and migration of fibroblasts. However, the net effect of ASCs is an overall remodeling of fibrotic tissue, resulting in a reduction in fibrosis [70, 190, 191].

As previously discussed, ASCs have proved efficacious in mitigating fibrotic factors following soft tissue injury. However, the efficacy of autologously versus allogenically derived ASCs remains a topic of interest. In the case of accidental irradiation to a patient, a patient will require high doses of MSCs following exposure. In cases of full exposure to radiation, such as in the case of radiation-related incidents, or acts of terrorism, a patient may be in need of allogeneic stem cells, owing to irradiative damage to one's own adipose tissue and ASCs. To compare the efficacies of different sources of ASCs, Riccobono et al. conducted a study using autologous, allogenic, and acellular forms of ASCs in a porcine model of cutaneous radiation syndrome. In the study, animals received local irradiation with a ^{60}Co gamma source at a dose of 50 Gy. The control group was treated with an acellular vehicle. Minipigs treated with autologous stem cells demonstrated improved wound healing without necrosis in comparison with the allogenic and acellular treatment groups. The allogeneic stem cell treatment demonstrated a clinical outcome that was not significantly different from that of the control group [192].

A large body of clinical evidence exists on the beneficial impact of autologous fat grafting in improvement of radiation-induced fibrotic changes in the breast, head and neck, and extremities. In a key study, Rigotti et al. fat-grafted 20 patients with severe symptoms or irreversible functional damage following irradiation of the breast, chest wall, or supraclavicular region, and all but one patient experienced improvement in pain, fibrosis, ulceration, atrophy, and retraction. Furthermore, by 4–6 months, the irradiated tissue appeared hydrated and well vascularized with mature adipocytes and regression of fibrosis that remained stable at 1-year follow-up. Interestingly, the injected lipoaspirate did not contain well-preserved adipocytes but did contain ASCs, suggesting that the latter may be responsible for the benefits of fat grafting in the irradiated tissue [193]. Karmali et al. recently published their experience of 119 consecutive patients undergoing autologous fat grafting for oncologic head and neck reconstruction, of which 69% had received radiotherapy. The authors showed that fat grafting enhanced the aesthetic outcomes either by complementing or replacing reconstructive flaps while having low complication rates and no evidence of cancer recurrence association [194]. More studies exist confirming the impact of fat grafting in irradiated tissue, but further research is needed to determine the true benefit of ASCs in human trials.

Future prospects for use of adipose-derived stem cells in wound healing and fibrosis

Stem-cell-based therapies may provide a novel treatment to facilitate more effective and faster healing of cutaneous wounds. The knowledge acquired from both in vitro and in vivo studies points to an indisputable potential of ASCs in mitigating wounds and fibrosis, although there remain limitations in their efficacy due to low survival rates of ASCs when transplanted at the site of injury [195]. To address this issue, researchers have investigated the utilization of ASCs in combination with biological materials and techniques in the promotion of wound healing [80]. Cell-sheet engineering is a method by which efficacy of cell transplantation may be enhanced. In an in vivo study, researchers engineered biological sheets containing ASCs from the epididymis adipose tissue of wild-type Lewis rats, using temperature-responsive culture dishes and normal culture medium containing ascorbic acid that they transplanted into Zucker diabetic fatty (ZDF) rats, a rat model of type 2 diabetes and obesity, known for diminished wound healing. ASC sheets were transplanted at the posterior cranial surface of the Zucker rats, where a full-thickness wound had been created, and improved wound healing was observed in ZDF rats that received ASC sheets compared with ZDF rats without the transplantation of ASC sheets [139]. Application of triple-layer ASC sheets of human origin resulted in significantly improved healing of full-thickness wounds in nude mice [196].

Researchers have been able to utilize an injectable, biocompatible, and thermosensitive hydrogel Pluronic F-127, which functioned as a 3D cell microenvironment in which allogeneic nondiabetic ASCs were encapsulated, and subsequently administered topically to an area of a full-thickness cutaneous wound in a streptozotocin-induced diabetic rat model. This treatment demonstrated enhanced angiogenesis as assessed by CD31 marker, and increased cell proliferation as assessed by Ki67 marker at the site of injury, in addition to facilitating regeneration of granulation tissue and accelerating wound closure. These changes were accompanied by increased transcription of VEGF and TGF-β1 in comparison with control treatment groups [195]. In another study of a streptozotocin-induced diabetic Sprague Dawley rat model, researchers tested the effect of ASC-seeded silk fibroin 50:50 (SF)/chitosan (CS) versus an SF/CS graft by itself on cutaneous wound healing. SF/CS is a synthetic porous scaffold, or matrix, that mimics the extracellular matrix on which cells can grow and multiply [197]. The SF/CS blend film was seeded with ASCs of the same genus. SF/CS film supports the delivery, engraftment, and secretion of stem cells, in addition to allowing for their differentiation into vascular and epithelial components. Cutaneous wounds treated with ASC-seeded SF/CS film have had significantly higher rates of wound closure in comparison with the control group and the SF/CS-film-only treatment. A significant increase in amounts of EGF, TGF-β, and

VEGF was observed in the wounds of mice treated with ASC-seeded SF/CS in comparison with the control group and the SF/CS-only treatment group [198]. In another study, Jongbeom Na et al. generated a transferrable free-stranding sheet of human ASCs by harvesting human ASCs using the photothermal method. shADSC sheets were implanted to wounded diabetic mice, and a 30% increase in wound closure rate compared with control groups was observed [199].

Another approach aimed at improving the survival of injected cells is the utilization of keratin, an extracellular matrix protein, in the transplantation of ASCs at the wound site. The transplantation of porcine ASCs on a keratin coating derived from human hair has been demonstrated to significantly improve ASC adhesion at the site of injury in addition to enhancing the proliferation, viability, and adipogenic and osteogenic differentiation potential of ASCs [80, 200]. Other therapeutic approaches, including the use of *Aloe vera* in combination with ASCs, have also been tested with regard to wound healing. Researchers evaluated the in vivo effects of *Aloe vera* hydrogel loaded with allogeneic ASCs and administered via intradermal injection in a rat model of burn wound while using demineralized bone matrix (DBM) as dressing in the experiments. Increased angiogenesis, reepithelialization, and levels of TGF-β1 in the wounds treated with *Aloe vera* ASCs with DBM in comparison with *Aloe vera* and DBM–A*loe vera* groups were observed 14 days post injury. The DBM–A*loe vera*/ASCs significantly enhanced wound healing, showing a marked decrease in levels of transforming growth factor-β1 (TGF-β1), and interleukin-1β markedly decreased at day 7 post injury in addition to reduced scar formation in comparison with other treatment groups [201].

Conclusions

The discovery of ASCs has contributed to an evolution of our understanding of the regenerative properties of adipose tissue. These stem cells have the advantage of being multipotent, easily accessible, and found in great quantity in the subcutaneous adipose tissue of patients. Chronic wounds and fibrosis have been an increasing burden to our society, with current treatment options falling short in various aspects. In recent years, researchers and clinicians have identified ASCs as a potentially valuable therapeutic tool in overcoming these shortcomings. An increasing level of preclinical evidence supports the beneficial impact of ASCs in wound healing and fibrosis. However, clinical research is still lacking, and further studies are necessary. The future is bright for regenerative research, and ASCs may represent the ideal conductor of a complex and fine-tuned wound healing orchestra.

References

[1] P.A. Zuk, et al., Multilineage cells from human adipose tissue: implications for cell-based therapies, Tissue Eng. 7 (2) (2001) 211–228.

[2] P.A. Zuk, et al., Human adipose tissue is a source of multipotent stem cells, Mol. Biol. Cell 13 (12) (2002) 4279–4295.

[3] M. Strioga, et al., Same or not the same? Comparison of adipose tissue-derived versus bone marrow-derived mesenchymal stem and stromal cells, Stem Cells Dev. 21 (14) (2012) 2724–2752.

[4] D. Breitkreutz, N. Mirancea, R. Nischt, Basement membranes in skin: unique matrix structures with diverse functions? Histochem. Cell Biol. 132 (1) (2009) 1–10.

[5] S.A. Gerber, et al., Interleukin-12 preserves the cutaneous physical and immunological barrier after radiation exposure, Radiat. Res. 183 (1) (2015) 72–81.
[6] M. Otsuka, G. Egawa, K. Kabashima, Uncovering the mysteries of Langerhans cells, inflammatory dendrtic epidermal cells (IDEC) and monocyte-derived Langerhans cells-like cells (LC-like cells) in the epidermis, Front. Immunol. 9 (2018) 1768.
[7] D.E.A. Komi, K. Khomtchouk, P.L. Santa Maria, A review of the contribution of mast cells in wound healing: Involved molecular and cellular mechanisms, Clin Rev Allergy Immunol (2019) 1–15.
[8] A.S. MacLeod, et al., Skin-resident T cells sense ultraviolet radiation–induced injury and contribute to DNA repair, J. Immunol. 192 (12) (2014) 5695–5702.
[9] K. Müller, V. Meineke, Radiation-induced mast cell mediators differentially modulate chemokine release from dermal fibroblasts, J. Dermatol. Sci. 61 (3) (2011) 199–205.
[10] A.V. Nguyen, A.M. Soulika, The dynamics of the skin's immune system, Int. J. Mol. Sci. 20 (8) (2019).
[11] C. Noli, A. Miolo, The mast cell in wound healing, Vet. Dermatol. 12 (6) (2001) 303–313.
[12] A.C.O. de Gonzalez, et al., Wound healing—A literature review, An. Bras. Dermatol. 91 (5) (2016) 614–620.
[13] J. Ahamed, J. Laurence, Role of platelet-derived transforming growth factor-beta1 and reactive oxygen species in radiation-induced organ fibrosis, Antioxid. Redox Signal. 27 (13) (2017) 977–988.
[14] J.E. Janis, B. Harrison, Wound healing: part I. Basic science, Plast. Reconstr. Surg. 138 (3S) (2016) 9S–17S.
[15] A.M. Szpaderska, et al., The effect of thrombocytopenia on dermal wound healing, J. Investig. Dermatol. 120 (6) (2003) 1130–1137.
[16] S.A. Eming, T. Krieg, J.M. Davidson, Inflammation in wound repair: molecular and cellular mechanisms, J. Investig. Dermatol. 127 (3) (2007) 514–525.
[17] A. Ejaz, J.S. Greenberger, P.J. Rubin, Understanding the mechanism of radiation induced fibrosis and therapy options, Pharmacol. Ther. (2019) 107399.
[18] H.A. de Carvalho, R.C. Villar, Radiotherapy and immune response: The systemic effects of a local treatment, Clinics (Sao Paulo) 73 (suppl 1) (2018) e557s.
[19] J. Deckers, H. Hammad, E. Hoste, Langerhans cells: sensing the environment in health and disease, Front. Immunol. 9 (2018) 93.
[20] E.R. Miller 3rd, et al., Meta-analysis: high-dosage vitamin E supplementation may increase all-cause mortality, Ann. Intern. Med. 142 (1) (2005) 37–46.
[21] T.D. Smith, et al., Harnessing macrophage plasticity for tissue regeneration, Adv. Drug Deliv. Rev. 114 (2017) 193–205.
[22] B.N. Brown, et al., Macrophage polarization: an opportunity for improved outcomes in biomaterials and regenerative medicine, Biomaterials 33 (15) (2012) 3792–3802.
[23] K.L. Spiller, et al., Sequential delivery of immunomodulatory cytokines to facilitate the M1-to-M2 transition of macrophages and enhance vascularization of bone scaffolds, Biomaterials 37 (2015) 194–207.
[24] Z. Julier, et al., Promoting tissue regeneration by modulating the immune system, Acta Biomater. 53 (2017) 13–28.
[25] P. Laurent, et al., Immune-mediated repair: a matter of plasticity, Front. Immunol. 8 (2017) 454.
[26] S. Leibovich, R. Ross, The role of the macrophage in wound repair. A study with hydrocortisone and antimacrophage serum, Am. J. Pathol. 78 (1) (1975) 71.
[27] M. Juhas, et al., Incorporation of macrophages into engineered skeletal muscle enables enhanced muscle regeneration, Nat. Biomed. Eng. 2 (12) (2018) 942.
[28] A. Young, C.-E. McNaught, The physiology of wound healing, Surgery (Oxford) 29 (10) (2011) 475–479.
[29] H. Grubbs, B. Manna, Wound physiology, in: StatPearls [Internet], StatPearls Publishing, 2018.
[30] A.J. Singer, R.A. Clark, Cutaneous wound healing, N. Engl. J. Med. 341 (10) (1999) 738–746.
[31] M.G. Tonnesen, X. Feng, R.A. Clark, Angiogenesis in wound healing, in: Journal of Investigative Dermatology Symposium Proceedings, Elsevier, 2000.
[32] K.S. Midwood, L.V. Williams, J.E. Schwarzbauer, Tissue repair and the dynamics of the extracellular matrix, Int. J. Biochem. Cell Biol. 36 (6) (2004) 1031–1037.
[33] D.W. Thomas, et al., Cutaneous wound healing: A current perspective, J. Oral Maxillofac. Surg. 53 (4) (1995) 442–447.
[34] M.B. Witte, A. Barbul, General principles of wound healing, Surg. Clin. N. Am. 77 (3) (1997) 509–528.
[35] P. Goodarzi, et al., Adipose tissue-derived stromal cells for wound healing, in: Cell Biology and Translational Medicine, vol 4, Springer, 2018, pp. 133–149.

[36] N.B. Nardi, L. da Silva Meirelles, Mesenchymal stem cells: isolation, in vitro expansion and characterization, in: Stem Cells, Springer, 2008, pp. 249–282.
[37] D. Baksh, L. Song, R. Tuan, Adult mesenchymal stem cells: characterization, differentiation, and application in cell and gene therapy, J. Cell. Mol. Med. 8 (3) (2004) 301–316.
[38] S. Maxson, et al., Concise review: role of mesenchymal stem cells in wound repair, Stem Cells Transl. Med. 1 (2) (2012) 142–149.
[39] C. Gu, et al., Angiogenic effect of mesenchymal stem cells as a therapeutic target for enhancing diabetic wound healing, Int. J. Low. Extrem. Wounds 13 (2) (2014) 88–93.
[40] R.S. Waterman, et al., A new mesenchymal stem cell (MSC) paradigm: polarization into a pro-inflammatory MSC1 or an immunosuppressive MSC2 phenotype, PLoS One 5 (4) (2010), e10088.
[41] V. Volarevic, M. Lako, M. Stojkovic, Stem cells, inflammation, and fibrosis, Stem Cells Int. 2016 (2016).
[42] S. Galland, I. Stamenkovic, Mesenchymal stromal cells in cancer: a review of their immunomodulatory functions and dual effects on tumor progression, J. Pathol. 250 (2019) 555–572.
[43] S.C. Wu, W. Marston, D.G. Armstrong, Wound care: the role of advanced wound healing technologies, J. Vasc. Surg. 52 (3) (2010) 59S–66S.
[44] H. Mizuno, M. Tobita, A.C. Uysal, Concise review: adipose-derived stem cells as a novel tool for future regenerative medicine, Stem Cells 30 (5) (2012) 804–810.
[45] O.S. Beane, et al., Impact of aging on the regenerative properties of bone marrow-, muscle-, and adipose-derived mesenchymal stem/stromal cells, PLoS One 9 (12) (2014), e115963.
[46] S. Kern, et al., Comparative analysis of mesenchymal stem cells from bone marrow, umbilical cord blood, or adipose tissue, Stem Cells 24 (5) (2006) 1294–1301.
[47] D.A. De Ugarte, et al., Comparison of multi-lineage cells from human adipose tissue and bone marrow, Cells Tissues Organs 174 (3) (2003) 101–109.
[48] T. Utsunomiya, et al., Human adipose-derived stem cells: potential clinical applications in surgery, Surg. Today 41 (1) (2011) 18–23.
[49] J. Fraser, et al., Differences in stem and progenitor cell yield in different subcutaneous adipose tissue depots, Cytotherapy 9 (5) (2007) 459–467.
[50] W. Tsuji, J.P. Rubin, K.G. Marra, Adipose-derived stem cells: implications in tissue regeneration, World J. Stem Cells 6 (3) (2014) 312–321.
[51] Y. Tang, et al., A comparative assessment of adipose-derived stem cells from subcutaneous and visceral fat as a potential cell source for knee osteoarthritis treatment, J. Cell. Mol. Med. 21 (9) (2017) 2153–2162.
[52] M.E. Zwierzina, et al., Characterization of DLK1 (PREF1)+/CD34+ cells in vascular stroma of human white adipose tissue, Stem Cell Res. 15 (2) (2015) 403–418.
[53] A. Ejaz, M. Mattesich, W. Zwerschke, Silencing of the small GTPase DIRAS3 induces cellular senescence in human white adipose stromal/progenitor cells, Aging (Albany NY) 9 (3) (2017) 860.
[54] A. Ejaz, et al., Weight loss upregulates the small GTPase DIRAS3 in human white adipose progenitor cells, which negatively regulates adipogenesis and activates autophagy via Akt–mTOR inhibition, EBioMedicine 6 (2016) 149–161.
[55] D. Minteer, K.G. Marra, J.P. Rubin, Adipose-derived mesenchymal stem cells: biology and potential applications, Adv. Biochem. Eng. Biotechnol. 129 (2013) 59–71.
[56] M.J. Varma, et al., Phenotypical and functional characterization of freshly isolated adipose tissue-derived stem cells, Stem Cells Dev. 16 (1) (2007) 91–104.
[57] L. Zimmerlin, et al., Mesenchymal markers on human adipose stem/progenitor cells, Cytometry A 83 (1) (2013) 134–140.
[58] S.K. Kang, et al., Expression of telomerase extends the lifespan and enhances osteogenic differentiation of adipose tissue–derived stromal cells, Stem Cells 22 (7) (2004) 1356–1372.
[59] K. Safford, et al., Characterization of neuronal/glial differentiation of murine adipose-derived adult stromal cells, Exp. Neurol. 187 (2) (2004) 319–328.
[60] P.J. Kingham, et al., Adipose-derived stem cells differentiate into a Schwann cell phenotype and promote neurite outgrowth in vitro, Exp. Neurol. 207 (2) (2007) 267–274.
[61] Y. Xu, et al., Myelin-forming ability of Schwann cell-like cells induced from rat adipose-derived stem cells in vitro, Brain Res. 1239 (2008) 49–55.

[62] V. Trottier, et al., IFATS collection: using human adipose-derived stem/stromal cells for the production of new skin substitutes, Stem Cells 26 (10) (2008) 2713–2723.
[63] H. Kajiyama, et al., Pdx1-transfected adipose tissue-derived stem cells differentiate into insulin-producing cells in vivo and reduce hyperglycemia in diabetic mice, Int. J. Dev. Biol. 54 (4) (2009) 699–705.
[64] R. Talens-Visconti, et al., Human mesenchymal stem cells from adipose tissue: differentiation into hepatic lineage, Toxicol. In Vitro 21 (2) (2007) 324–329.
[65] Z. Si, et al., Adipose-derived stem cells: sources, potency, and implications for regenerative therapies, Biomed. Pharmacother. 114 (2019) 108765.
[66] R. Berry, et al., Imaging of adipose tissue, Methods Enzymol. 537 (2014) 47–73.
[67] R. Berry, M.S. Rodeheffer, Characterization of the adipocyte cellular lineage in vivo, Nat. Cell Biol. 15 (3) (2013) 302–308.
[68] N. Bertozzi, et al., The biological and clinical basis for the use of adipose-derived stem cells in the field of wound healing, Ann Med Surg (Lond). 20 (2017) 41–48.
[69] A. Ejaz, et al., Adipose-derived stem cell therapy ameliorates ionizing irradiation fibrosis via hepatocyte growth factor-mediated transforming growth factor-β downregulation and recruitment of bone marrow cells, Stem Cells 37 (6) (2019) 791–802.
[70] W.-S. Kim, et al., Wound healing effect of adipose-derived stem cells: a critical role of secretory factors on human dermal fibroblasts, J. Dermatol. Sci. 48 (1) (2007) 15–24.
[71] H. Mizuno, et al., In vivo adipose tissue regeneration by adipose-derived stromal cells isolated from GFP transgenic mice, Cells Tissues Organs 187 (3) (2008) 177–185.
[72] B.J. Philips, K.G. Marra, J.P. Rubin, Adipose stem cell-based soft tissue regeneration, Expert Opin. Biol. Ther. 12 (2) (2012) 155–163.
[73] B.M. Strem, et al., Multipotential differentiation of adipose tissue-derived stem cells, Keio J. Med. 54 (3) (2005) 132–141.
[74] U. Sivan, K. Jayakumar, L.K. Krishnan, Constitution of fibrin-based niche for in vitro differentiation of adipose-derived mesenchymal stem cells to keratinocytes, Biores Open Access 3 (6) (2014) 339–347.
[75] T.G. Ebrahimian, et al., Cell therapy based on adipose tissue-derived stromal cells promotes physiological and pathological wound healing, Arterioscler. Thromb. Vasc. Biol. 29 (4) (2009) 503–510.
[76] G. Pelizzo, et al., Mesenchymal stromal cells for cutaneous wound healing in a rabbit model: pre-clinical study applicable in the pediatric surgical setting, J. Transl. Med. 13 (1) (2015) 219.
[77] S.J. Hong, et al., Topically delivered adipose derived stem cells show an activated-fibroblast phenotype and enhance granulation tissue formation in skin wounds, PLoS One 8 (1) (2013), e55640.
[78] R. Hu, et al., Fibroblast-like cells differentiated from adipose-derived mesenchymal stem cells for vocal fold wound healing, PLoS One 9 (3) (2014), e92676.
[79] S.K. Bhatia, Tissue engineering for clinical applications, Biotechnol. J. 5 (12) (2010) 1309–1323.
[80] P. Li, X. Guo, A review: therapeutic potential of adipose-derived stem cells in cutaneous wound healing and regeneration, Stem Cell Res. Ther. 9 (1) (2018) 302.
[81] J. Rehman, et al., Secretion of angiogenic and antiapoptotic factors by human adipose stromal cells, Circulation 109 (10) (2004) 1292–1298.
[82] J. Zhao, et al., The effects of cytokines in adipose stem cell-conditioned medium on the migration and proliferation of skin fibroblasts in vitro, Biomed. Res. Int. 2013 (2013) 11.
[83] S.C. Heo, et al., Tumor necrosis factor-α-activated human adipose tissue–derived mesenchymal stem cells accelerate cutaneous wound healing through paracrine mechanisms, J. Investig. Dermatol. 131 (7) (2011) 1559–1567.
[84] J.R. Edgar, Q&A: what are exosomes, exactly? BMC Biol. 14 (1) (2016) 46.
[85] L. Wang, et al., Exosomes secreted by human adipose mesenchymal stem cells promote scarless cutaneous repair by regulating extracellular matrix remodelling, Sci. Rep. 7 (1) (2017) 13321.
[86] L. Hu, et al., Exosomes derived from human adipose mensenchymal stem cells accelerates cutaneous wound healing via optimizing the characteristics of fibroblasts, Sci. Rep. 6 (1) (2016) 32993.
[87] S. Mou, et al., Extracellular vesicles from human adipose derived stem cells for the improvement of angiogenesis and fat grafting application, Plast. Reconstr. Surg. 144 (2019) 869–880.
[88] A.A. Borovikova, et al., Adipose-derived tissue in the treatment of dermal fibrosis: Antifibrotic effects of adipose-derived stem cells, Ann. Plast. Surg. 80 (3) (2018) 297–307.

[89] F. Bussolino, et al., In vitro and in vivo activation of endothelial cells by colony-stimulating factors, J. Clin. Invest. 87 (3) (1991) 986–995.
[90] E.Y. Lee, et al., Hypoxia-enhanced wound-healing function of adipose-derived stem cells: increase in stem cell proliferation and up-regulation of VEGF and bFGF, Wound Repair Regen. 17 (4) (2009) 540–547.
[91] G.D. Yancopoulos, et al., Vascular-specific growth factors and blood vessel formation, Nature 407 (6801) (2000) 242–248.
[92] B.M. Borena, et al., Regenerative skin wound healing in mammals: state-of-the-art on growth factor and stem cell based treatments, Cell. Physiol. Biochem. 36 (1) (2015) 1–23.
[93] K.E. Johnson, T.A. Wilgus, Vascular endothelial growth factor and angiogenesis in the regulation of cutaneous wound repair, Adv. Wound Care 3 (10) (2014) 647–661.
[94] S. Gehmert, et al., Angiogenesis: the role of PDGF-BB on adipose-tissue derived stem cells (ASCs), Clin. Hemorheol. Microcirc. 48 (1) (2011) 5–13.
[95] Y. Cao, et al., Human adipose tissue-derived stem cells differentiate into endothelial cells in vitro and improve postnatal neovascularization in vivo, Biochem. Biophys. Res. Commun. 332 (2) (2005) 370–379.
[96] F. Lu, et al., Improved viability of random pattern skin flaps through the use of adipose-derived stem cells, Plast. Reconstr. Surg. 121 (1) (2008) 50–58.
[97] L. Cai, et al., Suppression of hepatocyte growth factor production impairs the ability of adipose-derived stem cells to promote ischemic tissue revascularization, Stem Cells 25 (12) (2007) 3234–3243.
[98] T.A. Mendel, et al., Pericytes derived from adipose-derived stem cells protect against retinal vasculopathy, PLoS One 8 (5) (2013), e65691.
[99] G.J. Hausman, R.L. Richardson, Adipose tissue angiogenesis1,2, J. Anim. Sci. 82 (3) (2004) 925–934.
[100] G.E. Kilroy, et al., Cytokine profile of human adipose-derived stem cells: expression of angiogenic, hematopoietic, and pro-inflammatory factors, J. Cell. Physiol. 212 (3) (2007) 702–709.
[101] S. Liu, et al., Synergistic angiogenesis promoting effects of extracellular matrix scaffolds and adipose-derived stem cells during wound repair, Tissue Eng. A 17 (5–6) (2010) 725–739.
[102] K. Matsuda, et al., Adipose-derived stem cells promote angiogenesis and tissue formation for in vivo tissue engineering, Tissue Eng. A 19 (11 – 12) (2013) 1327–1335.
[103] B.J. Philips, et al., Prevalence of endogenous CD34+ adipose stem cells predicts human fat graft retention in a xenograft model, Plast. Reconstr. Surg. 132 (4) (2013) 845–858.
[104] H. Suga, et al., Paracrine mechanism of angiogenesis in adipose-derived stem cell transplantation, Ann. Plast. Surg. 72 (2) (2014) 234–241.
[105] W.-S. Kim, et al., Evidence supporting antioxidant action of adipose-derived stem cells: protection of human dermal fibroblasts from oxidative stress, J. Dermatol. Sci. 49 (2) (2008) 133–142.
[106] W.-S. Kim, B.-S. Park, J.-H. Sung, The wound-healing and antioxidant effects of adipose-derived stem cells, Expert Opin. Biol. Ther. 9 (7) (2009) 879–887.
[107] C.H. Pinheiro, et al., Local injections of adipose-derived mesenchymal stem cells modulate inflammation and increase angiogenesis ameliorating the dystrophic phenotype in dystrophin-deficient skeletal muscle, Stem Cell Rev. 8 (2) (2012) 363–374.
[108] B. Puissant, et al., Immunomodulatory effect of human adipose tissue-derived adult stem cells: comparison with bone marrow mesenchymal stem cells, Br. J. Haematol. 129 (1) (2005) 118–129.
[109] M.A. Gonzalez, et al., Adipose-derived mesenchymal stem cells alleviate experimental colitis by inhibiting inflammatory and autoimmune responses, Gastroenterology 136 (3) (2009) 978–989.
[110] K.A. Keyser, K.E. Beagles, H.P. Kiem, Comparison of mesenchymal stem cells from different tissues to suppress T-cell activation, Cell Transplant. 16 (5) (2007) 555–562.
[111] K. McIntosh, et al., The immunogenicity of human adipose-derived cells: temporal changes in vitro, Stem Cells 24 (5) (2006) 1246–1253.
[112] J.A. Plock, et al., The influence of timing and frequency of adipose-derived mesenchymal stem cell therapy on immunomodulation outcomes after vascularized composite allotransplantation, Transplantation 101 (1) (2017) e1–e11.
[113] J.A. Plock, et al., Adipose- and bone marrow-derived mesenchymal stem cells prolong graft survival in vascularized composite Allotransplantation, Transplantation 99 (9) (2015) 1765–1773.
[114] G.C. Prendergast, et al., Indoleamine 2,3-dioxygenase pathways of pathogenic inflammation and immune escape in cancer, Cancer Immunol. Immunother. 63 (7) (2014) 721–735.

[115] O. DelaRosa, et al., Requirement of IFN-γ–mediated indoleamine 2, 3-dioxygenase expression in the modulation of lymphocyte proliferation by human adipose–derived stem cells, Tissue Eng. A 15 (10) (2009) 2795–2806.
[116] Q. Shang, et al., Delivery of adipose-derived stem cells attenuates adipose tissue inflammation and insulin resistance in obese mice through remodeling macrophage phenotypes, Stem Cells Dev. 24 (17) (2015) 2052–2064.
[117] A.S. Klar, J. Zimoch, T. Biedermann, Skin tissue engineering: application of adipose-derived stem cells, Biomed. Res. Int. 2017 (2017) 9747010.
[118] K.D. Phipps, et al., Alternatively activated M2 macrophages improve autologous fat graft survival in a mouse model through induction of angiogenesis, Plast. Reconstr. Surg. 135 (1) (2015) 140–149.
[119] T. Kotani, et al., Anti-inflammatory and anti-fibrotic effects of intravenous adipose-derived stem cell transplantation in a mouse model of bleomycin-induced interstitial pneumonia, Sci. Rep. 7 (1) (2017) 14608.
[120] L.E. Kokai, K. Marra, J.P. Rubin, Adipose stem cells: biology and clinical applications for tissue repair and regeneration, Transl. Res. 163 (4) (2014) 399–408.
[121] G.B.D. Disease, I. Injury, C. Prevalence, Global, regional, and national incidence, prevalence, and years lived with disability for 310 diseases and injuries, 1990–2015: a systematic analysis for the Global Burden of Disease Study 2015, Lancet (London, England) 388 (10053) (2016) 1545–1602.
[122] L. Branski, D. Herndon, R. Barrow, A Brief History of Acute Burn Care Management. Total Burn Care, Elsevier, Amsterdam, Netherlands, 2012.
[123] American Burn Association, Burn Incidence and Treatment in the United States: 2012 Fact Sheet. 2012, 2013. http://www.ameriburn.org/resources_factsheet.php.
[124] American Burn Association, Burn Incidence and Treatment in the United States: 2016, 2016. http://www.ameriburn.org/resources_factsheet.php.
[125] A. Shpichka, et al., Skin tissue regeneration for burn injury, Stem Cell Res. Ther. 10 (1) (2019) 94.
[126] J.A. Haagsma, et al., The global burden of injury: incidence, mortality, disability-adjusted life years and time trends from the global burden of Disease study 2013, Inj. Prev. 22 (1) (2016) 3.
[127] C.H. Lin, et al., Autologous adipose-derived stem cells reduce burn-induced neuropathic pain in a rat model, Int. J. Mol. Sci. 19 (1) (2017).
[128] C.L. Franck, et al., Influence of adipose tissue-derived stem cells on the burn wound healing process, Stem Cells Int. 2019 (2019) 2340725.
[129] P. Foubert, et al., Adipose-derived regenerative cell therapy for burn wound healing: a comparison of two delivery methods, Adv. Wound Care 5 (7) (2015) 288–298.
[130] N.S. Piccolo, M.S. Piccolo, M.T.S. Piccolo, Fat grafting for treatment of burns, burn scars, and other difficult wounds, Clin. Plast. Surg. 42 (2) (2015) 263–283.
[131] S. Gal, J.I. Ramirez, P. Maguina, Autologous fat grafting does not improve burn scar appearance: a prospective, randomized, double-blinded, placebo-controlled, pilot study, Burns 43 (3) (2017) 486–489.
[132] S.J. Schmieder, S.J. Ferrer-Bruker, Hypertrophic Scarring, StatPearls Publishing, 2019.
[133] J. Deng, et al., Inhibition of pathological phenotype of hypertrophic scar fibroblasts via coculture with adipose-derived stem cells, Tissue Eng. Part A 24 (5–6) (2018) 382–393.
[134] M. Turns, Diabetic foot ulcer management: the podiatrist's perspective, Br. J. Community Nurs. (Suppl) (2013) S14. s16–9.
[135] A. Raghav, et al., Financial burden of diabetic foot ulcers to world: a progressive topic to discuss always, Ther. Adv. Endocrinol. Metab. 9 (1) (2018) 29–31.
[136] M. Vaalamo, T. Leivo, U. Saarialho-Kere, Differential expression of tissue inhibitors of metalloproteinases (TIMP-1, −2, −3, and −4) in normal and aberrant wound healing, Hum. Pathol. 30 (7) (1999) 795–802.
[137] A.B. Wysocki, L. Staiano-Coico, F. Grinnell, Wound fluid from chronic leg ulcers contains elevated levels of metalloproteinases MMP-2 and MMP-9, J. Invest. Dermatol. 101 (1) (1993) 64–68.
[138] S. McLennan, D. Yue, S. Twigg, Molecular aspects of wound healing in diabetes, Primary Intention: Aust J. Wound Manag. 14 (1) (2006) 8.
[139] Y. Au-Kato, et al., Creation and transplantation of an adipose-derived stem cell (ASC) sheet in a diabetic wound-healing model, J. Vis. Exp. 126 (2017), e54539.
[140] R.F. Irons, et al., Acceleration of diabetic wound healing with adipose-derived stem cells, endothelial-differentiated stem cells, and topical conditioned medium therapy in a swine model, J. Vasc. Surg. 68 (6s) (2018) 115s–125s.

[141] M. Nambu, et al., Accelerated wound healing in healing-impaired db/db mice by autologous adipose tissue-derived stromal cells combined with atelocollagen matrix, Ann. Plast. Surg. 62 (3) (2009).
[142] H. Ghaneialvar, et al., Adipose derived mesenchymal stem cells improve diabetic wound healing in mouse animal model: extracellular matrix remodeling maybe a potential therapeutic usage of stem cells, Biomed. Res. 28 (8) (2017) 3672–3679.
[143] H.C. Lee, et al., Safety and effect of adipose tissue-derived stem cell implantation in patients with critical limb ischemia, Circ. J. (2012) 1204091686.
[144] A. Bura, et al., Phase I trial: the use of autologous cultured adipose-derived stroma/stem cells to treat patients with non-revascularizable critical limb ischemia, Cytotherapy 16 (2) (2014) 245–257.
[145] G. Marino, et al., Therapy with autologous adipose-derived regenerative cells for the care of chronic ulcer of lower limbs in patients with peripheral arterial disease, J. Surg. Res. 185 (1) (2013) 36–44.
[146] K. Miller, R. Siegel, A. Jemal, Cancer Treatment & Survivorship Facts & Figures 2016–2017, Corporate Center: American Cancer Society Inc, Atlanta, GA, 2016.
[147] R. Govindan, V.T. DeVita, Hellman, and Rosenberg's Cancer: Principles & Practice of Oncology Review, Lippincott Williams & Wilkins, 2009.
[148] D.E. Citrin, Recent developments in radiotherapy, N. Engl. J. Med. 377 (11) (2017) 1065–1075.
[149] B.R. Panel, Cancer Moonshot Blue Ribbon Panel Report, vol. 2016, 2016.
[150] D. De Ruysscher, et al., Radiotherapy toxicity, Nat. Rev. Dis. Primers. 5 (1) (2019) 13.
[151] G.C. Barnett, et al., Normal tissue reactions to radiotherapy: towards tailoring treatment dose by genotype, Nat. Rev. Cancer 9 (2) (2009) 134.
[152] J.L. Ryan, Ionizing radiation: the good, the bad, and the ugly, J. Investig. Dermatol. 132 (3) (2012) 985–993.
[153] F.N. Bray, et al., Acute and chronic cutaneous reactions to ionizing radiation therapy, Dermatol. Ther. 6 (2) (2016) 185–206.
[154] N. Salvo, et al., Prophylaxis and management of acute radiation-induced skin reactions: a systematic review of the literature, Curr. Oncol. 17 (4) (2010) 94.
[155] J.L. Harper, et al., Skin toxicity during breast irradiation: pathophysiology and management, South. Med. J. 97 (10) (2004) 989–994.
[156] F.A. Mendelsohn, et al., Wound care after radiation therapy, Adv. Skin Wound Care 15 (5) (2002) 216–224.
[157] M.D. Stubblefield, Radiation fibrosis syndrome: neuromuscular and musculoskeletal complications in cancer survivors, PM&R 3 (11) (2011) 1041–1054.
[158] B. Zackrisson, et al., A systematic overview of radiation therapy effects in head and neck cancer, Acta Oncol. 42 (5–6) (2003) 443–461.
[159] R.E. Jones, et al., Wound healing and fibrosis: current stem cell therapies, Transfusion 59 (S1) (2019) 884–892.
[160] K. Müller, V. Meineke, Radiation-induced alterations in cytokine production by skin cells, Exp. Hematol. 35 (4) (2007) 96–104.
[161] K. Randall, J.E. Coggle, Expression of transforming growth factor-beta 1 in mouse skin during the acute phase of radiation damage, Int. J. Radiat. Biol. 68 (3) (1995) 301–309.
[162] C. Li, et al., TGF-beta1 levels in pre-treatment plasma identify breast cancer patients at risk of developing post-radiotherapy fibrosis, Int. J. Cancer 84 (2) (1999) 155–159.
[163] M. Martin, J. Lefaix, S. Delanian, TGF-beta1 and radiation fibrosis: a master switch and a specific therapeutic target? Int. J. Radiat. Oncol. Biol. Phys. 47 (2) (2000) 277–290.
[164] M.O. Li, et al., Transforming growth factor-beta regulation of immune responses, Annu. Rev. Immunol. 24 (2006) 99–146.
[165] T. Avraham, et al., Radiation therapy causes loss of dermal lymphatic vessels and interferes with lymphatic function by TGF-beta1-mediated tissue fibrosis, Am. J. Physiol. Cell Physiol. 299 (3) (2010) C589–C605.
[166] J.W. Lee, et al., Inhibition of Smad3 expression in radiation-induced fibrosis using a novel method for topical transcutaneous gene therapy, Arch. Otolaryngol. Head Neck Surg. 136 (7) (2010) 714–719.
[167] J.H. Park, et al., SKI2162, an inhibitor of the TGF-beta type I receptor (ALK5), inhibits radiation-induced fibrosis in mice, Oncotarget 6 (6) (2015) 4171–4179.
[168] T. Wynn, Cellular and molecular mechanisms of fibrosis, J. Pathol. 214 (2) (2008) 199–210.
[169] B. Hinz, et al., The myofibroblast: one function, multiple origins, Am. J. Pathol. 170 (6) (2007) 1807–1816.
[170] J.J. Tomasek, et al., Myofibroblasts and mechano-regulation of connective tissue remodelling, Nat. Rev. Mol. Cell Biol. 3 (5) (2002) 349–363.

[171] J. Yarnold, M.-C.V. Brotons, Pathogenetic mechanisms in radiation fibrosis, Radiother. Oncol. 97 (1) (2010) 149–161.
[172] C. Bastianutto, et al., Local radiotherapy induces homing of hematopoietic stem cells to the irradiated bone marrow, Cancer Res. 67 (21) (2007) 10112–10116.
[173] D.A. Banyard, et al., Phenotypic analysis of stromal vascular fraction after mechanical shear reveals stress-induced progenitor populations, Plast. Reconstr. Surg. 138 (2) (2016) 237e–247e.
[174] I.S. Yun, et al., Effect of human adipose derived stem cells on scar formation and remodeling in a pig model: a pilot study, Dermatol. Surg. 38 (10) (2012) 1678–1688.
[175] W. Sun, et al., Adipose-derived stem cells alleviate radiation-induced muscular fibrosis by suppressing the expression of TGF-beta1, Stem Cells Int. 2016 (2016) 5638204.
[176] W. Sun, et al., Adipose-derived stem cells alleviate radiation-induced muscular fibrosis by suppressing the expression of TGF-β1, Stem Cells Int. 2016 (2016) 5638204.
[177] P. Butala, et al., 6: augmentation of fat graft survival with progenitor cell mobilization, Plast. Reconstr. Surg. 125 (6) (2010) 12.
[178] A. Ejaz, et al., Adipose-derived stem cell therapy ameliorates ionizing irradiation fibrosis via hepatocyte growth factor-mediated transforming growth factor-beta downregulation and recruitment of bone marrow cells, Stem Cells 37 (6) (2019) 791–802.
[179] M. Spiekman, et al., Adipose tissue–derived stromal cells inhibit TGF-β1–induced differentiation of human dermal fibroblasts and keloid scar–derived fibroblasts in a paracrine fashion, Plast. Reconstr. Surg. 134 (4) (2014) 699–712.
[180] Y. Liu, Hepatocyte growth factor in kidney fibrosis: therapeutic potential and mechanisms of action, Am. J. Physiol. Renal Physiol. 287 (1) (2004) F7–F16.
[181] Y. Liu, J. Yang, Hepatocyte growth factor: new arsenal in the fights against renal fibrosis? Kidney Int. 70 (2) (2006) 238–240.
[182] E.F. Cahill, et al., Hepatocyte growth factor is required for mesenchymal stromal cell protection against bleomycin-induced pulmonary fibrosis, Stem Cells Transl. Med. 5 (10) (2016) 1307–1318.
[183] H. Chen, et al., Mesenchymal stem cells combined with hepatocyte growth factor therapy for attenuating ischaemic myocardial fibrosis: assessment using multimodal molecular imaging, Sci. Rep. 6 (2016) 33700.
[184] K. Iekushi, et al., Hepatocyte growth factor attenuates renal fibrosis through TGF-β1 suppression by apoptosis of myofibroblasts, J. Hypertens. 28 (12) (2010) 2454–2461.
[185] H. Ogaly, et al., Hepatocyte growth factor mediates the antifibrogenic action of Ocimum bacilicum essential oil against CCl4-induced liver fibrosis in rats, Molecules 20 (8) (2015) 13518–13535.
[186] H. Suga, et al., IFATS collection: fibroblast growth factor-2-induced hepatocyte growth factor secretion by adipose-derived stromal cells inhibits postinjury fibrogenesis through ac-Jun N-terminal kinase-dependent mechanism, Stem Cells 27 (1) (2009) 238–249.
[187] N. Funato, et al., Basic fibroblast growth factor induces apoptosis in myofibroblastic cells isolated from rat palatal mucosa, Biochem. Biophys. Res. Commun. 240 (1) (1997) 21–26.
[188] H.-X. Shi, et al., The anti-scar effects of basic fibroblast growth factor on the wound repair in vitro and in vivo, PLoS One 8 (4) (2013), e59966.
[189] J.H. Kim, et al., Mechanisms of Radiation-Induced Skin Injury and Implications for Future Clinical Trials, Taylor & Francis., 2013.
[190] R. Kumar, et al., Lipotransfer for radiation-induced skin fibrosis, Br. J. Surg. 103 (8) (2016) 950–961.
[191] S.H. Lee, et al., Paracrine effects of adipose-derived stem cells on keratinocytes and dermal fibroblasts, Ann. Dermatol. 24 (2) (2012) 136–143.
[192] D. Riccobono, et al., Application of adipocyte-derived stem cells in treatment of cutaneous radiation syndrome, Health Phys. 103 (2) (2012) 120–126.
[193] G. Rigotti, et al., Clinical treatment of radiotherapy tissue damage by lipoaspirate transplant: a healing process mediated by adipose-derived adult stem cells, Plast. Reconstr. Surg. 119 (5) (2007) 1409–1422.
[194] R.J. Karmali, et al., Outcomes following autologous fat grafting for oncologic head and neck reconstruction, Plast. Reconstr. Surg. 142 (3) (2018) 771–780.
[195] L. Kaisang, et al., Adipose-derived stem cells seeded in Pluronic F-127 hydrogel promotes diabetic wound healing, J. Surg. Res. 217 (2017) 63–74.

[196] Y.-C. Lin, et al., Evaluation of a multi-layer adipose-derived stem cell sheet in a full-thickness wound healing model, Acta Biomater. 9 (2) (2013) 5243–5250.
[197] A.S. Gobin, V.E. Froude, A.B. Mathur, Structural and mechanical characteristics of silk fibroin and chitosan blend scaffolds for tissue regeneration, J. Biomed. Mater. Res. A 74 (3) (2005) 465–473.
[198] Y.-Y. Wu, et al., Experimental study on effects of adipose-derived stem cell–seeded silk fibroin chitosan film on wound healing of a diabetic rat model, Ann. Plast. Surg. 80 (5) (2018).
[199] J. Na, et al., Protein-engineered large area adipose-derived stem cell sheets for wound healing, Sci. Rep. 8 (1) (2018) 15869.
[200] Y.L. Wu, et al., Modulation of keratin in adhesion, proliferation, adipogenic, and osteogenic differentiation of porcine adipose-derived stem cells, J. Biomed. Mater. Res. B Appl. Biomater. 105 (1) (2017) 180–192.
[201] A. Oryan, et al., Healing potential of injectable Aloe vera hydrogel loaded by adipose-derived stem cell in skin tissue-engineering in a rat burn wound model, Cell Tissue Res. 377 (2) (2019) 215–227.

CHAPTER

13

Oncologic safety of adipose-derived stem cell application

Hakan Orbay[a] *and David E. Sahar*[b]

[a]Department of Surgery, Crozer Chester Medical Center, Upland, PA, United States [b]University of California Davis Medical Center, Sacramento, CA, United States

Fat grafting is a commonly used method in postmastectomy breast reconstruction for the treatment of contour irregularities and volume deficits. The procedure is relatively short, safe, can be performed in an outpatient setting, and has high patient satisfaction [1]. The discovery of ASCs and their regenerative potential has sparked further interest in this procedure [2, 3]. ASCs are particularly attractive for breast reconstruction as they prevent resorption of the fat grafts, preferentially differentiate into adipocytes, and help with tissue repair of injury due to radiation treatment [4, 5]. Overall, there is interest in using autologous tissue over implants, owing to inherent risks of prosthetic reconstruction. However, the oncological safety of fat grafting for postmastectomy breast reconstruction has been controversial [6–8]. This controversy stems from studies that documented stimulatory effects of ASCs on breast cancer cells in both in vitro and in vivo animal experiments [6, 9–11]. However, clinical studies did not show an increased oncologic risk after postmastectomy fat grafting [12].

Interaction of adipose-derived stem cells and breast cancer at a cellular level

ASCs secrete a range of angiogenic/growth factors that may stimulate residual tumor cells and trigger tumorigenesis, tumor progression, and tumor recurrence or metastasis [8, 13, 14] (Fig. 1). Moreover, adipocytes and ASCs are a part of the tumor microenvironment that closely influences the behavior of the tumor cells since breast cancer grows in close anatomical proximity to adipose tissue. A subset of adipocytes known as "cancer-associated adipocytes" have been shown to play an active role in tumor progression and metastasis. ASCs can differentiate into cancer-associated adipocytes [8]. Several genes involved in cell growth, extracellular matrix deposition/remodeling, and angiogenesis are expressed at higher levels in ASCs than in breast adipose tissue [15]. ASCs play a direct role in extracellular

https://doi.org/10.1016/B978-0-12-819376-1.00007-X

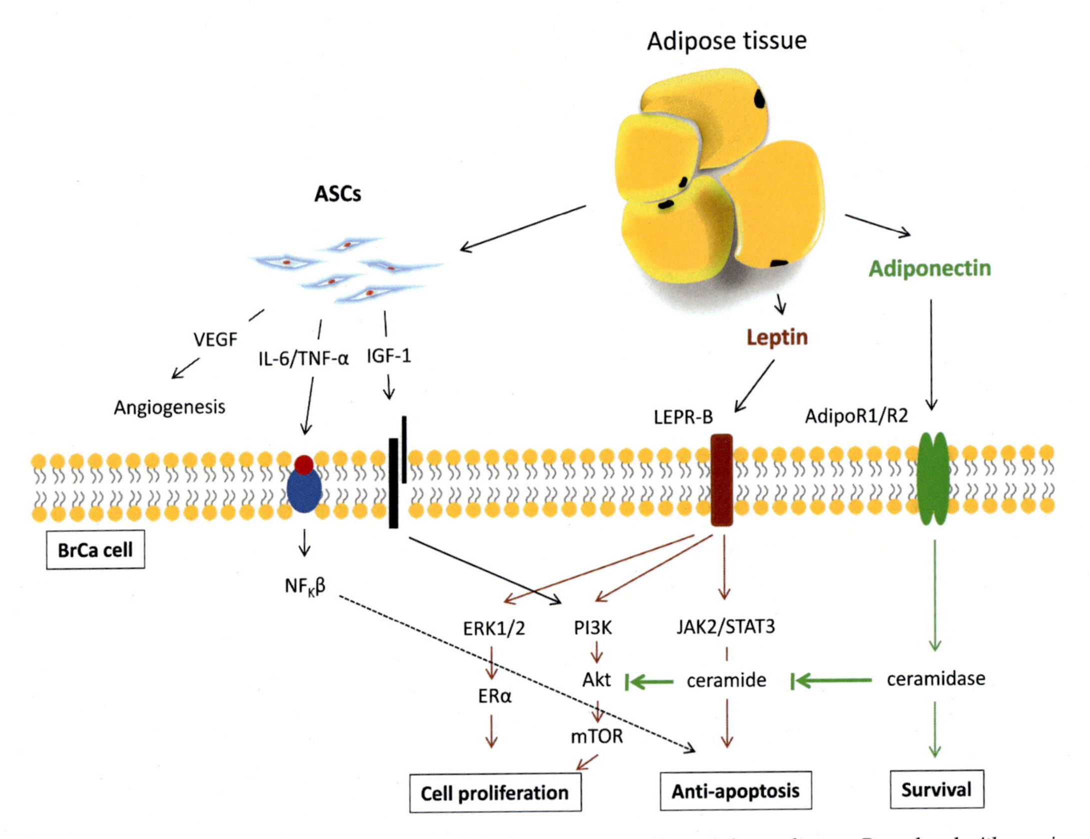

FIG. 1 Complex interaction between the ASCs, the breast cancer cells, and the mediators. *Reproduced with permission from Orbay, H., et al., Fat graft safety after oncologic surgery: addressing the contradiction between in vitro and clinical studies. Plast. Reconstr. Surg., 2018. 142(6): p. 1489–1499.*

matrix (ECM) remodeling occurring during tumor growth [16]. ASCs also trigger a desmoplastic reaction within the tumors, resulting in myofibroblast recruitment. These interactions within the tumor microenvironment explain how ASCs may create a favorable milieu for tumor growth [8].

Clinical studies on oncologic safety of fat grafting

There are several large cohort and matched cohort studies examining the long-term cancer recurrence risk after postmastectomy breast reconstruction with fat grafting (Table 1) [12, 17–28, 30]. Contrary to the experimental data mentioned above, none of these clinical studies showed increased breast cancer recurrence risk with postmastectomy fat grafting. In a multicenter study, Petit et al. retrospectively reviewed 646 patients who had lipofilling to the breast after breast-conserving surgery or mastectomy [27]. They reported a small increase in locoregional recurrence rates in the breast-conserving surgery group after lipofilling

TABLE 1 Significant clinical studies on the oncologic safety of fat grafting.

				Treatment		Histology		
Study	Type of study	Groups	*n*	Mastectomy	BCS	Invasive	CIS	Conclusion
Cohen et al. [17]	Cohort	AFT	162	162	0	111	51	This was a large, single-institution study that supported the oncologic safety of fat grafting
		Control	414	414	0	331	83	
Fertsch et al. [18]	Matched cohort	AFT	100	100	0	91	9	No general increased recurrence risk was observed between the lipofilling and control group
		Control	100	100	0	91	9	
Gale et al. [19]	Matched cohort	AFT	211	176	35	184	27	This study found no evidence of increased oncologic risk associated with fat grafting in women previously treated for breast cancer
		Control	422	358	64	368	54	
Kim et al. [20]	Cohort	AFT	102	102	0	60	42	Authors proposed autologous fat graft is an efficient and safe technique for secondary breast reconstruction
		Control	449	449	0	–	–	
Krastev et al. [13]	Matched cohort	AFT	282	161	139	254	46	No significant differences were observed in the locoregional recurrence rates between the AFT and control groups after 5 years of follow-up
		Control	300	150	150	259	41	
Kronowitz et al. [21]	Cohort	AFT	660	581	79	552	108	The study results showed no increase in locoregional recurrence, systemic recurrence, or secondary breast cancer incidence
		Control	609	536	73	548	61	
Masia et al. [22]	Cohort	AFT	100	107	0	91	16	There were no significant differences in recurrence between patients who underwent fat grafting and those who did not after mastectomy
		Control	107	107	0	93	14	
Petit et al. [23]	Matched cohort	AFT[2]	321	196	125	284	37	The lipofilling group had a higher risk of local events when the analysis was limited to intraepithelial neoplasia, but overall lipofilling was a safe procedure in breast cancer patients
		Control	642	392	250	568	74	

Continued

TABLE 1 Significant clinical studies on the oncologic safety of fat grafting.—cont'd

Study	Type of study	Groups	*n*	Treatment		Histology		Conclusion
				Mastectomy	BCS	Invasive	CIS	
Petit et al. [24]	Matched cohort	AFT	59	47	12	0	59	Higher risk of local recurrence was observed in intraepithelial neoplasia patients following lipofilling. This was a subgroup cohort analysis of the previous study
		Control	118	94	24	0	118	
Petit et al. [16]	Matched cohort	AFT	322	0	322	322	0	Fat grafting seemed to be a safe procedure after breast conservative treatment for breast cancer patients
		Control	322	0	322	322	0	
Petit et al. [25]	Multicenter/ retrospective	AFT	646	370	143	405	108	AFT after breast cancer treatment did not affect radiologic follow-up after breast-conserving surgery; however, a cautious oncologic follow-up protocol was advised
Seth et al. [26]	Cohort	AFT	67	67	0	50	17	Fat grafting after breast reconstruction did not adversely affect local tumor recurrence or survival on long-term follow-up
		Control	763	763	0	587	176	
Silva-Vergara et al. [27]	Matched cohort	AFT	205	147	58	161	44	An increased risk of locoregional recurrence was detected when lipofilling took place within the first 36 months after cancer surgery. Overall, fat grafting was a safe procedure from an oncologic standpoint
		Control	410	286	124	335	75	
Myckathyn et al. [28]	Case-cohort	AFT	64	64	0	64	0	Fat transfer was not associated with an increased probability of breast cancer recurrence in this multicenter case-cohort study
		Control	972	972	0	972	0	
Perez-Cano et al. [29]	Prospective trial	Enriched fat grafting	67	–	67	–	–	The authors claimed that enriched fat grafting is safe from an oncologic standpoint for breast reconstruction after BCS[a]

[a] *This study had a short follow-up period, and there was no matched control group.*

n, *number of breasts;* AFT, *autologous fat transfer;* BCS, *breast-conserving surgery.*

Adapted from Lohsiriwat, V., et al., Autologous fat transplantation in patients with breast cancer: "silencing" or "fueling" cancer recurrence? Breast, 2011. 20(4): p. 351–7.

(0.4% per year versus 2.07% per year), but there was no difference for the mastectomy group in comparison with the data from the European Institute of Oncology. Only 12 new radiologic lesions appeared during the follow-up of their patients, and only 2 of these were found to be local recurrence in further work-up. However, this study could not provide definitive proof regarding the safety of lipofilling in terms of cancer recurrence or distant metastasis owing to the lack of a matched control group. In two consecutive matched cohort studies, the same group again showed a significant increase in the locoregional recurrence rates only in patients with intraepithelial neoplasms after fat grafting [24, 25].

In a large matched cohort study by Kronowitz et al., the authors examined breast cancer recurrence rates in 719 breasts treated with fat grafting after oncologic resection [17]. The study results showed no increase in locoregional recurrence, systemic recurrence, or secondary breast cancer incidence. Several other well-designed clinical studies as seen in Table 1 repeatedly showed clinical safety of postmastectomy fat grafting. Finally, a meta-analysis including 59 studies and a total of 4292 patients showed that autologous fat transfer did not result in an increased rate of locoregional recurrence in patients with breast cancer [14]. Of note, the radiographic changes caused by lipofilling have been well studied in the literature, and usually the changes resulting from fat injection are easily distinguished from the neoplastic images [31, 32]. Therefore, there has been no evidence that fat grafting to the breast interferes with breast cancer detection either [27].

Oncologic safety of enriched fat grafting to the breast

Despite the growing body of data showing oncologic safety of postmastectomy fat grafting, there is still a scarcity of studies on the oncologic safety of enriched fat grafting. The RESTORE-2 trial prospectively assessed the oncological safety of ASC-enriched fat grafting in patients undergoing breast-conserving surgery. The study population was small, the follow-up period was short (12 months), and there was no matched control group; however, there was no incidence of local recurrence, and the authors concluded that enriched fat grafting is safe in the wake of recent oncologic resection. A definitive conclusion on the safety of enriched fat grafting cannot be drawn from this study. Therefore, we recommend caution when using this technique in patients seeking postmastectomy breast reconstruction [29].

Guidelines and recommendations on fat grafting

Recommendations by the American Society of Plastic Surgeons (ASPS) on fat grafting evolved over the years on the basis of the available scientific data. In 1987, ASPS banned autologous fat grafting to breasts primarily due to concerns over future cancer surveillance in the setting of fat necrosis, and secondarily because of inconsistency in graft retention [33]. In 2007, the ASPS established a task force to reevaluate the potential hazards and benefits of breast fat grafting. The task force concluded that current radiographic technology can distinguish grafted fat from potentially dangerous lesions with acceptable risk; therefore, autologous fat grafting to the breasts may be useful and safe, but lacks standardization [34]. In 2009, the ASPS lifted the ban on autologous fat grafting owing to lack of evidence; however, the ASPS Fat Graft Task Force stated that surgeons should exercise "caution when considering fat grafting procedure in patients at high risk for breast cancer" [6, 35]. More recently, The Plastic Surgery Foundation formed a fat graft registry as a quality improvement initiative [36].

In December 2014, the Food and Drug Administration (FDA) published a draft guidance for the use of human cells, tissues, and cellular and tissue-based products (HCT/P) from adipose tissue: regulatory considerations, labeled 21 CFR [37]. Per this publication, adipose tissue must meet the following criteria for clinical use: (1) minimal manipulation; (2) homologous use only; (3) no combination of cells or tissues with another article, except for water, crystalloids, or a sterilizing, preserving, or storage agent, provided that the addition does not raise new clinical safety concerns; and (4) adipose tissue cannot have a systemic effect and be dependent upon the metabolic activity of living cells for its primary function, unless for autologous use, allogeneic use in a first-degree or second-degree blood relative, or reproductive use. Despite recognizing autologous fat grafting to the breast as a nonhomologous use and therefore not compliant with its regulations, the FDA made an exception for certain surgical techniques that allow intraoperative harvest and injection, including the Coleman technique, which utilizes an intraoperative centrifuge [6]. In simpler terms, autologous fat grafting processed in the standard manner can be safely performed in breasts for reconstructive or aesthetic indications without US FDA premarket approval. However, enriched fat grafting is not an approved procedure per these guidelines and should only be used on human subjects as part of a clinical trial [38].

In addition to US regulations, in 2015, French health authorities released their official recommendations on fat grafting to the breast to improve the safety of the procedure. According to these recommendations, *fat grafting to a normal breast is contraindicated* when patients have unreasonable expectations (e.g., major breast augmentation); insufficient body fat reserves, weight instability; familial, histologic, or genetic (*BRCA1* and *BRCA2*, *PTEN*, and *P53*) risk factors for breast cancer; or abnormality on preoperative radiologic evaluation (American College of Radiology Appropriateness Criteria rating of 3, 4, 5, or 6). In preparation for fat grafting, French recommendations include that patients <30 years get a breast ultrasound, while patients >30 years get a mammogram and breast ultrasound. The postoperative follow-up is recommended at 1 year and must be identical to the preoperative assessment [39].

According to the guidelines released by French autorities the *contraindications to breast reconstruction following breast conserving surgery* include: local recurrence on clinical or radiologic examination; metastatic disease; incomplete tumor resection; less than 2 years since local treatment; and incomplete conservative treatments, whether surgical, radiologic, or hormone therapy [39].

The *contraindications for breast reconstruction after mastectomy* are local recurrence detected on clinical examination and/or abnormal radiologic evaluation of the contralateral breast; and metastatic disease, less than 2 years after treatment when there is a high risk of local neoplastic recurrence (inflammatory breast cancer, high-grade in situ carcinoma or sarcoma in a young woman) [40].

Conclusion

The earlier experimental data reporting increased growth and metastasis of breast cancer cells with coinjection of fat grafts were contradicted by later, more clinically relevant studies.

These later studies showed that the injection of fat grafts and ASCs *alone* did not increase the growth of breast cancer cells in vivo [41, 42]. These data combined with clinical outcomes of large cohorts document the safety of postmastectomy fat grafting. However, there are insufficient clinical data to provide the same degree of reassurance regarding the oncologic safety of ASC-enriched fat grafts compared with traditional fat grafting [38]. This is partially because it is not an FDA-approved clinical application in the United States [37]. Moreover, recent animal studies showed increased growth of breast cancer xenografts when coinjected with enriched fat grafts [41]. Precaution is recommended when using ASC-enriched fat grafting for postmastectomy breast reconstruction until more data are available.

References

[1] A.W.W. Brown, et al., Patient reported outcomes of autologous fat grafting after breast cancer surgery, Breast 35 (2017) 14–20.

[2] P.A. Zuk, et al., Human adipose tissue is a source of multipotent stem cells, Mol. Biol. Cell 13 (12) (2002) 4279–4295.

[3] P.A. Zuk, et al., Multilineage cells from human adipose tissue: implications for cell-based therapies, Tissue Eng. 7 (2) (2001) 211–228.

[4] G. Rigotti, et al., Clinical treatment of radiotherapy tissue damage by lipoaspirate transplant: a healing process mediated by adipose-derived adult stem cells, Plast. Reconstr. Surg. 119 (5) (2007) 1409–1422 (discussion 1423-4).

[5] S. Kuno, K. Yoshimura, Condensation of tissue and stem cells for fat grafting, Clin. Plast. Surg. 42 (2) (2015) 191–197.

[6] H.J. Charvet, et al., The oncologic safety of breast fat grafting and contradictions between basic science and clinical studies: a systematic review of the recent literature, Ann. Plast. Surg. 75 (4) (2015) 471–479.

[7] R.A. Agha, et al., Use of autologous fat grafting for breast reconstruction: a systematic review with meta-analysis of oncological outcomes, J. Plast. Reconstr. Aesthet. Surg. 68 (2) (2015) 143–161.

[8] N. O'Halloran, et al., Adipose-derived stem cells in novel approaches to breast reconstruction: their suitability for tissue engineering and oncological safety, Breast Cancer (Auckl.) 11 (2017), 1178223417726777.

[9] H.J. Charvet, et al., In vitro effects of adipose-derived stem cells on breast Cancer cells harvested from the same patient, Ann. Plast. Surg. 76 (Suppl 3) (2016) S241–S245.

[10] S. Gebremeskel, et al., Promotion of primary murine breast cancer growth and metastasis by adipose-derived stem cells is reduced in the presence of autologous fat graft, Plast. Reconstr. Surg. 143 (1) (2019) 137–147.

[11] R.A. Pearl, S.J. Leedham, M.D. Pacifico, The safety of autologous fat transfer in breast cancer: lessons from stem cell biology, J. Plast. Reconstr. Aesthet. Surg. 65 (3) (2012) 283–288.

[12] T. Krastev, et al., Long-term follow-up of autologous fat transfer vs conventional breast reconstruction and association with cancer relapse in patients with breast cancer, JAMA Surg 154 (1) (2019) 56–63.

[13] V. Lohsiriwat, et al., Autologous fat transplantation in patients with breast cancer: "silencing" or "fueling" cancer recurrence? Breast 20 (4) (2011) 351–357.

[14] T.K. Krastev, Y. Jonasse, M. Kon, Oncological safety of autologous lipoaspirate grafting in breast cancer patients: a systematic review, Ann. Surg. Oncol. 20 (1) (2013) 111–119.

[15] M. Zhao, et al., Multipotent adipose stromal cells and breast cancer development: think globally, act locally, Mol. Carcinog. 49 (11) (2010) 923–927.

[16] R. Hass, A. Otte, Mesenchymal stem cells as all-round supporters in a normal and neoplastic microenvironment, Cell Commun Signal 10 (1) (2012) 26.

[17] S.J. Kronowitz, et al., Lipofilling of the breast does not increase the risk of recurrence of breast cancer: a matched controlled study, Plast. Reconstr. Surg. 137 (2) (2016) 385–393.

[18] T.M. Myckatyn, et al., Cancer risk after fat transfer: a multicenter case-cohort study, Plast. Reconstr. Surg. 139 (1) (2017) 11–18.

[19] C. Silva-Vergara, et al., Breast cancer recurrence is not increased with lipofilling reconstruction: a case-controlled study, Ann. Plast. Surg. 79 (3) (2017) 243–248.

[20] S. Fertsch, et al., Increased risk of recurrence associated with certain risk factors in breast cancer patients after DIEP-flap reconstruction and lipofilling-a matched cohort study with 200 patients, Gland Surg 6 (4) (2017) 315–323.

[21] K.L. Gale, et al., A case-controlled study of the oncologic safety of fat grafting, Plast. Reconstr. Surg. 135 (5) (2015) 1263–1275.
[22] J. Masia, et al., Oncological safety of breast cancer patients undergoing free-flap reconstruction and lipofilling, Eur. J. Surg. Oncol. 41 (5) (2015) 612–616.
[23] J.Y. Petit, et al., Evaluation of fat grafting safety in patients with intraepithelial neoplasia: a matched-cohort study, Ann. Oncol. 24 (6) (2013) 1479–1484.
[24] J.Y. Petit, et al., Fat grafting after invasive breast cancer: a matched case-control study, Plast. Reconstr. Surg. 139 (6) (2017) 1292–1296.
[25] J.Y. Petit, et al., Locoregional recurrence risk after lipofilling in breast cancer patients, Ann. Oncol. 23 (3) (2012) 582–588.
[26] A.K. Seth, et al., Long-term outcomes following fat grafting in prosthetic breast reconstruction: a comparative analysis, Plast. Reconstr. Surg. 130 (5) (2012) 984–990.
[27] J.Y. Petit, et al., The oncologic outcome and immediate surgical complications of lipofilling in breast cancer patients: a multicenter study—Milan-Paris-Lyon experience of 646 lipofilling procedures, Plast. Reconstr. Surg. 128 (2) (2011) 341–346.
[28] O. Cohen, et al., Determining the oncologic safety of autologous fat grafting as a reconstructive modality: an institutional review of breast cancer recurrence rates and surgical outcomes, Plast Reconstr Surg 140 (3) (2017) 382e–392e.
[29] R. Perez-Cano, et al., Prospective trial of adipose-derived regenerative cell (ADRC)-enriched fat grafting for partial mastectomy defects: the RESTORE-2 trial, Eur. J. Surg. Oncol. 38 (5) (2012) 382–389.
[30] H.Y. Kim, et al., Autologous fat graft in the reconstructed breast: fat absorption rate and safety based on sonographic identification, Arch. Plast. Surg. 41 (6) (2014) 740–747.
[31] S.R. Pulagam, T. Poulton, E.P. Mamounas, Long-term clinical and radiologic results with autologous fat transplantation for breast augmentation: case reports and review of the literature, Breast J. 12 (1) (2006) 63–65.
[32] R.P. Parikh, et al., Differentiating fat necrosis from recurrent malignancy in fat-grafted breasts: an imaging classification system to guide management, Plast. Reconstr. Surg. 130 (4) (2012) 761–772.
[33] Report on autologous fat transplantation. ASPRS Ad-Hoc Committee on New Procedures, Plast Surg Nurs 7 (4) (1987) 140–141. September 30, 1987.
[34] K.A. Gutowski, Current applications and safety of autologous fat grafts: a report of the ASPS fat graft task force, Plast. Reconstr. Surg. 124 (1) (2009) 272–280.
[35] Fat Transfer/Fat Graft and Fat Injection:ASPS Guiding Principles, 2009. Available at: http://www.plasticsurgery.org/Documents/medical-professionals/health-policy/guiding-principles/ASPS-Fat-Transfer-Graft-Guiding-Principles.pdf.
[36] GRAFT, The Plastic Surgery Foundation, Available at https://www.thepsf.org/research/clinical-impact/graft.html.
[37] FDA Human Cells, Tissues, and Cellular and Tissue-Based Products (HCT/Ps) from Adipose Tissue: Regulatory Considerations; Draft Guidance, 2014. Available at http://www.fda.gov/BiologicsBloodVaccines/GuidanceComplianceRegulatoryInformation/Guidances/Tissue/ucm427795.html.
[38] R.J. Rohrich, D. Wan, Making sense of stem cells and fat grafting in plastic surgery: the hype, evidence, and evolving U.S. food and drug administration regulations, Plast Reconstr Surg 143 (2) (2019) 417e–424e.
[39] B. Chaput, et al., Recurrence of an invasive ductal breast carcinoma 4 months after autologous fat grafting, Plast Reconstr Surg 131 (1) (2013), 123e-4e.
[40] B. Chaput, et al., For the first time, a national health authority provides official recommendations for autologous fat grafting in the breast, Plast Reconstr Surg 136 (5) (2015), 713e-4e.
[41] H. Orbay, et al., Fat graft safety after oncologic surgery: addressing the contradiction between in vitro and clinical studies, Plast. Reconstr. Surg. 142 (6) (2018) 1489–1499.
[42] M.M.A. Silva, et al., Oncologic safety of fat grafting for autologous breast reconstruction in an animal model of residual breast Cancer, Plast. Reconstr. Surg. 143 (1) (2019) 103–112.

CHAPTER

14

FDA regulation of adipose cell use in clinical trials and clinical translation

Mary Ann Chirba[a,b,c], *Veronica Morgan Jones*[d], *Patsy Simon*[e], *and Adam J. Katz*[f]

[a]Boston College Law School, Newton Centre, MA, United States [b]Tufts Medical School, Boston, MA, United States [c]New York University Law School, New York, NY, United States [d]Department of Surgery, Wake Forest, School of Medicine, Winston-Salem, NC, United States [e]Department of Plastic Surgery, Regulatory and Clinical Affairs, Center for Innovation in Restorative Medicine, University of Pittsburgh, Pittsburgh, PA, United States [f]Department of Plastic Surgery, Wake Forest University, Winston-Salem, NC, United States

We're at the beginning of a paradigm change in medicine ... this field is dynamic and complex ... [bringing] unique challenges to researchers, health care providers, and the FDA .... ***Scott Gottlieb, M.D., Former FDA Commissioner (2017)***

The ages live in history through their anachronisms. ***Oscar Wilde (1894)***

As the world of information explodes at exponential pace, it is not uncommon for terms and concepts of the past to continue in the present without realizing (or at least acknowledging) that yesterday's "facts" have become today's anachronisms. This has certainly been true for the terms commonly used to describe regenerative products derived from adipose—terms that, today, reflect an outdated and inaccurate understanding of fat's nature, function, and composition.

Unfortunately, in regulating adipose-derived cell therapies, the FDA has too often relied on anachronistic classification of adipose strictly as a structural tissue for the purpose of applying the human cells, tissues, and cellular and tissue-based products (HCT/P) regulatory

Scientific Principles of Adipose Stem Cells
https://doi.org/10.1016/B978-0-12-819376-1.00003-2

framework while steadily resisting calls to update its thinking and revise its regulatory strategy. While the FDA now acknowledges that adipose has both nonstructural and structural functions, as a practical matter, if the agency is classifying adipose as such to protect patients from rogue stem cell clinics, it is not working. Regulating adipose on the basis of factual inaccuracies is unlikely to hamper those who reject the FDA's fundamental authority to regulate cell therapies and serves only to defeat or at least distort the meaningful assessment of risk. Obviously, this undermines the FDA's stated objective of tailoring the regulation of cell therapies to the risks they present.

As a legal matter, the FDA's risk-based approach to cell therapies is potentially risky as regulatory decisions rooted in factual inaccuracies may not withstand litigation and could instead aid unscrupulous actors already intent on undermining the FDA's regulatory authority on the subject. In the June 2019 case of *Kisor v. Wilkie,* the US Supreme Court emphasized that (a) agencies are bound by the regulations they promulgate; (b) courts should not defer to, or uphold, an agency's interpretation of a regulation if that regulation is not "genuinely ambiguous," even in areas of great scientific complexity; and (c) when a regulation is ambiguous, an agency's interpretation of it must reflect the agency's "authoritative, expertise-based, and fair and considered judgment[.]" For adipose, the FDA has promulgated—and is bound by—regulations that are either quite clear or, if ambiguous, leave no room for ignoring the basic biology of what adipose is and how it functions.

> Put simply, it is quite reasonable to view the foundation upon which adipose-based therapies are currently regulated as *inconsistent* with modern scientific understanding of adipose tissue and its many functions—not to mention its primary function (energy storage).

This chapter provides an overview of the FDA, its beginnings and its general risk-based approach to regulating cell and tissue therapies, and the basics of law and regulation as applied to regenerative medicine. It then discusses the FDA's regulation of adipose-derived cell therapies, providing a close look at inconsistencies between the regulatory framework and adipose biology. It concludes with an overview of recent developments in methods and mechanisms for obtaining approval of regenerative products and biologics in general, particularly with regard to real-world evidence (RWE) and innovative trial design.

Legal disclaimer

- This chapter offers a basic review of the laws, regulations, and interpretive guidance underlying FDA oversight of regenerative products, particularly those derived from adipose. It is not intended to provide specific legal advice concerning particular practices, procedures, or products.
- Because working with regenerative products raises complex issues of science, medicine, and law, readers are strongly encouraged to discuss any questions or concerns with an attorney having expertise in this area. Moreover, in light of the FDA's recent escalation of enforcement efforts, it is usually most cost-effective to seek legal counsel sooner rather than later.

Adipose: Composition and functions

Histology textbooks commonly refer to four main tissue types that make up the human body: muscle, epithelial, connective, and nervous [1, 2]. A *tissue* is traditionally defined as a group of cells working together, while an *organ* is defined as several tissues working together.

- *Anachronism no. 1*: Histology texts commonly characterize adipose as a loose connective "tissue" [3, 4]. However, adipose "tissue" is actually composed of three different tissue types: connective tissue, muscle tissue (smooth muscle in the veins and arteries within adipose), and nervous tissue (the sympathetic nerves that innervate adipose and control lipolysis and use of stored energy) [5, 6]. Thus, given its histology, adipose is more accurately classified as an *organ*—i.e., composed of several tissues working together.

All organs can further be described as consisting of two tissue partitions: parenchymal and stromal. *Parenchyma* is commonly defined as "the functional tissue of an organ as distinguished from the connective and supporting tissue" [7]. For example, the muscle cells of the heart (cardiomyocytes), the insulin-making and secreting cells of the pancreas (beta cells), and the hepatocyte cells of the liver that process toxins and make blood-clotting proteins are all considered part of the parenchymal, or functioning, part of their respective organs. The *stromal* portion of an organ is essentially everything else, and is often defined as "the supportive tissue of an organ, consisting of connective tissue and blood vessels" [7]. Due to its supportive role within an organ, stromal tissue has until recently been perceived as subordinate to the parenchyma at best, and unimportant and superfluous at worst.

- *Anachronism no. 2*: It is now recognized that automatically viewing stroma as merely supportive of, and subordinate to, parenchyma is wrong, making the "primacy of parenchyma" paradigm another anachronism. The proper and effective function of parenchyma is (inter)dependent on the presence and function of the formerly "nondescript" stromal compartment. The stroma directly impacts the health, growth, repair, and function of parenchymal cells [8] (and therefore the health, growth, repair, and function of an organ). It does so by playing a central, controlling role in processes such as inflammation (which impacts repair/destruction) [9] and/or (neo)vascularization (the supply of blood, oxygen, nutrients, etc.) [10].

Consequently, the function of a cell population isolated from a given tissue/organ is highly dependent on whether those cells are derived from the parenchymal component of the organ or the stromal component. Generally speaking, the primary function of a given organ is defined (and differentiated) by its parenchymal cell component, whereas the supportive functions of stromal tissue are relatively conserved across all organs and tissue types.

- *Anachronism no. 3*: As explained in greater detail, *infra*, the FDA's current framework for regulating and thus guiding the therapeutic use of cells from a tissue/organ does not consider distinctions in parenchymal versus stromal sourcing or evaluate a source's potential impact on a therapeutic application. Given what science now understands about the interrelationship of an organ's parenchymal and stromal components, a cell's locational source should be considered to understand a cell product's functions in the donor and recipient, and meaningfully assess the risks of its clinical applications.

Like other organs/tissues, adipose is composed of parenchyma and stroma. The parenchymal cells of adipose—"adipocytes"—are cells whose primary and differentiating function is the ability to store energy in the form of triglycerides [11, 12]. Indeed, the "signet ring" appearance of adipocytes on routine histology is entirely due to its storage of lipid and is the *sine qua non* of the adipose organ [13]. Histologically (and functionally to a considerable degree), the stromal component of adipose is often regarded as nondescript and without any significant defining or differentiating characteristics from the stroma of a majority of many other organs.

As discussed in greater depth later in this chapter, the FDA regulates cell and tissue products in large part on the basis of whether it considers the product's "basic function" to be structural or nonstructural. This chapter will also demonstrate how and why the FDA's determinations of basic functions cannot always be squared with biological reality. For now, it is sufficient to recognize that adipose is ubiquitous in the human body and can differ in composition and function depending on its location/depot. At some locations, adipose primarily serves structural functions by providing cushioning, padding, and physical support; in most instances, however, the functions of adipose are predominantly nonstructural because they involve endocrine, metabolic, and other systemic activities.

In certain but limited areas of the body, adipose acts as a cushion and shock absorber—functions that the FDA classifies as "structural." This occurs in the balls and heels of the feet, within the orbit, around viscera, and along neurovascular bundles [8]. Adipose depots of this kind will not atrophy or diminish with caloric deprivation [14, 15].

In contrast, subcutaneous adipose tissue—the most common source of tissue/cells for therapeutic applications—is very different in terms of its chemistry, histology, and function [16–18]. Instead of serving the localized and structural functions of providing cushioning and physical support, subcutaneous adipose is most important for its *nonstructural* endocrine-related activities in storing energy and regulating metabolism. Indeed, the ability of adipocytes to store energy in the form of lipids is what quite literally defines adipose as adipose. The efficient energy storage of triglycerides in adipocytes also serves as an excellent insulator of the human body, giving adipose an important role in controlling and maintaining body temperature. Furthermore, it is now well established that adipocytes and the adipose *organ* secrete a wide variety of systemically active factors—including hormones, growth factors, and cytokines (also known as *adipokines*). All of these purposes and activities fit the FDA's definition of "nonstructural" functions.

In Table 1, a number of important adipokines are identified with their systemic, nonstructural functions [19–22]. Given the breadth of adipokines' functional versatility, this list is in no way exhaustive.

Again, despite its detail, this list provides an incomplete picture of what adipose is and does. It nevertheless shows that adipose does much more than serve the nonsystemic structural functions of cushioning and support in limited areas of the body. Given its diffuse location and diverse activities throughout the body, adipose is more accurately categorized as a dynamic endocrine and immunologic organ, with a systemic role in the control of energy storage/use and metabolism [23–27].

Accordingly, attempts to define adipose as either predominantly "structural" or "nonstructural," terms used by the FDA when regulating cell and tissue therapies, creates a stumbling block as both are accurate. For many Americans, adipose is a major structural component of the body. As a counterargument, very lean people do fine without it.

TABLE 1 Example adipokines with endocrine (e.g., systemic) function.

Factor	Function
11-Hydroxysteroid dehydrogenase type 1 (11-HSD type 1)	Converts inactive cortisone to active cortisol, eventually leading to increased insulin resistance
Adiponectin	Modulates endothelial adhesion molecules; inhibits inflammatory responses; link to atherosclerosis, which may increase cardiovascular risk; antidiabetic properties
Adipsin	Activates alternative complement pathway
Acylation-stimulating protein	Postprandial clearance of TAG
Angiotensinogen	Overactive (RAAS) releasing aldosterone from adrenal cortex; blood pressure regulation
Apolipoprotein E	Mediates lipid transfer between circulating lipoproteins and tissues; binds inflammatory components to clear pathogenic agents (innate immunity)
Aromatase	Increases conversion of androgens to estrogen/estrone
Cholesteryl ester transfer protein (CETP)	Involved in the transfer of neutral lipids, including cholesteryl ester and triglyceride, among lipoprotein particles
Deiodinase	Increases conversion of T4 to T3
Fasting-induced adipose factor (FIAF)	Possibly signaling molecule working with leptin
Growth-hormone-binding proteins	Increases proportion of bound GH; GH is inversely related to BMI/visceral fat
Interleukin-6	Implicated in acute-phase reactant
Leptin	Starvation signal; channels FA into adipose; limits TAG deposit in other tissue; releases GnRH from hypothalamus; regulates TRH from hypothalamus; releases aldosterone
Metallothionein	Protects FA from oxidation
Plasminogen activator inhibitor-1 (PAI-1) and tissue factor	Inhibits plasminogen
Resistin	Induces insulin resistance
Retinol-binding protein	
Tissue factor	Initiates coagulation cascade; cell surface receptor to activate factor VII
Transforming growth factor beta (TGF-β)	Regulates preadipocyte proliferation, differentiation, and apoptosis
Tumor necrosis factor alpha (TNF-α)	Systemic inflammation; development of insulin resistance

FA, fatty acid; *TRH*, thyrotropin-releasing hormone; *TAG*, triacylglyerols; *GnRH*, gonadotropin-releasing hormone; *RAAS*, renin-angiotensin-aldosterone system.

As noted, subcutaneous adipose is by far the most frequently utilized depot for cell therapy/regenerative medicine applications. When used for this purpose, subcutaneous adipose is more important for its nonstructural metabolic and biochemical functions relative to its structural/cushioning importance, which is also illustrated by the various, albeit rare inheritable [28] and acquired [29] lipodystrophy syndromes involving selective loss of adipose from various regions of the body [30]. The extent of fat loss in these syndromes can range from very small areas on one part of the body to near total absence of adipose in the entire body [31]. Congenital generalized lipodystrophy (CGL), a rare autosomal recessive disorder, gives some interesting insights into the primary role of adipose. Characterized by *near total loss of body fat*, patients with CGL often exhibit and suffer from many clinical findings consistent with other endocrine disorders, such as changes in skin color, liver disease [32], hirsutism, reproductive inability [33], muscular hypertrophy or atrophy [33], and diabetes and hypertriglyceridemia [34], that predispose to coronary artery disease, generalized atherosclerosis [35], and pancreatitis [31,36]. Thus, the near-total lack of adipose is not associated with serious structural complications, but rather serious metabolic and biochemical abnormalities that have systemic repercussions [37]. These are properly described as "nonstructural" according to the FDA's definitional rubric.

To summarize, the primary and most critical of the many functions of the adipose *organ*—especially the subcutaneous adipose *organ*—is *not* structural in nature. As detailed below, the FDA now acknowledges that adipose has both nonstructural and structural functions, but nevertheless persists in regulating "adipose tissue" as a "a structural tissue (and only a structural tissue) for the purpose of applying the HCT/P regulatory framework." This does not reflect biological reality and cannot be reconciled with the extensive body of scientific research that currently exists. Rather, regulating on the basis of anachronistic terms and concepts creates needless dilemmas for patients and clinicians, and raises serious concerns about the legal viability of the FDA's solely structural characterization if properly challenged in court.

A quick primer: Legislation, regulations and subregulatory guidance

Legislation: Federal agencies implement and enforce statutes enacted by Congress. The FDA is an agency within the US Department of Health and Human Services and is charged with implementing numerous statutes, giving it broad jurisdiction over a large and varied group of products. This chapter will focus primarily on the FDA's interpretation, implementation, and enforcement of the Federal Food, Drug, and Cosmetic Act (FDCA) and Public Health Services Act (PHSA). It will also describe certain FDA initiatives pursuant to the 21st Century Cures Act that affect regenerative medicine.

Regulations: Administrative agencies implement statutes by promulgating regulations. Under the federal Administrative Procedure Act, 5 United States Code (USC) § 553(b)–(d), regulations that impose substantive obligations on regulated actors (as opposed to interim or ministerial/administrative regulations) must be developed through the process of "notice and comment" rulemaking. The agency begins by publishing a Notice of Proposed Rulemaking in the Federal Register and inviting the public to submit comments for an identified period (usually 30 days). At the close of the "notice and comment" period, the agency can take its time in deciding whether and how to finalize the proposed regulation. When finalizing a regulation or "rule," the agency must consider public comments, but there is no need to adopt or specifically respond to them. When final, the regulation will be codified in the Code of Federal Regulations (CFR) and appear in the Federal Register with an accompanying statement of reasoning and conclusions. Thus, an agency has discretion in deciding when and how to regulate. Once finalized, a regulation has the full force of law unless revised or rescinded by the agency or invalidated by a court. An agency also has discretion in determining whether and how to implement and enforce a regulation. Enforcement methods can include inspection, warnings, civil litigation for monetary or injunctive relief, and in extreme cases can extend to criminal prosecution seeking monetary penalties or imprisonment.

Administrative guidance: At any time after finalizing a rule, an agency may issue subregulatory "guidance" to assist regulated actors in understanding and complying with their legal obligations. In theory, agency "guidance" differs from the regulations they explain in two significant respects. First, guidance interprets and explains existing regulatory requirements; it cannot impose new regulatory obligations. As such, it need not be subject to formal notice and comment procedures, although an agency should seek public input before finalizing a draft guidance. Second, because guidance should be interpretive only and lacks the force of law enjoyed by statutes and regulations, it is supposed to be nonbinding. The following language serves as the standard preface for FDA guidance documents:

> FDA's guidance documents, including this guidance, do not establish legally enforceable responsibilities. Instead, guidances describe the FDA's current thinking on a topic and should be viewed only as recommendations, unless specific regulatory or statutory requirements are cited. The use of the word "should" in FDA's guidances means that something is suggested or recommended, but not required.

While agency guidance may stop short of a legally binding regulation, it nevertheless explains what will violate the underlying regulation and indicate what will warrant enforcement proceedings. Thus, despite a guidance's disclaimer of its nonbinding effect, its substantive content is dismissed at significant peril. Nevertheless, as merely interpretive and advisory, final guidance rarely qualifies for judicial review unless it is found to be an impermissible effort to create new legally binding obligations without employing proper rulemaking process.

Judicial review of regulations and guidance

Like any administrative agency, the FDA has broad discretion in its interpretation of a statute to promulgate regulations to implement and enforce it. It also enjoys discretion in developing guidance to explain the agency's "thinking" about its regulatory apparatus and enforcement priorities. Courts are especially willing to grant broad deference to an agency's actions when regulating matters involving technical and scientific expertise—as is clearly the case with most matters within the FDA's scope of responsibilities, particularly cell and tissue therapies.

Judicial deference has its limits, however. Typically, judicial review is reserved for an agency's "final action"—such as an enforcement action for an alleged violation of a specific regulation in a concrete "case or controversy" between the agency and a party directly affected and aggrieved by an enforcement action. Untitled and even warning letters cannot be challenged in court since they fall short of any final action and in no way bind the agency to act on its warnings. The threat of seizure, shutdown, or even criminal prosecution is not enough as long as it is just a warning or threat.

When a final action by the agency has occurred, a regulation can be challenged as invalid on its face or as implemented. A regulation can be invalidated for imposing new substantive obligations without adhering to proper notice and comment procedures, exceeding the agency's jurisdiction under the statute being enforced or conflicting with the statute's language or other clear expression of Congress' intent in enacting the statute.

- **Judicial "Chevron" deference to an agency's interpretation of a *statute*:**
  - So-called "Chevron" deference derives from the US Supreme Court's decision in *Chevron USA, Inc. v. Natural Resources Defense Council, Inc.*, 467 US 837 (1984). As previously explained, Congress enacts statutes, which administrative agencies interpret and implement by promulgating and enforcing regulations. Under *Chevron*, a court should defer to an agency's interpretation of that statute as long as the agency's regulation represents a reasonable interpretation, even if there are policy or other reasons to prefer alternative readings of that statute.
  - Conversely, a regulation that is not a reasonable interpretation because it lacks logical and factual support will be judicially invalidated as arbitrary and capricious.
- **Judicial "Auer" deference to an agency's interpretation of a *regulation*:**
  - A somewhat different form of judicial deference to a regulatory agency's decision is known as "Auer" deference, articulated by the US Supreme Court in *Auer v. Robbins*, 519 US 452 (1997). Unlike *Chevron* deference to an agency's interpretation of an ambiguous *statute* authored by Congress, *Auer* deference applies to an agency's interpretation of an ambiguous regulation promulgated by the agency itself. According to *Auer*, if a *regulation* is genuinely ambiguous, a court will normally defer to the promulgating agency's interpretation of that regulation (often in the form of interpretive guidance) as long as that interpretation is reasonable, even if there are policy or other reasons to prefer alternative readings. *Auer v. Robbins*, 519 US 452 (1997).
  - Again, though, judicial deference to agency decision making has limits, and the US Supreme Court is increasingly focused on clarifying and enforcing those limits. Most pertinent for this discussion are *Auer's* requirements that the regulation be genuinely ambiguous, the agency's reading of it must be reasonable, and it must "reflect an agency's authoritative, expertise-based, 'fair and considered judgment'." *Kisor v. Wilkie*, 139 S. Ct. 2400, 2413–2414 (2019).
  - As discussed later, this may not be the case with certain aspects of FDA regulation of adipose cell and tissue therapies, given the FDA's occasional preference for anachronistic views about adipose's form and function as opposed to fully informed and updated expertise-based judgment informed by contemporary science.

Because judicial review is reserved for challenges to an agency's final actions, it is not normally available for guidance. A guidance may be "final" in the sense of the agency finalizing its explanation of a given topic, regulation, or other issue, but it is still guidance. As such, it is advisory only, and not the kind of final action that might warrant judicial review. However, should a court conclude that a guidance effectively imposes new substantive and legally binding obligations instead of interpreting existing regulations, it will review the document as a de facto final regulation. The purported guidance can be invalidated for violating notice and comment requirements, exceeding the agency's authority under the controlling statute, or being arbitrary and capricious for lacking any basis in reason and fact.

The Food and Drug Administration: Administrative structure and role

The FDA has jurisdiction over a large and diverse set of products, ranging from veterinary products, cosmetics, and vaping devices to food, drugs, and gene therapies. The FDA exercises its broad responsibilities through category-specific "Centers." The Center for Biologics Evaluation and Research (CBER) has primary oversight authority for cell and tissue products, including those derived from adipose. The Center for Devices and Radiologic Health (CDRH) is involved in regulating devices used for regenerative products, and the Center for Drug Evaluation and Research (CDER) may share responsibility for combination products. While the centers and products may differ, the FDA's mission in overseeing all of them is consistent: regulate in a manner that safeguards the public health by balancing a product's risks and benefits. The overriding goal is to promote innovation while protecting patients from products that are unsafe and/or ineffective for their intended use. In theory at least, "risk-based regulation" informs all areas of FDA oversight to tailor the extent and type of regulation to a product's real or potential risks and benefits.

Under the FDCA, certain medical drugs and devices must obtain premarket approval before being marketed and sold to the public. During early development of such products, the FDA must approve an Investigational New Drug application (IND) application to permit testing on human subjects, typically for the purpose of conducting three-phase clinical trials (with randomized and double-blinded being the gold standard) and submitting results for approval of a New Drug Application (NDA). A similar sequence pertains to biologics in need of premarket approval under § 351 of the Public Health Service Act (PHSA), which requires an IND for human testing and, ultimately, an approved Biologics License Application (BLA).

Practice note: The investigational new drug application and biologics license application

The Investigational New Drug Application (IND) requests FDA authorization to administer an investigational drug or biological product to humans. See 21 CFR 312.

An approved IND must be secured prior to interstate shipment and administration of any new drug or biological product that is not the subject of an approved NDA or BLA.

Applicants for an IND will need to submit:

- **Form FDA 1571** Cover Sheet
- Applicant information
- Introductory statement and investigational plan
- Investigator's brochure
- Clinical protocol
- Chemistry, Manufacturing, and Control (CMC) information

Continued

- Pharmacology and toxicology information
- Previous human experience with the investigational drug/biologic
- Labeling
- FDA-requested relevant information
- Additional information, as applicable

The Biologics License Application (BLA) process is regulated under 21 CFR 600–680 pursuant to 42 USC § 262(k).

A BLA is a request for permission to introduce, or deliver for introduction, a biologic product into interstate commerce. See 21 CFR 601.2. A BLA can be submitted by any legal person or entity who is engaged in manufacturing the product or takes responsibility for compliance with product and establishment standards.

Form 356h specifies the requirements for a BLA, which include:

- Applicant information
- Product/manufacturing information
- Preclinical studies
- Clinical studies demonstrating that the product is safe and effective for its intended use
- Proposed labeling information

If satisfied by the foregoing information, the FDA will issue a biologics license to signify that a product has been approved for marketing.

Issuance of a biologics license further demonstrates that the product, the manufacturing process, and the manufacturing facilities have met all applicable requirements to ensure the continued safety, purity, and potency of the product.

Products that require an approved BLA or NDA may not be marketed, sold, or used without one. Doing so renders the product adulterated and misbranded and may lead the FDA to exercise one of its enforcement options, which include an advisory Untitled Letter, a formal Warning Letter, temporary or permanent shutdown/injunction of some or all of a facility's operations, confiscation of property, and imposition of civil fines, with attendant civil litigation. The FDA also has power to initiate criminal prosecution, which, if successful, can result in criminal financial penalties or imprisonment.

The three-phase, randomized controlled clinical trial (RCT) remains the gold standard for demonstrating a product's safety and efficacy when seeking premarket approval under the FDCA, PHSA, and 21st Century Cures Act. As briefly described toward the end of this chapter, the FDA has taken a number of steps to streamline and update the review process and bring new products to market more quickly. In addition to developing several expedited approval pathways and designations, the agency will consider new forms of study design and evidence, at least for certain types of regulatory approvals. It also encourages investigators to seek agency input early in the development process (even before a pre-IND meeting) as a way of optimizing a study's likelihood of meeting the agency's expectations, particularly when the study involves an innovative design or relies on "real world evidence" from sources other than an RCT.

FDA regulation of cell and tissue products

The FDA's authority to regulate: FDA regulation of medical devices, drugs, and biologics is well accepted. In contrast, even though the FDA has asserted its authority to regulate cell therapies for close to two decades, its ability to do this is widely challenged, if not overtly derided.

This is especially true for autologous stem cell therapies, i.e., treating a patient with one's own cells. To this day, whether through website proclamations or public comments submitted in response to the FDA's proposed regulations or guidance, patients and providers routinely (and often indignantly) ask *how can the government regulate my own cells as a drug?* This question seems to make perfect sense given the colloquial understanding of a "drug" as a chemically synthesized and commercially manufactured product as opposed to the cells and tissues which comprise the human body. In this instance, though, the common usage of the term "drug" is not relevant. Rather, the FDA is legally required to regulate products that fit the technical definitions of "biologics" and/or "drugs" as used, respectively, by the PHSA and the FDCA—and stem cells qualify as both.

Stem cells are FDA-regulated "biologics": In 42 USC § 262(i)(1), the PHSA defines a "biological product" as "a virus, therapeutic serum, toxin, antitoxin, vaccine, blood, blood component or derivative, allergenic product, protein (except any chemically synthesized polypeptide), or analogous product ... applicable to the prevention, treatment, or cure of a disease or condition of human beings." A stem cell therapy is therefore a "biological product" since it is "an analogous product" to prevent, treat, or cure a person's disease and condition.

Stem cells are FDA-regulated "drugs": The FDCA defines a "drug" in 21 USC § 321(g)(1) as an "article" that is "intended for use in the diagnosis, cure, mitigation, treatment, or prevention of disease in man or other animals ... [and/or] intended to affect the structure or any function of the body of man or other animal ...", according to 21 USC § 321(g)(1). Accordingly, a stem cell product is an "article" that qualifies as a "drug" when its use is intended to diagnose, cure, treat, etcetera, a disease.

The PHSA directs the FDA to determine whether the product is "pure and potent"; under the FDCA, the FDA must decide whether a product must be "safe and effective" for its intended use. The FDA will evaluate a product's intended use on the basis of objective manifestations of the manufacturer's intent as indicated by labeling claims, advertising materials, oral or written statements, or other representations, including statements made by a company's agents or representatives.

Practice note: FDA authority to regulate cells

Before examining the specifics of how the FDA regulates stem cell therapies, it is important to address two common controversies about, and challenges to, the FDA's basic ability to regulate such products at all:

Controversy no. 1: "FDA regulation of cell therapies is illegal because cells are not drugs." While this statement may have some appeal based on the colloquial use of "cells" and "drugs," the plain text of the FDA's enabling statutes—text that binds both regulators and courts—reveals this statement to be inaccurate.

- ***The FDA has statutory authority to regulate a patient's cells even when those cells are used to treat that same patient.*** Cell and tissue products satisfy the controlling statutes' technical definitions of both "drugs" and "biologics," meaning that the FDA can regulate them. Rejecting the FDA's authority to do so is akin to rejecting the Internal Revenue Service's (IRS's) authority to impose penalties for nonpayment of taxes. Both agencies are obligated to enforce their enabling statutes as written; both have significant discretion in promulgating and implementing regulations to do so; and both can make regulated actors pay dearly for noncompliance. For these reasons, courts to date have rejected challenges to the FDA's jurisdiction to regulate cells, tissues, and cell- and tissue-derived products.

Continued

Controversy no. 2: "FDA regulation of cell therapies is illegal because it interferes with the Practice of Medicine." This is another popular but nevertheless inaccurate perception of the respective functions of federal and state oversight of cell products and the providers who use them.

- ***FDA regulation of biologics, drugs and devices routinely affects, influences—and, in many instances, effectively restricts—the practice of medicine*** by determining whether and under what conditions a regulated product can be used. This is equally true for regulated cell and tissue products. In all areas of medical practice and clinical investigation, a physician carries multiple obligations under both federal and state law. While it is true that state law has primary oversight responsibility for the practice of medicine, this does not limit or displace federal law. Should the two conflict, federal law will preempt state law.

- ***Relying on "the practice of medicine" to resist FDA regulation of cell therapies (including their permissible intended uses)*** is not a valid defense and exposes the clinician to serious enforcement sanctions, which typically come with significant financial costs, including legal fees.

FDA regulation of stem cell therapies invariably raises complex and interrelated issues of science, medicine, and law. As previously noted, when facing questions or concerns about compliance obligations, it is typically wisest and most cost-effective to be proactive in consulting an attorney with expertise in FDA regulation of regenerative medicine. Doing so sooner rather than later can avert the sizeable expenses of having to respond to enforcement measures that could have been anticipated and avoided or at least mitigated.

The HCT/P framework

In 2005, the FDA formally acknowledged that regulating cell and tissue products as both "safe and effective drugs" under the FDCA and "pure and potent biologics" under the PHSA was potentially inefficient and confusing. Consequently, it announced a new framework to regulate "human cells, tissues and cellular and tissue-based products" (HCT/Ps) as "biological product[s]" subject to oversight under sections 351 and 361 of the PHSA as codified by 42 USC §§ 262 and 264, respectively.

Preliminary definitions: Before examining the HCT/P framework more closely, it is helpful to review its core concepts—all of which are defined by regulation 21 CFR § 1271.3.

- *HCT/P* "means articles containing or consisting of human cells or tissues that are intended for implantation, transplantation, infusion, or transfer into a human recipient."
- *Autologous use* "means the implantation, transplantation, infusion, or transfer of human cells or tissue back into the individual from whom the cells or tissue were recovered."
- *Homologous use* "means the repair, reconstruction, replacement, or supplementation of a recipient's cells or tissues with an HCT/P that performs the same basic function or functions in the recipient as in the donor."
  - Whether an HCT/P is intended for homologous use will be determined on the basis of the labeling, advertising, or other indicia of the manufacturer's objective intent.
  - An HCT/P is intended for homologous use if it is intended to serve the same function in the recipient as originally performed in the donor. This is true even if the HCT/P's ultimate location in the recipient differs from its original location in the donor.

- *Minimal manipulation* means:
  - "*For structural tissue*, processing that does not alter the original relevant characteristics of the tissue relating to the tissue's utility for reconstruction, repair, or replacement."
  - "*For cells or* nonstructural *tissues*, processing that does not alter the relevant biological characteristics of cells or tissues."
- *Processing* "means any activity performed on an HCT/P, other than recovery, donor screening, donor testing, storage, labeling, packaging, or distribution, such as testing for microorganisms, preparation, sterilization, steps to inactivate or remove adventitious agents, preservation for storage, and removal from storage."

Again, all of these definitions can be found in 21 CFR § 1271.3 and, as formal regulations, are binding on regulated actors—as well as on the FDA as regulator.

The framework's risk-based structure: The HCT/P framework consists of three tiers of escalating oversight that, in theory, gear the nature and extent of regulation to a product's risk of infection or transmission of an infectious disease.

Tier 1: No oversight
- The FDA exempts certain products as "not considered HCT/Ps"—meaning that, biologically, they fall within the definition of an HCT/P but are expressly excluded from §§ 361 and 351 oversight.
- Examples of 351–361 exempt products include vascularized human organs for transplantation, whole blood and blood components, "secreted or extracted human products" such as milk or collagen, and "[m]inimally manipulated bone marrow for homologous use and not combined with another article[.]" Such products may nevertheless be subject to federal regulation outside the HCT/P framework.
- These terms and concepts are again defined in 21 CFR § 1271.3.

Tier 2: Minimal oversight under Section 361
- Section 361 products require domestic and foreign establishments that manufacture HCT/Ps to register with the FDA, submit a list identifying each HCT/P manufactured, and follow good tissue practices. There is no need to fulfill § 351's requirements of premarketing review (including an approved IND and BLA or NDA) and current good manufacturing practices (cGMPs).
- To qualify as a § 361 product under 21 CFR § 1271.10, an HCT/P must be *all* of the following:

  (1) Minimally manipulated

  (i) *For structural tissue*: minimal manipulation means processing that does not alter the original relevant characteristics of the tissue relating to the tissue's utility for reconstruction, repair, or replacement.

  (ii) *For cells or nonstructural tissues*: minimal manipulation means processing that does not alter the relevant biological characteristics of cells or tissues.

  (2) Intended for homologous use only, as reflected by the labeling, advertising, or other indications of the manufacturer's objective intent;

  (i) Homologous use means the repair, reconstruction, replacement, or supplementation of a recipient's cells or tissues with an HCT/P that performs the same basic function or functions in the recipient as in the donor.

(3) **Involve no combination with another article,** except for water, crystalloids, or a sterilizing, preserving, or storage agent, provided that the addition of water, crystalloids, or the sterilizing, preserving, or storage agent does not raise new clinical safety concerns with respect to the HCT/P; *and*

(4) **Either**:

  (i) **Has no systemic effect or dependence on metabolic activity** of living cells for its primary function; *or*

  (ii) **Has a systemic effect or depends on metabolic activity** of living cells for its primary function, *and* its intended use is one of the following:
  (a)autologous,
  (b)allogeneic in first or second degree blood relative, or
  (c)reproductive.

Tier 3: Section 351 premarket approval

- Failing any one or more of § 361's four requirements signals that, at least in the FDA's view, the product poses greater risks that warrant § 351 premarket review, placing a solo physician or small clinic on the same regulatory pathway traveled by large pharmaceutical manufacturers when bringing commercially manufactured products to the mass market.
- As previously explained, the clinician would need to file an IND; conduct expensive and time-consuming three-phase clinical trials to establish safety, purity, potency, efficacy, and stability; and pay a substantial user fee for the FDA to review the product with no guarantee of ultimate approval.
- A physician or clinician engaged in recovering, screening, testing, processing, storing, labeling, packaging, or distributing a § 351 HCT/P also qualifies as a "manufacturer" and, therefore, must conform to cGMPs requirements as set forth in 21 CFR § 207.3.

Tier 3: "Same Surgical Procedure" exception to § 351 regulation

- The narrow "Same Surgical Procedure" (SSP) exception under 21 CFR § 1271.15(b) permits a § 351 product to be used without having to meet § 351's rigorous requirements. The exception applies to certain procedures that, in the agency's view, present no greater risks of contamination or transmission of communicable disease beyond those typically associated with surgery.
- To qualify, the HCT/Ps must be removed from the patient and implanted back into that same patient "during the same surgical procedure," undergo "no intervening processing steps beyond rinsing, cleansing, sizing, or shaping," and remain in its "original form."
- A "same surgical procedure" can be a single surgery, although harvest and return can be on separate dates as long as the HCT/Ps are not processed beyond being "rinsed or cleaned and temporarily stored after being labeled, pending implantation … [with] no other processing steps and no other manufacturing steps beyond labeling and storage" and remain in their "original form."
- ***Unlike § 361, the § 351 SSP exception does not require homologous use.***
- However, even minimal manipulation may entail too much processing to remain within the SSP exception's narrow limits. If so, the product would need to satisfy all four of § 361's

criteria (including its requirement of homologous use) or obtain premarket approval, etcetera, under § 351.

- For additional information, see US FDA, Same Surgical Procedure Exception under 21 CFR 1271.15(b): Questions and Answers Regarding the Scope of the Exception—Guidance for Injury (November 2017) (hereafter "Final SSP Guidance").

Determining regulatory status under the HCT/P framework

The HCT/P framework effectively recognizes four categories of increasingly regulated cell and tissue products (Table 2):

(a) Non-HCT/Ps that are expressly exempt from regulation under this framework
(b) Section 351 SSP products that are not restricted to homologous use but face tight restrictions in the way of processing
(c) Section 361's minimal requirements for registration and current good tissue practices (cGTPs), assuming minimal manipulation, homologous use, no combination with other articles, and *either* no systemic effect or dependence on metabolic activity for its primary function *or* is otherwise intended for autologous, first or second degree allogeneic or reproductive use
(d) Full regulation under § 351, including premarket approval and cGMPs for products failing to qualify for any of the foregoing categories

Basic decision tree: Anticipating how an adipose-derived HCT/P will be regulated can be broken down into five basic steps:

1. Consider whether the product is expressly exempt from HCT/P regulation. *If not,*
2. Assess whether the "§ 351 Same Surgical Procedure" applies. *If not,*
3. Determine whether § 361 applies, by first determining whether the product is structural or nonstructural based on its basic functions(s). (Note: Although biology would disagree, the FDA insists that it cannot be both.)
4. Given the HCT/P's structural or nonstructural character, evaluate whether the product satisfies *all* of § 361's four requirements. *If not,*
5. Pursue premarket approval under § 351, but consider availability of recent measures to adapt § 351 criteria to regenerative therapies such as early FDA consultation, flexible trial design, multicenter studies, RWE, or eligibility for accelerated review through a regenerative medicine advanced therapy (RMAT) designation or other product designation or expedited pathway under 21 USC § 356 (further detailed below).

To assist in determining how a particular use of an HCT/P will be regulated under the HCT/P framework, the Final MM/HU Guidance provides the advisory decision tree in Fig. 1.

FDA regulation of adipose cell and tissue therapies: Definitions and dilemmas

For many clinicians, having to satisfy § 351's requirements for premarket review, cGMPs and the like can present obstacles of time and expense that may prove insurmountable. In such cases, § 351 classification effectively truncates further development of, and patient access to, a therapy even if it involves using a patient's own cells to deliver superior results with

TABLE 2 A detailed explanation of determining regulatory status.

21 CFR 1271	Category	Criteria	Compliance
1271.3 (d)	Non-HCT/P → *Expressly exempt based on listed cell/tissue types*	Step 1: Consult non-HCT/P list • Blood/blood products (e.g., PRP) • Vascularized human organs for transplant • Bone marrow if minimally manipulated and homologous use → *If product not listed, go to Step 2*	No § 361 or § 351 oversight
1271.15 (b)(3)	§ 351 SSP → *Risks no greater than typically associated with surgery* → *HCT/Ps must retain original form*	Step 2: Satisfy SSP criteria? • Autologous • Single establishment • Single procedure (intervening days permitted) • No processing other than donor screening and testing; rinsing, cleansing, sizing, shaping → *If fail any, go to Step 3*	No § 361 or § 351 oversight
1271.10	§ 361 HCT/P → *Prevent introduction, transmission, or spread of communicable diseases*	Step 3, Part 1: Structural/nonstructural classification regarding minimal manipulation 1. *Minimal manipulation*: based on impact of manufacturing/processing on HCT/P's original relevant characteristics "as the HCT/P exists in the donor, and not based on the intended use of the HCT/P in the recipient" • *Structural*: No alteration of original relevant structural characteristics, e.g., "strength, flexibility, cushioning, covering, compressibility, and response to friction and shear" • *Nonstructural*: No alteration of original relevant biological characteristics, e.g., "differentiation and activation state, proliferation potential, and metabolic activity" Step 3, Part 2: If minimal manipulation, satisfy remaining § 361 criteria? 2. *Homologous use*: Intended to perform same basic (structural/nonstructural) function[s] in recipient as it did in donor before harvest or processing	Minimal § 361 Oversight • Establishment Registration & Listing • Donor screening and testing • cGTPs

		3. No combination with other articles (but does permit combination with water, crystalloids, and agents for sterilization, preservation, or storage as long as no additional safety concerns) 4. EITHER: • No systemic effect and primary function does not depend on metabolic activity of living cells, *OR* • Systemic effect or depends on metabolic activity for primary function, AND – Autologous, or – Allogenic—first/second degree blood relative, OR – Reproductive use → *If not § 351 SSP or § 361, go to Step 4*	
1271.20	§ 351 HCT/P → *Must establish product is safe, effective, pure, and potent for intended use*	Step 4: Submit IND; seek premarket approval and BLA	§ 351 Premarketing approval and cGMPs, etc. Consider availability of • Early FDA "INTERACT" consult • Flexible trial design • RWE • Expedited paths/designations, including RMAT

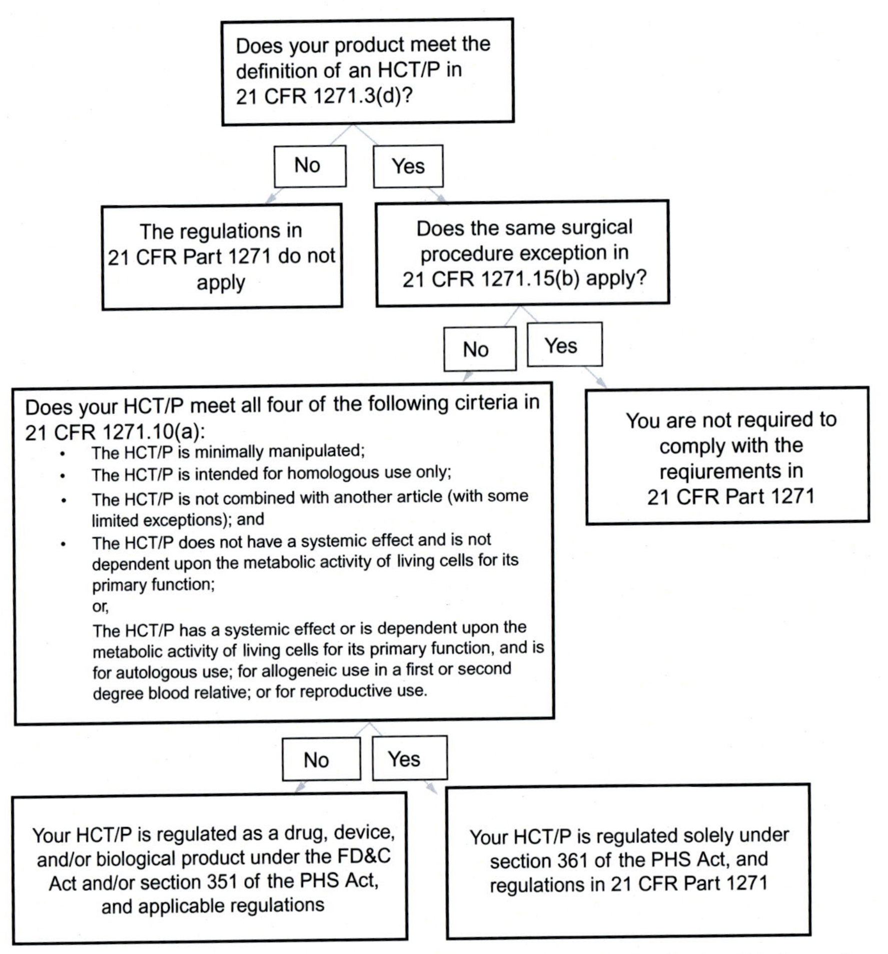

FIG. 1 Flowchart to illustrate how to apply the criteria in 21 CFR 1271.15(b) and 1271.10(a). Image from https://www.fda.gov/media/109176/download, p. 5 (Accessed February 2021).

fewer risks than traditional treatment options. At the same time, § 351 oversight is essential for therapies that genuinely pose greater risks due the HCT/P's characteristics, mechanism(s) of action, and/or circumstances of use. Despite the HCT/P framework's basic premise of fitting the extent of oversight to the extent of risk, the FDA's rationale for subjecting certain clinical applications of adipose to § 351 oversight is questionable and arguably anachronistic in terms of basic biology. This can make it even more difficult to understand and fulfill one's compliance obligations under the HCT/P framework.

At its core, the framework uses a series of interrelated regulations to assess risk based on the HCT/P's source and function in the host, the degree of any processing postharvest and its intended function and use in the recipient. To avoid § 351 premarket approval requirements, a cell product must be specifically exempt as a "non-HCT/P," qualify for § 351's SSP exception (and thus maintain its original form from point of harvest to ultimate implantation), or fall within § 361 (and thus be intended for homologous use and no more than minimally manipulated). As creatures of regulation, these concepts and criteria have the force of law, binding both the regulated and the regulator. And, as previously explained, an agency's authority to write a regulation may come with significant discretion to interpret it, but any interpretation must still reflect a reasonable reading of the regulation, particularly when it comes to concrete definitions. Thus, any explanations of, or statements about, the HCT/P framework of specific parts thereof from the FDA via agency guidance, public statements, enforcement efforts, etc., must reflect a reasonable interpretation of the plain text of these regulations.

Where a product falls under the HCT/P framework will depend in the first instance on whether its "basic function" is "structural" or "nonstructural" since, in various ways, this will influence the degree of processing and manipulation as well as the existence of homologous use that are central to categorization under § 361 or § 351's SSP exception. For adipose HCT/Ps, it is at this starting point that the FDA has made certain choices and relied on anachronistic views that are questionable at best because they are difficult to square with any reasonable interpretation of the regulations' concrete definitions of key terms or contemporary understanding of what adipose is and does.

Regulating based on an HCT/P's "basic function or functions"

Background: The FDA's regulation of an HCT/P will depend in large part on how the agency views its "basic function." Its current strategy for regulating adipose is perceived by many as inaccurate and not reflective of current understanding of adipose biology. This perception may be particularly frustrating to many since the FDA's current approach derives from the agency's multiyear effort to clarify § 361 and the § 351 SSP exception as applied to HCT/Ps, particularly adipose.

Between 2013 and 2016, the FDA published a suite of four interrelated draft guidances to explain § 351's same surgical procedure exception (October 2014); § 361's minimal manipulation (December 2014) and homologous use (October 2015) requirements; and particular issues specific to adipose tissue (December 2014). Though these documents were well intended, they were immediately criticized for compounding instead of allaying confusion and, in some instances, propagating factual inaccuracies, particularly with regard to adipose. Especially controversial was the Draft Guidance on Minimal Manipulation's effort to clarify the meaning of a structural or nonstructural "basic function" by introducing the new and undefined concept of "main function."

Using "main function" to explain "basic function" was criticized for being vague and for attempting to create new regulatory obligations without following notice and comment requirements.

In finalizing the two separate draft guidances on minimal manipulation and homologous use into a single Guidance on Minimal Manipulation and Homologous Use ("Final MM/HU Guidance"), the FDA abandoned its attempt to explain the critical concept of "basic function," declaring instead that there was actually nothing to clarify. Specifically:

> The basic function of an HCT/P is what it does from a biological/physiological point of view, or is capable of doing when in its native state. By "basic" we mean the function or functions that *are commonly attributed to the HCT/P as it exists in the donor. Basic functions are well understood;* it should not be necessary to perform laboratory,

preclinical, or clinical studies to demonstrate a basic function or functions for the purpose of applying the HCT/P regulatory framework. Also, clinical effects of the HCT/P in the recipient that are not basic function or functions of the HCT/P in the donor would generally not be considered basic function or functions of the HCT/P for the purpose of applying the definition of homologous use.

By deeming basic functions to be "well understood," the FDA chose to overlook numerous criticisms about its treatment of the basic functions of several HCT/Ps, including adipose, amnion, and skin. The supposition that basic function is obvious and clear is also at odds with public acknowledgments by FDA leadership that HCT/Ps are "remarkably complex biologic entities," [38] and "much more complex than traditional products that FDA regulates[.]" (Bauer, S.R., as quoted in FDA, *Consumer Update: Adult Stem Cell Research Shows Promise*, April 2104).

Distinguishing structural from nonstructural functions

The regulations that form the HCT/P framework and the Final MM/HU Guidance recognize that an HCT/P can have more than one "basic function or functions." Moreover, phrases such as "basic function or functions," "basic functions," and "multiple functions" permeate the Final MM/HU Guidance for defining homologous use, and for the broader "purpose of applying the HCT/P regulatory framework." Notwithstanding these repeated references to an HCT/P's potential for having more than one "basic function," the Final MM/HU Guidance does not clarify the meaning of "basic functions" beyond stating that they "are well understood." The FDA also chose to determine the actual specific function (note the singular) of specific types of multifunctional HCT/Ps, particularly those derived from adipose.

A closer examination of the FDA's treatment of adipose illustrates the difficulty of regulating a product as either structural or nonstructural in function regardless of a multifunction product's specific intended use. In theory, an either/or model offers the advantage of administrative ease. As applied to adipose and other multifunction HCT/Ps, however, a solely structural categorization risks being anything but easy if it cannot be aligned with the basic biology of a product's nonstructural applications.

This regulatory dilemma is rooted in the FDA's own regulations and guidance that recognize that a given HCT/P can have a "basic function or basic functions." The Final MM/HU Guidance appropriately affirms that some HCT/Ps perform multiple functions. For the purpose of applying the HCT/P framework, however, the Final MM/HU Guidance states that the FDA will evaluate an HCT/P as either structural or nonstructural, but not both. It provides no rationale for, on the one hand, acknowledging the biological reality of multiple functions for certain HCT/Ps, and on the other, refusing to consider this reality regardless of an HCT/P's original function(s) in the donor and intended use(s) in the recipient.

Classifying an HCT/P as either structural or nonstructural is the initial step of evaluating whether an HCT/P can satisfy § 361's requirement of minimal manipulation. Understanding what an HCT/P is and does also influences the determination of homologous use under § 361 and "remaining such HCT/P" under § 351's SSP exception. It is therefore critical to get this initial determination right. In the course of explaining what will and will not qualify as minimal manipulation, the Final MM/HU Guidance goes into great detail about the differences between structural and nonstructural HCT/Ps. The basics are as follows:

1. Applicable regulations recognize that some HCT/Ps have more than one "basic function," and this is again affirmed in the Final MM/HU Guidance.
2. Nevertheless, the Final MM/HU Guidance makes clear that, for the purpose of determining minimal manipulation and applying the HCT/P framework in general, the FDA will evaluate an HCT/P on the basis of its function *in the donor* prior to harvest and processing, and will categorize it as *either* structural *or* cell/nonstructural—but not both.
3. According to the Final MM/HU Guidance, basic functions "are well understood."
4. *Structural HCT/Ps* physically support or serve as a barrier or conduit, or connect, cover, or cushion *in the donor*.

(a) *Examples of structural HCT/Ps [according to the FDA]*:

- Bone
- Skin
- Amniotic membrane and umbilical cord
- Blood vessel
- *Adipose* tissue
  - *[Note: This is the FDA categorizing adipose as structural despite its predominantly nonstructural activities in the body.]*
- Articular *cartilage*
- Nonarticular *cartilage*
- Tendon or *ligament*

(b) *Relevant structural characteristics* have a meaningful bearing on the tissue's utility for reconstruction, repair, or replacement *in the donor*. Examples:

- Strength
- Flexibility
- Cushioning
- Covering
- Compressibility
- Response to friction and shear

5. *Cells/nonstructural HCT/Ps* serve metabolic or other biochemical roles in the body such as hematopoietic, immune, and endocrine functions *in the donor*.

[Note: Even though many of adipose's diverse and most important functions are properly described as endocrine, metabolic, biological, and therefore nonstructural, the FDA refuses to consider adipose as such when applying the HCT/P regulatory framework.]

(a) *Examples of cells/nonstructural HCT/Ps*:

- Reproductive cells or tissues (e.g., oocytes)
- Hematopoietic stem/progenitor cells (e.g., cord blood)
- Lymph nodes and thymus
- Parathyroid glands

- Peripheral nerve
- Pancreatic tissue

(b) *Relevant biological characteristics of cells or nonstructural tissues* generally include the properties *in the donor* that contribute to the cells or tissue's function or functions. Examples:
- Differentiation and activation state
- Proliferation potential
- Metabolic activity

The basic functions of adipose HCT/Ps: Nonstructural (and structural)

According to the Final MM/HU Guidance, "basic functions are well understood." According to the FDA's own definitions of structural and nonstructural functions as applied to well-established facts, adipose clearly serves both, but, as explained at the beginning of this chapter and elsewhere in this publication, adipose is predominantly cellular/nonstructural. Dispersed throughout the body, adipose varies in function and form depending on where it is found. Its role in providing structural cushioning and support is limited and localized. In contrast, the list of subcutaneous fat's nonstructural biological functions is long and continues to grow with ongoing advances in adipose science.

The Final MM/HU Guidance gets many of these facts right when describing what adipose *is*, only to reject (or overrule?) basic biology when it comes to what adipose *does*. For instance:

> What adipose is: [Citing *Junqueira's Basic Histology: Text & Atlas, 13th Ed. (despite more recent editions)*] "Adipose tissue is typically defined as a connective tissue composed of clusters of cells (adipocytes) surrounded by a reticular fiber network and interspersed small blood vessels, divided into lobes and lobules by connective tissue septa. Additionally, adipose tissue contains other cells, including preadipocytes, fibroblasts, vascular endothelial cells, and macrophages."
>
> What adipose *does*: "Adipose tissue provides cushioning and support for other tissues, including the skin and internal organs, stores energy in the form of lipids, and insulates the body, among other functions. While adipose tissue has multiple functions, because it is predominantly composed of adipocytes and surrounding connective tissues that provide cushioning and support to the body, FDA considers adipose tissue to be a structural tissue for the purpose of applying the HCT/P regulatory framework."

It is worth noting that there is no readily apparent logical or biological basis for acknowledging the presence of many cell types in describing what adipose *is*, and then implying that most of them are superfluous in describing what adipose *does*. Further, in terms of what adipose *is* and *does*, it is misleading to omit any mention of subcutaneous fat's many types of adipokines and their many nonstructural functions. There is also no apparent rationale for emphasizing the physical prevalence of adipocytes and their importance in the (nonstructural) functions of storing energy and providing insulation, only to invoke them as a key reason for regulating adipose as solely structural. And finally, it is somewhat perplexing that the FDA cites a single (outdated) histology textbook in support of its solely structural characterization when that text expressly states that adipose is a nonstructural specialized connective tissue—a fact that was repeatedly brought to the agency's attention during public testimony and written comments in response to its draft adipose guidance.

Compounding these problems is the Final MM/HU Guidance's instructions for evaluating an adipose-derived HCT/P's ability to satisfy § 361's requirement of undergoing no more than minimal manipulation (emphasis added):

- "To evaluate whether processing of adipose tissue would meet the regulatory definition of minimal manipulation, you should consider whether the processing alters the original relevant characteristics of the adipose tissue *relating to its utility to provide cushioning and support*."
- Original relevant characteristics of structural tissues generally include the properties of that tissue *in the donor* that contribute to ... the tissue's *utility for reconstruction, repair, or replacement*. ... Examples of relevant characteristics of structural tissues include *strength, flexibility, cushioning, covering, compressibility, and response to friction and shear.*

Consequently, unless the § 351 SSP exception applies, using adipose-derived HCT/Ps for any of their myriad and predominantly nonstructural functions is automatically disqualified from § 361 oversight for the illogical and troubling reason that the FDA simply refuses to provide any mechanism for assessing risk of nonstructural use short of full-blown premarket review.

Given the FDA's draft and final HCT/P guidances on minimal manipulation, homologous use, the SSP exception, and adipose, the FDA's basic approach to regulating adipose-derived HCT/Ps can be summarized as follows:

- The FDA will interpret and implement the HCT/P regulations based on the concept of "basic function."
- Basic functions are "well understood" and will be categorized as either structural or cell/nonstructural.
- An HCT/P can have more than one basic function.
- The FDA will evaluate and regulate a multifunction HCT/P as *either* structural *or* nonstructural, but *not both.*
- The FDA's categorization of a multifunction HCT/P as either structural or nonstructural—but not both—will hold for *all* regulatory decisions, regardless of varying functions in the donor or potential applications in the recipients (even when autologous).
- Adipose has various structural and nonstructural components, and serves more than one basic function.
- The FDA will nevertheless evaluate and regulate *all* adipose regenerative products as structural tissue serving only structural basic functions.
- In effect, the FDA has endowed itself with the sole and final authority to:
  - Discount any need to define the "well understood" concept of "basic function or functions";
  - Evaluate a multifunction HCT/P as either structural or nonstructural for all purposes, but never both—regardless of the particular characteristics of that HCT/P's histology, source, or function in the donor or intended uses/therapeutic applications in the recipient (even when autologous); and therefore,
  - Reshape biology to fit its administrative needs.

- Although the Final MM/HU Guidance might indicate that the FDA apparently has final say in determining whether an HCT/P is structural or nonstructural, a court would rely on basic biology because it cannot defer to *factual errors*.

Agency guidance documents are supposed to make it easier for regulated actors to understand and fulfill their regulatory compliance obligations and predict the consequences of noncompliance. When it comes to adipose, however, the Final MM/HU and SSP Guidances' reliance on biological anachronisms in terms of what adipose is and does undermines the agency's important goals of improving regulatory clarity and consistency. The net result is an overall strategy that manages to be fairly aggressive in evaluating and regulating the risks of structural adipose therapies while simultaneously declining to evaluate the benefits and risks of adipose's nonstructural uses, thereby depriving patients of access to the large majority of adipose's basic functions in the body. In comments and public testimony, comparable concerns were raised about the FDA's single functional classification of other multifunction HCT/Ps, including those derived from dermis and amnion.

In terms of the basics of administrative practice, the FDA's insistence on regulating adipose as solely structural jeopardizes the agency's core mission of balancing risks and benefits in order to promote innovation while protecting patients against unsafe and ineffective products. In terms of the basics of administrative law, regulating on the basis of biologic inaccuracies is an increasingly risky strategy given the US Supreme Court's June 2019 decision in *Kisor v. Wilkie*. In *Kisor*, the Court exhorted lower courts to exercise significant caution when reviewing an agency's interpretation of its own regulations, even when the interpretation is a reasonable one. For adipose HCT/Ps, it is hard to see how a solely structural categorization can be deemed reasonable when holistic review of adipose biology shows that a majority of adipose function falls within the Final MM/HU Guidance's definitions that bifurcate functions into those that are nonstructural (i.e., "a metabolic or biochemical functions, such as, hematopoietic, immune, and endocrine functions") as opposed to those that are structural ("to physically support or serve as a barrier or conduit, or connect, cover, or cushion"). For many working in the field, this distinction—particularly given the FDA's resistance to consider evaluating adipose-derived HCT/Ps for their nonstructural properties—is a continuing source of frustration and confusion. In the meantime, however, regulated actors should not take it upon themselves to ignore current regulations and guidance. Unless and until a rule or interpretation is revised, replaced, or rescinded, it should be viewed as having its intended effect.

Examples of § 351 SSP-exempt versus § 361 versus § 351 clinical application of adipose HCT/Ps

Adipose and the § 351 SSP exemption

The final guidance on the Same Surgical Procedure Exception to § 351 premarket approval provides the following examples of when adipose will and will not qualify to avoid § 351 premarket review requirements as provided by 21 CFR § 1271.15(b).

- *Satisfies the § 351 SSP exception*: Adipose tissue is recovered by tumescent liposuction and then centrifuged to remove debris and extracellular fluid.

- Beyond minimal handling, there is no further processing (via enzymatic digestion or mechanical disruption) to isolate stem cells (often referred to as stromal vascular fraction, or SVF), and nothing more is added.
  - Note: the cleansed adipose tissue may be aliquoted, temporarily stored, and injected during a predetermined, limited number of subsequent operations to achieve the desired effect (common for certain kinds of plastic surgery and dermatologic procedures).
- Its use is autologous because it is returned to the donor.
- The adipose tissue remains "such HCT/P" because it retains its original form as a connective tissue composed of clusters of adipocytes and other cells surrounded by a reticular fiber network and interspersed small blood vessels.
- Because this qualifies for the § 351 SSP exception, its use need not be homologous.

- *Fails the § 351 SSP exception*: Adipose tissue is recovered by tumescent liposuction and processed to isolate SVF.
  - SVF is often obtained by enzymatic digestion or mechanical disruption.
  - Methods of cell isolation would typically cause the adipose tissue to no longer be "such HCT/P."
  - This disqualifies the product from regulation under § 351's SSP exception.
  - NOTE: As explained below, the FDA also considers SVF to involve more than minimal manipulation, which disqualifies it for regulation as a § 361 product.
  - Thus, SVF products will need to undergo formal premarket review and meet regulatory requirements for an IND, BLA, and cGMPs.

Adipose and § 361

To qualify for regulation as a § 361 product under 21 CFR § 1271.10 and thereby avoid § 351 premarket approval requirements, an HCT/P must be:

(1) No more than minimally manipulated
- *For structural tissue*: minimal manipulation is processing that does not alter the original relevant characteristics of the tissue relating to the tissue's utility for reconstruction, repair, or replacement.
- *For cells or nonstructural tissues*: minimal manipulation is processing that does not alter the relevant biological characteristics of cells or tissues.

(2) Intended for homologous use (i.e., it will perform the same basic function or functions in the recipient as in the donor, although its eventual location in the recipient may differ from the original locus of harvest in the donor).
(3) Combined with no other articles (other than water, crystalloids, or a sterilizing, preserving, or storage agent that pose no additional safety risks).
(4) Either has no systemic effect or dependence on metabolic activities for its primary function or, if it does, be either autologous, first or second degree allogeneic, or for reproductive use.

The FDA's insistence on evaluating adipose, a complex and multifunctional organ, as a tissue whose basic functions are solely structural significantly restricts an adipose HCT/P's ability to satisfy § 361. This is especially problematic when it comes to § 361's minimal manipulation and homologous use criteria as demonstrated by the following examples, all of which are provided by the FDA in the Final MM/HU Guidance.

Adipose, § 361 and minimal manipulation

- *Fails § 361 owing to more than minimal manipulation* via decellularization:

  A manufacturer removes cells from adipose with the intent of using the remaining decellularized extracellular matrix for a structural purpose.

  - According to the FDA, this qualifies as *more than minimal manipulation* since removing adipocytes, etc., alters adipose's original relevant structural characteristics for providing cushioning and support.
  - On this point, the FDA specifically emphasizes that the original characteristics relevant to structural functions being altered include the removal of adipose's bulk and lipid storage capacity—even though lipid storage is a nonstructural function.
  - In contrast, removing cells from dermis to leave a decellularized extracellular matrix with the intent of using it for a structural purpose does qualify as minimal manipulation since it does not alter the dermal HCT/P's relevant structural characteristics.

- *Fails § 361 owing to more than minimal manipulation* via SVF:
  - A manufacturer recovers adipose tissue by tumescent liposuction. The lipoaspirate is processed through enzymatic digestion or mechanical disruption to isolate cellular components as SVF for the purpose of obtaining adipose-derived stromal/stem cells.
  - According to the FDA, this involves *more than minimal manipulation* because the processing breaks down and eliminates the adipocytes and the surrounding structural components that provide cushioning and support.
  - On this point, the FDA explains that this alters adipose's original relevant "structural" characteristics of the HCT/P relating to its utility for reconstruction, repair, or replacement—even though reconstruction and repair are often accomplished through the nonstructural functions of adipose's component cells.

Adipose, § 361, and homologous use

Because the FDA evaluates adipose as a solely structural tissue for the purpose of § 361, its intended use in the recipient must involve "repair, reconstruction, replacement, or supplementation of adipose tissue" to qualify as a homologous use.

- *Satisfies § 361's requirement of homologous use* because basic function is structural (but still needs to meet remaining requirements for § 361 regulation):
  - Adipose tissue is used to fill voids in the face or hands (e.g., for cosmetic reasons). This is homologous use because providing cushioning and support is a basic function of adipose tissue.

- *Satisfies § 361's requirement of homologous use* because basic function is structural (but again, would still need to meet remaining § 361 requirements).
  - Adipose tissue is used for transplantation into the subcutaneous areas of breast for reconstruction or augmentation procedures. This is homologous use because providing cushioning and support is a basic function of adipose tissue.
  - NOTE: In earlier draft guidance, the FDA characterized this particular application as nonhomologous because adipose cannot restore lactation, which, according to the FDA, is the basic function of the breast.
    - After being roundly criticized for this position, the agency reversed course, but this episode underscores the dangers of a government agency placing itself in charge of defining biology.

- *Fails § 361's requirement of homologous use*: Using adipose to treat a degenerative, inflammatory, or demyelinating disorder.
  - This is generally not considered a homologous use because limiting the autoimmune reaction and promoting remyelination are not basic functions of adipose tissue.
  - According to the FDA, using adipose-derived HCT/Ps to treat "a degenerative, inflammatory, or demyelinating disorder" would generally be considered a nonhomologous use—even though, as a matter of basic biology, adipose's many cell types serve important nonstructural functions, some of which are involved in regulating inflammation, repairing scarring, and regenerating damaged tissue.
  - Currently, any use of an adipose-derived HCT/P for any of adipose's myriad nonstructural functions will automatically be classified as nonhomologous use and disqualify the HCT/P from § 361 oversight.
    - NOTE: A nonstructural application might still be eligible for the § 351 SSP exception, which does not mandate homologous use. If not, the product will need to undergo § 351 premarket review.
- *Fails § 361's requirement of homologous use*: An HCT/P from adipose tissue is used to treat musculoskeletal conditions such as arthritis or tendonitis by regenerating or promoting the regeneration of articular cartilage or tendon.
  - This is generally not considered a homologous use because regenerating or promoting the regeneration of cartilage or tendon is not a basic function of adipose tissue.

The FDA's current enforcement strategy for adipose HCT/Ps

When the FDA announced its comprehensive framework for regulating regenerative medicine in November 2017, it also launched a 3-year period of "enforcement discretion" or lenience in order to encourage developers to seek early agency input on an HCT/Ps status as a § 351 SSP-exempt, § 361, or § 351 product. (FDA's Comprehensive New Policy Approach to Facilitate the Development of Innovative Regenerative Medicine Products to Improve Human Health, 2017.) This period ended in November 2020 (FDA Regenerative Medicine Update: Compliance and Enforcement, December 2018). While in effect, the agency emphasized that enforcement lenience should not be viewed as abstinence. Instead, it would be stepping up its enforcement efforts against those clinics that it deemed to present especially high risks for patients due to the nature of their HCT/P products and/or their mode of administration. Since then, the FDA has proceeded on a clinic-by-clinic basis in conducting inspections and sending out Untitled and Warning Letters. Many have involved adipose products, although umbilical and amniotic HCT/Ps are also common targets.

Common concerns include:

- Cell culture disqualified under § 361 owing to more than minimal manipulation and combination with other articles
- SVF disqualified as § 351 SSP product owing to excessive processing and not remaining "such HCT/P" on reimplantation
- SVF also fails § 361's requirements of minimal manipulation, homologous use, and no combination with other articles
- Moving HCT/P between distinct locations may disqualify product from § 351 SSP exception or § 361
- Deviations from CGTPs and, if § 351 applies, requirements for premarket review with IND, BLA, and CGMPs

Some of those letters elicited assurances that the recipients would review their operations and improve compliance. Others ignored such missives and continue to do business as usual.

When warnings go unheeded, the FDA can sue, as it did in May 2018 in Florida (*United States v. US Stem Cell Clinics, LLC*) and California (*United States. v. California Stem Cell Treatment Center, Inc.*). Both cases involve adipose SVF, albeit in different ways. In each case, the defendants zealously defended their work. Early correspondence and press coverage revealed a variety of defense theories, ranging from the FDA having no authority to regulate cells (it does) or improperly interfering with the practice of medicine (it can). As the litigation progressed, all parties dug more deeply into the fine technicalities of what adipose is and does when subjected to certain forms of processing and used for certain purposes. As of this writing, both cases remain pending at various stages of the trial and appeals process. A final disposition of either case is likely several years away owing to the legal and scientific complexities at issue, with additional delays due to COVID-19.

With or without a pandemic, though, the FDA's basic strategy of initiating complex litigation against one clinic or chain of clinics is extremely inefficient and consequently ineffective. The sad reality is that there are thousands of clinics that are not playing by the rules. Perhaps these actors do not understand or simply do not respect technically complicated rules built on questionable outcomes and imprecise logic. Or perhaps clinic operators understand quite well that the odds of being singled out and sued by the federal government are minimal. For the FDA, the most important "perhaps" might be this: if challenged in the right case in the right way under *Kisor v. Wilkie*, the FDA's current approach to regulating adipose products based on questionable interpretation of biology could collapse as a result of tenuous facts and arguably shaky logic.

In the meantime, those who work with adipose-derived HCT/Ps should play by the FDA's rules. The FDA encourages early and ongoing consultation to ensure that those working with HCT/Ps understand how their products will be regulated. The agency's INTERACT initiative is summarized below. Seeking individualized legal advice from an attorney who is experienced in dealing with the FDA and well-versed in HCTP regulations in particular is typically a worthwhile investment. If unanticipated, later problems almost inevitably carry much larger legal tabs.

Advocating for larger changes to the FDA's current approach to regulating adipose should be nonadversarial and pursued outside the judicial system if at all possible. Regulation is extremely complicated, but litigation is even more so. For any litigant, bringing or defending against a lawsuit is an enormously stressful, time-consuming, and financially expensive endeavor—even if one ultimately prevails.

CBER's INTERACT program

Recognizing that investigating HCT/Ps and other biological products often involve novel issues of safety, rapidly developing manufacturing technologies, and innovative testing methods, the FDA has taken a number of steps to encourage sponsors to seek agency input early in the development process. The **IN**itial **T**argeted **E**ngagement for **R**egulatory **A**dvice on CBER produc**T**s (INTERACT) program encourages sponsors to meet with CBER for an informal consultation in the early stage of development, even before a pre-IND meeting. An INTERACT consultation does not replace a pre-IND meeting. It simply offers a way to improve the efficiency of product development and review by giving the sponsor nonbinding advice as to what will and will not satisfy agency requirements later in the agency's process of reviewing the particular process.

Notably, an INTERACT meeting is not available to discuss generalized, open-ended questions that do not pertain to a particular product. Rather, the sponsor must present a specific investigational product or product derivation strategy to be evaluated in a clinical study. Through an INTERACT consultation (usually a single meeting), CBER representatives can provide *nonbinding*, preliminary advice to:

(1) assist sponsors conducting early product characterization and preclinical proof-of-concept studies
(2) initiate discussion for new delivery devices
(3) inform sponsors about overall early-phase clinical two-trial design elements
(4) identify critical issues or deficiencies for sponsors to address in the development of innovative products

For instance, a pre-IND INTERACT consultation can be helpful when—in the agency's view—an innovative product "introduces new safety concerns due to the unknown safety profiles resulting from the use of complex manufacturing technologies, innovative devices, or cutting-edge testing methodologies." An INTERACT meeting is not indicated when the sponsor has already received agency input or where the product development has advanced to the point of qualifying for a pre-IND meeting.

Additional components of the FDA's comprehensive regulation of regenerative medicine

Anyone working with therapeutic applications of adipose-derived HCT/Ps in a clinical setting must pay careful attention to the previously discussed final guidances on (1) minimal manipulation and homologous use, and (2) § 351's Same Surgical Procedure Exception. While this chapter focuses primarily on the FDA's framework for regulating adipose products as either § 351, § 351-exempt or § 361 HCT/Ps, this is only part of the agency's "comprehensive framework" for regulating regenerative medicine.

Two additional final guidances now complete the picture: one dealing with "Evaluation of *Devices* Used with Regenerative Medicine Advanced Therapies" (which is beyond the scope of this discussion but is nevertheless included in this chapter's references), and another on "Expedited Approval for Regenerative Products." Before passage of the 21st Century Cures Act in December 2016, a regenerative product, like any medical drug or biologic in need of premarket approval, could seek expedited review if it met the FDA's requirements for priority review, accelerated approval, breakthrough therapy designation, and fast track designation—all of which were originally implemented through agency guidance. The Cures Act codified these mechanisms and directed the FDA to create a new accelerated approval mechanism for what is now known as the "regenerative medicine advanced therapy" or "RMAT" designation. See 21 USC § 356(g), as added by § 3033 of the 21st Century Cures Act.

Thus, following enactment of the Cures Act, 21 USC § 356 now recognizes five basic methods for obtaining some form of expedited review:

1. Priority review
2. Accelerated approval
3. Fast track designation
4. Breakthrough therapy designation
5. Regenerative medicine advance therapy (RMAT) designation

Distinguishing between the different mechanisms for expedited review is complicated and often confusing. For instance, fast track designation, breakthrough therapy designation, and RMAT designation are separate and distinct, and they have different (although conceptually related) requirements. A product can receive more than one designation, but this requires submitting separate applications for each one (even if each application includes identical information). Despite their individual and collective complexity, the three different designation programs as well as priority review and accelerated approval mechanisms share certain core objectives and terminology.

Core objectives:

1. All focus primarily, if not exclusively, on products that target a serious or life-threatening condition.
2. All show a potential to provide a significant benefit over available therapies and/or meet an unmet need.
3. All endorse flexibility in basing approval decisions on evidence other than the gold standard of randomized clinical trials.
4. While all permit increased flexibility in the regulatory approval process, none change regulatory approval standards.
5. Thus, all maintain the need for an IND, NDA or BLA at or before seeking expedited review.

Key terms

- **Serious disease or condition**
  - Life threatening;
  - Will become more serious if left untreated; or
  - Has a substantial impact on day-to-day functioning;
  - The condition need not be irreversible if it is persistent or recurrent, but it should be more than short-term or self-limiting.
- **Unmet medical need**
  - Available therapies are not adequate to diagnose or treat. Includes:
    - No available therapy
    - Alternative for those who cannot tolerate available therapy
    - An immediate need for a defined population (i.e., to treat a serious condition with no or limited treatment)
    - A longer-term need for society (e.g., to address the development of resistance to antibacterial drugs)
- **Intended to treat a serious condition**
  - Improving diagnosis
  - Improving outcome
  - Preventing/mitigating a serious treatment-related effect or adverse outcomes
  - Preventing/reducing progression to a more serious condition or advanced stage of the disease
- **Available/existing therapy**
  - Already licensed or approved for use in the United States for the same indication for which the new drug is offered
  - Reflects current standard of care for the specific indication and disease stage for which the product has been developed
- **Demonstrating the potential to meet an unmet need—FDA determines:**
  - Type of supporting evidence
  - Degree of effect or potential based on where the product is in the drug development process and what form of expedited method is requested
- **Surrogate endpoint (SEP)** (discussed in more detail later in this chapter)
  - A marker such as a laboratory measurement, radiographic image, physical sign, or other measure; *and*
  - Is thought to predict clinical benefit, but is *not* a measure of clinical benefit.
- **Intermediate clinical endpoint**
  - Measures a therapeutic effect before an effect on irreversible morbidity and mortality (IMM); *and*
  - Is considered reasonably likely to predict the drug's effect on IMM or other clinical benefit.

- **Clinically significant endpoint**
  - Measures an effect on IMM or on symptoms that represent serious consequences of a disease
  - Can also refer to findings that suggest an effect on IMM or serious symptom

The final guidance on Expedited Programs for Regenerative Medicine Therapies (February 2019) provides the information contained in Table 3, which compares the RMAT and breakthrough therapy designations:

Table 4 provides a more detailed comparison of the objectives and criteria for the FDA's five expedited review mechanisms.

Applying for the RMAT designation

- Submit a written request to CBER
  - Do so at the time of submitting a new IND or an IND amendment.
  - Requests cannot be submitted if the IND is inactive or on clinical hold.
  - Cover letter should indicate inclusion of RMAT request in bold, uppercase letters, e.g.:

TABLE 3 "Comparison of Breakthrough Therapy Designation and Regenerative Medicine Advanced Therapy Designation" from https://www.fda.gov/media/120267/download (Accessed February 2021).

	Breakthrough therapy designation	**Regenerative medicine advanced therapy designation**
Statute	Section 506(a) of the FD&C Act, as added by section 902 of the Food and Drug Administration Safety and Innovation Act of 2012 (FDASIA)	Section 506(g) of the FD&C Act, as added by section 3033 of the 21st Century Cures Act
Qualifying criteria	A drug that is intended to treat a serious condition, *and* preliminary clinical evidence indicates that the drug may demonstrate substantial improvement on a clinically significant endpoint(s) over available therapies	A drug is a regenerative medicine therapy, *and* the drug is intended to treat, modify, reverse, or cure a serious condition, *and* preliminary clinical evidence indicates that the drug has the potential to address unmet medical needs for such disease or condition
Features	• All fast-track designation features, including: – Actions to expedite development and review – Rolling review • Intensive guidance on efficient drug development, beginning as early as phase 1 • Organizational commitment involving senior managers	• All breakthrough therapy designation features, including early interactions to discuss any potential surrogate or intermediate endpoints • Statute addresses potential ways to support accelerated approval and satisfy postapproval requirements
When to submit	With the IND or after and, ideally, no later than the end-of-phase-2 meeting	
FDA response	Within 60 calendar days after receipt of request	
Designation rescission	Designation may be rescinded later in product development if the product no longer meets the designation-specific qualifying criteria	

TABLE 4 Comparison of options for expedited review (Authority: See generally, 21 USC § 356).

	Priority review	Accelerated approval	Fast-track designation	Breakthrough therapy designation	Regenerative advanced therapy designation
Structure	Product designation	Approval pathway	Product designation	Product designation	Product designation
Advantages	FDA prioritizes resources to complete faster review (average 6 months vs 10 months standard review)	Faster review of drugs with measurable promise where standard review would take years to complete	• Expedite development and review • Ongoing interaction with FDA: pre-IND, end-of-phase 1 and 2; discuss study design, dose response, use of biomarkers, extent of safety data • If supported by clinical data at time of BLA/NDA, or efficacy supplement, may be eligible for priority review • If preliminary clinical data of effectiveness, FDA may conduct rolling review of NDA/BLA as completed (rather than at completion of studies)	• Intensive guidance in development process • Rolling review • Organizational and senior FDA commitment • Other efforts to expedite	• Intensive guidance throughout development and approval process
Target #1	Serious condition	Same	Same	Same	Same
OR Target #2	Qualified infectious disease	XXX	Qualified infectious disease	XXX	XXX
OR Target #3	• Drug with priority review voucher • RPD PR (rare pediatric disease priority review)	XXX	XXX	XXX	XXX

Continued

TABLE 4 Comparison of options for expedited review (Authority: See generally, 21 USC § 356)—cont'd

	Priority review	Accelerated approval	Fast-track designation	Breakthrough therapy designation	Regenerative advanced therapy designation
OR Target #4	Supplemental proposed labeling change based on pediatric exclusivity study under 21 USC § 355a	XXX	XXX	XXX	XXX
Purpose re: unmet needs/ clinical benefit	• *Potential for significant improvement* in safety OR effectiveness over available treatments • Increased effectiveness in treatment, prevention, diagnosis. Examples: – Eliminate/ substantially reduce treatment limiting adverse reaction – Improve patient compliance expected to improve serious outcomes – Evidence of safety and effectiveness in new population	*Meaningful advantage* Over available therapies	Potential to address Unmet medical need *Theoretical*	*Potential for substantial improvement* on available treatments	*Potential to meet* Unmet need

Evidence	• Clinical trials • Other scientifically valid information	• Effect on surrogate endpoint reasonably likely to predict clinical benefit or • Effect on clinical intermediate endpoint that can be measured earlier than irreversible morbidity or mortality	• Nonclinical data supporting theoretical, mechanistic, pharmacologic rationale, or evidence of nonclinical activity • If clinical data at time of BLA/NDA or efficacy supplement, may qualify for priority review	• Preliminary clinical evidence (typically phase 1 or 2 trials) of treatment effect that, while not sufficient to establish safety and effectiveness for purpose of approval, may represent substantial improvement (ideally, a "clear advantage") over available therapies on one or more clinically significant endpoints • Should include sufficient # of patients to be credible • Nonclinical info supporting clinical evidence of drug activity	Nonclinical data
Endpoints	Primary/ intermediate/ surrogate endpoints	Clinical or surrogate endpoints		Clinically significant surrogate or intermediate endpoint—sponsor must justify clinical significance	
Timing	• Apply for priority review designation at time of original NDA/BLA/ efficacy supplement filing • Designation assigned at time of application	• Discuss with the FDA review division during drug development • Must submit promotional materials for FDA review before using or distributing	On/after IND submission, preferably before pre-NDA/ pre-BLA meeting; see 21 CFR 312.47(b)(2)	On or after IND submission (ideally no later than an end-of-phase-2 meeting, defined at 21 CFR § 312.47(b)(1))	On/after IND submission

Continued

TABLE 4 Comparison of options for expedited review (Authority: See generally, 21 USC § 356)—cont'd

	Priority review	Accelerated approval	Fast-track designation	Breakthrough therapy designation	Regenerative advanced therapy designation
	• FDA to respond within 60 days of receiving original BLA/NDA or efficacy supplement				
Other	Postapproval confirmatory trials *may* be required	• Postapproval confirmatory trials *required* • Subject to expedited withdrawal of approval	• Postapproval confirmatory trials *may* be required • May be rescinded if qualifications lapse	• Postapproval confirmatory trials *required* • May be rescinded if qualifications lapse	Postapproval confirmatory trials *may* be required

INITIAL [AMENDED] INVESTIGATIONAL NEW DRUG SUBMISSION and REQUEST FOR REGENERATIVE MEDICINE ADVANCED THERAPY DESIGNATION

- Application should include:
  - Concise summary of supporting information (including: 12 FDA's Standard Operating Policy and Procedure (SOPP) 8212, entitled "Management of Breakthrough Therapy Designated Products: Sponsor Interactions and Status Assessment Including Rescinding")
  - A description of the investigational product
  - Rationale for satisfying RMAT definition re: "intended to treat serious disease or condition"
  - Summary of risks and benefits associated with any currently available therapies
  - Description of specific unmet medical need
  - Description of preliminary clinical evidence of product's potential to address unmet need, including:
    - Conditions for product administration,
    - Outcome assessment, and
    - Patient monitoring
  - Description of patients and their outcomes, including
    - Number of patients who have received drug and
    - Design, conduct, and analyses of any clinical investigations.
- **RESPONSE:** CBER will provide a written response no later than 60 calendar days after receiving the RMAT designation request.
  - If the request is not approved, CBER will explain its rationale.
  - An approved RMAT designation is subject to later rescission if, during continued development, the product no longer satisfies RMAT criteria.

Practice note: Expedited approval versus expanded access

Expedited approval methods should not be confused with the FDA's "Expanded Access" program.

- ***Expedited approval*** is just that—approval to market a product sooner than conventional § 351 review would normally permit by pursuant priority review, accelerated approval, fast track designation, breakthrough therapy designation, or regenerative medicine advanced therapy designation.
- ***Expedited approval seeks approval to market*** and use the product for a large population. Even when it targets a smaller group, approval pertains to a general category of patients as opposed to specifically identified individuals.
- ***Expanded access permits immediate treatment*** with an *unapproved* product. Also known as "compassionate use," the process seeks *access* to treatment. It does not request *approval* of that treatment or authorization to investigate that treatment. Expanded access is typically sought on behalf of an individual patient, but it may also be made available to treat larger numbers.

The Cures Act, real-world evidence, surrogate endpoints, and more: Impact on FDA approval of HCT/Ps

With the goal of making the approval process for drugs and biologics more efficient and effective, the 21st Century Cures Act requires the FDA to establish a "program framework" "to evaluate the potential use of RWE." 21 USC §355 g(a). This framework will evaluate a previously approved drug for a new indication, a required postapproval study, or other circumstances as identified by the FDA. The statute simply directs the FDA to permit sponsors to use RWE more often and develop guidelines for doing so. It does not require sponsors to use RWE nor does it obligate the FDA to accept proffered RWE in every circumstance.

Prior to the Cures Act's enactment in December 2016, the FDA had already taken a number of steps to improve the speed and efficiency of every phase of the product approval life cycle, including initial research and development, product review and approval for marketing, and postapproval oversight of product performance. The agency was also interested in increasing patient input and expanding patient access to innovative products. Thus, by passing the Cures Act, Congress was requiring the FDA to continue what it had already begun, including providing assistance to product sponsors in developing novel trial designs and determining how to use RWE in evaluating whether a product is safe and effective for its intended use and making other regulatory decisions. However, the Cures Act does not reduce or otherwise alter the FDA's longstanding standards for determining whether a sponsor's submitted evidence is adequate to support a particular regulatory decision. RWE, innovative trial design designs, and the like have value, but the RCT continues to be the gold standard or best evidence when seeking to establish a product's basic safety and efficacy. The Cures Act simply codifies what had already been emerging at the FDA: a hierarchy of different kinds of evidence, with RCT at the top, followed by various kinds of evidence derived from sources other than RCTs and originally generated for purposes other than premarket approval. See 21 USC §§ 355g, 356.

As of this writing, the FDA is still in the process of finalizing guidance to regularize the submission of RWE and real-world data (RWD) to support regulatory decisions (e.g., applications for an IND/NDA/BLA, approval of a new indication, etc.). However, given the FDA's obligations under the Cures Act and its own pre-Cures initiatives, the agency has already demonstrated its willingness to consider RWE and the results of innovative clinical studies when assessing a product's safety and effectiveness for a particular intended use.

Clarifying terms

The FDA has been equally determined to clarify terms to enable developers to be on the same page as the agency with regard to compliance and enforcement. Defining terminology is critical because certain core terms are often understood to be synonymous and interchangeable when they are actually quite different from each other. The most commonly confused pairings are (a) biomarkers and SEPs, and (b) RWD and RWE. As defined by the FDA, each has a distinct meaning; none are interchangeable, but all are interrelated.

Biomarkers versus surrogate endpoints

Although often used synonymously, *biomarkers are not SEPs* although SEPs frequently depend on biomarkers.

Biomarkers: *A biomarker indicates a health characteristic* (usually biologic or physiologic in nature) in terms of its presence, absence, or risk. There is nothing new about biomarkers since RCTs typically evaluate direct relationships between a biomarker and the primary endpoint of interest, i.e., the disease or condition being studied.

FDA-qualified biomarkers: The FDA recognizes seven types of biomarkers based on the type of health characteristic indicated:

1. *Susceptibility* biomarkers (where presence of disease is unknown) indicate the potential for, or risk of, developing a specific disease or condition when that disease or condition is not known or apparent in the individual.
2. *Diagnostic* biomarkers (where presence of disease is in question) detect or confirm the presence of disease or condition, or a subtype of the same.
3. *Prognostic* biomarkers (where presence of disease is known) can identify the likelihood of the presence, progress, or recurrence of a specific clinical event.
4. *Predictive biomarkers* (a type of outcome) can signal an increased risk of a favorable or unfavorable outcome following exposure to a specific agent or event.
5. *Pharmacodynamic or response biomarkers* evidence a biological response to an exposure. It does not establish a product or treatment's effectiveness but may be helpful with proof of concept.
6. *Monitoring biomarkers* (where presence of disease is known) are captured by repeated measurements to assess status of disease, condition, or evidence of exposure.
7. *Safety biomarkers* reflect serial measurements that attempt to assess the likelihood, presence, or extent of toxicity other than an adverse event.

Surrogate endpoints: *A SEP indicates a correlation (existing or potential) between a biomarker and specific clinical response to a treatment,* intervention, or exposure of interest (hereafter, collectively referred to as "treatment"). The strength of this correlation can significantly affect a SEP's evidentiary value. The utility of a SEP in obtaining product approval will depend on the strength of this relationship, the stage of product review, and the availability of alternative evidence.

FDA-qualified surrogate endpoints: The FDA recognizes three types of SEPs based on the strength of information establishing the SEP's ability to predict a specific clinical response.

(a) *Validated SEP*: *Strong evidence of ability to predict* based on a sufficiently demonstrated correlation between the treatment and a specific clinical benefit or adverse reaction.
(b) *Reasonably likely SEP*: *Reasonable to expect a predictive correlation* between a treatment and specific clinical benefit or adverse reaction due to a strong mechanistic or epidemiologic supporting rationale.
(c) *Candidate SEP*: *Remains under evaluation for a sufficient correlation indicating ability to predict* a specific clinical response to a treatment.

Real-world evidence versus real-world data

Even before the 2016 Cures Act, the FDA had acknowledged that, in an age of big data and instantaneous quantitative analysis, modernizing the product approval process must reach beyond the constraints of the randomized clinical trial to other "real world" sources of information. This helps to "close the evidence gap" between the kind of information traditionally used for regulatory decisions and the wide variety and long-available information used by the medical community, payers, etc. Doing so promotes the FDA's objectives of improving patient voice and moving toward a process of continuous evaluation of product performance, from initial development to broad clinical use.

The terms "real world data" (RWD) and "real world evidence" (RWE) are not synonymous. In essence, RWE represents RWD subject to analysis and evaluation. RWD can be valuable if collected systematically and routinely from sources and for purposes other than RCTs. RWD typically record patient health status or delivery/consumption of health care. Sources of RWD include electronic health records, medical claims and billing records, product and disease registries, mobile device/apps data, and patient-generated data including those generated at home or other decentralized settings. RWE is clinical evidence derived from analyzing RWD with the goal of understanding the uses or potential risks and benefits of a particular medical product, treatment, or other exposure.

Using RWD and RWE in support of regulatory decisions

The FDA's willingness to consider RWD and RWE reflects its recognition that, while still the evidentiary gold standard, RCTs as the sole source of information has its limitations. For example:

- For legitimate and important reasons, RCTs employ tight controls of sample selection and product use that may end up excluding what RWE can provide: significant information about the product's performance in day-to-day clinical practice.
- RWE can also facilitate evaluation of products that are difficult to study through RCTs owing to low prevalence, ethical constraints, or other obstacles.
- Conversely, evaluating large sets of RWD can provide information on a wider patient population that would be difficult and potentially impossible to obtain if relying solely on RCTs.

Other than circumstances where collecting RCT data is infeasible, RWE will not replace the use of clinical trials; it will instead expand, supplement, and literally inform the FDA's regulatory decisions. On its own, RWD can be used to improve the efficiency of RCTs with respect to:

- Generating hypotheses for testing
- Identifying drug development tools (including biomarker identification)
- Assessing trial feasibility by examining the impact of planned inclusion/exclusion criteria
- Informing prior probability distributions in Bayesian statistical models
- Identifying prognostic indicators or patient baseline characteristics for enrichment or stratification
- Assembling geographically distributed research cohorts (e.g., in drug development for rare diseases or targeted therapeutics)

If systematically collected, RWD and RWE may be sufficiently relevant and reliable to support a wide range of regulatory determinations and objectives, including:

- Control arm for pivotal clinical study
- Effectiveness or safety for a new product approval (e.g., collecting information about effectiveness or safety outcomes from an RWD source in a randomized clinical trial)

- Developing objective performance criteria (OPC) and performance goals (PGs)
- Labeling changes for an approved product, including:
  - Additional or modified indications
  - Changes in dose, dose regimen, or routes of administration
  - Use in a new population
  - Adding comparative effectiveness information
  - Adding safety information
  - Feasibility studies
- Additional postmarketing evidence of safety/effectiveness
- Natural history studies for developing clinical outcomes assessments or biomarkers
- Postmarketing surveillance and adverse event reporting

While the FDA has proactively encouraged the use of RWD and RWE for certain purposes, it has moved slowly and deliberately in evaluating whether and how to use RWE during initial premarket review and approval of a product as safe and effective for its intended use. As compared with RCTs, RWE may offer the benefit of more closely reflecting reality, but reality is messy. Consequently, despite its promise, RWE carries risks that RCTs are designed to avoid. For example, the validity and reliability of RWE may be compromised by irregular use, inaccurate measurement, improper or inconsistent methods of collection, and known and unknown biases and confounders. Such information may be more likely to impair rather than improve the decisions of patients and their providers.

Given the broad spectrum of potential RWD and rapid advances in how to generate and evaluate it, the agency currently has no plans to describe or require specific methods for developing RWD into RWE. It will instead conduct a case-by-case evaluation of whether the proffered RWE and underlying RWD are suitable and sufficient to support the specific regulatory decision in that particular case. Thus, a product's sponsor should seek early input from the FDA concerning study design and protocols, types of RWD, methods of generating and assessing RWE to assess whether that evidence will be sufficiently robust, relevant, and reliable given the regulatory question presented. Early consultation does not guarantee eventual approval, but it can certainly optimize the conditions for success.

The FDA is still in the process of developing guidance on the use of RWD and RWE for approval decisions involving HCT/Ps and other biologics. The FDA's August 2017 Final Guidance on "Use of Real-World Evidence to Support Regulatory Decision-Making for Medical Devices" gives a good picture of what the FDA will look for when evaluating the quality, integrity, and reliability of RWD and/or RWE for cell and tissue products.

1. *Methods of data accrual and collection*: To date, there are no hard and fast criteria, but evidence of data assurance, quality, and reliability may be found in:
   - Operational manuals
   - Qualified personnel
   - Site training and support
   - Common data capture form
   - Common definitional framework
   - Specified sources and technical methods for data element capture
   - Patient selection and enrollment criteria to minimize bias and ensure representation of real-world population

- Institutional review board (IRB)-reviewed and FDA-compliant methods for protecting patient privacy and obtaining informed consent

2. *Common problems with data quality, consistency, accuracy, and reliability*:
- Missing data
- Variable methods for extracting, recording, and tracking data
- Limited information technology capabilities
- Data captured by clinicians
- Comorbidities
- Different clinical endpoints
- Different methods of collection, reliability, recording precision in electronic health records (EHRs); lack of interoperability
- Data identifiable? Anonymized?
- Potential impact of, or on, informed consent requirements

Protecting data integrity is nothing new, but it is a serious concern when dealing with RWD, RWE, SEPS, and innovative or blended study designs. Problems with data integrity can do more than compromise the data's utility. The FDA will view them as potential signs of larger problems with the design and conduct of the overall study and adherence to CGMPs in general. Thus, developers should consult 21 CFR Part 211 to ensure compliance with cGMPs, and seek additional FDA input as needed, especially when it comes to:

- Record-keeping
- Installation and operational qualifications for computer systems
- Areas needing restricted and secure access
- Backup systems
- Protocols for managing metadata, ensuring audit trails, validating workflow, and much more

3. *Enforcement*: A major share of CGMP warning letters are triggered by data integrity issues, particularly with regard to:
- Inadequate controls over access to computerized systems
- Noncontemporaneous record-keeping and deletion
- Falsification, alteration, or other forms of intentional or inadvertent data manipulation or contamination

As with any enforcement matter, the FDA can use an array of increasingly severe methods to enforce data integrity and other CGMP requirements, ranging from inspections, warning letters, and regulatory meetings to injunctions, fines, seizures, consent decrees, and, in extreme cases, criminal prosecution.

Conclusion

Regulating a multifunction HCT/P product as *either* structural *or* nonstructural for *every* application of that product (but *never* both), regardless of its potentially varied functions in the donor and uses in the recipient, is frustratingly limiting. A more logical approach would determine the function of interest based on its intended use in the recipient, compare it with that function as originally performed in the donor, and reason from there. For adipose, it is also difficult to understand why, if all adipose-derived HCT/Ps must be categorized as either structural or nonstructural for all purposes, the singular characterization of structural was chosen even though adipose is much more important for its nonstructural activities. From a biological standpoint, this is anachronistic or at least confusing, and therefore, from a legal standpoint, this is worrisome. The FDA should be mindful of what a court cannot overlook: the agency may have jurisdiction to regulate biology, but it has no authority to rewrite it.

One possible rationale for treating adipose as solely structural is that, as a woefully underresourced agency, the FDA needs every ounce of administrative efficiency and convenience it can muster as it confronts the unmanageable and growing problem of what it has termed "unscrupulous" clinics that are using easily harvested adipose to market high-risk therapies to unsuspecting patients [39]. Viewed in this context, perhaps such a rigid approach to regulating adipose makes sense, even if it glosses over basic biology and automatically disqualifies all nonstructural adipose from § 361 oversight.

Regulating on the basis of biological inaccuracies does nothing to reduce risks and serves only to heighten them. Foreclosing meaningful assessment of risk for recognized and emerging therapies that rely on adipose's predominantly nonstructural functions does not protect patients in need of better treatment options. It will not discourage dubious clinics from offering dubious products backed by dubious promises. The US Supreme Court has repeatedly emphasized—as it recently did in *Kisor v. Wilkie* (June 2019)—that regulatory decisions rooted in factual inaccuracies, "convenient litigating positions," and "posthoc rationalizations" cannot stand. The unfortunate outcome is that, instead of exercising more leverage over rogue clinics, the FDA is simply giving those entities more to work with should they get sued.

Hopefully, the FDA will soon acknowledge that relying on selective interpretations of biology to curtail irresponsible clinics is not working and, indeed, can never do enough. The FDA should instead (a) revise its approach to the complex adipose organ to accurately reflect what its tissue and cell components actually are and do, and (b) work with other state and federal agencies (particularly those with jurisdiction over unfair and deceptive trade practices such as the Federal Trade Commission) to build a comprehensive oversight and enforcement strategy based on shared resources and complementary expertise.

In the meantime, those who are subject to existing regulations should take their compliance obligations seriously—whether or not they agree with the FDA's choices and underlying rationale. Questions or concerns regarding the applicability, meaning, and requirements of the complex HCT/P framework should seek early input from the FDA or an experienced attorney—or better yet, both.

References

[1] L.P. Gartner, Textbook of Histology, fourth ed., Elsevier, Philadelphia, PA, 2016.

[2] W.K. Ovalle, P.C. Nahirney, Netter's Essential Histology, Saunders, Philadelphia, 2013.

[3] W.K. Ovalle, P.C. Nahirney, Connective tissue, in: Netter's Essential Histology, Saunders, Philadelphia, 2013, pp. 51–70.

[4] L.P. Gartner, J.L. Hiatt, Connective tissue, in: Concise Histology, Saunders, Philadelphia, 2011, pp. 62–73.

[5] T.J. Bartness, et al., Neural innervation of white adipose tissue and the control of lipolysis, Front. Neuroendocrinol. 35 (2014) 473–493.
[6] F. Kreier, H.P. Sauerwein, R.M. Buijs, Selective parasympathetic innervation of subcutaneous and intra-abdominal fat: functional implications, J. Clin. Invest. (2002) 1243–1250.
[7] A.L. Kierszenbaum, L.L. Tres, Connective tissue, in: Histology and Cell Biology: An Introduction to Pathology, Elsevier, Philadelphia, 2019, pp. 135–175.
[8] B. Young, G. O'Dowd, P. Woodford, Supporting/connective tissues, in: Wheater's Functional Histology, Churchill Livingstone, Philadelphia, 2014, pp. 65–81.
[9] A. Naylor, A. Filer, C. Buckley, The role of stormal cells in the persistence of chronic inflammation, Clin. Exp. Immunol. 171 (2012) 30–35.
[10] B. Ghesquiere, et al., Metabolism of stromal and immune cells in health and disease, Nature 511 (7508) (2014) 167–176.
[11] M.F. White, K.D. Copps, The mechanisms of insulin action, in: Endocrinology: Adult and Pediatric, Saunders, Philadelphia, 2016, pp. 556–585.
[12] J.E. Hall, Lipid metabolism, in: Guyton and Hall Textbook of Medical Physiology, Elsevier, Philadelphia, 2016, pp. 863–874.
[13] S. Standring, Integrating cells into tissues, in: Gray's Anatomy: The Anatomical Basis of Clinical Practice, Elsevier, Philadelphia, 2016, pp. 28–41.
[14] C. Merlotti, et al., Subcutaneous fat loss is greater than visceral fat loss with diet and exercise, weight-loss promoting drugs and bariatric surgery: a critical review and meta-analysis, Int. J. Obes. 41 (2017) 672–682.
[15] T. Chaston, J. Dixon, Factors associated with percent change in visceral versus subcutaneous abdominal fat during weight loss: findings from a systematic review, Int. J. Obes. 32 (2008) 619–628.
[16] M.M. Ibrahim, Subcutaneous and visceral adipose tissue: structural and functional differences, Obes. Rev. 11 (2010) 11–18.
[17] M. Lafontan, J. Girard, Impact of visceral adipose tissue on liver metabolism part I: heterogeneity of adipose tissue and functional properties of visceral adipose tissue, Diabetes Metab. 34 (2008) 317–327.
[18] S.R. Smith, et al., Contributions of total body fat, abdominal subcutaneous adipose tissue departments, and visceral adipose tissue to the metabolic complications of obesity, Metabolism 4 (2001) 425–435.
[19] P. Trayhurn, J. Beattie, Physiological role of adipose tissue: white adipose tissue as an endocrine and secretory organ, Proc. Nutr. Soc. 60 (3) (2001) 329–339.
[20] M. Poddar, Y. Chetty, V.T. Chetty, How does obesity affect the endocrine system? A narrative review, Clin. Obes. 7 (2016) 136–144.
[21] A. Rodriguez, et al., Revisiting the adipocyte: a model for integration of cytokine signaling in the regulation of energy metabolism, Am. J. Phsyiol. Endocrinol. Metab. 309 (8) (2015) E691–E714.
[22] P. Huebbe, G. Rimbach, Evolution of human apolipoprotein E (APOE) isoforms: gene structure, protein function and interaction with dietary factors, Ageing Res. Rev. 37 (2017) 146–161.
[23] R.J. Sulston, W.P. Cawthorn, Bone marrow adipose tissue as an endocrine organ: close to the bone? Horm. Mol. Biol. Clin. Invest. 28 (1) (2016) 21–38.
[24] F. Villarroya, et al., Brown adipose tissue as a secretory organ, Nat. Rev. Endocrinol. 13 (2017) 26–35.
[25] R.W. Grant, V.D. Dixit, Adipose tissue as an immunological organ, Obesity 23 (3) (2015) 512–518.
[26] Z.L. Sebo, M.S. Rodeheffer, Assembling the adipose organ: adipocyte lineage segregation and adipogenesis in vivo, Development 146 (7) (2019) 1–11.
[27] C. Bouch, Anaesthesia for the obese patient, in: Smith and Aitkenhead's Textbook of Anaesthesia, Elseveir, Philadelphia, 2019, pp. 656–665.
[28] W.E. Kim, Diseases of subcutaneous tissue, in: Nelson Textbook of Pediatrics, Elsevier, Philadelphia, 2020, pp. 3531–3535.
[29] C.R. Kahn, H. Ferris, B. O'neill, Pathophysiology of type 2 diabetes mellitus, in: Williams Textbook of Endocrinology, Elsevier, Philadelphia, 2016, pp. 1349–1370.
[30] A. Garg, Lipodystrophies, Am. J. Med. 108 (2000) 143–152.
[31] M.A. Tsoukas, C.S. Mantzoros, Lipodystrophy syndromes, in: Endocrinology: Adult and Pediatric, Saunders, Philadelphia, 2016, pp. 648–661.
[32] B. Akinci, A. Garg, Natural history of congenital generalized lipodystrophy: a nationwide study from Turkey, J. Clin. Endocrinol. Metab. 101 (7) (2016) 2759–2767.
[33] S. Tan, M. Tang, H. Tey, Lipodystrophies, in: Dermatology, Elsevier, Philadelphia, 2018, pp. 1758–1774.

[34] I. Hussain, N. Patni, A. Garg, Lipodystrophies, dyslipidaemias and atherosclerotic cardiovascular disease, Pathology 51 (2) (2019) 202–212.
[35] J. Lima, et al., Clinical and laboratory data of a large series of patients with congenital generalized lipodystrophy, Diabetol. Metab. Syndr. 8 (23) (2016) 1–7.
[36] P. Smith, et al., Lipodystrophy, pancreatitis, and eosinophilia, Gut 16 (1975) 230–234.
[37] B. Luzar, E. Calonje, Inflammatory disease of the subcutaneous fat, in: McKee's Pathology of the Skin With Clinical Correlations, Elsevier, 2020, pp. 351–388.
[38] P. Marks, S. Gottlieb, Balancing safety and innovation for cell-based regenerative medicine, New Engl. J. Med. 378 (10) (2018) 954–959.
[39] FDA, Statement from FDA Commissioner Scott Gottlieb, M.D. on the FDA's new policy steps and enforcement efforts to ensure proper oversight of stem cell therapies and regenerative medicine (August 28), FDA, 2017. https://www.fda.gov/news-events/press-announcements/statement-fda-commissioner-scott-gottlieb-md-fdas-new-policy-steps-and-enforcement-efforts-ensure.

Further reading

Legal

Public Health Service Act, 42 U.S.·C § 262, et seq. (2019). https://www.govinfo.gov/app/details/USCODE-2010-title42/USCODE-2010-title42-chap6A-subchapII-partF-subpart1-sec262.
Food, Drug and Cosmetics Act, 21 U.S.C. § 321, et seq. (2018). https://uscode.house.gov/browse/prelim@title21&edition=prelim.
21st Century Cures Act – Title III, Pub. L. 114–255 (2016). https://www.congress.gov/114/plaws/publ255/PLAW-114publ255.pdf.
Administrative Procedure Act, 5 U.S.C. § 553 ("Rulemaking"). https://www.govinfo.gov/content/pkg/USCODE-2011-title5/pdf/USCODE-2011-title5-partI-chap5-subchapII-sec553.pdf.
Code of Federal Regulations Part 1271: Human Cells, Tissues, and Cellular and Tissue-Based Products. www.govregs.com/regulations/title21_chapterI_part1271.
C.F.R. Part 211: Current Good Manufacturing Practice for Finished Pharmaceuticals. https://www.accessdata.fda.gov/scripts/cdrh/cfdocs/cfcfr/CFRSearch.cfm?CFRPart=211.
Kisor v. Wilkie, 139 S. Ct. 2400, 2413–2414 (2019). https://www.supremecourt.gov/opinions/18pdf/18-15_9p6b.pdf.
Auer v. Robbins, 519 U.S. 452 (1997). https://www.oyez.org/cases/1996/95-897.
Chevron U.S.A., Inc. v. Natural Resources Defense Council, Inc., 467 U.S. 837 (1984). https://www.oyez.org/cases/1983/82-1005.

FDA Publications: HCT/P and/or Adipose Specific

U.S. Food & Drug Admin., Regulatory Considerations for Human Cells, Tissues, and Cellular and Tissue-Based Products: Minimal Manipulation and Homologous Use Guidance for Industry and Food and Drug Administration Staff (July 2020) (*Final MM/HU Guidance*). https://www.fda.gov/media/109176/download.
U.S. Food & Drug Admin., Expedited Programs for Regenerative Medicine Therapies for Serious Conditions Guidance for Industry (February 2019). https://www.fda.gov/media/120267/download.
U.S. Food & Drug Admin., Same Surgical Procedure Exception: Questions and Answers Regarding the Scope of the Exception: Guidance for Industry (November 2017) (*Final SSP Guidance*). https://www.fda.gov/media/89920/download.
U.S. Food & Drug Admin., Regenerative Medicine Update: Information for Industry on FDA's Compliance and Enforcement Policy Regarding Certain Regulatory Requirements (December 2018). https://www.fda.gov/vaccines-blood-biologics/cellular-gene-therapy-products/regenerative-medicine-update-information-industry-fdas-compliance-and-enforcement-policy-regarding.
Statement by FDA Commissioner Scott Gottlieb, M.D., and Biologics Center Director Peter Marks, M.D., Ph.D. on FDA's Continued Efforts to Stop Stem Cell Clinics and Manufacturers From Marketing Unapproved Products That Put Patients at Risk, (April 19, 2019). https://www.fda.gov/news-events/press-announcements/statement-fda-commissioner-scott-gottlieb-md-and-biologics-center-director-peter-marks-md-phd-fdas.

U.S. Food & Drug Admin., Statement From FDA Commissioner Scott Gottlieb, M.D. on FDA's Comprehensive New Policy Approach to Facilitating the Development of Innovative Regenerative Medicine Products to Improve Human Health (November 16, 2017). https://www.fda.gov/newsevents/newsroom/pressannouncements/ucm585342.htm.

Bauer, S.R., as quoted in U.S. Food & Drug Admin., Consumer Update: Adult Stem Cell Research Shows Promise (April 2104). https://www.fda.gov/forconsumers/consumerupdates/ucm393030.htm.

FDA Publications: General Information

U.S. Food & Drug Admin. SOPP 8212: Management of Breakthrough Therapy-Designated Products: Sponsor Interactions and Status Assessment Including Rescinding (June 2016). https://www.fda.gov/downloads/BiologicsBloodVaccines/GuidanceComplianceRegulatoryInformation/ProceduresSOPPs/UCM506016.pdf.

U.S. Food & Drug Admin., Laws, Regulations, Policies and Procedures for Drug Applications (December 2014). www.fda.gov/drugs/development-approval-process-drugs/laws-regulations-policies-and-procedures-drug-applications.

U.S. Food & Drug Admin., Guidance, Compliance & Regulatory Information (Biologics). www.fda.gov/vaccines-blood-biologics/guidance-compliance-regulatory-information-biologics.

U.S. Food & Drug Admin., Investigational New Drug Application (October 2017). www.fda.gov/drugs/types-applications/investigational-new-drug-ind-application.

U.S. Food & Drug Admin., Instructions for Filling Out Form FDA 356h – Application to Market a New or Abbreviated New Drug or Biologic for Human Use. www.fda.gov/media/84223/download.

U.S. Food & Drug Admin., Tissue and Tissue Products (July 2019). www.fda.gov/vaccines-blood-biologics/tissue-tissue-products.

U.S. Food & Drug Admin., Guidance For Industry Expedited Programs for Serious Conditions – Drugs and Biologics (May 2014). https://www.fda.gov/media/86377/download.

U.S. Food & Drug Admin., INTERACT Meetings (Initial Targeted Engagement for Regulatory Advice on CBER products) (May 2019). https://www.fda.gov/vaccines-blood-biologics/industry-biologics/interact-meetings-initial-targeted-engagement-regulatory-advice-cber-products.

U.S. Food & Drug Admin., Submitting Documents Using Real-World Data and Real-World Evidence to FDA for Drugs and Biologics: DRAFT Guidance for Industry (May 2019). https://www.fda.gov/media/124795/download.

U.S. Food & Drug Admin., Framework for FDA's Real-World Evidence Program (May 2018). https://www.fda.gov/media/120060/download.

U.S. Food & Drug Admin., Use of Real-World Evidence to Support Regulatory Decision-Making for Medical Devices: Guidance for Industry and Food and Drug Administration Staff (August 2017). https://www.fda.gov/media/99447/download.

U.S. Food & Drug Admin., Speech: Gottlieb, S. Remarks to the National Academy of Sciences on the Impact of Real World Evidence on Medical Product Development (September 26, 2017). https://www.fda.gov/news-events/speeches-fda-officials/remarks-national-academy-sciences-impact-real-world-evidence-medical-product-development-09262017.

PART 4

Engineering with adipose stem cell

CHAPTER

15

Biomaterial control of adipose-derived stem/stromal cell differentiation

John Walker[a] *and Lauren Flynn*[a,b]

[a]Department of Anatomy and Cell Biology, Schulich School of Medicine and Dentistry, The University of Western Ontario, London, ON, Canada [b]Department of Chemical and Biochemical Engineering, Thompson Engineering Building, The University of Western Ontario, London, ON, Canada

Part 1: General introduction

In this chapter, we will begin with a brief introduction to the field of biomaterials and the benefits of combining adipose-derived stem/stromal cells (ASCs) with biomaterials. The remainder of the chapter will focus on two key application areas that link these fields: (i) the use of mesenchymal stem/stromal cells (MSCs) as in vitro tools to assess the inductive properties of biomaterials on cell differentiation and (ii) the use of biomaterials as delivery systems to improve MSC retention and tissue regeneration in vivo (Fig. 1). Discussion of the first aspect will highlight the development of biomaterial scaffolds and controlled delivery systems that have been explored as platforms to direct MSC differentiation toward the osteogenic, chondrogenic, or adipogenic lineages. For the latter, we will focus specifically on strategies involving ASCs. However, as much of the collective understanding of the inductive properties of biomaterials on cell differentiation derives from studies using a variety MSC populations, we will highlight key examples involving other MSC sources where appropriate.

Introduction to biomaterials

The definition of biomaterials given by the American National Institutes of Health (NIH) is as follows [1]:

Scientific Principles of Adipose Stem Cells
https://doi.org/10.1016/B978-0-12-819376-1.00013-5

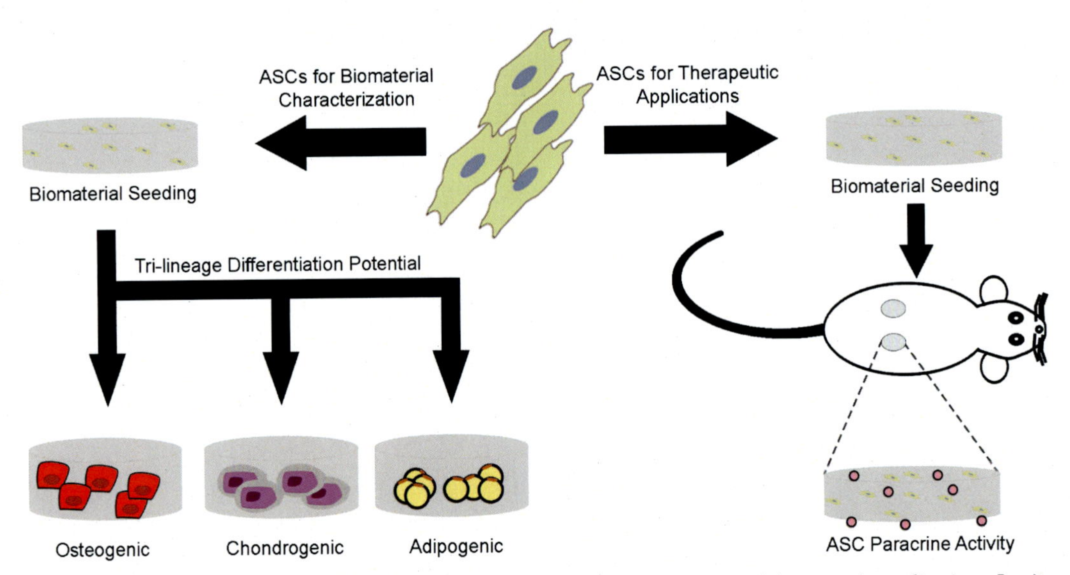

FIG. 1 Integration of ASCs in the development of biomaterials for cell culture and therapeutic applications. In vitro testing with ASCs can provide valuable information on the capacity of biomaterial scaffolds to direct the lineage-specific differentiation of MSCs or other progenitor cell populations. Further, biomaterials can be used to develop tissue-specific cell delivery platforms to enhance in vivo cell retention, survival, and proregenerative paracrine functionality.

> Any matter, surface, or construct that interacts with biological systems. Biomaterials can be derived from nature or synthesized in the laboratory using metallic components, polymers, ceramics, or composite materials. Medical devices made of biomaterials are often used to replace or augment a natural function.

This definition reflects the broad nature of biomaterials and their diverse range of forms and functions. For this chapter, we will focus on two main strategies that apply biomaterials for the differentiation of ASCs: (i) scaffold-based biomaterials that mimic the extracellular matrix (ECM), and (ii) biomaterials designed for the controlled release of bioactive payloads (Fig. 2). The key difference between these two approaches is that ECM mimetics are designed to provide structural support to seeded and/or infiltrating cell populations, while biomaterials designed for controlled release focus on the localized delivery of soluble factors over prolonged periods. However, these designs are not mutually exclusive, and many biomaterials integrate features associated with both classifications. Regardless, at their core, these categories represent two distinct mechanisms through which exogenous materials can modulate cell function.

Historically, many biomaterials have been designed as potential therapeutics for in vivo applications in tissue engineering and regenerative medicine, which will be the main focus of this chapter. However, another growing area of interest is in the use of biomaterials to develop improved 3D culture models that more closely replicate in vivo environments for therapeutic testing [2–4]. It is also worth noting that a key consideration for developing biomaterials is the dichotomy between complexity and cost effectiveness [5]. While increasing biomaterial complexity may serve to better recapitulate the complex microenvironment

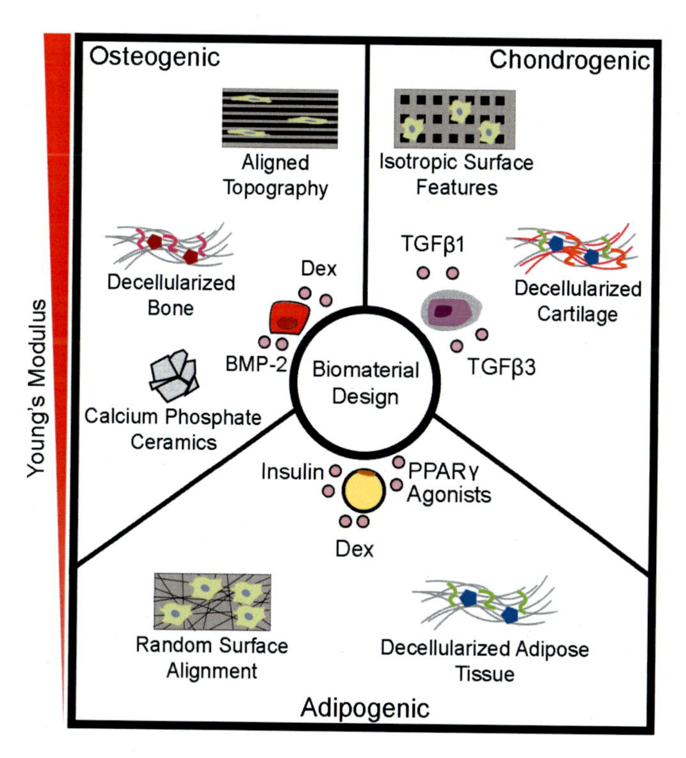

FIG. 2 Biomaterial properties for guided differentiation. Directed differentiation of MSCs can be achieved by altering the biochemical, mechanical, and structural properties of biomaterials. Diverse structural and soluble biochemical cues have proven to be effective at stimulating differentiation. Modifying the physical properties of biomaterials to mimic those in the microenvironment of native tissues can further direct MSC differentiation.

within native tissues, the cost of making such systems may limit their utility. In developing biomaterials, most strategies can be classified as either "bottom-up" approaches that apply simple biomaterials created from individual polymers (natural or synthetic) or inorganic materials as a base to which other bioactive components can be added, or alternatively "top-down" strategies that use methods such as tissue decellularization to create cell-free complex ECM systems [6–8]. In addition, the complexity of any biomaterial system can be increased by incorporating a cellular component to create responsive cellular biomaterials that may be applied as therapeutics. These concepts for biomaterial design will be further discussed throughout this chapter.

Mechanisms of ASC-mediated regeneration and their role in biomaterials design

Broadly speaking, MSCs can be applied in either "cell-replacement" or "cell-empowerment" strategies to promote tissue regeneration [9]. For example, tissue engineering approaches seek to harness the capacity of MSCs to differentiate into more specialized cell types and replace cell populations that have been damaged or lost owing to injury or disease [9, 10]. In contrast, cell-empowerment approaches strive to apply MSCs more transiently, exploiting their ability to secrete exosomes, cytokines, and growth factors, which can function in a paracrine manner to direct the response of host cell populations and help to establish a microenvironment within damaged tissues that induces regeneration [9, 10]. Limitations to both approaches include the poor retention and survival of cell populations injected into injured tissues, as well as a lack of understanding of how to effectively control differentiation and proregenerative factor production. Regardless of the mechanisms of regeneration, biomaterials offer the potential to provide a cell-supportive environment that directs cell function both in vitro and in vivo.

MSC differentiation has classically been studied by culturing cells on tissue culture polystyrene (TCPS) in the presence of potent soluble factors specific for each lineage. Although the differentiation cocktails employed can vary between groups, the overall goal is to stimulate expression and activation of lineage-specific transcription factors to promote differentiation. While multiple transcription factors regulate the differentiation of ASCs toward a specific lineage, the following are often targeted as early regulators of these processes: runt-related transcription factor-2 (Runx2) for osteogenic differentiation; SRY-related HMG-box (Sox9) for chondrogenic differentiation; and peroxisome proliferator-activated receptor-γ (PPARγ) for adipogenic differentiation [11].

Despite their capacity to be induced toward the osteogenic, chondrogenic, and adipogenic lineages in culture, the in vivo relevance of the trilineage differentiation potential possessed by MSCs is debated. A pivotal study by Crisan et al. identified the characteristic properties of MSCs, including trilineage differentiation capacity, in isolated and cultured perivascular cells from several organs [12]. This finding led to the attractive hypothesis that endogenous MSC populations exist within the perivascular niche of many organs, ready to initiate tissue repair upon injury. More recently, this has been challenged using lineage tracing strategies to monitor these populations in vivo over time [13]. Specifically, Guimarães-Camboa et al. identified that perivascular cells exhibited characteristic MSC properties upon isolation and expansion, yet they remained as perivascular cells in vivo following induced cardiac and skeletal muscle fibrosis, as well as in aged animals [13]. While there may be endogenous MSC populations that reside outside of the perivascular niche, this study suggests that cell populations can gain unnatural plasticity through culture. Thus, trilineage differentiation capacity, which is used to characterize MSCs in vitro [14], may not actually be a property associated with endogenous populations. Moreover, a growing body of evidence from studies applying MSCs in vivo in cell-based therapies for a range of applications suggests that there is minimal long-term engraftment or differentiation observed despite evidence of positive therapeutic effects [15–17].

Despite this knowledge, cultured MSCs are useful when designing cell-instructive biomaterial platforms by functioning as a well-characterized surrogate for endogenous progenitor cell populations enabling assessment of biomaterial capacity to promote differentiation. This is supported by previous studies by our group investigating the proadipogenic properties of decellularized adipose tissue (DAT). DAT scaffolds were shown to support adipogenic differentiation of human ASCs in vitro [18, 19], while host-derived adipogenesis was also observed within the DAT scaffolds in vivo [20]. Similarly, Li et al. developed a controlled release scaffold system to deliver bone morphogenetic protein-2 (BMP-2) and dexamethasone for osteogenic differentiation [21]. Investigation of this system showed that it induced osteogenesis in rat bone-marrow-derived MSCs in vitro. Further, in vivo characterization of the cell-free scaffolds in a rat calvarial defect model showed that they promoted new bone formation relative to control scaffolds without BMP-2 or dexamethasone. These systems, which will be discussed in more detail later, support the link between the capacity of a material to promote MSC differentiation in vitro and its ability to support tissue-specific regeneration in vivo.

Regardless of their differentiation capacity, the cell-empowerment properties of MSCs remain a valuable resource for cell therapy [9, 10]. Harnessing the potent secretory behavior of MSCs represents a highly promising strategy for inducing host tissue regeneration for a diverse array of clinical applications [15]. Key amongst the proregenerative factors secreted by MSCs are those with angiogenic, antiapoptotic, and immunomodulatory properties [22]. In this context, their plastic nature may still be a useful property as researchers are beginning to uncover how their differentiation status affects their secretory profile [23, 24]. Overall, although the understanding of MSC biology has changed considerably since their initial discovery, their potential as therapeutics remains strong. This is especially true of ASCs, as will be discussed in more detail in the following section.

Cell-empowerment effects of ASCs

The immunomodulatory capacity of ASCs has been reported to be similar or potentially even stronger than that of bone-marrow-derived MSCs [25–28]. For example, Yoo et al. noted similar responses between human bone marrow- and adipose-derived MSC populations when cocultured with activated human peripheral blood mononuclear cells (PBMCs) [25]. Specifically, both populations were able to greatly reduce PBMC proliferation and expression of the proinflammatory cytokines interferon-γ (IFNγ) and tumor necrosis factor-α (TNFα). Within their system, similar cytokine expression levels were observed in both MSC populations. In another study using age-matched MSC donors, Melief et al. observed that human ASCs were more effective than their bone-marrow-derived counterparts at reducing the proliferation of PBMCs, potentially mediated by the increased expression of interleukin-6 (IL-6) and IL-10, two potent immunosuppressive cytokines [26]. Interestingly, these two factors have also been shown to be expressed at higher levels in murine ASCs compared with murine bone-marrow-derived MSCs [27].

Additionally, of critical importance for tissue regeneration, are the well described proangiogenic effects of ASCs, which are linked to their secretion of factors including vascular endothelial growth factor (VEGF), fibroblast growth factor-2 (FGF-2), and hepatocyte growth factor (HGF) [29–32]. Expression of these factors can also be enhanced in ASCs through culture under low oxygen tension [29–34].

Part 2: Properties of biomaterial scaffolds for guided differentiation

Introduction to the ECM

When designing scaffold-based biomaterials, it is helpful to have an understanding of the ECM that forms the natural scaffold within tissues. The ECM has a tissue-specific composition that comprises fibrous proteins and interfibrillar components [35, 36]. The fibrous proteins consist primarily of varying types of collagen and impart structure, strength, and resilience to the tissues [35, 37]. The interfibrillar molecules are varied in nature and include proteoglycans, glycoproteins, glycosaminoglycans (GAGs), and matricellular proteins [36]. These components have diverse roles in cell signaling and cellular organization, as well as tissue structure and mechanical properties [35, 36]. Spatially, the ECM is segregated into the interstitial matrix and the basement membranes that separate the stromal cells from the epithelium and endothelium [38]. The composition and organization of the ECM within each of these regions plays important roles in directing cell function and defining the overall tissue properties [38].

The composition of the ECM is highly complex, exemplified by proteomic characterizations by our group [39] and others, including pioneering work by Naba et al. [40] (further reviewed in Refs. [41, 42]). In addition to the expected structural components, the ECM can also contain a diverse range of sequestered soluble factors that function as important biological mediators [39]. Temporally, the ECM can change with age [43] and disease status [44], and during regular reproductive cycling [45]. For tissue engineering, it is not feasible to develop biomaterials that replicate the full complexity of the native ECM. In general, for biomaterials design, the goal is to replicate certain critical factors, in a cost-effective manner, and develop scaffolds that support cell infiltration and differentiation toward a desirable cell phenotype that promotes constructive remodeling and recapitulates the native tissues.

Cell-ECM interactions

Stromal cells interact with the ECM primarily through integrin receptors, allowing for bidirectional communication using both physical and biochemical signals [46, 47]. Integrins are heterodimeric proteins that consist of an α and a β subunit, which together recognize specific motifs present within the ECM [46, 47]. This specificity arises from 24 unique integrin configurations derived from the combination of 18 α and 8 β subunits [48]. Integrins form larger adhesion complexes, adhesomes, that transmit force between the cell and the ECM, converting mechanical forces into biochemical signals that are unique to individual integrin pairs [46, 49]. Depending on the forces applied, the integrin adhesomes can change in composition and mechanical properties, modulating force transmission and biochemical signaling [49, 50]. Although the mechanisms of this response are not fully elucidated, one of the ways in which the adhesome proteins can be altered by external force is through conformational changes. For example, when Talin, a protein that links integrins to the cytoskeleton, is stretched, a vinculin binding site is exposed, allowing for the recruitment of vinculin, which stabilizes the focal adhesion [51, 52].

Once mechanical forces are transmitted into the cell, gene expression can be altered via mechanosensing transcription factors. Two of these pathways that have been well studied are the myocardin-related transcription factor (MRTF)/serum response factor (SRF) pathway, and the Yes-associated protein (YAP) and transcriptional coactivator with PDZ-binding motif (TAZ) mediated pathway [53–56]. In the absence of mechanical stimuli, MRTF is sequestered within the cytoplasm via interactions with actin monomers. Upon integrin-mediated Rho activation [57], subsequent actin polymerization releases MRTF and allows translocation into the nucleus, enabling downstream transcription through its association with SRF [58]. Conversely, YAP and TAZ, two highly related transcription factors, are excluded from the nucleus through phosphorylation mediated by large tumor suppressor kinase 1 and 2 (LATS1/2) [59]. Under mechanical tension, integrin-mediated signaling impairs phosphorylation via LATS1/2, promoting translocation of YAP and TAZ, a process that may be further enhanced via stretch-responsive pores within the nuclear membrane [59–61]. These mechanisms provide a direct means of mechanosensing within cells; however, recent data also suggest that surface stiffness can modulate other signaling pathways [62, 63]. For example, ASCs displayed stiffness-dependent responses to platelet-derived growth factor-BB (PDGF-BB) and BMP-2 [63]. These effects may be mediated by growth factor receptor clustering associated with integrin activation [64, 65]. Thus, even though there are specific signaling pathways responsible for sensing the mechanical properties of the microenvironment, substrate stiffness may have more subtle effects on a range of pathways.

The specificity of integrins for unique ECM motifs also allows cells to sense the biochemical composition of their surroundings [47]. Cells bind collagen fibers directly via $\alpha_1\beta_1$, $\alpha_2\beta_1$, $\alpha_{10}\beta_1$, and $\alpha_{11}\beta_1$ integrins [66], and can also access cryptic asparagine-glycine-aspartate (RGD) motifs upon collagen denaturation [67]. These RGD sites are classical cell adhesion sites present on cell adhesive glycoproteins including fibronectin [68]. Additional adhesive motifs have also been identified on several other ECM proteins such as isoleucine-lysine-valine-alanine-valine (IKVAV) sites on laminin and glycine-phenylalanine-hydroxyproline-glycine-glutamate-arginine (GFOGER) sites on collagen [69]. Importantly, for the design of biomaterials, incorporation of these sequences offers a means to induce cell adhesion and modulate cell function through integrin activation on polymers that do not innately support cell attachment [69, 70]. This strategy can enhance the bioactivity and cell-supportive nature of synthetic polymers that have been designed with mechanical properties and/or degradation kinetics targeted for specific applications [69, 70]. Further, additional components, such as growth factors, can be added to further stimulate specific signaling pathways and direct cell functions including proliferation, migration, or differentiation.

The ECM contains a wealth of information within its biochemical, mechanical, and structural properties, which cells can sense and respond to through ECM synthesis and remodeling, creating a bidirectional signaling loop. The cell-instructive potential of the ECM can be harnessed when designing biomaterials for ASC culture and delivery by mimicking selected properties within engineered platforms. However, when seeking to elucidate the roles of each of these factors, it is important to recognize that scaffold composition, biomechanics, and structure are often interdependent [71]. For example, altering a material's stiffness via cross-linking or by changing the material density can impact parameters including surface chemistry and/or microstructure that confound the interpretation of the findings [71]. Nevertheless, as the field continues to develop, an increasing number of studies using tightly controlled systems will help to identify cause-effect relationships for specific properties that can be integrated in the rational design of scaffolds for ASC culture and delivery.

Biochemical composition for guided differentiation

The biochemical composition of a scaffold is one of the most important factors to consider when designing biomaterials to guide cell proliferation and differentiation. In this section, we will provide a brief introduction into three classes of biomaterials that have been commonly used for ASC differentiation and tissue regeneration. The two main designs that will be discussed are bottom-up approaches in which specific building blocks are systematically combined to generate a well-defined material, and top-down approaches in which biologically derived samples are processed to yield more complex ECM mimetics. In addition, we will provide a brief introduction into inorganic biomaterials, which, although synthesized in a bottom-up approach, are considerably different than their organic counterparts. It is important to note that all of these designs have shown clinical utility. For example, organic bottom-up and top-down approaches have been widely used for wound healing applications [72, 73], whereas inorganic biomaterials are commonly used for bone replacements [74, 75].

Bottom-up organic biomaterials

Bottom-up designs for biomaterials are developed using synthetic or natural polymers as the structural backbone [76]. An increasing variety of synthetic polymers are available that offer great versatility, allowing for alternative polymerization techniques, better control over the addition of functional groups, and enhanced tunability of physical properties to meet application-specific requirements [76]. Due to their innate bioactivity, biomaterials derived from natural polymers, such as collagen, fibrin, hyaluronic acid, silk fibroin, alginate, and chitosan remain widely studied [77, 78].

In bottom-up approaches, these base structures are modified with selected components that provide biological signals to induce tissue-specific differentiation [7]. For example, these strategies can integrate the controlled release of various biological factors that stimulate differentiation, which will be discussed in detail in the section "Part 3: Controlled release systems for guided differentiation." One advantage of using bottom-up approaches is that the synthesis of these biomaterials can be done in a highly reproducible manner with a high level of quality control. Further, due to this reproducibility and the overall simplicity in the designs, it is easier to characterize the individual and combined effects of each component on the cellular response. With respect to their inductive potential, bottom-up approaches typically rely on the delivery of biological factors, or tunability of mechanical and structural properties to promote tissue-specific differentiation. These topics will be discussed in more detail later.

Top-down organic biomaterials

In top-down designs, biomaterials are typically derived from complex biological samples using a process known as decellularization. These biological samples can be sourced from ECM generated through cell culture or from native tissues [79, 80]. The general goal in decellularization is to reduce immunogenicity associated with cells and cellular components, while simultaneously preserving as much of the ECM as possible to mimic the native microenvironment [79]. Proteomic investigations of decellularized materials underscore the complexity of these systems, which can contain many hundreds of proteins [39–42].

Complex, decellularized biomaterials have innate bioactivity, and a wide variety have been reported to have inductive properties for directing cell differentiation. For example, Hashimoto et al. observed increased alkaline phosphatase (ALP) activity in rat bone-marrow-derived MSCs cultured on decellularized porcine bone compared with TCPS after 21 days in culture in osteogenic differentiation medium lacking dexamethasone [81]. Likewise, human ASCs grown on decellularized porcine cartilage differentiated along the chondrogenic lineage in differentiation medium lacking exogenous growth factors [82]. In this study, ASCs displayed enhanced gene expression of aggrecan and type II collagen by 2 weeks relative to initial expression, and the mechanical properties of the constructs approached native cartilage by 42 days in culture [82]. Similarly, DAT was shown to be adipo-inductive for human ASCs cultured in adipogenic differentiation medium or proliferation medium [18, 19]. More specifically, gene expression of the adipogenic markers lipoprotein lipase (LPL), peroxisome proliferator activated receptor gamma (PPARγ), and CCAAT/enhancer-binding protein-α

(CEBPα); activity of glycerol-3-phosphate dehydrogenase (GPDH), a marker of lipid metabolism; and lipid accumulation were all enhanced when the ASCs were cultured on the DAT, even in samples cultured in proliferation medium that would normally suppress adipogenesis [18–20]. Others have shown that adipose-derived ECM can be further engineered to better promote adipogenesis through the incorporation of additional factors. Specifically, Zhang et al. isolated murine DAT, which on its own promoted minimal fat formation. In contrast, the addition of FGF-2 into the DAT led to a potent adipogenic response [83].

From these findings, it is evident that decellularized materials can be used to generate biomaterials designed to promote lineage-specific differentiation and have the potential to be further modified to enhance these effects. Thus, even though decellularized biomaterials have recognized limitations including batch-to-batch variation, their naturally inductive properties make them attractive for potential tissue replacement and regeneration strategies.

Inorganic biomaterials

Inorganic biomaterials, including those derived from metals and ceramics, have great utility for orthopedic implants and are often studied for their osteoinductive properties [74, 75, 84]. Commonly used metallic biomaterials include stainless steel, titanium alloys, and cobalt-chromium alloys [75]. Various ceramics have also been studied, including alumina, zirconia, and calcium phosphate-based materials. Calcium phosphates are of particular interest for the development of orthopedic implants given their similarities to the inorganic component of mineralized bone and osteoinductive nature [74, 75].

While the mechanical properties of these inorganic materials enable their use in load-bearing applications, their inorganic nature also has limitations. For example, metallic biomaterials can corrode over time, releasing cytotoxic metal ions into the surrounding tissues, leading to the degradation of the implant [85]. Alternatively, ceramics are susceptible to fracture [75, 86]. Ongoing efforts focus on developing biodegradable and bioactive ceramics and metal alloys that may better support tissue remodeling and repair [87].

Although titanium was classically considered a biologically inert material, the use of chemical and physical means to create porous titanium meshes was shown to impart osteoinductive properties [88]. Implanting both porous and nonporous titanium into the dorsal muscles of dogs, Fujibayashi et al. observed new bone tissue formation in the porous constructs, but not within the native titanium [88]. Similarly, human and rat ASCs, as well as human umbilical-cord-derived MSCs all displayed enhanced osteogenic differentiation when cultured on roughened titanium compared with those grown on smooth titanium [89–91]. Thus, while titanium may not innately provide a strong osteogenic stimulus, these studies support that surface modifications can enhance this property.

A range of calcium phosphate-based ceramics exist with varying structural and biochemical properties, and biphasic calcium phosphate ceramics are widely used in the clinic for a range of orthopedic applications [92]. Two commonly studied variants are hydroxyapatite and β-tricalcium phosphate [74], which differ in their rate of degradation. The more soluble β-tricalcium phosphate degrades more quickly than hydroxyapatite [74, 93], releasing calcium and phosphate ions, which are potent inducers of osteogenesis [74, 93, 94]. To achieve a balance between mechanical stability and osteoinduction, studies have combined the two into biphasic ceramics. In an intramuscular implant model in mice, the composites yielded the greatest de novo bone formation compared with either ceramic alone [95]. Similar findings have been observed in vitro, where rat MSCs seeded onto biphasic hydroxyapatite/β-tricalcium phosphate ceramics showed enhanced osteogenic differentiation relative to those seeded on either ceramic alone [84].

Mechanical properties for guided differentiation

In addition to the biochemical composition, the mechanical properties of a material can have a profound influence on MSC differentiation [96]. A variety of model systems have been developed to investigate the influence of substrate mechanics on lineage-specific differentiation [96]. The most common parameter explored is the effects of varying the Young's modulus of an underlying cell-adhesive substrate, measured through compressive testing.

One useful system that researchers have designed is the micropost assay, in which cells are grown on patterned polymer pillars of differing lengths on a micrometer scale [71, 97]. In this assay, the length of the pillars is inversely proportional to substrate rigidity, while allowing for control of the surface chemistry and surface area available for adhesion. Using this assay, Fu et al. demonstrated that the osteogenic differentiation of human bone-marrow-derived MSCs was enhanced on more rigid pillars (spring constant $k = \sim 1500\,\text{nN}/\mu\text{m}$), whereas adipogenic differentiation was augmented on the more compliant pillars ($k = \sim 2\,\text{nN}/\mu\text{m}$) [71].

In another approach, Wen et al. used polyacrylamide hydrogels synthesized using different concentrations of acrylamide monomer or bis-acrylamide cross-linker to make different hydrogel preparations with unique porosity-stiffness combinations [98]. Using this methodology, it was determined that ASC differentiation toward the adipogenic lineage was favored on softer gels with a Young's modulus of 4 kPa regardless of gel porosity, whereas osteogenic differentiation was favored on stiffer gels with a Young's modulus of 30 kPa, again independent of porosity [98].

Thus, although other properties may be altered by changing the material stiffness, these studies highlight that stiffness is a potent modulator of MSC differentiation. Moreover, these findings support the hypothesis that the lineage-specific differentiation of MSCs is favored when the Young's modulus of the substrate mimics that of the native microenvironment of the desired tissue type [99–102]. In living tissues, these values range from ~1 kPa in soft tissues such as brain [103] to greater than 1 GPa for trabecular bone [104, 105]. For comparison, the modulus of adipose tissue is in the range of 2–4 kPa [106], cartilage ranges from 20 kPa to 6 MPa, depending on the zone measured [107], and nonmineralized osteoid surrounding bone tissue in which osteoblasts reside measures ~35 kPa [108], while TCPS measures in the order of 1 GPa [109].

As the immunomodulatory secretory behavior of ASCs is increasingly being recognized as a primary mechanism for in vivo regeneration [15], it is also important to understand how mechanical properties can modulate the ASC secretome. This is an emerging area of research, and although stiffness-sensitive expression of secreted proteins has been observed, there is a need for further investigation. A recent study supports that integrin signaling can modulate the secretory profile of MSCs [110]. More specifically, Wan et al. observed that cultured ASCs showed decreased expression of the immunomodulatory cytokines, cyclooxygenase-2

(Cox-2), TNFα-induced protein-6 (TSG-6), interleukin-1 receptor antagonist (IL-1ra), and monocyte chemoattractant protein-1 (MCP-1) when cultured with an inhibitor of focal adhesion kinase (FAK), an important downstream mediator of integrin signaling [110]. Similarly, investigating differences between MSCs grown on alginate hydrogels with Young's moduli of 3 kPa or 18 kPa, Darnell et al. observed differences in the fraction of cells expressing the immunomodulatory proteins IDO-1, Cox-2, Osteopontin, and SDF-1 [111].

A major limitation of these studies is that the biological implications of these expression changes are difficult to assess. Recognizing this challenge, Murphy et al. probed the effects of the human bone-marrow-derived MSC secretome on murine macrophages [112]. Using a fibrin gel system, their team observed that VEGF expression was enhanced on stiffer gels with a Young's modulus of 40 kPa, whereas PGE_2 expression was favored on softer gels with a modulus of 5 kPa. In contrast, when the conditioned medium was transferred to a culture of macrophages, medium from MSCs grown on the softer matrix promoted greater VEGF expression in the macrophage population [112]. Thus, the interaction between MSCs and macrophages resulted in the opposite response compared with MSCs alone. Taken together, this highlights the challenge of studying immunomodulatory effects in a controlled in vitro system and emphasizes the need for further studies in more complex models, including coculture systems and in vivo models, to advance in the rational design of materials that harness the capacity of substrate biomechanics to direct the desired cellular response.

Microstructural parameters for guided differentiation

In addition to their biochemical and mechanical properties, biomaterials can be engineered to have diverse structural properties. The macrostructural form a biomaterial takes, such as an injectable particulate, 3D porous foam, hydrogel, or surface coating, is typically chosen based on the desired application. However, the microstructure, including topography, porosity, and pore size, can vary widely depending on material processing and can have a significant impact on the cellular response. For example, scaffold microstructure has been shown to influence cell polarity, binding site availability, nutrient exchange, migration, and biodegradation [113–117]. In this section, common structural modifications will be discussed with regard to their effects on differentiation.

Topography

Surface topography is a well-studied structural property of materials as it can be easily modified in a variety of ways. For example, using lithography, etching-based approaches, electrospinning, pattern transfer, or surface roughening methods, nano- to microscale topographies can be created that mimic structural patterns found within the native ECM [118]. With the exception of some novel methods designed to increase the porosity of electrospun scaffolds to improve cell infiltration and provide a true 3D environment [119], these techniques are primarily used to modify 2D surfaces, giving them pseudo 3D properties. Clinically, these modifications have been useful for improving the integration of inorganic implants for orthopedic or dental applications [120, 121].

Electrospun scaffolds can be created to have uniformly or randomly aligned fibers by manipulating the speed of the collecting mandrel during production. MSCs, including ASCs, cultured on aligned fibers display enhanced FAK and extracellular signal-regulated kinases (ERK) signaling over their randomly aligned counterparts [110, 122, 123]. Downstream FAK activation is associated with osteogenic differentiation, although most studies suggest that osteogenic induction on aligned fibers is modest [123–128]. Investigating the osteo- and adipoinductive properties of 3-hydroxybutyrate-*co*-3-hydroxyhexanoate (PHBHHx) fiber alignment on rat bone-marrow-derived MSCs, Wang et al. reported increased osteogenic differentiation on aligned fibers and increased adipogenic differentiation on random fibers [123]. However, they also noted that the mechanical properties of their scaffolds differed, with the aligned fibers showing increased stiffness, which confounds the interpretation of the specific effects of fiber alignment [123]. Nevertheless, all of these studies report changes in cell morphology with fiber alignment, with more elongated cells on aligned fibers and greater cell spreading on randomly aligned fibers [123–128]. Culturing human bone marrow MSCs on patterned adhesive islands on an otherwise nonadhesive substrate, Kilian et al. demonstrated that the shape of individual MSCs could be altered [129]. Importantly, elongated cells preferentially differentiated toward the osteogenic lineage, while cells that had a more rounded shape displayed preferential differentiation toward the adipogenic lineage [129]. Taken together, it is possible that fiber alignment does have a minor effect on MSC differentiation, potentially through a mechanism mediated by changes in cell shape.

Altering topographies of otherwise planar surfaces through additive or subtractive methods offers another technique to modify MSC differentiation [130, 131]. Unfortunately, variations in experimental design, including the specific materials employed, diversity of surface features, and wide range of outcome measures, makes it challenging to identify specific topographies that show the greatest potential for directing cell differentiation. At present, the most well studied and best supported use for modified surfaces is for the osteogenic differentiation of MSCs compared with planar surfaces. In general, rough surfaces with dense surface features have been reported to promote the osteogenic differentiation of MSCs [132–135]. These surfaces have diverse features ranging in depth from $\sim$0.1 to 5 μm and have typically displayed only a modest effect on osteogenic differentiation [132–136]. While there are few studies investigating surface topography on chondrogenesis, Wu et al. showed that nanoscale isotropic patterns (pillars and holes) promoted the chondrogenic differentiation of human bone-marrow-derived MSCs as compared with those cultured on either grooves or planar surfaces [137]. Similarly, while there is evidence to support that surface modifications can enhance adipogenic differentiation in cells cultured in differentiation medium, there is limited evidence to support that surface topographies alone have inductive effects for this lineage [135, 138]. In general, the soft, porous biomaterials that would be most appropriate for regenerating soft tissues are not well suited to retaining structural features, limiting the exploration of topographies for chondrogenic and adipogenic differentiation.

Three-dimensional structure

Whereas surface topography can be altered to modulate cellular interactions on biomaterials with a 2D geometry, 3D structural properties can similarly be modified in biomaterials designed to support cellular infiltration. These biomaterials have the potential to be applied as proregenerative scaffolds for in vivo applications, as well as in 3D in vitro cell expansion and differentiation platforms [139]. In general, MSCs cultured on TCPS display reduced proliferation and differentiation potential over time in culture [140]. Culturing in 3D systems may better conserve or potentially even enhance the proregenerative capacity of the MSCs during expansion [139]. Specifically, mouse bone-marrow-derived MSCs cultured using a hanging drop culture method, in which the cells form aggregated spheroids that promote cell-cell and cell-ECM interactions, displayed enhanced adipogenic and osteogenic differentiation compared with controls grown on TCPS [139]. Moreover, human umbilical-cord-derived MSCs displayed enhanced immunomodulatory properties when cultured on 3D substrates compared with those cultured on TCPS [141]. Interestingly, these differences were independent of the biochemistry of the 3D substrates, which were synthesized from collagen, chitosan, or poly(lactic-*co*-glycolic acid) (PLGA) [141]. Thus, 3D biomaterials hold great promise for the design of culture platforms seeking to harness both ASC differentiation capacity and proregenerative paracrine functionality.

Unlike 2D geometries, 3D structures are designed to promote the influx of cells into the material core, as well as the diffusion of gases and nutrients to support those cells. While these factors can be controlled externally, such as by using bioreactor systems [142, 143], they can also be modified by altering intrinsic biomaterial properties such as porosity, pore size, and overall construct size [116, 144, 145].

With respect to MSC differentiation, pore sizes have been suggested to impact both chondrogenic and osteogenic differentiation. Investigations of both rat bone-marrow-derived MSCs and human ASCs have shown improved chondrogenic differentiation in biomaterials with larger pore sizes (300–400 μm) as compared with smaller pore sizes (~100 μm) [146, 147]. In contrast, human bone-marrow-derived MSCs cultured on β-tricalcium phosphate scaffolds with pore sizes of ~40 μm and 65% porosity displayed modestly improved osteogenic differentiation as compared with scaffolds with a pore size of ~135 μm and 75% porosity [148]. Similarly, rat bone-marrow-derived MSCs cultured on polyethylene terephthalate fiber scaffolds displayed improved osteogenic differentiation when porosity was decreased from 97% to 93% [149]. These studies suggest that larger pore sizes may promote chondrogenic differentiation, whereas smaller pore sizes and reduced porosity may be favorable for osteogenic differentiation. It is important to note the relationship between porosity and surface area for cell attachment in 3D scaffolds. These properties can influence cell clustering within the scaffolds, which may also have profound effects on the differentiation response. For example, a high degree of cell-cell contact may be favorable for MSC differentiation toward the chondrogenic and adipogenic lineages [150, 151].

Part 3: Controlled release systems for guided differentiation

In addition to the scaffold-based biomaterials that have been discussed thus far, controlled release systems offer an alternative and potentially complementary strategy to stimulate ASC differentiation and promote tissue regeneration. Whereas scaffold biomaterials are designed to mimic the ECM through their biochemical, mechanical, and structural properties, implantable controlled release systems are designed to deliver soluble factors locally and, ideally, stably over prolonged periods [152]. Further, the systemic delivery of controlled release systems has also been explored using various approaches to try to target specific tissues [153]. These systems strive to overcome the limitations of systemic drug delivery, including temporal variability and the potential for undesired systemic side effects [152–156].

Controlled release systems may also have utility in cell culture models. For example, Hegde et al. developed hydrogel disks that allowed for the continuous release of glucose or protein hydrolysate to feed cells at steady state rather than intermittently through medium changes [157]. However, with respect to MSC differentiation in vitro, the delivery of differentiation factors via controlled release systems has shown comparable levels of differentiation as compared with periodic replenishment via medium changes, as will be highlighted below. Regardless, in vitro assessment offers a platform to test and refine these systems to improve their therapeutic capacity prior to in vivo testing, which is required to assess the true potential of the systems.

Several of the systems explored below could also be characterized as scaffold-based biomaterials. However, the systems discussed in this section have been shown to specifically support MSC differentiation through the delivery of inductive factors in a controlled and localized manner.

Encapsulation-based controlled release

Encapsulation-based controlled release systems rely on the rates of passive diffusion and degradation of polymer carriers, or the fusion of membrane-based carriers with the recipient cells. When introduced systemically in vivo, their distribution is primarily determined by anatomy, particle surface charge, and particle size [153, 158]. The addition of targeting ligands can enhance cell fusion for membranous particles, but has not been found to typically promote tissue localization [158]. In this section, three types of systems that rely on encapsulation will be discussed with respect to their effects on MSC differentiation and tissue regeneration, including polymer microspheres, synthetic core-shell structures, and biologically derived vesicles.

Microspheres

Polymer microspheres are solid particles with diameters ranging within the micrometer scale that can be synthesized from natural or synthetic polymers [156, 159]. These biomaterials incorporate the factors of interest into the polymer network. Parameters including the polymer chemistry, molecular weight, and particle size can be tuned to control the rates of degradation and diffusion [156, 159]. Larger particles can act as substrates for cell adhesion, whereas smaller microspheres have been shown to be able to cross the cell membrane, allowing for intracellular delivery [160]. Intracellular accumulation has also been observed with particles in the nanometer scale [161]. Microspheres of varying sizes have been successfully applied for MSC differentiation.

Miao et al. investigated alginate microspheres loaded with VEGF-A with diameters of ~1 μm for applications in osteogenic differentiation [160]. Although VEGF-A primarily signals through its cognate cell surface receptors, previous work has identified a potential noncanonical pathway for intracellular VEGF-A that promotes osteogenic differentiation of MSCs [162]. Miao et al. provided evidence of intracellular accumulation of VEGF-A using confocal microscopy, concomitant with enhanced osteogenic differentiation. These findings demonstrate the capacity of their system to deliver a payload intracellularly, which may be beneficial for future strategies.

Several microsphere formulations have been tested for their capacity to induce the chondrogenic differentiation of MSCs, primarily by controlling the release of TGFβ1 or TGFβ3. By incorporating PLGA microspheres loaded with TGFβ1 into cell pellets, Solorio et al. observed higher levels and more uniform expression of GAGs in human bone-marrow-derived MSCs when higher concentrations of the microspheres were added [163]. In vitro, TGFβ1 release was observed from free microspheres over 21 days, although ~50% was released within the first 2 days [163]. However, it is possible that the microspheres incorporated within the cell pellets would have more desirable release kinetics, similar to the reduced initial burst release observed when microspheres are encapsulated within hydrogels as compared with free floating [164]. In a follow-up study investigating TGFβ1-loaded gelatin microspheres, Solorio et al. explored how microsphere properties affected GAG accumulation when the MSCs were cultured in chondrogenic differentiation media lacking TGFβ1 [165]. Specifically, the level of gelatin cross-linking was shown to have no impact on the level of GAG production [165], despite a reduced initial burst release that would enable more uniform delivery over time [166]. In contrast, the amount of TGFβ1 delivered, controlled by either the amount of TGFβ1 loaded per microsphere or the total number of microspheres delivered, had the greatest positive impact on GAG production [165]. Using a similar experimental design involving 3D pellet cultures of human bone-marrow-derived MSCs incorporating TGFβ3-loaded collagen microspheres, Mathieu et al. showed that chondrogenic gene expression and qualitative histological assessments were similar between cells grown with TGFβ3-loaded microspheres as compared with TGFβ3 delivery through the medium over a 3-week period [167]. When the microspheres were implanted subcutaneously into severe combined immunodeficiency (SCID) mice in combination with human bone-marrow-derived MSCs, the TGFβ3 loaded microspheres yielded greater accumulation of cartilage-like tissue as compared with unloaded controls, supporting the potential of localized delivery to enhance in vivo differentiation [167]. In another study, human ASCs encapsulated within *N*-methacrylate glycol chitosan hydrogels containing microspheres loaded with TGFβ3 and BMP-6, displayed slightly improved GAG accumulation in vitro at 28 days relative to delivery of the growth factors in culture medium [168]. However, further validation of the system through in vivo testing would be required.

Intracellular delivery via microspheres may also hold promise for chondrogenic differentiation. For example, Park et al. loaded PLGA microspheres with Sox9, a transcription factor associated with chondrogenic differentiation of MSCs, and coated them with TGFβ3 as a strategy to induce the chondrogenic differentiation of human bone-marrow-derived MSCs [169]. Using this system, they showed that chondrogenic gene expression and GAG content were induced by either TGFβ3 coated or Sox9 loaded microspheres, with an augmented response when both factors were delivered in combination. Similarly, subcutaneous implantation of microspheres incorporating both TGFβ3 and Sox9 into mice resulted in the greatest chondrogenic gene expression and GAG production [169]. Although the addition of Sox9 was supportive of chondrogenic differentiation, the mechanism through which it was internalized by the cell or the efficiency of this process were not reported. Unlike the mechanism described by Miao et al. [160] in which the small microspheres (~1 μm diameter) were taken up by the cells, these microspheres were too large, with diameters between 50 and 80 μm, suggesting that Sox9 was traversing the membrane after release from the microspheres or that it was acting through an unforeseen pathway.

Controlled release systems have also been explored for the adipogenic differentiation of MSCs. For example, delivery of both dexamethasone and insulin in PLGA microspheres was shown to enhance intracellular lipid accumulation in human ASCs [170]. Other examples have been shown to induce adipogenesis in vivo. While both FGF-1 and FGF-2 were found to induce a potent angiogenic response when delivered subcutaneously in microsphere systems in animal models, only FGF-2 induced an adipogenic response [171, 172]. More specifically, FGF-2 was incorporated into gelatin microspheres that were embedded into a collagen matrix and implanted subcutaneously into mice using a chamber system [172]. Although delivery of FGF-2 within the collagen matrix alone led to a limited amount of new fat formation, FGF-2 loaded into the gelatin microspheres resulted in a potent adipogenic response, potentially due to more favorable release kinetics [172]. Finally, although not a microsphere system, the delivery of pioglitazone, a PPARγ agonist, by incorporation into a noninductive collagen scaffold, was shown to stimulate de novo fat formation when implanted subcutaneously in a murine model [173].

Synthetic core-shell delivery vehicles

Core-shell systems represent another promising approach for the controlled release of bioactive payloads. These systems have an outer core designed to prolong factor release and prevent the initial burst release relative to single polymer systems [159]. The most common methods of synthesizing core-shell delivery vehicles are through the use of coaxial electrospraying or electrospinning to create spheroid structures or fibers, respectively. Using these methodologies, several systems have been developed for the osteogenic and chondrogenic differentiation of MSCs.

Delivery of BMP-2 has been commonly explored, both alone and in combination with dexamethasone, to develop materials for osteogenic differentiation. BMP-2, is a member of the TGFβ superfamily, and a key factor involved in cartilage development and bone fracture healing [174, 175]. Importantly, it has been shown to be a potent mediator of both osteogenic and chondrogenic differentiation of MSCs in vitro [176]. Dexamethasone induces the gene expression of the osteogenic transcription factor *Runx2*, and further promotes its activity [177], and has been shown to have an additive effect with BMP-2 in ectopic bone formation in vivo [178]. Using these differentiation factors, Choi et al. synthesized PLGA microspheres coated in an alginate shell for osteogenic differentiation of rat bone-marrow-derived MSCs [179]. Microspheres were synthesized with either dexamethasone in the core and BMP-2 in the shell or vice versa. Localization of either factor within the core enabled more uniform release with a reduced initial burst effect as compared with localization within the shell [179]. Using end-point RT-PCR, it was shown that a core loaded with dexamethasone and a shell loaded with BMP-2 resulted in the greatest osteogenic gene expression in the cultured MSCs, with significantly enhanced gene expression of type I collagen, osteocalcin, and osteopontin as compared with the control group.

These findings have been further supported by several other groups investigating the delivery of BMP-2 alone or in combination with dexamethasone for the purpose of inducing osteogenic differentiation and tissue repair. Using core-shell fibrous delivery systems to release both factors, Su et al. and Li et al. observed enhanced osteogenic differentiation of human and rat bone-marrow-derived MSCs, respectively, in culture, as evidenced through increased ALP activity [21, 180]. Further, in a rat calvarial defect model, enhanced tissue repair was observed when scaffolds delivering both BMP-2 and dexamethasone were implanted [21]. Similar findings have been observed when BMP-2 was delivered alone [181, 182]. Using nanofibrous systems to deliver BMP-2, Srouji et al. and Zhu et al. both noted improved osteogenic differentiation of human bone-marrow-derived MSCs as measured by increased ALP activity [181], and a slight increase in osteogenic gene expression [182]. Most notably, both of these systems greatly improved tissue repair in a rat calvarial defect model [181, 182].

As has been observed with microspheres, core-shell delivery of TGFβ isoforms has been shown to induce chondrogenic differentiation in culture. In particular, synthetic polymer-based core-shell systems have been successfully applied to deliver TGFβ1 to induce the differentiation of rat [183], and human [184], bone-marrow-derived MSCs in vitro. Specifically, Man et al. investigated chondrogenic gene expression and GAG content to assess chondrogenic induction [183]. Using core-shell fibers to deliver TGFβ1, increased chondrogenic differentiation of MSCs was observed as compared with unloaded fibers, but this was slightly reduced relative to unloaded fibers with TGFβ1 included in the medium. Zhu et al. used an interesting 3D printing strategy to generate patterned gels containing TGFβ1-loaded core-shell nanoparticles [184]. Human MSCs incorporated into the gels during synthesis showed increased gene expression of type II collagen, Sox9, and Aggrecan, compared with controls without TGFβ1-loaded nanoparticles. Overall, further validation of these platforms through in vivo testing is warranted to more fully assess their potential.

Biologically derived vesicles

A growing area of interest is the delivery of exosomes, which are cell-secreted membranous vesicles with diameters ranging from 30 to 100nm [158, 185]. Due to their size, surface biochemistry, biocompatibility, and ability to be modified, exosomes have shown promise for drug delivery [158, 185]. Within the MSC field, exosomes are thought to be responsible for at least some of the potent secretory behavior of MSCs [186]. MSC-derived exosomes have been shown to have both immunomodulatory and angiogenic properties [187, 188]. Interestingly, the osteogenic differentiation of human bone-marrow-derived MSCs was enhanced by treatment with exosomes isolated from MSCs previously induced toward the osteogenic lineage as compared with treatment with exosomes isolated from MSCs cultured in growth medium [23]. It has been subsequently shown that the functionality of exosomes secreted by MSCs is dependent on the stage of differentiation, with more mature osteogenic populations secreting more potent osteogenic exosomes [24]. Overall, exosomes for cell differentiation are only in the early stages of investigation. However, it is evident that they may offer novel mechanisms for modifying MSC differentiation with applications both in vivo and in vitro.

Affinity-based controlled release

A recent advance in the controlled release field is the development of affinity-based systems. The major difference between affinity-based and encapsulation-based methods is that affinity-based systems use specific, defined molecular interactions to control release [189]. These affinity-based systems have made use of both natural and synthetic binding partners to trap ligands within biomaterials for delivery [189]. For example, heparin can be used for its innate ability to bind multiple growth factors [190]. Alternatively, histidine-tagged proteins have been linked to scaffolds through interactions with nickel and copper ions recruited by metal-chelating ligands [191]. More recent technologies have enabled design of protein- and DNA-based binding partners, such as antibody mimetics and aptamers, respectively, with optimized binding efficiencies to target a vast array of ligands for delivery [189]. Importantly, these technologies allow for tighter control and more predictable release through modification of the dissociation constant between the binding pair [189].

Another major advantage of affinity-based systems over classical encapsulation-based methods is that the factor to be delivered can be efficiently loaded in neutral aqueous buffers, which is especially important for protein delivery so as to maintain an appropriate conformation [189]. To our knowledge, engineered binding partners have not yet been explored for MSC differentiation; however, using a natural binding partner, Re'em et al. developed constructs for chondrogenic differentiation [192]. Specifically, alginate sulfate scaffolds were explored for their binding affinity with TGFβ1. The response was compared with alginate scaffolds lacking the sulfate group, which, when loaded with TGFβ1, were observed to release 90% of the loaded protein within the first day, as compared with 90% release by day 7 for the alginate sulfate scaffolds [192]. Notably, the affinity-based system led to qualitatively improved chondrogenic differentiation of human bone marrow MSCs relative to the unmodified alginate scaffolds. Overall, affinity-based controlled release systems may offer benefits over the traditional encapsulation-based delivery vehicles; however, further exploration is required. As technologies to design proteins and nucleic acids with specified binding affinities become more accessible, these methods may offer widespread benefits for the development of controlled release systems.

Part 4: Combining ASCs with biomaterials for cell-based therapies

In the previous sections, we discussed the utility of trilineage differentiation as a tool to assess the inductive properties of biomaterials in vitro. However, the secretome is increasingly being realized as the primary mediator of the therapeutic effects of MSC populations, including ASCs [15, 193]. This change in perspective from cell replacement to cell empowerment applications is exemplified in an opinion letter by Arnold Caplan, who originally coined the term "Mesenchymal Stem Cells" [194]. In this letter, Dr. Caplan proposed changing the name of this population to "Medicinal Signaling Cells" to better represent their current perceived clinical utility. In this section, we will focus on this aspect of MSC biology, with an emphasis specifically on ASCs and their applications when combined with biomaterials.

The potent secretory behavior of ASCs makes them a strong therapeutic candidate for regenerative medicine; however, cell retention after delivery is poor, and ASCs alone are insufficient for conditions where there is a substantial loss of tissue volume that needs to be replaced [195–198]. Both of these limitations can potentially be addressed by delivering the cells within biomaterials designed to harness their proregenerative effects.

Relative to delivery in suspension, the use of biomaterials has been shown to enhance ASC retention and efficacy when implanted in vivo for a number of applications [195–197]. As an example, in a retrospective study investigating ASC delivery to osteoarthritic knees of patients to promote cartilage regeneration, Kim et al. found that delivery of autologous human ASCs in fibrin glue resulted in significantly better outcomes than injection of cells in saline

[197]. Similarly, rat ASCs implanted into ischemic hearts following induced myocardial infarction in a rat model were retained at higher levels when delivered in a collagen gel as compared with delivery in medium [196]. Notably, this was only investigated at 24h post implantation, when the collagen delivery group had already declined to ~25% retention, relative to ~5% in medium. In another study using a dorsal subcutaneous implant model, an RGD-containing hydrogel was shown to increase retention of human ASCs compared with saline delivery up to 10 days postimplantation [195]. However, by 14 days postimplantation, less than 5% of cells were retained in either the hydrogel or the saline delivery groups [195]. While these studies highlight the improved retention of ASCs when delivered using biomaterials, they also suggest that further research is required to develop systems that better support the long-term retention of viable cells in vivo. Regardless, combining ASCs with biomaterials has been a promising strategy to promote bone, cartilage and adipose tissue regeneration in animal models, as discussed in the next sections.

ASC-seeded biomaterials for bone regeneration

ASCs have shown strong potential to support the development of new bone tissue at sites of injury. In an early study, Pieri et al. investigated the dose dependence of autologous ASC delivery on the repair of calvarial defects in rabbits [199]. Using a commercial decellularized bovine bone scaffold, they investigated ASC doses of 3×10^5, 3×10^6, and 3×10^7 cells per scaffold. While all ASC-seeded groups outperformed the scaffold alone, there was a dose-dependent relationship, with improved histological outcomes and new bone formation in the scaffolds incorporating higher densities of ASCs [199].

Similar results have also been observed in other species using different scaffolds for delivery. For example, Carvalho et al. delivered human ASCs into calvarial defects in nude mice using scaffolds consisting of a mixture of polycaprolactone and starch [200]. In this model, the scaffolds alone led to improvements in healing, reaching ~40% mineralized bone tissue within the defect by 8 weeks as measured by microcomputed tomography (μCT), but the highest response was observed when they were combined with the ASCs, reaching ~70% mineralized bone tissue within the same time frame [200]. Similarly, Liu et al. delivered allogeneic ASCs using a coral-derived scaffold, consisting of >97% calcium carbonate, into calvarial defects in canines [201]. This model also showed significantly enhanced bone formation in the scaffolds seeded with the ASCs as compared with the coral scaffold alone. Notably, this study also explored the systemic immune response by assaying the blood composition. Interestingly, no differences were noted in the ratio of CD4 to CD8 T-cell populations, or in the levels of several cytokines including IFNγ and IL-10 [201]. In general, these studies support the potential of seeding scaffolds with autologous or allogeneic ASCs for the treatment of bone defects.

Recognizing the proregenerative paracrine capacity of ASCs and the potential for using conditioned media as a cell-free therapy, Linero et al. investigated its use for the repair of surgical defects within a rabbit mandible, compared with delivery of ASCs [32]. More specifically, human ASCs or ASC-conditioned medium was loaded into human blood plasma hydrogels for delivery. Interestingly, both conditioned medium and ASCs displayed similar results, outperforming the hydrogel alone with respect to the amount of new bone tissue formed within the defect [32]. While these findings suggest that conditioned medium may be a promising alternative to cell delivery, the lifespan of the ASCs in the immunocompetent rabbit model may have been limited by their xenogeneic sourcing, which could have impacted their efficacy. Thus, going forward, further exploration using allogeneic populations of ASCs and/or testing in immunocompromised animals is warranted.

In another interesting approach, Lin et al. engineered ASCs to express high levels of BMP-2 or TGFβ3 and assessed their potential for repairing rabbit calvarial defects [202]. The ASCs expressing either BMP-2 or TGFβ3 were delivered into defects using PLGA scaffolds coated with hydroxyapatite or gelatin scaffolds. The findings indicated that the gelatin scaffolds outperformed the PLGA scaffolds in terms of bone formation, and that increasing the expression of BMP-2 in the ASCs was more effective than augmenting TGFβ3 expression. Overall, the delivery of engineered cells within biomaterial scaffolds could hold potential for enhancing tissue-specific regeneration.

Finally, to assess the potential of ASC-seeded scaffolds in humans, Sándor et al. developed a system incorporating several inductive strategies. Specifically, osteoinductive β-tricalcium phosphate granules were loaded with BMP-2 and seeded with autologous ASCs prepared following Good Manufacturing Practice (GMP) guidelines [203]. In this case study, a single patient with a 10-cm anterior mandibular defect received the biomaterial implant and showed a positive surgical outcome up to a final 10-month follow-up period. Further, bone cores taken at this time showed integration of the β-tricalcium phosphate granules with newly formed bone tissue, as well as vascular ingrowth [203].

ASC-seeded biomaterials for cartilage regeneration

The delivery of ASCs using biomaterials has also shown promise for healing cartilage defects; however, studies for this application have primarily focused on predifferentiating ASCs prior to implantation. For example, seeding a variety of scaffolds with autologous ASCs predifferentiated in chondrogenic medium has shown positive results on the repair of surgically induced defects in the femurs of rabbits [204–206], and pigs [207]. For example, Cui et al. isolated porcine ASCs for autologous transplantation and seeded them onto PLGA scaffolds. The constructs were then differentiated for 2 weeks prior to implantation into surgical defects made on the trochlear surface of the femur. At 3 and 6 months postimplantation, the biomaterial treatment outperformed empty defects, and by 6 months the treated groups reached ~90% of the GAG content, and ~90% of the elastic modulus of uninjured cartilage [207].

Similarly, Wei et al. and Zhang et al. isolated rabbit ASCs for autologous transplantation using PLGA and PLGA/chitosan scaffolds, respectively [204, 205]. The ASCs were predifferentiated on the scaffolds in vitro for 2 weeks in chondrogenic medium prior to implantation. In both studies, the delivery of the ASC-seeded scaffolds improved histological scores relative to the scaffolds alone. Zhang et al. also noted that the GAG and type II collagen content, and the Young's modulus of the regenerated cartilage, reached approximately 75%–85% of the values measured in normal cartilage by 12 weeks [204, 205]. Similar findings have also been observed using a decellularized cartilage matrix for delivery of ASCs [206]. In this study, Kang et al. implanted the ASCs in combination with the decellularized ECM into a femur defect model following 5 days of predifferentiation in culture. Improved histological scores were observed compared with delivery of the decellularized scaffold alone, with the ASC-seeded group showing GAG and type II collagen content, as well as stiffness, that was similar to uninjured cartilage at 6 months [206].

ASC-seeded biomaterials for adipose tissue regeneration

There is strong interest in the delivery of ASCs using scaffolds for adipose tissue regeneration for applications in plastic and reconstructive surgery. For example, in an early study, Mauney et al. delivered human ASCs on silk fibroin scaffolds intramuscularly in nude mice [208]. While only one of four mice in the unseeded group showed evidence of adipogenesis, fat formation was observed in three of four mice in the ASC-seeded group, supporting the potential of the cell-based approach to enhance adipose tissue regeneration [208].

As previously discussed, biomaterials derived from DAT have shown promising adipo-inductive effects [18–20]. Further, numerous studies support that adipose tissue regeneration within DAT scaffolds can be enhanced by seeding the scaffolds with ASCs [209–211]. For example, Choi et al. explored the delivery of human ASCs on milled human DAT powders [209]. Suspensions containing DAT powders with or without ASCs were implanted subcutaneously into nude mice. While the DAT on its own had a modest effect on new fat formation at 8 weeks, this was greatly increased when the DAT was delivered in combination with the ASCs [209]. Similarly, our laboratory found that allogeneic rat ASCs greatly improved adipose tissue formation within DAT scaffolds following subcutaneous implantation in an immunocompetent rat model [210]. More specifically, ASC seeding significantly enhanced fat formation within the DAT, with remodeling of ~55% of the implant cross-sectional area into new fat observed at 12 weeks as compared with <5% in the unseeded group. While the donor ASCs were detected within the implant area up to 8 weeks, the ASCs were not observed at 12 weeks, suggesting that the ASCs promoted adipogenesis via the recruitment of endogenous adipogenic progenitor cells [210]. In another approach, Cheung et al. explored a composite system incorporating DAT particles within tunable hydrogels [211]. No fat formation was observed at 12 weeks postimplantation when cell-free hydrogels were delivered in an immunocompetent rat model. However, the addition of allogeneic ASCs was sufficient to promote de novo fat formation [211], again supporting the benefits of a cell-based approach for adipose regeneration.

Predifferentiated ASCs have also been investigated for adipose regeneration [212]. Specifically, Itoi et al. explored adipogenesis in scaffolds synthesized from type I collagen, polyglycolic acid (PGA), and hyaluronic acid both alone and in combination with noninduced ASCs or ASCs that were cultured in adipogenic induction media for 2 weeks prior to seeding scaffolds [212]. They applied ASCs derived from green fluorescent protein (GFP) transgenic mice and implanted the scaffolds subcutaneously into athymic mice. Interestingly, only the addition of predifferentiated ASCs seeded on collagen scaffolds resulted in substantial fat formation by 8 weeks post implantation, characterized by ~65% of the cross-sectional area remodeled into adipose tissue [212]. In contrast, seeding with undifferentiated ASCs on any scaffold or predifferentiated ASCs on PGA or hyaluronic acid scaffolds resulted in <15% adipose remodeling. Investigated only in the collagen scaffold, donor ASCs were retained in the 8-week tissue, but it was not clear whether these cells were differentiated adipocytes. Overall, these findings support further investigation of the effects of delivering undifferentiated versus predifferentiated ASCs within a variety of biomaterial scaffolds to support stable adipose tissue regeneration.

Part 5: Discussion and conclusions

In this chapter, we have discussed two main uses of MSCs, including ASCs, for the design and development of inductive biomaterials for applications in cell culture and delivery. These include (i) the use of ASCs as a tool to assess differentiation in vitro, and (ii) the use of ASCs to enhance biomaterial-induced tissue regeneration. To summarize, biomaterial-guided differentiation relies on the stimulation of ASCs through lineage-specific pathways. These pathways can be targeted through biochemical means, such as through the release of soluble factors, as with calcium phosphate ceramics or controlled release systems, or through complex interactions with the ECM, as with decellularized materials. Alternatively, biomaterials can promote differentiation via mechanical or structural means, by mimicking the physical state of tissue-specific ECM. For in vivo tissue regeneration strategies, ASCs are a useful tool to augment the regenerative capacity of biomaterials for the regeneration of mesenchymal tissues. In this final section, we will briefly discuss some of the major challenges facing the field moving forward, as well as potential opportunities to meet these challenges.

First, it is important to discuss the property of "inductivity" and how this terminology lacks standardization both in vitro and in vivo. The first issue with this terminology is that the magnitude of the response is often not considered when discussing whether a material is inductive. Thus, two materials with drastically different capacities to stimulate differentiation can both be considered "inductive." This issue primarily arises from a lack of agreement on standardized assays that should be used to assess lineage-specific potential in vitro and in vivo. However, even more problematic is the use of suboptimal controls for comparison. While controlled release systems are consistently compared with unloaded carriers as a control, selecting appropriate controls for scaffold-based biomaterials is often more challenging. As a field, there is a need to select well-characterized reference samples with which to compare novel materials. For example, a novel complex 3D scaffold may be compared with TCPS in one study and with a synthetic hydrogel in another. Assuming TCPS and the hydrogel have different inductive potentials, that novel 3D scaffold may be said to be inductive in one study and noninductive in the other. In vivo, noninductive reference samples for scaffold-based biomaterials are rarely included, likely due to a lack of consensus in the field as to which controls are appropriate. As the field continues to grow with ever more complex biomaterial designs, an inability to compare different designs is limiting. There is an increasing need to identify appropriate controls for each application to better standardize research outcomes, and to create a more directed research effort aimed at improving clinical outcomes.

As discussed in the section "Part 2: Properties of biomaterial scaffolds for guided differentiation," the cellular response on biomaterial scaffolds can be impacted by the biochemical, mechanical, and structural properties of the scaffold. Developing strategies to assess the effects of these individual parameters is challenging but important for advancing in the rational design of instructive biomaterials for MSC culture and delivery. More specifically, changes in one property, such as composition, will often impact both the scaffold structure and biomechanics. Recognizing this problem, our group has recently published work in which we compared DAT scaffolds with scaffolds derived from purified type I collagen [213]. Importantly, the two systems were designed to have similar starting mechanical and structural properties, to be able to more specifically assess the effects of scaffold composition on the cellular response [213]. In general, these types of controlled comparative studies are needed to help understand the benefits of specific material choices and scaffold designs.

One area of research that we predict will be highly influential moving forward is the application of biomaterials as a means of harnessing and controlling MSC paracrine function. As the secretion of paracrine factors is increasingly being realized as the major therapeutic role of MSCs upon implantation in vivo [15, 193], there is a need to explore how the secretory profile can be modified with regenerative biomaterials, and whether this is an effective strategy to improve tissue repair. Above, we highlighted studies that investigated the effect of mechanical properties on paracrine function; however, the potential for this is only beginning to be realized.

Predifferentiation of ASCs is another strategy to modify the secretome, which is in the early stages of investigation [23, 24]. It is evident that changes in cell state, including differentiation, can alter the inductive potential of secreted exosomes [23, 24]. Furthermore, predifferentiation of ASCs has been useful in vivo to promote tissue regeneration, especially in cartilage as highlighted above. However, there is not strong evidence that predifferentiated ASCs can persist longer than noninduced ASCs in immunocompetent animals. In general, there is a need for greater integration of technologies that enable long-term tracking of donor cell populations in living animals to be able to more fully assess cell retention and the mechanisms of the observed regenerative responses. Regardless, the findings suggest that predifferentiated populations may also primarily stimulate regeneration through cell-empowerment mechanisms.

Alternative strategies for modulating ASC secretion have also been investigated (reviewed in Ref. [214]). These include a range of priming regimens, including cytokine stimulation, hypoxic conditioning, or genetic manipulation, to promote desirable outcomes when implanted, such as enhanced angiogenesis, immune modulation, or tissue-specific regeneration [214]. Moving forward, we hypothesize that strategies to direct the secretory profile of ASCs through biomaterial design, in combination with induced differentiation or cell priming will enable more potent systems for guided tissue regeneration.

Additionally, there remains a need for further understanding of the fundamental biology of the interactions between MSCs delivered within biomaterials and host cell populations. Upon implantation, biomaterials interact with a variety of cell types, including endothelial cells, stromal cells, and immune cells, altering their complex signaling cascades and, ideally, leading to tissue regeneration. Macrophages specifically are early responders to tissue damage and mediators of long-term tissue remodeling [215], and their responses can be guided by biomaterial design [216], providing additional means of dictating in vivo outcomes. Further, despite strong evidence for tissue-specific regeneration following biomaterial implantation, as discussed throughout this chapter, little is known about the endogenous progenitor populations that give rise to these new tissues and the signals required to activate these populations. A better understanding of the interactions between seeded biomaterials and the host populations will be highly influential for the design of the next generations of biomaterials.

In conclusion, the interplay between ASCs and biomaterials is an expanding area of research, and one that holds great potential for therapeutic applications. In this chapter, we have described parameters to consider when developing novel biomaterials for scaffold-based and controlled delivery applications. Further, we have provided evidence of the potent effects of ASCs to support biomaterial-induced tissue regeneration strategies. Finally, we have suggested some areas for improvement that should be considered in future experimental design, as well as active areas of research that we perceive as being highly influential moving forward.

References

[1] Nataional Institute of Biomedical Imaging and Bioengineering, Biomaterials, (n.d.). https://www.nibib.nih.gov/science-education/science-topics/biomaterials (Accessed 5 February 2020).

[2] A. Nyga, U. Cheema, M. Loizidou, 3D tumour models: novel in vitro approaches to cancer studies, J. Cell Commun. Signal. 5 (2011) 239–248, https://doi.org/10.1007/s12079-011-0132-4.

[3] M. Jannasch, S. Gaetzner, F. Groeber, T. Weigel, H. Walles, T. Schmitz, J. Hansmann, An in vitro model mimics the contact of biomaterials to blood components and the reaction of surrounding soft tissue, Acta Biomater. 89 (2019) 227–241, https://doi.org/10.1016/j.actbio.2019.03.029.

[4] Y. Zhao, N.T. Feric, N. Thavandiran, S.S. Nunes, M. Radisic, The role of tissue engineering and biomaterials in cardiac regenerative medicine, Can. J. Cardiol. 30 (2014) 1307–1322, https://doi.org/10.1016/j.cjca.2014.08.027.

[5] E.S. Place, N.D. Evans, M.M. Stevens, Complexity in biomaterials for tissue engineering, Nat. Mater. 8 (2009) 457–470, https://doi.org/10.1038/nmat2441.

[6] S. Zhang, Fabrication of novel biomaterials through molecular self-assembly, Nat. Biotechnol. 21 (2003) 1171–1178, https://doi.org/10.1038/nbt874.

[7] M.S. Weisenberger, T.L. Deans, Bottom-up approaches in synthetic biology and biomaterials for tissue engineering applications, J. Ind. Microbiol. Biotechnol. 45 (2018) 599–614, https://doi.org/10.1007/s10295-018-2027-3.

[8] P.M. Crapo, T.W. Gilbert, S.F. Badylak, An overview of tissue and whole organ decellularization processes, Biomaterials 32 (2011) 3233–3243, https://doi.org/10.1016/j.biomaterials.2011.01.057.

[9] Y. Wang, X. Chen, W. Cao, Y. Shi, Plasticity of mesenchymal stem cells in immunomodulation: pathological and therapeutic implications, Nat. Immunol. 15 (2014) 1009–1016, https://doi.org/10.1038/ni.3002.

[10] M.F. Pittenger, D.E. Discher, B.M. Péault, D.G. Phinney, J.M. Hare, A.I. Caplan, Mesenchymal stem cell perspective: cell biology to clinical progress, NPJ Regen. Med. 4 (2019), https://doi.org/10.1038/s41536-019-0083-6.

[11] S.G. Almalki, D.K. Agrawal, Key transcription factors in the differentiation of mesenchymal stem cells, Differentiation 92 (2016) 41–51, https://doi.org/10.1016/j.diff.2016.02.005.

[12] M. Crisan, S. Yap, L. Casteilla, C.W. Chen, M. Corselli, T.S. Park, G. Andriolo, B. Sun, B. Zheng, L. Zhang, C. Norotte, P.N. Teng, J. Traas, R. Schugar, B.M. Deasy, S. Badylak, H.J. Buhring, J.P. Giacobino, L. Lazzari, J. Huard, B. Péault, A perivascular origin for mesenchymal stem cells in multiple human organs, Cell Stem Cell 3 (2008) 301–313, https://doi.org/10.1016/j.stem.2008.07.003.

[13] N. Guimarães-Camboa, P. Cattaneo, Y. Sun, T. Moore-Morris, Y. Gu, N.D. Dalton, E. Rockenstein, E. Masliah, K.L. Peterson, W.B. Stallcup, J. Chen, S.M. Evans, Pericytes of multiple organs do not behave as mesenchymal stem cells in vivo, Cell Stem Cell 20 (2017) 345–359. e5 https://doi.org/10.1016/j.stem.2016.12.006.

[14] M. Dominici, K. Le Blanc, I. Mueller, I. Slaper-Cortenbach, F.C. Marini, D.S. Krause, R.J. Deans, A. Keating, D.J. Prockop, E.M. Horwitz, Minimal criteria for defining multipotent mesenchymal stromal cells. The International Society for Cellular Therapy position statement, Cytotherapy 8 (2006) 315–317, https://doi.org/10.1080/14653240600855905.

[15] D.J. Prockop, "Stemness" does not explain the repair of many tissues by mesenchymal stem/multipotent stromal cells (MSCs), Clin. Pharmacol. Ther. 82 (2007) 241–243, https://doi.org/10.1038/sj.clpt.6100313.

[16] Y. Iso, J.L. Spees, C. Serrano, B. Bakondi, R. Pochampally, Y.H. Song, B.E. Sobel, P. Delafontaine, D.J. Prockop, Multipotent human stromal cells improve cardiac function after myocardial infarction in mice without long-term engraftment, Biochem. Biophys. Res. Commun. 354 (2007) 700–706, https://doi.org/10.1016/j.bbrc.2007.01.045.

[17] E.M. Horwitz, P.L. Gordon, W.K.K. Koo, J.C. Marx, M.D. Neel, R.Y. McNall, L. Muul, T. Hofmann, Isolated allogeneic bone marrow-derived mesenchymal cells engraft and stimulate growth in children with osteogenesis imperfecta: implications for cell therapy of bone, Proc. Natl. Acad. Sci. U. S. A. 99 (2002) 8932–8937, https://doi.org/10.1073/pnas.132252399.

[18] L.E. Flynn, The use of decellularized adipose tissue to provide an inductive microenvironment for the adipogenic differentiation of human adipose-derived stem cells, Biomaterials 31 (2010) 4715–4724, https://doi.org/10.1016/j.biomaterials.2010.02.046.

[19] A.E.B. Turner, C. Yu, J. Bianco, J.F. Watkins, L.E. Flynn, The performance of decellularized adipose tissue microcarriers as an inductive substrate for human adipose-derived stem cells, Biomaterials 33 (2012) 4490–4499, https://doi.org/10.1016/j.biomaterials.2012.03.026.

[20] C. Yu, J. Bianco, C. Brown, L. Fuetterer, J.F. Watkins, A. Samani, L.E. Flynn, Porous decellularized adipose tissue foams for soft tissue regeneration, Biomaterials 34 (2013) 3290–3302, https://doi.org/10.1016/j.biomaterials.2013.01.056.

[21] L. Li, G. Zhou, Y. Wang, G. Yang, S. Ding, S. Zhou, Controlled dual delivery of BMP-2 and dexamethasone by nanoparticle-embedded electrospun nanofibers for the efficient repair of critical-sized rat calvarial defect, Biomaterials 37 (2015) 218–229, https://doi.org/10.1016/j.biomaterials.2014.10.015.

[22] V.B.R. Konala, M.K. Mamidi, R. Bhonde, A.K. Das, R. Pochampally, R. Pal, The current landscape of the mesenchymal stromal cell secretome: a new paradigm for cell-free regeneration, Cytotherapy 18 (2016) 13–24, https://doi.org/10.1016/j.jcyt.2015.10.008.

[23] R. Narayanan, C. Huang, S. Ravindran, Hijacking the cellular mail: exosome mediated differentiation of mesenchymal stem cells, Stem Cells Int. (2016) 3808674, https://doi.org/10.1155/2016/3808674.

[24] X. Wang, O. Omar, F. Vazirisani, P. Thomsen, K. Ekström, Mesenchymal stem cell-derived exosomes have altered microRNA profiles and induce osteogenic differentiation depending on the stage of differentiation, PLoS One 13 (2018) 4–6, https://doi.org/10.1371/journal.pone.0193059.

[25] K.H. Yoo, I.K. Jang, M.W. Lee, H.E. Kim, M.S. Yang, Y. Eom, J.E. Lee, Y.J. Kim, S.K. Yang, H.L. Jung, K.W. Sung, C.W. Kim, H.H. Koo, Comparison of immunomodulatory properties of mesenchymal stem cells derived from adult human tissues, Cell. Immunol. 259 (2009) 150–156, https://doi.org/10.1016/j.cellimm.2009.06.010.

[26] S.M. Melief, J.J. Zwaginga, W.E. Fibbe, H. Roelofs, Adipose tissue-derived multipotent stromal cells have a higher immunomodulatory capacity than their bone marrow-derived counterparts, Stem Cells Transl. Med. 2 (2013) 455–463, https://doi.org/10.5966/sctm.2012-0184.

[27] M. El-Sayed, M.A. El-Feky, M.I. El-Amir, A.S. Hasan, M. Tag-Adeen, Y. Urata, S. Goto, L. Luo, C. Yan, T.S. Li, Immunomodulatory effect of mesenchymal stem cells: cell origin and cell quality variations, Mol. Biol. Rep. 46 (2019) 1157–1165, https://doi.org/10.1007/s11033-018-04582-w.

[28] M. Strioga, S. Viswanathan, A. Darinskas, O. Slaby, J. Michalek, Same or not the same? Comparison of adipose tissue-derived versus bone marrow-derived mesenchymal stem and stromal cells, Stem Cells Dev. 21 (2012) 2724–2752, https://doi.org/10.1089/scd.2011.0722.

[29] K. Rubina, N. Kalinina, A. Efimenko, T. Lopatina, V. Melikhova, Z. Tsokolaeva, V. Sysoeva, V. Tkachuk, Y. Parfyonova, Adipose stromal cells stimulate angiogenesis via promoting progenitor cell differentiation, secretion of angiogenic factors, and enhancing vessel maturation, Tissue Eng. Part A 15 (2009) 2039–2050, https://doi.org/10.1089/ten.tea.2008.0359.

[30] H. Nakagami, K. Maeda, R. Morishita, S. Iguchi, T. Nishikawa, Y. Takami, Y. Kikuchi, Y. Saito, K. Tamai, T. Ogihara, Y. Kaneda, Novel autologous cell therapy in ischemic limb disease through growth factor secretion by cultured adipose tissue-derived stromal cells, Arterioscler. Thromb. Vasc. Biol. 25 (2005) 2542–2547, https://doi.org/10.1161/01.ATV.0000190701.92007.6d.

[31] H. Suga, J.P. Glotzbach, M. Sorkin, M.T. Longaker, G.C. Gurtner, Paracrine mechanism of angiogenesis in adipose-derived stem cell transplantation, Ann. Plast. Surg. 72 (2014) 234–241, https://doi.org/10.1097/SAP.0b013e318264fd6a.

[32] I. Linero, O. Chaparro, Paracrine effect of mesenchymal stem cells derived from human adipose tissue in bone regeneration, PLoS One 9 (2014) 1–12, https://doi.org/10.1371/journal.pone.0107001.

[33] S.T. Hsiao, Z. Lokmic, H. Peshavariya, K.M. Abberton, G.J. Dusting, S.Y. Lim, R.J. Dilley, Hypoxic conditioning enhances the angiogenic paracrine activity of human adipose-derived stem cells, Stem Cells Dev. 22 (2013) 1614–1623, https://doi.org/10.1089/scd.2012.0602.

[34] M. Sumi, M. Sata, N. Toya, K. Yanaga, T. Ohki, R. Nagai, Transplantation of adipose stromal cells, but not mature adipocytes, augments ischemia-induced angiogenesis, Life Sci. 80 (2007) 559–565, https://doi.org/10.1016/j.lfs.2006.10.020.

[35] C. Frantz, K.M. Stewart, V.M. Weaver, The extracellular matrix at a glance, J. Cell Sci. 123 (2010) 4195–4200, https://doi.org/10.1242/jcs.023820.

[36] H. Järveläinen, A. Sainio, M. Koulu, T.N. Wight, R. Penttinen, Extracellular matrix molecules: potential targets in pharmacotherapy, Pharmacol. Rev. 61 (2009) 198–223, https://doi.org/10.1124/pr.109.001289.

[37] R.O. Hynes, A. Naba, Overview of the matrisome—an inventory of extracellular matrix constituents and functions, Cold Spring Harb. Perspect. Biol. 4 (2012) 1–16, https://doi.org/10.1101/cshperspect.a004903.

[38] R. Sekiguchi, K.M. Yamada, Basement Membranes in Development and Disease, first ed., Elsevier Inc., 2018, https://doi.org/10.1016/bs.ctdb.2018.02.005.

[39] M. Kuljanin, C.F.C. Brown, M.J. Raleigh, G.A. Lajoie, L.E. Flynn, Collagenase treatment enhances proteomic coverage of low-abundance proteins in decellularized matrix bioscaffolds, Biomaterials 144 (2017) 130–143, https://doi.org/10.1016/j.biomaterials.2017.08.012.

[40] A. Naba, K.R. Clauser, S. Hoersch, H. Liu, S.A. Carr, R.O. Hynes, The matrisome: *in silico* definition and *in vivo* characterization by proteomics of normal and tumor extracellular matrices, Mol. Cell. Proteomics 11 (2012), https://doi.org/10.1074/mcp.M111.014647. M111.014647.

[41] A. Byron, J.D. Humphries, M.J. Humphries, Defining the extracellular matrix using proteomics, Int. J. Exp. Pathol. 94 (2013) 75–92, https://doi.org/10.1111/iep.12011.

[42] T.J. McKee, G. Perlman, M. Morris, S.V. Komarova, Extracellular matrix composition of connective tissues: a systematic review and meta-analysis, Sci. Rep. 9 (2019) 1–15, https://doi.org/10.1038/s41598-019-46896-0.

[43] A.J. Turlo, Y. Ashraf Kharaz, P.D. Clegg, J. Anderson, M.J. Peffers, Donor age affects proteome composition of tenocyte-derived engineered tendon, BMC Biotechnol. 18 (2018) 1–11, https://doi.org/10.1186/s12896-018-0414-5.

[44] A. Naba, O.M.T. Pearce, A. Del Rosario, D. Ma, H. Ding, V. Rajeeve, P.R. Cutillas, F.R. Balkwill, R.O. Hynes, Characterization of the extracellular matrix of normal and diseased tissues using proteomics, J. Proteome Res. 16 (2017) 3083–3091, https://doi.org/10.1021/acs.jproteome.7b00191.

[45] E.T. Goddard, R.C. Hill, A. Barrett, C. Betts, Q. Guo, O. Maller, V.F. Borges, K.C. Hansen, P. Schedin, Quantitative extracellular matrix proteomics to study mammary and liver tissue microenvironments, Int. J. Biochem. Cell Biol. 81 (2016) 223–232, https://doi.org/10.1016/j.biocel.2016.10.014.

[46] Z. Sun, S.S. Guo, R. Fässler, Integrin-mediated mechanotransduction, J. Cell Biol. 215 (2016) 445–456.

[47] A.J. García, Get a grip: integrins in cell-biomaterial interactions, Biomaterials 26 (2005) 7525–7529, https://doi.org/10.1016/j.biomaterials.2005.05.029.

[48] Y. Takada, X. Ye, S. Simon, The integrins, Genome Biol. 8 (2007), https://doi.org/10.1186/gb-2007-8-5-215.

[49] S.E. Winograd-katz, B. Geiger, The integrin adhesome: from genes and proteins to human disease, Nat. Rev. Mol. Cell Biol. 15 (2014) 273–288, https://doi.org/10.1038/nrm3769.

[50] B.D. Hoffman, C. Grashoff, M.A. Schwartz, Dynamic molecular processes mediate cellular mechanotransduction, Nature 475 (2011) 316–323, https://doi.org/10.1038/nature10316.

[51] S.E. Lee, R.D. Kamm, M.R.K. Mofrad, Force-induced activation of Talin and its possible role in focal adhesion mechanotransduction, J. Biomech. 40 (2007) 2096–2106, https://doi.org/10.1016/j.jbiomech.2007.04.006.

[52] A. Del Rio, R. Perez-Jimenez, R. Liu, P. Roca-Cusachs, J.M. Fernandez, M.P. Sheetz, Stretching single talin rod molecules activates vinculin binding, Science (80-.) 323 (2009) 638–641, https://doi.org/10.1126/science.1162912.

[53] S. Dupont, L. Morsut, M. Aragona, E. Enzo, S. Giulitti, M. Cordenonsi, F. Zanconato, J. Le Digabel, M. Forcato, S. Bicciato, N. Elvassore, S. Piccolo, Role of YAP/TAZ in mechanotransduction, Nature 474 (2011) 179–183, https://doi.org/10.1038/nature10137.

[54] G. Brusatin, T. Panciera, A. Gandin, A. Citron, S. Piccolo, Biomaterials and engineered microenvironments to control YAP/TAZ-dependent cell behaviour, Nat. Mater. 17 (2018) 1063–1075, https://doi.org/10.1038/s41563-018-0180-8.

[55] M. Finch-edmondson, M. Sudol, Framework to function: mechanosensitive regulators of gene transcription, Cell. Mol. Biol. Lett. (2016) 1–23, https://doi.org/10.1186/s11658-016-0028-7.

[56] A. Totaro, T. Panciera, S. Piccolo, YAP/TAZ upstream signals and downstream responses, Nat. Cell Biol. 20 (2018) 888–899, https://doi.org/10.1038/s41556-018-0142-z.

[57] C.D. Lawson, K. Burridge, The on-off relationship of Rho and Rac during integrin-mediated adhesion and cell migration, Small GTPases 5 (2014) e27958, https://doi.org/10.4161/sgtp.27958.

[58] F. Miralles, G. Posern, A. Zaromytidou, R. Treisman, Actin dynamics control SRF activity by regulation of its coactivator MAL, Cell 113 (2003) 329–342.

[59] O. Dobrokhotov, M. Samsonov, M. Sokabe, H. Hirata, Mechanoregulation and pathology of YAP/TAZ via Hippo and non-Hippo mechanisms, Clin. Transl. Med. (2018) 1–14, https://doi.org/10.1186/s40169-018-0202-9.

[60] A. Elosegui-artola, I. Andreu, A.E.M. Beedle, D. Navajas, S. Garcia-manyes, A. Elosegui-artola, I. Andreu, A.-E.M. Beedle, A. Lezamiz, M. Uroz, A.J. Kosmalska, Force triggers YAP nuclear entry by regulating transport across nuclear pores, Cell 171 (2017) 1397–1410. e14 https://doi.org/10.1016/j.cell.2017.10.008.
[61] L. Boeri, D. Albani, M.T. Raimondi, E. Jacchetti, Mechanical regulation of nucleocytoplasmic translocation in mesenchymal stem cells: characterization and methods for investigation, Biophys. Rev. 11 (2019) 817–831, https://doi.org/10.1007/s12551-019-00594-3.
[62] H. Safaee, M.A. Bakooshli, S. Davoudi, R.Y. Cheng, A.J. Martowirogo, E.W. Li, C.A. Simmons, P.M. Gilbert, Tethered jagged-1 synergizes with culture substrate stiffness to modulate Notch-induced myogenic progenitor differentiation, Cell. Mol. Bioeng. 10 (2017) 501–513, https://doi.org/10.1007/s12195-017-0506-7.
[63] J.M. Banks, L.C. Mozdzen, B.A.C. Harley, R.C. Bailey, The combined effects of matrix stiffness and growth factor immobilization on the bioactivity and differentiation capabilities of adipose-derived stem cells, Biomaterials 35 (2014) 8951–8959, https://doi.org/10.1016/j.biomaterials.2014.07.012.
[64] A. Becchetti, G. Petroni, A. Arcangeli, Ion channel conformations regulate integrin-dependent signaling, Trends Cell Biol. 29 (2019) 298–307, https://doi.org/10.1016/j.tcb.2018.12.005.
[65] H. Hamidi, J. Ivaska, Every step of the way: integrins in cancer progression and metastasis, Nat. Rev. Cancer (2018) 1–16, https://doi.org/10.1038/s41568-018-0038-z.
[66] D.J. White, S. Puranen, M.S. Johnson, J. Heino, The collagen receptor subfamily of the integrins, Int. J. Biochem. Cell Biol. 36 (2004) 1405–1410, https://doi.org/10.1016/j.biocel.2003.08.016.
[67] A.V. Taubenberger, M.A. Woodruff, H. Bai, D.J. Muller, D.W. Hutmacher, The effect of unlocking RGD-motifs in collagen I on pre-osteoblast adhesion and differentiation, Biomaterials 31 (2010) 2827–2835, https://doi.org/10.1016/j.biomaterials.2009.12.051.
[68] E.F. Plow, T.A. Haas, L. Zhang, J. Loftus, J.W. Smith, Ligand binding to integrins, J. Biol. Chem. 275 (2000) 21785–21788, https://doi.org/10.1074/jbc.R000003200.
[69] M.B. Rahmany, M. Van Dyke, Biomimetic approaches to modulate cellular adhesion in biomaterials: a review, Acta Biomater. 9 (2013) 5431–5437, https://doi.org/10.1016/j.actbio.2012.11.019.
[70] U. Hersel, C. Dahmen, H. Kessler, RGD modified polymers: biomaterials for stimulated cell adhesion and beyond, Biomaterials 24 (2003) 4385–4415, https://doi.org/10.1016/S0142-9612(03)00343-0.
[71] J. Fu, Y. Wang, M.T. Yang, R.A. Desai, X. Yu, Z. Liu, C.S. Chen, Mechanical regulation of cell function with geometrically modulated elastomeric substrates, Nat. Methods 7 (2010) 733–736, https://doi.org/10.1038/nmeth.1487.
[72] M. Mir, M.N. Ali, A. Barakullah, A. Gulzar, M. Arshad, S. Fatima, M. Asad, Synthetic polymeric biomaterials for wound healing: a review, Prog. Biomater. 7 (2018) 1–21, https://doi.org/10.1007/s40204-018-0083-4.
[73] D.L. Snyder, N. Sullivan, K.M. Schoelles, Skin Substitutes for Treating Chronic Wounds – Technology Assessment, Agency Healthc. Res. Qual., 2012, p. 89. http://www.ncbi.nlm.nih.gov/pubmed/25356454.
[74] Z. Tang, X. Li, Y. Tan, H. Fan, X. Zhang, The material and biological characteristics of osteoinductive calcium phosphate ceramics, Regen. Biomater. 5 (2018) 43–59, https://doi.org/10.1093/rb/rbx024.
[75] K.S. Katti, Biomaterials in total joint replacement, Colloids Surf. B Biointerfaces 39 (2004) 133–142, https://doi.org/10.1016/j.colsurfb.2003.12.002.
[76] Y. Liang, L. Li, R.A. Scott, K.L. Kiick, 50th Anniversary perspective: polymeric biomaterials: diverse functions enabled by advances in macromolecular chemistry, Macromolecules 50 (2017) 483–502, https://doi.org/10.1021/acs.macromol.6b02389.
[77] C. Houacine, S.S. Yousaf, I. Khan, R.K. Khurana, K.K. Singh, Potential of natural biomaterials in nano-scale drug delivery, Curr. Pharm. Des. 24 (2019) 5188–5206, https://doi.org/10.2174/1381612825666190118153057.
[78] V.V. Hiew, S.F.B. Simat, P.L. Teoh, The advancement of biomaterials in regulating stem cell fate, Stem Cell Rev. Rep. 14 (2018) 43–57, https://doi.org/10.1007/s12015-017-9764-y.
[79] T.J. Keane, I.T. Swinehart, S.F. Badylak, Methods of tissue decellularization used for preparation of biologic scaffolds and in vivo relevance, Methods 84 (2015) 25–34, https://doi.org/10.1016/j.ymeth.2015.03.005.
[80] C.Y. Gao, Z.H. Huang, W. Jing, P.F. Wei, L. Jin, X.H. Zhang, Q. Cai, X.L. Deng, X.P. Yang, Directing osteogenic differentiation of BMSCs by cell-secreted decellularized extracellular matrixes from different cell types, J. Mater. Chem. B 6 (2018) 7471–7485, https://doi.org/10.1039/c8tb01785a.
[81] Y. Hashimoto, S. Funamoto, T. Kimura, K. Nam, T. Fujisato, A. Kishida, The effect of decellularized bone/bone marrow produced by high-hydrostatic pressurization on the osteogenic differentiation of mesenchymal stem cells, Biomaterials 32 (2011) 7060–7067, https://doi.org/10.1016/j.biomaterials.2011.06.008.
[82] N.C. Cheng, B.T. Estes, H.A. Awad, F. Guilak, Chondrogenic differentiation of adipose-derived adult stem cells by a porous scaffold derived from native articular cartilage extracellular matrix, Tissue Eng. Part A 15 (2009) 231–241, https://doi.org/10.1089/ten.tea.2008.0253.

[83] S. Zhang, Q. Lu, T. Cao, W.S. Toh, Adipose tissue and extracellular matrix development by injectable decellularized adipose matrix loaded with basic fibroblast growth factor, Plast. Reconstr. Surg. 137 (2016) 1171–1180, https://doi.org/10.1097/PRS.0000000000002019.

[84] Y. Li, T. Jiang, L. Zheng, J. Zhao, Osteogenic differentiation of mesenchymal stem cells (MSCs) induced by three calcium phosphate ceramic (CaP) powders: a comparative study, Mater. Sci. Eng. C 80 (2017) 296–300, https://doi.org/10.1016/j.msec.2017.05.145.

[85] N.S. Manam, W.S.W. Harun, D.N.A. Shri, S.A.C. Ghani, T. Kurniawan, M.H. Ismail, M.H.I. Ibrahim, Study of corrosion in biocompatible metals for implants: a review, J. Alloys Compd. 701 (2017) 698–715, https://doi.org/10.1016/j.jallcom.2017.01.196.

[86] F. Traina, M. De Fine, A. Di Martino, C. Faldini, Fracture of ceramic bearing surfaces following total hip replacement: a systematic review, Biomed. Res. Int. 2013 (2013), https://doi.org/10.1155/2013/157247.

[87] C. Gao, S. Peng, P. Feng, C. Shuai, Bone biomaterials and interactions with stem cells, Bone Res. 5 (2017) 1–33, https://doi.org/10.1038/boneres.2017.59.

[88] S. Fujibayashi, M. Neo, H.M. Kim, T. Kokubo, T. Nakamura, Osteoinduction of porous bioactive titanium metal, Biomaterials 25 (2004) 443–450, https://doi.org/10.1016/S0142-9612(03)00551-9.

[89] T.A.B. Bressel, J.D.F. De Queiroz, S.M. Gomes Moreira, J.T. Da Fonseca, E.A. Filho, A.C. Guastaldi, S.R. Batistuzzo De Medeiros, Laser-modified titanium surfaces enhance the osteogenic differentiation of human mesenchymal stem cells, Stem Cell Res. Ther. 8 (2017) 1–11, https://doi.org/10.1186/s13287-017-0717-9.

[90] J. Stepanovska, R. Matejka, M. Otahal, J. Rosina, L. Bacakova, The effect of various surface treatments of Ti_6Al_4V on the growth and osteogenic differentiation of adipose tissue-derived stem cells, Coatings 10 (2020) 1–23, https://doi.org/10.3390/COATINGS10080762.

[91] F. Benazzo, L. Botta, M.F. Scaffino, L. Caliogna, M. Marullo, S. Fusi, G. Gastaldi, Trabecular titanium can induce in vitro osteogenic differentiation of human adipose derived stem cells without osteogenic factors, J. Biomed. Mater. Res. Part A 102 (2014) 2061–2071, https://doi.org/10.1002/jbm.a.34875.

[92] J.M. Bouler, P. Pilet, O. Gauthier, E. Verron, Biphasic calcium phosphate ceramics for bone reconstruction: a review of biological response, Acta Biomater. 53 (2017) 1–12, https://doi.org/10.1016/j.actbio.2017.01.076.

[93] A.M.C. Barradas, V. Monticone, M. Hulsman, C. Danoux, H. Fernandes, Z. Tahmasebi Birgani, F. Barrère-De Groot, H. Yuan, M. Reinders, P. Habibovic, C. Van Blitterswijk, J. De Boer, Molecular mechanisms of biomaterial-driven osteogenic differentiation in human mesenchymal stromal cells, Integr. Biol. (United Kingdom) 5 (2013) 920–931, https://doi.org/10.1039/c3ib40027a.

[94] A.M.C. Barradas, H.A.M. Fernandes, N. Groen, Y.C. Chai, J. Schrooten, J. Van de Peppel, J.P.T.M. Van Leeuwen, C.A. Van Blitterswijk, J. De Boer, A calcium-induced signaling cascade leading to osteogenic differentiation of human bone marrow-derived mesenchymal stromal cells, Biomaterials 33 (2012) 3205–3215, https://doi.org/10.1016/j.biomaterials.2012.01.020.

[95] J. Wang, Y. Chen, X. Zhu, T. Yuan, Y. Tan, Y. Fan, X. Zhang, Effect of phase composition on protein adsorption and osteoinduction of porous calcium phosphate ceramics in mice, J. Biomed. Mater. Res. Part A 102 (2014) 4234–4243, https://doi.org/10.1002/jbm.a.35102.

[96] A.J. Steward, D.J. Kelly, Mechanical regulation of mesenchymal stem cell differentiation, J. Anat. 227 (2015) 717–731, https://doi.org/10.1111/joa.12243.

[97] A. Saez, A. Buguin, P. Silberzan, B. Ladoux, Is the mechanical activity of epithelial cells controlled by deformations or forces? Biophys. J. (2005) 52–54, https://doi.org/10.1529/biophysj.105.071217.

[98] J.H. Wen, L.G. Vincent, A. Fuhrmann, Y.S. Choi, K.C. Hribar, H. Taylor-Weiner, S. Chen, A.J. Engler, Interplay of matrix stiffness and protein tethering in stem cell differentiation, Nat. Mater. 13 (2014) 979–987, https://doi.org/10.1038/nmat4051.

[99] A.J. Engler, S. Sen, H.L. Sweeney, D.E. Discher, Matrix elasticity directs stem cell lineage specification, Cell 126 (2006) 677–689, https://doi.org/10.1016/j.cell.2006.06.044.

[100] D.A. Young, Y.S. Choi, A.J. Engler, K.L. Christman, Stimulation of adipogenesis of adult adipose-derived stem cells using substrates that mimic the stiffness of adipose tissue, Biomaterials 34 (2013) 8581–8588, https://doi.org/10.1016/j.biomaterials.2013.07.103.

[101] M. Sun, G. Chi, P. Li, S. Lv, J. Xu, Z. Xu, Y. Xia, Y. Tan, J. Xu, L. Li, Y. Li, Effects of matrix stiffness on the morphology, adhesion, proliferation and osteogenic differentiation of mesenchymal stem cells, Int. J. Med. Sci. 15 (2018) 257–268, https://doi.org/10.7150/ijms.21620.

[102] N. Huebsch, P.R. Arany, A.S. Mao, D. Shvartsman, O.A. Ali, S.A. Bencherif, J. Rivera-Feliciano, D.J. Mooney, Harnessing traction-mediated manipulation of the cell/matrix interface to control stem-cell fate, Nat. Mater. 9 (2010) 518–526, https://doi.org/10.1038/nmat2732.

[103] S. Budday, R. Nay, R. de Rooij, P. Steinmann, T. Wyrobek, T.C. Ovaert, E. Kuhl, Mechanical properties of gray and white matter brain tissue by indentation, J. Mech. Behav. Biomed. Mater. 46 (2015) 318–330, https://doi.org/10.1016/j.jmbbm.2015.02.024.
[104] D.T. Butcher, T. Alliston, V.M. Weaver, A tense situation: forcing tumour progression, Nat. Rev. Cancer 9 (2009) 108–122, https://doi.org/10.1038/nrc2544.
[105] D. Wu, P. Isaksson, S.J. Ferguson, C. Persson, Young's modulus of trabecular bone at the tissue level: a review, Acta Biomater. 78 (2018) 1–12, https://doi.org/10.1016/j.actbio.2018.08.001.
[106] N. Alkhouli, J. Mansfield, E. Green, J. Bel, B. Knight, N. Liversedge, J.C. Tham, R. Welbourn, A.C. Shore, K. Kos, C.P. Winlove, The mechanical properties of human adipose tissues and their relationships to the structure and composition of the extracellular matrix, Am. J. Physiol. Endocrinol. Metab. 305 (2013) 1427–1435, https://doi.org/10.1152/ajpendo.00111.2013.
[107] J. Antons, M.G.M. Marascio, J. Nohava, R. Martin, L.A. Applegate, P.E. Bourban, D.P. Pioletti, Zone-dependent mechanical properties of human articular cartilage obtained by indentation measurements, J. Mater. Sci. Mater. Med. 29 (2018), https://doi.org/10.1007/s10856-018-6066-0.
[108] A. Buxboim, I.L. Ivanovska, D.E. Discher, Matrix elasticity, cytoskeletal forces and physics of the nucleus: how deeply do cells "feel" outside and in? J. Cell Sci. 123 (2010) 297–308, https://doi.org/10.1242/jcs.041186.
[109] I. Levental, P.C. Georges, P.A. Janmey, Soft biological materials and their impact on cell function, Soft Matter 3 (2007) 299–306, https://doi.org/10.1039/b610522j.
[110] S. Wan, X. Fu, Y. Ji, M. Li, X. Shi, Y. Wang, FAK- and YAP/TAZ dependent mechanotransduction pathways are required for enhanced immunomodulatory properties of adipose-derived mesenchymal stem cells induced by aligned fi brous scaffolds, Biomaterials 171 (2018) 107–117, https://doi.org/10.1016/j.biomaterials.2018.04.035.
[111] M. Darnell, L. Gu, D. Mooney, RNA-seq reveals diverse effects of substrate stiffness on mesenchymal stem cells, Biomaterials 181 (2018) 182–188, https://doi.org/10.1016/j.biomaterials.2018.07.039.
[112] K.C. Murphy, J. Whitehead, D. Zhou, S.S. Ho, J.K. Leach, Engineering fibrin hydrogels to promote the wound healing potential of mesenchymal stem cell spheroids, Acta Biomater. 64 (2017) 176–186, https://doi.org/10.1016/j.actbio.2017.10.007.
[113] S. Khetan, M. Guvendiren, W.R. Legant, D.M. Cohen, C.S. Chen, J.A. Burdick, Degradation-mediated cellular traction directs stem cell fate in covalently crosslinked three-dimensional hydrogels, Nat. Mater. 12 (2013) 458–465, https://doi.org/10.1038/nmat3586.
[114] C. Loebel, R.L. Mauck, J.A. Burdick, Local nascent protein deposition and remodelling guide mesenchymal stromal cell mechanosensing and fate in three-dimensional hydrogels, Nat. Mater. 18 (2019) 883–891, https://doi.org/10.1038/s41563-019-0307-6.
[115] V.N. Goral, Y.C. Hsieh, O.N. Petzold, J.S. Clark, P.K. Yuen, R.A. Faris, Perfusion-based microfluidic device for three-dimensional dynamic primary human hepatocyte cell culture in the absence of biological or synthetic matrices or coagulants, Lab Chip 10 (2010) 3380–3386, https://doi.org/10.1039/c0lc00135j.
[116] J. Rouwkema, B.F.J.M. Koopman, C.A.V. Blitterswijk, W.J.A. Dhert, J. Malda, Supply of nutrients to cells in engineered tissues, Biotechnol. Genet. Eng. Rev. 26 (2009) 163–178, https://doi.org/10.5661/bger-26-163.
[117] H. Miyoshi, T. Adachi, Topography design concept of a tissue engineering scaffold for controlling cell function and fate through actin cytoskeletal modulation, Tissue Eng. Part B Rev. 20 (2014) 609–627, https://doi.org/10.1089/ten.teb.2013.0728.
[118] W. Chen, Y. Shao, X. Li, G. Zhao, J. Fu, Nanotopographical surfaces for stem cell fate control: engineering mechanobiology from the bottom, Nano Today 9 (2014) 759–784, https://doi.org/10.1016/j.nantod.2014.12.002.
[119] J. Wu, Y. Hong, Enhancing cell infiltration of electrospun fibrous scaffolds in tissue regeneration, Bioact. Mater. 1 (2016) 56–64, https://doi.org/10.1016/j.bioactmat.2016.07.001.
[120] W. Liu, S. Liu, L. Wang, Surface modification of biomedical titanium alloy: micromorphology, microstructure evolution and biomedical applications, Coatings 9 (2019) 249, https://doi.org/10.3390/coatings9040249.
[121] A. Jemat, M.J. Ghazali, M. Razali, Y. Otsuka, Surface modifications and their effects on titanium dental implants, Biomed. Res. Int. 2015 (2015), https://doi.org/10.1155/2015/791725.
[122] M.N. Andalib, J.S. Lee, L. Ha, Y. Dzenis, J.Y. Lim, Focal adhesion kinase regulation in stem cell alignment and spreading on nanofibers, Biochem. Biophys. Res. Commun. 473 (2016) 920–925, https://doi.org/10.1016/j.bbrc.2016.03.151.
[123] Y. Wang, R. Gao, P.P. Wang, J. Jian, X.L. Jiang, C. Yan, X. Lin, L. Wu, G.Q. Chen, Q. Wu, The differential effects of aligned electrospun PHBHHx fibers on adipogenic and osteogenic potential of MSCs through the regulation of PPARγ signaling, Biomaterials 33 (2012) 485–493, https://doi.org/10.1016/j.biomaterials.2011.09.089.

[124] S. Wang, S. Zhong, C.T. Lim, H. Nie, Effects of fiber alignment on stem cells-fibrous scaffold interactions, J. Mater. Chem. B 3 (2015) 3358–3366, https://doi.org/10.1039/c5tb00026b.

[125] P. Newman, J.L. Galenano-Ninõ, P. Graney, J.M. Razal, A.I. Minett, J. Ribas, R. Ovalle-Robles, M. Biro, H. Zreiqat, Relationship between nanotopographical alignment and stem cell fate with live imaging and shape analysis, Sci. Rep. 6 (2016) 1–12, https://doi.org/10.1038/srep37909.

[126] A. Ince Yardimci, O. Baskan, S. Yilmaz, G. Mese, E. Ozcivici, Y. Selamet, Osteogenic differentiation of mesenchymal stem cells on random and aligned PAN/PPy nanofibrous scaffolds, J. Biomater. Appl. 34 (2019) 640–650, https://doi.org/10.1177/0885328219865068.

[127] Z. Yin, X. Chen, H.X. Song, J.J. Hu, Q.M. Tang, T. Zhu, W.L. Shen, J.L. Chen, H. Liu, B.C. Heng, H.W. Ouyang, Electrospun scaffolds for multiple tissues regeneration invivo through topography dependent induction of lineage specific differentiation, Biomaterials 44 (2015) 173–185, https://doi.org/10.1016/j.biomaterials.2014.12.027.

[128] D. Olvera, B.N. Sathy, S.F. Carroll, D.J. Kelly, Modulating microfibrillar alignment and growth factor stimulation to regulate mesenchymal stem cell differentiation, Acta Biomater. 64 (2017) 148–160, https://doi.org/10.1016/j.actbio.2017.10.010.

[129] K.A. Kilian, B. Bugarija, B.T. Lahn, M. Mrksich, Geometric cues for directing the differentiation of mesenchymal stem cells, Proc. Natl. Acad. Sci. U. S. A. 107 (2010) 4872–4877, https://doi.org/10.1073/pnas.0903269107.

[130] R.J. McMurray, N. Gadegaard, P.M. Tsimbouri, K.V. Burgess, L.E. McNamara, R. Tare, K. Murawski, E. Kingham, R.O.C. Oreffo, M.J. Dalby, Nanoscale surfaces for the long-term maintenance of mesenchymal stem cell phenotype and multipotency, Nat. Mater. 10 (2011) 637–644, https://doi.org/10.1038/nmat3058.

[131] J.K. Chaudhary, P.C. Rath, Microgrooved-surface topography enhances cellular division and proliferation of mouse bone marrow-derived mesenchymal stem cells, PLoS One 12 (2017) 1–22, https://doi.org/10.1371/journal.pone.0182128.

[132] A.B. Faia-Torres, M. Charnley, T. Goren, S. Guimond-Lischer, M. Rottmar, K. Maniura-Weber, N.D. Spencer, R.-L. Reis, M. Textor, N.M. Neves, Osteogenic differentiation of human mesenchymal stem cells in the absence of osteogenic supplements: a surface-roughness gradient study, Acta Biomater. 28 (2015) 64–75, https://doi.org/10.1016/j.actbio.2015.09.028.

[133] W. Qian, L. Gong, X. Cui, Z. Zhang, A. Bajpai, C. Liu, A.B. Castillo, J.C.M. Teo, W. Chen, Nanotopographic regulation of human mesenchymal stem cell osteogenesis, ACS Appl. Mater. Interfaces 9 (2017) 41794–41806, https://doi.org/10.1021/acsami.7b16314.

[134] M.J. Davison, R.J. McMurray, C.A. Smith, M.J. Dalby, R.D. Meek, Nanopit-induced osteoprogenitor cell differentiation: the effect of nanopit depth, J. Tissue Eng. 7 (2016), https://doi.org/10.1177/2041731416652778.

[135] G. Abagnale, M. Steger, V.H. Nguyen, N. Hersch, A. Sechi, S. Joussen, B. Denecke, R. Merkel, B. Hoffmann, A. Dreser, U. Schnakenberg, A. Gillner, W. Wagner, Surface topography enhances differentiation of mesenchymal stem cells towards osteogenic and adipogenic lineages, Biomaterials 61 (2015) 316–326, https://doi.org/10.1016/j.biomaterials.2015.05.030.

[136] M.J. Dalby, N. Gadegaard, R. Tare, A. Andar, M.O. Riehle, P. Herzyk, C.D.W. Wilkinson, R.O.C. Oreffo, The control of human mesenchymal cell differentiation using nanoscale symmetry and disorder, Nat. Mater. 6 (2007) 997–1003, https://doi.org/10.1038/nmat2013.

[137] Y.N. Wu, J.B.K. Law, A.Y. He, H.Y. Low, J.H.P. Hui, C.T. Lim, Z. Yang, E.H. Lee, Substrate topography determines the fate of chondrogenesis from human mesenchymal stem cells resulting in specific cartilage phenotype formation, Nanomedicine 10 (2014) 1507–1516, https://doi.org/10.1016/j.nano.2014.04.002.

[138] R. Ortiz, I. Aurrekoetxea-Rodríguez, M. Rommel, I. Quintana, M.M. Vivanco, J.L. Toca-Herrera, Laser surface microstructuring of a bio-resorbable polymer to anchor stem cells, control adipocyte morphology, and promote osteogenesis, Polymers (Basel) 10 (2018), https://doi.org/10.3390/polym10121337.

[139] Y.J. Bae, Y.R. Kwon, H.J. Kim, S. Lee, Y.J. Kim, Enhanced differentiation of mesenchymal stromal cells by three-dimensional culture and azacitidine, Blood Res. 52 (2017) 18–24, https://doi.org/10.5045/br.2017.52.1.18.

[140] M.M. Bonab, K. Alimoghaddam, F. Talebian, S.H. Ghaffari, A. Ghavamzadeh, B. Nikbin, Aging of mesenchymal stem cell in vitro, BMC Cell Biol. 7 (2006) 1–7, https://doi.org/10.1186/1471-2121-7-14.

[141] J. Li, T. Chen, X. Huang, Y. Zhao, B. Wang, Y. Yin, Y. Cui, Y. Zhao, R. Zhang, X. Wang, Y. Wang, J. Dai, Substrate-independent immunomodulatory characteristics of mesenchymal stem cells in three-dimensional culture, PLoS One 13 (2018) 1–17, https://doi.org/10.1371/journal.pone.0206811.

[142] D.A. Gaspar, V. Gomide, F.J. Monteiro, The role of perfusion bioreactors in bone tissue engineering, Biomatter 2 (2012) 167–175, https://doi.org/10.4161/biom.22170.

[143] F.W. Janssen, J. Oostra, A. Van Oorschot, C.A. Van Blitterswijk, A perfusion bioreactor system capable of producing clinically relevant volumes of tissue-engineered bone: in vivo bone formation showing proof of concept, Biomaterials 27 (2006) 315–323, https://doi.org/10.1016/j.biomaterials.2005.07.044.
[144] Q.L. Loh, C. Choong, Three-dimensional scaffolds for tissue engineering applications: role of porosity and pore size, Tissue Eng. Part B Rev. 19 (2013) 485–502, https://doi.org/10.1089/ten.teb.2012.0437.
[145] F.P.W. Melchels, B. Tonnarelli, A.L. Olivares, I. Martin, D. Lacroix, J. Feijen, D.J. Wendt, D.W. Grijpma, The influence of the scaffold design on the distribution of adhering cells after perfusion cell seeding, Biomaterials 32 (2011) 2878–2884, https://doi.org/10.1016/j.biomaterials.2011.01.023.
[146] A. Matsiko, J.P. Gleeson, F.J. O'Brien, Scaffold mean pore size influences mesenchymal stem cell chondrogenic differentiation and matrix deposition, Tissue Eng. Part A 21 (2015) 486–497, https://doi.org/10.1089/ten.tea.2013.0545.
[147] S.H. Oh, T.H. Kim, G. Il Im, J.H. Lee, Investigation of pore size effect on chondrogenic differentiation of adipose stem cells using a pore size gradient scaffold, Biomacromolecules 11 (2010) 1948–1955, https://doi.org/10.1021/bm100199m.
[148] P. Kasten, I. Beyen, P. Niemeyer, R. Luginbühl, M. Bohner, W. Richter, Porosity and pore size of β-tricalcium phosphate scaffold can influence protein production and osteogenic differentiation of human mesenchymal stem cells: an in vitro and in vivo study, Acta Biomater. 4 (2008) 1904–1915, https://doi.org/10.1016/j.actbio.2008.05.017.
[149] Y. Takahashi, Y. Tabata, Effect of the fiber diameter and porosity of non-woven PET fabrics on the osteogenic differentiation of mesenchymal stem cells, J. Biomater. Sci. Polym. Ed. 15 (2004) 41–57, https://doi.org/10.1163/156856204322752228.
[150] S. Gobaa, S. Hoehnel, M. Roccio, A. Negro, S. Kobel, M.P. Lutolf, Artificial niche microarrays for probing single stem cell fate in high throughput, Nat. Methods 8 (2011) 949–955, https://doi.org/10.1038/nmeth.1732.
[151] B. Cao, Z. Li, R. Peng, J. Ding, Effects of cell-cell contact and oxygen tension on chondrogenic differentiation of stem cells, Biomaterials 64 (2015) 21–32, https://doi.org/10.1016/j.biomaterials.2015.06.018.
[152] J. Li, D.J. Mooney, Designing hydrogels for controlled drug delivery, Nat. Rev. Mater. 1 (2016), https://doi.org/10.1038/natrevmats.2016.71.
[153] J.R. Weiser, W.M. Saltzman, Controlled release for local delivery of drugs: barriers and models, J. Control. Release 190 (2014) 664–673, https://doi.org/10.1016/j.jconrel.2014.04.048.
[154] Z. Wang, Z. Wang, W.W. Lu, W. Zhen, D. Yang, S. Peng, Novel biomaterial strategies for controlled growth factor delivery for biomedical applications, NPG Asia Mater. 9 (2017), https://doi.org/10.1038/am.2017.171. e435-17.
[155] S. Vrignaud, J.P. Benoit, P. Saulnier, Strategies for the nanoencapsulation of hydrophilic molecules in polymer-based nanoparticles, Biomaterials 32 (2011) 8593–8604, https://doi.org/10.1016/j.biomaterials.2011.07.057.
[156] N.K. Varde, D.W. Pack, Microspheres for controlled release drug delivery, Expert. Opin. Biol. Ther. 4 (2004) 35–51, https://doi.org/10.1517/14712598.4.1.35.
[157] S. Hegde, T. Pant, K. Pradhan, M. Badiger, M. Gadgil, Controlled release of nutrients to mammalian cells cultured in shake flasks, Biotechnol. Prog. 28 (2012) 188–195, https://doi.org/10.1002/btpr.729.
[158] S.A.A. Kooijmans, R.M. Schiffelers, N. Zarovni, R. Vago, Modulation of tissue tropism and biological activity of exosomes and other extracellular vesicles: new nanotools for cancer treatment, Pharmacol. Res. 111 (2016) 487–500, https://doi.org/10.1016/j.phrs.2016.07.006.
[159] S. Freiberg, X.X. Zhu, Polymer microspheres for controlled drug release, Int. J. Pharm. 282 (2004) 1–18, https://doi.org/10.1016/j.ijpharm.2004.04.013.
[160] T. Miao, K.S. Rao, J.L. Spees, R.A. Oldinski, Osteogenic differentiation of human mesenchymal stem cells through alginate-graft-poly (ethylene glycol) microsphere-mediated intracellular growth factor delivery, J. Control. Release 192 (2014) 57–66, https://doi.org/10.1016/j.jconrel.2014.06.029.
[161] S.L. Fenn, T. Miao, R.M. Scherrer, R.A. Oldinski, Dual-cross-linked methacrylated alginate sub-microspheres for intracellular chemotherapeutic delivery, ACS Appl. Mater. Interfaces 8 (2016) 17775–17783, https://doi.org/10.1021/acsami.6b03245.
[162] Y. Liu, A.D. Berendsen, S. Jia, S. Lotinun, R. Baron, N. Ferrara, B.R. Olsen, Intracellular VEGF regulates the balance between osteoblast and adipocyte differentiation, J. Clin. Invest. 122 (2012) 3101–3113, https://doi.org/10.1172/JCI61209.
[163] L.D. Solorio, A.S. Fu, R. Hernández-Irizarry, E. Alsberg, Chondrogenic differentiation of human mesenchymal stem cell aggregates via controlled release of TGF-β1 from incorporated polymer microspheres, J. Biomed. Mater. Res. Part A 92 (2010) 1139–1144, https://doi.org/10.1002/jbm.a.32440.

[164] A.J. DeFail, C.R. Chu, N. Izzo, K.G. Marra, Controlled release of bioactive TGF-β1 from microspheres embedded within biodegradable hydrogels, Biomaterials 27 (2006) 1579–1585, https://doi.org/10.1016/j.biomaterials.2005.08.013.

[165] L.D. Solorio, C.D. Dhami, P.N. Dang, E.L. Vieregge, E. Alsberg, Spatiotemporal regulation of chondrogenic differentiation with controlled delivery of transforming growth factor-β1 from gelatin microspheres in mesenchymal stem cell aggregates, Stem Cells Transl. Med. 1 (2012) 632–639, https://doi.org/10.5966/sctm.2012-0039.

[166] L. Solorio, C. Zwolinski, A.W. Lund, M.J. Farrell, J.P. Stegemann, Gelatin microspheres crosslinked with genipin for local delivery of growth factors, J. Tissue Eng. Regen. Med. 4 (2010) 514–523, https://doi.org/10.1002/term.267.

[167] M. Mathieu, S. Vigier, M.N. Labour, C. Jorgensen, E. Belamie, D. Noël, Induction of mesenchymal stem cell differentiation and cartilage formation by cross-linker-free collagen microspheres, Eur. Cells Mater. 28 (2014) 82–97. https://doi.org/10.22203/eCM.v028a07.

[168] A. Sukarto, C. Yu, L.E. Flynn, B.G. Amsden, Co-delivery of adipose-derived stem cells and growth factor-loaded microspheres in RGD-grafted N-methacrylate glycol chitosan gels for focal chondral repair, Biomacromolecules 13 (2012) 2490–2502, https://doi.org/10.1021/bm300733n.

[169] J.S. Park, H.J. Lim, S.W. Yi, K.H. Park, Stem cell differentiation-related protein-loaded PLGA microspheres as a novel platform micro-typed scaffold for chondrogenesis, Biomed. Mater. 11 (2016), https://doi.org/10.1088/1748-6041/11/5/055003.

[170] J.P. Rubin, A. DeFail, N. Rajendran, K.G. Marra, Encapsulation of adipogenic factors to promote differentiation of adipose-derived stem cells, J. Drug Target. 17 (2009) 207–215, https://doi.org/10.1080/10611860802669231.

[171] M.L. Moya, M.H. Cheng, J.J. Huang, M.E. Francis-Sedlak, S.W. Kao, E.C. Opara, E.M. Brey, The effect of FGF-1 loaded alginate microbeads on neovascularization and adipogenesis in a vascular pedicle model of adipose tissue engineering, Biomaterials 31 (2010) 2816–2826, https://doi.org/10.1016/j.biomaterials.2009.12.053.

[172] A.V. Vashi, K.M. Abberton, G.P. Thomas, W.A. Morrison, A.J. O'Connor, J.J. Cooper-White, E.W. Thompson, Adipose tissue engineering based on the controlled release of fibroblast growth factor-2 in a collagen matrix, Tissue Eng. 12 (2006) 3035–3043, https://doi.org/10.1089/ten.2006.12.3035.

[173] M. Yazawa, T. Mori, Y. Nakayama, K. Kishi, Basic study of soft tissue augmentation by adipose-inductive biomaterial, J. Biomed. Mater. Res. Part B Appl. Biomater. 103 (2015) 92–96, https://doi.org/10.1002/jbm.b.33180.

[174] B. Shu, M. Zhang, R. Xie, M. Wang, H. Jin, W. Hou, D. Tang, S.E. Harris, Y. Mishina, R.J. O'Keefe, M.J. Hilton, Y. Wang, D. Chen, BMP2, but not BMP4, is crucial for chondrocyte proliferation and maturation during endochondral bone development, J. Cell Sci. 124 (2011) 3428–3440, https://doi.org/10.1242/jcs.083659.

[175] K. Tsuji, A. Bandyopadhyay, B.D. Harfe, K. Cox, S. Kakar, L. Gerstenfeld, T. Einhorn, C.J. Tabin, V. Rosen, BMP2 activity, although dispensable for bone formation, is required for the initiation of fracture healing, Nat. Genet. 38 (2006) 1424–1429, https://doi.org/10.1038/ng1916.

[176] S. Scarfì, Use of bone morphogenetic proteins in mesenchymal stem cell stimulation of cartilage and bone repair, World J. Stem Cells 8 (2016) 1–12, https://doi.org/10.4252/wjsc.v8.i1.1.

[177] F. Langenbach, J. Handschel, Effects of dexamethasone, ascorbic acid and beta-glycerophosphate on the osteogenic differentiation of stem cells in vitro, Stem Cell Res. Ther. 4 (2013) 117, https://doi.org/10.1186/scrt328.

[178] M. Yuasa, T. Yamada, T. Taniyama, T. Masaoka, W. Xuetao, T. Yoshii, M. Horie, H. Yasuda, T. Uemura, A. Okawa, S. Sotome, Dexamethasone enhances osteogenic differentiation of bone marrow-and muscle-derived stromal cells and augments ectopic bone formation induced by bone morphogenetic protein-2, PLoS One 10 (2015) 1–23, https://doi.org/10.1371/journal.pone.0116462.

[179] D.H. Choi, C.H. Park, I.H. Kim, H.J. Chun, K. Park, D.K. Han, Fabrication of core-shell microcapsules using PLGA and alginate for dual growth factor delivery system, J. Control. Release 147 (2010) 193–201, https://doi.org/10.1016/j.jconrel.2010.07.103.

[180] Y. Su, Q. Su, W. Liu, M. Lim, J.R. Venugopal, X. Mo, S. Ramakrishna, S.S. Al-Deyab, M. El-Newehy, Controlled release of bone morphogenetic protein 2 and dexamethasone loaded in core-shell PLLACL-collagen fibers for use in bone tissue engineering, Acta Biomater. 8 (2012) 763–771, https://doi.org/10.1016/j.actbio.2011.11.002.

[181] S. Srouji, D. Ben-David, R. Lotan, E. Livne, R. Avrahami, E. Zussman, Slow-release human recombinant bone morphogenetic protein-2 embedded within electrospun scaffolds for regeneration of bone defect: In vitro and in vivo evaluation, Tissue Eng. Part A 17 (2011) 269–277, https://doi.org/10.1089/ten.tea.2010.0250.

[182] H. Zhu, D. Yu, Y. Zhou, C. Wang, M. Gao, H. Jiang, H. Wang, Biological activity of a nanofibrous barrier membrane containing bone morphogenetic protein formed by core-shell electrospinning as a sustained delivery vehicle, J. Biomed. Mater. Res. Part B Appl. Biomater. 101 (2013) 541–552, https://doi.org/10.1002/jbm.b.32854.

[183] Z. Man, L. Yin, Z. Shao, X. Zhang, X. Hu, J. Zhu, L. Dai, H. Huang, L. Yuan, C. Zhou, H. Chen, Y. Ao, The effects of co-delivery of BMSC-affinity peptide and rhTGF-β1 from coaxial electrospun scaffolds on chondrogenic differentiation, Biomaterials 35 (2014) 5250–5260, https://doi.org/10.1016/j.biomaterials.2014.03.031.
[184] W. Zhu, H. Cui, B. Boualam, F. Masood, E. Flynn, R.D. Rao, Z.Y. Zhang, L.G. Zhang, 3D bioprinting mesenchymal stem cell-laden construct with core-shell nanospheres for cartilage tissue engineering, Nanotechnology 29 (2018), https://doi.org/10.1088/1361-6528/aaafa1.
[185] G. Kibria, E.K. Ramos, Y. Wan, D.R. Gius, H. Liu, Exosomes as a drug delivery system in cancer therapy: potential and challenges, Mol. Pharm. 15 (2018) 3625–3633, https://doi.org/10.1021/acs.molpharmaceut.8b00277.
[186] B. Crivelli, T. Chlapanidas, S. Perteghella, E. Lucarelli, L. Pascucci, A.T. Brini, I. Ferrero, M. Marazzi, A. Pessina, M.L. Torre, Mesenchymal stem/stromal cell extracellular vesicles: from active principle to next generation drug delivery system, J. Control. Release 262 (2017) 104–117, https://doi.org/10.1016/j.jconrel.2017.07.023.
[187] N. Heldring, I. Mäger, M.J.A. Wood, K. Le Blanc, S.E.L. Andaloussi, Therapeutic potential of multipotent mesenchymal stromal cells and their extracellular vesicles, Hum. Gene Ther. 26 (2015) 506–517, https://doi.org/10.1089/hum.2015.072.
[188] D.G. Phinney, M.F. Pittenger, Concise review: MSC-derived exosomes for cell-free therapy, Stem Cells 35 (2017) 851–858, https://doi.org/10.1002/stem.2575.
[189] M.M. Pakulska, S. Miersch, M.S. Shoichet, Designer protein delivery: from natural to engineered affinity-controlled release systems, Science (80-.) 351 (2016), https://doi.org/10.1126/science.aac4750.
[190] S.E. Sakiyama-Elbert, Incorporation of heparin into biomaterials, Acta Biomater. 10 (2014) 1581–1587, https://doi.org/10.1016/j.actbio.2013.08.045.
[191] C.-C. Lin, A.T. Metters, Metal-chelating affinity hydrogels for sustained protein release, J. Biomed. Mater. Res. Part A 83A (2007) 954–964, https://doi.org/10.1002/jbm.a.31282.
[192] T. Re'em, Y. Kaminer-Israeli, E. Ruvinov, S. Cohen, Chondrogenesis of hMSC in affinity-bound TGF-beta scaffolds, Biomaterials 33 (2012) 751–761, https://doi.org/10.1016/j.biomaterials.2011.10.007.
[193] A.J. Salgado, R.L. Reis, N. Sousa, J.M. Gimble, Adipose tissue derived stem cells secretome: soluble factors and their roles in regenerative medicine, Curr. Stem Cell Res. Ther. 5 (2010) 103–110. http://www.eurekaselect.com/openurl/content.php?genre=article&issn=1574-888X&volume=5&issue=2&spage=103.
[194] A.I. Caplan, What's in a name? Tissue Eng. Part A 16 (2010) 2415–2417, https://doi.org/10.1089/ten.tea.2010.0216.
[195] A. Parisi-Amon, W. Mulyasasmita, C. Chung, S.C. Heilshorn, Protein-engineered injectable hydrogel to improve retention of transplanted adipose-derived stem cells, Adv. Healthc. Mater. 2 (2013) 428–432, https://doi.org/10.1002/adhm.201200293.
[196] M.E. Danoviz, J.S. Nakamuta, F.L.N. Marques, L. dos Santos, E.C. Alvarenga, A.A. dos Santos, E.L. Antonio, I.T. Schettert, P.J. Tucci, J.E. Krieger, Rat adipose tissue-derived stem cells transplantation attenuates cardiac dysfunction post infarction and biopolymers enhance cell retention, PLoS One 5 (2010) 1–9, https://doi.org/10.1371/journal.pone.0012077.
[197] Y.S. Kim, Y.J. Choi, D.S. Suh, D.B. Heo, Y. Il Kim, J.S. Ryu, Y.G. Koh, Mesenchymal stem cell implantation in osteoarthritic knees: is fibrin glue effective as a scaffold? Am. J. Sports Med. 43 (2015) 176–185, https://doi.org/10.1177/0363546514554190.
[198] C.M. Mahoney, C. Imbarlina, C.C. Yates, K.G. Marra, Current therapeutic strategies for adipose tissue defects/repair using engineered biomaterials and biomolecule formulations, Front. Pharmacol. 9 (2018) 1–12, https://doi.org/10.3389/fphar.2018.00507.
[199] F. Pieri, E. Lucarelli, G. Corinaldesi, N.N. Aldini, M. Fini, A. Parrilli, B. Dozza, D. Donati, C. Marchetti, Dose-dependent effect of adipose-derived adult stem cells on vertical bone regeneration in rabbit calvarium, Biomaterials 31 (2010) 3527–3535, https://doi.org/10.1016/j.biomaterials.2010.01.066.
[200] P.P. Carvalho, I.B. Leonor, B.J. Smith, I.R. Dias, R.L. Reis, J.M. Gimble, M.E. Gomes, Undifferentiated human adipose-derived stromal/stem cells loaded onto wet-spun starch-polycaprolactone scaffolds enhance bone regeneration: nude mice calvarial defect in vivo study, J. Biomed. Mater. Res. Part A 102 (2014) 3102–3111, https://doi.org/10.1002/jbm.a.34983.
[201] G. Liu, Y. Zhang, B. Liu, J. Sun, W. Li, L. Cui, Bone regeneration in a canine cranial model using allogeneic adipose derived stem cells and coral scaffold, Biomaterials 34 (2013) 2655–2664, https://doi.org/10.1016/j.biomaterials.2013.01.004.
[202] C.Y. Lin, Y.H. Chang, K.C. Li, C.H. Lu, L.Y. Sung, C.L. Yeh, K.J. Lin, S.F. Huang, T.C. Yen, Y.C. Hu, The use of ASCs engineered to express BMP2 or TGF-β3 within scaffold constructs to promote calvarial bone repair, Biomaterials 34 (2013) 9401–9412, https://doi.org/10.1016/j.biomaterials.2013.08.051.

[203] G.K. Sándor, V.J. Tuovinen, J. Wolff, M. Patrikoski, J. Jokinen, E. Nieminen, B. Mannerström, O.P. Lappalainen, R. Seppänen, S. Miettinen, Adipose stem cell tissue-engineered construct used to treat large anterior mandibular defect: a case report and review of the clinical application of good manufacturing practice-level adipose stem cells for bone regeneration, J. Oral Maxillofac. Surg. 71 (2013) 938–950, https://doi.org/10.1016/j.joms.2012.11.014.

[204] Y. Wei, N. Hu, H. Wang, Y. Wu, L. Deng, J. Qi, Cartilage regeneration of adipose-derived stem cells in a hybrid scaffold from fibrin-modified PLGA, Cell Transplant. 18 (2009) 159–170, https://doi.org/10.3727/096368909788341261.

[205] K. Zhang, Y. Zhang, S. Yan, L. Gong, J. Wang, X. Chen, L. Cui, J. Yin, Repair of an articular cartilage defect using adipose-derived stem cells loaded on a polyelectrolyte complex scaffold based on poly(l-glutamic acid) and chitosan, Acta Biomater. 9 (2013) 7276–7288, https://doi.org/10.1016/j.actbio.2013.03.025.

[206] H. Kang, J. Peng, S. Lu, S. Liu, L. Zhang, J. Huang, X. Sui, B. Zhao, A. Wang, W. Xu, Z. Luo, Q. Guo, In vivo cartilage repair using adipose-derived stem cell-loaded decellularized cartilage ECM scaffolds, J. Tissue Eng. Regen. Med. 8 (2014) 442–453, https://doi.org/10.1002/term.1538.

[207] L. Cui, Y. Wu, L. Cen, H. Zhou, S. Yin, G. Liu, W. Liu, Y. Cao, Repair of articular cartilage defect in non-weight bearing areas using adipose derived stem cells loaded polyglycolic acid mesh, Biomaterials 30 (2009) 2683–2693, https://doi.org/10.1016/j.biomaterials.2009.01.045.

[208] J.R. Mauney, T. Nguyen, K. Gillen, C. Kirker-Head, J.M. Gimble, D.L. Kaplan, Engineering adipose-like tissue in vitro and in vivo utilizing human bone marrow and adipose-derived mesenchymal stem cells with silk fibroin 3D scaffolds, Biomaterials 28 (2007) 5280–5290, https://doi.org/10.1016/j.biomaterials.2007.08.017.

[209] J.S. Choi, H. Yang, B.S. Kim, J.D. Kim, J.Y. Kim, B. Yoo, K. Park, H.Y. Lee, Y.W. Cho, Human extracellular matrix (ECM) powders for injectable cell delivery and adipose tissue engineering, J. Control. Release 139 (2009) 2–7, https://doi.org/10.1016/j.jconrel.2009.05.034.

[210] T.T.Y. Han, S. Toutounji, B.G. Amsden, L.E. Flynn, Adipose-derived stromal cells mediate in vivo adipogenesis, angiogenesis and inflammation in decellularized adipose tissue bioscaffolds, Biomaterials 72 (2015) 125–137, https://doi.org/10.1016/j.biomaterials.2015.08.053.

[211] H.K. Cheung, T.T.Y. Han, D.M. Marecak, J.F. Watkins, B.G. Amsden, L.E. Flynn, Composite hydrogel scaffolds incorporating decellularized adipose tissue for soft tissue engineering with adipose-derived stem cells, Biomaterials 35 (2014) 1914–1923, https://doi.org/10.1016/j.biomaterials.2013.11.067.

[212] Y. Itoi, M. Takatori, H. Hyakusoku, H. Mizuno, Comparison of readily available scaffolds for adipose tissue engineering using adipose-derived stem cells, J. Plast. Reconstr. Aesthetic Surg. 63 (2010) 858–864, https://doi.org/10.1016/j.bjps.2009.01.069.

[213] P. Morissette Martin, A. Grant, D.W. Hamilton, L.E. Flynn, Matrix composition in 3-D collagenous bioscaffolds modulates the survival and angiogenic phenotype of human chronic wound dermal fibroblasts, Acta Biomater. 83 (2019) 199–210, https://doi.org/10.1016/j.actbio.2018.10.042.

[214] Y. Seo, T.H. Shin, H.S. Kim, Current strategies to enhance adipose stem cell function: an update, Int. J. Mol. Sci. 20 (2019), https://doi.org/10.3390/ijms20153827.

[215] A. Mantovani, S.K. Biswas, M.R. Galdiero, A. Sica, M. Locati, Macrophage plasticity and polarization in tissue repair and remodelling, J. Pathol. 229 (2013) 176–185, https://doi.org/10.1002/path.4133.

[216] G.S.A. Boersema, N. Grotenhuis, Y. Bayon, J.F. Lange, Y.M. Bastiaansen-Jenniskens, The effect of biomaterials used for tissue regeneration purposes on polarization of macrophages, Biores. Open Access. 5 (2016) 6–14, https://doi.org/10.1089/biores.2015.0041.

CHAPTER

16

Genetic modification of adipose-derived stem cells for bone regeneration

Harsh N. Shah[a,b], Abra H. Shen[a], Sandeep Adem[a], Ankit Salhotra[a,b], Michael T. Longaker[a,b], and Derrick C. Wan[a]

[a]Department of Surgery, Division of Plastic and Reconstructive Surgery, Stanford University School of Medicine, Stanford, CA, United States [b]Institute for Stem Cell Biology and Regenerative Medicine, Stanford University School of Medicine, Stanford, CA, United States

Introduction

ASCs are an abundant heterogenous cell population that is an autologous cell source for tissue regenerative strategies. ASCs may be isolated in large numbers from small amounts of adipose tissue and lipoaspirate with low donor site morbidity, and they have significant therapeutic potential as they possess the ability to differentiate into osteogenic, chondrogenic, adipogenic, and myogenic phenotypes. In 2001, Zuk et al. established a method to direct the differentiation of ASCs along various lineages through addition of various supplements to the cell culture medium [1]. For example, the addition of ascorbate-2-phosphate in concert with specific ex vivo culture conditions could promote acquisition of a chondrogenic fate, while the addition of β-glycerophosphate directed cells along an osteogenic pathway. Though this landmark study did not include functional assays, the differentiated cells were shown to express specific markers indicative of chondrocytes and osteoblasts, respectively.

Since 2001, investigations differentiating ASCs toward an osteogenic lineage have blossomed from altering culture medium conditions to using genomic editing. As such, this chapter will focus on the various research techniques applied to direct ASCs toward an osteoblastic fate. The addition of specific factors important in propagating BMP, HH, and WNT signaling pathways has been shown to aid in the in vitro differentiation of ASC cultures toward

Scientific Principles of Adipose Stem Cells
https://doi.org/10.1016/B978-0-12-819376-1.00010-X

osteogenesis. These studies have served to develop a foundational understanding as to how ASCs may be genetically manipulated to promote osteogenesis. To alter the genetic landscape of the innate ASC genome, electroporation, viral vectors, minicircles—nonviral DNA vectors—and nanoparticles have been used to introduce essential genes required for osteogenesis. Additionally, specific editing of the genome using emerging CRISPR technology can assist the efficacy of ASC differentiation to osteogenic cells. Finally, this chapter will cover the implementation of differentiated ASCs in the clinical setting to remedy bone defects.

In vitro differentiation

Bone morphogenetic protein signaling pathway

BMPs are a group of growth factors belonging to the larger transforming growth factor-β superfamily that function through a classical signal transduction pathway beginning with BMP binding to BMP receptors (BMPRs) on the cell surface. BMPRs are transmembrane proteins of the serine/threonine kinase family, and form a heterotetrameric complex consisting of two type 1 and two type 2 receptors [2–4]. The interaction between BMPs and their receptors stabilizes the complex, resulting in the phosphorylation of the type 1 receptor by the type 2 receptor [2–4]. The phosphorylated type 1 receptors propagate intracellular signaling through the phosphorylation of various receptor mothers against decapentaplegic (SMADs), which then localize to the nucleus and bind/activate target genes [2–5].

Previous work has shown that many of the 13 classified BMPs have key roles in osteoblast differentiation of various mesenchymal cell types [6, 7]. For example, exogenous BMP-2 has been shown to promote the expression of markers related to osteoblast differentiation in *Runx-2*-deficient cell lines [8, 9]. BMP-2 has also been shown to interact with transcription factors associated with osteoblast differentiation such as osterix and *Runx-2* during osteoblast differentiation, indicating BMP-2 may be important in the signaling pathway of cells acquiring an osteoblastic fate [6–9]. As such, these studies have shown that BMP-2 may use an *Runx-2*-independent pathway to promote expression of osteoblast differentiation markers, though interactions with *Runx-2* may still be important for terminal osteoblast differentiation. Furthermore, exogenous BMP-7 treatment was found to promote osteoblastic differentiation in a myogenic C2C12 cell line. Furthermore BMP-7 treatment resulted in elevated mRNA expression of osteocalcin and collagen type 1 [10]. Finally, osteoblastic differentiation could be driven in C3H10T1/2 mesenchymal cells through in vitro culture with BMP-9 [11, 12]. As various BMPs have been shown to be successful in inducing osteoblastic differentiation in a variety of mesenchymal cell lines, the same rationale can be applied to ASC differentiation into osteoblasts.

Modification of human-derived ASCs with BMP has been shown to successfully direct cellular differentiation toward an osteogenic fate. Culturing human ASCs in medium supplemented with exogenous BMP-2 has been found to promote production of a mineralized extracellular matrix containing hydroxyapatite. This has also been shown to correlate with upregulated transcription of runx2 and osteocalcin, both of which are well-recognized markers for osteogenesis [13]. Use of exogenous BMPs to induce ASC osteogenesis has also been extended for reconstruction of in vivo bone defects. For example, studies have utilized human-derived ASCs cultured in BMP-2 prior to introduction into a murine calvarial defect and demonstrated enhanced ability to form bone with significantly reduced defect size through activation of the BMP signaling pathway [14]. Collectively, these findings highlight the fundamental capacity to manipulate ASCs with growth factors such as BMPs and serve as a foundation for development of strategies to genetically modify these cells for the purpose of bone regeneration (Fig. 1).

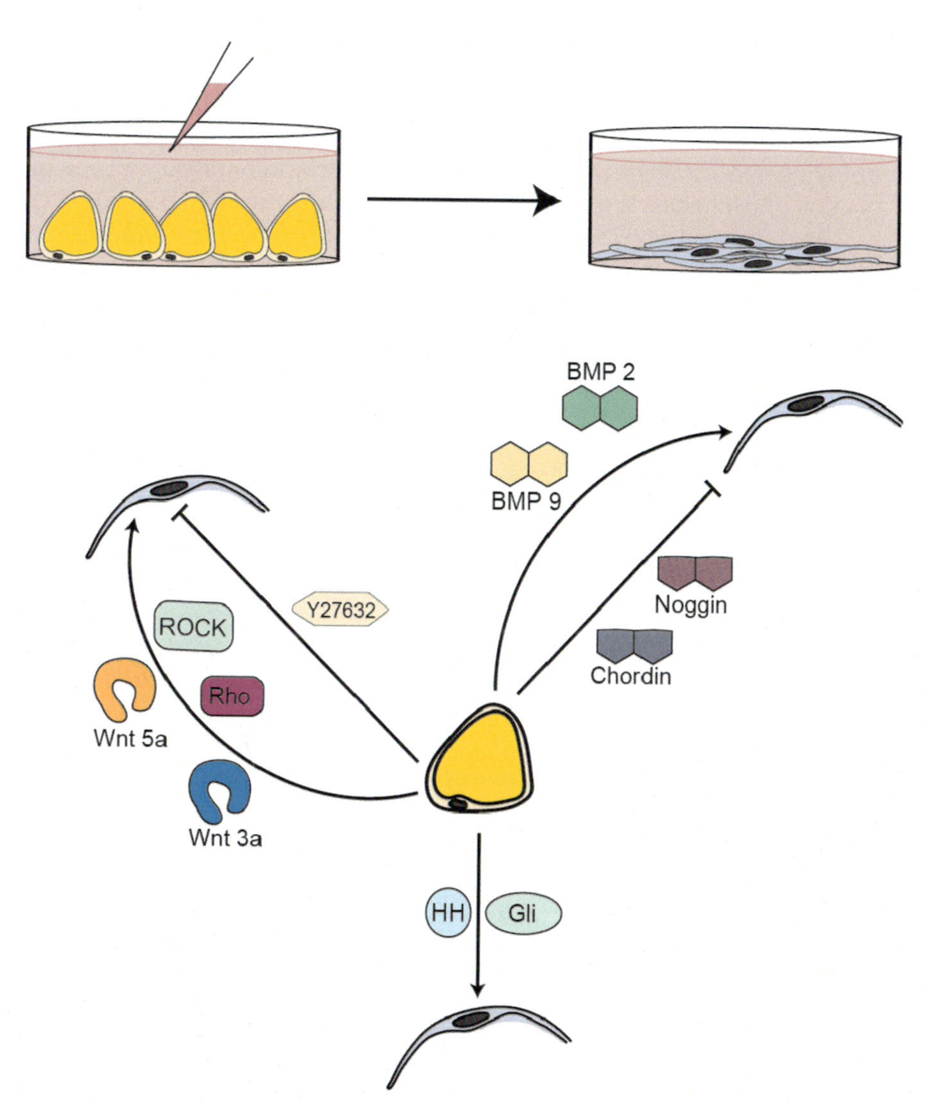

FIG. 1 In vitro differentiation of ASC cell culture toward osteogenic cells. In the ASC cell culture, various proteins can be added into the culture media to drive differentiation toward osteogenesis. Beginning with the ASC in the middle, the presence of various proteins in the culture media can drive or hinder osteogenesis. For example, increased concentrations of BMP2 and BMP9 in the culture media will drive osteogenic differentiation of the ASC. However, the presence of Chordin or Noggin, which are BMP antagonists, will arrest the propagation of the BMP pathway, thereby, reducing the osteogenic differentiation potential of the ASC.

As an alternative approach, rather than exogenous delivery of BMP to direct ASC differentiation fate toward osteogenesis, suppression of BMP antagonists such as noggin and chordin can be used to achieve a similar desired outcome. Such a strategy could mitigate known complications of excess BMP signaling, including inflammation, seroma formation, and heterotopic ossification. Both noggin and chordin are secreted polypeptide proteins that antagonize endogenous BMP signaling [2–5], and knockdown of noggin has been shown to potentiate the functional effects of BMP and promote osteogenic differentiation. Specifically, human-derived ASCs transfected with short hairpin RNA (shRNA) targeting noggin transcript was shown to increase the ability of cells to acquire an osteogenic fate and facilitate repair of critical-sized murine calvarial defects [15, 16]. With chordin, a similar gene knockdown approach has been reported in ASCs using RNA interference [17, 18]. Studies employing this strategy reported improved matrix mineralization, calcium content, and alkaline phosphatase activity following chordin suppression [19, 20]. These studies indicate that decreased levels of natural BMP antagonists can increase endogenous levels of BMP activity, thereby enhancing commitment of human ASCs along an osteogenic fate.

Hedgehog signaling pathway

While much of the literature has focused on manipulation of BMP signaling in ASCs to promote bone formation, other pathways have been found to be similarly capable of guiding osteogenic lineage commitment in mesenchymal cells. HH is a secreted protein essential for organismal development and organogenesis, regulation of cell cycle and differentiation, and cellular survival. Furthermore, HH signaling functions as a morphogen, and in the absence of HH ligand, Patched (PTC) inhibition of Smoothened (SMO) is removed [21–24], and downstream activation of transcriptional gene targets such as the *Gli* family of transcription factors has been observed [25–27]. *Gli* transcription factors, specifically *Gli2*, mediate HH signaling and activate *Bmp-2* gene expression, which are required for normal osteoblast differentiation and subsequent bone formation [28]. HH signaling has been demonstrated to promote osteogenesis while concurrently repressing adipogenesis [21–23]. Therefore, manipulating HH signaling in an in vitro system presents another potential method of directing fate acquisition of ASCs and facilitate bone formation [21–23].

Previous research has introduced recombinant Sonic hedgehog (SHH), the best-studied ligand of the HH signaling pathway, into ASC culture and demonstrated increased osteogenesis [29, 30]. Specifically, exogenous recombinant SHH was reported to increase extracellular matrix calcium deposition by mouse-derived ASCs, with associated upregulation of runx2 and col1 transcription [31]. Interestingly, these cells also expressed higher levels of BMP, which paralleled prior reports demonstrating SHH as capable of altering cellular response to BMP signaling. Conversely, ASCs grown in differentiation medium supplemented with recombinant BMP-2 have also shown higher expression levels of Shh signaling pathway markers such as *Shh*, *Ptc*, and *Gli1* [31], suggesting a bidirectional interaction between HH and BMP signaling to promote osteogenic differentiation in ASCs. These findings also highlight the need for further research to elucidate a mechanism connecting both pathways in driving differentiation of ASCs along an osteogenic lineage (Fig. 1).

In concert with these in vitro studies, HH-modified ASCs delivered on a scaffold have been shown to accelerate calvarial defect healing in mice. Previous studies have cultured ASCs ex vivo in medium supplemented with SHH and confirmed RUNX2 and osteocalcin expression [14, 16]. By seeding these pretreated cells onto a scaffold for implantation, they promoted bone regeneration in critical-sized mouse calvarial defects. These reports thus support a role for HH signaling in ASC osteogenic differentiation and open another avenue for genetic modification to promote bone formation.

Wingless signaling pathway

The WNT signaling pathway may be employed for nearby cell-cell communication in paracrine fashion or by the same cell in autocrine fashion, and WNT signaling has been implicated in embryonic development controlling body axis patterning, cell proliferation, and cell migration [32, 33]. Canonical WNT signaling pathway begins with WNT protein binding to a G-protein-coupled receptor, Frizzled (FZ), and a coreceptor, lipoprotein receptor-related protein (LRP) 5/6 [32–34], and ultimately leads to β-catenin accumulation and nuclear translocation where it acts as a transcriptional coactivator for target genes. Canonical WNT signaling has been shown to be important for normal skeletal development, with studies demonstrating downstream activation of RUNX2 and osteocalcin expression following administration of WNT3A to human ASCs. Inhibition of WNT signaling with NOTCH-1 overexpression was also found to inhibit mesenchymal cell osteogenesis.

Alternatively, the noncanonical WNT signaling pathway does not involve β-catenin or the LRP5/6 coreceptor [34–37]. In this pathway, WNT proteins still bind to the FZ receptor, but downstream signaling results in Rho-associated kinase (ROCK) activation and cytoskeletal rearrangement [35–37]. Importantly, changes in cell shape have been shown to help regulate commitment of mesenchymal cells, and shape-dependent control of lineage commitment has been found to be mediated by RhoA activity. WNT5A, a noncanonical WNT pathway ligand, has been studied as a specific WNT protein capable of inducing osteogenic differentiation of human ASCs [34, 38]. Human ASCs cultured with WNT5A were found to upregulate expression of RUNX2 and osteocalcin, along with increase mineralization of the extracellular matrix [34, 38] (Fig. 1).

Finally, rather than administering WNT ligand to promote osteogenesis in ASCs, manipulation of the downstream signaling pathway can similarly enhance bone formation by these cells [39, 40]. For example, administration of ROCK-activity inhibitor Y27632 to human ASCs was found to significantly impair osteogenic differentiation [41]. Forced expression of a dominant-negative *RhoA-N19* was also reported to inhibit mesenchymal cell osteogenesis and instead directed a commitment switch to an adipogenic fate. These findings raise the potential for alternative strategies to direct ASCs toward osteogenesis through enhancing the activity of downstream WNT signaling proteins such as RHO or ROCK. These alterations can be potentially accomplished through administration of exogenous small-molecule activators or genetic modification to regulate cellular processes such as bone differentiation that may involve the actin cytoskeleton (Fig. 1).

Approaches for gene alterations

Electroporation

Electroporation applies an electric field to facilitate delivery of nucleic acids and other impermeable molecules to a cell. When exposed to an electric field, the hydrophobic bilayer of the plasma membrane builds up charge, creating a transmembrane potential. The transmembrane potential results in the distribution of proteins and carbohydrates within the cell dependent on their interaction with the transmembrane potential. Once the induced transmembrane potential exceeds 500 mV, the membrane will form hydrophobic pores allowing water and ion movement [42]. The greater the time the cell is exposed to the electric field, the bigger the hydrophobic pores become, allowing larger impermeable molecules to enter and leave the cell [41]. The two wave forms used in electroporation are square wave and exponential decay, which can be optimized for each cell type. Square-wave electroporators deliver pulses to the cells with a set voltage for a defined amount of time, with the pulses typically lasting between 100 μs and 100 ms [42,43]. Conversely, exponential-decay electroporators deliver a peak energy that dissipates in an exponential manner [42,44]. Though both wave-type electrophoreses are effective in creating pores, square-wave electrophoresis has been shown to be safer and less damaging to the cell. Such an understanding of hydrophobic pore formation, using an electric field, provides a unique technique to introduce genetic material into the ASC to genetically modify the cell to guide osteogenic differentiation.

For gene delivery to occur through electroporation, two key entities need to be present in the vicinity of the cell: nucleic acids and the electric field. DNA needs to be present during the moment of the electric field pulsation due to the size of the molecule [45]. As the cell membrane destabilizes and the permeation begins, smaller pores coalesce into larger pores. In addition, the rate of larger pore formation must be greater than the rate of the small pores closing. The pulsing of the electric field needs to be optimized for each cell type to ensure a substantial amount of large-pore formation for DNA to enter the cell. The mechanism by which DNA enters the cell via electroporation remains unknown. One mechanism suggests DNA directly enters the cytosol, with experiments indicating that electroporation can cause direct entry of DNA across the membrane [46]. Another mechanism suggests DNA must be trapped within lipid vesicles at the membrane since spots of DNA accumulation cannot be removed upon polarity reversal [47]. Once in the cytosol, DNA must be translocated into the nucleus to enact an effect on the cell. As such, DNA must rely upon cytoskeletal elements such as actin microfilaments and microtubules for directed movement [44]. DNA binds to proteins, such as dynein and kinesin, forming a large protein-DNA complex. Previous research has shown that DNA moves along microtubules with kinetics and dynamics similar to microtubule-based movement of organelles and proteins [48, 49]. Once the DNA has been brought into the cell through electroporation, it must enter the nucleus for appropriate gene expression. The DNA contains sequences called DNA nuclear targeting sequences (DTS), which form complexes with specific nuclear localization signaling containing proteins. These nuclear localization complexes bind to importins to enable entry into the nucleus through the nuclear pore [50, 51].

Electroporation has been researched for the genetic modification of ASCs to guide the cells toward an osteogenic fate. Lee et al. determined that transfection of ASCs with *Runx2* or

Osterix through electroporation resulted in increased bone formation [52]. The cells transfected with *Runx2* gene expressed significantly higher mRNA and proteins levels of RUNX2 and osterix. Also, the cell transfected with the *Osterix* gene expressed significantly higher mRNA and proteins levels of osterix. For gene transfection of ASC, the study applied electroporation at 1400 V with two pulses lasting 20 ms each [52]. Regarding transfection efficiency, 53% of the ASCs were transduced with the *Runx2* gene, and 42% of the ASCs were transduced with *Osterix* gene [52]. Upon 14 days of culturing, genetically manipulated ASCs underwent osteogenic differentiation as determined by gene expression analysis and functional assays. At day 14, mRNA expression of osteocalcin and collagen type 1, alpha 1, both essential in bone formation, was significantly higher in the transfected cells [52]. Furthermore, western blot analysis indicated a significant increase in protein levels for both osteocalcin and collagen type 1, alpha 1 in the differentiated transfected cells [52]. Finally, functional assays measuring mineralization showed increased mineralization of the transfected cells over the 14-day period compared with the control group [52].

The study by Lee et al. provided insight into applying the electroporation technique as a potential method to genetically modify ASCs toward an osteogenic fate; however, the overall applicability for human treatment is yet to be determined. First, in the study, applying 1400 V has the potential to cause physiological, irreversible damage to the cells. To put the voltage into perspective, the voltage in electrical outlets reaches a maximum of 240 V; therefore, the ASC is receiving 5.8 times the electrical outlet voltage for gene transfection. Furthermore, cellular physiology depends on charge separation, such as ATP production in the electron transport chain. As such, applying an exogenous electrical stimulus may disrupt normal cellular physiology and function. Second, the current method of electroporation relies on a buffer containing high concentration of salts such as potassium chloride (KCl) and magnesium chloride ($MgCl_2$) [42]. These high concentrations enable the propagation of the electrical current through the culture system; however, the concentrations are not physiologically appropriate. Since the cells are metabolically active during electroporation, the high salt concentrations can further affect their physiological activity. Overall, in vitro and in vivo studies have shown the increased efficiency of gene transfer by electroporation and the ability for the technique to genetically modify ASCs for an osteogenic fate. However, further research needs to be conducted to determine the potential harm the technique inflicts on the cells and the applicability for treatments.

Viral transduction

The use of viral constructs to genetically modify cells has long been employed by molecular biologists to target delivery of genetic material. A variety of viral vectors exist, and the process of using a virus to transfer genetic material into a cell is termed transduction. A multitude of factors must be taken into consideration when designing a virus vector, with transgene selection one of the most important considerations [53, 54]. Specifically, genes involved in cell-cycle regulation or apoptosis pose a particular challenge for clinical therapeutic applications because these genes may also be involved in multiple signaling pathways. Other factors to consider when choosing the transgene is whether the gene of interest is expressed ubiquitously or expressed only in specific cells and tissues [53, 54]. Furthermore, proper promoter selection is important to drive transgene expression at desirable levels [53]. Addition of selectable markers can also aid in the identification of specific cells that have undergone successful viral infection [53]. Finally, safety considerations need to be considered when constructing viral vectors for clinical applications [53]. The activity of promoters and enhancer elements or aberrant splicing events within the vector can lead to the activation of genes flanking the integration site, commonly referred to as insertional mutagenesis [55–57]. If genes near the integration site are proto-oncogenes, then insertional mutagenesis has the potential to drive clonal proliferation of cells [56]. Therefore, viral vectors must be selected carefully when designing potential clinical strategies. Many different types of viral vectors exist for use depending on the therapy of interest, though with genetic manipulation of ASCs, the most commonly used viral vectors include retroviruses/lentiviruses and adenoviruses.

Retroviruses are RNA viruses possessing the ability to reverse transcribe an RNA genome into a double-stranded DNA intermediate. The DNA intermediate can stably integrate into the host cell genome for further replication. A few examples of retroviruses used in cellular transfection include gammaretrovirus, Moloney murine leukemia virus, and lentivirus. The lentivirus is the most common viral vector used within the group, particularly for ASCs. Lentiviruses contain a *gag* gene encoding for viral structure proteins, a *pol* gene encoding for reverse transcriptase, and a *pro* gene, encoding for protease [58, 59]. In addition to these major genes, the viral genome contains the regulatory genes *tat* and *rev*, which are necessary for lentiviral replication [60, 61]. Retroviral vectors transduce a wide variety of cell types, including ASCs, while containing a relatively low toxicity profile. The integration of genetic information into the host genome is a unique characteristic of these viruses that allows for higher and more durable levels of transgene expression. Nonetheless, with introduction of a newly incorporated gene comes the potential for insertional mutagenesis, as discussed above.

In contrast to lentiviruses, adenoviruses carry their genetic material in the form of double-stranded DNA [62]. Once the viral DNA makes its way into the nucleus, it exists in a free-floating form and remains unincorporated in the host's genome [62]. The host's cellular machinery is utilized for transcription alongside the host's own genetic material. This allows for transient high expression of genes by the infected cell with decreased risk for host cell mutagenesis. Most adenovirus vectors are genetically modified versions of adenovirus serotype 5, which consists of two types: replication-defective and replication-competent [63]. The replication-defective vectors delete the essential E1A and E1B genes and replace them with an expression cassette containing a high activity promoter to drive expression of the foreign transgene [62, 64]. The E1A genes encodes for proteins essential for adenovirus replication, while the E1B gene encodes for proteins inhibiting the host cell apoptotic response to the infection [63]. Vectors with the deleted E1A and E1B genes are considered leaky, resulting in transduced cells being eliminated by the host T-cell response, thereby eventually extinguishing transgene expression. As such, a new replication-defective adenovirus vector has been created called the "helper-dependent" adenovirus vector. In this type of viral vector, most of the adenoviral genome has been deleted, but the origins of DNA replication, at each end of the genome, and approximately 500 base pairs required to package the genome into virions, are maintained [62, 64]. These vectors have a high cloning capacity, enabling them to retain multiple transgenes. Finally, perhaps the greatest challenge for adenovirus transfection is their ability to induce a strong immunogenic response in the host. The adenovirus capsid is

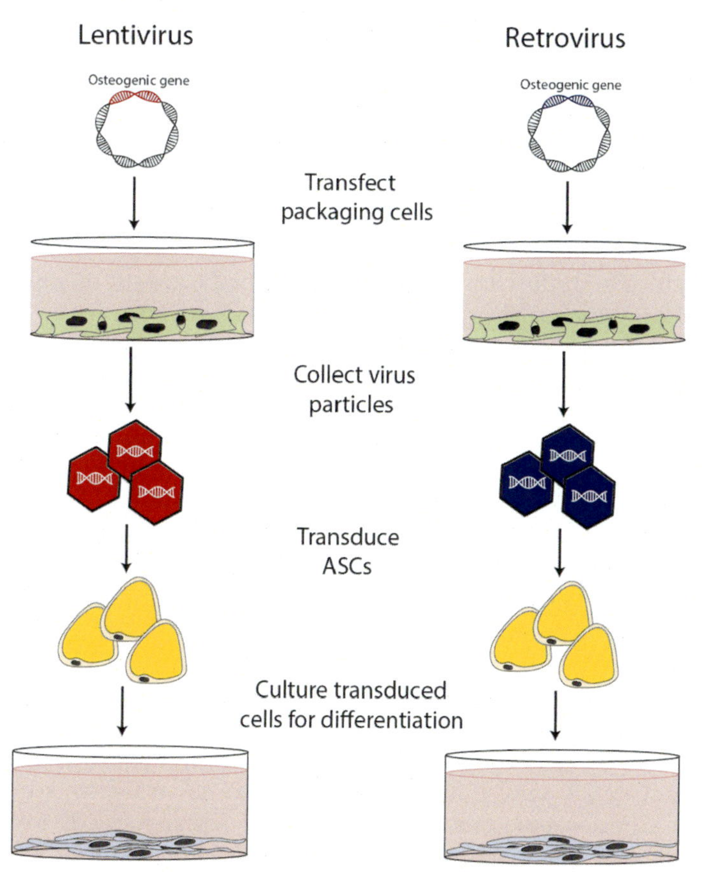

FIG. 2 Viral transfection of ASC cell culture for osteogenic differentiation. In both lentiviral and retroviral transfection, the viral particle will contain a pro-osteogenic gene such as Runx2 or Osterix along with a selection marker such as GFP or kanamycin resistant gene. The viral particles will initially transfect a packaging cell line such as 293T. Using the property of the selection marker through the green fluorescence or survival in the presence of kanamycin antibiotic, the viral particles are purified from the cells with successful viral infection. This initial step enables exponential production of the viral particle using a cell line that is not scientific interest. Once the viral particles have been purified, these particles will be used to transfect the ASC. With the presence of the osteogenic gene and appropriate cell culturing, the viral particle will use the host transcription machinery to overexpress the osteogenic gene. This overexpression will enable the differentiation of the ASC into an osteogenic cell.

the most important antigen, which elicits a strong innate immune response through the activation of $CD4^+$ and $CbD8^+$ T cells [62, 65]. As such, overcoming the immunogenicity of the adenovirus vector is necessary for clinical applications of gene therapy and ASC transfection in patients (Fig. 2).

Although both retrovirus systems have respective advantages and disadvantages, the use of viral systems to deliver targeted genes holds great therapeutic promise, especially in regenerative bone biology. With the accessibility of ASCs, scientists have employed retrovirus systems to drive expression of pro-osteogenic genes such as BMPs in these cells to enhance bone differentiation. Studies in small animal models have used a lentiviral system to induce expression of BMP-2 in ASCs, and these reports have shown potential for accelerating in vitro bone formation. Another study in rats showed successful bone regeneration after the delivery of ASCs transduced with BMP-2 to femoral defects [13]. Microcomputed tomography revealed higher bone volume in comparison with controls with the use of infected cells. Similarly, transducing ASCs with BMP-2 via an adenovirus was found to increase bone formation at the site of the defect [66]. Both viral vectors have demonstrated promising therapeutic potential that can be employed to promote bone regeneration of ASCs through viral transduction with BMP.

Few viral transduction studies have been conducted for the genetic manipulation of ASCs in regard to the WNT and HH pathways. Lentiviral vectors have been created to alter the WNT signaling pathway. One specific vector constitutively activates β-catenin, which is the downstream target of the pathway that will be translocated into the nucleus to act as a transcriptional coactivator of target genes such as osteocalcin and *Runx2* [67]. Furthermore, as discussed in the section on the WNT signaling pathway, WNT is necessary for normal skeletal development. Currently, research has not been conducted to understand the effect of viral transfection of WNT or other downstream signaling targets on the genetic modification of ASCs toward an osteogenic fate. Therefore, applying the technology in ASCs with the existing lentiviral vectors can provide another method of modulating the WNT pathway and guiding the cells toward an osteogenic differentiation fate resulting in bone formation. Similarly, only one study has been conducted on the gene-enhanced bone regeneration capacity of ASCs through the retroviral transfection of the Shh gene [68]. The study used a murine leukemia retrovirus as the transfection vector to introduce human SHH cDNA into the rat ASCs. The transfected cells were then assembled onto an alginate-collagen matrix, and the matrix construct was introduced into rabbit calvarial defects, which resulted in the formation of full-thickness bone in the SHH-transduced group. Overall, viral transfection technology is an effective method to modify the ASCs toward an osteogenic fate; however, further research needs to be conducted on modulating important signaling pathways such as WNT, HH, and BMP.

Minicircles

While extensive research has been performed on gene therapy using viral vectors, and despite its effectiveness and significant promise for inducing expression of genes within infected cells, safety remains a significant concern. Ideally, a gene delivery system should have high transfection rates and specificity, have low toxicity, and be biologically compatible to effectively deliver desired genes. To address these disadvantages, nonviral gene vectors, such as minicircles, have been introduced to serve an alternative to the traditional viral vector approaches owing to their promising properties such as targeted tissue delivery, low detrimental effects, and low chances for immunogenicity [69].

Conventional plasmid transgene delivery systems contain prokaryotic DNA that carries a risk of triggering an immune response when transfecting eukaryotic cells. As the ultimate goal of plasmid transfections of ASCs directing cell fate toward an osteogenic lineage is its clinical application, safety precautions need to be developed for this technology. Currently, plasmid vectors frequently contain a prokaryotic origin of replication and an antibiotic resistance marker, which are both undesirable constructs for clinical applications [70–72].

Furthermore, mammalian cells have developed resistance toward prokaryotic genetic material, resulting in undesirable side effects. For example, short immunostimulatory DNA sequences in the prokaryotic plasmid backbone encode for antigens that result in the mammalian host mounting an immunogenic response. While this may be desirable for applications where an immunogenic response is necessary, such as with cancer therapeutics, in the context of ASC genetic modification, the response is unnecessary and potentially detrimental to the recipient's health. To address these limitations, a minicircle vector has been developed to overcome the challenges stated above [69].

Minicircles are small circular nonviral DNA vectors that are used as common transgene carriers for the genetic alteration of mammalian cells. Minicircles are a product of site-specific recombination between the two *attP* and *attB* sequences driven by the *Escherichia coli* bacteriophage λ integrase. In vivo excision enables insertion of a therapeutic expression cassette [69, 73]. Furthermore, the minicircles do not contain plasmid backbone sequences, thereby reducing the deleterious and immunogenic effects of the prokaryotic sequences [69, 73]. As such, minicircles are devoid of all prokaryotic DNA, meaning they are less likely to be perceived as foreign and destroyed by host cells, and may thus provide a higher transfection efficiency compared with typical plasmid DNA [74].

The preparation of minicircle vectors begins with *E. coli* plasmids that are loaded with eukaryotic genes of interest in the form of vectors to become a recombinant parental plasmid. The prokaryotic backbone within parental plasmid is then excised by a recombinase enzyme, resulting in the generation of a minicircle containing only the eukaryotic vector components [75, 76]. One additional advantage of nonviral minicircle transfection is that it minimizes the risk of insertional mutagenesis and host rejection posed by traditional viral transduction. The nonviral characteristic of minicircles allows for repeated transfection of a gene of interest without integrating into the host DNA, as with viral-based methods [77] (Fig. 3).

Minicircles have been used extensively in research relevant to ASCs and bone regeneration. Studies done by Jia and Narsinh have shown this method to be successful for reprogramming human ASCs into human-induced pluripotent stem cells (iPSCs) by minicircle-mediated transfection of pluripotent factors (OCT4, SOX2, LIN28, NANOG) [78]. Of note, the reprogramming efficiency using nonviral minicircles was found to be about half that of viral transduction; however, conventional plasmid-based transfection techniques were not effective at producing any iPSCs [78, 79]. Once created from ASCs, minicircle-generated iPSCs were then found to be capable of in vivo calvarial bone regeneration with guidance from BMP-2 [80]. Minicircles have also been used to genetically manipulate ASCs to directly promote bone formation. Levi and colleagues demonstrated that knockdown of the BMP-2 antagonist noggin using minicircle transfection of shRNA in human ASCs enhanced in vitro differentiation and in vivo regeneration of calvarial bone [16]. Osteogenic gene expression in human ASCs following minicircle-mediated Noggin knockdown was significantly higher than in control ASCs, and repair of mouse critical-sized calvarial defects with these genetically modified ASCs was accelerated.

The ability for ASCs to participate in bone regeneration is also dependent on survival postimplantation. To promote ASC viability, Hyun and colleagues transfected ASCs with minicircles to induce BCL-2 overexpression [81]. They found not only enhanced survivability of cells following placement in critical-sized mouse calvarial defects, but also enhanced capacity for bone regeneration by these cells. Another study by Im et al. demonstrated a role of minicircle

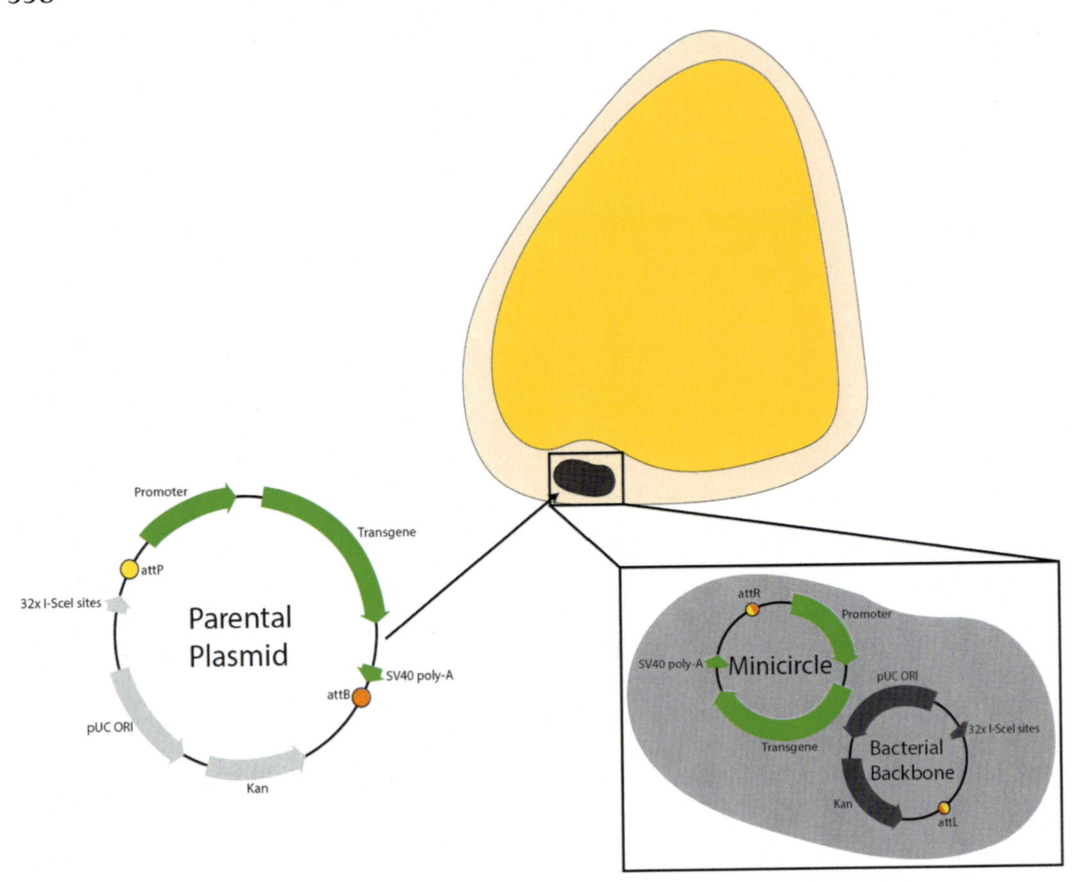

FIG. 3 Minicircle genetic manipulation in the ASC driving osteogenesis. The parental plasmid consists of a bacterial backbone containing selection markers that can be used to identify the cells which have successfully incorporated the plasmid into their nucleus. The minicircle portion of the parental plasmid contains the promoter and transgene of interest, which will continually be expressed using the host cell's transcriptional machinery. Upon insertion of the parental plasmid into the host cell nucleus, cleavage at the attP and attB sites will separate the minicircle and bacterial backbone. This will enable the proper function of the minicircle through the transgene overexpression. The minicircle can be engineered to contain any transgene; however, for the genetic modification of ASC for osteogenic differentiation, optimal transgenes would include: Runx2, Osterix, and Osteocalcin.

transfection in ASCs to improve chondrogenesis. Minicircles containing SOX-6 and SOX-9 transcription factors were found to enhance the chondrogenic potential of ASCs in vitro and also improved the regenerative ability of minicircle-transfected ASCs to form cartilage in both mouse and goat knee injury models [82]. Chondrogenic and osteogenic differentiation potential of ASCs transfected with SOX-9 minicircles have been similarly tested on dog and horse models, with studies showing greater transfection efficacy and differentiation capacity than for ASCs transfected with conventional plasmids [83]. These studies thus illustrate the therapeutic potential of a strategy employing minicircles to genetically manipulate ASCs for osteogenic purposes.

Minicircles also facilitate synergistic processes through combinatorial gene transfection where more than one gene of interest is inserted into a minicircle [84]. This may be applicable in cases such as osteonecrosis of the femoral head, where impaired blood supply limits the ability of osteogenic cells to successful regenerate bone. Current therapies include total hip arthroplasty, even in young patients, resulting in the need for subsequent future surgeries to replace hardware as the patient ages. As such, minicircles have been employed to transfect ASCs with both vascular endothelial growth factor (VEGF) and BMP-2. In concert with the angiogenic role of VEGF to reestablish lost blood supply to the niche, BMP-2 can direct cell fate differentiation of the ASCs toward osteogenesis to replace the necrotic bone. In a study by Lee and colleagues, simultaneous activation of osteogenesis and angiogenesis was observed in BMP-2/VEGF minicircle-transfected ASCs [85] that were capable of healing critical-sized calvarial defects and long-bone segmental defects in immunocompromised mice. Over a 56-day period, implanted cells in Matrigel did not migrate outside the site of implantation, and the presence of transfected cells enhanced both angiogenesis and osteogenesis, contributing to accelerated bone regeneration [85]. Specifically, transcriptomic analysis revealed elevated levels of RUNX2, BMP-2, and VEGF, as well as increased RNA transcription of YAP/TAZ and downstream signaling pathway molecules such as ANRKD1 [85]. All of these targets have been implicated in the activation of both osteogenesis and angiogenesis.

Overall, minicircle transfection is an appealing technique owing to its large loading capacity for genes, ease of scalable target gene production, and low integration risk to the host genome through nonviral delivery. Although superior to conventional plasmid DNA, the transfection efficacy of minicircles needs to be further optimized to achieve clinical relevance. Strategies such as incorporating the properties of highly branched poly (β-amino ester) have been shown to improve minicircle transfection efficiency while preserving the target cell viability [86]. Ultimately, minicircle transfection is a promising strategy for genetic manipulation of ASCs and has an encouraging future for widespread applications in stem cell research.

CRISPR

The clustered regularly interspaced short palindromic repeat (CRISPR) system is a revolutionary research tool in molecular biology that allows for highly precise genome modification. CRISPR was first identified in 1987 in the *E. coli* genome as a cluster of repetitive palindromic DNA sequences interspaced with nonidentical spacer DNA [87]. CRISPR-associated (Cas) proteins were also identified that helped the host organism defend itself from viral attack [88]. These Cas proteins are nucleases and helicases that cleave foreign viral genomic material, and the spacer DNA represents a historical account of old viral infections that enabled the bacteria to resist similar subsequent infections more readily. The CRISPR-Cas system is thus an innate prokaryotic defense system [89].

Researchers have since identified Cas9 in *Streptococcus pyogenes* [90], which consists of two primary components, an endonuclease that cleaves double-stranded DNA and a guide RNA (gRNA) stand that contains a target sequence of interest to be cleaved. Importantly, the gRNA strand can be easily replaced by synthetic oligonucleotides that are customized to contain a target sequence of choice for highly specific gene silencing, editing, or replacement [91]. The use of the CRISPR/Cas9 system has proven to be an effective gene editing strategy in vitro in

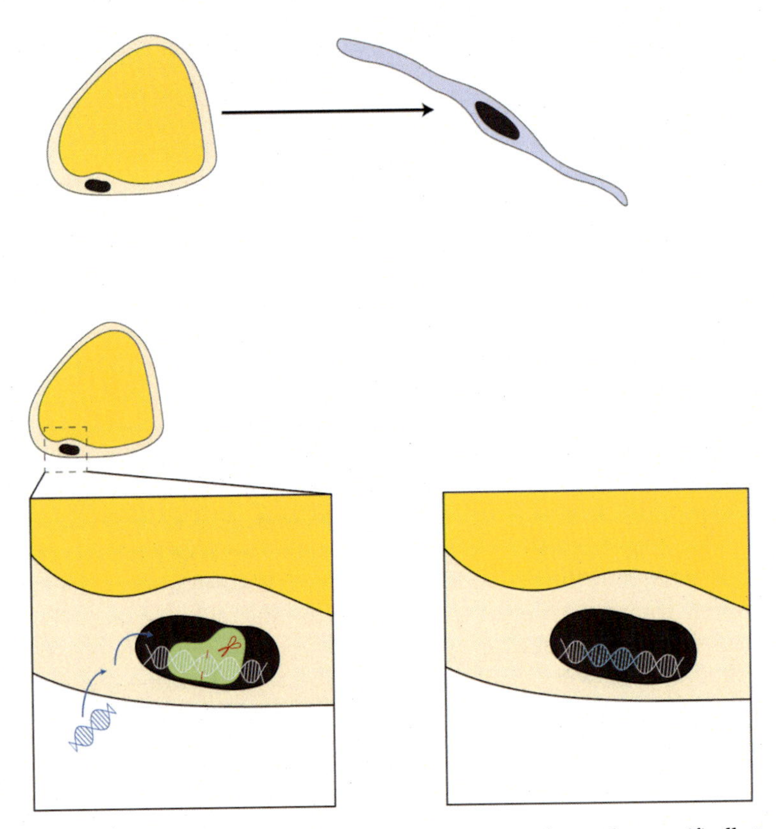

FIG. 4 CRISPR-Cas9 genetic modification of the ASC. CRISPR-Cas9 can be used to specifically target the integration of a gene into the genetic material of the ASC. The Cas9 enzyme will direct specific base pair cuts on each strand of the ASC genetic material using a guide RNA that has been engineered through CRISPR. The guide RNA targets the location on the genetic material that is of interest enabling the attached Cas9 to cut the DNA at specific sequences. Specifically, the guide RNA should be created to target base pairs of a gene located near a highly active promoter. Once the double strand breaks are made, a vector containing the osteogenic gene is introduced into the system to incorporate the gene in the genetic material. Through homology-directed repair, a mechanism to repair double-stranded DNA break, the osteogenic gene will be permanently incorporated into the cell's genetic material for further transcription.

various eukaryotic cell types and has been shown to have multiplexing capabilities [92, 93]. Further advances in CRISPR research have led to modification of the system by removing endonuclease activity and coupling transcriptional activators/inhibitors (dCas9) that can be utilized to control specific gene expression [94] (Fig. 4).

With these recent discoveries, the CRISPR/Cas9 system has been heavily investigated for gene therapy in stem cells. For example, Gerace and colleagues have highlighted a potential clinical application to treat type 1 diabetes by using CRISPR to alter the genome of autologous mesenchymal stem cells (MSCs) to differentiate into insulin-producing β-cells through dCas9 upregulation of islet cell transcription factors [95]. Another study by Farhang demonstrated the utilization of dCas9 in human ASCs to downregulate the expression of inflammatory cytokine cell receptors TNFR1 and IL1R1 [96]. After CRISPR modulation of the genome, these ASCs demonstrated an improvement of chondrogenic differentiation capacity and cell survival in an inflammatory environment, indicating a potential antiinflammatory treatment for various musculoskeletal diseases. Additionally, studies done by Hsu and colleagues in rats demonstrated that activation of key neurotrophic factor genes facilitated by dCas9 in ASCs could stimulate neural proliferation in vitro and peripheral nerve regeneration in vivo following sciatic nerve injury [97].

Another important application of CRISPR is in the study of mutations through gene knockout. In a report by Zhao et al., CRISPR/Cas9 was used to knock out BRCA1, a tumor suppressive gene, in ASCs [98], which were then cultured with a breast cancer cell line. The BRCA1-knockout ASCs were observed to eventually reach a senescent state in which they secreted inflammatory cytokines, leading to more aggressive tumor proliferation of the surrounding breast cancer cells. Findings from this study have allowed clinicians to gain greater insight into evaluating patients with BRCA1 gene mutation expression pathways and mechanisms involved in breast cancer progression.

In the context of promoting osteogenesis by ASCs, CRISPR-based research has been limited thus far. A possible method in which CRISPR can be implemented is by delivering dCas9 transcriptional activators and gRNAs to target the upregulation of genes such as TGFβ, RUNX2, and SP7 that induce osteogenic differentiation in ASCs [99, 100]. Importantly, some limitations of this methodology may include off-target effects of Cas9, the editing efficiency of CRISPR, rejection of stem cell transplants, and the risk of tumorigenesis with ASC manipulation; thus, rigorous experimentation is essential before clinical implementation [101, 102]. Nevertheless, the CRISPR/Cas9 system is an exceptional technique that holds remarkable promise for novel stem cell research and therapies in the future.

Nanoparticles

Nanoparticles are another method to introduce genes into a target cell population to alter endogenous expression. Naked genetic material cannot be readily internalized by the target cell population owing to phagocyte uptake and immune response stimulation. Nanotechnology for gene delivery has been explored to address this limitation. Important aspects of a clinically viable nanomolecule used for gene delivery include encapsulation efficiency, nanoparticle stability, endocytosis by target cell, and toxicity of pharmacology.

Various materials have been investigated for use as delivery vehicles for genetic material, though poly(lactic-*co*-glycolic acid) (PLGA), a biodegradable polymer, may hold particular promise. Already approved by the United States Food and Drug Administration (FDA) for clinical use as a drug delivery device, PLGA has also been explored for use as a gene vector delivery system owing to its stable property and ability to protect DNA from degradation in vivo [103, 104]. As such, DNA and RNA can be encapsulated in PLGA nanoparticles through a double-emulsion solvent evaporation method [103, 104]. Furthermore, to increased cellular endocytosis, PLGA nanoparticles have been modified with biocompatible cationic components such as chitosan, a naturally occurring polycation [103, 104]. The addition of chitosan to PLGA nanoparticles provides a positive charge to achieve higher transfection efficiency and increases the biocompatibility of nanoparticles within targeted cells. However, while the positive charge may solve certain drawbacks in nanoparticle biology, it also creates new challenges such as aggregation and nonspecific incorporation by off-target cells [103–105]. One approach to limit undesired aggregation with serum proteins in the bloodstream is through PEGylation of nanocarriers. Without PEGylation, large nanoparticle aggregates have been shown to form and be cleared by phagocytic cells and the reticuloendothelial system [103–105].

Another important consideration for nanoparticle development in gene delivery is size, as studies have shown that particle size significantly affects cellular uptake [106, 107]. In particular, PLGA nanoparticle-DNA constructs fractionated using a 100-nm pore-size membrane were evaluated for cellular uptake, and smaller-sized particles with a mean diameter of 97nm were found to have a 27-fold higher transfection rate compared with larger-sized particles [108]. This enhanced efficiency of smaller-sized nanoparticles has been attributed to improved tissue diffusion and less hindrance to cellular incorporation (Fig. 5).

To overcome limitations in target specificity with nanoparticles, a specific ligand-receptor-mediated targeting strategy has been explored [103–105]. With this strategy, conjugated ligands, such as an antibody, protein, peptide, or aptamer, interact with receptors on target cells, which then allow for specific uptake of the nanoparticle-DNA construct. This approach depends on the use of a ligand-receptor pair with strong binding affinity and the presence of distinct receptors on the target cell [103–105]. To further improve cellular internalization, cell-penetrating peptides with membrane translocation sequences or protein transduction domains have also been added

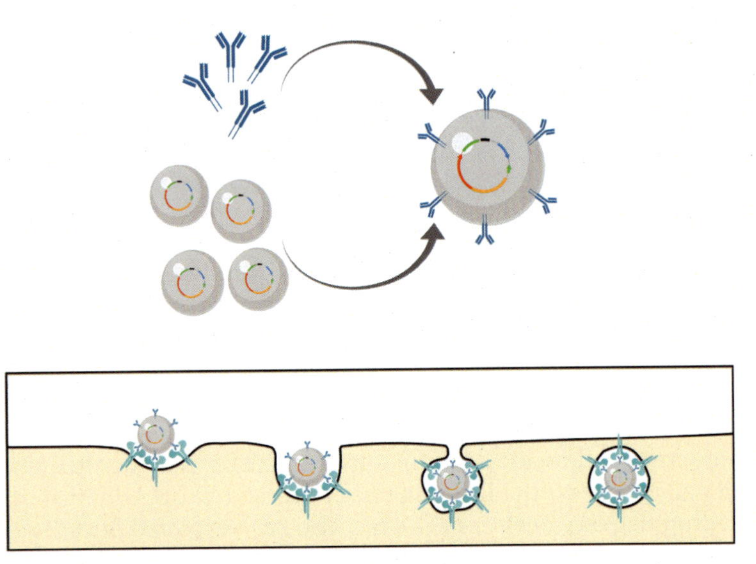

FIG. 5 Nanoparticle application for the introduction of osteogenic genes. Using synthetic material to form the nanoparticle, the osteogenic plasmid RNA (e.g., Runx2, Osterix, etc.) can be encapsulated within the nanoparticle. The surface of the nanoparticle will be encased with immunoglobins which can bind to the receptors on the surface of the ASC. The binding of the immunoglobulin to the receptor will cause the endocytosis of the nanoparticle into the cell. Once inside the cell, the plasmid DNA would be translocated into the nucleus to undergo transcription for eventual translation to protein for osteogenic differentiation.

to the surface of nanoparticles [103–105]. These surface peptides enable the quick cellular uptake of the nanoparticle–gene construct through direct endocytosis.

With encouraging developments in nanoparticle-mediated manipulation of cells, emerging studies over the past 5 years have employed this strategy with ASCs to guide bone formation. As previously discussed, BMP-2 is among the most well-investigated factors related to bone regeneration, and nanoparticles have been used to stimulate endogenous signaling activity mimicking this growth factor to promote osteogenesis in ASCs. By incorporating Phenamil, a small molecule that can induce osteogenic differentiation through activation of BMP's signaling pathway, as well as BMP-2 into a slow-release construct consisting of PLGA, ASCs exposed to this matrix were observed to have significantly enhanced in vitro bone formation [109]. Furthermore, functional assays conducted using ASCs treated with nanoparticle-delivered Phenamil and BMP-2 demonstrated increased alkaline phosphatase and alizarin red staining [109]. As such, PLGA particles incorporating Phenamil were found to augment the activity of BMP-2 on the induction of ASCs toward osteogenic cells, resulting in increased bone formation.

Another study using bioactive glass nanoparticles was found to similarly enhance osteogenic differentiation in ASCs. These nanoparticles are specifically designed to possess biomimetic micro/nanoscale topological structures with high surface area to enhance biomineralization ability for bone tissue regeneration [110]. Following uptake by ASCs, bioactive glass nanoparticles dispersed throughout the cytoplasm and nucleus, and, when introduced into cells cultured with osteoinductive medium, enhanced expression of osteogenic markers was observed and the bone-forming potential of ASCs was found to significantly increase [110]. Though the clinical use of bioactive glass nanoparticles may be limited owing to their nonbiodegradable nature, these findings reveal an important concept with strategies to genetically modify ASCs for bone regeneration. To maximize osteogenic differentiation potential of ASCs, multiple experimental strategies may be applied in concert to capitalize on the synergistic strengths of each individual approach.

Interestingly, a study incorporated the nanoparticle technology with minicircles to drive the osteogenic potential of ASCs for bone regeneration in mice calvarial defects [111]. Initially, the magnetic nanoparticles and minicircles were created separately. The magnetic nanoparticles were synthesized using an iron oxide core coated with PEI-B. In addition, the minicircles were synthesized by ligating the parental plasmid to the amplifying B-cell lymphoma 2 (BCL-2) gene, which has been shown to inhibit apoptosis in implanted cells [81]. Furthermore, the minicircle expressed GFP for downstream determination of the integration into the cells. Both the magnetic nanoparticles and minicircles were complexed together, and the complex particles were isolated from solution through a magnetic pull-down. A scaffold was constructed containing the complexed nanoparticle-minicircles. The scaffold was placed in the calvarial defects in the orientation where the nanoparticles were in contact with the dura mater. The opposing side of the scaffold, with no nanoparticles present, were seeded with the isolated ASCs, which were isolated from human abdominal lipoaspirate. Finally, a magnet was placed over the scaffold to attract the complexes toward the ASCs for magnetofection. This methodology enables temporal and spatial control of the genetic modification of ASCs and their eventual osteogenic fate. The study determined that enhancing the survival of the ASCs through the presence of the BCL-2 gene increased the bone healing potential in a critical-size calvarial defect. This particular study combined two different technologies to accomplish the goal of genetically changing ASCs, allowing for greater control over the ASC ability for osteogenic differentiation. In addition to introducing the BCL-2 gene, an osteogenic-specific gene such as RUNX2 or osterix could also be introduced through the nanoparticle-minicircle complex to further enhance the osteogenic potential of ASCs.

Clinical applications

Rather than relying on the gold standard of autologous grafts for bone regeneration, autologous ASCs harvested from the patient provide an effective alternative to the traditional approaches. Current clinical applications regarding the genetic modification of ASCs toward an osteogenic fate have resulted in a multitude of case reports demonstrating the successful repair of bony defects in the craniomaxillofacial region. In case reports, ASCs were harvested from the patient, either from abdominal fat or iliac crest, and expanded in vitro. During the expansion, the cells were exogenously treated with recombinant human BMP-2 (rhBMP-2). The treated cells were implanted into the bony defect to promote bone regeneration as the ASCs had been genetically modified with rhBMP-2 to guide the cells toward an osteogenic fate. One example of such an application is a case study where a 65-year-old male had undergone a hemimaxillectomy for a recurrent keratocyst [112]. The cyst removal had been complicated by oronasal and oroantral fistula formation. During the reconstructive surgery, surgeons removed ASCs from the patient's subcutaneous abdominal fat and genetically modified the cells as described above. Over the course of 8 months, skeletal scintigraphy confirmed the presence of mature bone structure at the site where the genetically modified ASCs were transplanted.

Furthermore, a few clinical trials have been initiated to determine the therapeutic implications of using genetically modified ASCs for bone regeneration. Specifically, a phase I/II clinical trial in Israel has completed its evaluation into the efficacy of BonoFill, a biological bone filler using autologous ASCs to regenerate bone in the maxillofacial region [113]. Though the results are not published, there are encouraging efforts to determine the applicability of ASCs augmented with biologics for patient care. Similar to case reports, ASCs have been modified through the addition of exogenous factors and in vitro differentiation. Though this is one method of guiding ASCs toward an osteogenic fate, the technique is not the most effective owing to the uneven distribution of the exogenous factors in the culture system resulting in various stages of osteogenic differentiation. Additionally, ASCs have been manipulated outside of the body; therefore, additional precautions need to be taken to ensure sterility of the cells along with providing expansion conditions similar to their innate environment. Translational research for genetic manipulation of ASCs toward an osteogenic fate is currently in its infancy. The addition of exogenous factors to an ASC culture system is one technique of many, discussed throughout the chapter, that enables the cells to express bone-forming genes and characteristics. As such, these initial clinical trials provide an opportunity for further genetic manipulation of ASCs to increase the effectiveness of their osteogenic fate differentiation through viral transfection and CRISPR.

Future directions and conclusions

Advances in our ability to genetically manipulate ASCs have created a new field of research with exceptional potential for bone regenerative therapeutics. The accessibility of ASCs from patient lipoaspirate makes this cell an extremely favorable building block for bone regeneration. Since ASCs are native to the patient, the risk of rejection may be minimized relative to other synthetic bone graft materials. Furthermore, coupling the osteogenic potential of ASCs with the native regenerative capabilities of bone throughout adulthood makes therapeutic possibilities transcending multiple specialties limitless. However, prior to realizing the therapeutic potential of ASCs, the safest and most effective method for promoting ASC osteogenic differentiation needs to be determined. Over the past 20 years, various experimental methodologies to drive osteogenic lineage commitment have been created, ranging from in vitro differentiation protocols to targeted gene manipulation with CRISPR technology. With the variety of potential options, certain challenges need to be addressed regarding the genetic manipulation of ASCs toward an osteogenic fate.

Since the first description of ASCs, our understanding of their functional heterogeneity has continued to improve, and thus, isolation of a purer population of osteogenic ASCs would be ideal for use in bone regenerative strategies. Regardless of the method used to drive osteogenic differentiation, enrichment of ASCs for osteogenic progenitors would potentially enhance functional outcomes. Having a purer population of cells at the onset will ensure patients receive the optimal cell population for bone regeneration. With emerging studies identifying cell surface markers for these more bone-lineage-committed ASCs, our ability to prospectively isolate these cells by flow cytometry has increasingly become a reality. With ASC enrichment, strategies to genetically manipulate these cells for bone regeneration using viruses, minicircle, CRISPR, or nanoparticle technologies may also become even more effective.

Gaps in our knowledge of ASCs with respect to their genetic/transcriptional similarity to endogenous osteoblasts also exist. While current research has extensively shown similarity between genetically manipulated ASCs and osteoblasts, with secretion of a mineralized matrix and expression of various traditional markers including alkaline phosphatase, runx2, osteopontin, and osteocalcin, questions still remain regarding how closely these ASCs mimic the full osteoblast transcriptome. With technological advances in analytic tools, one strategy to address this question would be through use of single-cell analysis and bioinformatic approaches to determine cluster proximity and interrogate relationships between the two cell populations through certain genes such as SOX9, osteocalcin, gremlin, BMP-2, and sclerostin. Furthermore, cluster visualization will aid in determining homogeneity of ASC-modified osteoblast differentiation and effectiveness of different strategies to genetically manipulate ASCs.

A future direction for modifying ASCs to promote osteogenesis would be to understand the ability of these cells to support a bone niche. Physiologically, bone interacts with immune cells, nerves, and the hematopoietic system to various degrees. As such, factors released by cells from other tissue types can influence bone homeostasis through osteoblasts and vice versa. In light of this, the question remains whether genetically manipulated ASCs directed along an osteogenic lineage have the ability to support a hematopoietic stem cell niche. Furthermore, during injury, the recruitment of immune cells can alter the local bone environment, and understanding the functional response of ASC-modified osteoblasts to secreted factors from immune cells will determine the functional capabilities of the differentiated cell. Finally, during regeneration, peripheral nerves secrete neurotrophic factors to regrow damaged axons. These factors also promote expansion of the skeletal stem cell and downstream osteoblasts to facilitate bone repair; therefore, determining the response of ASC-modified osteoblasts to these same neurotrophic factors will be essential. Elucidation of these open-ended questions may begin through use of novel ex vivo three-dimensional culture systems, as multiple cell populations can be cultured together to better understand the niche interactions.

Finally, when protocols are evaluated to genetically manipulate ASCs for bone regeneration, it is necessary to consider their potential for translational success. Though viral transfections may be very effective, the prospects of such an approach being approved for patient treatment may be limited, given the risk of off-target genetic modifications. Given this consideration, newer technologies for genetic modification using minicircles, CRISPR, or biocompatible/biodegradable nanoparticles may find use in clinical applications sooner. And though incorporation of ASC-modified osteoblasts into therapeutic strategies for patient care is a goal of regenerative medicine, further research needs to be conducted before initiation of more clinical trials. Nonetheless, great strides in genetically modifying ASCs for osteogenic differentiation have been made over the past decade. With the advent of numerous molecular technologies and biomedical techniques, the ability to meet the proposed challenges to ensure safe and effective genetic modification of ASCs is within our grasp.

Financial disclosure statement

The authors have no relevant affiliations or financial involvement with any organization or entity with a financial interest in or financial conflict with the subject matter or materials discussed in this manuscript.

References

[1] P.A. Zuk, et al., Multilineage cells from human adipose tissue: implications for cell-based therapies, Tissue Eng 7 (2) (2001) 211–228.

[2] B. Schmierer, C.S. Hill, TGFbeta-SMAD signal transduction: molecular specificity and functional flexibility, Nat. Rev. Mol. Cell Biol. 8 (12) (2007) 970–982.

[3] V.S. Salazar, L.W. Gamer, V. Rosen, BMP signalling in skeletal development, disease and repair, Nat. Rev. Endocrinol. 12 (4) (2016) 203–221.

[4] M. Wu, G. Chen, Y.P. Li, TGF-β and BMP signaling in osteoblast, skeletal development, and bone formation, homeostasis and disease, Bone Res. 4 (2016) 16009–16030.

[5] T. Katagiri, T. Watabe, Bone morphogenetic proteins, Cold Spring Harb. Perspect. Biol. 8 (6) (2016) a021899.

[6] X. Zhang, J. Guo, Y. Zhou, G. Wu, The roles of bone morphogenetic proteins and their signaling in the osteogenesis of adipose-derived stem cells, Tissue Eng. Part B Rev. 20 (1) (2014) 84–92.

[7] B.E. Grottkau, Y. Lin, Osteogenesis of adipose-derived stem cells, Bone Res. 1 (2013) 133–145.

[8] T. Matsubara, et al., BMP2 regulates osterix through Msx2 and Runx2 during osteoblast differentiation, J. Biol. Chem. 283 (43) (2008) 29119–29125.

[9] T. Liu, et al., BMP-2 promotes differentiation of osteoblasts and chondroblasts in Runx2-deficient cell lines, J. Cell. Physiol. 211 (3) (2007) 728–735.

[10] L.C.C. Yeh, A.D. Tsai, J.C. Lee, Osteogenic protein-1 (OP-1, BMP-7) induces osteoblastic cell differentiation of the pluripotent mesenchymal cell line C2C12, J. Cell. Biochem. 87 (3) (2002) 292–304.

[11] R. Zhang, et al., BMP9-induced osteogenic differentiation is partially inhibited by miR-30a in the mesenchymal stem cell line C3H10T1/2, J. Mol. Histol. 46 (4–5) (2015) 399–407.

[12] M. Mie, H. Ohgushi, Y. Yanagida, T. Haruyama, E. Kobatake, M. Aizawa, Osteogenesis coordinated in C3H10T1/2 cells by adipogenesis-dependent BMP-2 expression system, Tissue Eng. 6 (1) (2000) 9–18.

[13] S. Bougioukli, et al., Gene therapy for bone repair using human cells: superior osteogenic potential of bone morphogenetic protein 2-transduced mesenchymal stem cells derived from adipose tissue compared to bone marrow, Hum. Gene Ther. 29 (4) (2018) 507–519.
[14] M.P. Murphy, N. Quarto, M.T. Longaker, D.C. Wan, Calvarial defects: cell-based reconstructive strategies in the murine model, Tissue Eng. Part C Methods 23 (12) (2017) 971–981.
[15] J.W. Lowery, B. Brookshire, V. Rosen, A survey of strategies to modulate the bone morphogenetic protein signaling pathway: current and future perspectives, Stem Cells Int. 2016 (2016) 7290686.
[16] B. Levi, et al., Nonintegrating knockdown and customized scaffold design enhances human adipose-derived stem cells in skeletal repair, Stem Cells 29 (12) (2011) 2018–2029.
[17] E. Bernstein, A.A. Caudy, S.M. Hammond, G.J. Hannon, Role for a bidentate ribonuclease in the initiation step of RNA interference, Nature 409 (6818) (2001) 363–366.
[18] G.S. Mack, MicroRNA gets down to business, Nat. Biotechnol. 25 (6) (2007) 631–638.
[19] H. Schneider, B. Sedaghati, A. Naumann, M.C. Hacker, M. Schulz-Siegmund, Gene silencing of chordin improves BMP-2 effects on osteogenic differentiation of human adipose tissue-derived stromal cells, Tissue Eng. Part A 20 (1–2) (2014) 335–345.
[20] G. Tardif, J.P. Pelletier, D. Hum, C. Boileau, N. Duval, J. Martel-Pelletier, Differential regulation of the bone morphogenic protein antagonist chordin in human normal and osteoarthritic chondrocytes, Ann. Rheum. Dis. 65 (2) (2006) 261–264.
[21] H. Strutt, et al., Mutations in the sterol-sensing domain of patched suggest a role for vesicular trafficking in smoothened regulation, Curr. Biol. 11 (8) (2001) 608–613.
[22] R.B. Corcoran, M.P. Scott, Oxysterols stimulate sonic hedgehog signal transduction and proliferation of medulloblastoma cells, Proc. Natl. Acad. Sci. USA 103 (22) (2006) 8408–8413.
[23] D. Carpenter, et al., Characterization of two patched receptors for the vertebrate hedgehog protein family, Proc. Natl. Acad. Sci. USA 95 (23) (1998) 13630–13634.
[24] E.H. Villavicencio, D.O. Walterhouse, P.M. Iannaccone, The sonic hedgehog-Patched-Gli pathway in human development and disease, Am. J. Hum. Genet. 67 (5) (2000) 1047–1054.
[25] F. Rahnama, et al., Inhibition of GLI1 gene activation by Patched1, Biochem. J. 394 (1) (2006) 19–26.
[26] T. Nakamura, et al., Induction of osteogenic differentiation by hedgehog proteins, Biochem. Biophys. Res. Commun. 237 (2) (1997) 465–469.
[27] M. Kasper, G. Regl, A.M. Frischauf, F. Aberger, GLI transcription factors: mediators of oncogenic hedgehog signalling, Eur. J. Cancer 42 (4) (2006) 437–445.
[28] M. Zhao, M. Qiao, S.E. Harris, D. Chen, B.O. Oyajobi, G.R. Mundy, The zinc finger transcription factor Gli2 mediates bone morphogenetic protein 2 expression in osteoblasts in response to hedgehog signaling, Mol. Cell. Biol. 26 (16) (2006) 6197–6208.
[29] Q. Chen, et al., Fate decision of mesenchymal stem cells: adipocytes or osteoblasts? Cell Death Differ. 23 (7) (2016) 1128–1139.
[30] A.W. James, et al., Sonic hedgehog influences the balance of osteogenesis and adipogenesis in mouse adipose-derived stromal cells, Tissue Eng. Part A 16 (8) (2010) 2605–2616.
[31] T. Yuasa, et al., Sonic hedgehog is involved in osteoblast differentiation by cooperating with BMP-2, J. Cell. Physiol. 193 (2) (2002) 225–232.
[32] B.T. MacDonald, K. Tamai, X. He, Wnt/β-catenin signaling: components, mechanisms, and diseases, Dev. Cell 17 (1) (2009) 9–26.
[33] T.P. Rao, M. Kühl, An updated overview on wnt signaling pathways: a prelude for more, Circ. Res. 106 (12) (2010) 1798–1806.
[34] L. Grumolato, et al., Canonical and noncanonical Wnts use a common mechanism to activate completely unrelated coreceptors, Genes Dev. 24 (22) (2010) 2517–2530.
[35] Y. Komiya, R. Habas, Wnt signal transduction pathways, Organogenesis 4 (2) (2008) 68–75.
[36] R. Sugimura, L. Li, Noncanonical Wnt signaling in vertebrate development, stem cells, and diseases, Birth Defects Res. C Embryo Today 90 (4) (2010) 243–256.
[37] M.D. Gordon, R. Nusse, Wnt signaling: multiple pathways, multiple receptors, and multiple transcription factors, J. Biol. Chem. 281 (32) (2006) 22429–22433.
[38] X. Liu, et al., Wnt signaling in bone formation and its therapeutic potential for bone diseases, Ther Adv Musculoskelet Dis 5 (1) (2013) 13–31.
[39] R. Van Amerongen, C. Fuerer, M. Mizutani, R. Nusse, Wnt5a can both activate and repress Wnt/B-catenin signaling during mouse embryonic development, Dev. Biol. 369 (1) (2012) 101–114.

[40] A. Santos, A.D. Bakker, J.M.A. De Blieck-Hogervorst, J. Klein-Nulend, WNT5A induces osteogenic differentiation of human adipose stem cells via rho-associated kinase Rock, Cytotherapy 12 (7) (2010) 924–932.

[41] E. Neumann, M. Schaefer-Ridder, Y. Wang, P.H. Hofschneider, Gene transfer into mouse lyoma cells by electroporation in high electric fields, EMBO J. 1 (7) (1982) 841–845.

[42] M. Flanagan, et al., Competitive electroporation formulation for cell therapy, Cancer Gene Ther. 18 (8) (2011) 579–586.

[43] T. Kotnik, L.M. Mir, K. Flisar, M. Puc, D. Miklavcic, Cell membrane electropermeabilization by symmetrical bipolar rectangular pulses. Part I. Increased efficiency of permeabilization, Bioelectrochemistry 54 (1) (2001) 83–90.

[44] J.L. Young, D.A. Dean, Electroporation-mediated gene delivery, Adv. Genet. 89 (2015) 49–88.

[45] L.M. Mir, Nucleic acids electrotransfer-based gene therapy (electrogenetherapy): past, current, and future, Mol. Biotechnol. 43 (2) (2009) 167–176.

[46] N.I. Hristova, I. Tsoneva, E. Neumann, Sphingosine-mediated electroporative DNA transfer through lipid bilayers, FEBS Lett. 415 (1) (1997) 81–86.

[47] M. Golzio, J. Teissié, M.P. Rols, Direct visualization at the single-cell level of electrically mediated gene delivery, Proc. Natl. Acad. Sci. USA 99 (3) (2002) 1292–1297.

[48] E.E. Vaughan, D.A. Dean, Intracellular trafficking of plasmids during transfection is mediated by microtubules, Mol. Ther. 13 (2) (2006) 422–428.

[49] M.A. Badding, D.A. Dean, Highly acetylated tubulin permits enhanced interactions with and trafficking of plasmids along microtubules, Gene Ther. 20 (6) (2013) 616–624.

[50] J. Vacik, B.S. Dean, W.E. Zimmer, D.A. Dean, Cell-specific nuclear import of plasmid DNA, Gene Ther. 6 (6) (1999) 1006–1014.

[51] D.A. Dean, B.S. Dean, S. Muller, L.C. Smith, Sequence requirements for plasmid nuclear import, Exp. Cell Res. 253 (2) (1999) 713–722.

[52] J.S. Lee, J.M. Lee, G. Il Im, Electroporation-mediated transfer of Runx2 and Osterix genes to enhance osteogenesis of adipose stem cells, Biomaterials 32 (3) (2011) 760–768.

[53] S. Cooray, S.J. Howe, A.J. Thrasher, Retrovirus and lentivirus vector design and methods of cell conditioning, Methods Enzymol. 507 (2012) 29–57.

[54] M. Flasshove, et al., Type and position of promoter elements in retroviral vectors have substantial effects on the expression level of an enhanced green fluorescent protein reporter gene, J. Cancer Res. Clin. Oncol. 126 (7) (2000) 391–399.

[55] V. Sandrin, et al., Lentiviral vectors pseudotyped with a modified RD114 envelope glycoprotein show increased stability in sera and augmented transduction of primary lymphocytes and CD34+ cells derived from human and nonhuman primates, Blood 100 (3) (2002) 823–832.

[56] Y. Yang, et al., Inducible, high-level production of infectious murine leukemia retroviral vector particles pseudotyped with vesicular stomatitis virus G envelope protein, Hum. Gene Ther. 6 (9) (1995) 1203–1213.

[57] A. Schambach, H. Wodrich, M. Hildinger, J. Bohne, H.G. Kräusslich, C. Baum, Context dependence of different modules for posttranscriptional enhancement of gene expression from retroviral vectors, Mol. Ther. 2 (5) (2000) 435–445.

[58] A. Miller, Development and applications of retroviral vectors, in: J.M. Coffin, S.H. Hughes, H.E. Varmus (Eds.), Retroviruses, Cold Spring Harbor Laboratory Press, Cold Spring Harbor, NY, 1997.

[59] Y. Ikeda, Y. Takeuchi, F. Martin, F.L. Cosset, K. Mitrophanous, M. Collins, Continuous high-titer HIV-1 vector production, Nat. Biotechnol. 21 (5) (2003) 569–572.

[60] V. Vogt, Retroviral virions and genomes, in: J.M. Coffin, S.H. Hughes, H.E. Varmus (Eds.), Retroviruses, Cold Spring Harbor Laboratory Press, Cold Spring Harbor, NY, 1997.

[61] G. Pal, C. Parolin, Y. Takeuchi, M. Pizzato, Progress with retroviral gene vectors, Rev. Med. Virol. 10 (3) (2000) 185–202.

[62] W. Wold, K. Toth, Adenovirus vectors for gene therapy, vaccination and cancer gene therapy, Curr. Gene Ther. 13 (6) (2014) 421–433.

[63] A. Berk, Adenoviridae, Lippincott Williams & Wilkins, Philadelphia, PA, 2013.

[64] N. Brunetti-Pierri, P. Ng, Helper-dependent adenoviral vectors for liver-directed gene therapy, Hum. Mol. Genet. 20 (2011) R7–R13.

[65] C. Serangeli, et al., Ex vivo detection of adenovirus specific CD4+ T-cell responses to HLA-DR-epitopes of the hexon protein show a contracted specificity of THELPER cells following stem cell transplantation, Virology 397 (2) (2010) 277–284.
[66] S.Y. Park, K.H. Kim, S. Kim, Y.M. Lee, Y.J. Seol, BMP-2 gene delivery-based bone regeneration in dentistry, Pharmaceutics 11 (8) (2019) 393–419.
[67] C. Fuerer, R. Nusse, Lentiviral vectors to probe and manipulate the Wnt signaling pathway, PLoS One 5 (2) (2010) e9370.
[68] P.C. Edwards, et al., Sonic hedgehog gene-enhanced tissue engineering for bone regeneration, Gene Ther. 12 (1) (2005) 75–86.
[69] A.M. Darquet, B. Cameron, P. Wils, D. Scherman, J. Crouzet, A new DNA vehicle for nonviral gene delivery: supercoiled minicircle, Gene Ther 4 (12) (1997) 1341–1349.
[70] A. Valera, J.C. Perales, M. Hatzoglou, F. Bosch, Expression of the neomycin-resistance (neo) gene induces alterations in gene expression and metabolism, Hum. Gene Ther. 5 (4) (1994) 449–456.
[71] J. Hartikka, et al., An improved plasmid DNA expression vector for direct injection into skeletal muscle, Hum. Gene Ther. 7 (10) (1996) 1205–1217.
[72] Y. Sato, et al., Immunostimulatory DNA sequences necessary for effective intradermal gene immunization, Science 273 (5273) (1996) 352–354.
[73] H. Yin, R.L. Kanasty, A.A. Eltoukhy, A.J. Vegas, J.R. Dorkin, D.G. Anderson, Non-viral vectors for gene-based therapy, Nat. Rev. Genet. 15 (8) (2014) 541–555.
[74] Z.Y. Chen, C.Y. He, A. Ehrhardt, M.A. Kay, Minicircle DNA vectors devoid of bacterial DNA result in persistent and high-level transgene expression in vivo, Mol. Ther. 8 (3) (2003) 495–500.
[75] M.A. Kay, C.Y. He, Z.Y. Chen, A robust system for production of minicircle DNA vectors, Nat. Biotechnol. 28 (12) (2010) 1287–1289.
[76] K. Nehlsen, S. Broll, J. Bode, Replicating minicircles: generation of nonviral episomes for the efficient modification of dividing cells, Gene Ther. Mol. Biol. 10 (1) (2006) 233–244.
[77] C.L. Hardee, L.M. Arévalo-Soliz, B.D. Hornstein, L. Zechiedrich, Advances in non-viral DNA vectors for gene therapy, Genes 8 (2) (2017) 65–87.
[78] F. Jia, et al., A nonviral minicircle vector for deriving human iPS cells, Nat. Methods 7 (3) (2010) 197–199.
[79] K.H. Narsinh, F. Jia, R.C. Robbins, M.A. Kay, M.T. Longaker, J.C. Wu, Generation of adult human induced pluripotent stem cells using nonviral minicircle DNA vectors, Nat. Protoc. 6 (1) (2011) 78–88.
[80] B. Levi, et al., In vivo directed differentiation of pluripotent stem cells for skeletal regeneration, Proc. Natl. Acad. Sci. USA 109 (50) (2012) 20379–20384.
[81] J. Hyun, et al., Enhancing in vivo survival of adipose-derived stromal cells through Bcl-2 overexpression using a minicircle vector, Stem Cells Transl. Med. 2 (9) (2013) 690–702.
[82] J.-Y. Ko, J. Lee, J. Lee, Y.H. Ryu, G.-I. Im, SOX-6, 9-transfected adipose stem cells to treat surgically-induced osteoarthritis in goats, Tissue Eng. Part A 25 (13–14) (2019) 990–1000.
[83] N. Tidd, J. Michelsen, B. Hilbert, J.C. Quinn, Minicircle mediated gene delivery to canine and equine mesenchymal stem cells, Int. J. Mol. Sci. 18 (4) (2017) 819–833.
[84] H. Feichtinger, G. Hacobian, A. Hofmann, A.T. Wassermann, K. Zimmermann, M. van Griendven, M. Redl, Constitutive and inducible co-expression for non-viral osteoinductive gene therapy, Eur. Cell. Mater. 27 (2014) 166–184.
[85] E. Lee, J.Y. Ko, J. Kim, J.W. Park, S. Lee, G. Il Im, Osteogenesis and angiogenesis are simultaneously enhanced in BMP2-/VEGF-transfected adipose stem cells through activation of the YAP/TAZ signaling pathway, Biomater. Sci. 7 (11) (2019) 4588–4602.
[86] S. Liu, et al., Highly branched poly(β-amino ester) delivery of minicircle DNA for transfection of neurodegenerative disease related cells, Nat. Commun. 10 (1) (2019) 3307–3321.
[87] Y. Ishino, H. Shinagawa, K. Makino, M. Amemura, A. Nakatura, Nucleotide sequence of the iap gene, responsible for alkaline phosphatase isoenzyme conversion in Escherichia coli, and identification of the gene product, J. Bacteriol. 169 (12) (1987) 5429–5433.
[88] K.S. Makarova, N.V. Grishin, S.A. Shabalina, Y.I. Wolf, E.V. Koonin, A putative RNA-interference-based immune system in prokaryotes: computational analysis of the predicted enzymatic machinery, functional analogies with eukaryotic RNAi, and hypothetical mechanisms of action, Biol. Direct 1 (2006) 7–33.
[89] P. Horvath, R. Barrangou, CRISPR/Cas, the immune system of bacteria and archaea, Science 327 (5962) (2010) 167–170.

[90] M. Jinek, K. Chylinski, I. Fonfara, M. Hauer, J.A. Doudna, E. Charpentier, A programmable dual-RNA-guided DNA endonuclease in adaptive bacterial immunity, Science 337 (6096) (2012) 816–821.
[91] F.A.F.A. Ran, et al., XOne-step generation of mice carrying reporter and conditional alleles by CRISPR/cas-mediated genome engineering, Cell 154 (6) (2013) 1370–1379.
[92] P. Mali, et al., RNA-guided human genome engineering via Cas9, Science 339 (6121) (2013) 823–826.
[93] L. Cong, et al., Multiplex genome engineering using CRISPR/Cas systems, Science 339 (6121) (2013) 819–823.
[94] L.S. Qi, et al., Repurposing CRISPR as an RNA-γuided platform for sequence-specific control of gene expression, Cell 152 (5) (2013) 1173–1183.
[95] D. Gerace, R. Martiniello-Wilks, N.T. Nassif, S. Lal, R. Steptoe, A.M. Simpson, CRISPR-targeted genome editing of mesenchymal stem cell-derived therapies for type 1 diabetes: a path to clinical success? Stem Cell Res. Ther. 8 (1) (2017) 62–72.
[96] N. Farhang, et al., CRISPR-based epigenome editing of cytokine receptors for the promotion of cell survival and tissue deposition in inflammatory environments, Tissue Eng. Part A 23 (15–16) (2017) 738–749.
[97] M.-N. Hsu, et al., CRISPR-based activation of endogenous neurotrophic genes in adipose stem cell sheets to stimulate peripheral nerve regeneration, Theranostics 9 (21) (2019) 6099–6111.
[98] R. Zhao, et al., CRISPR/Cas9-mediated BRCA1 knockdown adipose stem cells promote breast cancer progression, Plast. Reconstr. Surg. 143 (3) (2019) 747–756.
[99] M. Deng, et al., TGFβ3 recruits endogenous mesenchymal stem cells to initiate bone regeneration, Stem Cell Res. Ther. 8 (1) (2017) 258–270.
[100] A. Infante, C.I. Rodríguez, Osteogenesis and aging: lessons from mesenchymal stem cells, Stem Cell Res. Ther. 9 (1) (2018) 244.
[101] S.W. Cho, et al., Analysis of off-target effects of CRISPR/Cas-derived RNA-guided endonucleases and nickases, Genome Res. 24 (1) (2014) 132–141.
[102] F. Clément, E. Grockowiak, F. Zylbersztejn, G. Fossard, S. Gobert, V. Maguer-Satta, Stem cell manipulation, gene therapy and the risk of cancer stem cell emergence, Stem Cell Investig. 4 (7) (2017) 67–82.
[103] C. Liu, N. Zhang, Nanoparticles in gene therapy: principles, prospects, and challenges, Prog. Mol. Biol. Transl. Sci. 104 (2011) 509–562.
[104] J. Chen, Z. Guo, H. Tian, X. Chen, Production and clinical development of nanoparticles for gene delivery, Mol. Ther. Methods Clin. Dev. 3 (2016) 16023.
[105] M.K. Riley, W. Vermerris, Recent advances in nanomaterials for gene delivery—a review, Nanomaterials 7 (5) (2017) 94–113.
[106] W. Zauner, N.A. Farrow, A.M.R. Haines, In vitro uptake of polystyrene microspheres: effect of particle size, cell line and cell density, J. Control. Release 71 (1) (2001) 39–51.
[107] M.P. Desai, V. Labhasetwar, E. Walter, R.J. Levy, G.L. Amidon, The mechanism of uptake of biodegradable microparticles in Caco-2 cells is size dependent, Pharm. Res. 14 (11) (1997) 1568–1573.
[108] S. Prabha, W.Z. Zhou, J. Panyam, V. Labhasetwar, Size-dependency of nanoparticle-mediated gene transfection: studies with fractionated nanoparticles, Int. J. Pharm. 244 (1–2) (2002) 105–115.
[109] J. Fan, et al., Delivery of phenamil enhances BMP-2-induced osteogenic differentiation of adipose-derived stem cells and bone formation in calvarial defects, Tissue Eng. Part A 21 (13–14) (2015) 2053–2065.
[110] Y. Guo, et al., Monodispersed bioactive glass nanoparticles enhance the osteogenic differentiation of adipose-derived stem cells through activating TGF-Beta/Smad3 signaling pathway, Part. Part. Syst. Charact. 35 (7) (2018) 1800087.
[111] E. Brett, et al., Magnetic nanoparticle-based upregulation of B-cell lymphoma 2 enhances bone regeneration, Stem Cells Transl. Med. 6 (1) (2017) 151–160.
[112] K. Mesimäki, et al., Novel maxillary reconstruction with ectopic bone formation by GMP adipose stem cells, Int. J. Oral Maxillofac. Surg. 38 (3) (2009) 201–209.
[113] BonusBio, Filling Bone Defects/Voids With Autologous BonoFill for Maxillofacial Bone Regeneration, *ClinicalTrails.gov*, 2014.

CHAPTER

17

Adipose-derived stromal/stem cells for bone tissue engineering applications

Nathalie Faucheux[a,b], *Fabien Kawecki*[c,d], *Jessica Jann*[a,b], *François A. Auger*[c,d], *Roberto D. Fanganiello*[e,*], *and Julie Fradette*[c,d,*]

[a]Laboratory of Cell-Biomaterial Biohybrid Systems, Department of Chemical and Biotechnological Engineering, University of Sherbrooke, Sherbrooke, QC, Canada [b]CHUS Research Center, Sherbrooke, QC, Canada [c]The Tissue Engineering Laboratory (LOEX), a Laval University Research Center, and Regenerative Medicine Division, CHU of Quebec— Laval University Research Center, Quebec, QC, Canada [d]Department of Surgery, Faculty of Medicine, Laval University, Quebec, QC, Canada [e]Group of Oral Ecology Research (GREB), Faculty of Dentistry, Laval University, Quebec, QC, Canada

Enhancing bone wound healing: From native grafts, to material- and cell-based tissue engineering approaches

Bones are the basic mechanical support of the locomotor system. Bones protect vital organs, act as an important reservoir of minerals, and contain bone marrow, which is the main site of hematopoiesis. Bone tissue is a mineralized, vascularized, and innervated tissue that can remodel and self-renew through a process involving two main cell types: the osteoclasts, which resorb the bone matrix, and the osteoblasts, which form a new extracellular matrix (ECM) [1, 2]. An imbalance between the rates of resorption and bone formation leads to bone disorders and pathologies, such as osteoporosis, resulting in significant changes in bone mass and architecture. Bone disorders can be a source of fractures and massive bone loss, in

*Co-senior authors.

https://doi.org/10.1016/B978-0-12-819376-1.00018-4

addition to defects caused by traumatic injuries, infections, tumor resections, or congenital malformations. It is essential to design improved strategies to accelerate bone formation and healing for all these situations that necessitate skeletal tissue repair and/or reconstruction.

Therapeutic options relying on bone grafting

Autologous bone grafting, also called autografting or autogenous bone grafting, is the most-used therapeutic option for bone tissue reconstruction, and is considered as the gold standard clinical procedure in this field [3]. This procedure involves harvesting bone from the same patient receiving the graft. Bone tissue is obtained during a separate surgery, from a "donor site," which is usually the iliac crest, the fibula, the ribs, the distal portion of the femoral bone, or specific parts of the skull, such as the mandibular symphysis or the mandible's coronoid process [4]. For example, cases of mandibular continuity defects smaller than 6cm in length are customarily treated with nonvascularized corticocancellous autologous bone grafts collected from the posterior iliac crest. Larger mandibular nonunion defects are often treated with vascularized autologous grafts isolated from the fibula [5].

A triad of key processes associated with guided bone regeneration is osteoinduction, osteoconduction, and osteogenesis (Fig. 1).

Osteoinduction comprises the dynamics through which osteoprogenitor cells or other types of undifferentiated cells are induced, by biochemical signals and/or physical cues from

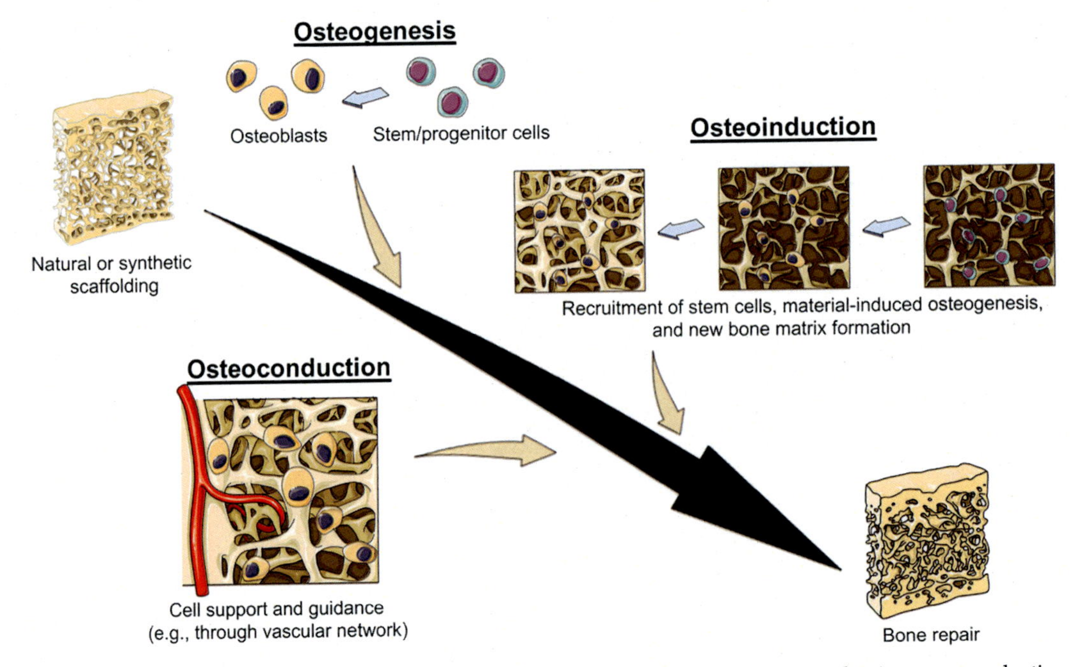

FIG. 1 Schematic of the key processes associated with guided bone regeneration: osteoinduction, osteoconduction, and osteogenesis. A part of the figure was created using Servier Medical Art. https://smart.servier.com.

the environment, to differentiate toward the osteoblastic lineage [6]. Osteoconduction is the provision of a physical framework that supports and promotes the attachment, migration, proliferation, and differentiation of osteoprogenitor cells and osteoblasts. Osteogenesis is the actual differentiation of osteoprogenitor cells into osteoblasts and osteocytes, yielding new bone tissue. Autografts act on these three main components necessary for optimal bone repair by providing osteogenic cells and osteoinductive molecules, as well as a framework (matrix) for bone regeneration. Autografts also have a low graft and immune rejection. On the other hand, autografts are associated with several disadvantages such as postoperative pain, donor site morbidity, limitations in the availability of bone that can be harvested, and differences in shape and architecture between the bone tissue used as a graft and the bone defect to be repaired [7–10]. A study by Younger and Chapman with 239 patients that received autografts from the iliac crest reported a major complication rate in 8.6% of the cases, including infection, wound drainage, large hematomas, reoperation, and sensory loss, as well as minor complications in about 20% of the cases [11]. High and unpredictable resorption rates have also been reported in other studies [12, 13].

An alternative to bone autografting is allogeneic grafting (also called allografting), for which bone tissue is obtained either from cadavers or from living donors matching the individual receiving the graft, commonly garnered from bone banks. Before storage, in both cases, bone blocks are explanted, split, and treated so that all living cells and potential contaminants are removed. Decellularization of allografts removes the need for immunosuppression. The use of allografts eliminates the need for additional surgery on the patient, but is associated with the constraints of tissue matching, batch variability, potential of insufficient blood supply, the possibility of disease transmission, immunologic rejection, and graft-versus-host disease. Challenges related to the stability and integration of the implant with the surrounding tissue are also paramount since they are usually acellular at the time of the implantation [14].

Materials for bone tissue engineering

Unfavorable outcomes associated with the limitations of autografts and allografts have prompted scientists and clinicians to develop biomaterials that can be used as scaffolds for guided bone regeneration. Engineered bone substitutes that mimic the osteogenic potential of bone grafts reduce disease transmission risks and the number of surgical procedures required, decrease immunogenicity issues, and provide a much larger supply of homogeneous implants. Research and development in the field of biomaterial sciences has moved from adapting inert, industrial materials amenable to applications in biomedical fields, to the design of structural elements with defined physical and chemical features that are called "scaffolds" [15, 16].

An ideal biomaterial scaffold should be osteoinductive, osteoconductive, and osteogenic, mediating formation of bone that is morphologically, physiologically, and functionally similar to the patient's bone previous to the injury. For this reason, one of the main concepts in biomaterials design for bone tissue engineering is the devising of porous three-dimensional (3D) scaffolds that can give physical and structural support, while providing an optimal microenvironment to guide cellular behavior with spatiotemporal cues during bone regeneration [17, 18]. Other pivotal features that a model scaffold should have are the ability to support angiogenic processes, biomechanical stability, and biodegradability within a time frame that matches bone substitution and regeneration [16].

There is a myriad of commercially available materials for bone regeneration. The most common materials are based on alpha- or beta-tricalcium phosphate (TCP), amorphous calcium phosphates, calcium phosphate granules, blocks and cements, ceramics, such as hydroxyapatites (HA) or glass ceramics, among others (reviewed in Ref. [19]). Synthetic polymers such as polylactic acid (PLA) and polyglycolic acid (PGA), as well as naturally derived materials, such as collagens, gelatins, and hyaluronic acid, are also available. Some composite compounds containing chitosan, gelatin, or collagen associated with HA nanoparticles have shown impressive results on the differentiation of mesenchymal stromal/stem cells (MSCs) [20]. These materials are designed to mimic the organic and/or the inorganic components of bone, and display good osteointegrative features, with different degrees of osteoinductive and osteoconductive properties. Yet, none of these materials are originally osteogenic since they lack a cellular component.

Cell-based strategies for bone engineering

Over the past decades, bone tissue engineering has become very promising, especially for the repair of massive bone tissue losses. Methods for bone tissue engineering have constantly evolved owing to the discovery of new cell sources and the design of innovative biomaterials and scaffold types. Strategies aiming at bone formation and repair must consider the physiology of this tissue and its mechanical function. Different techniques now associate one or more scaffolding elements with regenerative cells having the potential to differentiate toward an osteoblastic lineage under the action of various biochemical cues, as well as structural and mechanical factors [21].

Many cell types can be associated with these biomaterial scaffolds to elicit osteogenic properties. The two main categories of cells exhibiting osteogenic differentiation potential that are used for bone tissue engineering are osteoblasts and MSCs. Osteoblasts are already primed and committed toward the osteogenic lineage, but their harvest is associated with donor site morbidity. Furthermore, they are limited in number (cell yield at extraction), and they also have limited in vitro proliferation potential [22, 23].

MSCs can be isolated from many different tissues (bone marrow, adipose and muscle tissues, among many others) and have the dual ability of extensive cell replication as well as multilineage differentiation potential. Under appropriate in vitro conditions and proper biochemical signals, MSCs can be directed to differentiate toward specific cell types, including the osteogenic lineage. Among other cell sources amenable to bone tissue engineering applications, embryonic stem cells (ESCs) offer interesting characteristics, in particular regarding self-renewal and pluripotency [24]. The literature describes numerous stimuli, such as supplementation with osteogenic factors, that were used to induce ESC differentiation toward the osteoblastic lineage [25]. However, due to ESCs' sourcing, applications in tissue engineering are limited by regulatory and ethical issues. Induced pluripotent stem cells (iPSCs) are expected to offer considerable potential for regenerative medicine in the coming years. These cells display pluripotency similarly to ESCs, but they are obtained by genetic reprogramming of differentiated adult cells, such as blood cells, fibroblasts of the skin, epithelial cells, or adult mesenchymal cells. Reprogramming of adult differentiated cells into iPSCs consists of overexpression of four essential genes (Oct3/4, Sox2, c-Myc, and Klf4) that activate transcriptomic and biochemical pathways of dedifferentiation and proliferation [26]. Protocols inducing their differentiation toward the osteoblastic lineage have recently been published [27], but many technological challenges have to be overcome before iPSCs are routinely used for bone tissue engineering. Currently, adult MSCs remain one of the most promising sources of cells for clinical applications.

Bone repair using bone marrow or adipose-derived mesenchymal stromal/stem cells

MSCs have been isolated and characterized from various tissues, including bone marrow, adipose tissue (fat depots), muscle, umbilical cord, and placenta [28]. The most-studied MSCs for bone regeneration are the bone marrow stromal/stem cells (BMSCs), isolated from the bone marrow stroma. BMSCs are capable of undergoing osteogenic, chondrogenic, and adipogenic differentiation, along with differentiation into endothelial lineage cells [29, 30]. It has been demonstrated that BMSCs can be induced in vitro to differentiate into osteoblasts and osteocytes after treatment with an osteogenic cocktail consisting of dexamethasone, ascorbic acid, and β-glycerophosphate [30]. In addition, BMSCs have been seeded onto different biomaterials and used for intramembranous and endochondral in vivo bone regeneration, in animal models [31–33]. Adipose tissue is one of the most abundant and ubiquitous tissue types in mammals, and ASCs have attracted great attention for regenerative medicine in general, and especially for cartilage and bone applications [34–37].

In vitro osteogenic differentiation conditions and dynamics have been thoroughly reported in the literature [38–44]. The behavior of ASCs and BMSCs, as well as the panel of cell surface markers expressed during in vitro osteogenic differentiation, is highly similar [28, 37, 45]. Cytokines and growth factors can also be used to induce ASCs' osteogenic differentiation and the roles of bone morphogenetic proteins (BMPs) [46–49], insulin growth factors 1 and 2 (IGFs 1 and 2) [50, 51], sonic hedgehog [52–54], and valproic acid [55], among others, have been reported.

In addition to their potential toward osteogenic differentiation, ASCs' secretome, defined as the panel of molecules secreted by ASCs, has been shown to contain bone-inducing (e.g., BMP-2) and angiogenic (e.g., vascular endothelial growth factor [VEGF]) factors that can act in a paracrine manner [56–58]. These bioactive factors can be secreted as soluble growth factors or within exosomes, and increasing evidence supports the contribution of exosome-derived micro-ribonucleic acids (miRNAs) to enhanced osteogenic processes (reviewed in Ref. [57]).

Interestingly, Overman et al. reported differences in ASCs' secretome when cultured in association with a 3D biphasic ceramic scaffold (60% HA, 40% β-TCP) in comparison to 2D monolayers, on plastic surfaces [59]. When cultured in association with the biomaterial, ASCs had a more than twofold increase of approximately 20 growth factors, cytokines, and adhesion molecules, such as interleukin (IL)-6, fibroblast growth factor (FGF)-7, and VCAM1 [59].

The expression profile of several ASC-derived factors can be modulated by stressors (e.g., hypoxia) [60–62], as well as by the regulation of specific biochemical pathways such as transforming growth factor-beta (TGF-β)-1 and hypoxia-inducible factor 1-alpha (HIF-1-alpha) [63, 64]. For example, Frazier et al. reported a more than twofold decrease in the expression of ECM elements related to osteogenesis, such as fibronectin 1, TGF-β1-induced protein, osteonectin, and collagen alpha-1 (1) and alpha-2 (1) chains when human ASCs (hASCs) were cultured under hypoxia (low levels of oxygen, usually under 5%) [60]. The hypothesis that remodeling of ECM is a pivotal biological process linked to ASCs under hypoxia was validated by Riis and colleagues [62]. Proteomic profiling by mass spectrometry of ASCs exposed to low oxygen levels was performed, and nearly 10% of the proteome was altered. Most of the differentially expressed proteins under these conditions were connected with ECM remodeling and control of cellular metabolism [62].

Finally, ASCs are also very efficient for production, secretion, and assembly of endogenous ECM components upon stimulation with serum and ascorbic acid [21, 65, 66]. This property is the basis of a method called the self-assembly approach of tissue engineering leading to the production of matrix-enriched cell sheets by mesenchymal cells [67–69]. Using a modified protocol, hASCs were differentiated into the osteoblastic lineage in parallel to ECM matrix stimulation, allowing the formation of osseous cell sheets after 21 days [70] or 35 days [71] of culture in vitro. The resulting cell sheets were further customized by stacking multiple cell sheets to obtain thicker constructs featuring osteogenic differentiation and mineralization, without the requirement of exogenous biomaterials.

Assessing adipose-derived stromal/stem cell osteogenic properties in vitro and in vivo

Osteogenic differentiation can be modulated by the properties of the scaffolds/biomaterials/matrices [59, 72, 73], and many research teams are investigating the osteogenic differentiation potential of ASCs in vitro when used in combination with different types of biomaterials. Differentiation along the osteogenic lineage has been reported when ASCs were associated with calcium phosphate ceramics of different porosities and surface areas [74], with titanium alloys [75, 76], with natural, semisynthetic, and synthetic polymers [77–80], with bioactive glasses [81], and with a myriad of combinations of two or more different biomaterials [82–87].

An important step to achieve translation of in vitro findings into clinical applications is to validate ASCs' osteogenic potential in an in vivo setting. Experimental designs to test ASCs' osteogenic potential in vivo can vary widely. ASCs can easily be isolated from different organisms, usually mammals, such as rats, mice, or humans, and from subcutaneous adipose depots from different anatomical sites. Cells can be associated or not with a biomaterial before implantation, and can also be previously primed toward osteogenic differentiation through treatment with osteoinductive factors, or not. Common animal models used to test ASCs' potential for bone formation and regeneration are mice, rats, and rabbits, and less common ones are dogs, pigs, and sheep. The anatomical sites where cells or 3D constructs are transplanted can also vary, from subcutaneous or intramuscular sites, where ectopic bone formation is expected to occur, to bone defects created by a surgical procedure [88, 89].

Ectopic bone formation

Subcutaneous or intramuscular implantation is used to probe bone formation at these ectopic sites and to assess the direct effects of osteogenic induction and/or of biomaterials on

ASCs without the interference of the endogenous host's osteogenic niche. A number of studies adopted ectopic bone formation experiments to study the osteogenic potential of ASCs in vivo. In Lee et al., ASCs were isolated from epididymal adipose tissue of Lewis rats, expanded in vitro, and induced toward osteogenic differentiation with cell culture medium supplemented with a cocktail of ascorbic acid, β-glycerophosphate, and dexamethasone [90]. ASCs were seeded upon resorbable PGA grafts and implanted in subcutaneous pockets in Lewis rats. The authors reported bone formation after 8 weeks, as assessed by hematoxylin and eosin (H&E) staining and immunohistochemical staining of osteocalcin [90]. When in vivo transplantation is xenogeneic in nature, the use of nude, athymic, or immunosuppressed animals is strategic since these animals display weakened inflammatory and immune rejection responses. For example, Leong et al. implanted hASCs seeded upon a medical-grade polycaprolactone (PCL)-TCP biomaterial subcutaneously in the dorsal region of athymic nude rats [91]. The study compared uninduced hASCs with hASCs that were either induced with osteogenic factors for 2 weeks before seeding into scaffolds, or induced with the same factors after seeding into scaffolds. The authors reported widespread foci of mineralization and deposition of ECM composed of collagen type I, osteopontin, and osteonectin in the first two groups, 6 or 12 weeks after implantation. Yet, no mature and organized bone tissue organization was visible [91].

Bone repair models

Surgical models of bone defects produced under controlled procedures are necessary for clinical translation (Table 1). Defects can be created in calvarial bones, intramembranous bones of the skull, or in long, endochondral bones, such as femur, tibia, ulna, or radius

TABLE 1 Main studies reporting osteogenic potential of adipose-derived stromal/stem cells (ASCs) using animal models.

Surgical site	Source of ASCs	Material	Animal model	References
Tibia	Rabbit	HA	Rabbit	1
	Rabbit	None (injection)	Rabbit	2
	Human	PRP	Dog	3
	Mouse	None (injection)	Mouse	4
	Mouse	Demineralized bone	Sheep	5
Ulna	Rabbit	PLGA	Rabbit	6
	Rabbit	Demineralized bone	Rabbit	7
	Rat	Demineralized bone	Rat	8
Radius	Dog	β-TCP	Dog	9
	Rabbit	PLA and PCL	Rabbit	10
	Rabbit	HA and PCL and collagen gel	Rabbit	10

Continued

TABLE 1 Main studies reporting osteogenic potential of adipose-derived stromal/stem cells (ASCs) using animal models—cont'd

Surgical site	Source of ASCs	Material	Animal model	References
Femur	Mouse	None (injection)	Mouse	11
	Rat	Fibrin	Rat	12
	Rat	β-TCP	Rat	13
	Human	Collagen	Rat	14
	Human	PLDLA	Rat	15
	Sheep	Titanium	Sheep	16
	Human	Collagen and ceramic material	Rat	17
	Rat	Collagen	Rat	18
	Rabbit	None (injection)	Rabbit	19
	Rabbit	HA/α-TCP	Rabbit	20
Vertebrae	Pig	None (injection)	Mouse	21
	Pig	Cancellous bone	Rat	22
Mandible	Pig	None (injection)	Pig	23
	Human	HA and collagen	Rat	24
	Mouse	Chitosan and CS	Rat	25
	Pig	Demineralized bone	Pig	26
Alveolar bone	Rat	PLGA	Rat	27
Calvarial bones	Human	PLGA	Rat	28
	Rabbit	HA and PLGA and collagen	Rabbit	29
	Human	PLGA	Mouse	30
	Rat	β-TCP	Rat	31
	Human	HA/PLGA	Mouse	32
	Mouse	None (injection)	Mouse	33
	Human	None (injection)	Mouse	34
	Rat	Demineralized bone and PLA	Rat	35
	Human	PLGA and PCL	Rat	36
	Human	HA and β-TCP	Rat	37
	Human	PLGA	Rat	38
	Dog	Coral	Dog	39
	Dog	Coral	Dog	40

TABLE 1 Main studies reporting osteogenic potential of adipose-derived stromal/stem cells (ASCs) using animal models—cont'd

Surgical site	Source of ASCs	Material	Animal model	References
	Pig	Collagen	Pig	41
	Rat	Demineralized bone	Rat	42
	Human	PCL and PLGA and β-TCP	Rat	43
	Human	PLGA	Mouse	44
	Rabbit	Collagen	Rabbit	45
	Human	HA and PLGA	Mouse	46
	Human	None (injection)	Rat	47
	Rat	PRP	Rat	48
	Rabbit	PLA	Rabbit	49
	Rat	PLA	Rat	50
	Rabbit	Bioactive glass/β-TCP	Rabbit	51
	Human	BioOss™ and collagen	Rat	52
	Dog	PCL and β-TCP	Dog	53
	Rabbit	Bioactive glass/TCP	Rabbit	54
	Human	PLGA	Mouse	55
	Rat	Chitosan and HA	Rat	56
	Human	Decellularized tendon/HA	Mouse	57
	Mouse	Bioactive glass	Rat	58
	Mouse	Starch-polycaprolactone	Mouse	59

Abbreviations: ASCs, *adipose-derived stromal/stem cells;* α-TCP, *alpha-tricalcium phosphate;* β-TCP, *beta-tricalcium phosphate;* CS, *chondroitin sulfate;* HA, *hydroxyapatite;* PCL, *polycaprolactone;* PLA, *polylactic acid;* PLDLA, *poly(*L*-lactide-co-*DL*-lactide);* PLGA, *poly(lactic-co-glycolic acid);* PRP, *platelet-rich plasma.*

[1] *Arrigoni E, de Girolamo L, Di Giancamillo A, Stanco D, Dellavia C, Carnelli D, Campagnol M, et al. Adipose-derived stem cells and rabbit bone regeneration: histomorphometric, immunohistochemical and mechanical characterization.* Journal of Orthopaedic Science: Official Journal of the Japanese Orthopaedic Association. ***2013****; 18:331–339.*

[2] *Sunay O, Can G, Cakir Z, Denek Z, Kozanoglu I, Erbil G, Yilmaz M, Baran Y. Autologous rabbit adipose tissue-derived mesenchymal stromal cells for the treatment of bone injuries with distraction osteogenesis.* Cytotherapy. ***2013****; 15:690–702.*

[3] *Cruz ACC, Caon T, Menin Á, Granato R, Boabaid F, Simões CMO. Adipose-derived stem cells incorporated into platelet-rich plasma improved bone regeneration and maturation in vivo.* Dental Traumatology: Official Publication of International Association for Dental Traumatology. ***2015****; 31:42–48.*

[4] *Wallner C, Abraham S, Wagner JM, Harati K, Ismer B, Kessler L, Zollner H, et al. Local Application of Isogenic Adipose-Derived Stem Cells Restores Bone Healing Capacity in a Type 2 Diabetes Model.* Stem Cells Translational Medicine. ***2016****; 5(6):836–844.*

[5] *Hernandez-Hurtado AA, Borrego-Soto G, Marino-Martinez IA, Lara-Arias J, Romero-Diaz VJ, Abrego-Guerra A, Vilchez-Cavazos JF, et al. Implant Composed of Demineralized Bone and Mesenchymal Stem Cells Genetically Modified with AdBMP2/AdBMP7 for the Regeneration of Bone Fractures in* Ovis aries. Stem Cells International. ***2016****; 2016:7403890.*

[6] *Kim A, Kim D-H, Song H-R, Kang W-H, Kim H-J, Lim H-C, Cho D-W, Bae J-H. Repair of rabbit ulna segmental bone defect using freshly isolated adipose-derived stromal vascular fraction.* Cytotherapy. ***2012****; 14:296–305.*

[7] *Gu H, Xiong Z, Yin X, Li B, Mei N, Li G, Wang C. Bone regeneration in a rabbit ulna defect model: use of allogeneic adipose-derivedstem cells with low immunogenicity.* Cell and Tissue Research. ***2014****; 358:453–464.*

[8] *Wen C, Yan H, Fu S, Qian Y, Wang D, Wang C. Allogeneic adipose-derived stem cells regenerate bone in a critical-sized ulna segmental defect.* Experimental Biology and Medicine (Maywood, NJ). ***2016****; 241:1401–1409.*

[9] *Kang BJ, Ryu HH, Park SS, Koyama Y, Kikuchi M, Woo HM, Kim WH, Kweon OK. Comparing the osteogenic potential of canine mesenchymal stem cells derived from adipose tissues, bone marrow, umbilical cord blood, and Wharton's jelly for treating bone defects.* Journal of Veterinary Science. **2012**; *13:299–310.*
[10] *Hao W, Dong J, Jiang M, Wu J, Cui F, Zhou D. Enhanced bone formation in large segmental radial defects by combining adipose-derived stem cells expressing bone morphogenetic protein 2 with nHA/RHLC/PLA scaffold.* International Orthopaedics. **2010**; *34:1341–1349.*
[11] *Lee S-W, Padmanabhan P, Ray P, Gambhir SS, Doyle T, Contag C, Goodman SB, Biswal S. Stem cell-mediated accelerated bone healing observed with in vivo molecular and small animal imaging technologies in a model of skeletal injury.* Journal of Orthopaedic Research: Official Publication of the Orthopaedic Research Society. **2009**; *27:295–302.*
[12] *Keibl C, Fügl A, Zanoni G, Tangl S, Wolbank S, Redl H, van Griensven M. Human adipose derived stem cells reduce callus volume upon BMP-2 administration in bone regeneration.* Injury. **2011**; *42:814–820.*
[13] *Qing W, Guang-Xing C, Lin G, Liu Y. The osteogenic study of tissue engineering bone with BMP2 and BMP7 gene-modified rat adipose-derived stem cell.* Journal of Biomedicine & Biotechnology. **2012**; *2012:410879.*
[14] *Shoji T, Ii M, Mifune Y, Matsumoto T, Kawamoto A, Kwon S-M, Kuroda T, et al. Local transplantation of human multipotent adipose-derived stem cells accelerates fracture healing via enhanced osteogenesis and angiogenesis.* Laboratory Investigation; a Journal of Technical Methods and Pathology. **2010**; *90:637–649.*
[15] *Chou Y-F, Zuk PA, Chang T-L, Benhaim P, Wu BM. Adipose-derived stem cells and BMP2: part 1. BMP2-treated adipose-derived stem cells do not improve repair of segmental femoral defects.* Connective Tissue Research. **2011**; *52:109–118.*
[16] *Godoy Zanicotti D, Coates DE, Duncan WJ. In vivo bone regeneration on titanium devices using serum-free grown adipose-derived stem cells, in a sheep femur model.* Clinical Oral Implants Research. **2017**; *28:64–75.*
[17] *Peterson B, Zhang J, Iglesias R, Kabo M, Hedrick M, Benhaim P, Lieberman JR. Healing of critically sized femoral defects, using genetically modified mesenchymal stem cells from human adipose tissue.* Tissue Engineering. **2005**; *11:120–129.*
[18] *Nomura I, Watanabe K, Matsubara H, Hayashi K, Sugimoto N, Tsuchiya H. Uncultured autogenous adipose-derived regenerative cells promote bone formation during distraction osteogenesis in rats.* Clinical Orthopaedics and Related Research. **2014**; *472:3798–3806.*
[19] *Cao Z, Hou S, Sun D, Wang X, Tang J. Osteochondral regeneration by a bilayered construct in a cell-free or cell-based approach.* Biotechnology Letters. **2012**; *34:1151–1157.*
[20] *Fernandez FB, Shenoy S, Suresh Babu S, Varma HK, John A. Short-term studies using ceramic scaffolds in lapine model for osteochondral defect amelioration.* Biomedical Materials (Bristol, England). **2012**; *7:035005.*
[21] *Sheyn D, Pelled G, Zilberman Y, Talasazan F, Frank JM, Gazit D, Gazit Z. Nonvirally Engineered Porcine Adipose Tissue-Derived Stem Cells: Use in Posterior Spinal Fusion.* Stem Cells. **2008**; *26:1056–1064.*
[22] *Sheyn D, Kallai I, Tawackoli W, Cohn Yakubovich D, Oh A, Su S, Da X, et al. Gene-modified adult stem cells regenerate vertebral bone defect in a rat model.* Molecular Pharmaceutics. **2011**; *8:1592–1601.*
[23] *Wilson SM, Goldwasser MS, Clark SG, Monaco E, Bionaz M, Hurley WL, Rodriguez-Zas S, et al. Adipose-derived mesenchymal stem cells enhance healing of mandibular defects in the ramus of swine.* Journal of Oral and Maxillofacial Surgery: Official Journal of the American Association of Oral and Maxillofacial Surgeons. **2012**; *70:e193–203.*
[24] *Parrilla C, Saulnier N, Bernardini C, Patti R, Tartaglione T, Fetoni AR, Pola E, et al. Undifferentiated human adipose tissue-derived stromal cells induce mandibular bone healing in rats.* Archives of Otolaryngology–Head & Neck Surgery. **2011**; *137:463–470.*
[25] *Fan J, Park H, Lee MK, Bezouglaia O, Fartash A, Kim J, Aghaloo T, Lee M. Adipose-derived stem cells and BMP-2 delivery in chitosan-based 3D constructs to enhance bone regeneration in a rat mandibular defect model.* Tissue Engineering - Part A. **2014**; *20:2169–2179.*
[26] *Bhumiratana S, Bernhard JC, Alfi DM, Yeager K, Eton RE, Bova J, Shah F, et al. Tissue-engineered autologous grafts for facial bone reconstruction.* Science Translational Medicine. **2016**; *8.*
[27] *Akita D, Kano K, Saito-Tamura Y, Mashimo T, Sato-Shionome M, Tsurumachi N, Yamanaka K, et al. Use of Rat Mature Adipocyte-Derived Dedifferentiated Fat Cells as a Cell Source for Periodontal Tissue Regeneration.* Frontiers in Physiology. **2016**; *7:50.*
[28] *Wang C-Z, Chen S-M, Chen C-H, Wang C-K, Wang G-J, Chang J-K, Ho M-L. The effect of the local delivery of alendronate on human adipose-derived stem cell-based bone regeneration.* Biomaterials. **2010**; *31:8674–8683.*
[29] *Lin C-Y, Chang Y-H, Li K-C, Lu C-H, Sung L-Y, Yeh C-L, Lin K-J, et al. The use of ASCs engineered to express BMP2 or TGF-β3 within scaffold constructs to promote calvarial bone repair.* Biomaterials. **2013**; *34:9401–9412.*
[30] *Levi B, Nelson ER, Li S, James AW, Hyun JS, Montoro DT, Lee M, et al. Dura mater stimulates human adipose-derived stromal cells to undergo bone formation in mouse calvarial defects.* Stem Cells (Dayton, Ohio). **2011**; *29:1241–1255.*
[31] *Deng Y, Zhou H, Zou D, Xie Q, Bi X, Gu P, Fan X. The role of miR-31-modified adipose tissue-derived stem cells in repairing rat critical-sized calvarial defects.* Biomaterials. **2013**; *34:6717–6728.*
[32] *Lo DD, Hyun JS, Chung MT, Montoro DT, Zimmermann A, Grova MM, Lee M, et al. Repair of a critical-sized calvarial defect model using adipose-derived stromal cells harvested from lipoaspirate.* Journal of Visualized Experiments: JoVE. **2012**.
[33] *Levi B, James AW, Nelson ER, Hu S, Sun N, Peng M, Wu J, Longaker MT. Studies in adipose-derived stromal cells: migration and participation in repair of cranial injury after systemic injection.* Plastic and Reconstructive Surgery **2011**; *127(3):1130–1140.*
[34] *Levi B, James AW, Nelson ER, Peng M, Wan DC, Commons GW, Lee M, et al. Acute skeletal injury is necessary for human adipose-derived stromal cell-mediated calvarial regeneration.* Plastic and Reconstructive Surgery. **2011**; *127:1118–1129.*
[35] *Rhee SC, Ji Y-h, Gharibjanian NA, Dhong ES, Park SH, Yoon E-S. In vivo evaluation of mixtures of uncultured freshly isolated adipose-derived stem cells and demineralized bone matrix for bone regeneration in a rat critically sized calvarial defect model.* Stem Cells and Development. **2011**; *20:233–242.*

[36] *Hong JM, Kim BJ, Shim J-H, Kang KS, Kim K-J, Rhie JW, Cha HJ, Cho D-W. Enhancement of bone regeneration through facile surface functionalization of solid freeform fabrication-based three-dimensional scaffolds using mussel adhesive proteins.* Acta Biomaterialia. **2012**; *8:2578–2586.*
[37] *Jo CH, Yoon PW, Kim H, Kang KS, Yoon KS. Comparative evaluation of in vivo osteogenic differentiation of fetal and adult mesenchymal stem cell in rat critical-sized femoral defect model.* Cell and Tissue Research. **2013**; *353:41–52.*
[38] *Yoon E, Dhar S, Chun DE, Gharibjanian NA, Evans GRD. In vivo osteogenic potential of human adipose-derived stem cells/poly lactide-co-glycolic acid constructs for bone regeneration in a rat critical-sized calvarial defect model.* Tissue Engineering. **2007**; *13:619–627.*
[39] *Cui L, Liu B, Liu G, Zhang W, Cen L, Sun J, Yin S, et al. Repair of cranial bone defects with adipose derived stem cells and coral scaffold in a canine model.* Biomaterials. **2007**; *28:5477–5486.*
[40] *Liu G, Zhang Y, Liu B, Sun J, Li W, Cui L. Bone regeneration in a canine cranial model using allogeneic adipose derived stem cells and coral scaffold.* Biomaterials. **2013**; *34:2655–2664.*
[41] *Stockmann P, Park J, von Wilmowsky C, Nkenke E, Felszeghy E, Dehner J-F, Schmitt C, et al. Guided bone regeneration in pig calvarial bone defects using autologous mesenchymal stem/progenitor cells - a comparison of different tissue sources.* Journal of Cranio-Maxillo-Facial Surgery: Official Publication of the European Association for Cranio-Maxillo-Facial Surgery. **2012**; *40:310–320.*
[42] *Kim HP, Ji Y-H, Rhee SC, Dhong ES, Park SH, Yoon E-S. Enhancement of bone regeneration using osteogenic-induced adipose-derived stem cells combined with demineralized bone matrix in a rat critically-sized calvarial defect model.* Current Stem Cell Research & Therapy. **2012**; *7:165–172.*
[43] *Kim JY, Jin GZ, Park IS, Kim JN, Chun SY, Park EK, Kim SY, et al. Evaluation of solid free-form fabrication-based scaffolds seeded with osteoblasts and human umbilical vein endothelial cells for use in vivo osteogenesis.* Tissue Engineering - Part A. **2010**; *16:2229–2236.*
[44] *Ko E, Yang K, Shin J, Cho S-W. Polydopamine-assisted osteoinductive peptide immobilization of polymer scaffolds for enhanced bone regeneration by human adipose-derived stem cells.* Biomacromolecules. **2013**; *14:3202–3213.*
[45] *Smith DM, Cooper GM, Afifi AM, Mooney MP, Cray J, Rubin JP, Marra KG, Losee JE. Regenerative surgery in cranioplasty revisited: the role of adipose-derived stem cells and BMP-2.* Plastic and Reconstructive Surgery. **2011**; *128:1053–1060.*
[46] *Levi B, James AW, Nelson ER, Li S, Peng M, Commons GW, Lee M, et al. Human adipose-derived stromal cells stimulate autogenous skeletal repair via paracrine Hedgehog signaling with calvarial osteoblasts.* Stem Cells and Development. **2011**; *20:243–257.*
[47] *Behr B, Tang C, Germann G, Longaker MT, Quarto N. Locally applied vascular endothelial growth factor A increases the osteogenic healing capacity of human adipose-derived stem cells by promoting osteogenic and endothelial differentiation.* Stem Cells (Dayton, Ohio). **2011**; *29:286–296.*
[48] *Tajima S, Tobita M, Orbay H, Hyakusoku H, Mizuno H. Direct and indirect effects of a combination of adipose-derived stem cells and platelet-rich plasma on bone regeneration.* Tissue Engineering - Part A. **2015**; *21:895–905.*
[49] *Di Bella C, Farlie P, Penington AJ. Bone regeneration in a rabbit critical-sized skull defect using autologous adipose-derived cells.* Tissue Engineering -Part A. **2008**; *14:483–490.*
[50] *Shah AR, Cornejo A, Guda T, Sahar DE, Stephenson SM, Chang S, Krishnegowda NK, et al. Differentiated adipose-derived stem cell cocultures for bone regeneration in polymer scaffolds in vivo.* The Journal of Craniofacial Surgery. **2014**; *25:1504–1509.*
[51] *Lappalainen OP, Karhula S, Haapea M, Kyllönen L, Haimi S, Miettinen S, Saarakkala S, et al. Bone healing in rabbit calvarial critical-sized defects filled with stem cells and growth factors combined with granular or solid scaffolds.* Child's Nervous System. **2016**; *32:681–688.*
[52] *Daei-Farshbaf N, Ardeshirylajimi A, Seyedjafari E, Piryaei A, Fadaei Fathabady F, Hedayati M, Salehi M, et al. Bioceramic-collagen scaffolds loaded with human adipose-tissue derived stem cells for bone tissue engineering.* Molecular Biology Reports. **2014**; *41:741–749.*
[53] *Kim Y, Lee SH, Kang B-J, Kim WH, Yun H-S, Kweon O-K. Comparison of Osteogenesis between Adipose-Derived Mesenchymal Stem Cells and Their Sheets on Poly-ε-Caprolactone/β-Tricalcium Phosphate Composite Scaffolds in Canine Bone Defects.* Stem Cells International. **2016**; *2016:8414715.*
[54] *Lappalainen O-P, Haapea M, Serpi R, Lehtonen S, Ylikontiola L, Korpi J, Serlo W, Sándor GKB. Iron-labeled adipose stem cells and neovascularization in rabbit calvarial critical-sized defects.* Oral Surgery, Oral Medicine, Oral Pathology and Oral Radiology. **2016**; *121: e104–110.*
[55] *Liao YH, Chang YH, Sung LY, Li KC, Yeh CL, Yen TC, Hwang SM, et al. Osteogenic differentiation of adipose-derived stem cells and calvarial defect repair using baculovirus-mediated co-expression of BMP-2 and miR-148b.* Biomaterials. **2014**; *35(18):4901–4910.*
[56] *Calis M, Demirtas TT, Atilla P, Tatar I, Ersoy O, Irmak G, Celik HH, et al. Estrogen as a novel agent for induction of adipose-derived mesenchymal stem cells for osteogenic differentiation: in vivo bone tissue-engineering study.* Plastic and Reconstructive Surgery. **2014**; *133:499e-510e.*
[57] *Ko E, Alberti K, Lee JS, Yang K, Jin Y, Shin J, Yang HS, et al. Nanostructured Tendon-Derived Scaffolds for Enhanced Bone Regeneration by Human Adipose-Derived Stem Cells.* ACS Applied Materials & Interfaces. **2016**; *8:22819–22,829.*
[58] *Saçak B, Certel F, Akdeniz ZD, Karademir B, Ercan F, Özkan N, Akpinar İN, Çelebiler Ö. Repair of critical size defects using bioactive glass seeded with adipose-derived mesenchymal stem cells.* Journal of Biomedical Materials Research - Part B Applied Biomaterials. **2017**; *105:1002–1008.*
[59] *Carvalho PP, Leonor IB, Smith BJ, Dias IR, Reis RL, Gimble JM, Gomes ME. Undifferentiated human adipose-derived stromal/stem cells loaded onto wet-spun starch-polycaprolactone scaffolds enhance bone regeneration: nude mice calvarial defect in vivo study.* Journal of Biomedical Materials Research Part A. **2014**; *102:3102–3111.*

Based on M. Barba, G. Di Taranto, W. Lattanzi, Adipose-derived stem cell therapies for bone regeneration, Expert. Opin. Biol. Ther. 17(6) (2017) 677–689.

[88, 89]. Vertebral bone models to evaluate ASCs' potential uses in posterior spinal fusion are less common but have also been explored [88, 92, 93].

Critical-sized bone defects do not heal spontaneously. Critical-sized defects in the calvarial bones of the skull are relatively easy to create and reproduce since the calvarial plate is practically a bidimensional shell, and fixation of engineered constructs is facilitated since most of the flanking bone tissue can be kept intact. Calvarial defects also have the advantage of bearing minimal mechanical loads, which contributes to the stabilization of the implant during the surgical procedure and during the postsurgical period of recovery, which can last for weeks or months.

In a study by Cowan et al., ASCs were isolated from the subcutaneous anterior abdominal wall of Friend leukemia virus B (FVB) albino mice, seeded into osteoinductive apatite-coated poly(lactic-*co*-glycolic acid) (PLGA) grafts, and implanted in calvarial critical size defects produced in the parietal bone of these animals [94]. Bone regeneration was reported by X-ray, histology, and molecular imaging as early as 2 weeks after graft implantation. Complete repair of the defects was shown 12 weeks after the surgical procedure. In this study, the authors also seeded the same scaffolds with BMSCs isolated from the same animals, and reported similar rates of bone formation mediated by BMSCs and ASCs [94]. In another study, Yoon et al. reported the isolation of hASCs after a liposuction procedure, precultivation with osteogenic medium, and association with PLGA for implantation in critical-size calvarial defects in Rowett Nude rats [95]. According to the timing of exposure of hASCs to osteogenic medium before their association with PLGA scaffolds, four groups were formed: three experimental groups, where hASCs were induced in vitro toward osteogenic differentiation for 1, 7, or 14 days, and the control group, for which hASCs were associated with the scaffolds without previous exposure to osteogenic medium. Radiodensitometric and histomorphometric analyses showed that, although all experimental groups displayed some extent of bone regeneration at the defect sites, hASCs that had been predifferentiated for 14 days in vitro contributed with more robust bone formation [95]. Levi et al. seeded freshly derived undifferentiated cells isolated from human lipoaspirate material onto apatite-coated PLGA and implanted the constructs in critical-sized calvarial defects in the parietal bones of nude mice [96]. Newly formed bone was detected as early as 2 weeks after the surgeries, by histology, in situ hybridization, and histomorphometry. Nearly complete healing of the defects was achieved after 4 weeks of implantation.

Segmental critical-sized defects in long bones can be generated to study the effects of load bearing and of external physical forces. For example, defects can be produced in the femoral bone when the interference of these forces is being fully considered. Otherwise, if the effects of external forces and impacts need to be reduced to a minimum, segmental defects in the radius can be generated, where the ulna absorbs some of these forces. A detailed list of in vivo studies that have been performed using ASCs of different species (including hASCs) in various experimental settings is included in Table 1.

Comparison with dental pulp cells

Another source of MSCs with high osteogenic potential that can be adopted as an alternative to BMSCs is stromal/stem cells from dental pulp tissues, either termed dental pulp stem cells (DPSCs), when isolated from permanent teeth, or stem cells from human exfoliated deciduous teeth (SHED), when obtained from baby teeth [97–99]. Similar to hASCs, DPSCs and

SHED bear a high potential to be used in cell-based regenerative therapies because of minimally invasive process of isolation, and differentiation plasticity even after several passages in vitro [100–104]. DPSCs and SHED have been used extensively in research involving animal models, such as for reconstruction of cranial defects and trabecular bone defects in mice, and to guide regeneration of critical-sized bone defects in rodents [102, 105–107]. These cells have also been used in association with biomaterials for bone regeneration in human clinical trials [108, 109].

The osteogenic potential of MSCs obtained from different tissue sources is heterogeneous and can vary greatly, depending on parameters such as subjects' age and health, protocols and reagents used for isolation, in vitro expansion, osteoinduction, and cryopreservation, among many others. There is a paucity of studies devised to compare directly the osteogenic potential of different MSCs under the same standardized conditions. Studies by Fanganiello and colleagues reported that SHED display higher in vitro osteogenic potential when compared with hASCs and with BMSCs under the same controlled conditions, although from different donors [51, 110]. It was also indicated that the osteogenic potential of these MSCs can be optimized with treatment with exogenous insulin growth factor 2 (IGF2); with preselection of subpopulations with higher IGF2 expression or lower CD105 (endoglin) expression, a type I transmembrane glycoprotein and TGF-β1 coreceptor; or with forced downregulation of CD105 by the micro-RNA hsa-miR-1287 [51, 110]. This last finding is in accordance with the work of Levi et al. that first suggested that CD105 expression could be used as an important predictor of osteopotential in hASCs, that TGF-β1/Smad2 signaling in hASCs' populations selected for low CD105 expression is associated with its high osteogenic potential, and that this pathway could be activated to enhance osteogenic potential in hASCs [111]. By the same reasoning that subpopulations of hASCs with higher osteogenic potential can be selected by fluorescence-activated cell sorting (FACS) based on differential expression of transmembrane proteins, Chung et al. showed that CD90 (Thy-1) can be used to select subpopulations of hASCs with higher osteogenic potential [112]. McArdle et al. showed that BMP receptor type-IB (BMPR-IB) could be used for the same purpose [113].

Refining engineering strategies for optimal osteogenesis and functionality

Apart from the cell-based engineering strategies described in the previous sections, several approaches have been tested to further improve functional outcomes, including the combined use of cells with BMPs and the use of mechanical stimulation to favor osteogenic differentiation processes.

Native bone tissue is continuously exposed to several types and degrees of mechanical stress. Different devices, such as bioreactors, have been developed to mimic these mechanical constraints in vitro [114, 115]. For many years, osteocytes were considered as the only mechanosensitive cells, but it was also revealed that MSCs respond to different mechanical stimuli, such as compression, stretching, or shear stress [116]. The use of bioreactors aims to ensure the dynamic perfusion of 3D matrices, to supply the core of the structure with nutrients and oxygen, and to eliminate cellular wastes [114, 115]. Dynamic culture can have significant consequences on the proliferation and the differentiation of MSCs [117] and, consequently, on bone formation and vascularization of newly formed bone tissues after implantation [118]. Indeed, Jagodzinski et al. demonstrated that a mechanical compression at a

10% amplitude (0.05 Hz) applied for 3 weeks on bovine acellular matrix (Tutobone disc, Tutogen Medical Inc.) loaded with human BMSCs (hBMSCs) significantly increased cell proliferation, osteoblastic *RUNX2* gene expression, and osteocalcin secretion compared with static culture condition [117].

Finally, to ensure optimal survival and viability of tissue-engineered bone substitutes upon implantation, several strategies have addressed ways to provide adequate vascular support [119]. For example, delivery of proangiogenic factors, either by direct supplementation, delivery by viral vectors, or scaffold functionalization, has been performed to promote angiogenesis. The vascularization process can also be stimulated by seeding endothelial cells or their progenitors in association with MSCs, including ASCs, in 3D scaffolds [120, 121]. This approach takes advantage of the capacity of endothelial cells to assemble into structures mimicking the capillary system, resulting in prevascularization of the constructs in vitro. Then, upon in vivo implantation, the microvasculature from the host bed can readily connect with the preformed capillaries, ensuring a faster vascular supply to the constructs [122]. Numerous studies have described the benefits of coculturing endothelial cells and mesenchymal lineage progenitors to promote osteoendothelial crosstalk and to stimulate both in vivo osteogenesis and angiogenesis in newly formed tissues [21, 123]. However, great care should be taken to fully characterize each newly developed prevascularized model since the cell source and species, as well as the nature of the chosen culture model, can influence MSCs/endothelial cells' interactions [124].

Enhancing bone formation with bone morphogenetic proteins

Bone morphogenetic protein family and its signal transduction

As mentioned before, bones are continuously being remodeled and can self-repair by primary bone repair and endochondral bone formation [125]. However, the potential for bone healing and regeneration is severely reduced in the aging population. A fracture that does not heal normally is a major clinical problem that can lead to patient disability and significant financial burden [126, 127]. Several cytokines and members of the TGF-β superfamily such as BMPs have been identified as key players in the fracture healing process and bone remodeling.

In 1965, Urist confirmed that new bone formation can occur after implantation of demineralized bone components into rabbit muscles. He identified BMP as responsible for this ectopic bone formation [128]. However, further extraction procedures were later required to better characterize and purify these BMPs [129, 130]. Such studies have permitted the isolation of complementary deoxyribonucleic acid (cDNA) clones, encoding several BMPs such as BMP-2 and BMP-7, paving the way to the production of recombinant human BMPs (rhBMPs) and their clinical use [131, 132]. To date, around 20 BMPs have been identified, but not all of them have osteoinductive properties [133].

BMP family

BMPs, except BMP-1, are members of the TGF-β superfamily, characterized by a common cystine knot domain of six highly conserved cysteines stabilizing the monomer by three intramolecular disulfide bonds [134]. Both TGF-β and BMPs can bind specific kinase receptors that activate a Smad protein-based canonical signaling pathway (Smad 2/3 for TGF-β and Smad 1/5/8 for BMPs) (Table 2). However, several recent studies suggest antagonist effects of TGF-β and BMPs signaling in bone formation and disease progression such as cancer (as reviewed by Ref. [135]).

BMPs are classified into four main subgroups based on their sequence homology: (1) the *Drosophila* decapentaplegic (dpp) subgroup (BMP-2/-4); (2) the 60A subgroup (BMP-5, -6, -7, -8, -8b); (3) the BMP-9 and BMP-10 subgroup; and (4) the BMP-12, -13, -14 subgroup [132, 136–138] (Table 2). For example, members of the dpp and 60A subgroups share between 55% and 61% amino acid identities [137]. Zhang et al. have recently proposed another classification of 14 BMPs based on expression pattern profiles of 519 genes in multipotent C3H10T1/2 cells transduced with adenoviral vectors encoding BMPs (AdBMPs) [139]. They classified these 14 BMPs into three subclusters: (1) BMP-2/BMP-4, BMP-6/BMP-7, and BMP-9; (2) BMP-5, BMP-11, BMP-12/BMP-13/BMP-14, and BMP-15; and (3) BMP-3, BMP-8, and BMP-10 [139]. The members of cluster 1 are well known to induce multilineage (osteogenic, chondrogenic, or adipogenic) differentiation of MSCs depending on the cell culture conditions, while the members of the second cluster are tenogenic factors. In contrast, cluster 3 regroups BMPs having a very different effect on cells. BMP-3 is a BMP antagonist, while BMP-10 controls heart development and vascular remodeling [133, 140].

Active BMPs are obtained through homodimerization or heterodimerization (e.g., BMP-2/BMP-7) via a disulfide bond, involving a seventh conserved cysteine within each monomer [141]. Israel et al. found that both BMP-2/BMP-7 and BMP-2/BMP-6 heterodimers are more potent than BMP-2 homodimers to induce bone formation in vivo [142]. Several studies have also found that some BMPs are more potent osteoinductive factors than others [133, 143–145]. Using direct intramuscular injection of AdBMPs (10^9 plaque forming units) into the quadriceps of athymic mice for 5 weeks, Luu et al. found that, among 14 BMPs, BMP-2 and BMP-7 induced chondrogenesis, while ectopic bone formation was observed in the BMP-9-treated mice [133].

The use of knockout mice also revealed the important role played by BMPs subfamilies, not only in the regulation of bone formation and remodeling but also in embryogenesis and vascular development (Table 2) [146]. For example, while loss of BMP-5 leads to skeletal malformation [147], homozygous mutant embryos for BMP-2 and BMP-4 die between day 6.5 (E6.5) and 10.5 (E10.5) with abnormal heart development and lack of mesodermal differentiation, respectively [148, 149]. BMP-7 knockout mice die shortly after birth because of kidney malformation [150]. In contrast, BMP-3 knockout mice revealed that this BMP acts as a negative regulator of bone density, preventing bone formation and growth [151].

Bone morphogenetic protein signaling pathways and their regulation

The Smad canonical pathway

Active BMPs act on cells by binding with different affinity to type I and type II Ser/Thr kinase receptors, leading to the stimulation of the canonical Smad or mitogen-activated protein kinases (MAPK) signaling pathways (Figs. 2 and 3) [141, 152]. In fact, the canonical Smad pathway is activated when the binding of BMPs induces receptor oligomerization, while their interaction with preassembled receptor complexes induces the MAPK pathway activation

TABLE 2 Bone morphogenetic protein (BMP) family and their receptors.

BMP subgroups	% Amino acid identity (versus BMP-2)	Knockout mouse phenotype	Type I receptor	Type II receptor	Smad 1/5/8 activation (time/dose of BMP/cells)
BMP-2	100	Embryonically lethal (between E7.5 and E9.0), malformation in cardiac development[1]; spontaneous fractures in limbs and incomplete fracture healing[2]; severe disorganization of chondrocytes during endochondral bone development[3]	BMPRIa, BMPRIb	BMPRII, ActRIIA, ActRIIB	2h of incubation with BMP-2 (100ng/mL)—stem cells from the apical papilla (SCAP)[4]; 30min of incubation with BMP-2 (0.38nM)—MC3T3-E1 murine preosteoblasts[5]; 10min of incubation with BMP2 (10nM)—C3H10 T1/2 murine cells[6]
BMP-4	92	Embryonically lethal (between 6.5 and 9.5days), defect in mesoderm formation: disorganized posterior structures, reduction in extraembryonic mesoderm[7]; defects in cardiac development[8]; impaired insulin sensitivity and enlarged white adipocyte morphology[9]; defects in tracheal formation[10]; defects in outflow-tract (OFT) myocardium development[11]; abnormal development of inner ear[12]; defects in mandibular development[13]			30min of incubation with BMP-4 (10ng/mL)—central nervous system (CNS) stem cells[14]; 7days of incubation with BMP-4—neural stem cells[15]
BMP-3	49	Enhanced bone density[16]	ActRIb	ActRIIA, ActRIIB	No Smad1/5/8 activation[17]
BMP-5	61	Short ear phenotype[18]; long bones: smaller and frailer[19]	Alk2, BMPRIa, BMPRIb	ActRIIA, ActRIIB, BMPRII	18h of incubation with BMP-5 (100ng/mL)—human pluripotent stem cells[20]
BMP-6	61	Defects in sternum ossification[21]; long bones: smaller[22]; decreased fertility[23]			Incubation with BMP-6 (500ng/mL)—C2C12 mouse myoblasts and MC3T3-E1 murine preosteoblasts[24]; 18h of incubation with BMP-6 (100ng/mL)—human pluripotent stem cells[20]

BMP-7	60	Death after birth (first day): renal dysplasia and anophthalmia[25]; defects in nephrogenesis, eye development and skeletal patterning (rib cage, skull, hindlimbs)[26]; defects in corticogenesis[27]; abnormal brown adipogenesis and energy expenditure[28]; limits Langerhans cell proliferation[29]			30 min of incubation with BMP-7 (25 ng/mL)—L6E9 rat myoblasts[30]; 18 h of incubation with BMP-7 (100 ng/mL)—human pluripotent stem cells[20]
BMP-8	55	Abnormal spermatogenesis: germ cell degeneration leading to infertility[31,32]			Incubation with BMP-8A (3 nM)—HEK-293T cells and P19 mouse embryonic carcinoma[33]
BMP-9 (GDF2)	NA, but notable changes in residues in receptor binding regions between BMP-2 and BMP-9: acidic to basic Asn59 (BMP-2) versus Lys53 (BMP-9) and basic to acidic His54 (BMP-2) versus Asp58 (BMP-9)[34]	Abnormal lymphatic development[35,36]	Alk1, Alk2	BMPRII, ActRIIA, ActRIIB	15 min of incubation with BMP-9 (10 ng/mL)—C3H10T1/2 murine cells[37]; 30 min of incubation with BMP-9 (150 ng/mL)—human osteoclasts[38]; 4 h of incubation with BMP-9 (0.38 nM)—MC3T3-E1 murine preosteoblasts[5]
BMP-10	NA, but shares the highest sequence homology with BMP-9 (65%)[39]	Embryonically lethal (between E9.5 and E10.5) defects in cardiac development[40]			18 h of incubation with BMP-10 (100 ng/mL)—human pluripotent stem cells[20]
BMP-15	34	Viable mice with minimal alterations in follicle development[41]	BMPRIb	BMPRII	30 min and 1 h of incubation with BMP-15 (100 ng/mL)—human granulosa cells[42]

[1] *Zhang H, Bradley A. Mice deficient for BMP2 are nonviable and have defects in amnion/chorion and cardiac development.* Development. ***1996***; *122(10):2977–2986.*
[2] *Tsuji K, Bandyopadhyay A, Harfe BD, Cox K, Kakar S, Gerstenfeld L, Einhorn T, et al. BMP2 activity, although dispensable for bone formation, is required for the initiation of fracture healing.* Nature Genetics. ***2006***; *38(12):1424–1429.*
[3] *Shu B, Zhang M, Xie R, Wang M, Jin H, Hou W, Tang D, et al. BMP2, but not BMP4, is crucial for chondrocyte proliferation and maturation during endochondral bone development.* Journal of Cell Science. ***2011***; *124(Pt 20):3428–3440.*
[4] *Xiao M, Yao B, Zhang BD, Bai Y, Sui W, Wang W, Yu Q. Stromal-derived Factor-1alpha signaling is involved in bone morphogenetic protein-2-induced odontogenic differentiation of stem cells from apical papilla via the Smad and Erk signaling pathways.* Experimental Cell Research. ***2019***; *381(1):39–49.*
[5] *Drevelle O, Daviau A, Lauzon MA, Faucheux N. Effect of BMP-2 and/or BMP-9 on preosteoblasts attached to polycaprolactone functionalized by adhesive peptides derived from bone sialoprotein.* Biomaterials. ***2013***; *34(4):1051–1062.*
[6] *Berasi SP, Varadarajan U, Archambault J, Cain M, Souza TA, Abouzeid A, Li J, et al. Divergent activities of osteogenic BMP2, and tenogenic BMP12 and BMP13 independent of receptor binding affinities.* Growth Factors. ***2011***; *29(4):128–139.*
[7] *Winnier G, Blessing M, Labosky PA, Hogan BL. Bone morphogenetic protein-4 is required for mesoderm formation and patterning in the mouse.* Genes & Development. ***1995***; *9 (17):2105–2116.*

[8] *Jiao K, Kulessa H, Tompkins K, Zhou Y, Batts L, Baldwin HS, Hogan BL. An essential role of Bmp4 in the atrioventricular septation of the mouse heart.* Genes & Development. **2003**; *17(19):2362–2367.*

[9] *Qian SW, Tang Y, Li X, Liu Y, Zhang YY, Huang HY, Xue RD, et al. BMP4-mediated brown fat-like changes in white adipose tissue alter glucose and energy homeostasis.* Proceedings of the National Academy of Sciences of the United States of America. **2013**; *110(9):E798–807.*

[10] *Li Y, Gordon J, Manley NR, Litingtung Y, Chiang C. Bmp4 is required for tracheal formation: a novel mouse model for tracheal agenesis.* Developmental Biology. **2008**; *322(1):145–155.*

[11] *Liu W, Selever J, Wang D, Lu MF, Moses KA, Schwartz RJ, Martin JF. Bmp4 signaling is required for outflow-tract septation and branchial-arch artery remodeling.* Proceedings of the National Academy of Sciences of the United States of America. **2004**; *101(13):4489–4494.*

[12] *Liu W, Selever J, Murali D, Sun X, Brugger SM, Ma L, Schwartz RJ, et al. Threshold-specific requirements for Bmp4 in mandibular development.* Developmental Biology. **2005**; *283 (2):282–293.*

[13] *Chang W, Lin Z, Kulessa H, Hebert J, Hogan BL, Wu DK. Bmp4 is essential for the formation of the vestibular apparatus that detects angular head movements.* PLoS Genetics. **2008**; *4(4): e1000050.*

[14] *Rajan P, Panchision DM, Newell LF, McKay RD. BMPs signal alternately through a SMAD or FRAP-STAT pathway to regulate fate choice in CNS stem cells.* The Journal of Cell Biology. **2003**; *161(5):911–921.*

[15] *Fang H, Song P, Shen C, Liu X, Li H. Bone mesenchymal stem cell-conditioned medium induces the upregulation of Smad6, which inhibits the BMP-4/Smad1/5/8 signaling pathway.* Neurological Research. **2016**; *38(11):965–972.*

[16] *Daluiski A, Engstrand T, Bahamonde ME, Gamer LW, Agius E, Stevenson SL, Cox K, et al. Bone morphogenetic protein-3 is a negative regulator of bone density.* Nature Genetics. **2001**; *27(1):84–88.*

[17] *Miyazono K, Kamiya Y, Morikawa M. Bone morphogenetic protein receptors and signal transduction.* Journal of Biochemistry. **2010**; *147(1):35–51.*

[18] *Kingsley DM, Bland AE, Grubber JM, Marker PC, Russell LB, Copeland NG, Jenkins NA. The mouse short ear skeletal morphogenesis locus is associated with defects in a bone morphogenetic member of the TGFβ superfamily.* Cell. **1992**; *71(3):399–410.*

[19] *Mikic B, Van Der Meulen MCH, Kingsley DM, Carter DR. Long bone geometry and strength in adult BMP-5 deficient mice.* EMC - Cardiologie-Angiologie. **1995**; *16(4):445–454.*

[20] *Lichtner B, Knaus P, Lehrach H, Adjaye J. BMP10 as a potent inducer of trophoblast differentiation in human embryonic and induced pluripotent stem cells.* Biomaterials. **2013**; *34(38):9789–9802.*

[21] *Solloway MJ, Dudley AT, Bikoff EK, Lyons KM, Hogan BLM, Robertson EJ. Mice lackingBmp6 function.* Developmental Genetics. **1998**; *22(4):321–339.*

[22] *Perry MJ, McDougall KE, Hou SC, Tobias JH. Impaired growth plate function in bmp-6 null mice.* Bone. **2008**; *42(1):216–225.*

[23] *Sugiura K, Su YQ, Eppig JJ. Does bone morphogenetic protein 6 (BMP6) affect female fertility in the mouse?* Biology of Reproduction. **2010**; *83(6):997–1004.*

[24] *Ebisawa T, Tada K, Kitajima I, Tojo K, Sampath TK, Kawabata M, Miyazono K, Imamura T. Characterization of bone morphogenetic protein-6 signaling pathways in osteoblast differentiation.* Journal of Cell Science. **1999**; *112(20):3519–3527.*

[25] *Dudley AT, Lyons KM, Robertson EJ. A requirement for bone morphogenetic protein-7 during development of the mammalian kidney and eye.* Genes & Development. **1995**; *9 (22):2795–2807.*

[26] *Luo G, Hofmann C, Bronckers AL, Sohocki M, Bradley A, Karsenty G. BMP-7 is an inducer of nephrogenesis, and is also required for eye development and skeletal patterning.* Genes & Development. **1995**; *9(22):2808–2820.*

[27] *Segklia A, Seuntjens E, Elkouris M, Tsalavos S, Stappers E, Mitsiadis TA, Huylebroeck D, et al. Bmp7 regulates the survival, proliferation, and neurogenic properties of neural progenitor cells during corticogenesis in the mouse.* PLoS One. **2012**; *7(3):e34088.*

[28] *Tseng YH, Kokkotou E, Schulz TJ, Huang TL, Winnay JN, Taniguchi CM, Tran TT, et al. New role of bone morphogenetic protein 7 in brown adipogenesis and energy expenditure.* Nature. **2008**; *454(7207):1000–1004.*

[29] *Yasmin N, Bauer T, Modak M, Wagner K, Schuster C, Koffel R, Seyerl M, et al. Identification of bone morphogenetic protein 7 (BMP7) as an instructive factor for human epidermal Langerhans cell differentiation.* The Journal of Experimental Medicine. **2013**; *210(12):2597–2610.*

[30] *Meurer SK, Esser M, Tihaa L, Weiskirchen R. BMP-7/TGF-beta1 signaling in myoblasts: components involved in signaling and BMP-7-dependent blockage of TGF-beta-mediated CTGF expression.* European Journal of Cell Biology. **2012**; *91(6–7):450–463.*

[31] *Zhao GQ, Deng K, Labosky PA, Liaw L, Hogan BL. The gene encoding bone morphogenetic protein 8B is required for the initiation and maintenance of spermatogenesis in the mouse.* Genes & Development. **1996**; *10(13):1657–1669.*

[32] *Zhao GQ, Liaw L, Hogan BL. Bone morphogenetic protein 8A plays a role in the maintenance of spermatogenesis and the integrity of the epididymis.* Development. **1998**; *125(6):1103–1112.*

[33] *Wu FJ, Lin TY, Sung LY, Chang WF, Wu PC, Luo CW. BMP8A sustains spermatogenesis by activating both SMAD1/5/8 and SMAD2/3 in spermatogonia.* Science Signaling. **2017**; *10(477).*

[34] *Brown MA, Zhao Q, Baker KA, Naik C, Chen C, Pukac L, Singh M, et al. Crystal structure of BMP-9 and functional interactions with pro-region and receptors.* The Journal of Biological Chemistry. **2005**; *280(26):25111–25,118.*

[35] *Levet S, Ciais D, Merdzhanova G, Mallet C, Zimmers TA, Lee SJ, Navarro FP, et al. Bone morphogenetic protein 9 (BMP9) controls lymphatic vessel maturation and valve formation.* Blood. **2013**; *122(4):598–607.*

[36] *Yoshimatsu Y, Lee YG, Akatsu Y, Taguchi L, Suzuki HI, Cunha SI, Maruyama K, et al. Bone morphogenetic protein-9 inhibits lymphatic vessel formation via activin receptor-like kinase 1 during development and cancer progression.* Proceedings of the National Academy of Sciences of the United States of America. **2013**; *110(47):18940–18,945.*

[37] *Cheng A, Gustafson AR, Schaner Tooley CE, Zhang M. BMP-9 dependent pathways required for the chondrogenic differentiation of pluripotent stem cells.* Differentiation. **2016**; *92(5):298–305.*

[38] *Fong D, Bisson M, Laberge G, McManus S, Grenier G, Faucheux N, Roux S. Bone morphogenetic protein-9 activates Smad and ERK pathways and supports human osteoclast function and survival in vitro.* Cell Signal. **2013**; *25(4):717–728.*

[39] *David L, Mallet C, Mazerbourg S, Feige JJ, Bailly S. Identification of BMP9 and BMP10 as functional activators of the orphan activin receptor-like kinase 1 (ALK1) in endothelial cells.* Blood. **2007**; *109(5):1953–1961.*

[40] *Chen H, Shi S, Acosta L, Li W, Lu J, Bao S, Chen Z, et al. BMP10 is essential for maintaining cardiac growth during murine cardiogenesis.* Development. **2004**; *131(9):2219–2231.*

[41] *Yan C, Wang P, DeMayo J, DeMayo FJ, Elvin JA, Carino C, Prasad SV, et al. Synergistic roles of bone morphogenetic protein 15 and growth differentiation factor 9 in ovarian function.* Molecular Endocrinology. **2001**; *15(6):854–866.*

[42] *Chang HM, Cheng JC, Klausen C, Leung PC. BMP15 suppresses progesterone production by down-regulating StAR via ALK3 in human granulosa cells.* Molecular Endocrinology. **2013**; *27(12):2093–2104.*

NA, *not available.*

Based on D.M. Kingsley, The TGF-beta superfamily: new members, new receptors, and new genetic tests of function in different organisms, Genes Dev. 8(2) (1994) 133–146; R.N. Wang, J. Green, Z. Wang, Y. Deng, M. Qiao, M. Peabody, Q. Zhang, et al., Bone morphogenetic protein (BMP) signaling in development and human diseases, Genes Dis. 1(1) (2014) 87–105; D. Yadin, P. Knaus, T.D. Mueller, Structural insights into BMP receptors: specificity, activation and inhibition, Cytokine Growth Factor Rev. 27 (2016) 13–34; M.A. Lauzon, E. Bergeron, B. Marcos, N. Faucheux, Bone repair: new developments in growth factor delivery systems and their mathematical modeling, J Control. Release 162(3) (2012) 502–520.

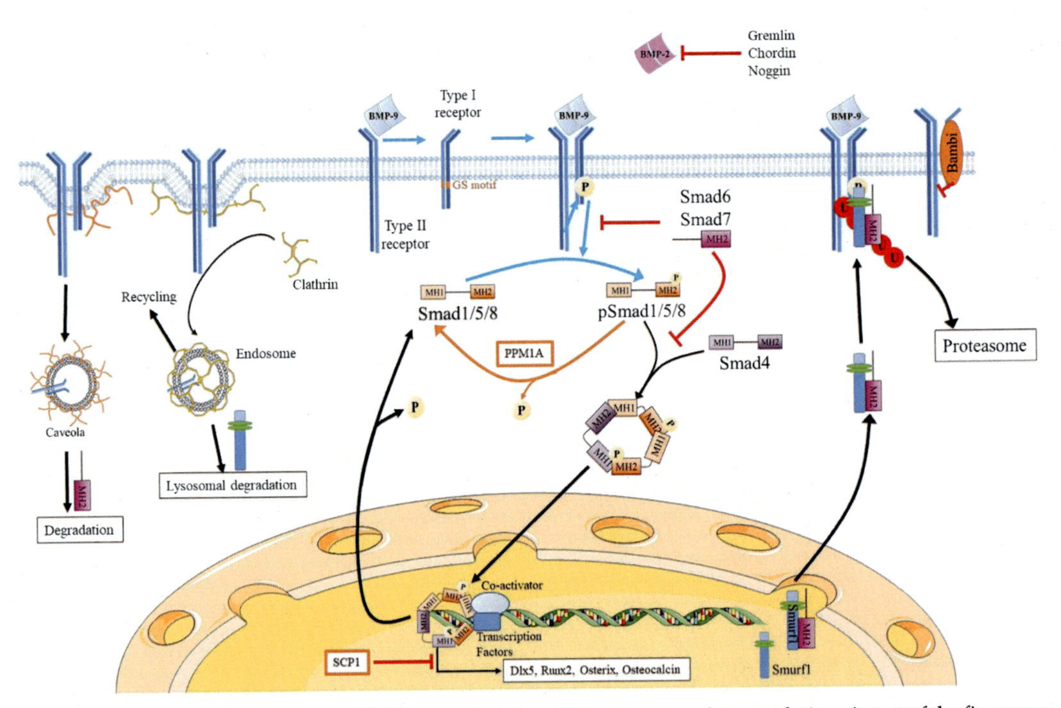

FIG. 2 Bone morphogenetic protein (BMP) canonical signaling pathway and its regulation. A part of the figure was created using Servier Medical Art. https://smart.servier.com. *Modified from M. Ehrlich, Endocytosis and trafficking of BMP receptors: regulatory mechanisms for fine-tuning the signaling response in different cellular contexts, Cytokine Growth Factor Rev. 27 (2016) 35–42; D. Yadin, P. Knaus, T.D. Mueller, Structural insights into BMP receptors: specificity, activation and inhibition, Cytokine Growth Factor Rev. 27 (2016) 13–34; S. Kokabu, T. Katagiri, T. Yoda, V. Rosen, Role of Smad phosphatases in BMP-Smad signaling axis-induced osteoblast differentiation, J. Oral Biosci. 54(2) (2012) 73–78; H. Hassanisaber, L. Rouleau, N. Faucheux, Effect of BMP-9 on endothelial cells and its role in atherosclerosis, Front. Biosci. 24(6) (2019) 994–1023; K. Miyazono, S. Maeda, T. Imamura, BMP receptor signaling: transcriptional targets, regulation of signals, and signaling cross-talk, Cytokine Growth Factor Rev. 16(3) (2005) 251–263; S. Beauvais, O. Drevelle, J. Jann, M.A. Lauzon, M. Foruzanmehr, G. Grenier, S. Roux, N. Faucheux, Interactions between bone cells and biomaterials: an update, Front. Biosci. 8 (2016) 227–263; J. Jann, S. Gascon, S. Roux, N. Faucheux, Influence of the TGF-beta superfamily on osteoclasts/osteoblasts balance in physiological and pathological bone conditions, Int. J. Mol. Sci. 21(20) (2020) 7597–7651.*

through p38 phosphorylation [153, 154]. BMP dimers bind to the Ser/Thr kinase type I receptors with their wrist epitopes, while their knuckle epitopes are recognized by the Ser/Thr kinase type II receptors [138, 155]. BMPs signal transduction can therefore be regulated by the internalization of these cell-surface receptors through clathrin-dependent and independent mechanisms [154]. Currently, seven type I [activin receptor-like kinases 1 to 7 (ALK1–7)] and five type II receptors [BMP receptor type II (BMPRII), activin receptor type IIA (ActRIIA), activin receptor type IIB (ActRIIB), TGF-β receptor type II (TbRII), and anti-Mullerian hormone receptor type II (AMHRII)] have been identified [156–158]. For example, human ASCs express several type I receptors, including ALK2 (ACVR1, ACTRI) and ALK-6 (BMPR-IB) as well as BMPRII type II receptor [113, 159, 160]. Three of the type I receptors, ActRIB/ALK4, TbRI/ALK5, and ALK7, can initiate TGF-β signaling, while ALK1, ALK2, BMPRIA/ALK3, and BMPRIB/ALK6 can trigger BMP signaling. Interestingly, BMP-9

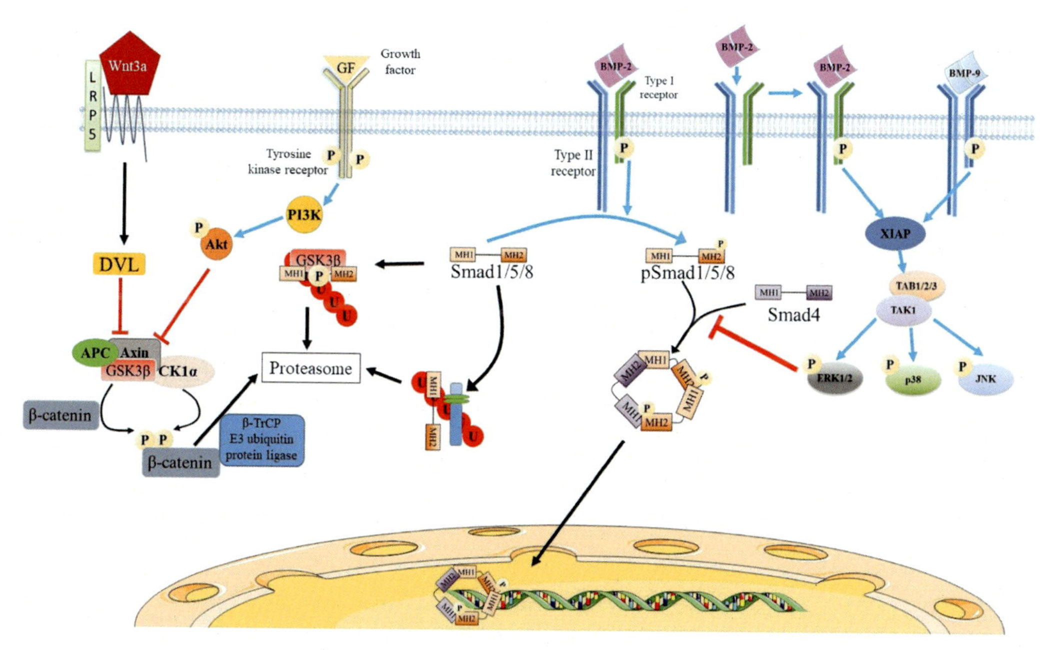

FIG. 3 Bone morphogenetic protein (BMP) noncanonical signaling pathways and their effect on Smad activation. A part of the figure was created using Servier Medical Art. https://smart.servier.com. *Modified from H. Hassanisaber, L. Rouleau, N. Faucheux, Effect of BMP-9 on endothelial cells and its role in atherosclerosis, Front. Biosci. 24(6) (2019) 994–1023; S. Beauvais, O. Drevelle, J. Jann, M.A. Lauzon, M. Foruzanmehr, G. Grenier, S. Roux, N. Faucheux, Interactions between bone cells and biomaterials: an update, Front. Biosci. 8 (2016) 227–263; J. Jann, S. Gascon, S. Roux, N. Faucheux, Influence of the TGF-beta superfamily on osteoclasts/osteoblasts balance in physiological and pathological bone conditions, Int. J. Mol. Sci. 21(20) (2020) 7597–7651.*

mainly interacts with type I receptors ALK1 and ALK2, while BMP-2/-4 signaling is mediated via ALK3 and ALK6 [161]. Furthermore, the Ser/Thr kinase type II receptors, TbRII and AMHRII, only recognize TGF-β and anti-Mullerian hormone, respectively [155], while BMP-2/-4/-5/-6/-7/-8 bind BMPRII, ActRIIA, and ActRIIB [162].

Upon BMP binding, the constitutively active Ser/Thr kinase type II receptors phosphorylate the type I receptors at their Gly/Ser rich motif (GS motif, Fig. 2). The activated type I receptors phosphorylate, in turn, the Smad 1/5/8, also called receptor-regulated Smad proteins (R-Smad), on two Ser residues located in their C-terminal Ser-X-Ser (SXS) motif [163]. Phosphorylated R-Smad can then interact with Smad 4, their common mediator (Co-Smad), to form heterodimeric complexes. These complexes are translocated with other transcription factors into the nucleus to regulate the expression of genes encoding proteins involved in the osteogenic differentiation of stem cells [164]. For example, BMP-2/BMP-4 and BMP-9 induce the expression of the transcription factor Runx2 in C3H10T1/2 transduced by AdBMPs [139]. The expression of another transcription factor, Osterix (Osx), was activated in C3H10T1/2 cells stimulated for 4 days with BMP-2 at 300 ng/mL [165].

The BMP pathways can be inhibited by inactive membrane receptors lacking the intracellular Ser/Thr kinase domain (decoy-receptor BMP and activin membrane-bound protein,

BAMBI), or extracellular molecules such as Noggin and Chordin (Fig. 2). Such molecules can bind several BMPs, such as BMP-2, BMP-4, and BMP-7, preventing their interaction with type I and type II Ser/Thr kinase receptors [157, 166, 167]. The canonical Smad pathway can also be blocked by intracellular molecules such as Smad 6/7, also called I-Smad. The binding of Smad7 to the intracellular domains of Ser/Thr kinase type I receptors prevents the phosphorylation of R-Smad, which blocks signal transduction [168]. In addition, Smad6 C-domain can interact with Smad1, preventing the complex formation between R-Smads and Smad 4 [169]. BMPs regulate their own signaling by promoting or repressing the expression of I-Smad. For example, using C3H10T1/2 cells transduced with AdBMPs, Zhang et al. have recently found that BMP-2/BMP-4, BMP-7, BMP-8, and BMP-9 induced the expression of the genes encoding Smad6 and Smad7 after 36h [139]. BMP-7 induced the highest messenger RNA (mRNA) levels of Smad6 and Smad7, while BMP-13 blocked I-Smad gene expression [139]. Various intracellular phosphatases such as small C-terminal domain phosphatase 1 (SCP-1) and protein phosphatase magnesium-dependent 1A (PPM1A) can also inactivate R-Smad by dephosphorylating its C-terminal SXS motif [170–172]. In addition, these R-Smad proteins are broken down by the proteasome after their ubiquitination by a E3 ubiquitin ligase called Smad ubiquitin regulatory factor 1 (Smurf1) [173]. Murakami and Etlinger have recently found that Smurf1 also regulates basal BMPR endocytosis and lysosomal degradation [174].

Noncanonical pathways

BMPs also act on cells by activating the MAPK pathways. Upon BMP binding, the type I Ser/Thr kinase receptors interact with the X chromosome-linked inhibitor of apoptosis (XIAP), TGF-β-activated kinase 1/MAP3K7 binding protein 1 (TAB1), and TGF-β-activated kinase 1 (TAK1), leading to the activation of the three MAPK cascades, ERK1/2, p38, and JNK [175] (Fig. 3). Recently, Zhang et al. found that BMP-2/BMP-4 and BMP-7 upregulate the mRNA levels of several members of the Smad-independent MAP kinase pathway in C3H10T1/2 cells transduced by AdBMPs [139]. The MAPK pathways can also antagonize the canonical Smad cascade by phosphorylating the linker region of R-Smad, thus preventing their nuclear localization [176]. Other signaling pathways such as the canonical Wnt and PI3K/Akt pathways regulate BMP-induced osteogenesis [177, 178]. Indeed, the Wnt/beta-catenin pathway is important for determining the fate of stem cells, promoting osteogenic differentiation while preventing adipogenic differentiation [179].

BMP effects in vitro and in vivo

Roles played by bone morphogenetic proteins in bone fracture healing

Even though fractured bones can self-repair, approximately 10% to 15% do not heal properly, resulting in a nonunion fracture. The fractured bone healing process has four complex and dynamic temporal reactive phases [180–182], which are initiated by the inflammatory phase and characterized by the formation of a hematoma due to vascular disruption, made of platelets and a fibrin network in the fracture site. The hematoma ensures the availability of a provisional matrix for neutrophils, macrophages to clear cellular debris, osteoclasts, and undifferentiated MSCs [183]. MSCs are recruited by released growth factors from activated platelets such as TGF-β and platelet-derived growth factor (PDGF). During the inflammatory phase, several cytokines, including tumor necrosis factor alpha (TNF-alpha), IL-1β, IL-6, VEGF, and angiopoietin-1, are also released by neutrophils or macrophages. These cytokines enhance the recruitment of other inflammatory cells and endothelial progenitor cells involved in angiogenesis. After the inflammatory phase, a second phase called fibrocartilage or soft callus formation occurs. During this phase, chondrocytes derived from MSCs proliferate and synthesize a cartilaginous matrix that progressively replaces the granulation tissue. TGF-β3 and FGF-2 control MSC differentiation into chondrocytes and their proliferation [184, 185]. Then, the cartilaginous matrix is calcified by hypertrophic chondrocytes that subsequently undergo apoptosis. The third phase of the bone fracture healing process is called the hard callus formation, allowing MSCs to differentiate into osteoblasts to synthesize woven bone on the matrix left by the chondrocytes. Finally, the woven bone that has been formed is remodeled by osteoclasts. This last phase of remodeling leads to regeneration of the bone structure and function [180–182].

BMPs play a crucial role in the fracture healing process since they are involved in chondrogenic and osteogenic differentiation of MSCs. BMPs, which are produced during fracture healing from macrophages, MSCs, chondrocytes, and osteoblasts, show independent or synergistic activities with each other [186–188]. Their effects are also coordinated with those of other growth factors and ECM proteins such as collagen type I [188–190]. BMPs show different temporal patterns of expression during the fracture healing process depending on the animal model used [191, 192]. Cho et al. have followed the mRNA levels of BMPs in fractured mouse tibias for 4 weeks [191]. They found that *BMP-2* was the earliest activated gene (day 1), while the genes encoding BMP-4, BMP-7, and BMP-8 were expressed between weeks 2 and 3. Interestingly, BMP-5 and BMP-6 mRNA levels were elevated throughout the 4 weeks of the healing process [191]. In contrast, Cottrell et al. found that BMP-2 and BMP-4 mRNA levels in the fractured femur calluses of female Sprague-Dawley rats were higher after 3 weeks [192]. Indeed, the BMP-2 mRNA level at 3 weeks was almost fourfold higher than that at day 2. Furthermore, only mRNA levels for BMP-2 and BMP-4 varied over time [192].

Osteogenic differentiation of mesenchymal stromal/stem cells induced by bone morphogenetic proteins

Since BMPs such as BMP-2 are required in the bone healing process, several research groups have studied the multipotent differentiation of BMSCs and ASCs following BMP treatment in vitro and in vivo (Table 3) [193–197]. Song et al. found that BMP-2 at the physiological dose of 10 ng/mL induced the osteogenic differentiation of hASCs isolated from lipoaspirates, as shown by an increase in alkaline phosphatase (ALP) activity after 1 week, and mineralization after 3 weeks, compared with the control. The effect of BMP-2 on this differentiation, as shown by mineralization, was dose and time dependent. A synergistic effect of BMP-2 and vitamin D3 was also observed on the osteogenic differentiation of hASCs. However, Song et al. did not verify the effect of BMP-2 on the commitment of hASCs into chondrogenic or adipogenic lineage [193]. Banka et al. found that BMP-2 (300 ng/mL) induced the chondrogenic differentiation of hASCs, as shown by increased aggrecan gene expression in week 1, 2, and 3 and sulfated glycosaminoglycans synthesis after 3 weeks [194]. BMP-2 at 300 ng/mL did not induce the osteogenic differentiation of hASCs. The levels of mRNA for Runx2, Osterix, and osteocalcin were similar to the control [194]. Zuk et al. have also observed a lack of effect of BMP-2 on the osteogenic capacity of hASCs [42]. However,

TABLE 3 Effect of bone morphogenetic proteins (BMPs) on stem cell fate.

Cell (type, species)	Source and extraction procedure	BMP	Operating conditions (serum, dose of BMP, time)	In vitro differentiation	In vivo effects	References
hASCs	Liposuction from three patients (~32 years old); digestion with 1.5 mg/mL of collagenase; erythrocyte lysis buffer; filtration (100 μm nylon mesh); erythrocyte lysis buffer	BMP-2	BMP-2 (10, 50, 100 ng/mL) and/or vitamin D_3 (10^{-8}, 10^{-7}, 10^{-6} M) with 10% of fetal bovine serum (FBS)	BMP-2 (10 ng/mL) increased ALP activity (7 days) and mineralization (14 and 21 days); dose response effect of BMP-2 on mineralization at 14 days; vitamin D_3 dose-dependently increased the osteogenic effect (ALP staining and mineralization at 14 days) of BMP-2 (50 ng/mL)	NA	1
hASCs	hASCs obtained by liposuction from Invitrogen (STEMPRO) were grown in MesenPRO RS medium; hASCs seeded on type-I-collagen-coated tissue culture plate for experiments	BMP-2 or BMP-7	300 ng/mL in osteogenic differentiation medium containing ascorbate, L-glutamine, dexamethasone, β-glycerophosphate	BMP-2 or BMP-7 induced chondrogenic differentiation: gene expression of aggrecan (7, 14, and 21 days) and sulfated glycosaminoglycan synthesis (21 days)	NA	2
				No effect on osteogenic differentiation (mRNA levels for Runx2, Osterix, and Osteocalcin, similar to CTL)	NA	
hASCs	Liposuction from 10 patients (~44 years old); digestion with 0.075% collagenase; filtration (100 μm nylon mesh)	rhBMP-2	25 ng/mL with 5% of FBS	No effect on matrix mineralization (21 days) or expression of several osteogenic markers (mRNA of Runx2, ALP, and osteocalcin after 7 days)	NA	3
Canine ASCs	Liposuction from adult beagle dogs; digestion with 0.075% collagenase for 45 min at 37°C; centrifugation and filtration (150 μm nylon mesh)	hBMP-2 adenoviral vector	Culture in specific osteogenic (dexamethasone, vitamin D3, ascorbate-2-phosphate, β-glycerophosphate) or adipogenic (dexamethasone, insulin, indomethacin, and isobutyl-methylxanthine) DMEM medium, both with 10% FBS	Verified the osteogenic differentiation of ASCs in osteogenic medium by an increase in ALP activity (14 days) and mineralization (von Kossa staining 28 days); verified the adipogenic differentiation of ASCs in adipogenic medium by an increase in lipid droplets (Oil red O staining 21 days)	ASCs transduced by AdBMP2 combined with β-TCF carrier: promoted repair of osteoperiosteal bone defects in dogs (4 weeks)	4

hASCs	hASCs from Zen-Bio, Inc. (hASC-ZB) or the Établissement Français du Sang (hASC-EFS)	BMP-9	100ng/mL	Highest positive contribution to ALP expression (staining after 7days)	NA	5
hASCs	hASCs from subcutaneous tissue obtained during surgical procedure of a human iliac bone graft harvest; digestion for 30min at 37°C with 0.075% collagenase type II; erythrocyte lysis buffer for 10min at room temperature; filtration (100μm nylon filter) and centrifugation	BMP-2, BMP-6, BMP-9 +IGF2, and BMP-2, -6, -9	Days 1–2: 50ng/mL; days 3–6: 100ng/mL; days 7–10: 200ng/mL; culture in osteogenic medium (OM) with 10% FBS	BMP-9+IGF2 or combination of BMP-2, -6, -9 in osteogenic OM: increased ALP expression (10days) but no change in osteocalcin expression (10 and 18days)	NA	6
hBMSCs	hBMSCs from STEMCELL Technologies	BMP-2, -4, -5, -6, -7, or -9	0 to 500ng/mL with 10% FBS	BMP-6 (500ng/mL), BMP-7 (100–500ng/mL), and BMP-9 (100–500ng/mL); increased ALP activity (3days); highest osteogenic activity of BMP-6 and BMP-9	NA	7
hBMSCs	hBMSCs from American Type Culture Collection (ATCC)	BMP-2 versus BMP-9	Transfected by chemically modified RNA encoding BMP-2 or BMP-9	Increased ALP mRNA level (3days) and matrix mineralization (14days) by BMP-9 relative to BMP-2	Similar critical size calvarial bone defect (BMP-9 versus BMP-2 after 4weeks); increased bone formation by BMP-9	8
C2C12	Murine myoblasts	14 types of BMPs	Using recombinant adenoviruses expressing 14 types of BMPs (AdBMPs)	BMP-9, BMP-6, BMP-2, BMP-7, and BMP-4 (decreasing order of osteogenic activity) enhanced ALP activity (after 40h)	BMP-6 and BMP-9 induced the most robust and mature ossification (radiographic and histological evaluation after 3 and 5weeks)	9

Continued

TABLE 3 Effect of bone morphogenetic proteins (BMPs) on stem cell fate.—cont'd

Cell (type, species)	Source and extraction procedure	BMP	Operating conditions (serum, dose of BMP, time)	In vitro differentiation	In vivo effects	References
Porcine ASCs and BMSCs	ASCs and BMSCs were isolated from subcutaneous adipose tissues and costal BM of pigs, respectively; digestion with 0.075% collagenase at 37°C for 1h; layered on lymphocyte separation medium; centrifugation (900*g* for 30min at 30°C)	rhBMP-6 versus rhBMP-2	Nucleofected with rhBMP-6 or rhBMP-2	BMSCs nucleofected with rhBMP-6 had greater osteogenic differentiation (ALP activity and matrix mineralization after 7days)	Faster and higher bone formation with rhBMP-6 versus rhBMP-2 (after 2, 4, and 6weeks)	10

[1] *Song I, Kim BS, Kim CS, Im GI. Effects of BMP-2 and vitamin D3 on the osteogenic differentiation of adipose stem cells.* Biochemical and Biophysical Research Communications. **2011**; *408(1):126–131.*
[2] *Banka S, Mukudai Y, Yoshihama Y, Shirota T, Kondo S, Shintani S. A combination of chemical and mechanical stimuli enhances not only osteo- but also chondro-differentiation in adipose-derived stem cells.* Journal of Oral Biosciences. **2012**; *54(4):188–195.*
[3] *Zuk P, Chou YF, Mussano F, Benhaim P, Wu BM. Adipose-derived stem cells and BMP2: part 2. BMP2 may not influence the osteogenic fate of human adipose-derived stem cells.* Connective Tissue Research. **2011**; *52(2):119–132.*
[4] *Li H, Dai K, Tang T, Zhang X, Yan M, Lou J. Bone regeneration by implantation of adipose-derived stromal cells expressing BMP-2.* Biochemical and Biophysical Research Communications. **2007**; *356(4):836–842.*
[5] *Kuterbekov M, Machillot P, Baillet F, Jonas AM, Glinel K, Picart C. Design of experiments to assess the effect of culture parameters on the osteogenic differentiation of human adipose stromal cells.* Stem Cell Research & Therapy. **2019**; *10(1):256.*
[6] *Acil Y, Ghoniem AA, Gulses A, Kisch T, Stang F, Wiltfang J, Gierloff M. Suppression of osteoblast-related genes during osteogenic differentiation of adipose tissue derived stromal cells.* Journal of Cranio-Maxillo-Facial Surgery. **2017**; *45(1):33–38.*
[7] *Rivera JC, Strohbach CA, Wenke JC, Rathbone CR. Beyond osteogenesis: an in vitro comparison of the potentials of six bone morphogenetic proteins.* Frontiers in Pharmacology. **2013**; *4:125.*
[8] *Khorsand B, Elangovan S, Hong L, Dewerth A, Kormann MS, Salem AK. A Comparative Study of the Bone Regenerative Effect of Chemically Modified RNA Encoding BMP-2 or BMP-9.* The AAPS Journal. **2017**; *19(2):438–446.*
[9] *Kang Q, Sun MH, Cheng H, Peng Y, Montag AG, Deyrup AT, Jiang W, et al. Characterization of the distinct orthotopic bone-forming activity of 14 BMPs using recombinant adenovirus-mediated gene delivery.* Gene Therapy. **2004**; *11(17):1312–1320.*
[10] *Mizrahi O, Sheyn D, Tawackoli W, Kallai I, Oh A, Su S, Da X, et al. BMP-6 is more efficient in bone formation than BMP-2 when overexpressed in mesenchymal stem cells.* Gene Therapy. **2013**; *20(4):370–377.*

NA, *not available.*

others have found an effect of BMP-2 on bone formation in vivo by using ASCs transduced by vectors expressing BMP-2 [195, 198]. For example, Dragoo et al. isolated hASCs from liposuction material, established them in vitro and preinduced them toward osteogenic differentiation through transfection with an AdBMP-2. hASCs were seeded onto collagen type 1 sponges, and the grafts were implanted intramuscularly in hindlimbs of SCID mice. Abundant bone formation was reported after 6 weeks of implantation, as assessed through H&E staining [36]. In the same way, Li et al. showed that canine ASCs transduced by AdBMP-2 and injected at 5×10^6 cells into triceps surae musculature of nude mice promoted bone formation after 4 weeks compared with nontransduced or AdLacZ-transduced ASCs [198]. When combined with porous β-TCP ceramic scaffold (pore size of 400 μm with 60% porosity), ASCs transduced by AdBMP-2 promoted new woven bone in bilateral-ulnar segmental defects in dogs. Interestingly, ASCs transduced by AdBMP-2 plus TCP scaffolds were more efficient than TCP combined with osteogenic-induced ASCs. The osteogenic differentiation medium (OM) used was Dulbecco modified Eagle's minimal essential medium (DMEM) containing fetal bovine serum (FBS), dexamethasone, vitamin D3, ascorbate-2-phosphate, and β-glycerophosphate [198]. Bone healing of critical calvarial bone defect in nude mice by hASCs transduced with a baculovirus vector expressing BMP-2 was also promoted by coexpression of the miRNA miR-148b [195].

Other BMPs such as BMP-7, BMP-6, and BMP-9 also induced osteogenic differentiation of ASCs [178, 199–201]. Using the statistical method of "design of experiments" (DOE), Kuterbekov et al. have recently analyzed systematically and under the same controlled conditions the effect of eight parameters (BMP-9, ASCs supplied by different providers, seeding density, basal culture medium, serum source, dexamethasone, ascorbate-2-phosphate, and β-glycerophosphate) on osteogenic differentiation of ASCs [201]. BMP-9 (100 ng/mL), ASC source, seeding density, and dexamethasone showed a net positive contribution to ALP expression (ALP staining after 1 week). On the contrary, the medium, the serum source (i.e., 5% allogeneic human platelet lysate versus 10% xenogeneic FBS), ascorbate-2-phosphate, and β-glycerophosphate exerted a net negative contribution.

There are several studies indicating that BMP-9 may be a more osteoinductive factor than BMP-2 [196, 202]. For example, hBMSCs transfected by chemically modified RNAs encoding BMP-2 (cmRNA-BMP-2) or BMP-9 (cmRNA-BMP-9) had activated genes encoding early (ALP and Runx2) and late (osteocalcin) osteogenic markers after 3 days of incubation [196]. The ALP mRNA level was more important in hBMSCs transfected by cmRNA-BMP-9 compared with those transfected by cmRNA-BMP-2. These hBMSCs (cmRNA-BMP-9) also induced a higher matrix mineralization after 14 days. However, using calvarial bone defects (5 mm diameter) in rats, the authors did not find any difference in the fraction of regenerated bone volume to the total tissue volume between the cmRNA-BMP-9 and cmRNA-BMP-2 embedded collagen matrices after 4 weeks. Some histomorphometry analyses only revealed more bone tissue formation in the cmRNA-BMP-9 groups compared with cmRNA- BMP-2 groups or empty defects [196]. In the same way, our research group recently found that a dose of 0.05 nM BMP-9 was enough to induce a significant increase in ALP activity in vitro in murine $Sca1^+$ $CD31^-$ Lin^- muscle resident stromal cells compared with a dose of 1 nM BMP-2 [202]. This effect might depend on the presence of serum since we found that FBS can potentiate the effect of BMP-9 on osteoblastic differentiation [203]. Mizrahi et al. also found that

ASCs or BMSCs with rhBMP-6 were more efficient to induce bone formation compared with BMP-2 [200].

BMP-9 has also been used in combination with IGF2 (0.85 ng/mL) or BMP-2 and BMP-6 to improve the osteogenic differentiation of hASCs [204]. hASCs isolated from adipose tissue of healthy female donors were incubated with 50 ng/mL of each BMP between days 1 and 2, 100 ng/mL between days 3 and 6, and 200 ng/mL between days 7 and 10. The hASCs treated with OM supplemented with BMP-9 plus IGF2 or a combination of BMP-9, BMP-2, and BMP-6 showed an increase in ALP expression on day 10 compared with ASCs cultured with OM alone. Nevertheless, BMP-9 plus IGF2 or combinations of BMP-9, BMP-2, and BMP-6 in OM did not increase the osteocalcin mRNA levels at days 10 or 18, compared with OM alone. In this study, controls using BMP-9 $\pm$ OM were missing. The BMP-9 effect on hASC differentiation into osteoblast lineage in the presence of other growth factors would therefore be difficult to interpret [204].

To summarize, BMPs not only provide morphogenetic signals for skeletal development during embryogenesis, but also play a critical role during bone fracture healing and remodeling due to differentiation induction of skeletal stem cells into chondrocytes and osteoblasts. ASCs have generated great interest in the development of new strategies to enhance bone regeneration since these cells can differentiate into osteoblast and chondroblast lineages. Interestingly, BMPs such as BMP-2, BMP-6, BMP-7, and BMP-9 can act on ASC fate. In addition, among the osteoinductive members of the BMP family, only BMP-2 and BMP-7 have been used for clinical applications. However, due to the supraphysiological doses applied and off-label uses, these BMPs have induced severe complications and adverse effects. Thus, a lot of work remains to determine not only safe dosage of BMPs but also appropriate release kinetic profiles from delivery systems or scaffolds for improving outcomes of bone regeneration. New avenues in the development of biomimetic materials bearing peptides derived from BMPs could also replace the use of supraphysiological BMP dose and favor a more local delivery of the osteoinductive factors. Thus, ASCs appear as a promising alternative to BMSC for bone tissue engineering and clinical applications. Nevertheless, the ability of ASCs to differentiate into osteoblast lineage with or without BMPs strongly depends on the source, the extraction procedure, and the culture medium [201, 205]. Of course, all these parameters will affect clinical outcomes.

Clinical applications of adipose-derived stromal/stem cells and adipose-derived stromal/stem cell-based engineered products for bone repair

Although the results obtained with several cell-based tissue-engineered products are satisfactory in preclinical models [206], the number of clinical trials using MSCs in the field is still very limited [44, 207]. Such trials are indicated for a small number of patients suffering mainly from massive bone loss in the long bones [208], or for maxillofacial surgery applications [209, 210]. The major difficulties encountered for establishing clinical trials are related to the regulatory aspects of the development of these innovative therapy products under good manufacturing practice (GMP) conditions, to ensure optimal biosafety for the patient [211].

One of the first clinical applications of cells extracted from adipose tissue was reported by Lendeckel et al. [210]. In this case report, the authors have described the treatment of a

7-year-old girl suffering from severe head injuries resulting in multifragmented calvarial fractures. To improve regeneration, autologous cancellous bone from the dorsal iliac crest and fat tissue of the gluteal region were harvested concomitantly. The adipose tissue was digested enzymatically, and the resulting cells were not cultured in vitro or preinduced toward osteogenesis before association with the scaffolds. Autologous fibrin glue generated from the patient's plasma was added to the scaffolds, and the mixture was combined with the cancellous bone and a resorbable microporous mesh to stabilize the defects. A postoperative follow-up demonstrated symmetrical calvaria contour with no neurological deficits from the patient. In addition, ultrasound evaluations have shown a correct and stable position of the cancellous bone and the macroporous mesh after 2 and 6 weeks. Three-dimensional computed tomography (CT) scans were performed after 3 months and showed marked ossification in the defects [210]. This first case reported promising results for the use of autologous cells from adipose tissue in bone engineering for cranioplasty.

In 2009, Mesimäki et al. used a microvascular custom-made ectopic bone flap to regenerate a large maxillary bone defect in a 65-year-old patient who had undergone hemimaxillectomy due to a recurrent keratocyst [212]. These authors were the first to use autologous hASCs cultured and amplified in vitro for such repair under xenogen-free culture conditions and according to GMP standards. hASCs were then associated with β-TCP that was previously incubated in the presence of rhBMP-2 (12 mg) for 48 h, then encased into a titanium cartridge and placed into the left rectus abdominis muscle of the patient. After 8 months, the presence of mature bone tissue, with widespread vasculature, was verified, and the graft was placed in the maxillary defect. Primary stability of the maxillary newly formed bone was achieved 4 months after the implant placement. The authors also report regeneration of the palatal mucosa and dental rehabilitation with a temporary acrylic fixed bridge [212].

More recently, Khojasteh et al. evaluated the concomitant use of autologous bone and autologous ASCs extracted from the buccal fat pad (BFSCs) for alveolar cleft osteoplasty [213]. Ten patients suffering from congenital unilateral anterior maxillary cleft were included in this phase I clinical trial and divided into three groups. The first, second, and third group were respectively treated with anterior iliac crest (AIC) bone covered with a collagen membrane (AIC group; three patients), lateral ramus cortical bone plate (LRCP) associated with BFSCs loaded in natural bovine bone mineral (LRCP + BFSC group; three patients), or AIC bone associated with BFSCs loaded in natural bovine bone mineral covered with a collagen membrane (AIC + BFSC group; four patients). A 6-month postoperative evaluation using cone-beam CT showed an improvement of new bone formation for the groups containing BFSCs (LRCP + BFSC group: 75.0 ± 3.5% and AIC + BFSC group: 82.5 ± 6.45%) compared with the AIC group (70.0 ± 10.4%). This study supported the promising clinical potential of ASCs for the reconstruction of alveolar cleft bone defects.

Large frontal and frontotemporal craniofacial defects were also regenerated with the use of autologous hASCs, as reported in Thesleff et al., where cells seeded upon granular β-TCP were used to treat four patients undergoing cranioplasty [214]. CT scans showed ossification in all cases 3 months after the operation, with no signs of complications at this time point, and the bone density of the grafts, measured in Hounsfield units, tended to increase [214]. In Sándor et al., autologous hASCs were isolated, expanded, and associated with β-TCP (in granules or in pliable strips) or with Biogran® bioglass (Biomet 3i) to regenerate bone defects in the frontal sinus, craniofacial bone, mandible, or nasal septum of 13 patients. In some cases,

BMP-2 was added, and in others, the grafts were placed inside a titanium cage [215]. Follow-up varied from 12 to 52 months, and successful reconstruction of the defects was reported in 10 out of the 13 patients. Some patients for whom mandibular bone was repaired underwent successful dental implant procedures afterwards. On the other hand, the authors reported resorption rates higher than expected for grafts used to treat cranial defects and that were not encased in titanium [215].

In line with these findings, Thesleff et al. [216] performed a 6- to 7-year clinical follow-up study of the four cases reported in 2011, as well as of one of the cases reported in Sándor et al. [215]. They reported that three out of five patients needed reoperation owing to graft-related problems: two of them had marked resorption of the graft, and one developed an infection 7.3 years after the first surgery, which required an additional surgical intervention for graft removal [216]. As for the remaining two patients, one displayed a fully ossified graft but needed to be reoperated 2.2 years after the first surgery because of the recurrence of a meningioma, while the other had an uneventful clinical follow-up, even though X-ray images displayed bone hypodensity at the ridges of the implants. In light of these findings, this study concluded that in long-term follow-up, the cranioplasties were not satisfactory in most of the cases, either due to deficient ossification or because of infection or recurrence of the tumor [216].

In Sandor et al., and in Wolf et al., hASCs were used in association with β-TCP granules and BMP-2 to treat four cases of mandibular ameloblast resection defects. In three of the patients, dental implants were successfully placed months after mandibular regeneration [217, 218]. The largest cohort of bone tissue bioengineering with autologous hASCs involved 23 patients with large jaw defects and was reported in Sándor et al. [219]. Cells were associated with β-TCP or bioglass scaffolds and BMP-2, and successful results were reported in 85% of the cases, but no long-term follow-up study was performed. A more detailed list of clinical studies using adipose-derived cells to treat bone defects of human participants is included in Table 4.

TABLE 4 Published human clinical trials and experimental therapies using adipose-derived cells for bone regeneration.

Surgical site	Cell origin	Materials	Number of patients	References
Calvarial bone	Autologous[1]	Fragmented autologous bone from the iliac crest and fibrin glue	1	2
Maxilla	Autologous	β-TCP and BMP-2	1	3
Maxilla and mandible	Autologous	Biomatrix and PRP	8	4
Calvarial bone	Autologous	β-TCP	4	5
Calvarial bone	Autologous	Resorbable scaffolds with BMP-2	20	6
Mandible	Autologous	Titanium mesh with β-TCP and BMP-2	1	7

TABLE 4 Published human clinical trials and experimental therapies using adipose-derived cells for bone regeneration—cont'd

Surgical site	Cell origin	Materials	Number of patients	References
Mandible	Autologous	β-TCP	3	8
Frontal sinus, nasal septum, mandible, cranial bone	Autologous	Bioactive glass, β-TCP, and BMP-2	13	9
Hip	Autologous	PRP and hyaluronic acid, and calcium chloride	2	10
Femur	Autologous	PRP	2	11
Femur	Autologous	PRP	1	12
Humerus	Autologous	Fibrin gel	8	13
Femur, tibia, ulna, fibula, pelvic bone	Autologous	Human demineralized bone matrix	11	14

[1] *Uncultured stromal-vascular fraction (SVF) cells instead of cultured hASCs were used.*
[2] *Lendeckel S, Jödicke A, Christophis P, Heidinger K, Wolff J, Fraser JK, Hedrick MH, et al. Autologous stem cells (adipose) and fibrin glue used to treat widespread traumatic calvarial defects: case report.* Journal of Cranio-Maxillo-Facial Surgery: Official Publication of the European Association for Cranio-Maxillo-Facial Surgery. ***2004****; 32:370–373.*
[3] *Mesimäki K, Lindroos B, Törnwall J, Mauno J, Lindqvist C, Kontio R, Miettinen S, Suuronen R. Novel maxillary reconstruction with ectopic bone formation by GMP adipose stem cells.* International Journal of Oral and Maxillofacial Surgery. ***2009****; 38:201–209.*
[4] *Kulakov AA, Goldshtein DV, Grigoryan AS, Rzhaninova AA, Alekseeva IS, Arutyunyan IV, Volkov AV. Clinical study of the efficiency of combined cell transplant on the basis of multipotent mesenchymal stromal adipose tissue cells in patients with pronounced deficit of the maxillary and mandibulary bone tissue.* Bulletin of Experimental Biology and Medicine. ***2008****; 146:522–525.*
[5] *Thesleff T, Lehtimäki K, Niskakangas T, Mannerström B, Miettinen S, Suuronen R, Öhman J. Cranioplasty with adipose-derived stem cells and biomaterial: a novel method for cranial reconstruction.* Neurosurgery. ***2011****; 68:1535–1540.*
[6] *Sándor GKB. Tissue engineering of bone: Clinical observations with adipose-derived stem cells, resorbable scaffolds, and growth factors.* Annals of Maxillofacial Surgery. ***2012****; 2:8–11.*
[7] *Sándor GK, Tuovinen VJ, Wolff J, Patrikoski M, Jokinen J, Nieminen E, Mannerström B, et al. Adipose stem cell tissue-engineered construct used to treat large anterior mandibular defect: a case report and review of the clinical application of good manufacturing practice-level adipose stem cells for bone regeneration.* Journal of Oral and Maxillofacial Surgery: Official Journal of the American Association of Oral and Maxillofacial Surgeons. ***2013****; 71:938–950.*
[8] *Wolff J, Sándor GK, Miettinen A, Tuovinen VJ, Mannerström B, Patrikoski M, Miettinen S. GMP-level adipose stem cells combined with computer-aided manufacturing to reconstruct mandibular ameloblastoma resection defects: Experience with three cases.* Annals of Maxillofacial Surgery. ***2013****; 3:114–125.*
[9] *Sándor GK, Numminen J, Wolff J, Thesleff T, Miettinen A, Tuovinen VJ, Mannerström B, et al. Adipose stem cells used to reconstruct 13 cases with cranio-maxillofacial hard-tissue defects.* Stem Cells Translational Medicine. ***2014****; 3:530–540.*
[10] *Pak J. Regeneration of human bones in hip osteonecrosis and human cartilage in knee osteoarthritis with autologous adipose-tissue-derived stem cells: a case series.* Journal of Medical Case Reports. ***2011****; 5:296.*
[11] *Pak J. Autologous adipose tissue-derived stem cells induce persistent bone-like tissue in osteonecrotic femoral heads.* Pain Physician. ***2012****; 15(1):75–85.*
[12] *Pak J, Lee JH, Jeon JH, Lee SH. Complete resolution of avascular necrosis of the human femoral head treated with adipose tissue-derived stem cells and platelet-rich plasma.* The Journal of International Medical Research. ***2014****; 42(6):1353–1362.*
[13] *Saxer F, Scherberich A, Todorov A, Studer P, Miot S, Schreiner S, Guven S, et al. Implantation of Stromal Vascular Fraction Progenitors at Bone Fracture Sites: From a Rat Model to a First-in-Man Study.* Stem Cells. ***2016****; 34(12):2956–2966.*
[14] *Veriter S, Andre W, Aouassar N, Poirel HA, Lafosse A, Docquier PL, Dufrane D. Human Adipose-Derived Mesenchymal Stem Cells in Cell Therapy: Safety and Feasibility in Different "Hospital Exemption" Clinical Applications.* PLoS One. ***2015****; 10(10):e0139566.*
Abbreviations: BMP-2, *bone morphogenetic protein 2;* β-TCP, *beta-tricalcium phosphate;* PRP, *platelet-rich plasma.*
Based on F. Paduano, M. Marrelli, M. Amantea, C. Rengo, S. Rengo, M. Goldberg, G. Spagnuolo, M. Tatullo, Adipose tissue as a strategic source of mesenchymal stem cells in bone regeneration: a topical review on the most promising craniomaxillofacial applications, Int. J. Mol. Sci. 18(10) (2017); M. Torres-Torrillas, M. Rubio, E. Damia, B. Cuervo, A. Del Romero, P. Pelaez, D. Chicharro, et al., Adipose-derived mesenchymal stem cells: a promising tool in the treatment of musculoskeletal diseases, Int. J. Mol. Sci. 20(12) (2019).

Perspectives for next-generation bone tissue engineering: Biofabrication of personalized constructs

In the future, as technologies continue to evolve and become accessible to more researchers, we will likely see the emergence of new advanced and improved bone-like substitutes. For example, new techniques for the biofabrication of personalized and anatomically relevant living substitutes containing hASCs will further the field of bone tissue engineering. The use of bioprinting processes, such as laser-assisted, extrusion-based, inkjet, or acoustic bioprinting, will likely transform more traditional production of bone-like constructs [220, 221]. These technologies have been developed to address several limitations associated with the manufacturing of living tissues possessing suitable biological complexity and adequate mechanical properties [222, 223]. Based on a bottom-up (layer-by-layer) approach, additive manufacturing technology allows the in vitro production of small-to-large living constructs with great spatial control of the architecture during the production [223]. In addition, computer-assisted processes can considerably enhance the reproducibility and reduce the cost of production compared with traditional operator-dependent processes [224, 225].

In this context, Hung et al. developed a hybrid scaffold composed of decellularized bone matrix (DBM) particles combined with PCL using 3D printing extruder technology [226]. In that study, hASCs were seeded into the scaffolds after printing and cultured for 3 weeks in the absence of exogenous osteogenic factors. Despite the lack of chemical inducers in the culture medium, hASCs exhibited an upregulation of osteogenic genes when in contact with the hybrid scaffolds compared with pure PCL scaffolds. This result shows the important role of matrix composition and cell-matrix crosstalk during osteogenesis. The implantation of hASC-seeded hybrid scaffolds into critical-sized murine calvarial defects showed greater bone regeneration after 1 and 3 months compared with hASC-seeded pure PCL scaffolds [226]. This study highlights the potential of 3D printing combined with cell seeding in the field of bone tissue engineering, a fast-growing field of applications [227, 228].

The main advantages of bioprinting technologies are the high resolution of matrix and cell depositions that can be achieved, as well as the automation of the biofabrication process. Indeed, the evolution of computer-assisted robots has considerably improved the precision required for complex geometric tasks executed with high-accuracy motions. In the context of clinical applications, this technology can be associated with 3D imaging systems, such as CT or magnetic resonance imaging (MRI). Prior to the biofabrication of a personalized substitute, an initial step of 3D scanning for the anatomic features of the patient provides information on the structure of the injured tissue. Based on the images of the patient's anatomy, an additional step of computer-aided design and mathematical modeling is necessary to determine the volume of the anatomic tissue that must be reconstructed by the bioprinter [229].

Initially, the bioprinting process was developed to produce in vitro biological substitutes, which can then be potentially grafted. However, Keriquel et al. were among the first to use a different strategy and print biological material directly into the wounds using a live animal model [224]. In a first study, a laser-assisted bioprinter was used to deposit nano-HA within 4-mm mouse calvarial defects. X-ray microtomography and post mortem histology showed heterogeneous results in terms of bone regeneration after 1 week, 1 month, and 3 months at the printed site. This proof of concept established the possibility to use computer-assisted modality in the context of regenerative medicine through 3D bioprinting. More recently, Keriquel et al. repeated the experiment and upgraded their process of in situ bioprinting [227]. Bioprinted nano-HA material in combination with a collagen gel and murine BMSCs was evaluated on different cell printing geometries (ring or disk) to achieve guided calvarial bone defect repair in mice. The results showed a significant amount of newly formed bone within the defects containing disk-shaped cell pattern after 1 and 2 months. This promising work confirmed the potential of this new bioprinting strategy for bone tissue engineering.

Conclusion

Considering the extended lifespan and the increased prevalence of obesity of the population, the incidence of musculoskeletal system-associated disorders and injuries is expected to continue to grow. The repair of bone defects using tissue engineering approaches is considered highly relevant in the therapeutic realm. In fact, there is a sense of urgency with respect to the translation of findings in stem cell biology, tissue bioengineering, and biomaterials design into clinical applications. In vivo tests using many different animal models, as well as clinical trials and experimental therapies with human participants, have paved the way for a broader use of hASCs in clinical settings. Nevertheless, some important obstacles still need to be overcome to achieve optimal bench-to-bedside translation of regenerative therapies based on hASCs. Variations associated with the anatomical sources of hASCs, with patients' clinical histories and health conditions at the time of the regenerative medical procedure, and with the site and severity of the bone defects, as well as variability in the protocols used for hASC isolation and in vitro expansion, among others, make comparisons among different studies difficult. This is also reflected in difficulties to systematically replicate results reported in the literature by distinct groups. There is a need for consensus, guidelines, and standardized GMP protocols for utilization of different scaffolds in combination with hASCs and bone-inducing factors to ensure a more comprehensive translation of cell-based bioengineering constructs into clinical environments. Prospective randomized clinical trials, performed by multicentric research and clinical groups, with more homogeneous protocols, larger number of participants per set, and more stringent criteria to attest efficacy and the regeneration of fully functional bone tissue, and long-term follow-up studies are also necessary. Lastly, but not of less importance, the more we characterize and understand the molecular and cellular mechanisms behind each particular use of ASCs, the more likely it is that the results described in one study will be replicated in distinct experiments and, ultimately, translated efficiently and safely to clinical settings. Tackling these important challenges will undoubtedly advance the promising field of bone tissue engineering.

Acknowledgments

We address special thanks to Annie Banville for her proofreading of the manuscript and Dr. O. Drevelle for his help in the preparation of the Figures. This work was funded in part by the Quebec Cell, Tissue and Gene Therapy Network—Thé Cell (a thematic network supported by the Fonds de recherche du Québec-Santé).

References

[1] K. Ikeda, S. Takeshita, The role of osteoclast differentiation and function in skeletal homeostasis, J. Biochem. 159 (1) (2016) 1–8.
[2] M. Capulli, R. Paone, N. Rucci, Osteoblast and osteocyte: games without frontiers, Arch. Biochem. Biophys. 561 (2014) 3–12.
[3] W.G. De Long, T.A. Einhorn, K. Koval, M. McKee, W. Smith, R. Sanders, T. Watson, Bone grafts and bone graft substitutes in orthopaedic trauma surgery, J. Bone Joint Surg. 89 (2007) 649–658.
[4] S. Sittitavornwong, R. Gutta, Bone graft harvesting from regional sites, Oral Maxillofac. Surg. Clin. North Am. 22 (3) (2010) 317–330 (v–vi).
[5] D.A. Hidalgo, A. Rekow, A review of 60 consecutive fibula free flap mandible reconstructions, Plast. Reconstr. Surg. 96 (3) (1995) 585–596 (discussion 597–602).
[6] E. Garcia-Gareta, M.J. Coathup, G.W. Blunn, Osteoinduction of bone grafting materials for bone repair and regeneration, Bone 81 (2015) 112–121.
[7] E.D. Arrington, W.J. Smith, H.G. Chambers, A.L. Bucknell, N.A. Davino, Complications of iliac crest bone graft harvesting, Clin. Orthop. Relat. Res. (1996) 300–309.
[8] J.C. Banwart, M.A. Asher, R.S. Hassanein, Iliac crest bone graft harvest donor site morbidity: a statistical evaluation, Spine 20 (1995) 1055–1066.
[9] T.A. St John, A.R. Vaccaro, A.P. Sah, M. Schaefer, S.C. Berta, T. Albert, A. Hilibrand, Physical and monetary costs associated with autogenous bone graft harvesting, Am. J. Orthop. 32 (2003) 18–23.
[10] S.C. Ludwig, J.M. Kowalski, S.D. Boden, Osteoinductive bone graft substitutes, Eur. Spine J. 9 (2000) S119–S125.
[11] E.M. Younger, M.W. Chapman, Morbidity at bone graft donor sites, J. Orthop. Trauma 3 (1989) 192–195.
[12] R.M. Kline, S.A. Wolfe, Complications associated with the harvesting of cranial bone grafts, Plast. Reconstr. Surg. 95 (1995) 5–13.
[13] G.A. Grant, M. Jolley, R.G. Ellenbogen, T.S. Roberts, J.R. Gruss, J.D. Loeser, Failure of autologous bone-assisted cranioplasty following decompressive craniectomy in children and adolescents, J. Neurosurg. 100 (2004) 163–168.
[14] F.R. Kloss, V. Offermanns, A. Kloss-Brandstatter, Comparison of allogeneic and autogenous bone grafts for augmentation of alveolar ridge defects—a 12-month retrospective radiographic evaluation, Clin. Oral Implants Res. 29 (11) (2018) 1163–1175.
[15] D. Yang, J. Xiao, B. Wang, L. Li, X. Kong, J. Liao, The immune reaction and degradation fate of scaffold in cartilage/bone tissue engineering, Mater. Sci. Eng. C Mater. Biol. Appl. 104 (2019) 109927.
[16] M. Darnell, D.J. Mooney, Leveraging advances in biology to design biomaterials, Nat. Mater. 16 (12) (2017) 1178–1185.
[17] C. Hu, D. Ashok, D.R. Nisbet, V. Gautam, Bioinspired surface modification of orthopedic implants for bone tissue engineering, Biomaterials 219 (2019) 119366.
[18] H.F. Pereira, I.F. Cengiz, F.S. Silva, R.L. Reis, J.M. Oliveira, Scaffolds and coatings for bone regeneration, J. Mater. Sci. Mater. Med. 31 (3) (2020) 27.
[19] M.P. Ginebra, M. Espanol, Y. Maazouz, V. Bergez, D. Pastorino, Bioceramics and bone healing, EFORT Open Rev. 3 (5) (2018) 173–183.
[20] M. Swetha, K. Sahithi, A. Moorthi, N. Srinivasan, K. Ramasamy, N. Selvamurugan, Biocomposites containing natural polymers and hydroxyapatite for bone tissue engineering, Int. J. Biol. Macromol. 47 (2010) 1–4.
[21] F. Kawecki, W.P. Clafshenkel, M. Fortin, F.A. Auger, J. Fradette, Biomimetic tissue-engineered bone substitutes for maxillofacial and craniofacial repair: the potential of cell sheet technologies, Adv. Healthcare Mater. 7 (6) (2017). 1700919.
[22] Y. Xiao, H. Qian, W.G. Young, P.M. Bartold, Tissue engineering for bone regeneration using differentiated alveolar bone cells in collagen scaffolds, Tissue Eng. 9 (2003) 1167–1177.
[23] J.E. Fleming Jr., C.N. Cornell, G.F. Muschler, Bone cells and matrices in orthopedic tissue engineering, Orthop. Clin. North Am. 31 (3) (2000) 357–374.
[24] R.A. Young, Control of the embryonic stem cell state, Cell 144 (2011) 940–954.
[25] J.M. Jukes, C.A. Van Blitterswijk, J. De Boer, Skeletal tissue engineering using embryonic stem cells, J. Tissue Eng. Regen. Med. 4 (2010) 165–180.
[26] S. Yamanaka, Induction of pluripotent stem cells from mouse fibroblasts by four transcription factors, Cell Prolif. 41 (Suppl. 1) (2008) 51–56.

[27] F. Li, S. Bronson, C. Niyibizi, Derivation of murine induced pluripotent stem cells (iPS) and assessment of their differentiation toward osteogenic lineage, J. Cell. Biochem. 109 (2010) 643–652.
[28] S. Kern, H. Eichler, J. Stoeve, H. Klüter, K. Bieback, Comparative analysis of mesenchymal stem cells from bone marrow, umbilical cord blood, or adipose tissue, Stem Cells 24 (2006) 1294–1301.
[29] A.J. Friedenstein, R.K. Chailakhyan, U.V. Gerasimov, Bone marrow osteogenic stem cells: in vitro cultivation and transplantation in diffusion chambers, Cell Prolif. 20 (1987) 263–272.
[30] M.F. Pittenger, A.M. Mackay, S.C. Beck, R.K. Jaiswal, R. Douglas, J.D. Mosca, M.A. Moorman, D.W. Simonetti, S. Craig, D.R. Marshak, Multilineage potential of adult human mesenchymal stem cells, Science 284 (**1999**) 143–147.
[31] H. Ohgushi, V.M. Goldberg, A.I. Caplan, Repair of bone defects with marrow cells and porous ceramic. Experiments in rats, Acta Orthop. Scand. 60 (1989) 334–339.
[32] S.P. Bruder, A.A. Kurth, M. Shea, W.C. Hayes, N. Jaiswal, S. Kadiyala, Bone regeneration by implantation of purified, culture-expanded human mesenchymal stem cells, J. Orthop. Res. 16 (1998) 155–162.
[33] J.-T. Schantz, S.H. Teoh, T.C. Lim, M. Endres, C.X.F. Lam, D.W. Hutmacher, Repair of calvarial defects with customized tissue-engineered bone grafts I. Evaluation of osteogenesis in a three-dimensional culture system, Tissue Eng. (9 Suppl. 1) (**2003**) S113–S126.
[34] R. Osinga, N. Di Maggio, A. Todorov, N. Allafi, A. Barbero, F. Laurent, D.J. Schaefer, I. Martin, A. Scherberich, Generation of a bone organ by human adipose-derived stromal cells through endochondral ossification, Stem Cells Transl. Med. 5 (8) (2016) 1090–1097.
[35] Y.D.C. Halvorsen, D. Franklin, A.L. Bond, D.C. Hitt, C. Auchter, A.L. Boskey, E.P. Paschalis, W.O. Wilkison, J.M. Gimble, Extracellular matrix mineralization and osteoblast gene expression by human adipose tissue-derived stromal cells, Tissue Eng. 7 (2001) 729–741.
[36] J.L. Dragoo, J.Y. Choi, J.R. Lieberman, J. Huang, P.A. Zuk, J. Zhang, M.H. Hedrick, P. Benhaim, Bone induction by BMP-2 transduced stem cells derived from human fat, J. Orthop. Res. 21 (2003) 622–629.
[37] P.A. Zuk, M. Zhu, P. Ashjian, D.A. De Ugarte, J.I. Huang, H. Mizuno, Z.C. Alfonso, J.K. Fraser, P. Benhaim, M.H. Hedrick, Human adipose tissue is a source of multipotent stem cells, Mol. Biol. Cell 13 (2002) 4279–4295.
[38] P.A. Zuk, The adipose-derived stem cell: looking back and looking ahead, Mol. Biol. Cell 21 (2010) 1783–1787.
[39] A. Sterodimas, J. De Faria, B. Nicaretta, I. Pitanguy, Tissue engineering with adipose-derived stem cells (ADSCs): current and future applications, J. Plast. Reconstr. Aesthet. Surg. 63 (2010) 1886–1892.
[40] M. Locke, J. Windsor, P.R. Dunbar, Human adipose-derived stem cells: isolation, characterization and applications in surgery, ANZ J. Surg. 79 (2009) 235–244.
[41] B.M. Strem, K.C. Hicok, M. Zhu, I. Wulur, Z. Alfonso, R.E. Schreiber, J.K. Fraser, M.H. Hedrick, Multipotential differentiation of adipose tissue-derived stem cells, Keio J. Med. 54 (2005) 132–141.
[42] P. Zuk, Y.F. Chou, F. Mussano, P. Benhaim, B.M. Wu, Adipose-derived stem cells and BMP2: part 2. BMP2 may not influence the osteogenic fate of human adipose-derived stem cells, Connect. Tissue Res. 52 (2) (2011) 119–132.
[43] R.J. Kroeze, M. Knippenberg, M.N. Helder, Osteogenic differentiation strategies for adipose-derived mesenchymal stem cells, Methods Mol. Biol. 702 (2011) 233–248.
[44] G. Hutchings, L. Moncrieff, C. Dompe, K. Janowicz, R. Sibiak, A. Bryja, M. Jankowski, P. Mozdziak, D. Bukowska, P. Antosik, J.A. Shibli, M. Dyszkiewicz-Konwinska, M. Bruska, B. Kempisty, H. Piotrowska-Kempisty, Bone regeneration, reconstruction and use of osteogenic cells; from basic knowledge, animal models to clinical trials, J. Clin. Med. 9 (1) (2020) 139, https://doi.org/10.3390/jcm9010139.
[45] N. Saulnier, M.A. Puglisi, W. Lattanzi, L. Castellini, G. Pani, G. Leone, S. Alfieri, F. Michetti, A.C. Piscaglia, A. Gasbarrini, Gene profiling of bone marrow- and adipose tissue-derived stromal cells: a key role of Kruppel-like factor 4 in cell fate regulation, Cytotherapy 13 (2011) 329–340.
[46] D.C. Wan, Y.-Y. Shi, R.P. Nacamuli, N. Quarto, K.M. Lyons, M.T. Longaker, Osteogenic differentiation of mouse adipose-derived adult stromal cells requires retinoic acid and bone morphogenetic protein receptor type IB signaling, Proc. Natl Acad. Sci. USA 103 (2006) 12335–12340.
[47] J.R. Dudas, K.G. Marra, G.M. Cooper, V.M. Penascino, M.P. Mooney, S. Jiang, J.P. Rubin, J.E. Losee, The osteogenic potential of adipose-derived stem cells for the repair of rabbit calvarial defects, Ann. Plast. Surg. 56 (2006) 543–548.
[48] L. Lecoeur, J.P. Ouhayoun, In vitro induction of osteogenic differentiation from non-osteogenic mesenchymal cells, Biomaterials 18 (1997) 989–993.

[49] J.L. Dragoo, J.R. Lieberman, R.S. Lee, D.A. Deugarte, Y. Lee, P.A. Zuk, M.H. Hedrick, P. Benhaim, Tissue-engineered bone from BMP-2-transduced stem cells derived from human fat, Plast. Reconstr. Surg. 115 (2005) 1665–1673.
[50] B. Levi, A.W. James, D.C. Wan, J.P. Glotzbach, G.W. Commons, M.T. Longaker, Regulation of human adipose-derived stromal cell osteogenic differentiation by insulin-like growth factor-1 and platelet-derived growth factor-α, Plast. Reconstr. Surg. 126 (2010) 41–52.
[51] R.D. Fanganiello, F.A.A. Ishiy, G.S. Kobayashi, L. Alvizi, D.Y. Sunaga, M.R. Passos-Bueno, Increased in vitro osteopotential in SHED associated with higher IGF2 expression when compared with hASCs, Stem Cell Rev. Rep. 11 (2015) 635–644.
[52] B. Levi, A.W. James, E.R. Nelson, S. Li, M. Peng, G.W. Commons, M. Lee, B. Wu, M.T. Longaker, Human adipose-derived stromal cells stimulate autogenous skeletal repair via paracrine hedgehog signaling with calvarial osteoblasts, Stem Cells Dev. 20 (2011) 243–257.
[53] A.W. James, P. Leucht, B. Levi, A.L. Carre, Y. Xu, J.A. Helms, M.T. Longaker, Sonic hedgehog influences the balance of osteogenesis and adipogenesis in mouse adipose-derived stromal cells, Tissue Eng. Part A 16 (2010) 2605–2616.
[54] A.W. James, S. Pang, A. Askarinam, M. Corselli, J.N. Zara, R. Goyal, L. Chang, A. Pan, J. Shen, W. Yuan, D. Stoker, X. Zhang, J.S. Adams, K. Ting, C. Soo, Additive effects of sonic hedgehog and Nell-1 signaling in osteogenic versus adipogenic differentiation of human adipose-derived stromal cells, Stem Cells Dev. 21 (2012) 2170–2178.
[55] Y. Xu, K.E. Hammerick, A.W. James, A.L. Carre, P. Leucht, A.J. Giaccia, M.T. Longaker, Inhibition of histone deacetylase activity in reduced oxygen environment enhances the osteogenesis of mouse adipose-derived stromal cells, Tissue Eng. Part A 15 (2009) 3697–3707.
[56] P. Hong, H. Yang, Y. Wu, K. Li, Z. Tang, The functions and clinical application potential of exosomes derived from adipose mesenchymal stem cells: a comprehensive review, Stem Cell Res Ther 10 (1) (2019) 242.
[57] G. Storti, M.G. Scioli, B.S. Kim, A. Orlandi, V. Cervelli, Adipose-derived stem cells in bone tissue engineering: useful tools with new applications, Stem Cells Int. 2019 (2019) 3673857.
[58] M. Maredziak, K. Marycz, D. Lewandowski, A. Siudzinska, A. Smieszek, Static magnetic field enhances synthesis and secretion of membrane-derived microvesicles (MVs) rich in VEGF and BMP-2 in equine adipose-derived stromal cells (EqASCs)-a new approach in veterinary regenerative medicine, In Vitro Cell. Dev. Biol. Anim. 51 (3) (2015) 230–240.
[59] J.R. Overman, M.N. Helder, C.M. ten Bruggenkate, E.A.J.M. Schulten, J. Klein-Nulend, A.D. Bakker, Growth factor gene expression profiles of bone morphogenetic protein-2-treated human adipose stem cells seeded on calcium phosphate scaffolds in vitro, Biochimie 95 (2013) 2304–2313.
[60] T.P. Frazier, J.M. Gimble, I. Kheterpal, B.G. Rowan, Impact of low oxygen on the secretome of human adipose-derived stromal/stem cell primary cultures, Biochimie 95 (2013) 2286–2296.
[61] N. Kalinina, D. Kharlampieva, M. Loguinova, I. Butenko, O. Pobeguts, A. Efimenko, L. Ageeva, G. Sharonov, D. Ischenko, D. Alekseev, O. Grigorieva, V. Sysoeva, K. Rubina, V. Lazarev, V. Govorun, Characterization of secretomes provides evidence for adipose-derived mesenchymal stromal cells subtypes, Stem Cell Res Ther 6 (2015) 221.
[62] S. Riis, A. Stensballe, J. Emmersen, C.P. Pennisi, S. Birkelund, V. Zachar, T. Fink, Mass spectrometry analysis of adipose-derived stem cells reveals a significant effect of hypoxia on pathways regulating extracellular matrix, Stem Cell Res Ther 7 (2016) 52.
[63] A. Gómez-Aristizábal, A. Sharma, M.A. Bakooshli, M. Kapoor, P.M. Gilbert, S. Viswanathan, R. Gandhi, Stage-specific differences in secretory profile of mesenchymal stromal cells (MSCs) subjected to early- vs late-stage OA synovial fluid, Osteoarthr. Cartil. 25 (2017) 737–741.
[64] T.M. Rodríguez, A. Saldías, M. Irigo, J.V. Zamora, M.J. Perone, R.A. Dewey, Effect of TGF-β1 stimulation on the secretome of human adipose-derived mesenchymal stromal cells, Stem Cells Transl. Med. 4 (2015) 894–898.
[65] M. Vermette, V. Trottier, V. Menard, L. Saint-Pierre, A. Roy, J. Fradette, Production of a new tissue-engineered adipose substitute from human adipose-derived stromal cells, Biomaterials 28 (18) (2007) 2850–2860.
[66] G.M. Fortier, R. Gauvin, M. Proulx, M. Vallee, J. Fradette, Dynamic culture induces a cell type-dependent response impacting on the thickness of engineered connective tissues, J. Tissue Eng. Regen. Med. 7 (4) (2013) 292–301.
[67] D. Chan, S.R. Lamande, W.G. Cole, J.F. Bateman, Regulation of procollagen synthesis and processing during ascorbate-induced extracellular matrix accumulation in vitro, Biochem. J. 269 (1990) 175–181.

[68] J.M. Davidson, P.A. LuValle, O. Zoia, D. Quaglino, M. Giro, Ascorbate differentially regulates elastin and collagen biosynthesis in vascular smooth muscle cells and skin fibroblasts by pretranslational mechanisms, J. Biol. Chem. 272 (1997) 345–352.
[69] J. Geesin, Regulation of collagen synthesis in human dermal fibroblasts in contracted collagen gels by ascorbic acid, growth factors, and inhibitors of lipid peroxidation, Exp. Cell Res. 206 (1993) 283–290.
[70] F. Kawecki, W.P. Clafshenkel, F.A. Auger, J.-M. Bourget, J. Fradette, R. Devillard, Self-assembled human osseous cell sheets as living biopapers for the laser-assisted bioprinting of human endothelial cells, Biofabrication 10 (2018) 035006.
[71] T. Galbraith, W.P. Clafshenkel, F. Kawecki, C. Blanckaert, B. Labbé, M. Fortin, F.A. Auger, J. Fradette, A cell-based self-assembly approach for the production of human osseous tissues from adipose-derived stromal/stem cells, Adv. Healthcare Mater. 6 (4) (2017) 1600889, https://doi.org/10.1002/adhm.201600889.
[72] A. Shridhar, B.G. Amsden, E.R. Gillies, L.E. Flynn, Investigating the effects of tissue-specific extracellular matrix on the adipogenic and osteogenic differentiation of human adipose-derived stromal cells within composite hydrogel scaffolds, Front. Bioeng. Biotechnol. 7 (2019) 402.
[73] R.Z. LeGeros, Calcium phosphate-based osteoinductive materials, Chem. Rev. 108 (11) (2008) 4742–4753.
[74] X. Li, H. Liu, X. Niu, Y. Fan, Q. Feng, F.-Z. Cui, F. Watari, Osteogenic differentiation of human adipose-derived stem cells induced by osteoinductive calcium phosphate ceramics, J Biomed Mater Res B Appl Biomater 97 (2011) 10–19.
[75] I. Tognarini, S. Sorace, R. Zonefrati, G. Galli, A. Gozzini, S. Carbonell Sala, G.D.Z. Thyrion, A.M. Carossino, A. Tanini, C. Mavilia, C. Azzari, F. Sbaiz, A. Facchini, R. Capanna, M.L. Brandi, In vitro differentiation of human mesenchymal stem cells on Ti6Al4V surfaces, Biomaterials 29 (2008) 809–824.
[76] G. Gastaldi, A. Asti, M.F. Scaffino, L. Visai, E. Saino, A.M. Cometa, F. Benazzo, Human adipose-derived stem cells (hASCs) proliferate and differentiate in osteoblast-like cells on trabecular titanium scaffolds, J. Biomed. Mater. Res. A 94 (2010) 790–799.
[77] V. Nardone, S. Fabbri, F. Marini, R. Zonefrati, G. Galli, A. Carossino, A. Tanini, M.L. Brandi, Osteodifferentiation of human preadipocytes induced by strontium released from hydrogels, Int. J. Biomater. 2012 (2012) 865291.
[78] S. Kishimoto, M. Ishihara, Y. Mori, M. Takikawa, Y. Sumi, S. Nakamura, T. Sato, T. Kiyosawa, Three-dimensional expansion using plasma-medium gel with fragmin/protamine nanoparticles and fgf-2 to stimulate adipose-derived stromal cells and bone marrow-derived mesenchymal stem cells, BioResearch 1 (2012) 314–323.
[79] C. Correia, S. Bhumiratana, L.-P. Yan, A.L. Oliveira, J.M. Gimble, D. Rockwood, D.L. Kaplan, R.A. Sousa, R.L. Reis, G. Vunjak-Novakovic, Development of silk-based scaffolds for tissue engineering of bone from human adipose-derived stem cells, Acta Biomater. 8 (2012) 2483–2492.
[80] C. Zhou, Q. Shi, W. Guo, L. Terrell, A.T. Qureshi, D.J. Hayes, Q. Wu, Electrospun bio-nanocomposite scaffolds for bone tissue engineering by cellulose nanocrystals reinforcing maleic anhydride grafted PLA, ACS Appl. Mater. Interfaces 5 (2013) 3847–3854.
[81] S. Haimi, G. Gorianc, L. Moimas, B. Lindroos, H. Huhtala, S. Räty, H. Kuokkanen, G.K. Sándor, C. Schmid, S. Miettinen, R. Suuronen, Characterization of zinc-releasing three-dimensional bioactive glass scaffolds and their effect on human adipose stem cell proliferation and osteogenic differentiation, Acta Biomater. 5 (2009) 3122–3131.
[82] D. Leong, W. Nah, A. Gupta, D. Hutmacher, M. Woodruff, The osteogenic differentiation of adipose tissue-derived precursor cells in a 3D scaffold/matrix environment, Curr. Drug Discov. Technol. 5 (2008) 319–327.
[83] C.M. Haslauer, A.K. Moghe, J.A. Osborne, B.S. Gupta, E.G. Loboa, Collagen-PCL sheath-core bicomponent electrospun scaffolds increase osteogenic differentiation and calcium accretion of human adipose-derived stem cells, J. Biomater. Sci. Polym. Ed. 22 (2011) 1695–1712.
[84] M.C. Williams, Optimizing radiation worker protection: the practical application of risk analysis, Health Phys. 59 (1990) 925–929.
[85] A.M. Müller, M. Davenport, S. Verrier, R. Droeser, M. Alini, C. Bocelli-Tyndall, D.J. Schaefer, I. Martin, A. Scherberich, Platelet lysate as a serum substitute for 2D static and 3D perfusion culture of stromal vascular fraction cells from human adipose tissue, Tissue Eng. A 15 (2009) 869–875.
[86] M.M. Asli, B. Pourdeyhimi, E.G. Loboa, Release profiles of tricalcium phosphate nanoparticles from poly(L-lactic acid) electrospun scaffolds with single component, core-sheath, or porous fiber morphologies: effects on hASC viability and osteogenic differentiation, Macromol. Biosci. 12 (2012) 893–900.

[87] S.D. McCullen, Y. Zhu, S.H. Bernacki, R.J. Narayan, B. Pourdeyhimi, R.E. Gorga, E.G. Loboa, Electrospun composite poly(L-lactic acid)/tricalcium phosphate scaffolds induce proliferation and osteogenic differentiation of human adipose-derived stem cells, Biomed. Mater. 4 (2009) 035002.
[88] M. Barba, G. Di Taranto, W. Lattanzi, Adipose-derived stem cell therapies for bone regeneration, Expert. Opin. Biol. Ther. 17 (6) (2017) 677–689.
[89] D. Wang, J.R. Gilbert, X. Zhang, B. Zhao, D.F.E. Ker, G.M. Cooper, Calvarial versus Long bone: implications for tailoring skeletal tissue engineering, Tissue Eng. Part B Rev. 26 (1) (2020) 46–63.
[90] J.A. Lee, B.M. Parrett, J.A. Conejero, J. Laser, J. Chen, A.J. Kogon, D. Nanda, R.T. Grant, A.S. Breitbart, Biological alchemy: engineering bone and fat from fat-derived stem cells, Ann. Plast. Surg. 50 (2003) 610–617.
[91] D.T. Leong, M.C. Abraham, S.N. Rath, T.-C. Lim, F.T. Chew, D.W. Hutmacher, Investigating the effects of preinduction on human adipose-derived precursor cells in an athymic rat model, Differentiation 74 (2006) 519–529.
[92] E.H. Schemitsch, Size matters: defining critical in bone defect size! J. Orthop. Trauma 31 (Suppl. 5) (2017) S20–S22.
[93] R.J. Kroeze, T.H. Smit, P.P. Vergroesen, R.A. Bank, R. Stoop, B. van Rietbergen, B.J. van Royen, M.N. Helder, Spinal fusion using adipose stem cells seeded on a radiolucent cage filler: a feasibility study of a single surgical procedure in goats, Eur. Spine J. 24 (5) (2015) 1031–1042.
[94] C.M. Cowan, Y.Y. Shi, O.O. Aalami, Y.F. Chou, C. Mari, R. Thomas, N. Quarto, C.H. Contag, B. Wu, M.T. Longaker, Adipose-derived adult stromal cells heal critical-size mouse calvarial defects, Nat. Biotechnol. 22 (2004) 560–567.
[95] E. Yoon, S. Dhar, D.E. Chun, N.A. Gharibjanian, G.R.D. Evans, In vivo osteogenic potential of human adipose-derived stem cells/poly lactide-co-glycolic acid constructs for bone regeneration in a rat critical-sized calvarial defect model, Tissue Eng. 13 (2007) 619–627.
[96] B. Levi, A.W. James, E.R. Nelson, D. Vistnes, B. Wu, M. Lee, A. Gupta, M.T. Longaker, Human adipose derived stromal cells heal critical size mouse calvarial defects, PLoS One 5 (6) (2010) e11177.
[97] S. Gronthos, M. Mankani, J. Brahim, P.G. Robey, S. Shi, Postnatal human dental pulp stem cells (DPSCs) in vitro and in vivo, Proc. Natl Acad. Sci. USA 97 (25) (2000) 13625–13630.
[98] M. Miura, S. Gronthos, M. Zhao, B. Lu, L.W. Fisher, P.G. Robey, S. Shi, SHED: stem cells from human exfoliated deciduous teeth, Proc. Natl Acad. Sci. USA 100 (10) (2003) 5807–5812.
[99] X. Shi, J. Mao, Y. Liu, Concise review: pulp stem cells derived from human permanent and deciduous teeth: biological characteristics and therapeutic applications, Stem Cells Transl. Med. 9 (4) (2020) 445–464.
[100] L. Ma, Y. Makino, H. Yamaza, K. Akiyama, Y. Hoshino, G. Song, T. Kukita, K. Nonaka, S. Shi, T. Yamaza, Cryopreserved dental pulp tissues of exfoliated deciduous teeth is a feasible stem cell resource for regenerative medicine, PLoS One 7 (12) (2012) e51777.
[101] A. Graziano, R. d'Aquino, G. Laino, A. Proto, M.T. Giuliano, G. Pirozzi, A. De Rosa, D. Di Napoli, G. Papaccio, Human CD34+ stem cells produce bone nodules in vivo, Cell Prolif. 41 (1) (2008) 1–11.
[102] T. Yamaza, A. Kentaro, C. Chen, Y. Liu, Y. Shi, S. Gronthos, S. Wang, S. Shi, Immunomodulatory properties of stem cells from human exfoliated deciduous teeth, Stem Cell Res Ther 1 (1) (2010) 5.
[103] X. Wang, X.J. Sha, G.H. Li, F.S. Yang, K. Ji, L.Y. Wen, S.Y. Liu, L. Chen, Y. Ding, K. Xuan, Comparative characterization of stem cells from human exfoliated deciduous teeth and dental pulp stem cells, Arch. Oral Biol. 57 (9) (2012) 1231–1240.
[104] F. Xie, J. He, Y. Chen, Z. Hu, M. Qin, T. Hui, Multi-lineage differentiation and clinical application of stem cells from exfoliated deciduous teeth, Hum. Cell 33 (2) (2020) 295–302.
[105] B.M. Seo, W. Sonoyama, T. Yamaza, C. Coppe, T. Kikuiri, K. Akiyama, J.S. Lee, S. Shi, SHED repair critical-size calvarial defects in mice, Oral Dis. 14 (5) (2008) 428–434.
[106] C.A. de Mendonca, D.F. Bueno, M.T. Martins, I. Kerkis, A. Kerkis, R.D. Fanganiello, H. Cerruti, N. Alonso, M.R. Passos-Bueno, Reconstruction of large cranial defects in nonimmunosuppressed experimental design with human dental pulp stem cells, J. Craniofac. Surg. 19 (1) (2008) 204–210.
[107] A. Novais, J. Lesieur, J. Sadoine, L. Slimani, B. Baroukh, B. Saubamea, A. Schmitt, S. Vital, A. Poliard, C. Helary, G.Y. Rochefort, C. Chaussain, C. Gorin, Priming dental pulp stem cells from human exfoliated deciduous teeth with fibroblast growth factor-2 enhances mineralization within tissue-engineered constructs implanted in craniofacial bone defects, Stem Cells Transl. Med. 8 (8) (2019) 844–857.
[108] R. d'Aquino, A. De Rosa, V. Lanza, V. Tirino, L. Laino, A. Graziano, V. Desiderio, G. Laino, G. Papaccio, Human mandible bone defect repair by the grafting of dental pulp stem/progenitor cells and collagen sponge biocomplexes, Eur. Cell. Mater. 18 (2009) 75–83.

[109] A. Giuliani, A. Manescu, M. Langer, F. Rustichelli, V. Desiderio, F. Paino, A. De Rosa, L. Laino, R. d'Aquino, V. Tirino, G. Papaccio, Three years after transplants in human mandibles, histological and in-line holotomography revealed that stem cells regenerated a compact rather than a spongy bone: biological and clinical implications, Stem Cells Transl. Med. 2 (4) (2013) 316–324.

[110] F.A.A. Ishiy, R.D. Fanganiello, G.S. Kobayashi, E. Kague, P.S. Kuriki, M.R. Passos-Bueno, CD105 is regulated by hsa-miR-1287 and its expression is inversely correlated with osteopotential in SHED, Bone 106 (2018) 112–120.

[111] B. Levi, D.C. Wan, J.P. Glotzbach, J. Hyun, M. Januszyk, D. Montoro, M. Sorkin, A.W. James, E.R. Nelson, S. Li, N. Quarto, M. Lee, G.C. Gurtner, M.T. Longaker, CD105 protein depletion enhances human adipose-derived stromal cell osteogenesis through reduction of transforming growth factor beta1 (TGF-beta1) signaling, J. Biol. Chem. 286 (45) (2011) 39497–39509.

[112] M.T. Chung, C. Liu, J.S. Hyun, D.D. Lo, D.T. Montoro, M. Hasegawa, S. Li, M. Sorkin, R. Rennert, M. Keeney, F. Yang, N. Quarto, M.T. Longaker, D.C. Wan, CD90 (Thy-1)-positive selection enhances osteogenic capacity of human adipose-derived stromal cells, Tissue Eng. Part A 19 (7–8) (2013) 989–997.

[113] A. McArdle, M.T. Chung, K.J. Paik, C. Duldulao, C. Chan, R. Rennert, G.G. Walmsley, K. Senarath-Yapa, M. Hu, E. Seo, M. Lee, D.C. Wan, M.T. Longaker, Positive selection for bone morphogenetic protein receptor type-IB promotes differentiation and specification of human adipose-derived stromal cells toward an osteogenic lineage, Tissue Eng. Part A 20 (21–22) (2014) 3031–3040.

[114] H. Nokhbatolfoghahaei, M.R. Rad, M.-M. Khani, S. Shahriari, N. Nadjmi, A. Khojasteh, Application of bioreactors to improve functionality of bone tissue engineering constructs: a systematic review, Curr. Stem Cell Res. Ther. 12 (2017) 564–599.

[115] B. Birru, N.K. Mekala, S.R. Parcha, Mechanistic role of perfusion culture on bone regeneration, J. Biosci. 44 (1) (2019) 23.

[116] R.M. Delaine-Smith, G.C. Reilly, Mesenchymal stem cell responses to mechanical stimuli, Muscles Ligaments Tendons J. 2 (2012) 169–180.

[117] M. Jagodzinski, A. Breitbart, M. Wehmeier, E. Hesse, C. Haasper, C. Krettek, J. Zeichen, S. Hankemeier, Influence of perfusion and cyclic compression on proliferation and differentiation of bone marrow stromal cells in 3-dimensional culture, J. Biomech. 41 (9) (2008) 1885–1891.

[118] A. Scherberich, R. Galli, C. Jaquiery, J. Farhadi, I. Martin, Three-dimensional perfusion culture of human adipose tissue-derived endothelial and osteoblastic progenitors generates osteogenic constructs with intrinsic vascularization capacity, Stem Cells 25 (2007) 1823–1829.

[119] M. Lovett, K. Lee, A. Edwards, D.L. Kaplan, Vascularization strategies for tissue engineering, Tissue Eng. Part B Rev. 15 (3) (2009) 353–370.

[120] H. Yu, P.J. VandeVord, L. Mao, H.W. Matthew, P.H. Wooley, S.-Y. Yang, Improved tissue-engineered bone regeneration by endothelial cell mediated vascularization, Biomaterials 30 (2009) 508–517.

[121] F. Kawecki, T. Galbraith, W.P. Clafshenkel, M. Fortin, F.A. Auger, J. Fradette, In vitro prevascularization of self-assembled human bone-like tissues and preclinical assessment using a rat calvarial bone defect model, Materials 14 (8) (2021) 2023.

[122] M.W. Laschke, Y. Harder, M. Amon, I. Martin, J. Farhadi, A. Ring, N. Torio-Padron, R. Schramm, M. Rucker, D. Junker, J.M. Haufel, C. Carvalho, M. Heberer, G. Germann, B. Vollmar, M.D. Menger, Angiogenesis in tissue engineering: breathing life into constructed tissue substitutes, Tissue Eng. 12 (8) (2006) 2093–2104.

[123] M. Grellier, L. Bordenave, J. Amédée, Cell-to-cell communication between osteogenic and endothelial lineages: implications for tissue engineering, Trends Biotechnol. 27 (2009) 562–571.

[124] S.J. Marfy-Smith, C.E. Clarkin, Are mesenchymal stem cells so bloody great after all? Stem Cells Transl. Med. 6 (1) (2017) 3–6.

[125] F. Shapiro, Bone development and its relation to fracture repair. The role of mesenchymal osteoblasts and surface osteoblasts, Eur. Cell. Mater. 15 (2008) 53–76.

[126] I. Dumic-Cule, M. Peric, L. Kucko, L. Grgurevic, M. Pecina, S. Vukicevic, Bone morphogenetic proteins in fracture repair, Int. Orthop. 42 (11) (2018) 2619–2626.

[127] Y. Wang, M.R. Newman, D.S.W. Benoit, Development of controlled drug delivery systems for bone fracture-targeted therapeutic delivery: a review, Eur. J. Pharm. Biopharm. 127 (2018) 223–236.

[128] M.R. Urist, Bone: formation by autoinduction, Science 150 (3698) (1965) 893–899.

[129] T.K. Sampath, A.H. Reddi, Dissociative extraction and reconstitution of extracellular matrix components involved in local bone differentiation, Proc. Natl Acad. Sci. USA 78 (12) (1981) 7599–7603.

[130] T.K. Sampath, J.E. Coughlin, R.M. Whetstone, D. Banach, C. Corbett, R.J. Ridge, E. Ozkaynak, H. Oppermann, D.C. Rueger, Bovine osteogenic protein is composed of dimers of OP-1 and BMP-2A, two members of the transforming growth factor-beta superfamily, J. Biol. Chem. 265 (22) (1990) 13198–13205.
[131] A.J. Celeste, J.A. Iannazzi, R.C. Taylor, R.M. Hewick, V. Rosen, E.A. Wang, J.M. Wozney, Identification of transforming growth factor beta family members present in bone-inductive protein purified from bovine bone, Proc. Natl Acad. Sci. USA 87 (24) (1990) 9843–9847.
[132] E. Ozkaynak, D.C. Rueger, E.A. Drier, C. Corbett, R.J. Ridge, T.K. Sampath, H. Oppermann, OP-1 cDNA encodes an osteogenic protein in the TGF-beta family, EMBO J. 9 (7) (1990) 2085–2093.
[133] H.H. Luu, W.X. Song, X. Luo, D. Manning, J. Luo, Z.L. Deng, K.A. Sharff, A.G. Montag, R.C. Haydon, T.C. He, Distinct roles of bone morphogenetic proteins in osteogenic differentiation of mesenchymal stem cells, J. Orthop. Res. 25 (5) (2007) 665–677.
[134] U.A. Vitt, S.Y. Hsu, A.J. Hsueh, Evolution and classification of cystine knot-containing hormones and related extracellular signaling molecules, Mol. Endocrinol. 15 (5) (2001) 681–694.
[135] J. Ning, Y. Zhao, Y. Ye, J. Yu, Opposing roles and potential antagonistic mechanism between TGF-beta and BMP pathways: implications for cancer progression, EBioMedicine 41 (2019) 702–710.
[136] J.M. Wozney, V. Rosen, A.J. Celeste, L.M. Mitsock, M.J. Whitters, R.W. Kriz, R.M. Hewick, E.A. Wang, Novel regulators of bone formation: molecular clones and activities, Science 242 (4885) (1988) 1528–1534.
[137] D.M. Kingsley, The TGF-beta superfamily: new members, new receptors, and new genetic tests of function in different organisms, Genes Dev. 8 (2) (1994) 133–146.
[138] M.A. Brown, Q. Zhao, K.A. Baker, C. Naik, C. Chen, L. Pukac, M. Singh, T. Tsareva, Y. Parice, A. Mahoney, V. Roschke, I. Sanyal, S. Choe, Crystal structure of BMP-9 and functional interactions with pro-region and receptors, J. Biol. Chem. 280 (26) (2005) 25111–25118.
[139] L. Zhang, Q. Luo, Y. Shu, Z. Zeng, B. Huang, Y. Feng, B. Zhang, X. Wang, Y. Lei, Z. Ye, L. Zhao, D. Cao, L. Yang, X. Chen, B. Liu, W. Wagstaff, R.R. Reid, H.H. Luu, R.C. Haydon, M.J. Lee, J.M. Wolf, Z. Fu, T.-C. He, Q. Kang, Transcriptomic landscape regulated by the 14 types of bone morphogenetic proteins (BMPs) in lineage commitment and differentiation of mesenchymal stem cells (MSCs), Genes Dis. 6 (3) (2019) 258–275.
[140] E. Tillet, S. Bailly, Emerging roles of BMP9 and BMP10 in hereditary hemorrhagic telangiectasia, Front. Genet. 5 (2014) 456.
[141] C. Sieber, J. Kopf, C. Hiepen, P. Knaus, Recent advances in BMP receptor signaling, Cytokine Growth Factor Rev. 20 (5–6) (2009) 343–355.
[142] D.I. Israel, J. Nove, K.M. Kerns, R.J. Kaufman, V. Rosen, K.A. Cox, J.M. Wozney, Heterodimeric bone morphogenetic proteins show enhanced activity in vitro and in vivo, Growth Factors 13 (3–4) (1996) 291–300.
[143] H. Cheng, W. Jiang, F.M. Phillips, R.C. Haydon, Y. Peng, L. Zhou, H.H. Luu, N. An, B. Breyer, P. Vanichakarn, J.-P. Szatkowski, J.Y. Park, T.C. He, Osteogenic activity of the fourteen types of human bone morphogenetic proteins (BMPs), J. Bone Joint Surg. Am. 85 (8) (2003) 1544–1552.
[144] Q. Kang, M.H. Sun, H. Cheng, Y. Peng, A.G. Montag, A.T. Deyrup, W. Jiang, H.H. Luu, J. Luo, J.P. Szatkowski, P. Vanichakarn, J.Y. Park, Y. Li, R.C. Haydon, T.C. He, Characterization of the distinct orthotopic bone-forming activity of 14 BMPs using recombinant adenovirus-mediated gene delivery, Gene Ther. 11 (17) (2004) 1312–1320.
[145] S.D. Boden, K. McCuaig, G. Hair, M. Racine, L. Titus, J.M. Wozney, M.S. Nanes, Differential effects and glucocorticoid potentiation of bone morphogenetic protein action during rat osteoblast differentiation in vitro, Endocrinology 137 (8) (1996) 3401–3407.
[146] X. Cao, D. Chen, The BMP signaling and in vivo bone formation, Gene 357 (1) (2005) 1–8.
[147] J.A. King, P.C. Marker, K.J. Seung, D.M. Kingsley, BMP5 and the molecular, skeletal, and soft-tissue alterations in short ear mice, Dev. Biol. 166 (1) (1994) 112–122.
[148] H. Zhang, A. Bradley, Mice deficient for BMP2 are nonviable and have defects in amnion/chorion and cardiac development, Development 122 (10) (1996) 2977–2986.
[149] G. Winnier, M. Blessing, P.A. Labosky, B.L. Hogan, Bone morphogenetic protein-4 is required for mesoderm formation and patterning in the mouse, Genes Dev. 9 (17) (1995) 2105–2116.
[150] A.T. Dudley, K.M. Lyons, E.J. Robertson, A requirement for bone morphogenetic protein-7 during development of the mammalian kidney and eye, Genes Dev. 9 (22) (1995) 2795–2807.
[151] A. Daluiski, T. Engstrand, M.E. Bahamonde, L.W. Gamer, E. Agius, S.L. Stevenson, K. Cox, V. Rosen, K.M. Lyons, Bone morphogenetic protein-3 is a negative regulator of bone density, Nat. Genet. 27 (1) (2001) 84–88.

[152] J. Greenwald, J. Groppe, P. Gray, E. Wiater, W. Kwiatkowski, W. Vale, S. Choe, The BMP7/ActRII extracellular domain complex provides new insights into the cooperative nature of receptor assembly, Mol. Cell 11 (3) (2003) 605–617.

[153] A. Nohe, S. Hassel, M. Ehrlich, F. Neubauer, W. Sebald, Y.I. Henis, P. Knaus, The mode of bone morphogenetic protein (BMP) receptor oligomerization determines different BMP-2 signaling pathways, J. Biol. Chem. 277 (7) (2002) 5330–5338.

[154] M. Ehrlich, Endocytosis and trafficking of BMP receptors: regulatory mechanisms for fine-tuning the signaling response in different cellular contexts, Cytokine Growth Factor Rev. 27 (2016) 35–42.

[155] D. Yadin, P. Knaus, T.D. Mueller, Structural insights into BMP receptors: specificity, activation and inhibition, Cytokine Growth Factor Rev. 27 (2016) 13–34.

[156] B. Schmierer, C.S. Hill, TGFbeta-SMAD signal transduction: molecular specificity and functional flexibility, Nat. Rev. Mol. Cell Biol. 8 (12) (2007) 970–982.

[157] R.N. Wang, J. Green, Z. Wang, Y. Deng, M. Qiao, M. Peabody, Q. Zhang, J. Ye, Z. Yan, S. Denduluri, O. Idowu, M. Li, C. Shen, A. Hu, R.C. Haydon, R. Kang, J. Mok, M.J. Lee, H.L. Luu, L.L. Shi, Bone morphogenetic protein (BMP) signaling in development and human diseases, Genes Dis. 1 (1) (2014) 87–105.

[158] D.P. Brazil, R.H. Church, S. Surae, C. Godson, F. Martin, BMP signalling: agony and antagony in the family, Trends Cell Biol. 25 (5) (2015) 249–264.

[159] M. Karbiener, C. Neuhold, P. Opriessnig, A. Prokesch, J.G. Bogner-Strauss, M. Scheideler, MicroRNA-30c promotes human adipocyte differentiation and co-represses PAI-1 and ALK2, RNA Biol. 8 (5) (2011) 850–860.

[160] A.C. Cruz, M.L. Silva, T. Caon, C.M. Simoes, Addition of bone morphogenetic protein type 2 to ascorbate and beta-glycerophosphate supplementation did not enhance osteogenic differentiation of human adipose-derived stem cells, J. Appl. Oral Sci. 20 (6) (2012) 628–635.

[161] G. Luther, E.R. Wagner, G. Zhu, Q. Kang, Q. Luo, J. Lamplot, Y. Bi, X. Luo, J. Luo, C. Teven, Q. Shi, S.H. Kim, J.L. Gao, E. Huang, K. Yang, R. Rames, X. Liu, M. Li, N. Hu, H. Liu, Y. Su, L. Chen, B.C. He, G.W. Zuo, Z.L. Deng, R.-R. Reid, H.H. Luu, R.C. Haydon, T.C. He, BMP-9 induced osteogenic differentiation of mesenchymal stem cells: molecular mechanism and therapeutic potential, Curr. Gene Ther. 11 (3) (2011) 229–240.

[162] K. Heinecke, A. Seher, W. Schmitz, T.D. Mueller, W. Sebald, J. Nickel, Receptor oligomerization and beyond: a case study in bone morphogenetic proteins, BMC Biol. 7 (2009) 59.

[163] J. Zhang, L. Li, BMP signaling and stem cell regulation, Dev. Biol. 284 (1) (2005) 1–11.

[164] J. Massague, J. Seoane, D. Wotton, Smad transcription factors, Genes Dev. 19 (23) (2005) 2783–2810.

[165] T. Matsubara, K. Kida, A. Yamaguchi, K. Hata, F. Ichida, H. Meguro, H. Aburatani, R. Nishimura, T. Yoneda, BMP2 regulates osterix through Msx2 and Runx2 during osteoblast differentiation, J. Biol. Chem. 283 (43) (2008) 29119–29125.

[166] J.A. McMahon, S. Takada, L.B. Zimmerman, C.M. Fan, R.M. Harland, A.P. McMahon, Noggin-mediated antagonism of BMP signaling is required for growth and patterning of the neural tube and somite, Genes Dev. 12 (10) (1998) 1438–1452.

[167] P. Seemann, A. Brehm, J. Konig, C. Reissner, S. Stricker, P. Kuss, J. Haupt, S. Renninger, J. Nickel, W. Sebald, J.C. Groppe, F. Ploger, J. Pohl, M. Schmidt-von Kegler, M. Walther, I. Gassner, C. Rusu, A.R. Janecke, K. Dathe, S. Mundlos, Mutations in GDF5 reveal a key residue mediating BMP inhibition by NOGGIN, PLoS Genet. 5 (11) (2009) e1000747.

[168] F. Kugimiya, S. Ohba, K. Nakamura, H. Kawaguchi, U.I. Chung, Physiological role of bone morphogenetic proteins in osteogenesis, J. Bone Miner. Metab. 24 (2) (2006) 95–99.

[169] A. Hata, G. Lagna, J. Massague, A. Hemmati-Brivanlou, Smad6 inhibits BMP/Smad1 signaling by specifically competing with the Smad4 tumor suppressor, Genes Dev. 12 (2) (1998) 186–197.

[170] S. Kokabu, T. Katagiri, T. Yoda, V. Rosen, Role of Smad phosphatases in BMP-Smad signaling axis-induced osteoblast differentiation, J. Oral Biosci. 54 (2) (2012) 73–78.

[171] M. Knockaert, G. Sapkota, C. Alarcon, J. Massague, A.H. Brivanlou, Unique players in the BMP pathway: small C-terminal domain phosphatases dephosphorylate Smad1 to attenuate BMP signaling, Proc. Natl Acad. Sci. USA 103 (32) (2006) 11940–11945.

[172] X. Duan, Y.Y. Liang, X.H. Feng, X. Lin, Protein serine/threonine phosphatase PPM1A dephosphorylates Smad1 in the bone morphogenetic protein signaling pathway, J. Biol. Chem. 281 (48) (2006) 36526–36532.

[173] H. Zhu, P. Kavsak, S. Abdollah, J.L. Wrana, G.H. Thomsen, A SMAD ubiquitin ligase targets the BMP pathway and affects embryonic pattern formation, Nature 400 (6745) (1999) 687–693.

[174] K. Murakami, J.D. Etlinger, Role of SMURF1 ubiquitin ligase in BMP receptor trafficking and signaling, Cell. Signal. 54 (2019) 139–149.
[175] H. Gu, Z. Huang, X. Yin, J. Zhang, L. Gong, J. Chen, K. Rong, J. Xu, L. Lu, L. Cui, Role of c-Jun N-terminal kinase in the osteogenic and adipogenic differentiation of human adipose-derived mesenchymal stem cells, Exp. Cell Res. 339 (1) (2015) 112–121.
[176] M. Kretzschmar, J. Doody, J. Massague, Opposing BMP and EGF signalling pathways converge on the TGF-beta family mediator Smad1, Nature 389 (6651) (1997) 618–622.
[177] N. Tang, W.X. Song, J. Luo, X. Luo, J. Chen, K.A. Sharff, Y. Bi, B.C. He, J.Y. Huang, G.H. Zhu, Y.X. Su, W. Jiang, M. Tang, Y. He, Y. Wang, L. Chen, G.W. Zuo, J. Shen, X. Pan, R.R. Reid, H.H. Luu, R.C. Haydon, T.C. He, BMP-9-induced osteogenic differentiation of mesenchymal progenitors requires functional canonical Wnt/beta-catenin signalling, J. Cell. Mol. Med. 13 (8B) (2009) 2448–2464.
[178] C. Yuan, X. Gou, J. Deng, Z. Dong, P. Ye, Z. Hu, FAK and BMP-9 synergistically trigger osteogenic differentiation and bone formation of adipose derived stem cells through enhancing Wnt-beta-catenin signaling, Biomed. Pharmacother. 105 (2018) 753–757.
[179] T. Kubota, T. Michigami, K. Ozono, Wnt signaling in bone metabolism, J. Bone Miner. Metab. 27 (3) (2009) 265–271.
[180] M.A. Lauzon, E. Bergeron, B. Marcos, N. Faucheux, Bone repair: new developments in growth factor delivery systems and their mathematical modeling, J. Control. Release 162 (3) (2012) 502–520.
[181] A. Schindeler, M.M. McDonald, P. Bokko, D.G. Little, Bone remodeling during fracture repair: the cellular picture, Semin. Cell Dev. Biol. 19 (5) (2008) 459–466.
[182] J.A. Cottrell, J.C. Turner, T.L. Arinzeh, J.P. O'Connor, The biology of bone and ligament healing, Foot Ankle Clin. 21 (4) (2016) 739–761.
[183] D. Shirley, D. Marsh, G. Jordan, S. McQuaid, G. Li, Systemic recruitment of osteoblastic cells in fracture healing, J. Orthop. Res. 23 (5) (2005) 1013–1021.
[184] B. Shen, A. Wei, H. Tao, A.D. Diwan, D.D. Ma, BMP-2 enhances TGF-beta3-mediated chondrogenic differentiation of human bone marrow multipotent mesenchymal stromal cells in alginate bead culture, Tissue Eng. Part A 15 (6) (2009) 1311–1320.
[185] X. Du, Y. Xie, C.J. Xian, L. Chen, Role of FGFs/FGFRs in skeletal development and bone regeneration, J. Cell. Physiol. 227 (12) (2012) 3731–3743.
[186] C.M. Champagne, J. Takebe, S. Offenbacher, L.F. Cooper, Macrophage cell lines produce osteoinductive signals that include bone morphogenetic protein-2, Bone 30 (1) (2002) 26–31.
[187] Z.S. Ai-Aql, A.S. Alagl, D.T. Graves, L.C. Gerstenfeld, T.A. Einhorn, Molecular mechanisms controlling bone formation during fracture healing and distraction osteogenesis, J. Dent. Res. 87 (2) (2008) 107–118.
[188] S. Rasi Ghaemi, B. Delalat, X. Ceto, F.J. Harding, J. Tuke, N.H. Voelcker, Synergistic influence of collagen I and BMP 2 drives osteogenic differentiation of mesenchymal stem cells: a cell microarray analysis, Acta Biomater. 34 (2016) 41–52.
[189] M. Murata, B.-Z. Huang, T. Shibata, S. Imai, N. Nagal, M. Arisue, Bone augmentation by recombinant human BMP-2 and collagen on adult rat parietal bone, Int. J. Oral Maxillofac. Surg. 28 (3) (1999) 232–237.
[190] Y.E. Antebi, J.M. Linton, H. Klumpe, B. Bintu, M. Gong, C. Su, R. McCardell, M.B. Elowitz, Combinatorial signal perception in the BMP pathway, Cell 170 (6) (2017) 1184–1196. e1124.
[191] T.J. Cho, L.C. Gerstenfeld, T.A. Einhorn, Differential temporal expression of members of the transforming growth factor beta superfamily during murine fracture healing, J. Bone Miner. Res. 17 (3) (2002) 513–520.
[192] J.A. Cottrell, O. Keane, S. Sutton Lin, J.P. O'Connor, BMP-2 modulates expression of other growth factors in a rat fracture healing model, J. Appl. Biomed. 12 (3) (2014) 127–135.
[193] I. Song, B.S. Kim, C.S. Kim, G.I. Im, Effects of BMP-2 and vitamin D3 on the osteogenic differentiation of adipose stem cells, Biochem. Biophys. Res. Commun. 408 (1) (2011) 126–131.
[194] S. Banka, Y. Mukudai, Y. Yoshihama, T. Shirota, S. Kondo, S. Shintani, A combination of chemical and mechanical stimuli enhances not only osteo- but also chondro-differentiation in adipose-derived stem cells, J. Oral Biosci. 54 (4) (2012) 188–195.
[195] Y.H. Liao, Y.H. Chang, L.Y. Sung, K.C. Li, C.L. Yeh, T.C. Yen, S.M. Hwang, K.J. Lin, Y.C. Hu, Osteogenic differentiation of adipose-derived stem cells and calvarial defect repair using baculovirus-mediated co-expression of BMP-2 and miR-148b, Biomaterials 35 (18) (2014) 4901–4910.
[196] B. Khorsand, S. Elangovan, L. Hong, A. Dewerth, M.S. Kormann, A.K. Salem, A comparative study of the bone regenerative effect of chemically modified RNA encoding BMP-2 or BMP-9, AAPS J. 19 (2) (2017) 438–446.

[197] R. Yanai, F. Tetsuo, S. Ito, M. Itsumi, J. Yoshizumi, T. Maki, Y. Mori, Y. Kubota, S. Kajioka, Extracellular calcium stimulates osteogenic differentiation of human adipose-derived stem cells by enhancing bone morphogenetic protein-2 expression, Cell Calcium 83 (2019) 102058.
[198] H. Li, K. Dai, T. Tang, X. Zhang, M. Yan, J. Lou, Bone regeneration by implantation of adipose-derived stromal cells expressing BMP-2, Biochem. Biophys. Res. Commun. 356 (4) (2007) 836–842.
[199] M. Yang, Q.J. Ma, G.T. Dang, K. Ma, P. Chen, C.Y. Zhou, In vitro and in vivo induction of bone formation based on ex vivo gene therapy using rat adipose-derived adult stem cells expressing BMP-7, Cytotherapy 7 (3) (2005) 273–281.
[200] O. Mizrahi, D. Sheyn, W. Tawackoli, I. Kallai, A. Oh, S. Su, X. Da, P. Zarrini, G. Cook-Wiens, D. Gazit, Z. Gazit, BMP-6 is more efficient in bone formation than BMP-2 when overexpressed in mesenchymal stem cells, Gene Ther. 20 (4) (2013) 370–377.
[201] M. Kuterbekov, P. Machillot, F. Baillet, A.M. Jonas, K. Glinel, C. Picart, Design of experiments to assess the effect of culture parameters on the osteogenic differentiation of human adipose stromal cells, Stem Cell Res Ther 10 (1) (2019) 256.
[202] G. Drouin, V. Couture, M.A. Lauzon, F. Balg, N. Faucheux, G. Grenier, Muscle injury-induced hypoxia alters the proliferation and differentiation potentials of muscle resident stromal cells, Skelet. Muscle 9 (1) (2019) 18.
[203] M.A. Lauzon, A. Daviau, O. Drevelle, B. Marcos, N. Faucheux, Identification of a growth factor mimicking the synergistic effect of fetal bovine serum on BMP-9 cell response, Tissue Eng. Part A 20 (17–18) (2014) 2524–2535.
[204] Y. Acil, A.A. Ghoniem, A. Gulses, T. Kisch, F. Stang, J. Wiltfang, M. Gierloff, Suppression of osteoblast-related genes during osteogenic differentiation of adipose tissue derived stromal cells, J. Craniomaxillofac. Surg. 45 (1) (2017) 33–38.
[205] C.F. Markarian, G.Z. Frey, M.D. Silveira, E.M. Chem, A.R. Milani, P.B. Ely, A.P. Horn, N.B. Nardi, M. Camassola, Isolation of adipose-derived stem cells: a comparison among different methods, Biotechnol. Lett. 36 (4) (2014) 693–702.
[206] J.C. Reichert, S. Saifzadeh, M.E. Wullschleger, D.R. Epari, M.A. Schütz, G.N. Duda, H. Schell, M. van Griensven, H. Redl, D.W. Hutmacher, The challenge of establishing preclinical models for segmental bone defect research, Biomaterials 30 (2009) 2149–2163.
[207] A. Chatterjea, G. Meijer, C. Van Blitterswijk, J. De Boer, Clinical application of human mesenchymal stromal cells for bone tissue engineering, Stem Cells Int. 2010 (2010) 215625.
[208] R. Quarto, M. Mastrogiacomo, R. Cancedda, S.M. Kutepov, V. Mukhachev, A. Lavroukov, E. Kon, M. Marcacci, Repair of large bone defects with the use of autologous bone marrow stromal cells, N. Engl. J. Med. 344 (2001) 385–386.
[209] G.J. Meijer, J.D. de Bruijn, R. Koole, C.A. van Blitterswijk, Cell based bone tissue engineering in jaw defects, Biomaterials 29 (2008) 3053–3061.
[210] S. Lendeckel, A. Jödicke, P. Christophis, K. Heidinger, J. Wolff, J.K. Fraser, M.H. Hedrick, L. Berthold, H.-P. Howaldt, Autologous stem cells (adipose) and fibrin glue used to treat widespread traumatic calvarial defects: case report, J. Craniomaxillofac. Surg. 32 (2004) 370–373.
[211] L. Sensebé, M. Krampera, H. Schrezenmeier, P. Bourin, R. Giordano, Mesenchymal stem cells for clinical application, Vox Sang. 98 (2010) 93–107.
[212] K. Mesimäki, B. Lindroos, J. Törnwall, J. Mauno, C. Lindqvist, R. Kontio, S. Miettinen, R. Suuronen, Novel maxillary reconstruction with ectopic bone formation by GMP adipose stem cells, Int. J. Oral Maxillofac. Surg. 38 (2009) 201–209.
[213] A. Khojasteh, L. Kheiri, H. Behnia, A. Tehranchi, P. Nazeman, N. Nadjmi, M. Soleimani, Lateral ramus cortical bone plate in alveolar cleft osteoplasty with concomitant use of buccal fat pad derived cells and autogenous bone: phase I clinical trial, Biomed. Res. Int. 2017 (2017) 6560234.
[214] T. Thesleff, K. Lehtimäki, T. Niskakangas, B. Mannerström, S. Miettinen, R. Suuronen, J. Öhman, Cranioplasty with adipose-derived stem cells and biomaterial: a novel method for cranial reconstruction, Neurosurgery 68 (2011) 1535–1540.
[215] G.K. Sándor, J. Numminen, J. Wolff, T. Thesleff, A. Miettinen, V.J. Tuovinen, B. Mannerström, M. Patrikoski, R. Seppänen, S. Miettinen, M. Rautiainen, J. Öhman, Adipose stem cells used to reconstruct 13 cases with cranio-maxillofacial hard-tissue defects, Stem Cells Transl. Med. 3 (2014) 530–540.
[216] T. Thesleff, K. Lehtimäki, T. Niskakangas, S. Huovinen, B. Mannerström, S. Miettinen, R. Seppänen-Kaijansinkko, J. Öhman, Cranioplasty with adipose-derived stem cells, beta-tricalcium phosphate granules and supporting mesh: six-year clinical follow-up results, Stem Cells Transl. Med. 6 (2017) 1576–1582.

[217] G.K. Sándor, V.J. Tuovinen, J. Wolff, M. Patrikoski, J. Jokinen, E. Nieminen, B. Mannerström, O.-P. Lappalainen, R. Seppänen, S. Miettinen, Adipose stem cell tissue-engineered construct used to treat large anterior mandibular defect: a case report and review of the clinical application of good manufacturing practice-level adipose stem cells for bone regeneration, J. Oral Maxillofac. Surg. 71 (2013) 938–950.

[218] J. Wolff, G.K. Sándor, A. Miettinen, V.J. Tuovinen, B. Mannerström, M. Patrikoski, S. Miettinen, GMP-level adipose stem cells combined with computer-aided manufacturing to reconstruct mandibular ameloblastoma resection defects: experience with three cases, Ann. Maxillofac. Surg. 3 (2013) 114–125.

[219] G.K.B. Sándor, Tissue engineering of bone: clinical observations with adipose-derived stem cells, resorbable scaffolds, and growth factors, Ann. Maxillofac. Surg. 2 (2012) 8–11.

[220] Z. Wan, P. Zhang, Y. Liu, L. Lv, Y. Zhou, Four-dimensional bioprinting: current developments and applications in bone tissue engineering, Acta Biomater. 101 (2020) 26–42.

[221] K. Hölzl, S. Lin, L. Tytgat, S. Van Vlierberghe, L. Gu, A. Ovsianikov, Bioink properties before, during and after 3D bioprinting, Biofabrication 8 (2016) 032002.

[222] D. Choudhury, S. Anand, M.W. Naing, The arrival of commercial bioprinters—towards 3D bioprinting revolution! Int. J. Bioprinting 4 (2) (2018) 139.

[223] S.V. Murphy, A. Atala, 3D bioprinting of tissues and organs, Nat. Biotechnol. 32 (2014) 773–785.

[224] V. Keriquel, F. Guillemot, I. Arnault, B. Guillotin, S. Miraux, J. Amédée, J.-C. Fricain, S. Catros, In vivo bioprinting for computer- and robotic-assisted medical intervention: preliminary study in mice, Biofabrication 2 (2010) 014101.

[225] C. Mandrycky, Z. Wang, K. Kim, D.H. Kim, 3D bioprinting for engineering complex tissues, Biotechnol. Adv. 34 (2016) 422–434.

[226] B.P. Hung, B.A. Naved, E.L. Nyberg, M. Dias, C.A. Holmes, J.H. Elisseeff, A.H. Dorafshar, W.L. Grayson, Three-dimensional printing of bone extracellular matrix for craniofacial regeneration, ACS Biomater. Sci. Eng. 2 (2016) 1806–1816.

[227] V. Keriquel, H. Oliveira, M. Rémy, S. Ziane, S. Delmond, B. Rousseau, S. Rey, S. Catros, J. Amédée, F. Guillemot, J.-C. Fricain, In situ printing of mesenchymal stromal cells, by laser-assisted bioprinting, for in vivo bone regeneration applications, Sci. Rep. 7 (2017) 1778.

[228] S.V. Murphy, P. De Coppi, A. Atala, Opportunities and challenges of translational 3D bioprinting, Nat. Biomed. Eng. 4 (4) (2020) 370–380.

[229] W. Sun, P. Lal, Recent development on computer aided tissue engineering—a review, Comput. Methods Prog. Biomed. 67 (2) (2002) 85–103.

CHAPTER

18

The hematopoietic potential of stem cells from the adipose tissue

Béatrice Cousin and Louis Casteilla

RESTORE, Toulouse University, INSERM-1301, CNRS-5070, EFS, ENVT, University Toulouse III - Paul Sabatier (UPS), Toulouse, France

Introduction: adipose tissue as a reservoir for stromal and hematopoietic stem cells

AT was long considered merely as a filler tissue. The last decades have challenged this traditional concept of AT, transforming it from an inert storage depot into an important endocrine organ, secreting a number of adipokines and playing a pivotal role in controlling whole-body energy homeostasis. It has thus been largely studied for its involvement in metabolic diseases such as diabetes and obesity, considered as one of the most serious public health problems of the 21st century. In parallel, the use of AT transfer in plastic and reconstructive surgery is not new and has been the subject of numerous studies. Nowadays, AT engineering is a fast-developing field, both in terms of fundamental research and medical applications, addressing issues related to current clinical pathology or trauma management of soft tissue injuries in different body locations.

AT is heterogeneous, containing a stroma-vascular fraction (SVF) and specific mature adipocytes, and is present in adult mammals, including in humans. Alongside mature adipocytes, abundant and heterogeneous cell populations including endothelial cells, adipose stromal/stem cells (ASCs), and immune cells constitute the AT-SVF that has been largely studied. Indeed, adipocyte progenitors, firstly named preadipocytes, are present throughout adult life. They can proliferate and/or be recruited according to physiological or pathological situations and participate in the turnover of adipocytes. Recently, these progenitors have been renamed adipose stromal/stem cells, as they are able to give rise not only to adipocytes but also to osteoblasts, chondrocytes, and endothelial-like cells [1]. It must be noted that AT is very abundant in the body and easy to sample by liposuction. Additionally, the frequency of ASC in AT is about 100 to 500 times higher than that of MSC in the BM. This opened

Scientific Principles of Adipose Stem Cells
https://doi.org/10.1016/B978-0-12-819376-1.00006-8

intensive research in the regenerative field. Beside the therapeutic uses of ASC, questions on the physiological relevance of the presence of such cells in AT and a putative new role of AT as a physiological reservoir of regenerative cells are being raised.

In addition, AT hosts a large population of immune cells that strongly impacts AT physiology and its dysfunctions [2, 3], as well as some of their progenitors [4]. This raises the question about a putative immune/hematopoietic role of AT in mammals, as demonstrated in the fat body of insect [5]. Indeed, in *Drosophila*, the fat body assumes the functions of mammalian homologs of the liver and the AT, as well as the hematopoietic and immune systems, suggesting that AT in mammals may have evolved from an ancestral structure and retained hematopoietic/immune characteristics. Moreover, spontaneous emergence of functional cardiomyocytes in SVF cell cultures has led to the identification of cardiogenic progenitors that display electrophysiological properties and express cardiac-specific transcription factors [6]. Altogether, these results suggest the presence of functional immature cells with striking properties in AT.

Adipose-derived stromal stem cells

Definition and comparison with the bone marrow mesenchymal stem cells

ASCs belong to the large family of mesenchymal stem cells (MSCs) with some specificities. MSCs were firstly identified and characterized in bone marrow [7]. Both MSCs and ASCs represent a population of "non-conventional" stem cells that is a two-faced entity, being stem cells for mesenchymal tissues (e.g., bone, cartilage or AT) and cells of niches for others stem cells (e.g., HSCs). The term "stem cell" seems inappropriate since the self-renewal properties of these cells have not yet been fully established [8]. Although this point is still debated, one of the features of ASCs is their ability to form a fibroblastic colony-forming unit (CFU-F), showing that this population contains progenitor cells [9]. ASCs are also considered as supporting cells through their high secretory and immunomodulatory properties [10], but several differences have been described at genomic, proteomic, and functional levels [11, 12]. Human ASCs and BM-MSCs display distinct immunophenotypes based on cell surface expression and intensity [13]. For instance, in contrast to BM-MSCs, native ASCs highly express the cell surface marker CD34. In contrast to BM-MSC, most of the ASCs do not display pericytic position or express in vivo pericytic markers such as NG2, CD140b, or alpha-smooth muscle actin, which appear during the culture process [14, 15]. However, as this population is heterogeneous, a subpopulation of ASCs has been shown to be in pericytic position [15].

Concerning their differentiation potentials, BM-MSCs are probably more committed towards osteoblastic and chondrogenic lineages, whereas the adipogenic and angiogenic potentials are higher in ASCs than in MSCs [16, 17].

Supportive function for hematopoiesis

Several lines of evidence indicate that medullar MSCs are key components of the HSC niche in the BM [18, 19]. The hematopoietic niche provides a spatially limited environment for HSCs that regulates their activity via a combination of soluble and adhesion factors

[20]. In the niche, BM-MSCs are precursors of cellular components (such as osteoblasts and adipocytes), and they secrete extracellular matrix proteins of the BM stroma, thus providing a suitable microenvironment to support HSC function and control HSC proliferation and differentiation.

MSCs from other tissues such as AT have also been described to provide the support of hematopoietic stem and progenitor cells (HSPCs) and simulate the BM niche conditions, although the mechanisms are not yet fully understood, and functional differences between BM-MSC and ASCs have been described. In vitro, the hematopoietic supportive ability of stromal cells was first demonstrated in cell lines such as fibroblastic C3H10T1/2 cultured in adipogenic conditions. Coculture of murine HSC with these cell lines in the preadipocyte stage led to a marked increase in granulocyte–macrophage progenitor number compared with cocultures on fibroblasts. Although the mechanism was not detailed, it was proposed that the preadipose feeder layer promotes the proliferation and the differentiation of HSC through the production of matrix compounds necessary for cell-to-cell interactions and the secretion of growth factors and cytokines such as IL6, IL7, TGF, G-CSF, M-CSF, and GM-CSF, although none of the cell lines alone was able to secrete all of these factors [21–23]. When these stromal cells were differentiated into adipocytes, spontaneously or under the influence of adipogenic compounds, the expression of the extracellular matrix and cytokines was often modified and affected hematopoiesis. In particular, the production of factors that affect very immature progenitors, such as CSF-1, IL6 and LIF, was decreased.

These observations were then extended by using both rodent and human ASCs. In human, ASCs from subcutaneous fat support the in vitro complete differentiation of hematopoietic progenitors along myeloid and lymphoid lineages, and promote in vivo grafting of hematopoietic progenitors in lethally irradiated mice [24–28]. This hematopoietic supportive potential was attributed to a constant and abundant secretion of a large range of hematopoietic factors (granulocyte/monocyte, granulocyte, and macrophage colony stimulating factors, interleukin 6, 7, and 8, SDF1, stem cell factor [SCF]) that support proliferation, differentiation, and survival of HSCs [25, 29, 30]. In addition, recent studies revealed that coculture of human ASCs with cord blood hematopoietic progenitors leads to an upregulation of adhesion molecules and a downregulation of genes encoding the majority of extracellular matrix components and cell-matrix adhesion molecules in ASCs [31]. Interestingly, as for BM-MSCs [32], oxygen tension modifies HSC expansion in cocultures with ASCs, through modifications in the expression of cytokine, adhesion molecules, and remodeling proteins, with low O_2 levels inducing increased expansion of primitive HSCs [28, 33].

The molecular mechanisms underlying the hematopoietic supporting activity have not been clearly identified in the AT. However, ASCs exhibit a perivascular phenotype in vitro and in vivo and express significant levels of CXCL12 [34, 35], suggesting that this factor could be one of the mediators of hematopoiesis support activity, since CXCR4-CXCL12 signaling has been shown to play a pivotal role in the BM, in the regulation of HSPC homing and subsequent engraftment [36–38].

ASCs injected alone or with a scaffold have also been used as therapeutic agents to promote endogenous hematopoiesis. ASCs exert a myeloprotective effect via the secretion of the antiinflammatory factor TSG-6 (TNF-stimulated gene 6 that is a component of a negative feedback loop capable of downregulating the inflammatory response), by fully restoring the number of HSCs previously decreased upon cigarette smoking exposure, although ASCs do not localize in the BM [39]. Cultured on a hydroxyapatite scaffold and implanted in vivo, ASCs promote the regeneration of the BM and support HSC populations, suggesting that a functional BM niche is reconstituted [40].

The comparison of the hematopoietic supportive ability of ASCs and BM-MSCs has led to apparent conflicting results. In vitro cocultures and progenitor assays showed that ASCs were more efficient than BM-MSCs to support hematopoietic progenitor formation [26, 34, 41], whereas maintenance of a primitive HSC phenotype was significantly higher on BM-MSCs in comparison with ASCs [25, 42]. In addition, coinjection of murine ASCs instead of murine BM-MSCs together with human HSC in vivo improves mice survival [26, 34]. It thus seems that, if MSCs remain the uncontested leader in supporting primary hematopoiesis, ASCs appear to be more efficient in maintaining and promoting the differentiation of precursor cells, already engaged in a terminal differentiation pathway. It has been proposed that cytokine and chemokine secretion of the stromal feeder layers might account for the differential supportive hematopoietic activity between ASCs and BM-MSCs. However, even though some discrete differences in secretome profiles have been identified, none of them was clearly correlated with a specific potential to maintain primitive function of HSCs [43].

Adipose stromal cells as ancestral immune cells

The immunomodulatory properties of BM-MSCs have been extensively described. Indeed, in response to inflammatory molecules, MSCs produce immunomodulatory factors that reduce the progression of inflammation and thus help to prepare the microenvironment for tissue repair [44]. Based on comparison with BM-MSCs and specific studies, ASCs have also been shown to possess broad immunoregulatory abilities and are capable of influencing both adaptive and innate immune responses by means of direct cell-to-cell interactions and soluble factor secretion [45].

Specific studies on cultured human ASCs show that they inhibit lymphocyte proliferation induced by allogenic mononuclear cells or mitogens, block their polarization in Th1 and Th17, and induce functional Treg [10, 46–48]. ASCs also inhibit the proliferation and differentiation of B cells through both direct and indirect effects [49]. They can educate macrophages to adapt an antiinflammatory phenotype, through multiple mechanisms, including the production of exosomes [50–52], and are also able to inhibit neutrophil recruitment [53]. These immunomodulatory functions can be modulated in vitro by treating ASCs with IFNγ [54]. Interestingly, it is now proposed that ASCs express higher antiinflammatory and immunosuppressive molecules than BM-MSCs, making them a potential valuable alternative in therapeutic assays [55].

In addition, ASCs have been shown to express markers usually described on immune cells and, also in some specific conditions, to behave as immune cells. Indeed, ASCs express some receptors such as Toll-like receptors (TLR) involved in the recognition of danger signals, and respond to TLR ligands by secreting proinflammatory cytokines such as tumor necrosis factor (TNF)-, interleukin (IL)-6, and IL-8 [56, 57]. In addition, it has been shown that in both in vitro and inflammatory conditions, ASCs are able to upregulate MHC class II, and potentially act as antigen-presenting cells [58, 59]. This suggests that ASCs may behave as immune sensors and induce immune response. Transcriptional analyses reveal that ASCs share the expression of a great number of genes with macrophages, particularly those related to endocytosis, actin remodeling, and vesicle trafficking [60]. Murine ASCs transplanted in the peritoneal cavity acquire phenotypic and functional features of macrophages, partially through cell-to cell contact [61]. Moreover, they are able to internalize yeast *Candida albicans* and exhibit some microbicide activities in both mice and human [62, 63] and have therefore been proposed to act as macrophage-like cells [62]. The broad range of antibacterial action of ASCs is mediated through multiple and combined actions depending on the context and bacterial strain. ASCs may alter bacterial membrane integrity, leading to reduction of cell growth and viability, abnormality in cell division, and sensitization to antibiotics. ASCs may also use reactive oxygen species (ROS) production via NADH-dependent mechanisms and phagocytosis to trigger their antibacterial effects [64]. In parallel to their macrophage-like phenotype and function, murine and human ASCs have also been shown to differentiate in vitro into platelet-producing megakaryocytes, and this differentiation is regulated by ASC-secreted thrombopoietin [65–67]. Whether this differentiation occurs in vivo and whether it is of physiological interest remains an open question.

In vivo, human ASCs have been successfully used to dampen inflammation in a wide range of pathological conditions [11], such as graft vs host disease [68], systemic lupus erythematosus [69], Crohn's disease [70], autoimmune encephalomyelitis [71, 72], acute respiratory distress syndrome [53], and ulcer [73], among others. It is noteworthy that the only therapeutic product available on the market and validated by the regulatory authorities corresponds to ASCs for treating fistula of Crohn's disease [74]. Indeed, ASCs present many advantages over BM-MSCs, such as a high frequency and availability, ease of obtention with minor invasive process, and no ethical issues. In addition, they exhibit a higher proliferation rate than BM-MSCs, and a potentially less active telomere-based senescence mechanism [75, 76]. In addition, their genetic stability in culture is higher compared with BM-MSCs, and they have been shown to be safe in transplantation studies [77]. Altogether, these characteristics make ASCs suitable for cell-based therapies in the context of chronic diseases.

Hematopoietic stem cells

Differentiation, distribution

The formation of immune cells mainly relies on the hematopoietic process that has been studied for a long time and largely described, at least in BM. Indeed, the classical model of hematopoiesis is a hierarchical model based on the presence in BM of a HSC able to self-renew and differentiate into both myeloid and lymphoid cells, and functionally defined as capable of rescuing lethally irradiated animals by restoring the entire immune system [78, 79].

It is now clear that HSCs form a heterogeneous population with differential reconstitution abilities. Indeed, different subpopulations of HSCs have been identified and characterized by their specific functional properties, including repopulation kinetics or differentiation potential [80]. In addition, two different models of differentiation have been proposed: the hierarchical model that describes a classical differentiation through a series of stem and progenitor cells along a lineage tree, and a continuum model in which multipotent stem cells undergo direct differentiation into distinct mature cell types [81, 82]. In addition, the contribution of HSCs or their downstream progenitors to steady-state hematopoiesis, as well as the precise hierarchy of the branching points of each lineage commitment, has been a subject of controversy [83]. Although this debate could lead to a confusing view of the hematopoietic process, this also indicates that HSCs and the hematopoietic hierarchy are able to adapt to various physiological and pathological conditions.

Beside their canonical differentiation potential, and based on in vitro and in vivo studies of single-progenitor analyses, it has also been proposed that HSCs are able to differentiate towards adipocytes [84, 85]. Detailed analysis of adipocytes and stromal cells, deriving from BM progenitors in vivo, showed that these cells differ from conventional white adipocytes by lower leptin expression, mitochondrial/peroxisomal content, and oxidative capacity, and higher inflammatory cytokine production [86]. This hematopoietic-to-mesenchymal transition occurs both in mice and humans through the formation of myeloid progenitors and is controlled by integrin β1 signaling [87, 88]. Altogether, these data suggest that hematopoietic progenitors may constitute a reserve pool for adipogenesis [89] and confirm the close relationship between adipocyte and monocyte/macrophage lineages [60]. In addition to HSCs localized in BM, one can postulate that HSCs residing in AT might also serve as a resident source of white adipocytes.

HSC are localized in specific environments in BM called niches that control their fate and their function [18]. The BM niches exhibit complexity both in terms of composition and function. Indeed, they form different heterogeneous compartments composed, at least in part, of osteoblasts, MSCs, vascular cells, and/or adipocytes that interact with HSCs and regulate their proliferation, differentiation, and migration [90]. It has been proposed that the distinct niches play distinct roles in the maintenance of HSCs. For example, the endosteal niche hosts and controls early lineage committed progenitors via the secretion of CXCL12, in contrast to the vascular niche, which promotes HSC maintenance by producing SCF [91]. The diversity of the hematopoietic niche together with the heterogeneity of the HSC population leads, therefore, to a complex system of bidirectional interactions that controls the production of circulating immune cells and opens emerging questions. One of them concerns the potential existence of hematopoietic niches for HSC maintenance in other adult tissues, and their role.

In this context, some recent studies have identified hematopoietic progenitors in the lung, where they contribute to megakaryocyte and platelet production, although the physiological relevance of this process remains elusive [92]. In addition, extramedullary hematopoiesis can occur in the liver and the spleen in the context of hematopoietic stress or hematological disorders [90]. We and others have also revealed the presence of HSCs in extramedullary AT [93, 94], and highlighted the role of this extramedullary hematopoietic process in the control of tissue homeostasis [95].

Adipose-derived hematopoietic activity in physiological and pathological situations

The presence of an abundant and complex immune cell population in AT is well established and described because of its strong involvement in AT homeostasis and its dysfunctions. Up to now and based on in vivo reconstitution assays, it was thought that these immune cells derived from the BM. However, our previous studies have shown that AT exhibits specific hematopoietic activity exerted by an unknown subpopulation of SVF cells [96]. Indeed, we showed that intraperitoneal injection of SVF cells in lethally irradiated mice prevents mortality and supports the reconstitution of lymphoid and myeloid lineages, suggesting that AT could be an unexpected source of hematopoietic progenitors. We thus prospectively isolated a peculiar functional hematopoietic stem/progenitor cell (HSPC) population from murine AT. These cells are phenotypically similar to BM HSCs and are able to differentiate into both myeloid and lymphoid lineages in vitro [94, 97]. A similar population has been identified in human subcutaneous and visceral fat (B. Cousin, personal data). On long-term reconstitution assay, they present intrinsic potential to replenish all major hematopoietic lineages as well as to efficiently generate myeloid cells within the AT. They are also able to give rise to multilineage engraftment in both secondary recipients and in utero transplantation. Thus, innate immune cells present in the subcutaneous AT may be renewed in situ via AT-HSC differentiation [94, 97]. In addition, this AT-specific hematopoietic activity may be generalized to all adipose deposits in mice, although with specificity according to the fat pad location, indicating that local cues control this AT-dependent hematopoietic potential. Indeed, subcutaneous AT exhibits higher intrinsic hematopoietic activity than visceral deposits, due to the specificity of the microenvironment in each fat pad [98]. This suggests that immune cells in visceral and subcutaneous fat do not have the same origin, and therefore do not potentially assume the same functions or exhibit the same responsiveness. Altogether, these data support the idea that AT could be considered not only as an immunological but also as a hematopoietic organ, and positions the AT as a major site of inflammatory cell production that could play a role in inflammatory disorders such as obesity and diabetes [99].

In this context, we recently demonstrated that AT and BM hematopoiesis is differently altered in diabetic mice. Indeed, diabetic diet intensifies the differentiation of AT-HSCs towards macrophages with proinflammatory mode, without any significant changes in BM-derived macrophages. Furthermore, AT-HSCs sorted from diabetic animals transferred into healthy mice maintained on normal diet increase AT inflammation and trigger the development of a diabetic state in the recipient animals. This unexpected result reveals the pivotal and causative role of AT hematopoiesis dysfunction in type 2 diabetes development [95].

This specific hematopoietic activity includes different types of myeloid cells, including monocytes and mast cells (MC). Indeed, we identified the AT as a reservoir of hematopoietic progenitors committed to the MC lineage, and a site of MC differentiation. Transplantation studies revealed that MCs present in the AT do not arise from medullar progenitors, but are produced in situ [97]. In addition, AT-derived MCs are capable of homing to different organs, such as skin and gut, where they acquire properties of functional tissue MCs [97]. In agreement with this, in chimeric mice whose BM-derived cells were donated from green fluorescent protein (GFP) transgenic animals, cardiac MCs populating the ischemic milieu did not carry the GFP transgene, suggesting that MCs that home to the infarcted heart do not arise from bone marrow progenitors. After acute myocardial infarction, heart function and remodeling are associated with the presence of MCs deriving from these AT-HSCs and are not dependent on BM-derived MCs [100]. These results suggest that AT-derived myeloid cells are able to home to the cardiac tissue, where they can participate in regenerative and/or reparative processes.

Conclusion

The presence in the AT of these different immature and plastic cells raises the question of whether AT can be considered as a physiological reservoir of cells that can be recruited in the context of tissue injury or degeneration. The identification of immune cells from AT hematopoiesis in the infarcted heart [100], or of ASCs in the injured muscle [101], supports this hypothesis. Considering the amount of AT in adult organisms, this specific function could be of crucial importance in some specific physiopathological contexts that remain to be described. In addition, the juxtaposition in the AT of metabolic and hematopoietic functions, together with endocrine and immune properties, is reminiscent of the *Drosophila* fat body that fulfills the functions of its mammalian counterparts, namely the liver, the AT, and the hematopoietic and immune system [5]. The hematopoietic potential of the AT raises the question of a developmental heritage, suggesting a physiological importance that remains to be elucidated.

References

[1] J.M. Gimble, A.J. Katz, B.A. Bunnell, Adipose-derived stem cells for regenerative medicine, Circ. Res. 100 (9) (2007) 1249–1260.
[2] D. Mathis, S.E. Shoelson, Immunometabolism: an emerging frontier, Nat. Rev. Immunol. 11 (2) (2011) 81.
[3] A. Schaffler, J. Scholmerich, Innate immunity and adipose tissue biology, Trends Immunol. 31 (6) (2010) 228–235.
[4] B. Prunet-Marcassus, et al., From heterogeneity to plasticity in adipose tissues: site-specific differences, Exp. Cell Res. 312 (6) (2006) 727–736.
[5] G.S. Hotamisligil, Inflammation and metabolic disorders, Nature 444 (7121) (2006) 860–867.
[6] V. Planat-Benard, et al., Spontaneous cardiomyocyte differentiation from adipose tissue stroma cells, Circ. Res. 94 (2) (2004) 223–229.
[7] A.J. Friedenstein, Precursor cells of mechanocytes, Int. Rev. Cytol. 47 (1976) 327–359.
[8] M. Kassem, P. Bianco, Skeletal stem cells in space and time, Cell 160 (1–2) (2015) 17–19.
[9] P. Bourin, et al., Stromal cells from the adipose tissue-derived stromal vascular fraction and culture expanded adipose tissue-derived stromal/stem cells: a joint statement of the International Federation for Adipose Therapeutics and Science (IFATS) and the International Society for Cellular Therapy (ISCT), Cytotherapy 15 (6) (2013) 641–648.
[10] B. Puissant, et al., Immunomodulatory effect of human adipose tissue-derived adult stem cells: comparison with bone marrow mesenchymal stem cells, Br. J. Haematol. 129 (1) (2005) 118–129.
[11] L. Casteilla, et al., Adipose-derived stromal cells: their identity and uses in clinical trials, an update, World J. Stem Cells. 3 (4) (2011) 25–33.
[12] D.T. Covas, et al., Multipotent mesenchymal stromal cells obtained from diverse human tissues share functional properties and gene-expression profile with CD146+ perivascular cells and fibroblasts, Exp. Hematol. 36 (5) (2008) 642–654.
[13] G. Pachon-Pena, et al., Stromal stem cells from adipose tissue and bone marrow of age-matched female donors display distinct immunophenotypic profiles, J. Cell. Physiol. 226 (3) (2011) 843–851.
[14] M. Maumus, et al., Native human adipose stromal cells: localization, morphology and phenotype, Int. J. Obes. 35 (9) (2011) 1141–1153.
[15] M. Crisan, et al., A perivascular origin for mesenchymal stem cells in multiple human organs, Cell Stem Cell 3 (3) (2008) 301–313.
[16] D. Noel, et al., Cell specific differences between human adipose-derived and mesenchymal-stromal cells despite similar differentiation potentials, Exp. Cell Res. 314 (7) (2008) 1575–1584.
[17] V. Planat-Benard, et al., Plasticity of human adipose lineage cells toward endothelial cells: physiological and therapeutic perspectives, Circulation 109 (5) (2004) 656–663.
[18] M.J. Kiel, S.J. Morrison, Uncertainty in the niches that maintain haematopoietic stem cells, Nat. Rev. Immunol. 8 (4) (2008) 290–301.

[19] S.J. Morrison, A.C. Spradling, Stem cells and niches: mechanisms that promote stem cell maintenance throughout life, Cell 132 (4) (2008) 598–611.
[20] R. Schofield, The relationship between the spleen colony-forming cell and the haemopoietic stem cell, Blood Cells 4 (1–2) (1978) 7–25.
[21] H. Kodama, et al., In vitro hemopoiesis within a microenvironment created by MC3T3-G2/PA6 preadipocytes, J. Cell. Physiol. 118 (3) (1984) 233–240.
[22] M. Nishikawa, et al., Changes in hematopoiesis-supporting ability of C3H10T1/2 mouse embryo fibroblasts during differentiation, Blood 81 (5) (1993) 1184–1192.
[23] J.M. Gimble, et al., The function of adipocytes in the bone marrow stroma: an update, Bone 19 (5) (1996) 421–428.
[24] J. Corre, et al., Human subcutaneous adipose cells support complete differentiation but not self-renewal of hematopoietic progenitors, J. Cell. Physiol. 208 (2) (2006) 282–288.
[25] G.E. Kilroy, et al., Cytokine profile of human adipose-derived stem cells: expression of angiogenic, hematopoietic, and pro-inflammatory factors, J. Cell. Physiol. 212 (3) (2007) 702–709.
[26] F. De Toni, et al., Human adipose-derived stromal cells efficiently support hematopoiesis in vitro and in vivo: a key step for therapeutic studies, Stem Cells Dev. 20 (12) (2011) 2127–2138.
[27] E.V. Maslova, et al., Enrichment of umbilical cord blood mononuclears with hemopoietic precursors in co-culture with mesenchymal stromal cells from human adipose tissue, Bull. Exp. Biol. Med. 156 (4) (2014) 584–589.
[28] E. Andreeva, et al., Hematopoiesis-supportive function of growth-arrested human adipose-tissue stromal cells under physiological hypoxia, J. Biosci. Bioeng. 127 (5) (2019) 647–654.
[29] E. Montelatici, et al., Defining the identity of human adipose-derived mesenchymal stem cells, Biochem. Cell Biol. 93 (1) (2015) 74–82.
[30] S.K. Kapur, A.J. Katz, Review of the adipose derived stem cell secretome, Biochimie 95 (12) (2013) 2222–2228.
[31] L.B. Buravkova, et al., The differential expression of adhesion molecule and extracellular matrix genes in mesenchymal stromal cells after interaction with cord blood hematopoietic progenitors, Dokl Biochem. Biophys. 479 (1) (2018) 69–71.
[32] P.Z. Andrade, et al., Ex vivo expansion of cord blood haematopoietic stem/progenitor cells under physiological oxygen tensions: clear-cut effects on cell proliferation, differentiation and metabolism, J. Tissue Eng. Regen. Med. 9 (10) (2015) 1172–1181.
[33] E.R. Andreeva, et al., Response of adipose tissue-derived stromal cells in tissue-related O_2 microenvironment to short-term hypoxic stress, Cells Tissues Organs 200 (5) (2015) 307–315.
[34] N. Nakao, et al., Adipose tissue-derived mesenchymal stem cells facilitate hematopoiesis in vitro and in vivo: advantages over bone marrow-derived mesenchymal stem cells, Am. J. Pathol. 177 (2) (2010) 547–554.
[35] C. Sengenes, et al., Chemotaxis and differentiation of human adipose tissue CD34+/CD31- progenitor cells: role of stromal derived factor-1 released by adipose tissue capillary endothelial cells, Stem Cells 25 (9) (2007) 2269–2276.
[36] N. Van Overstraeten-Schlogel, Y. Beguin, A. Gothot, Role of stromal-derived factor-1 in the hematopoietic-supporting activity of human mesenchymal stem cells, Eur. J. Haematol. 76 (6) (2006) 488–493.
[37] J. Juarez, L. Bendall, SDF-1 and CXCR4 in normal and malignant hematopoiesis, Histol. Histopathol. 19 (1) (2004) 299–309.
[38] M. Sharma, et al., Stromal-derived factor-1/CXCR4 signaling: indispensable role in homing and engraftment of hematopoietic stem cells in bone marrow, Stem Cells Dev. 20 (6) (2011) 933–946.
[39] J. Xie, et al., Human adipose-derived stem cells ameliorate cigarette smoke-induced murine myelosuppression via secretion of TSG-6, Stem Cells 33 (2) (2015) 468–478.
[40] T. Ueda, et al., Adipose-derived stromal cells grown on a hydroxyapatite scaffold can support hematopoiesis in regenerated bone marrow in vivo, Cell Biol. Int. 38 (6) (2014) 790–798.
[41] S. Nishiwaki, et al., Efficacy and safety of human adipose tissue-derived mesenchymal stem cells for supporting hematopoiesis, Int. J. Hematol. 96 (3) (2012) 295–300.
[42] W. Wagner, et al., Molecular and secretory profiles of human mesenchymal stromal cells and their abilities to maintain primitive hematopoietic progenitors, Stem Cells 25 (10) (2007) 2638–2647.
[43] W. Wagner, R. Saffrich, A.D. Ho, The stromal activity of mesenchymal stromal cells, Transfus. Med. Hemother. 35 (3) (2008) 185–193.
[44] Y. Wang, et al., Plasticity of mesenchymal stem cells in immunomodulation: pathological and therapeutic implications, Nat. Immunol. 15 (11) (2014) 1009–1016.
[45] N. Bertheuil, et al., Adipose mesenchymal stromal cells: definition, immunomodulatory properties, mechanical isolation and interest for plastic surgery, Ann. Chir. Plast. Esthet. 64 (1) (2019) 1–10.

[46] M. Najar, et al., Mesenchymal stromal cells use PGE2 to modulate activation and proliferation of lymphocyte subsets: combined comparison of adipose tissue, Wharton's Jelly and bone marrow sources, Cell Immunol. 264 (2) (2010) 171–179.
[47] R.A. Larocca, et al., Adipose tissue-derived mesenchymal stem cells increase skin allograft survival and inhibit Th-17 immune response, PLoS One 8 (10) (2013), e76396.
[48] A.U. Engela, et al., Interaction between adipose tissue-derived mesenchymal stem cells and regulatory T-cells, Cell Transplant. 22 (1) (2013) 41–54.
[49] M. Franquesa, et al., Human adipose tissue-derived mesenchymal stem cells abrogate plasmablast formation and induce regulatory B cells independently of T helper cells, Stem Cells 33 (3) (2015) 880–891.
[50] M. Sun, et al., Induction of macrophage M2b/c polarization by adipose tissue-derived mesenchymal stem cells, J. Immunol. Res. 2019 (2019) 7059680.
[51] J.S. Heo, Y. Choi, H.O. Kim, Adipose-derived mesenchymal stem cells promote M2 macrophage phenotype through exosomes, Stem Cells Int. 2019 (2019) 7921760.
[52] R.B.O. Ozdemir, et al., The investigation of immunomodulatory effects of adipose tissue mesenchymal stem cell educated macrophages on the CD4 T cells, Immunobiology 224 (4) (2019) 585–594.
[53] Y.J. Jung, et al., The effect of human adipose-derived stem cells on lipopolysaccharide-induced acute respiratory distress syndrome in mice, Ann. Transl. Med. 7 (22) (2019) 674.
[54] T.R.T. Serejo, et al., Assessment of the immunosuppressive potential of INF-gamma licensed adipose mesenchymal stem cells, their secretome and extracellular vesicles, Cells 8 (1) (2019).
[55] C. Menard, et al., Integrated transcriptomic, phenotypic, and functional study reveals tissue-specific immune properties of mesenchymal stromal cells, Stem Cells 38 (1) (2020) 146–159.
[56] M.E. Bernardo, W.E. Fibbe, Mesenchymal stromal cells: sensors and switchers of inflammation, Cell Stem Cell 13 (4) (2013) 392–402.
[57] G. Raicevic, et al., The source of human mesenchymal stromal cells influences their TLR profile as well as their functional properties, Cell. Immunol. 270 (2) (2011) 207–216.
[58] B. Purandare, et al., Temporal HLA profiling and immunomodulatory effects of human adult bone marrow- and adipose-derived mesenchymal stem cells, Regen. Med. 9 (1) (2014) 67–79.
[59] P. Bobyleva, et al., Reciprocal modulation of cell functions upon direct interaction of adipose mesenchymal stromal and activated immune cells, Cell Biochem. Funct. 37 (4) (2019) 228–238.
[60] G.M. Charriere, et al., Macrophage characteristics of stem cells revealed by transcriptome profiling, Exp. Cell Res. 312 (17) (2006) 3205–3214.
[61] G. Charriere, et al., Preadipocyte conversion to macrophage. Evidence of plasticity, J. Biol. Chem. 278 (11) (2003) 9850–9855.
[62] B. Cousin, et al., A role for preadipocytes as macrophage-like cells, FASEB J. 13 (2) (1999) 305–312.
[63] C. Saillan-Barreau, et al., Human adipose cells as candidates in defense and tissue remodeling phenomena, Biochem. Biophys. Res. Commun. 309 (3) (2003) 502–505.
[64] P. Monsarrat, et al., Broad-spectrum antibacterial effects of human adipose-derived stromal cells, Stem Cells Int. 2019 (2019) 5389629.
[65] Y. Matsubara, et al., Generation of megakaryocytes and platelets from human subcutaneous adipose tissues, Biochem. Biophys. Res. Commun. 378 (4) (2009) 716–720.
[66] Y. Ono-Uruga, et al., Human adipose tissue-derived stromal cells can differentiate into megakaryocytes and platelets by secreting endogenous thrombopoietin, J. Thromb. Haemost. 14 (6) (2016) 1285–1297.
[67] K. Tozawa, et al., Megakaryocytes and platelets from a novel human adipose tissue-derived mesenchymal stem cell line, Blood 133 (7) (2019) 633–643.
[68] R. Yanez, et al., Adipose tissue-derived mesenchymal stem cells have in vivo immunosuppressive properties applicable for the control of the graft-versus-host disease, Stem Cells 24 (11) (2006) 2582–2591.
[69] E.W. Choi, et al., Reversal of serologic, immunologic, and histologic dysfunction in mice with systemic lupus erythematosus by long-term serial adipose tissue-derived mesenchymal stem cell transplantation, Arthritis Rheum. 64 (1) (2012) 243–253.
[70] C. Thoma, Autologous adipose tissue injection promising for treatment of perianal fistulas in Crohn's disease, Nat. Rev. Gastroenterol. Hepatol. 16 (4) (2019) 198.

[71] G. Constantin, et al., Adipose-derived mesenchymal stem cells ameliorate chronic experimental autoimmune encephalomyelitis, Stem Cells 27 (10) (2009) 2624–2635.
[72] F. Yousefi, et al., In vivo immunomodulatory effects of adipose-derived mesenchymal stem cells conditioned medium in experimental autoimmune encephalomyelitis, Immunol. Lett. 172 (2016) 94–105.
[73] J. Bukowska, et al., Safety and efficacy of human adipose-derived stromal/stem cell therapy in an immunocompetent murine pressure ulcer model, Stem Cells Dev. (2020).
[74] L. Bernardi, et al., Transplantation of adipose-derived mesenchymal stem cells in refractory Crohn's disease: systematic review, Arq. Bras. Cir. Dig. 32 (4) (2019), e1465.
[75] L.E. Kokai, K. Marra, J.P. Rubin, Adipose stem cells: biology and clinical applications for tissue repair and regeneration, Transl. Res. 163 (4) (2014) 399–408.
[76] A. Hassanshahi, et al., Adipose-derived stem cells for wound healing, J. Cell. Physiol. 234 (6) (2019) 7903–7914.
[77] A. Bajek, et al., Adipose-derived stem cells as a tool in cell-based therapies, Arch. Immunol. Ther. Exp. 64 (6) (2016) 443–454.
[78] M.F. de Bruijn, et al., Definitive hematopoietic stem cells first develop within the major arterial regions of the mouse embryo, EMBO J. 19 (11) (2000) 2465–2474.
[79] E. Gunsilius, G. Gastl, A.L. Petzer, Hematopoietic stem cells, Biomed. Pharmacother. 55 (4) (2001) 186–194.
[80] M. Crisan, E. Dzierzak, The many faces of hematopoietic stem cell heterogeneity, Development 143 (24) (2016) 4571–4581.
[81] K.K. Hirschi, S. Nicoli, K. Walsh, Hematopoiesis lineage tree uprooted: every cell is a rainbow, Dev. Cell 41 (1) (2017) 7–9.
[82] L. Velten, et al., Human haematopoietic stem cell lineage commitment is a continuous process, Nat. Cell Biol. 19 (4) (2017) 271–281.
[83] Y. Zhang, et al., Hematopoietic hierarchy—an updated roadmap, Trends Cell Biol. 28 (12) (2018) 976–986.
[84] S.M. Majka, Y. Barak, D.J. Klemm, Concise review: adipocyte origins: weighing the possibilities, Stem Cells 29 (7) (2011) 1034–1040.
[85] Y. Sera, et al., Hematopoietic stem cell origin of adipocytes, Exp. Hematol. 37 (9) (2009). 1108–20, 1120 e1–4.
[86] S.M. Majka, et al., De novo generation of white adipocytes from the myeloid lineage via mesenchymal intermediates is age, adipose depot, and gender specific, Proc. Natl. Acad. Sci. U. S. A. 107 (33) (2010) 14781–14786.
[87] K.M. Gavin, et al., Hematopoietic-to-mesenchymal transition of adipose tissue macrophages is regulated by integrin beta1 and fabricated fibrin matrices, Adipocyte 6 (3) (2017) 234–249.
[88] K.M. Gavin, et al., De novo generation of adipocytes from circulating progenitor cells in mouse and human adipose tissue, FASEB J. 30 (3) (2016) 1096–1108.
[89] P. Arner, M. Ryden, The contribution of bone marrow-derived cells to the human adipocyte pool, Adipocyte 6 (3) (2017) 187–192.
[90] G.M. Crane, E. Jeffery, S.J. Morrison, Adult haematopoietic stem cell niches, Nat Rev Immunol 17 (9) (2017) 573–590.
[91] S.J. Morrison, D.T. Scadden, The bone marrow niche for haematopoietic stem cells, Nature 505 (7483) (2014) 327–334.
[92] E. Lefrancais, et al., The lung is a site of platelet biogenesis and a reservoir for haematopoietic progenitors, Nature 544 (7648) (2017) 105–109.
[93] J. Han, et al., Adipose tissue is an extramedullary reservoir for functional hematopoietic stem and progenitor cells, Blood 115 (5) (2010) 957–964.
[94] S. Poglio, et al., In situ production of innate immune cells in murine white adipose tissue, Blood 120 (25) (2012) 4952–4962.
[95] E. Luche, et al., Corrupted adipose tissue endogenous myelopoiesis initiates diet-induced metabolic disease, Elife 6 (2017).
[96] B. Cousin, et al., Reconstitution of lethally irradiated mice by cells isolated from adipose tissue, Biochem. Biophys. Res. Commun. 301 (4) (2003) 1016–1022.
[97] S. Poglio, et al., Adipose tissue as a dedicated reservoir of functional mast cell progenitors, Stem Cells 28 (11) (2010) 2065–2072.

[98] E. Luche, et al., Differential hematopoietic activity in white adipose tissue depending on its localization, J. Cell. Physiol. 230 (12) (2015) 3076–3083.
[99] B. Cousin, et al., Immuno-metabolism and adipose tissue: the key role of hematopoietic stem cells, Biochimie 124 (2016) 21–26.
[100] A. Ngkelo, et al., Mast cells regulate myofilament calcium sensitization and heart function after myocardial infarction, J. Exp. Med. 213 (7) (2016) 1353–1374.
[101] A. Girousse, et al., The release of adipose stromal cells from subcutaneous adipose tissue regulates ectopic intramuscular adipocyte deposition, Cell Rep. 27 (2) (2019) 323–333 e5.

CHAPTER

19

Adipose stem cells for peripheral nerve engineering

Benjamin K. Schilling[a], *George E. Panagis*[b], *Jocelyn S. Baker*[a], *and Kacey Marra*[c,d]

[a]Department of Bioengineering, School of Engineering, University of Pittsburgh, Pittsburgh, PA, United States [b]Department of Plastic Surgery, School of Medicine, University of Pittsburgh, Pittsburgh, PA, United States [c]Departments of Plastic Surgery and Bioengineering, University of Pittsburgh, Pittsburgh, PA, United States [d]McGowan Institute for Regenerative Medicine, University of Pittsburgh, Pittsburgh, PA, United States

Introduction to adipose-derived stem cells in peripheral nerve injury

Adipose stem cells (ASCs) mark a potential repository for therapeutic applications targeting peripheral nerce injury (PNI) and repair. Indeed, ASCs are a highly convenient autologous cell type relative to other stromal cells [1], and because of this has received consistently increasing attention [2, 3] since ASC's initial description at the turn of the millennium [4, 5]. Beyond the ease of acquisition and lack of controversy when harvesting from human tissue, the ease of in vitro expansion [6, 7], promotion of angiogenesis [8–10], and immunomodulatory effects [11–13] have resulted in a holistic appeal to their clinical and therapeutic translation. Worldwide, more than 200 ASC-related clinical trials in phases I–III were registered between 2007 and 2018 for various applications, promoting their translatability as a viable cell therapy option [14, 15]. Similarly, cell therapies for PNI have been pursued from preclinical investigations in recent decades as well [16, 17]. Despite the large body of preclinical work, of the 71 active or completed clinical trials registered on ClinicalTrials.gov for the treatment of PNI, only one involves stem cells (NCT03999424: *Autologous Human Schwann Cells in Peripheral Nerve Repair*), indicative of the absent clinical crossover of ASCs to date. ASCs have, however, been proposed for clinical use in pain management for reflex sympathetic dystrophy (NCT02987855: *Use of autologous adult adipose-derived stem/stromal cells in reflex sympathetic dystrophy (RSD), complex regional pain syndrome (CRPS), and fibromyalgia*) and

Scientific Principles of Adipose Stem Cells
https://doi.org/10.1016/B978-0-12-819376-1.00008-1

have been used for pain management in amputees (NCT01645722: *Enriched autologous fat grafting for treating pain at amputation sites*), though the latter of these studies utilized ASCs combined with grafted whole adipose tissue [18]. Notwithstanding advances in the regenerative space inclusive of polymeric conduits and wraps, the standard of care in PNI remains surgical rejoining of severed nerves in smaller injuries or the use of autologous nerve in cases of larger injuries that require the removal of the injured nerve segment [19]. There is still significant need for innovation in the PNI niche suggested by the continued annual expenditure approximated at $1.5 billion, which accounts for the socioeconomic impact, where 41% of patients who experience PNI do not return to work 1 year after injury intervention owing to residual complications [20, 21]. As a standalone therapy or acting synergistically within currently employed techniques, ASCs may contribute significantly to the healing peripheral nerve. Here, an overview of the peripheral nervous system (PNS) is given along with rationale and investigations supporting the use of ASCs toward a clinical end for PNI interventions.

Composition of the peripheral nervous system and adipose-derived stem cell crossover

Gestational origin and composition of the mature peripheral nervous system

The PNS is dichotomized into the autonomic and somatic nervous systems, the former being responsible for innervating glands and smooth muscle, and the latter responsible for locomotion and sensation. While the vast majority of functions within the autonomic nervous system (ANS) are involuntary, i.e., activation of the fight-or-flight or rest-and-digest responses, the ANS can act in concert with the somatic nervous systems (SNS) owing to their similarities in cellular and anatomical composition in conjunction with their integration at several neuroeffector connection sites [22, 23]. From gestation, both the ANS and SNS, and thus, the PNS, arise from the neural ectoderm. This is inherently distinct from the adipose tissue and the ASC cell population as they arise from the mesoderm [24]. Notable, however, is that neural crest cells undergo a developmental transdifferentiation into mesodermal cells linking the ASC-PNS cell heritage [25]. It is the neural ectoderm that gives rise to this neural crest as well as the neural tube, which subsequently compartmentalize into the brain, spinal cord, and the network of the peripheral nerves. The neuroepithelial cells differentiate into neurons, oligodendrocytes/Schwann cells, astrocytes/satellite cells, and the neuroglial support cells that surround, myelinate, and sheath the neurons [26], forming the bodily network of peripheral nerve (Fig. 1). Though composed similarly at the cellular level, two distinct nerve types exist that are directionally characterized on the basis of electrical impulse transduction. Afferent nerves signal toward the brain, and efferent nerves signal away from the brain; these nerve types are also commonly referred to as sensory and motor nerves, respectively [27].

Types of cells and extracellular matrix that comprise the nervous tissues

The dynamic interplay between cells, extracellular matrix (ECM), electrochemical signaling, and chemokine-based signaling is responsible for the overall health and proper function of the system. Schwann cells surround axons to form the myelin sheath, which

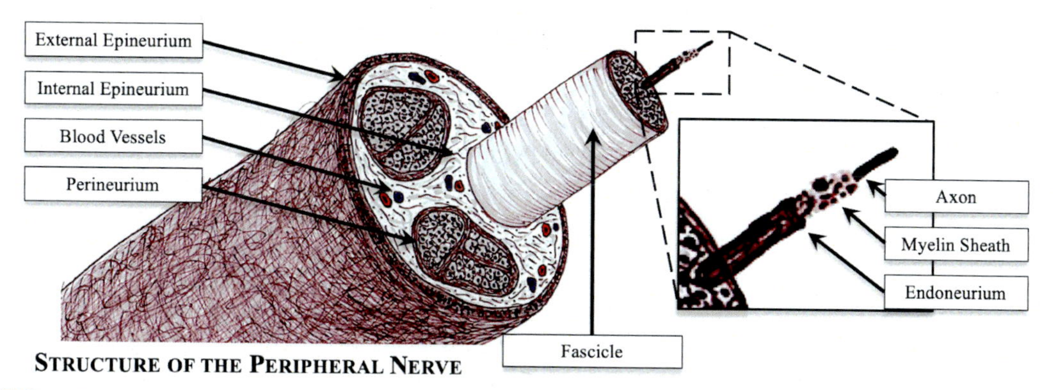

FIG. 1 Anatomy and structure of the peripheral nerve.

increases nerve conduction velocity. These cells also are responsible for secreting matrix proteins and soluble factors that are important in the growth, regeneration, and reinnervation of nerve fibers. These factors include NGF, GDNF, and BDNF [28, 29]. Astrocytes are integral cells in synaptogenesis, the development of a synapse. While the exact function of the astrocyte in injury is still unknown, there is speculation that these cells serve as reactionary agents in turnover and reconstruction, regulating the glial environment, the effects of which can be regenerative or detrimental [30, 31]. Both Schwann cells and astrocytes have responsibilities in the turnover and injury states for immune interaction through their respective paracrine effects [32]. As essential as the secretome may be, the neuronal ECM also plays a vital role in maintaining the intricate functions of the peripheral nerves. Basal lamina and basement membrane 1 proteins such as collagen IV, collagen VI, and laminins comprise the bulk of the ECM in neurons [33]. Laminin 2, laminin 8, and laminin 10 are particularly important for Schwann cell migration via the process of "radial migration," where damage to these proteins results in a host of peripheral neuropathies [34]. Proteoglycans surround the nodes of Ranvier, the composition of which is dependent on the function of the nerve [35]. The plasticity of the neurons is maintained by pentraxins and tenascin, with "long" pentraxin proteins doing so through their mediation of neural differentiation and synaptogenesis [36], and the ECM protein tenascin appearing to direct neurite outgrowth [37]. Astrocytes are primarily responsible for the production of thrombospondin, a key protein in the formation and health of the synapse [38]. Neurons terminate into the musculature at the neuromuscular junction (NMJ), which is composed of laminins, heparan sulfate proteoglycans, and various other glycoproteins, as well as collagen IV and the NMJ-specialized collagen Q (collagen-tail subunit of acetylcholinesterase). Notably, the integrity of collagen IV and its integration with laminin is essential in maintaining the presynaptic specialization at the NMJ [39].

Trauma-induced peripheral nerve injury and native healing mechanisms

Nerve injuries can be characterized broadly into three categories: (1) mechanical or traumatic, (2) vascular or ischemic, and (3) chemical or neurotoxic [40, 41]. Herein, we will focus primarily on mechanical and/or traumatic PNI with inclusion of neuronal ischemia pertinent to nerve injury and healing [42]. Approximately 20 million Americans are currently afflicted with some type of traumatic PNI [19]. Such injuries affect the various structures of the nerve and are assessed over five levels of severity in the Seddon and Sunderland grading system [43]. This system of grading was expanded to include a sixth (VI) classification by MacKinnon and Dellon, which includes combinations of grade III–V injuries along a damaged peripheral nerve segment [44]. These classifications are presented in Table 1. The severity and type of the PNI determine the potential and quality of native recovery or the need for surgical intervention. Injuries increasing in severity, i.e., past grade III, can result in serious complications and have been historically difficult to treat without long-term or permanent discomfort as well as diminished mobility and sensation. While the body has native repair mechanisms in place, this process is timely, and in the case of motor nerves, results in substantial muscle atrophy [28, 45]. In the post-PNI healing process, reinnervation can occur in two ways, by collateral branching and axonal regeneration. This dichotomy is based on severity of trauma, where collateral branching is induced in more minor instances, whereas axonal regeneration is activated in the event of moderate-to-severe PNI. In collateral branching, intact axons sprout toward a target organ or tissue. This is the primary form of recovery for minor injuries, when approximately 20%–30% of the axon has been damaged (grades I and II). This process begins 4 days after injury, and the healing process continues for several months, where an excessive amount of axonal branching results in some terminal ends not innervating the end terminus. These nonterminating, noninnervating ends will eventually be broken down naturally [46]. Notably, this process of collateral branching also occurs in nerve turnover, being an innate mechanism of PNS health [47].

More-severe injuries are quantified by significant damage to the axonal segment when damage to the endoneurium and beyond is imminent. In these instances, it is estimated that approximately 90% of axon damage has occurred, classified as grade III or beyond [48]. The process of regeneration occurs over two locations, i.e., the proximal and distal nerve segments, and in three phases, beginning with Wallerian degradation (primarily distal), followed by axonal regeneration (primarily proximal), and, finally, target reinnervation [49]. Following crush or transection injury, Wallerian degeneration takes place at the distal stump; without it, the recovery potential of the nerve is limited. This process involves clearing of axons and inhibitory cellular debris, including myelin, that is distal to the injury lesion. In turn, this degradation creates a favorable environment for regenerating axons to extend

TABLE 1 Nerve injury grading system of Seddon and Sunderland including the sixth (VI) classification added by MacKinnon and Dellon.

Injury		Myelin	Axon	Endoneurium	Perineurium	Epineurium
Neurapraxia	I	Demyelinated	Intact			
Axonotmesis (i.e., crush)	II		Damaged	Intact		
Axonotmesis/Neurotmesis (i.e., disconnection)	III			Damaged	Intact	
Axonotmesis/Neurotmesis	IV				Damaged	Intact
Axonotmesis/Neurotmesis	V					Transected
Mixed Injury*	VI			Varying levels of injury	Varying levels of injury	Varying levels of injury

toward their target reinnervation site. Triggering this regenerative degeneration is a rapid innate immune response, involving Schwann cells, fibroblasts, macrophages, endothelial cells, and their corresponding secretome [50–53]. Macrophages are recruited by several proinflammatory cytokines such as monocyte chemoattractant protein-1 (MCP-1, alternatively CCL2), macrophage inflammatory protein (MIP)-α, tissue necrosis factor (TNF)-α, and interleukin (IL)-1β, to phagocytose debris and to activate Schwann cells, which have two roles: assisting with phagocytosis and, later on, guiding regenerating axons [54, 55]. In distal degradation, macrophages and other phagocytes clear axonal debris, removing molecules that could inhibit future nerve regeneration. In the neural cell body region, macrophages trigger the conditioning lesion response, a process in which neurons increase their regeneration after a prior lesion [56]. In distal regeneration, Schwann cells proliferate to form the bands of Büngner and secrete neurotrophic factors that travel retrogradely (distal to proximal) to guide regenerating axons [57]. Recently, exosomal and extracellular vesicular content such as miR-132, being expressed by mature nerve tissue, and miR-3099, being overexpressed during injury and at the proximal stump, have been implicated as an activator of Schwann cells in PNI for inducing the recovery response [58, 59].

After PNI, axonal regeneration occurs at the proximal stump, where some Wallerian degeneration does occur, though in a retrograde fashion up to the first node of Ranvier and to a lesser extent relative to the distal stump [60]. Within a few hours, however, neuronal sprouts are formed, with terminal growth cones searching for neurotrophic factors secreted from the distal stump [61, 62]. A neuronal cell body post-PNI will undergo chromatolysis, which is a process where the cell body swells, its nucleus migrates peripherally with respect to the cell body, and the neuronal protein synthesis sites termed Nissl bodies deconstruct and disperse [63]. When the regenerating axons successfully reach the matrix of the distal stump, they grow within the bands of Büngner formed by Schwann cells [64]. Schwann cells continue congregating to form Büngner bands, lining the basal lamina, basement membrane, or connective tissue left behind after axonal and myelin degradation created distal to the injury site. This helps guide axons toward their target for reinnervation and eventually return to function. Throughout the time course of nerve regeneration post-PNI, approximately 40% of the dorsal root ganglions will undergo apoptosis [65]. Though some of this cell death is prospective and advantageous for reinnervation, deficient target-derived neurotrophic support may exacerbate apoptosis and lead to a dampening of distal signaling, resulting in scar formation [66].

Wallerian degeneration, axonal regeneration, and terminal reinnervation are all affected by the type of injury (i.e., crush, partial laceration, full transection), the physical length and duration of injury, and the quality of the Wallerian degeneration. Crush injuries do not tear apart connective tissue, while transection injuries do; severed axons regenerate after a crush injury but typically do not after a transection, as the transection requires complete turnover at the injury site involving axonal degeneration along with myelin sheath detachment and degradation [67]. The capacity for a Schwann cell to regenerate the injury site is limited [68, 69], where their decreased regenerative capacity parallels a decrease in neurotrophic factors as well as the reconstructive turnover of the basal lamina [70, 71]. Understanding these distinct sites of regeneration and the limitations of the sequential phenomena toward reinnervation may better assist the rationale of an ASC therapy in terms of both application site as well as the time of administration within the healing cascade.

Adipose-derived stem cells and the rationale for use in peripheral nerve injury

Similarities of adipose-derived stem cells and Schwann cells

Schwann cells are important contributors to nerve regeneration following traumatic injury through their role in clearing injured nerve tissue debris, secreting trophic factors, and guiding regenerating axons toward reinnervation. Perhaps unsurprisingly, Schwann cells have been widely shown to enhance axonal regeneration after PNI [72]. There exist, however, some impracticalities with the use of Schwann cells in procurement and expansion or manufacturing toward clinical therapies relative to ASCs. Key issues with Schwann cells for large-scale therapy are that they are difficult and relatively painful to harvest autologously; additionally, they are time consuming to expand in culture relative to ASCs [73, 74]. ASCs are non-tumorigenic in vivo [75], and compared with Schwann cells, ASCs possess not only a greater proliferative capacity but also retain their differentiation potential despite increasing passage numbers in vitro [76]. The proliferating ASC is responsible for secreting many factors relevant to nerve regeneration. These nerve-specific factors include NGF, GDNF, and BDNF, but also hepatocyte growth factor (HGF), insulin-like growth factor (IGF)-1, and angiopoietin (Ang)-1 [77], mimicking the factors that are secreted by cells within the PNS niche [78]. In the native PNS environment, NGF has been shown to enhance axonal regeneration [79], and BDNF has been shown to promote synapse plasticity and has also been shown to be required for myelination and regeneration of injured sciatic nerves [80, 81]. Recently, HGF has been shown to be a potent activator of Schwann cells with an upregulation of a proregenerative secretome [82]. In several studies, ASCs have been shown to secrete HGF [83, 84], and continue to do so even through the natural cascade of cell aging [85]. A comparison of the ASC and the Schwann cell is presented in Table 2.

Adipose-derived stem cells promote angiogenesis

The undifferentiated ASC may have further benefit in the context of secondary injury resulting from PNI, being ischemia induced from vascular damage as well as scar formation from a nerve that does not successfully reinnervate or is in a state of prolonged inflammation. Repair to the peripheral nerve must be done with exceptional care, ensuring that the repaired nerve is not under tension, though such repairs are not always successful [86–88]. Preclinical studies have shown that an 8% increase in tension on the nerve can have devastating consequences on the vasculature, inducing an up to 50% reduction in intraneural blood flow. Tensions exceeding 15% of the native resting tensions have been shown to induce an up to 80% reduction in blood flow [89]. Irrespective of tension, the wounded nerve site is inherently reduced in its intraneural vascular supply, which needs to be reestablished to facilitate complete nerve healing and functional regeneration [90–93]. The introduction of vascular endothelial growth factor (VEGF) has been shown to promote axon spouting and can improve the vasculature at the injury site. This vascular reconstruction is thought to be done by influencing vascular-constructing cell migration; it is also proposed that ASCs secreting VEGF mediate differentiation of local progenitor cells at the injury site toward various nerve cells or vasculature lineages. The ability of the ASCs to induce angiogenesis by HGF and VEGF, among other secretome factors [94, 95], may be essential to promoting rapid healing in the injured nerve, especially through matrix remodeling in situ [96]. ASC secretion of angiogenic

TABLE 2 Comparative characteristics between adipose-derived stem cells and Schwann cells.

	Secreted matrix proteins	Cytokines	Surface marker	In vitro doubling rate (days)	Morphology in culture
ASC	Collagens I, III, XVα, elastin, fibronectin, decorin, tenascin, tenomodulin	*Normal function*	*Positive markers:*	2–4	Spindle-shaped when anchored
		VEGF-A, FGF-9, PDGF	CD44, CD73, CD90,		
		Neuronal injury	*Negative markers:*		
		HGF, IGF-1, GDNF, BNF, NGF	CD31, CD45, CD235a		Spherical-shaped when detached
Schwann cells	Basil lamina proteins, collagen IV, VI, laminins, entactin, fibronectin	*Normal function*	*Common markers:* SOX10, S100	2	Bipolar morphologies with elongated processes containing a small number of thick bundles
		GDNF, BNF, NGF	*Myelinating Schwann cells*		
		Injury/ denervation	EGR2, MBP, MPZ		
		LIF, IL-6, IL-1	*Nonmyelinating Schwann cells*		
			GAP43, NCAM, P75NTR		

factors has also been shown to be increased when exposed to conditions of hypoxia [97], potentially enhancing their contribution of secreted factors to the nerve injury environment. Such ASC-mediated angiogenesis remains a controversial topic with respect to its tumorigenic nature but has largely been studied in areas of higher cancer incidence [98, 99], and in some areas has been reported to be a tumor-suppressing agent [100]. Despite its clinical importance, cancers of the PNS are relatively rare when compared with breast cancers, being the more common models in which ASCs are investigated for tumorigenic effects [101, 102]. This perhaps suggests a reduced likelihood that the angiogenic nature of ASCs in situ will result in tumorigenic effects if used in PNI, though such a hypothesis is, at this point, speculative.

Adipose-derived stem cells reduce scar formation

Notably, scar formation reduces the angiogenic potential of the injury site in PNI [103]. Neuronal scarring post-PNI typically results from prolonged local inflammation at the injury site; proinflammatory cytokines such as TNFα, transforming growth factor (TGF)-β, and IL-1α and IL-1β are upregulated in Schwann cells post-injury [104, 105] followed by deposition of collagen yielding fibrotic tissue [106, 107]. Prolonged inflammation causing scarring that is amply severe results in neuromuscular dysfunction, which may not be recoverable in the event that intervention is not rendered. ASCs have been shown to decrease inflammation at an injury site through immunomodulation. In concert with the secretion of angiogenic factors such as VEGF and HGF, it has been shown that ASCs secrete IL-6, IL-8, granulocyte-colony stimulating factor (G-CSF), granulocyte-macrophage colony-stimulating factor (GM-CSF), and MCP-1, thought to be responsible for attenuating the inflammatory cascade, promoting a more favorable environment for regeneration rather than fibrosis and scar formation [108].

In vivo studies: Adipose-derived stem cells in the peripheral nervous system

In vivo models of peripheral nerve injury

Due to the complex and synergistic factors of the mechanical environment, inflammatory response, vascular ingrowth, and electrochemical phenomena, in vitro assessments of PNI are inadequate. Further, the addition of ASCs, or any therapeutic modality, into the healing environment further limits the applicability of in vitro studies, given biocompatibility and host response are foregone [109]. To study the mechanisms of PNIs and potential therapeutics, reliable models of trauma-based PNI are required for experimentation. Such modeling requires the creation of an injury that is reproducible. The model should result in an expected level of impairment that is able to be effectively evaluated with established testing methods, and that is sustained for a predictable amount of time so that therapies can be accurately evaluated against controls.

Throughout the literature, many models across different species have been employed, each having unique advantages and disadvantages (Fig. 2A). The vast majority of PNI models are performed in rodents owing to their low cost and ease of husbandry relative to large animals.

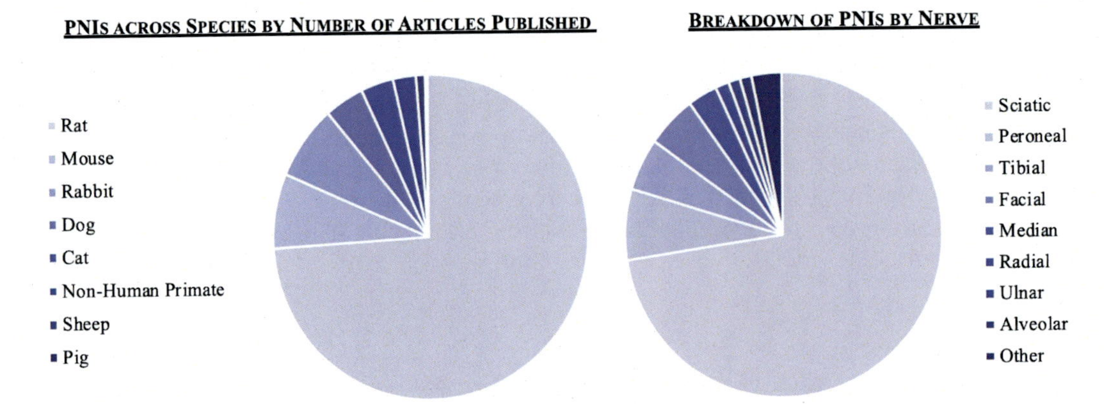

FIG. 2 Breakdown of animal model species in the published PNI literature (left) and a breakdown of the nerve injured in published literature (right).

Rats are typically preferred to mice in PNI modeling, given their larger overall size, and proportionally, their larger peripheral nerves. While the use of mice typically involves microsurgical skill, the mouse remains the predominant model for transgenic studies throughout the literature, PNI modeling included [110–114]. Larger animals such as rabbits, dogs, sheep, pigs, or nonhuman primates have also been used for PNI modeling, but have been less studied owing to their respective expenses relative to rodents, and all except for nonhuman primates are quadrupedal animals, thus adding to their approximation of the human condition [115]. Additionally, spontaneous axonal regeneration in humans does not match that of animals in terms of degree or rapidity after injury [116]. Therefore, the level of impairment normally observed in patients is understated in animal compression models of PNI. Despite this, the sciatic nerve has become well accepted and, as such, commonly utilized to model PNI in vivo [117]. The sciatic nerve in rats is relatively large in diameter, making it reasonably accessible for manipulation compared with its distal branches. While other nerves have been utilized in study as well (Fig. 2B), the sciatic nerve has seen the most usage in models of PNI [118]. A significant benefit of the sciatic nerve model is that surgeries are manageable with little or no microsurgical intervention necessary, owing to the relatively large diameter of the rat sciatic nerve. Distal to the sciatic nerve, the peroneal and tibial nerves can also be manipulated to create additional PNI, which results in a loss of muscular function. Despite the majority usage of the sciatic nerve, issues have been reported with autophagia in prolonged damage, and because of this, some in the field have moved toward using the distal branches of the sciatic nerve [119] or the median nerve. Injury to the median nerve results in partial impairment of the forelimb that improves animal well-being, resulting in fewer instances of self-mutilation. Further benefit of median nerve injury modeling is its clinical relevance; many patients experience injuries to the upper extremities, and the model lends itself well to modeling carpel tunnel syndrome [120].

Types of injuries and assessments for peripheral nerve injury in vivo models

The type of injury delivered to the nerve impacts the regeneration of the nerve as well as the functional return in the innervated muscle. A minimally detrimental traumatic injury model results from crushing or compression to a peripheral nerve. Crush or compression injuries can be achieved through bluntly crushing and/or compressing the nerve segment, or via ligation [121]. A crush or compression PNI involves applying force to the nerve to interrupt the continuity of the axons while leaving the connective scaffold intact, thereby maintaining the continuity of the nerve trunk. This continuity is particularly advantageous for providing an optimal pathway for axonal regrowth throughout the regeneration process, resulting in eventual reinnervation at the original target site [28]. Several surgical tools and devices have been developed to aid in the delivery of crush and/or compression injury, decreasing variability in animal injury models. Crush/compression surgical procedures only involve exposing the target nerve, which is a far more straightforward procedure than surgical anastomosis of two nerve segments, which may include additional therapies such as cuffs or conduits [122]. Once the nerve is exposed, however, the compression can be delivered by many methods, either via ligation or the preferred crush method (i.e., forceps, vasculature clamps, pressure-inducing conduits). Given the relative ease with which the crush injury can be induced and the common tools necessary for its creation, it has become a common model for the assessment of regrowth and recovery [123, 124]. Despite its attractiveness, the main disadvantage of the crush/compression model is its limited clinical translation, particularly due to the relative size between the rodent versus the human nerve in the context of a crush. While crushes are certainly a relevant phenomenon clinically, they are typically presented with polytrauma involving transection or laceration of the nerve [125].

PNI models can also be induced through total or partial transection/laceration of a nerve, or through creation of a gap yielding a proximal and distal stump. Such models are used to better mimic injury mechanics, surgical anatomies, and the longer-term healing and compensatory mechanisms that ensue post-injury. Common gap length typically does not exceed 20 mm, where the most common range for inducing a gap-based PNI is 6–10 mm [117]. Notably, clinical standards of practice can be mimicked in the laceration or gap creation models, anastomosing the nerve together after injury and/or suturing a cuff or conduit onto the site of injury. Notably, the clinical standard for repairing gapped segments is autografting, where the removed segment is addressed most commonly with the sural nerve, medial cutaneous antebrachial nerve, or saphenous nerve, and can be mismatched diametrically and/or in sensory versus motor function [126]. Due to regulations and ethical considerations for animal usage, autografts typically do not mimic the clinical scenarios owing to the inherent trauma of the donor site; it is typical of autografts in animal-based PNI models that the length of the nerve be rotated 180 degrees, creating a "reverse polarity" autograft for addressing the gap. Whether a more clinically relevant donor site will yield increasingly translatable information is, at this point, speculative [127].

Crush, transection, and gap PNI models result not only in damage to the affected nerve, but lead to denervation and the eventual atrophy of the target musculature. This process will begin immediately after injury is induced to the nerve, resulting, where the electrochemical signaling pathway to the target ceases, in deterioration of function, upregulation of inflammation, and potential fibrosis. Loss of function and sensory ability in the hind limb is a consequence of the rapid and aggressive muscle atrophy, which is associated with gap and transection nerve injuries [128, 129]. This effect of PNI on the musculature is often telling as the dysfunction can be measured using various functional tests that are well documented in the literature. Such tests typically include walking track analysis [130, 131], gait kinematics [132], sensory/stimuli response [133], nerve conduction velocity [134], and electrophysiological outputs such as the generation of isometric contraction force [135] or compound muscle action potential (CMAP) [136, 137]. There is speculation about the reliability of many commonly used tests that rely on monitoring and quantifying animal behavior, such as sciatic nerve indexing and stimuli-dependent sensory tests. Electrophysiological testing tends to produce more robust assessments of functionality and muscular health in PNI models, the cost of which is the highly invasive nature of such examinations [138]. Induction of median, ulnar, or radial nerve injury in bipedal models (i.e., nonhuman primates) specifically can yield assessments that better mimic functional return in humans. Such analyses include grasping strength as well as the fidelity and accuracy of digit motion. These tests are more translationally akin to human upper extremity function relative to the quadrupedal counterparts, but such modeling does not come without cost or controversy [139–141].

Histological and immunohistochemical evaluations

Schwann cells are essential in nerve health and the process of PNI healing. These cells can be identified with high accuracy via immunohistochemical (IHC) methods that are specific for binding to glial fibrillar acid protein (GFAP) and the S-100 protein [142]. Essential, too, is the maturation process of the axon. Axonal regrowth and repair can be identified through the investigation of neurotubules and neurofilaments. Both have similarly composed cytoskeletons, where growth-associated protein-43 (GAP-43) is specific of new axons not yet expressing neurofilaments, followed by staining for β-III tubulin and neurofilament as the axon matures [143]. Nerve fibers themselves can also offer specific and quantitative insights into the health and healing of the nerve in terms of fiber count, diameter and/or cross-sectional area, and density. The myelin also can be quantified for its thickness, followed by the calculation of the g-ratio, which is myelin thickness normalized to the axon diameter [144, 145]. With regard to IHC and cell/structure-specific staining, there is little crossover between the undifferentiated ASC and the nerve cells other than the progenitor marker CD34, which is lost with increasing cell maturity [146]. This lack of crossover may be of value for the identification of ASCs ectopically residing in the niche, and provides an option for cell tracing in addition to labeling prior to use as a therapeutic.

Adipose-derived stem cells in preclinical peripheral nerve injury studies

ASCs introduced into injury sites have been shown to promote tissue regeneration and minimize inflammatory response, and specifically, have been observed to recruit Schwann cells by secreting GDNF and BDNF as well as differentiate into Schwann cells [147]. Additionally, it has been shown that ASCs have the capacity to form myelin when under specific conditions [73]. Though the ASC can be coerced to a Schwann cell-like phenotype in vitro, there is debate on whether the ASC can transdifferentiate in vivo, and studies have presented mixed results. The potential for the ASC to be transdifferentiated, however, offers a unique benefit for its use.

Transdifferentiation of adipose-derived stem cells

Transdifferentiation of stem and stromal cells has become attractive for therapeutic intervention, and to do so with ASCs prior to use in PNI-relevant applications is no exception. While the undifferentiated ASC holds considerable potential in PNI with regard to its secretome, it possesses inherent differences relative to the Schwann cells, notably its lack of voltage gating and electrical transduction potential. Initial attempts at exploring the stemness of ASCs yielded about 10% conversion into neuronal-like phenotypes in culture [5]. To date, several additional protocols have been developed specifically for the transdifferentiation of ASCs into Schwann cells or the Schwann cell phenotype [73, 148–152]. Selected protocols and their reported results for transdifferentiation are summarized in Table 3. Though each protocol is unique, there are commonalities with respect to the duration required for the ASCs to begin expressing Schwann cell-like features. It seems that meaningful changes in the ASC phenotype toward that of the Schwann cell begins around days 7 to 10; the bipolar and tripolar spindle-like shape begins presenting at approximately the 1-week time point. After this, the Schwann cell-transdifferentiated

TABLE 3 Methods and results from multiple protocols for the transdifferentiation of adipose-derived stem cells into Schwann cell-like cells.

Method and reference	Protocol	Results
Intermittent induction	Sprague–Dawley rat ASCs (passage 2)	ASCs became morphologically similar to SCs under prolonged induction, exhibiting a narrow fusiform-like shape with a bipolar or tripolar structure;
	24h "pre-induction medium 1": serum-free Dulbecco's modified Eagle's medium (DMEM) containing 1mM β-mercaptoethanol (BME)	
	72h "pre-induction medium 2": DMEM, 10% fetal bovine serum (FBS), and 35ng/mL all trans-retinoic acid (RA)	
Sun et al. [144]	96h "complete induction medium": DMEM, 10% FBS, 5μM forskolin, 200ng/mL recombinant human heregulin-β1 (HRG), 10ng/mL bFGF, and 5ng/mL recombinant rat PDGF-AB	
	72h "incomplete induction medium": DMEM, 10% FBS, 5μM forskolin, and 200ng/mL HRG	Apoptosis of unsuccessfully differentiated ASCs occurred over time;
	72h "complete induction medium"	SC-ASCs expressed NGF, BDNF, and ciliary neurotrophic factor (CNTF)
OECCM+SB supplemented medium	Sprague–Dawley rat ASCs (passage 2)	Significant upregulation of markers S100 and GFAP and maintain phenotype after three passages post-transdifferentiation;
Xie et al. [145]	7Days: Olfactory ensheathing cell conditioned medium (OECCM), 20μM SB, 5μM forskolin, 5μM RA, 5μM BME, 10ng/mL bFGF, 5% FBS	Colocalization of Tuj-1 and BMP, suggestive that ASCs form myelin;
	Note: SB, or SB431542, is a small molecule and an inhibitor of the TGF-β/activin/nodal pathway	SC-ASCs secrete NGF, BDNF, and GDNF
SC-conditioned medium	Wistar rat ASCs (passage 4)	ADSCs changed from a monolayer of large cells with a flat morphology to a small number of bipolar or tripolar spindle-like shaped cells with progressively increasing cell densities
	24h: DMEM, 10% FBS, 1mM BME	
Fu et al. [146]	72h: DMEM/F12, 10% FBS, 35ng/mL all-trans-retinoic acid	SC-ASC expressed Schwann cell markers glial fibrillary acidic protein (GFAP), S100, and P75
	12days (SC-conditioned medium): DMEM, 10% FBS, 14μM forskolin, 5ng/mL PDGF-AA, 10ng/mL bFGF, 200ng/mL recombinant human HRG	
Her-beta	Human ASCs (passage 2)	Both undifferentiated and SC-ASCs expressed p75, suggesting this marker is not ideal as a differentiable characteristic marker;
	24h: DMEM, 10% FBS, 1mM 2-mercaptoethanol	Distinguishable spindle versus squamous at 14 days of incubation with expression of GFAP but *not* S100;

TABLE 3 Methods and results from multiple protocols for the transdifferentiation of adipose-derived stem cells into Schwann cell-like cells—cont'd

Method and reference	Protocol	Results
Tomita et al. [69]	72h: DMEM, 10% FBS, 35ng/mL all-trans-retinoic acid	
	14days (differentiation medium): DMEM, 10% FBS, 1mM 2-mercaptoethanol, 5ng/mL PDGF, 10ng/mL bFGF, 5.7μg/mL forskolin, 200ng/mL recombinant human HRG	SC-ASCs secreted markedly more NGF, GDNF, and BDNF relative to undifferentiated ASCs
Rat sciatic nerve conditioned medium	Sprague–Dawley rat ASCs (passage 3–5)	ASCs exhibited bipolar, spindle-like morphologies at 12h, and nearly all ASCs showed bi- or tripolar, elongated spindle shapes after 48h in induction medium
Liu et al. [148]	48h and prior to ASC culture: 1cm rat sciatic nerve fragments soaked in DMEM, 10% FBS, then filtered to remove fragments	SC-ASC developed thin cytoplasmic extensions and large nuclei (resembling Schwann cells)
	5 days: ASCs exposed to conditioned medium	SC-ASC presented GFAP and S100, not being present in undifferentiated ASCs

Abbreviations: ASC, adipose-derived stem cell; BDNF, brain-derived neurotrophic factor; BME, β-mercaptoethanol; CNTF, ciliary neurotrophic factor; DMEM, Dulbecco's modified Eagle's medium; GDNF, glial cell line-derived neurotrophic factor; GFAP, glial fibrillary acidic protein; FBS, fetal bovine serum; bFGF, fibroblast growth factor (basic); HRG, heregulin-β1; NGF, nerve growth factor; OECCM, olfactory ensheathing cell conditioned medium; PDGF, platelet-derived growth factor, PDGF-AB; RA, retinoic acid; SC, Schwann cell; SC-ASC, Schwann cell-transdifferentiated adipose-derived stem cell; TGF-β, transforming growth factor-b.

ASCs (SC-ASCs) begin to change their secretome, releasing NGF, BDNF, and ciliary neurotrophic factor (CNTF). Additionally, SC-ASCs begin expressing surface markers like those of Schwann cells, being glial fibrillary acidic protein (GFAP), S100, and P75, though P75 may be expressed even by undifferentiated ASCs when cultured in neuronal cell media. The ability of ASCs to transdifferentiate into the Schwann cell phenotype has not been met with absolute success; several papers describe a lack of in vivo transdifferentiation [153, 154]. Successful transdifferentiation protocols for ASC-to-Schwann cell conversion have been reported as a relatively time-intensive process that can take more than 3weeks to sufficiently expand and subsequently differentiate the amounts of cells necessary for achieving a therapeutic benefit [155, 156]. This does assume the use of ASCs from an autologous source, though any sort of cell therapy must be quick to expand, as the functional return post-PNI is inversely proportional to the time of denervation [157], and therefore a lengthy wait for a receiving a therapy can be at odds with its regenerative benefit.

Undifferentiated adipose-derived stem cells and Schwann-cell-transdifferentiated adipose-derived stem cells in peripheral nerve injury

The therapeutic potential of undifferentiated ASCs has been increasingly studied owing to their ability to secrete various factors that are advantageous in injury or disease state.

Additionally, their ease of harvest, expansion, and controllable differentiation capacity enhances their appeal for PNI-relevant applications. Since ASCs can be obtained as surgical waste from elective surgeries such as liposuction, body contouring, or weight loss procedures, they forego the ethical concerns of stem cell use while simultaneously mitigating the need for any significant donor site morbidity. Though the ease with which ASCs can be harvested and expanded in culture have led to their popular use in research, it is their success in the in vivo models for PNI that shows promise as potential therapies [158]. The preceding decade has resulted in significant in vivo studies of ASCs' effect on PNI. Common models for their assessment have included five peripheral nerves, being the cavernous [159–167], facial [168–171], laryngeal [172], peroneal [173, 174], and sciatic nerves [153, 154, 156, 175–202]. A summary of the models, their durations, and proposed mechanisms is presented in Table 4.

TABLE 4 Summarized models and results from in vivo modeling of adipose-derived stem cells (ASCs) and peripheral nerve injury (PNI), organized alphabetically by nerve injured and then by duration of the investigation.

Reference	Nerve	Injury type	Animal	Duration (weeks)	Results	Proposed mechanisms
Lin et al. [155]	Cavernous	Crush	Rat	1	Substantial recovery of erectile function when ASCs seeded into decellularized adipose	Downregulation of inflammation with nervous tissue ingrowth
Chen et al. [156]	Cavernous	Crush	Rat	4	ASC injection increased intracavernous pressure and presence of Schwann cells at dorsal stump	ASC-mediated increases in early pigment epithelium-derived factor (PEDF), followed by elevated nitric oxide synthase and phosphorylated Akt
Fandel et al. [157]	Cavernous	Crush	Rat	4	Intracavernous, but not perineural, ASC injection significantly improved erectile function	Injury-induced upregulation of stromal cell-derived factor-1 recruited injected ASCs, with neuroregenerative effects
Qui et al. [158]	Cavernous	Crush	Rat	4	Immediate and delayed ASC injection increased neuronal nitric oxide synthase and neurofilament; increased smooth muscle-to-collagen ratio	Increase of spontaneous nerve regeneration in ASC-treated conditions, producing prosurvival, antiapoptotic, and neurotrophic factors
Yang et al. [159]	Cavernous	Crush	Rat	4	ASC injection improved erectile function via nNOS-positive nerve regeneration	ASCs may promote neurite outgrowth without direct contact through the neurturin–GFRα2 pathway

TABLE 4 Summarized models and results from in vivo modeling of adipose-derived stem cells (ASCs) and peripheral nerve injury (PNI), organized alphabetically by nerve injured and then by duration of the investigation—cont'd

Reference	Nerve	Injury type	Animal	Duration (weeks)	Results	Proposed mechanisms
Ying et al. [160]	Cavernous	Crush	Rat	4	SC-ASC injection increased intracavernous pressure, myelinated axons, and dorsal neuronal nitric oxide synthase-positive fibers	SC-ASC-mediated production of neuronal growth factors; protection against hemorrhage-induced apoptosis
You et al. [161]	Cavernous	Crush	Rat	4	ASC injection significantly improved erectile function, though smooth muscle content was similar to controls	Possible mediation of neuronal nitric oxide synthase by injected ASCs
You et al. [162]	Cavernous	Crush	Rat	4	ASC injection increased maximal and mean intracavernous pressures, and increased erectile function	Increased expression of von Willebrand, thus, angiogenesis, may accelerate healing cascades
Ying et al. [163]	Cavernous	Crush	Rat	12	ASC injection increased mean arterial and intracavernous pressures; increased smooth muscle/collagen ratio in the corpus cavernosum	ASC secretome of VEGF, BDNF, and CXCL5 upregulated nNOS and acted on Janus kinase (JAK)/signal transducer and activator of transcription (STAT) pathway
Sun et al. [164]	Facial	Defect, 8mm	Rat	8	SC-ASCs in decellularized artery conduit were involved in the axonal regeneration and remyelination more so than ASCs	SC-ASCs interact with endogenous Schwan cells to stimulate nerve regeneration
Abbas et al. [165]	Facial	Transection	Rat	12	Significant ASC-mediated axonal regeneration quantified by electrophysiology and histology	SC-ASC transdifferentiation suspected
Ghoreishian et al. [166]	Facial	Defect, 7mm	Dog	12	ASCs in an ePTFE conduit increased nerve conduction velocity; no appreciable histological change between groups	Possible ASC-mediated acceleration of axonal regeneration through ePTFE tube

Continued

TABLE 4 Summarized models and results from in vivo modeling of adipose-derived stem cells (ASCs) and peripheral nerve injury (PNI), organized alphabetically by nerve injured and then by duration of the investigation—cont'd

Reference	Nerve	Injury type	Animal	Duration (weeks)	Results	Proposed mechanisms
Watanabe et al. [167]	Facial	Defect, 7mm	Rat	13	ASCs or SC-ASCs in collagen conduits performed to near-autologous metrics with facial palsy scoring system	Secretion of neurotrophic factors (NGF, BDNF, GDNF); hypoxia induced upregulation of angiogenic and antiapoptotic factors (VEGF, HGF, bFGF)
Li et al. [168]	Laryngeal	Crush	Rat	6	ASCs injection hastened vocal fold movement regeneration, compared to SC-ASCs and ECM injections	ASC secretion of trophic factors, stabilization of the microenvironment
Tomita et al. [169]	Peroneal	Defect, 10mm	Rat	10	SC-ASC increased nerve regeneration and motor functional recovery	SC-ASC associated with neurites and provided trophic support to existing and regenerating axons
Passipieri et al. [170]	Peroneal	Defect, 6mm	Rat	12	ASC in PCL conduit increased neurofilament and Schwann cell presence with enhanced recovery of muscular contraction	ASC-mediated repair at the distal stump promoted refunctionalization of neuromuscular junctions
Di Summa et al. [152]	Sciatic	Defect, 10mm	Rat	2	SC-ASC in fibrin conduit increased axonal regeneration and Schwann cell presence relative to sham	Likely similar mechanisms between SC-ASC and Schwann cells in regeneration
Erba et al. [171]	Sciatic	Defect, 10mm	Rat	2	ASCs in a conduit stimulate proximal axonal outgrowth; upregulate Schwann cell proliferation in distal stump	Regenerative paracrine secretome of ASCs; ASC downregulation of SRY gene
Kingham et al. [172]	Sciatic	Defect, 10mm	Rat	2	ASCs in fibrin conduit enhanced overall axon growth	ASCs increase intraneural angiogenesis, enhance GAP-43, and activate transcription factor (ATF)-3 expression
Suganuma et al. [173]	Sciatic	Defect, 10mm	Rat	2	ASCs in collagen conduit promoted significantly faster nerve regeneration per protein gene product 9.5	Likely secretion of some humoral factor such as Neu-1 and/or VEGF, with Schwann cell transdifferentiation not observed

TABLE 4 Summarized models and results from in vivo modeling of adipose-derived stem cells (ASCs) and peripheral nerve injury (PNI), organized alphabetically by nerve injured and then by duration of the investigation—cont'd

Reference	Nerve	Injury type	Animal	Duration (weeks)	Results	Proposed mechanisms
Marconi et al. [150]	Sciatic	Crush	Mouse	3	Significant improvement in fiber sprouting and the reduction of inflammatory infiltrates	ASCs induced a local production of GDNF through Schwann cell activation
Jiang et al. [174]	Sciatic	Defect, 20 mm	Rat	4	ASCs in decellularized nerve did not elicit an immune response	ASCs mediated $CD3^+$, $CD4^+$, and $CD8^+$ subsets in a manner similar to autograft
Kappos et al. [175]	Sciatic	Crush	Rat	4	ASC injection resulted in no significant differences in G-ratio or relative muscle mass compared with controls	Perhaps insufficiently severe model and/or insufficient stimulation by neuronal factors
Luo et al. [176]	Sciatic	Defect, 50 mm	Dog	4	ASCs with TGFβ in decellularized nerve ECM increased nerve regeneration more so than control	VEGF significantly increased in TGFβ-exposed ASCs; reduced inflammatory response; ASC reduction of apoptosis
Masgutov et al. [177]	Sciatic	Defect, 10 mm	Rat	4	ASCs increased survival of ganglia neurons, improved vascularization, increased distal myelination	ASC-mediated angiogenesis; possible SC-ASC transdifferentiation
Sánchez et al. [178]	Sciatic	Crush	Rat	4	ASC-mediated normalization of EMG amplitude and gait recovery, though no differences in muscle mass	Concentration of proregenerative molecules in the microenvironment such as BDNF, neurotrophin-3/4, VEGF, IGF1
Sowa et al. [179]	Sciatic	Defect, 5 mm	Mice	4	ASCs in gelatin conduits promoted axonal regeneration and myelin formation, comparable to Schwann cell control	No evidence of transdifferentiation suggesting naive mechanisms of ASC mediation
Tremp et al. [180]	Sciatic	Defect, 10 mm	Rat	4	Syngeneic ASCs in fibrin conduit increased axonal regenerated relative to xenogeneic ASCs	Transdifferentiation of transplanted ASC; immunosuppressive properties of inhibiting mixed lymphocyte proliferation

Continued

TABLE 4 Summarized models and results from in vivo modeling of adipose-derived stem cells (ASCs) and peripheral nerve injury (PNI), organized alphabetically by nerve injured and then by duration of the investigation—cont'd

Reference	Nerve	Injury type	Animal	Duration (weeks)	Results	Proposed mechanisms
Albright et al. [181]	Sciatic	Defect, 15mm	Rat	6	ASCs in poloxamer conduit increased axonal regrowth	Trophic and ASC-macrophage mediation promoted regeneration over inflammation
Hsieh et al. [182]	Sciatic	Defect, 10mm	Rat	6	ASCs on external wall of PLA conduit increased regenerated nerve and number of myelinated axons	ASC migration into conduit toward injury site acted as epineural-like support; possible local Schwann cell interactions
Hsueh et al. [183]	Sciatic	Defect, 10mm	Rat	6	ASCs in chitosan-coated silicone conduit increased myelinated axons density, muscle fiber diameter, and gait stride lengths	ASC-mediated inhibition of IL-1β and leukotriene B4 receptor-1; conversion into neurosphere morphology from ASCs
Dai et al. [184]	Sciatic	Defect 15mm	Rat	8	Combination of SCs and ASCs had the greatest functional gait recovery and nerve conduction velocity	Synergistic effects of occulting ASCs with Schwann cells, with emphasis on NGF production
Georgiou et al. [185]	Sciatic	Defect, 15mm	Rat	8	SC-ASCs in collagen conduit increased axon presence in distal, but not proximal, stump	SC-ASCs enhanced neurite outgrowth in situ; increased growth factor mRNAs possibly due to cell-ECM interaction
Liu et al. [186]	Sciatic	Defect, 10mm	Rat	8	ASCs in gelatin–ceramic conduit increased gait function and CMAP relative to autograft	Dampening of inflammatory and foreign body response, being both ASC and conduit mediated
Masgutov et al. [187]	Sciatic	Defect, 10mm	Rat	8	ASCs increased glial cell presence	ASC-mediated neurotrophic factor transport
Carriel et al. [188]	Sciatic	Defect, 10mm	Rat	12	ASC-filled conduit upregulated neurofilament and growth-associated protein (GAP-43)	Potential temporal regulation of nerve sprouting via neurofilament and GAP-43 mediation
Chato-Astrain et al. [189]	Sciatic	Defect, 10mm	Rat	12	ASCs increased nerve regeneration and functional recovery	Insufficient evidence for ASC-specific contribution when inside nerve conduit

TABLE 4 Summarized models and results from in vivo modeling of adipose-derived stem cells (ASCs) and peripheral nerve injury (PNI), organized alphabetically by nerve injured and then by duration of the investigation—cont'd

Reference	Nerve	Injury type	Animal	Duration (weeks)	Results	Proposed mechanisms
Hernandez-Cortes et al. [190]	Sciatic	Defect, 10mm	Rat	12	ASCs in a PCL conduit increased nerve area, myelin area, myelinated fibers	ASC-mediated enhancement of mid and distal regeneration zones
Kappos et al. [191]	Sciatic	Defect, 10mm	Rat	12	SC-ASCs in fibrin conduit showed less muscle atrophy and superior functional gait results	SC-ASC more closely mimicked the Schwann cell phenotype promoting the intrinsic regeneration cascade
Mohammadi et al. [192]	Sciatic	Defect, 10mm	Rat	12	ASCs in silicone conduit increased nerve fiber area, functional recovery, and gastrocnemius mass	Possible ASC-macrophage interactions promoting enhanced nerve regeneration via IL-1
Mohammadi et al. [193]	Sciatic	Defect, 10mm	Rat	12	ASCs in artery graft increased myelination, toe spread, and gastrocnemius muscle mass	ASC interaction with axons and Schwann cell-like cells assisting remyelination and structural recovery
Santiago et al. [149]	Sciatic	Defect, 6mm	Rat	12	ASCs in PCL conduit increased gastrocnemius muscle mass, but diminishing difference in gait relative to control	Likely ASC-mediated paracrine effects and not Schwann cell transdifferentiation at the sites of injury
Xu et al. [194]	Sciatic	Defect, 10mm	Rat	12	ASC and Schwann cells in silk-collagen conduit similar to autograft with considerably reduced local inflammation	ASCs transdifferentiate more quickly when in contact with Schwann cells, accelerating axonal growth
Saller et al. [195]	Sciatic	Defect, 20mm	Rat	16	ASCs applied at coaptation sites with reverse polarity autograft improved remyelination, axon ingrowth, and gait	ASC-mediated retrograde tracing of reestablished axonal tracts from the distal stump
Klein et al. [196]	Sciatic	Defect, 10mm	Rat	24	ASCs in collagen tube showed higher motor and sensory nerve conduction velocities, and increased Schwann cell presence	ASC-mediated organization of axonal arrangement throughout the conduit

Continued

TABLE 4 Summarized models and results from in vivo modeling of adipose-derived stem cells (ASCs) and peripheral nerve injury (PNI), organized alphabetically by nerve injured and then by duration of the investigation—cont'd

Reference	Nerve	Injury type	Animal	Duration (weeks)	Results	Proposed mechanisms
Orbay et al. [197]	Sciatic	Defect, 10mm	Rat	24	ASCs or SC-ASCs in silicone conduits increase myelination; ASCs in silicone conduit increased gait function	Paracrine effects of ASCs/SC-ASCs; native architecture (use of decellularized nerve) likely not relevant to myelination
Wei et al. [198]	Sciatic	Defect, 10mm	Rat	24	ASCs in chitosan/silk fibroin conduit enhanced nerve continuity and gait recovery, and better reinnervated the gastrocnemius	ASC-mediated and immediate prevention or reduction of axonal dieback

Abbreviations: ATF, activate transcription factor; ASC, adipose-derived stem cell; BDNF, brain-derived neurotrophic factor; CD, cluster of differentiation; CMAP, compound muscle action potential; CXCL5, C-X-C motif chemokine 5; ECM, extracellular matrix; FGFb, fibroblast growth factor (basic); GFRa2, GDNF family receptor alpha 2; GDNF, glial cell line-derived neurotrophic factor; (GAP-43), growth associate protein-43; HGF, hepatocyte growth factor; IL, interleukin; IGF-1, insulin-like growth factor-1; JAK, Janus kinase; NGF, nerve growth factor; iNOS, nitrous oxide synthase (inducible); nNOS, nitrous oxide synthase (neuronal); PEDF, pigment epithelium-derived factor; PCL, polycaprolactone; PLA, polylactic acid; ePTFE, polytetrafluoroethylene (expanded); Neu-1, sialidase 1 (lysosomal sialidase); SC-ASC, Schwann cell-transdifferentiation adipose-derived stem cell; STAT, signal transducer and activator of transcription; TFG-b, transforming growth factor-beta; VEGF, vascular endothelial growth factor.

The induction of PNIs commonly results from crush, laceration, or the creation and repair of a gap injury. While PNI to the cavernous nerve is induced almost exclusively by crush, injury to the sciatic nerve is more broadly characterized though crush injuries as well as gap resections with their subsequent repair. While these injuries are all addressed using ASCs, their administrations into the local site of injury are different, which appears to be determined by the type of injury; crush injuries typically receive ASC injections at the site of injury, whereas the options for therapeutics administration to sciatic nerve gaps and repair defects are more expansive. Nerve conduits are used clinically, and are often polymeric, i.e., poly(caprolactone) (PCL) and poly(lactic acid) (PLA), are decellularized from allogeneic nerve, are composed of purified collagens, or can be a composition of natural and synthetic materials [203, 204]. Building upon the current clinical foundation is the addition of ASCs into these conduit scaffolds. This has included embedding ASCs into a fibrin gel, seeding cells onto conduit walls, injecting cells into collagen, or injecting cells onto one of both of the coaptation sites. Despite the differences in models, injured nerves, and application techniques, the vast majority of the studies summarized here reported benefit when administered some sort of ASC therapy relative to control groups, where those controls typically did not involve a cell therapy. ASC interventions resulted in increased nerve conduction velocity, heightened electromyographic amplitude, and enhanced contractile strength of the innervated musculature. For the cavernous nerve specifically, ASC injection increased the maximal and mean intracavernous pressures and, in general, improved erectile function. Histologically, myelination and axonal regeneration appeared to be hastened by the presence of ASCs, with an increased ratio of smooth muscle to fibrous collagen. Mechanistically, there is still considerable investigation required to elucidate the means by which ASCs improved the PNI environment. There is a clear debate on the ability of ASCs to transdifferentiate to Schwann cells in vivo, and perhaps their ability to do so is influenced by the injury environment itself. Further, while the therapeutic potential of the ASC secretome is understood and is thought to contribute to antiinflammation and angiogenesis, its actual contribution to the PNI environment remains to be explained.

In vivo models and adipose-derived stem cell usage in nontraumatic pathologies

Peripheral neuropathies can occur when the axon or Schwann cells show dysfunction due to metabolic, toxic, infectious, or genetic causes. Current models for in vivo study of nontraumatic peripheral neuropathy pathologies include diabetic neuropathy (DN), chemotherapy, and immunodeficiency models. Diabetically induced neuropathies have been modeled in rodents with the use of streptozotocin injections. Streptozotocin is toxic to pancreatic beta cells and therefore can be used to closely model type 1 diabetes. Streptozotocin is effective in both mice and rats, which presents the opportunity for genetic variants as well as knockout models [205]. In recent years, ASC therapies specifically have seen only minimal investigation both in vitro and in vivo. In vitro, ASCs have been conditioned with the iron chelator deferoxamine to increase their expression of hypoxia inducible factor (HIF)-1α, VEGF and Ang-1, NGF, BDNF, and GDNF, as well as antiinflammatory cytokines such as IL-4 and IL-5 [206]. In vivo, a single intravenous injection of ASCs reverts neuropathic inflammation, restores skin innervation, and reduces peripheral immune activation [207]. Importantly, the axonal degradation in diabetic rodents rarely occurs in the same locations or with the same patterns as observed in humans with type 1 diabetic neuropathies. Reliable outcome measures are of concern as clinical outcome measures typically gauge levels of discomfort resulting from neuropathy. Subjective measurements such as pain can be particularly unreliable since these tests rely on observing animal behavior and correlating it with the histological outcomes of the affected nerve anatomy. Histological analyses have been of much importance to developing neuropathy injury models since high levels of injury pathology inconsistency have been reported, particularly with diabetic rat models [208].

Clinical trials and future considerations

Current clinical applications of adipose-derived stem cells addressing peripheral nerve injury

As of December 2019, there are nearly 400 investigations registered on ClinicalTrials.gov related to ASC therapies listed under the query "adipose stem cell." In stark contrast, there are no clinical trials currently listed when modifying the ASC query to include "peripheral nerve injury." Additionally, and as aforementioned, only one clinical trial involves stem cells seeking to treat PNIs (NCT03999424: *Autologous human Schwann cells in peripheral nerve repair*). Causalgia, or the burning sensation caused from nerve trauma, is among the pathologies sought to be addressed in this trial. This United States-based trial is described as an interventional study investigating the safety and efficacy of using adipose-derived cellular stromal vascular fraction, which includes ASCs in its contents. ASCs have been proposed, too, for clinical use in the pain management of reflex sympathetic dystrophy (RSD) (NCT02987855: *Use of autologous adult adipose-derived stem/stromal cells in reflex sympathetic dystrophy (RSD), complex regional pain syndrome (CRPS), and fibromyalgia*). While the pathophysiology is not fully understood, it is thought to involve atypical neuronal transmission that likely involves the peripheral nerve tissue [209]. While this study is still in recruitment, ASCs are autologously isolated from digested lipoaspirate, suspended in 500cc saline, and then deployed intravenously. Safety of the ASC therapy is to be the primary outcome measure, with secondary measures of change from baseline pain levels and any quality-of-life improvements. ASCs combined with grafted whole adipose tissue have been used for pain management in amputees as well (NCT01645722: *Enriched autologous fat grafting for treating pain at amputation sites*), where ASC-enriched fat grafting was injected at the amputation site. In this study, safety of the autologous therapy was established, though there was no significant improvement in pain scores at 2years, nor were there significant changes in disability indexes [18].

Additional relevant clinical investigations have been found in the published literature for applications in hemifacial atrophy as a result of Parry–Romberg disease, in erectile dysfunction after prostatectomy, and after conventional treatments were ineffective. Parry–Romberg disease is rare, resulting in facial atrophy, and can also be associated with congenital damage of the facial peripheral nerves. In this clinical investigation (Seoul, Korea), 1 million expanded ASCs were supplemented into an adipose tissue transfer and subsequently injected into the atrophic facial defect [210]. While this study was fundamentally cosmetic in nature, based on the mounting in vivo evidence, ASCs' paracrine effects on the local tissues, including the peripheral nervous tissue, should not be ignored, though any improvements in pain reported are inherently speculative and cannot be decoupled from the adipose tissue transfer itself. An investigation on stromal vascular fraction-mediated erectile dysfunction (Odense, Denmark) was a more direct investigation of a cellular therapy on penile nerve injury after undergoing a radical prostatectomy. Of the 17 patients receiving an injection of 8–32 million cells after isolation using the Cytori device (Lorem Cytori USA, Inc., San Diego, CA), 11 men regained the ability to have sexual intercourse, suggesting potential efficacy of cell therapy [211]. Again, however, the specific relevance of this investigation to the therapeutic potential of ASCs specifically cannot be decoupled from the other cells within stromal vascular fraction and their likely contribution to the nerve injury.

Conclusions and future directions

Whether used in PNI or in another facet of therapeutic intervention, the Office of Tissues and Advanced Therapies (OTAT) within the US Food and Drug Administration (FDA) would likely possess majority responsibility for the regulation of an ASC therapy. Depending on the indications for use or the process by which the cells are expanded, OTAT may defer to additional centers at the FDA such as the Center for Drug Evaluation and Research (CDER) or the Center for Biologics Evaluation and Research (CBER) to better understand the battery of requisite safety and efficacy testing prior to approval. While the vast majority of cell therapies that currently have FDA approval are blood cell derived, a shift from small molecules to cell therapies appears almost to be inevitable based on the pipeline of registered trials and published literature. The ease by which ASCs in particular can be harvested makes them an attractive prospect for cell therapy. Specifically relevant to PNI is the enduring annual expenditure that is burdensome both in the context of healthcare as well as in terms of lost employment. The clinical need remains for further development in PNI repair and, in particular, for addressing large defects where damaged nerve must be resected in gaps leaving only autografting as the viable option for any appreciable recovery. Indeed, the preclinical literature is as abundant as it is promising, with ASCs addressing applications across PNIs. It is, however, the elusive mechanisms of the ASCs' impact on the nervous tissue that, in part, hinder their clinical translatability. As with any therapy, there is an inherent danger of the unknown, which must be carefully balanced with the prospective benefit. Despite the current challenges and inherent risks, the potential use of ASCs in PNI is sound, given their ability to induce angiogenesis, downregulate inflammation, and mitigate fibrosis, as well as their ability to transdifferentiate into the Schwann cell-like lineage expressing relevant surface markers and soluble factors. Holistically, these attributes position ASC therapies as an ideal candidate for addressing a broad range of injuries to the various peripheral nerves.

References

[1] M. Locke, V. Feisst, P.R. Dunbar, Concise review: human adipose-derived stem cells: separating promise from clinical need, Stem Cells 29 (3) (2011) 404–411.

[2] V.V. Miana, E.A.P. Gonzalez, Adipose tissue stem cells in regenerative medicine, Ecancermedicalscience 12 (2018) 822.

[3] P.A. Zuk, The adipose-derived stem cell: looking back and looking ahead, Mol. Biol. Cell 21 (11) (2010) 1783–1787.

[4] P.A. Zuk, et al., Human adipose tissue is a source of multipotent stem cells, Mol. Biol. Cell 13 (12) (2002) 4279–4295.

[5] P.A. Zuk, et al., Multilineage cells from human adipose tissue: implications for cell-based therapies, Tissue Eng. 7 (2) (2001) 211–228.

[6] P.C. Baer, et al., Human adipose-derived mesenchymal stem cells in vitro: evaluation of an optimal expansion medium preserving stemness, Cytotherapy 12 (1) (2010) 96–106.

[7] P. Palumbo, et al., Methods of isolation, characterization and expansion of human adipose-derived stem cells (ASCs): an overview, Int. J. Mol. Sci. 19 (7) (2018).

[8] L. Zhao, T. Johnson, D. Liu, Therapeutic angiogenesis of adipose-derived stem cells for ischemic diseases, Stem Cell Res. Ther. 8 (1) (2017) 125.

[9] H. Suga, et al., Paracrine mechanism of angiogenesis in adipose-derived stem cell transplantation, Ann. Plast. Surg. 72 (2) (2014) 234–241.

[10] Z. Zhong, et al., GDNF secreted from adipose-derived stem cells stimulates VEGF-independent angiogenesis, Oncotarget 7 (24) (2016) 36829–36841.

[11] A.A. Leto Barone, et al., Immunomodulatory effects of adipose-derived stem cells: fact or fiction? Biomed. Res. Int. 2013 (2013) 383685.

[12] M. Pappalardo, et al., Immunomodulation in vascularized composite allotransplantation: what is the role for adipose-derived stem cells? Ann. Plast. Surg. 82 (2) (2019) 245–251.

[13] M. Waldner, et al., Characteristics and immunomodulating functions of adipose-derived and bone marrow-derived mesenchymal stem cells across defined human leukocyte antigen barriers, Front. Immunol. 9 (2018) 1642.

[14] M. Patrikoski, B. Mannerstrom, S. Miettinen, Perspectives for clinical translation of adipose stromal/stem cells, Stem Cells Int. 2019 (2019) 5858247.

[15] D.T. Chu, et al., Adipose tissue stem cells for therapy: an update on the progress of isolation, culture, storage, and clinical application, J. Clin. Med. 8 (7) (2019).
[16] C. O'Rourke, et al., An allogeneic 'off the shelf' therapeutic strategy for peripheral nerve tissue engineering using clinical grade human neural stem cells, Sci. Rep. 8 (1) (2018) 2951.
[17] S. Sayad Fathi, A. Zaminy, Stem cell therapy for nerve injury, World J. Stem Cells 9 (9) (2017) 144–151.
[18] D.A. Bourne, et al., Amputation-site soft-tissue restoration using adipose stem cell therapy, Plast. Reconstr. Surg. 142 (5) (2018) 1349–1352.
[19] D. Grinsell, C.P. Keating, Peripheral nerve reconstruction after injury: a review of clinical and experimental therapies, Biomed. Res. Int. 2014 (2014) 698256.
[20] C.N. Bruyns, et al., Predictors for return to work in patients with median and ulnar nerve injuries, J. Hand. Surg. Am. 28 (1) (2003) 28–34.
[21] J.B. Jaquet, et al., Median, ulnar, and combined median-ulnar nerve injuries: functional outcome and return to productivity, J. Trauma 51 (4) (2001) 687–692.
[22] K. Ondicova, B. Mravec, Multilevel interactions between the sympathetic and parasympathetic nervous systems: a minireview, Endocr. Regul. 44 (2) (2010) 69–75.
[23] M.J. Kenney, C.K. Ganta, Autonomic nervous system and immune system interactions, Compr. Physiol. 4 (3) (2014) 1177–1200.
[24] Z.L. Sebo, et al., A mesodermal fate map for adipose tissue, Development 145 (17) (2018).
[25] M.M. Smith, B.K. Hall, Development and evolutionary origins of vertebrate skeletogenic and odontogenic tissues, Biol. Rev. Camb. Philos. Soc. 65 (3) (1990) 277–373.
[26] M. Elshazzly, O. Caban, Embryology, central nervous system, StatPearls, Treasure Island (FL), 2019.
[27] P. Rigoard, et al., Anatomy and physiology of the peripheral nerve, Neurochirurgie 55 (Suppl 1) (2009) S3–12.
[28] R.M. Menorca, T.S. Fussell, J.C. Elfar, Nerve physiology: mechanisms of injury and recovery, Hand Clin. 29 (3) (2013) 317–330.
[29] Z. Wei, et al., Emerging role of Schwann cells in neuropathic pain: receptors, glial mediators and myelination, Front. Cell Neurosci. 13 (2019) 116.
[30] D. George, P. Ahrens, S. Lambert, Satellite glial cells represent a population of developmentally arrested Schwann cells, Glia 66 (7) (2018) 1496–1506.
[31] M.V. Sofroniew, Molecular dissection of reactive astrogliosis and glial scar formation, Trends Neurosci. 32 (12) (2009) 638–647.
[32] G.K. Tofaris, et al., Denervated Schwann cells attract macrophages by secretion of leukemia inhibitory factor (LIF) and monocyte chemoattractant protein-1 in a process regulated by interleukin-6 and LIF, J. Neurosci. 22 (15) (2002) 6696–6703.
[33] C.S. Barros, S.J. Franco, U. Muller, Extracellular matrix: functions in the nervous system, Cold Spring Harb. Perspect. Biol. 3 (1) (2011) a005108.
[34] M.E. Shy, Biology of peripheral inherited neuropathies: Schwann cell axonal interactions, Adv. Exp. Med. Biol. 652 (2009) 171–181.
[35] C. Melendez-Vasquez, et al., Differential expression of proteoglycans at central and peripheral nodes of Ranvier, Glia 52 (4) (2005) 301–308.
[36] Z. Wang, et al., The basic characteristics of the Pentraxin family and their functions in tumor progression, Front. Immunol. 11 (2020) 1757.
[37] B. Wehrle-Haller, M. Chiquet, Dual function of tenascin: simultaneous promotion of neurite growth and inhibition of glial migration, J. Cell Sci. 106 (Pt 2) (1993) 597–610.
[38] B. Stevens, Neuron-astrocyte signaling in the development and plasticity of neural circuits, Neurosignals 16 (4) (2008) 278–288.
[39] M.A. Fox, et al., Distinct target-derived signals organize formation, maturation, and maintenance of motor nerve terminals, Cell 129 (1) (2007) 179–193.
[40] R. Brull, et al., Pathophysiology and etiology of nerve injury following peripheral nerve blockade, Reg. Anesth. Pain Med. 40 (5) (2015) 479–490.
[41] B.S. Jortner, Mechanisms of toxic injury in the peripheral nervous system: neuropathologic considerations, Toxicol. Pathol. 28 (1) (2000) 54–69.
[42] T.K. Lim, et al., Peripheral nerve injury induces persistent vascular dysfunction and endoneurial hypoxia, contributing to the genesis of neuropathic pain, J. Neurosci. 35 (8) (2015) 3346–3359.

[43] S. Sunderland, A classification of peripheral nerve injuries producing loss of function, Brain 74 (4) (1951) 491–516.
[44] S.E. Mackinnon, A.L. Dellon, Surgery of the peripheral nerve, Thieme Medical Publishers, New York, Stuttgart, 1988. G. Thieme Verlag. xxiv, 638 p.
[45] A. Chhabra, et al., Peripheral nerve injury grading simplified on MR neurography: as referenced to Seddon and Sunderland classifications, Ind. J. Radiol. Imaging 24 (3) (2014) 217–224.
[46] D.A. Gibson, L. Ma, Developmental regulation of axon branching in the vertebrate nervous system, Development 138 (2) (2011) 183–195.
[47] G. Gallo, The cytoskeletal and signaling mechanisms of axon collateral branching, Dev. Neurobiol. 71 (3) (2011) 201–220.
[48] E.R. Lunn, M.C. Brown, V.H. Perry, The pattern of axonal degeneration in the peripheral nervous system varies with different types of lesion, Neuroscience 35 (1) (1990) 157–165.
[49] E.A. Huebner, S.M. Strittmatter, Axon regeneration in the peripheral and central nervous systems, Results Probl. Cell Differ. 48 (2009) 339–351.
[50] R. George, J.W. Griffin, The proximo-distal spread of axonal degeneration in the dorsal columns of the rat, J. Neurocytol. 23 (11) (1994) 657–667.
[51] A.D. Gaudet, P.G. Popovich, M.S. Ramer, Wallerian degeneration: gaining perspective on inflammatory events after peripheral nerve injury, J. Neuroinflammation 8 (2011) 110.
[52] S. Rotshenker, Wallerian degeneration: the innate-immune response to traumatic nerve injury, J. Neuroinflammation 8 (2011) 109.
[53] A. Llobet Rosell, L.J. Neukomm, Axon death signaling in Wallerian degeneration among species and in disease, Open Biol. 9 (8) (2019) 190118.
[54] K.R. Jessen, R. Mirsky, The repair Schwann cell and its function in regenerating nerves, J. Physiol. 594 (13) (2016) 3521–3531.
[55] J.A. Stratton, P.T. Shah, Macrophage polarization in nerve injury: do Schwann cells play a role? Neural Regen. Res. 11 (1) (2016) 53–57.
[56] R.E. Zigmond, F.D. Echevarria, Macrophage biology in the peripheral nervous system after injury, Prog. Neurobiol. 173 (2019) 102–121.
[57] S.P. Frostick, Q. Yin, G.J. Kemp, Schwann cells, neurotrophic factors, and peripheral nerve regeneration, Microsurgery 18 (7) (1998) 397–405.
[58] Q.Y. Liu, et al., Increased levels of miR-3099 induced by peripheral nerve injury promote Schwann cell proliferation and migration, Neural Regen. Res. 14 (3) (2019) 525–531.
[59] C. Yao, et al., Hypoxia-induced upregulation of miR-132 promotes Schwann cell migration after sciatic nerve injury by targeting PRKAG3, Mol. Neurobiol. 53 (8) (2016) 5129–5139.
[60] M.G. Burnett, E.L. Zager, Pathophysiology of peripheral nerve injury: a brief review, Neurosurg. Focus 16 (5) (2004), E1.
[61] Y. Liu, et al., Ryk-mediated Wnt repulsion regulates posterior-directed growth of corticospinal tract, Nat. Neurosci. 8 (9) (2005) 1151–1159.
[62] K. Kalil, L. Li, B.I. Hutchins, Signaling mechanisms in cortical axon growth, guidance, and branching, Front. Neuroanat. 5 (2011) 62.
[63] G.R. Evans, Peripheral nerve injury: a review and approach to tissue engineered constructs, Anat. Rec. 263 (4) (2001) 396–404.
[64] M.B. Bunge, et al., Role of peripheral nerve extracellular matrix in Schwann cell function and in neurite regeneration, Dev. Neurosci. 11 (4–5) (1989) 348–360.
[65] C.E. Schmidt, J.B. Leach, Neural tissue engineering: strategies for repair and regeneration, Annu. Rev. Biomed. Eng. 5 (2003) 293–347.
[66] A. Lisa, et al., Painful scar neuropathy: principles of diagnosis and treatment, Plast. Aesthetic Res. 3 (2) (2016) 68.
[67] R. Alvites, et al., Peripheral nerve injury and axonotmesis: state of the art and recent advances, Cogent. Med. 5 (1) (2018) 1466404.
[68] T. Gordon, N. Tyreman, M.A. Raji, The basis for diminished functional recovery after delayed peripheral nerve repair, J. Neurosci. 31 (14) (2011) 5325–5334.

[69] A. Hoke, Mechanisms of disease: what factors limit the success of peripheral nerve regeneration in humans? Nat. Clin. Pract. Neurol. 2 (8) (2006) 448–454.
[70] G. Terenghi, Peripheral nerve regeneration and neurotrophic factors, J. Anat. 194 (Pt 1) (1999) 1–14.
[71] S.K. Walsh, et al., Skin-derived precursor cells enhance peripheral nerve regeneration following chronic denervation, Exp. Neurol. 223 (1) (2010) 221–228.
[72] T. Hadlock, et al., A polymer foam conduit seeded with Schwann cells promotes guided peripheral nerve regeneration, Tissue Eng. 6 (2) (2000) 119–127.
[73] K. Tomita, et al., Glial differentiation of human adipose-derived stem cells: implications for cell-based transplantation therapy, Neuroscience 236 (2013) 55–65.
[74] J.R. Hess, et al., Use of cold-preserved allografts seeded with autologous Schwann cells in the treatment of a long-gap peripheral nerve injury, Plast. Reconstr. Surg. 119 (1) (2007) 246–259.
[75] L. Luo, et al., Molecular mechanisms of transdifferentiation of adipose-derived stem cells into neural cells: current status and perspectives, Stem Cells Int. 2018 (2018) 5630802.
[76] K.L. Burrow, J.A. Hoyland, S.M. Richardson, Human adipose-derived stem cells exhibit enhanced proliferative capacity and retain multipotency longer than donor-matched bone marrow mesenchymal stem cells during expansion in vitro, Stem Cells Int. 2017 (2017) 2541275.
[77] A.J. Salgado, et al., Adipose tissue derived stem cells secretome: soluble factors and their roles in regenerative medicine, Curr. Stem Cell Res. Ther. 5 (2) (2010) 103–110.
[78] J.S. Taylor, E.T. Bampton, Factors secreted by Schwann cells stimulate the regeneration of neonatal retinal ganglion cells, J. Anat. 204 (1) (2004) 25–31.
[79] R.M. Lindsay, Nerve growth factors (NGF, BDNF) enhance axonal regeneration but are not required for survival of adult sensory neurons, J. Neurosci. 8 (7) (1988) 2394–2405.
[80] J.Y. Zhang, et al., Endogenous BDNF is required for myelination and regeneration of injured sciatic nerve in rodents, Eur. J. Neurosci. 12 (12) (2000) 4171–4180.
[81] C.E. McGregor, A.W. English, The role of BDNF in peripheral nerve regeneration: activity-dependent treatments and Val66Met, Front. Cell Neurosci. 12 (2018) 522.
[82] K.R. Ko, et al., Hepatocyte growth factor (HGF) promotes peripheral nerve regeneration by activating repair Schwann cells, Sci. Rep. 8 (1) (2018) 8316.
[83] R. El-Habta, et al., The adipose tissue stromal vascular fraction secretome enhances the proliferation but inhibits the differentiation of myoblasts, Stem Cell Res. Ther. 9 (1) (2018) 352.
[84] N.A. Dzhoyashvili, et al., Disturbed angiogenic activity of adipose-derived stromal cells obtained from patients with coronary artery disease and diabetes mellitus type 2, J. Transl. Med. 12 (2014) 337.
[85] C. Ding, et al., HGF and BFGF secretion by human adipose-derived stem cells improves ovarian function during natural aging via activation of the SIRT1/FOXO1 signaling pathway, Cell Physiol. Biochem. 45 (4) (2018) 1316–1332.
[86] F. Zhang, et al., Quantification of nerve tension after nerve repair: correlations with nerve defects and nerve regeneration, J. Reconstr. Microsurg. 17 (6) (2001) 445–451.
[87] X. Zhu, H. Wei, H. Zhu, Nerve wrap after end-to-end and tension-free neurorrhaphy attenuates neuropathic pain: a prospective study based on cohorts of digit replantation, Sci. Rep. 8 (1) (2018) 620.
[88] H.M. Howarth, et al., Redistribution of nerve strain enables end-to-end repair under tension without inhibiting nerve regeneration, Neural Regen. Res. 14 (7) (2019) 1280–1288.
[89] W.L. Clark, et al., Nerve tension and blood flow in a rat model of immediate and delayed repairs, J. Hand. Surg. Am. 17 (4) (1992) 677–687.
[90] M.I. Hobson, C.J. Green, G. Terenghi, VEGF enhances intraneural angiogenesis and improves nerve regeneration after axotomy, J. Anat. 197 (Pt 4) (2000) 591–605.
[91] C. Emanueli, et al., Nerve growth factor promotes angiogenesis and arteriogenesis in ischemic hindlimbs, Circulation 106 (17) (2002) 2257–2262.
[92] V.H. Guaiquil, et al., VEGF-B selectively regenerates injured peripheral neurons and restores sensory and trophic functions, Proc. Natl. Acad. Sci. U. S. A. 111 (48) (2014) 17272–17277.
[93] H. Wang, et al., Overlapping mechanisms of peripheral nerve regeneration and angiogenesis following sciatic nerve transection, Front. Cell Neurosci. 11 (2017) 323.
[94] V. Eterno, et al., Adipose-derived mesenchymal stem cells (ASCs) may favor breast cancer recurrence via HGF/c-Met signaling, Oncotarget 5 (3) (2014) 613–633.
[95] C.S. Lee, et al., Adipose stem cells can secrete angiogenic factors that inhibit hyaline cartilage regeneration, Stem Cell Res. Ther. 3 (4) (2012) 35.

[96] Y.H. Song, et al., Adipose-derived stem cells increase angiogenesis through matrix metalloproteinase-dependent collagen remodeling, Integr. Biol. (Camb) 8 (2) (2016) 205–215.
[97] J. Rehman, et al., Secretion of angiogenic and antiapoptotic factors by human adipose stromal cells, Circulation 109 (10) (2004) 1292–1298.
[98] C. Jotzu, et al., Adipose tissue-derived stem cells differentiate into carcinoma-associated fibroblast-like cells under the influence of tumor-derived factors, Anal. Cell Pathol. (Amst) 33 (2) (2010) 61–79.
[99] A.L. Strong, et al., Obesity enhances the conversion of adipose-derived stromal/stem cells into carcinoma-associated fibroblast leading to cancer cell proliferation and progression to an invasive phenotype, Stem Cells Int. 2017 (2017) 9216502.
[100] K. Meier, et al., Silencing of ASC in cutaneous squamous cell carcinoma, PLoS One 11 (10) (2016), e0164742.
[101] I.M. Ariel, Tumors of the peripheral nervous system, CA Cancer J. Clin. 33 (5) (1983) 282–299.
[102] M. Shuayb, R. Begum, Unusual primary breast cancer—malignant peripheral nerve sheath tumor: a case report and review of the literature, J. Med. Case Rep. 11 (1) (2017) 161.
[103] B.S. Boyd, et al., Strain and excursion in the rat sciatic nerve during a modified straight leg raise are altered after traumatic nerve injury, J. Orthop. Res. 23 (4) (2005) 764–770.
[104] F. Fregnan, et al., Role of inflammatory cytokines in peripheral nerve injury, Neural Regen. Res. 7 (29) (2012) 2259–2266.
[105] P. Dubovy, I. Klusakova, I. Hradilova Svizenska, Inflammatory profiling of Schwann cells in contact with growing axons distal to nerve injury, Biomed. Res. Int. 2014 (2014) 691041.
[106] A. Lemke, et al., A novel experimental rat model of peripheral nerve scarring that reliably mimics post-surgical complications and recurring adhesions, Dis. Model Mech. 10 (8) (2017) 1015–1025.
[107] G. Koopmans, B. Hasse, N. Sinis, Chapter 19: The role of collagen in peripheral nerve repair, Int. Rev. Neurobiol. 87 (2009) 363–379.
[108] A. Banas, et al., Rapid hepatic fate specification of adipose-derived stem cells and their therapeutic potential for liver failure, J. Gastroenterol. Hepatol. 24 (1) (2009) 70–77.
[109] D.G. Hazzard, et al., Selection of an appropriate animal model to study aging processes with special emphasis on the use of rat strains, J. Gerontol. 47 (3) (1992) B63–B64.
[110] F.E. Holmes, S.A. Mahoney, D. Wynick, Use of genetically engineered transgenic mice to investigate the role of galanin in the peripheral nervous system after injury, Neuropeptides 39 (3) (2005) 191–199.
[111] R.J. Pope, et al., Characterization of the nociceptive phenotype of suppressible galanin overexpressing transgenic mice, Mol. Pain 6 (2010) 67.
[112] S.J. Gladman, et al., Improved outcome after peripheral nerve injury in mice with increased levels of endogenous omega-3 polyunsaturated fatty acids, J. Neurosci. 32 (2) (2012) 563–571.
[113] S.W. Kemp, et al., Functional recovery following peripheral nerve injury in the transgenic Thy1-GFP rat, J. Peripher. Nerv. Syst. 18 (3) (2013) 220–231.
[114] C. Magill, et al., Transgenic models of nerve repair and nerve regeneration, Neurol. Res. 30 (10) (2008) 1023–1029.
[115] C.B. Mohanty, D.I. Bhat, B.I. Devi, Use of animal models in peripheral nerve surgery and research, Neurol. India 67 (Supplement) (2019). p. S100-S105.
[116] Z. He, Y. Jin, Intrinsic control of axon regeneration, Neuron 90 (3) (2016) 437–451.
[117] D. Angius, et al., A systematic review of animal models used to study nerve regeneration in tissue-engineered scaffolds, Biomaterials 33 (32) (2012) 8034–8039.
[118] J.P. Charles, et al., Musculoskeletal geometry, muscle architecture and functional specializations of the mouse hindlimb, PLoS One 11 (4) (2016), e0147669.
[119] J.A. Batt, J.R. Bain, Tibial nerve transection - a standardized model for denervation-induced skeletal muscle atrophy in mice, J. Vis. Exp. 81 (2013), e50657.
[120] B.D. Clark, et al., Median nerve trauma in a rat model of work-related musculoskeletal disorder, J. Neurotrauma 20 (7) (2003) 681–695.
[121] T. Dowdall, I. Robinson, T.F. Meert, Comparison of five different rat models of peripheral nerve injury, Pharmacol. Biochem. Behav. 80 (1) (2005) 93–108.
[122] Z.Y. Liu, Z.B. Chen, J.H. Chen, A novel chronic nerve compression model in the rat, Neural Regen. Res. 13 (8) (2018) 1477–1485.
[123] G. Ronchi, et al., Functional and morphological assessment of a standardized crush injury of the rat median nerve, J. Neurosci. Methods 179 (1) (2009) 51–57.

[124] G. Ronchi, et al., Standardized crush injury of the mouse median nerve, J. Neurosci. Methods 188 (1) (2010) 71–75.
[125] G. Ronchi, et al., The median nerve injury model in pre-clinical research—a critical review on benefits and limitations, Front. Cell Neurosci. 13 (2019) 288.
[126] D. Cinteza, et al., Peripheral nerve—an appraisal of the current treatment options, Maedica (Buchar) 10 (1) (2015) 65–68.
[127] S.E. Roberts, et al., To reverse or not to reverse? A systematic review of autograft polarity on functional outcomes following peripheral nerve repair surgery, Microsurgery 37 (2) (2017) 169–174.
[128] B.K. Schilling, et al., Adipose-derived stem cells delay muscle atrophy after peripheral nerve injury in the rodent model, Muscle Nerve 59 (5) (2019) 603–610.
[129] A. Afshari, et al., Assessment of the effect of autograft orientation on peripheral nerve regeneration using diffusion tensor imaging, Ann. Plast. Surg. 80 (4) (2018) 384–390.
[130] L. Sarikcioglu, B.M. Demirel, A. Utuk, Walking track analysis: an assessment method for functional recovery after sciatic nerve injury in the rat, Folia Morphol. (Warsz) 68 (1) (2009) 1–7.
[131] A.S. Varejao, et al., Functional evaluation of peripheral nerve regeneration in the rat: walking track analysis, J. Neurosci. Methods 108 (1) (2001) 1–9.
[132] E.A. Kappos, et al., Validity and reliability of the CatWalk system as a static and dynamic gait analysis tool for the assessment of functional nerve recovery in small animal models, Brain Behav. 7 (7) (2017), e00723.
[133] Y.K. Kim, et al., The quantitative sensory testing is an efficient objective method for assessment of nerve injury, Maxillofac. Plast. Reconstr. Surg. 37 (1) (2015) 13.
[134] M.E. Walsh, et al., Use of nerve conduction velocity to assess peripheral nerve health in aging mice, J. Gerontol. A Biol. Sci. Med. Sci. 70 (11) (2015) 1312–1319.
[135] E.K. Merritt, et al., Functional assessment of skeletal muscle regeneration utilizing homologous extracellular matrix as scaffolding, Tissue Eng. Part A 16 (4) (2010) 1395–1405.
[136] W.D. Arnold, et al., Electrophysiological motor unit number estimation (MUNE) measuring compound muscle action potential (CMAP) in mouse hindlimb muscles, J. Vis. Exp. 103 (2015).
[137] A. Mallik, A.I. Weir, Nerve conduction studies: essentials and pitfalls in practice, J. Neurol. Neurosurg. Psychiatr. 76 (suppl 2) (2005). p. ii23.
[138] S. Geuna, The sciatic nerve injury model in pre-clinical research, J. Neurosci. Methods 243 (2015) 39–46.
[139] H.E. Cabaud, W.G. Rodkey, H.R. McCarroll Jr., Peripheral nerve injuries: studies in higher nonhuman primates, J. Hand. Surg. Am. 5 (3) (1980) 201–206.
[140] D. Wang, et al., A simple model of radial nerve injury in the rhesus monkey to evaluate peripheral nerve repair, Neural Regen. Res. 9 (10) (2014) 1041–1046.
[141] B.M. Altevogt, et al. (Eds.), Chimpanzees in biomedical and behavioral research: assessing the necessity, 2011. Washington (DC).
[142] M. Siemionow, et al., Peripheral nerve defect repair with epineural tubes supported with bone marrow stromal cells: a preliminary report, Ann. Plast. Surg. 67 (1) (2011) 73–84.
[143] C.H. Ma, et al., Accelerating axonal growth promotes motor recovery after peripheral nerve injury in mice, J. Clin. Invest. 121 (11) (2011) 4332–4347.
[144] C.L. Vleggeert-Lankamp, The role of evaluation methods in the assessment of peripheral nerve regeneration through synthetic conduits: a systematic review. Laboratory investigation, J. Neurosurg. 107 (6) (2007) 1168–1189.
[145] V. Carriel, et al., Histological assessment in peripheral nerve tissue engineering, Neural. Regen. Res. 9 (18) (2014) 1657–1660.
[146] L.E. Sidney, et al., Concise review: evidence for CD34 as a common marker for diverse progenitors, Stem Cells 32 (6) (2014) 1380–1389.
[147] A.D. Widgerow, et al., Neuromodulatory nerve regeneration: adipose tissue-derived stem cells and neurotrophic mediation in peripheral nerve regeneration, J. Neurosci. Res. 91 (12) (2013) 1517–1524.
[148] X. Sun, et al., Differentiation of adipose-derived stem cells into Schwann cell-like cells through intermittent induction: potential advantage of cellular transient memory function, Stem Cell Res. Ther. 9 (1) (2018) 133.
[149] S. Xie, et al., Efficient generation of functional Schwann cells from adipose-derived stem cells in defined conditions, Cell Cycle 16 (9) (2017) 841–851.
[150] X. Fu, et al., Induction of adipose-derived stem cells into Schwann-like cells and observation of Schwann-like cell proliferation, Mol. Med. Rep. 14 (2) (2016) 1187–1193.

[151] G. Wang, et al., Human eyelid adipose tissue-derived Schwann cells promote regeneration of a transected sciatic nerve, Sci. Rep. 7 (2017) 43248.
[152] Y. Liu, et al., A new method for Schwann-like cell differentiation of adipose derived stem cells, Neurosci. Lett. 551 (2013) 79–83.
[153] L.Y. Santiago, et al., Delivery of adipose-derived precursor cells for peripheral nerve repair, Cell Transplant 18 (2) (2009) 145–158.
[154] S. Marconi, et al., Human adipose-derived mesenchymal stem cells systemically injected promote peripheral nerve regeneration in the mouse model of sciatic crush, Tissue Eng. Part A 18 (11 – 12) (2012) 1264–1272.
[155] P.J. Kingham, et al., Adipose-derived stem cells differentiate into a Schwann cell phenotype and promote neurite outgrowth in vitro, Exp. Neurol. 207 (2) (2007) 267–274.
[156] P.G. di Summa, et al., Adipose-derived stem cells enhance peripheral nerve regeneration, J. Plast. Reconstr. Aesthet. Surg. 63 (9) (2010) 1544–1552.
[157] S. Jonsson, et al., Effect of delayed peripheral nerve repair on nerve regeneration, Schwann cell function and target muscle recovery, PLoS One 8 (2) (2013), e56484.
[158] M.K. Kolar, P.J. Kingham, Regenerative effects of adipose-tissue-derived stem cells for treatment of peripheral nerve injuries, Biochem. Soc. Trans. 42 (3) (2014) 697–701.
[159] G. Lin, et al., Cavernous nerve repair with allogenic adipose matrix and autologous adipose-derived stem cells, Urology 77 (6) (2011). 1509 e1–8.
[160] X. Chen, et al., Neurotrophic effect of adipose tissue-derived stem cells on erectile function recovery by pigment epithelium-derived factor secretion in a rat model of cavernous nerve injury, Stem Cells Int. 2016 (2016) 5161248.
[161] T.M. Fandel, et al., Recruitment of intracavernously injected adipose-derived stem cells to the major pelvic ganglion improves erectile function in a rat model of cavernous nerve injury, Eur. Urol. 61 (1) (2012) 201–210.
[162] X. Qiu, et al., Both immediate and delayed intracavernous injection of autologous adipose-derived stromal vascular fraction enhances recovery of erectile function in a rat model of cavernous nerve injury, Eur. Urol. 62 (4) (2012) 720–727.
[163] R. Yang, et al., Adipose-derived stem cells ameliorate erectile dysfunction after cavernous nerve cryoinjury, Andrology 3 (4) (2015) 694–701.
[164] C.C. Ying, et al., Neural-like cells from adipose-derived stem cells for cavernous nerve injury in rats, Neural Regen. Res. 14 (6) (2019) 1085–1090.
[165] D. You, et al., Comparative analysis of periprostatic implantation and intracavernosal injection of human adipose tissue-derived stem cells for erectile function recovery in a rat model of cavernous nerve injury, Prostate 73 (3) (2013) 278–286.
[166] D. You, et al., Comparative study of autologous stromal vascular fraction and adipose-derived stem cells for erectile function recovery in a rat model of cavernous nerve injury, Stem Cells Transl. Med. 4 (4) (2015) 351–358.
[167] C. Ying, et al., Effects of intracavernous injection of adipose-derived stem cells on cavernous nerve regeneration in a rat model, Cell Mol. Neurobiol. 33 (2) (2013) 233–240.
[168] F. Sun, et al., Repair of facial nerve defects with decellularized artery allografts containing autologous adipose-derived stem cells in a rat model, Neurosci. Lett. 499 (2) (2011) 104–108.
[169] O.L. Abbas, et al., Adipose-derived stem cells enhance axonal regeneration through cross-facial nerve grafting in a rat model of facial paralysis, Plast Reconstr. Surg. 138 (2) (2016) 387–396.
[170] M. Ghoreishian, et al., Facial nerve repair with Gore-Tex tube and adipose-derived stem cells: an animal study in dogs, J. Oral Maxillofac. Surg. 71 (3) (2013) 577–587.
[171] Y. Watanabe, et al., Undifferentiated and differentiated adipose-derived stem cells improve nerve regeneration in a rat model of facial nerve defect, J. Tissue Eng. Regen. Med. 11 (2) (2017) 362–374.
[172] Y. Li, W. Xu, L.Y. Cheng, Adipose-derived mesenchymal stem cells accelerate nerve regeneration and functional recovery in a rat model of recurrent laryngeal nerve injury, Neural Regen. Res. 12 (9) (2017) 1544–1550.
[173] K. Tomita, et al., Differentiated adipose-derived stem cells promote myelination and enhance functional recovery in a rat model of chronic denervation, J. Neurosci. Res. 90 (7) (2012) 1392–1402.
[174] J.A. Passipieri, et al., Adipose stem cells enhance nerve regeneration and muscle function in a peroneal nerve ablation model, Tissue Eng. Part A (2019).
[175] P. Erba, et al., Regeneration potential and survival of transplanted undifferentiated adipose tissue-derived stem cells in peripheral nerve conduits, J. Plast. Reconstr. Aesthet. Surg. 63 (12) (2010) e811–e817.

[176] P.J. Kingham, et al., Stimulating the neurotrophic and angiogenic properties of human adipose-derived stem cells enhances nerve repair, Stem Cells Dev. 23 (7) (2014) 741–754.
[177] S. Suganuma, et al., Uncultured adipose-derived regenerative cells promote peripheral nerve regeneration, J. Orthop. Sci. 18 (1) (2013) 145–151.
[178] L.F. Jiang, et al., T lymphocyte subsets and cytokines in rats transplanted with adipose-derived mesenchymal stem cells and acellular nerve for repairing the nerve defects, J. Korean Neurosurg. Soc. 58 (2) (2015) 101–106.
[179] E.A. Kappos, et al., Epineural adipose-derived stem cell injection in a sciatic rodent model, Brain Behav. 8 (7) (2018), e01027.
[180] H. Luo, et al., The protection of MSCs from apoptosis in nerve regeneration by TGFbeta1 through reducing inflammation and promoting VEGF-dependent angiogenesis, Biomaterials 33 (17) (2012) 4277–4287.
[181] R. Masgutov, et al., Allogenic adipose derived stem cells transplantation improved sciatic nerve regeneration in rats: autologous nerve graft model, Front. Pharmacol. 9 (2018) 86.
[182] D.N. Rodriguez Sanchez, et al., Canine adipose-derived mesenchymal stromal cells enhance neuroregeneration in a rat model of sciatic nerve crush injury, Cell Transplant 28 (1) (2019) 47–54.
[183] Y. Sowa, et al., Adipose-derived stem cells produce factors enhancing peripheral nerve regeneration: influence of age and anatomic site of origin, Stem Cells Dev. 21 (11) (2012) 1852–1862.
[184] M. Tremp, et al., The regeneration potential after human and autologous stem cell transplantation in a rat sciatic nerve injury model can be monitored by MRI, Cell Transplant 24 (2) (2015) 203–211.
[185] K.O. Allbright, et al., Delivery of adipose-derived stem cells in poloxamer hydrogel improves peripheral nerve regeneration, Muscle Nerve (2018).
[186] S.C. Hsieh, et al., Effect of an epineurial-like biohybrid nerve conduit on nerve regeneration, Cell Transplant 25 (3) (2016) 559–574.
[187] Y.Y. Hsueh, et al., Functional recoveries of sciatic nerve regeneration by combining chitosan-coated conduit and neurosphere cells induced from adipose-derived stem cells, Biomaterials 35 (7) (2014) 2234–2244.
[188] L.G. Dai, G.S. Huang, S.H. Hsu, Sciatic nerve regeneration by cocultured Schwann cells and stem cells on microporous nerve conduits, Cell Transplant 22 (11) (2013) 2029–2039.
[189] M. Georgiou, et al., Engineered neural tissue with aligned, differentiated adipose-derived stem cells promotes peripheral nerve regeneration across a critical sized defect in rat sciatic nerve, Biomaterials 37 (2015) 242–251.
[190] B.S. Liu, Y.C. Yang, C.C. Shen, Regenerative effect of adipose tissue-derived stem cells transplantation using nerve conduit therapy on sciatic nerve injury in rats, J. Tissue Eng. Regen. Med. 8 (5) (2014) 337–350.
[191] R.F. Masgutov, et al., Human adipose-derived stem cells stimulate neuroregeneration, Clin. Exp. Med. 16 (3) (2016) 451–461.
[192] V. Carriel, et al., Differential expression of GAP-43 and neurofilament during peripheral nerve regeneration through bio-artificial conduits, J. Tissue Eng. Regen. Med. 11 (2) (2017) 553–563.
[193] J. Chato-Astrain, et al., In vivo evaluation of nanostructured fibrin-agarose hydrogels with mesenchymal stem cells for peripheral nerve repair, Front. Cell Neurosci. 12 (2018) 501.
[194] P. Hernandez-Cortes, et al., Ghrelin and adipose-derived mesenchymal stromal cells improve nerve regeneration in a rat model of epsilon-caprolactone conduit reconstruction, Histol. Histopathol. 32 (6) (2017) 627–637.
[195] E.A. Kappos, et al., Peripheral nerve repair: multimodal comparison of the long-term regenerative potential of adipose tissue-derived cells in a biodegradable conduit, Stem Cells Dev. 24 (18) (2015) 2127–2141.
[196] R. Mohammadi, S. Azizi, K. Amini, Effects of undifferentiated cultured omental adipose-derived stem cells on peripheral nerve regeneration, J. Surg. Res. 180 (2) (2013) e91–e97.
[197] R. Mohammadi, A. Asadollahi, K. Amini, Uncultured undifferentiated adipose-derived nucleated cell fractions combined with inside-out artery graft accelerate sciatic nerve regeneration and functional recovery, Int. J. Oral Maxillofac. Surg. 43 (9) (2014) 1161–1168.
[198] Y. Xu, et al., A silk fibroin/collagen nerve scaffold seeded with a Co-culture of Schwann cells and adipose-derived stem cells for sciatic nerve regeneration, PLoS One 11 (1) (2016), e0147184.
[199] M.M. Saller, et al., Validation of a novel animal model for sciatic nerve repair with an adipose-derived stem cell loaded fibrin conduit, Neural Regen. Res. 13 (5) (2018) 854–861.
[200] S.M. Klein, et al., Peripheral motor and sensory nerve conduction following transplantation of undifferentiated autologous adipose tissue-derived stem cells in a biodegradable U.S. Food and Drug Administration-Approved Nerve Conduit, Plast. Reconstr. Surg. 138 (1) (2016) 132–139.

[201] H. Orbay, et al., Differentiated and undifferentiated adipose-derived stem cells improve function in rats with peripheral nerve gaps, J. Plast. Reconstr. Aesthet. Surg. 65 (5) (2012) 657–664.

[202] Y. Wei, et al., Chitosan/silk fibroin-based tissue-engineered graft seeded with adipose-derived stem cells enhances nerve regeneration in a rat model, J. Mater. Sci. Mater. Med. 22 (8) (2011) 1947–1964.

[203] J. Nan, et al., Use of nerve conduits for peripheral nerve injury repair: a web of science-based literature analysis, Neural Regen. Res. 7 (35) (2012) 2826–2833.

[204] G.H. Han, et al., Therapeutic strategies for peripheral nerve injury: decellularized nerve conduits and Schwann cell transplantation, Neural Regen. Res. 14 (8) (2019) 1343–1351.

[205] K.K. Wu, Y. Huan, Streptozotocin-induced diabetic models in mice and rats, Curr. Protoc. Pharmacol. 5 (2008) 47. Chapter 5: p. Unit.

[206] C. Oses, et al., Preconditioning of adipose tissue-derived mesenchymal stem cells with deferoxamine increases the production of pro-angiogenic, neuroprotective and anti-inflammatory factors: potential application in the treatment of diabetic neuropathy, PLoS One 12 (5) (2017), e0178011.

[207] A.T. Brini, et al., Therapeutic effect of human adipose-derived stem cells and their secretome in experimental diabetic pain, Sci. Rep. 7 (1) (2017) 9904.

[208] A. Hoke, Animal models of peripheral neuropathies, Neurotherapeutics 9 (2) (2012) 262–269.

[209] K.B. Guthmiller, M. Varacallo, Pain, complex regional pain syndrome (Reflex Sympathetic Dystrophy, RSD, CRPS). StatPearls [Internet], StatPearls Publishing, 2018.

[210] K.S. Koh, et al., Clinical application of human adipose tissue-derived mesenchymal stem cells in progressive hemifacial atrophy (Parry-Romberg disease) with microfat grafting techniques using 3-dimensional computed tomography and 3-dimensional camera, Ann. Plast. Surg. 69 (3) (2012) 331–337.

[211] M.K. Haahr, et al., Safety and potential effect of a single intracavernous injection of autologous adipose-derived regenerative cells in patients with erectile dysfunction following radical prostatectomy: an open-label phase I clinical trial, EBioMedicine 5 (2016) 204–210.

Index

Note: Page numbers followed by *f* indicate figures and *t* indicate tables.

B

C

D

E

F

G

H

I

Printed i
by Baker &

n the United States
r Taylor Publisher Services